# HIGH-FIELD ELECTRODYNAMICS

# CRC SERIES *in* PURE *and* APPLIED PHYSICS

**Dipak Basu**
*Editor-in-Chief*

## FORTHCOMING TITLES

**Introduction to Molecular Biophysics**
Jack Tuszynski

**Physics of Semiconductor Electron Devices**
Pratul K. Ajmera

**Fundamentals and Applications of Ultrasonic Waves**
David Cheeke

**Handbook of Particle Physics**
M. K. Sundaresan

# HIGH-FIELD ELECTRODYNAMICS

**Frederic V. Hartemann**
*Institute for Laser Science and Applications*
*Physics and Advanced Technologies Directorate*
*Lawrence Livermore National Laboratory*
*Livermore, CA, USA*

CRC PRESS
Boca Raton London New York Washington, D.C.

**Library of Congress Cataloging-in-Publication Data**

Hartemann, Frederic V.
High-field electrodynamics/Frederic V. Hartemann.
p. cm.—(CRC series in pure and applied physics)
Includes bibliographical references and index.
ISBN 0-8493-2378-9 (alk. paper)
1. Electrodynamics. I. Title. II. series.

QC631 .H37 2001
537.6—dc21 2001043670


International Standard Book Number 0-8493-2378-9
Library of Congress Card Number 2001043670
Printed in the United States of America 1 2 3 4 5 6 7 8 9 0
Printed on acid-free paper

# *Foreword*

In 1905, arguably one of the greatest years in physics, Albert Einstein discovered the concept of stimulated emission by carefully analyzing blackbody radiation in terms of the light quanta first postulated by Max Planck in 1900; this led to the invention of the laser in 1960. In the same year, Einstein's interpretation of the photoelectric effect in terms of photons gave a physical reality to Planck's quantum hypothesis. Paradoxically, the laser can be described as producing a classical electromagnetic wave, with an indefinite number of photons, and a well-defined phase. Thus, the ideas that Einstein applied to electrodynamics have come full circle, emphasizing the tight conceptual structure underlying classical and quantum electrodynamics. The last aspect of Einstein's trilogy is the celebrated theoretical formulation of special relativity; all three subjects are near the core of the present book. On May 16, 1960, the first laser, using ruby as its gain medium, was operated, and eventually produced peak powers in excess of a few kW; since that time, the peak power achieved by lasers has increased steadily, reaching the petawatt ($10^{15}$ W) level almost a quarter century later, on May 23, 1996, at Lawrence Livermore National Laboratory (LLNL). Currently, plans to reach one exawatt ($10^{18}$ W) are under consideration in various laboratories worldwide. In parallel, and starting with the discovery of the electron by Sir Joseph John Thomson in 1897, relativistic electron beams with energies now exceeding 50 GeV, have been produced, with extremely well-defined phase-space characteristics. Finally, convincing evidence of the particle nature of electromagnetic radiation was found in 1922 by the American physicist Arthur Holly Compton, who started the field of study of the interaction between relativistic electrons and photons; stimulated Compton scattering, first proposed by Paul Adrien Maurice Dirac, eventually led to the development of free-electron lasers.

Within this historical context, a new field of physics has recently emerged: high-field electrodynamics, where relativistic charged leptons interact with intense coherent electromagnetic radiation fields, classically or quantum mechanically, to produce a rich spectrum of phenomena, ranging from laser acceleration and nonlinear Compton scattering, to vacuum nonlinearities and the Casimir effect. This novel discipline has fundamental implications in terms of electrodynamics, and important applications in modern physics.

Although the literature is replete with excellent textbooks, both classic and modern, describing classical and quantum electrodynamics in great detail, as well as high-intensity lasers, relativistic electron beams, and accelerator physics, the confluence of these research areas, as embodied by the innovative discipline of high-field electrodynamics, has not yet been addressed in any comprehensive manner. It is, therefore, my pleasure to introduce Fred Hartemann's new book on this subject, which will fill this gap, nicely complementing the aforementioned works, and providing a wide variety of in-depth discussions of some of the fundamental interactions between relativistic electrons and positrons with ultrahigh intensity coherent photon fields. This book should prove an extremely valuable addition to the collection of any researcher in these fields, and to the graduate students pursuing a degree at the forefront of modern physics.

**Arthur K. Kerman**
Massachusetts Institute of Technology
Cambridge, MA
October 2001

# *Preface*

Since its inception, electrodynamics has evolved through sweeping changes brought about by Maxwell's unification of electricity and magnetism, Einstein's relativity, Dirac's quantum electrodynamics, and the successful renormalization program led by Feynman, Schwinger, Tomonaga, and Dyson, resulting in the most accurate scientific theory to date, QED. Modern gauge field theories have been profoundly influenced by the exquisite theoretical architecture of electrodynamics and the fundamental symmetries underlying its structure. Momentous discoveries, including gauge transformations, charge conservation, antimatter, spin, superconductivity, x-rays, radars, semiconductors, the laser, and optical communications have emerged, in large part, from electrodynamics. Through this extraordinary evolution, mirroring the progress of all modern physics, electrodynamics has retained an unmatched elegance and a magnificent economy of form, while remaining a profoundly important path to the fundamental understanding of the laws of nature. Students, teachers, engineers, and scientists alike should be well versed in electrodynamics, as it provides a paradigm for all scientific theories.

Recent developments in novel technologies, such as chirped-pulse amplification, high-brightness, relativistic electron sources, and femtosecond optics and diagnostics, have opened a new field of research: high-field electrodynamics. The systematic study of the interaction of relativistic electrons or positrons with coherent electromagnetic radiation, including ultra high-intensity laser pulses, forms the core of this novel discipline. The purpose of this book is therefore to provide a detailed introduction to the subject, beginning with the foundations, and progressing toward modern applications, including coherent synchrotron radiation, nonlinear Compton scattering, free-electron lasers, and laser acceleration. Because the field is new and evolving rapidly, I have striven to provide the reader with as broad a collection of theoretical techniques as possible; I have also endeavored to discuss a number of important ramifications of high-field electrodynamics, ranging from quantum optics, squeezed states, solitons, and the Einstein–Podolsky–Rosen paradox, to rotating charged black holes, Yang–Mills and non-Abelian gauge field theories, and the Bohm–Aharanov effect. Finally, where possible, a number of derivations are approached by at least two different routes to yield deeper physical insight and show how theoretical flexibility and different viewpoints can help gain a better understanding of the physics involved and a broader perspective on the nature of the scientific method.

In closing, and before acknowledging the individuals who each played an important, if indirect, role in the writing of this book, I would like to mention that I read with great interest the books entitled *The End of Science*, by John

Horgan, and *The End of Physics*, by David Lindley. Although similar in title, these books offer rather different views on the future of physics in particular and of science in general. In the first text, a rather pessimistic perspective is offered: the end of physics is close because, either a "theory of everything" (ToE) is at hand, and/or the human mind cannot comprehend things any further than our current grasp. I find myself firmly and decisively in the other camp: physics is likely to continue to the asymptotic forever: "There are more things in heaven and earth, Horatio, than are dreamt of in your philosophy" (*Hamlet, Prince of Denmark*, Act I, Scene 5).

During the writing of this book, Nora Konopka, my editor at CRC Press, has helped along the way with patience and enthusiasm, and I am very grateful for her unwavering guidance. I offer special thanks to Tony Troha, who is responsible for most of the illustrations and artwork included in this book. I also wish to acknowledge exceptionally stimulating interactions over my years on the path to physics, and on both sides of the Atlantic, with students, teachers, colleagues, and friends: Rokaya Al-Ayat, William Allis, Frédéric André, Tom Antonsen, Alain Aspect, Patrick Audebert, Hector Baldis, Chris Barty, Jacques Bauche, George Bekefi, Abe Bers, Monica Blank, Vladimir Bratman, Michel Bres, Scott Burns, Jean-Max Buzzi, Swapan Chattopadhyay, Chiping Chen, Pisin Chen, Shien-Chi Chen, Ray Chiao, Kwo-Ray Chu, Sam Chu, Chris Clayton, Yves Cohen, Thomas Compère-Morel, Jacques Cousteau, Tom Cowan, Bruce Danly, Ron Davidson, Jean-Loup Delcroix, Grigorij Denisov, Todd Ditmire, Calvin Domier, Henri Doucet, Harold Edgerton, Eric Esarey, Shaoul Ezekiel, Maurice Fabre, George Faillon, Joel Fajans, Roger Falcone, Marc Fitaire, Scott Fochs, Rick Freeman, Henry Freund, Miguel Furman, Pascal Garin, Nathalie Gerbelot, David Gibson, Bill Goldstein, Avi Gover, Phillipe Guidée, William Guss, Alan Guth, Vincent Hartz, Herman Haus, Jake Haimson, Jean-Claude Hartemann, Marc-Antoine Hartemann, Olivier Hartemann, Terry Hauptman, Andy Hazi, Jonathan Heritage, Heinrich Hora, Sylvie Joli, Chan Joshi, Erich Ippen, George Johnston, Daniel Kleppner, Kwang-Je Kim, Ken Kreisher, Ray Kurtzweil, Horatio Lamella-Riveira, Wim Leemans, Arnaud Le Foll, Jean Le Gaq, Peter Legourburu, Jon Leinaas, Greg Le Sage, Alain Magneville, Michel Marchal, Ivan Mastowski, Dave McDermott, Will Menninger, David Meyerhoffer, Howard Milchberg, Félix Mirabel, George Mourier, Gérard Mourou, Patrick Muggli, Bill Mulligan, Gregory Nusinovitch, Lynn Parayo, Claudio Pellegrini, Jean Perez Y Jorba, Mike Perry, Stefania Petracca, Yves Petroff, Denis Ranque, Judith Repetti, Jamie Rosenzweig, Bernhard Rupp, Alberto Santoni, Louis Sarliève, Livio Scarsi, Carl Schroeder, Gennady Schvets, Phil Sprangle, Paul Springer, John Swegle, Toshi Tajima, Valery Telnov, Richard Temkin, Laurent Terray, Joel Thiollier, Zeno Toffano, Thierry Trémeau, Abe Szoke, Han Uhm, Don Umstadter, Bill Unruh, Karl van Bibber, L. Van Hove, James van Meter, Toon Verhoeven, Arnold Vlieks, Glen Westenskow, John Woodworth, Jonathan Wurtele, Ming Xie, Kongyi Xu, David Yu, Alexander Zholents, and Max Zolotorev. I have learned exciting physics from each of them, sometimes through intense discussions and constructive criticism, always at a high level, and it has been a privilege to interact, over the years, with such gifted individuals.

Finally, I dedicate this book to three exceptional mentors: Victor Bogros, Arthur Kerman, and Robert Hauptman, as well as to my parents and Debbie Santa Maria, who have been with me throughout the creation of this book.

**Frederic V. Hartemann**
Kailua-Kona, HI
June 2001

# Contents

## Part 1 Foundations

# Part I

# Foundations

# 1

# *Overview*

## 1.1 Introduction

Electrodynamics is the branch of physics concerned with the interaction of charged particles and electromagnetic fields. Most macroscopic phenomena fall under this area of scientific knowledge, including optics, from lasers to astronomy; chemistry, from biomolecular physics to inorganic compounds; and solid-state physics, from semiconductors to superconductivity. The other three known interactions, namely gravitation and the weak and strong interactions, are somewhat more limited in scope. They apply either to very large-scale systems, such as galaxies, star clusters, and the topology of the universe, or to processes involving subatomic particles, including quarks, neutrinos, and charged leptons. The latter is exemplified by nuclear fusion reactions of elements lighter than iron in stars, for the strong force, and $\beta$-decay and the recently observed neutrino oscillations, for the weak interaction.

Recently, with the experimental discovery of the $W^+$, $W^-$, and $Z_0$ bosons, electrodynamics and the weak interaction have been unified into the electro-weak interaction, which governs the interaction of leptons, such as the electron, muon, and tau, and their antiparticles; vector bosons such as the photon, and the intermediate gauge bosons mentioned above; and the three generations of neutrinos associated with these leptons.

Within this context, Maxwell's equations play a major role, as they describe the behavior of the electromagnetic field both in vacuum and in the presence of sources. Historically, these equations were first discovered through the work of Ampère, Coulomb, Faraday, Gauss, and Laplace, to name a few, and were written in integral form. Maxwell unified these equations in a single set, gave their expression in differential form, and modified Ampère's theorem by introducing the concept of displacement current, as required for reasons of symmetry. Maxwell's prediction was first supported experimentally by Hertz, who demonstrated the propagation of electromagnetic waves in vacuum. The theory of electromagnetism unified electricity and magnetism under a common formalism, allowing the identification of light waves with electromagnetic waves, and ultimately instigated a radical reevaluation of the concept of space–time, which resulted in the theory of special relativity as formulated by Einstein.

## 1.2 The Relativistic Intensity Regime

Because the subject matter of this book is high-field electrodynamics, it is interesting to introduce two different fundamental scales of field in this first chapter before returning to Maxwell's equations. These two scales will play a critical role in a number of phenomena analyzed in detail in this work. The first scale is purely classical and corresponds to what is often called the normalized vector potential associated with an electromagnetic field distribution: this is the characteristic vector potential, $A = |\mathbf{A}|$, measured in units of $m_0c/e$, or "electron units," where $m_0 = 9.109\,389\,7(54) \times 10^{-31}$ kg is the rest mass of the electron, while $-e = -1.602\,177\,33(49) \times 10^{-19}$ C is the charge of the electron, and $c = 2.997\,924\,58 \times 10^{8}$ ms$^{-1}$ (exact) is the speed of light in vacuum. Since $eA$ has the dimension of a momentum, as can be seen by considering the Hamiltonian, $\mathcal{H} = c\sqrt{(\mathbf{p} - e\mathbf{A})^2 + m_0^2c^2} + e\varphi$, of a charged particle in an electromagnetic field, the normalized vector potential, $eA/m_0c$, is a dimensionless quantity. Here, $\mathbf{p} = m\mathbf{v} = \gamma m_0\mathbf{v} = \gamma\beta m_0 c$, is the momentum of the particle, while $e$ is its charge, and $\varphi$ is the scalar potential; $\boldsymbol{\beta} = \mathbf{v}/c$ is the velocity measured in units of $c$, and $\gamma = (1 - \beta^2)^{-1/2}$ is the relativistic mass factor, or temporal component of the four-velocity. These concepts will be discussed more extensively in Chapter 2, where the Lorentz transform and four-vector formalism are introduced. For an electromagnetic wave in vacuum, the normalized vector potential can be related to the electric field strength, $E$, by considering the frequency, $\omega$, of the wave: in the temporal gauge, where the scalar potential is set equal to zero, the relation between the electric field and the vector potential reduces to $\mathbf{E} = -\partial_t\mathbf{A}$; for a monochromatic wave of frequency $\omega$, this yields $eE/\omega m_0 c$; furthermore, in this case, the physical interpretation of the normalized potential is straightforward, as it corresponds to the normalized transverse momentum, or transverse component of the four-velocity, acquired by an electron submitted to the field of the wave. In other words, $p_\perp = eA_\perp$; $\gamma\beta_\perp = eA_\perp/m_0c$. Since the transverse vector potential is also a relativistic invariant, the normalized vector potential is a fundamental, covariant quantity characterizing the classical strength of the electromagnetic field. For normalized potentials exceeding unity, the transverse motion of the electron in the wave becomes relativistic. In the case of a laser operating at visible wavelengths, the required intensity is approximately $10^{17}$ W/cm$^2$, a number that can now readily be achieved using chirped-pulse amplification (CPA), where energies near one Joule can be obtained with a pulse duration well below 100 femtoseconds. This relativistic intensity regime is now being probed at a number of laboratories worldwide, and has already yielded a number of important and fundamental new experimental results, which are discussed in this book.

Finally, we note that the photon density can be conveniently expressed in terms of the rms normalized vector potential, $\mathcal{A} = eA/m_0c\sqrt{2}$, by considering the energy density, $\varepsilon_0 E^2$, the photon energy, $\hbar\omega$, and the relation $E = -\partial_t A$, to obtain the elegantly simple answer: $\mathcal{A}^2/r_0\lambda\!\!\!^{-}_c\lambda$. This result is eminently interesting, as it combines the classical and quantum scales: indeed, $r_0 = e^2/4\pi\varepsilon_0 m_0 c^2$ is the classical electron radius, while $\lambda\!\!\!^{-}_c = \hbar/m_0c$ is its Compton wavelength; $\lambda$ is the characteristic wavelength of the radiation under consideration.

## 1.3 The Schwinger Critical Field

The second fundamental scale of electric field strength is a quantum electrodynamical quantity, often referred to as the Schwinger critical field. In this case, the field is so intense that it can tunnel virtual electron–positron pairs, out of the quantum electrodynamical vacuum, into real pairs. The value of the critical field can be estimated by considering that the work produced by the field on an electron over a Compton wavelength must be equal to its rest mass: $eE^*\lambda\!\!\!^{-}_c = m_0c^2$; since $\lambda\!\!\!^{-}_c = \hbar/m_0c = 3.861\,593\,(22) \times 10^{-13}$ m, we find that $E^* = m_0^2c^3/e\hbar \simeq 1.323 \times 10^{18}$ V/m. Although such a high field cannot be directly produced, one can take advantage of the relativistic transform of the electromagnetic field tensor, which essentially multiplies the electric field by a factor $\gamma$. Therefore, starting from the peak field produced by a CPA laser, which corresponds to approximately $m_0c^2/e\lambda$, where $\lambda$ is the laser wavelength, and using a 50 GeV electron beam, with $\gamma \simeq 10^5$, researchers at the Stanford Linear Accelerator Center (SLAC) have recently started to test electrodynamics near the Schwinger critical field, with $\Upsilon = E/E^* \approx \gamma m_0c^2/e\lambda \approx 10^{-2}$. Near the high-upsilon regime, pair production and light-by-light scattering processes become significant, as well as other quantum electrodynamical effects.

## 1.4 Maxwell's Equations

We now return to the basic introduction to the fundamental equations of electrodynamics. Maxwell's equations can be divided into two sets: on the one hand, the source-free equations

$$\nabla \times \mathbf{E} + \partial_t \mathbf{B} = \mathbf{0}, \tag{1.1}$$

which is the law of induction, and

$$\nabla \cdot \mathbf{B} = 0, \tag{1.2}$$

which corresponds to the fact that magnetic monopoles are not observed; on the other hand, the set with sources,

$$\nabla \cdot \mathbf{E} = \frac{1}{\varepsilon_0}\rho, \tag{1.3}$$

which corresponds to Gauss' law, and

$$\nabla \times \mathbf{B} - \frac{1}{c^2}\partial_t \mathbf{E} = \mu_0 \mathbf{j}, \tag{1.4}$$

which is equivalent to Ampère's theorem.

In these equations, $\mathbf{E}$ is the electric field, which is, in general, a function of space and time: $\mathbf{E} = \mathbf{E}(\mathbf{x}, t)$; $\mathbf{B}(\mathbf{x}, t)$ is the magnetic induction, while $\mu_0 = 4\pi \times 10^{-7}$ Hm$^{-1}$ is the permeability of vacuum; $\varepsilon_0 = 8.854\,187817\cdots \times 10^{-12}$ Fm$^{-1}$ is the permittivity of vacuum; $\rho(\mathbf{x}, t)$ is the charge density of the source; and $\mathbf{j}(\mathbf{x}, t) = \rho(\mathbf{x}, t)\mathbf{v}(\mathbf{x}, t)$ is the current density of the source, as expressed in terms of the charge density and velocity field, $\mathbf{v}(\mathbf{x}, t)$. The differential operators applied to the electric field and magnetic induction are the partial derivative with respect to time, $\partial_t \equiv \frac{\partial}{\partial t}$; the divergence,

$$\nabla \cdot \mathbf{V} = \partial_x V_x + \partial_y V_y + \partial_z V_z; \tag{1.5}$$

and the curl,

$$\nabla \times \mathbf{V} = (\partial_y V_z - \partial_z V_y)\hat{x} + (\partial_z V_x - \partial_x V_z)\hat{y} + (\partial_x V_y - \partial_y V_x)\hat{z}, \tag{1.6}$$

as expressed in Cartesian coordinates, where $\mathbf{V}(\mathbf{x}, t) = [V_x(\mathbf{x}, t), V_y(\mathbf{x}, t), V_z(\mathbf{x}, t)]$ is a vector field, and where the spatial position is given by $\mathbf{x} = (x, y, z)$. Within this context, the gradient operator can be introduced: $\nabla \equiv (\partial_x, \partial_y, \partial_z)$. For cylindrical and spherical coordinates systems, which are also commonly used, we refer the reader to the notes at the end of this chapter.

The permeability and permittivity of vacuum are related to the speed of light:

$$\varepsilon_0 \mu_0 c^2 = 1, \tag{1.7}$$

where $c$ is the speed of light in vacuum.

This very important relation, and the fact that it holds in any reference frame, is one of the core concepts of special relativity; it will be discussed briefly at the end of this overview. Further discussions of this idea will be the focus of Chapter 2.

## 1.5 Fields and Inductions, the Minkowski Formalism

At this point, we note that the interaction of electromagnetic waves with matter can be described according to two different theoretical formulations. On the one hand, the electromagnetic properties of the medium may be defined by introducing relations between the fields and inductions; this approach is usually referred to as the Minkowski formalism. Generally, these so-called constitutive relations are nonlinear. The other formulation describes the reaction of the medium in terms of an induced four-vector current density, including both the charge and current densities resulting from the influence of the external electromagnetic fields on the medium. As long as the theoretical analysis of the interaction of electromagnetic radiation with matter is performed in the rest frame of the medium under consideration, these two formulations are equivalent. However, whereas the four-vector current density approach can lead to a covariant description of the electrodynamics of nonlinear media, the relations between fields and inductions become very complicated in any reference frame where the medium is in motion. A detailed comparison of the two formalisms will be given in Chapter 7. Still, it should be noted that in the rest frame of the medium, the constitutive relations describing its electromagnetic properties, which are generally derived from quantum mechanics and group theory, directly reflect the underlying spatial symmetries of the medium, and therefore are usually the preferred formulation in classical nonlinear optics. In view of this, it is easy to understand that, in the relativistic case, the difficulty arises from the fact that the Lorentz group conserves space–time symmetries rather than spatial symmetries. For example, it is possible to transform a tetragonal lattice into a cubic one through the Lorentz transform; its refractive and magnetic properties will then appear different in the two different frames.

Within the Minkowski formalism, one introduces the electric induction

$$\mathbf{D} = \varepsilon \mathbf{E}, \tag{1.8}$$

and the magnetic field

$$\mathbf{H} = \frac{1}{\mu}\mathbf{B}, \tag{1.9}$$

where $\varepsilon$ is the permittivity, and $\mu$ is the permeability of the medium in question.

As discussed above, these quantities can be quite complex, as they can acquire a tensorial form in an anisotropic medium, and can depend in a nonlocal and nonlinear way on the applied fields; this is discussed in Chapters 5 through 7. In addition, they do not transform relativistically in a simple way, as will be discussed extensively in Chapter 7. It should also be noted that an external electric field can induce magnetic phenomena, and that the magnetic field can produce an electric polarization of the medium; thus, cross-interactions are also possible. In general, these quantities represent, in a space- and time-averaged manner, the atomic reaction of a medium submitted to electromagnetic fields; these reactions result into the electric and magnetic inductions. Within the Minkowski formalism, the equations with sources take the simple form

$$\nabla \cdot \mathbf{D} = \rho, \tag{1.10}$$

and

$$\nabla \times \mathbf{H} - \partial_t \mathbf{D} = \mathbf{j}, \tag{1.11}$$

where $\rho$ and $\mathbf{j}$ now represent sources external to the medium under consideration.

In addition, the source-free equations now read

$$\nabla \times \left( \frac{1}{\varepsilon} \mathbf{D} \right) + \partial_t (\mu \mathbf{H}) = \mathbf{0}, \tag{1.12}$$

and

$$\nabla \cdot (\mu \mathbf{H}) = 0. \tag{1.13}$$

## 1.6 Potentials, Gauge Condition, and Wave Equation

At this point, we note that the source-free Equations 1.1 and 1.2 suggest the introduction of a vector potential, $\mathbf{A}(\mathbf{x}, t)$, and a scalar potential, $\varphi(\mathbf{x}, t)$, defined such that the magnetic induction

$$\mathbf{B} = \nabla \times \mathbf{A}, \tag{1.14}$$

in which case Equation 1.2 is automatically satisfied, and

$$\mathbf{E} = -\nabla \varphi - \partial_t \mathbf{A}, \tag{1.15}$$

which yields

$$\nabla \times \mathbf{E} + \partial_t \mathbf{B} = \nabla \times (-\nabla \varphi - \partial_t \mathbf{A}) + \partial_t (\nabla \times \mathbf{A}), \tag{1.16}$$

which is, indeed, identically equal to zero because the curl of a gradient is always equal to zero, as shown in the notes at the end of this chapter, and because the partial derivative with respect to time and the curl operator commute.

Having introduced the vector and scalar potentials, we can now reexamine the equations with sources. We first have

$$\nabla \cdot \mathbf{E} = \nabla \cdot (-\nabla\varphi - \partial_t \mathbf{A}) = -\Delta\varphi - \partial_t(\nabla \cdot \mathbf{A}) = \frac{1}{\varepsilon_0}\rho; \tag{1.17}$$

while Ampère's theorem now takes the form

$$\begin{aligned}\nabla \times \mathbf{B} - \frac{1}{c^2}\partial_t \mathbf{E} &= \nabla \times (\nabla \times \mathbf{A}) + \frac{1}{c^2}\partial_t(\nabla\varphi + \partial_t \mathbf{A}) \\ &= \nabla(\nabla \cdot \mathbf{A}) - \Delta\mathbf{A} + \frac{1}{c^2}\partial_t^2 \mathbf{A} + \nabla\left(\frac{1}{c^2}\partial_t \varphi\right) \\ &= \mu_0 \mathbf{j},\end{aligned} \tag{1.18}$$

where we have used the fact that the expression $\nabla \times (\nabla \times \mathbf{A})$ can be treated as a regular double cross-product, with $\mathbf{a} \times (\mathbf{b} \times \mathbf{c}) = (\mathbf{a} \cdot \mathbf{c})\mathbf{b} - (\mathbf{a} \cdot \mathbf{b})\mathbf{c}$, to yield

$$\nabla \times (\nabla \times \mathbf{A}) = \nabla(\nabla \cdot \mathbf{A}) - \nabla^2 \mathbf{A} = \nabla(\nabla \cdot \mathbf{A}) - \Delta \mathbf{A}. \tag{1.19}$$

Here, we have also introduced the Laplacian operator, $\Delta \equiv \nabla^2 \equiv \partial_x^2 + \partial_y^2 + \partial_z^2$, as expressed in Cartesian coordinates.

It proves insightful to rewrite Equation 1.18 as follows:

$$\left(\Delta - \frac{1}{c^2}\partial_t^2\right)\mathbf{A} + \mu_0 \mathbf{j} = \Box\, \mathbf{A} + \mu_0 \mathbf{j} = \nabla\left(\frac{1}{c^2}\partial_t \varphi + \nabla \cdot \mathbf{A}\right), \tag{1.20}$$

where we have simply grouped terms together, and introduced the d'Alembertian operator, also referred to as the photon propagator,

$$\Box \equiv \left(\Delta - \frac{1}{c^2}\partial_t^2\right). \tag{1.21}$$

In Cartesian coordinates,

$$\Box \equiv \left(\partial_x^2 + \partial_y^2 + \partial_z^2 - \frac{1}{c^2}\partial_t^2\right); \tag{1.22}$$

the expression of the d'Alembertian in cylindrical and spherical coordinates is discussed in the notes to Chapter 1. The quantity on the left-hand side of Equation 1.20 corresponds to the wave equation for the vector potential, driven by the source current density; the quantity on the right-hand side can be used in Gauss' law to rewrite Equation 1.17 as

$$\left(\Delta - \frac{1}{c^2}\partial_t^2\right)\varphi + \frac{1}{\varepsilon_0}\rho = \Box\varphi + \frac{1}{\varepsilon_0}\rho = -\partial_t\left(\frac{1}{c^2}\partial_t\varphi + \nabla\cdot\mathbf{A}\right), \tag{1.23}$$

where we have simply added the second derivative of the scalar potential with respect to time, $\frac{1}{c^2}\partial_t^2\varphi$, on each side of Equation 1.17.

The left-hand side of Equation 1.23 now corresponds to the wave equation for the scalar potential, driven by the source charge density. Both Equations 1.20 and 1.23 contain the common expression $\frac{1}{c^2}\partial_t\varphi + \nabla\cdot\mathbf{A}$, which is often set equal to zero:

$$\frac{1}{c^2}\partial_t\varphi + \nabla\cdot\mathbf{A} = 0. \tag{1.24}$$

This condition on the potentials is called the Lorentz gauge condition; it is discussed extensively in Chapter 3, along with other commonly used gauge conditions. One of the fundamental aspects of the gauge condition is that it is directly related to charge conservation, as shown in Chapter 3. Provided that the condition in Equation 1.24 is satisfied, or that the gradient and time derivative of $\frac{1}{c^2}\partial_t\varphi + \nabla\cdot\mathbf{A}$ are equal to zero, which are less restrictive conditions on the potentials, we recover the well-known driven wave equations

$$\Box\mathbf{A} + \mu_0\mathbf{j} = \mathbf{0}, \tag{1.25}$$

and

$$\Box\varphi + \frac{1}{\varepsilon_0}\rho = 0. \tag{1.26}$$

Equations 1.25 and 1.26, together with the gauge condition, completely describe the behavior of electromagnetic waves in the presence of sources.

## 1.7 The Coulomb Potential and Plane Waves

Finally, two interesting problems can be considered to illustrate briefly these equations. First, the Coulomb potential of a point electron at rest in vacuum, and second, the propagation of electromagnetic waves in a linear, homogeneous, time-independent, isotropic medium. In the first case, the sources take the form

$$\rho(\mathbf{x}, t) = -e\delta^3(\mathbf{x}-\mathbf{x}_0), \quad \mathbf{j}(\mathbf{x}, t) = \mathbf{v}(\mathbf{x}, t)\rho(\mathbf{x}, t) \tag{1.27}$$

where $\delta^3(\mathbf{x} - \mathbf{x}_0) = \delta(x - x_0)\delta(y - y_0)\delta(z - z_0)$ is the three-dimensional Dirac delta-function.

Note that, strictly speaking, $\delta(x)$ is a distribution; this implies that, in some cases, special care must be taken when manipulating mathematical expressions involving the Dirac delta-function. The static Coulomb problem can be solved in at least three different ways. First, one can use the divergence theorem and express the flux of the radial Coulomb electric field through a sphere centered around the charge as a function of the total charge enclosed in the volume delimited by the spherical boundary. Second, one can treat the problem as a Green function problem for an electrostatic potential. Finally, one can solve the problem for an extended charge distribution and take the limit when the distribution approaches a delta-function. The first two approaches are closely related and will be discussed in Chapter 5; here, we will use the third method and briefly verify that our result agrees with the application of the divergence theorem.

Since the problem is spherically symmetrical and time-independent, we use the divergence and Laplacian operators, as expressed in spherical coordinates, to rewrite Poisson's equation as

$$\frac{1}{r^2}\partial_r[r^2E_r(r)] = \frac{1}{r^2}d_r[r^2E_r(r)] = -\frac{1}{r}d_r^2[r\varphi(r)] = \frac{1}{\varepsilon_0}\rho(r), \tag{1.28}$$

where the charge density distribution satisfies the normalization condition

$$\int_0^\infty 4\pi r^2\rho(r)dr = -e. \tag{1.29}$$

Here, $-e$ is the elementary charge of the electron, and the differential element $4\pi r^2 dr$ results from the spherical symmetry of the problem.

Also note that because the scalar potential depends only on the radius, $r$, we can identify the partial derivative, $\partial_r \equiv \partial/\partial r$, with the total derivative, $d_r \equiv d/dr$. The electric field is purely radial, as required by the spherical symmetry, and is related to the scalar potential by

$$E_r(r) = -\partial_r\varphi(r). \tag{1.30}$$

As the approach used here is to first consider an extended charge distribution and then take the limit where the distribution tends to a delta-function, a simple model is provided by a Gaussian distribution, where

$$\rho(r) = \frac{-e}{2\pi^{3/2}r_0r^2}\exp\left[-\left(\frac{r}{r_0}\right)^2\right]. \tag{1.31}$$

First, we can verify that the total charge is equal to $-e$:

$$\int_0^\infty 4\pi r^2\rho(r)dr = \frac{-2e}{\sqrt{\pi}}\int_0^\infty e^{-x^2}dx, \tag{1.32}$$

where we have introduced the normalized radius, $x = \frac{r}{r_0}$.

The integral on the right-hand side of Equation 1.32 is

$$\int_0^\infty e^{-x^2}\,dx = \frac{\sqrt{\pi}}{2}; \tag{1.33}$$

using this result into Equation 1.32 yields the desired result. Next, the normalized Gaussian distribution tends to a spherical three-dimensional delta-function in the limit where its scale tends to zero; this is shown in detail in the notes at the end of this chapter.

Poisson's equation now takes the form

$$\frac{1}{r^2}d_r[r^2E_r(r)] = \frac{-e}{\varepsilon_0 2\pi^{3/2} r_0 r^2}\exp\left[-\left(\frac{r}{r_0}\right)^2\right]. \tag{1.34}$$

This differential equation can be resolved by introducing the function

$$f(r) = r^2E_r(r) = f_0\frac{2}{\sqrt{\pi}}\int_0^{r/r_0} e^{-x^2}\,dx = f_0\Phi\left(\frac{r}{r_0}\right), \tag{1.35}$$

where $\Phi$ is the probability integral, also known as the error function.

Indeed, we have

$$d_r f(r) = \frac{1}{r_0}f_0\frac{2}{\sqrt{\pi}}\exp\left[-\left(\frac{r}{r_0}\right)^2\right]; \tag{1.36}$$

using Equation 1.36 into Poisson's equation (Equation 1.34), we find that $f_0 = \frac{-e}{4\pi\varepsilon_0}$. The corresponding solution for the purely radial electric field is

$$E_r(r) = \frac{-e}{4\pi\varepsilon_0 r^2}\Phi\left(\frac{r}{r_0}\right); \tag{1.37}$$

now, the limit of the probability integral when the scale tends to zero is the unit step-function: $\lim_{r_0\to 0}[\Phi(\frac{r}{r_0})] = 1$, $r > 0$; therefore, when the charge distribution tends to a Dirac delta-function, we recover the well-known Coulomb field, $E_r(r) = -\frac{e}{4\pi\varepsilon_0 r^2} = -\partial_r\varphi(r)$, and potential $\varphi(r) = \frac{-e}{4\pi\varepsilon_0 r}$.

This is illustrated in Figure 1.1, where the normalized charge distribution and radial electric field are shown, as well as the probability function vs. the normalized radius, $r/r_0$. The convergence of the Gaussian charge density distribution toward a Dirac delta-function is shown in Figure 1.2, where normalized Gaussians are plotted for different values of the scale parameter, $r_0$.

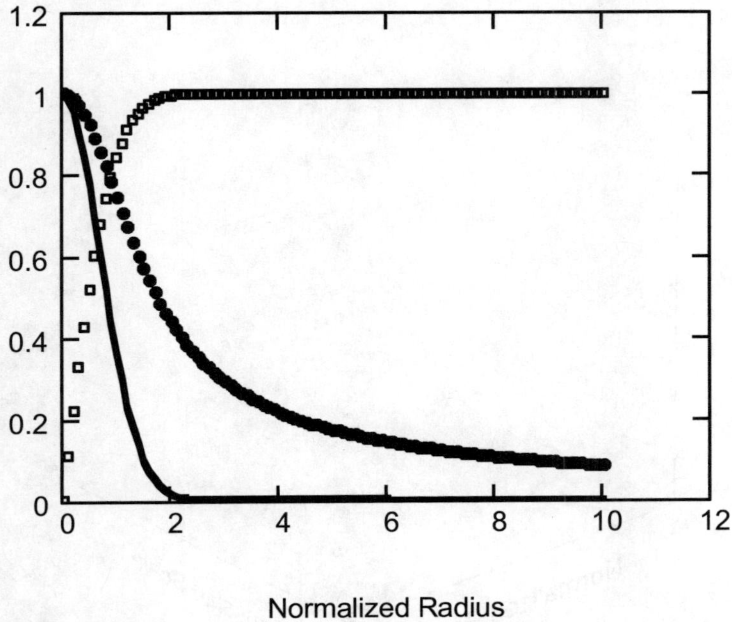


**FIGURE 1.1**
Gaussian, $\exp(-r^2)$ (solid line), error function, $\Phi(r) = 2\int_0^r e^{-x^2}dx/\sqrt{\pi}$ (squares), and normalized potential, $\Phi(r)/r$ (dots).

We now briefly compare the result derived above with a simple application of the divergence theorem, which states that

$$\iiint_{\mathcal{V}} \nabla \cdot \mathbf{U} dv = \iint_{S} \mathbf{n} \cdot \mathbf{U} ds: \tag{1.38}$$

the integral of the divergence of a vector field $\mathbf{U}$ over a given volume, $\mathcal{V}$, is equal to the flux $\mathbf{n} \cdot \mathbf{U}$ of that vector field through the surface $S$ delimiting that volume. In the case of a point charge, a sphere of radius $r$ can be chosen as the surface used to evaluate the flux of the radial electric field produced by the electron; we then have

$$\iiint_{\mathcal{V}} \nabla \cdot \mathbf{E}\, dv = \iiint_{\mathcal{V}} -\frac{e}{\varepsilon_0}\delta^3(\mathbf{x})d^3\mathbf{x} = -\frac{e}{\varepsilon_0}\iint_{S} \mathbf{n} \cdot \mathbf{E}\, ds = 4\pi r^2 E_r(r), \tag{1.39}$$

and it is easy to verify that, indeed,

$$E_r(r) = \frac{-e}{4\pi\varepsilon_0 r^2} = -\partial_r[\varphi(r)].$$

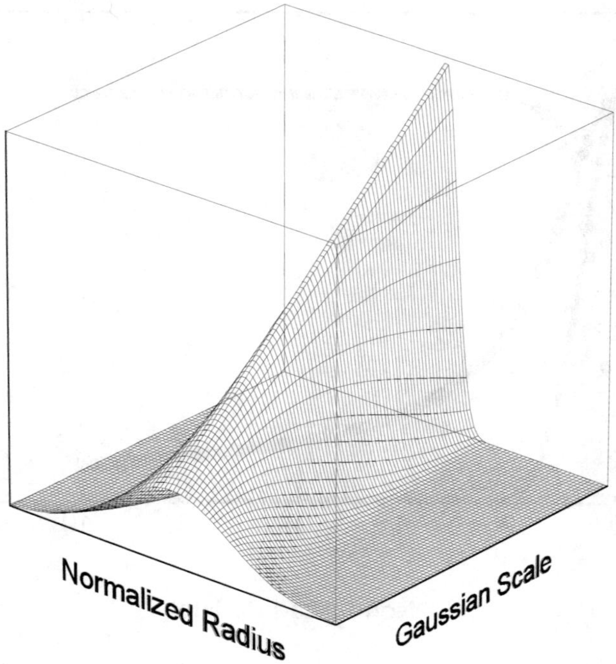


**FIGURE 1.2**
Evolution of a normalized Gaussian toward a Dirac delta-function.

Finally, we turn our attention to the propagation of electromagnetic waves in a linear, homogeneous, time-independent, isotropic medium, and in the absence of external sources; in this case, we start by taking the curl of Equation 1.11:

$$\nabla \times (\nabla \times \mathbf{H} - \partial_t \mathbf{D}) = \nabla(\nabla \cdot \mathbf{H}) - \Delta \mathbf{H} - \partial_t(\nabla \times \mathbf{D}) = \mathbf{0}. \tag{1.40}$$

Because the medium is linear and homogeneous, Equation 1.13 can be written as

$$\mu \nabla \cdot \mathbf{H} = 0; \tag{1.41}$$

the fact that the electromagnetic properties of the medium under consideration are also time-independent yields the additional condition

$$\frac{1}{\varepsilon} \nabla \times \mathbf{D} = -\mu \partial_t \mathbf{H}, \tag{1.42}$$

which easily derived from Equation 1.12. Using Equations 1.41 and 1.42 into Equation 1.40, we now have

$$(\Delta - \varepsilon\mu\partial_t^2)\mathbf{H} = \mathbf{0}, \tag{1.43}$$

which is the sought-after wave equation for electromagnetic waves propagating in a linear, homogeneous, time-independent, isotropic medium. While a general solution to this equation is beyond the scope of this introductory chapter, so-called "plane wave" solutions to Equation 1.43 play an important and ubiquitous role in classical and quantum electrodynamics and will therefore be discussed briefly here: in this case, we consider a wave that depends both on time and on a single spatial component; furthermore, we rotate our coordinate system so that the spatial component coincides with the z-coordinate: let us then show that

$$\mathbf{H}(z, t) = \mathbf{H}_0 + \mathbf{H}_+\left(z + \frac{1}{\sqrt{\varepsilon\mu}}t\right) + \mathbf{H}_-\left(z - \frac{1}{\sqrt{\varepsilon\mu}}t\right) \tag{1.44}$$

is a solution to Equation 1.43. Here, $\mathbf{H}_0$ is a constant three-vector, and $\mathbf{H}_+$ and $\mathbf{H}_-$ are arbitrary vector functions of $z \pm (t/\sqrt{\varepsilon\mu})$. The physics underlying this result is that electromagnetic waves propagate in the medium under consideration with the velocity $1/\sqrt{\varepsilon\mu}$; however, it is crucial to note that there are two distinct types of solutions: waves propagating forward in time and satisfying causality, where the field is a function of $z - (t/\sqrt{\varepsilon\mu})$; and waves propagating backward in time, in which case the solution is referred to as an advanced wave. These solutions reflect a fundamental symmetry of electrodynamics, and it will be shown in Chapter 10 that the advanced solutions play a major role, both in the physics of antiparticles and in the physics of electromagnetic renormalization.

To verify that Equation 1.44 is indeed a solution to the wave equation 1.43, we first define the new variables $z_\pm = z \pm (t/\sqrt{\varepsilon\mu})$, and note that by applying the chain rule to an arbitrary function $f(z_\pm)$, we have

$$\begin{aligned}[\Delta - \varepsilon\mu\partial_t^2]f(z_\pm) &= \nabla \cdot [\nabla f(z_\pm)] - \varepsilon\mu\partial_t^2 f(z_\pm) \\ &= \left(\frac{\partial^2 z_\pm}{\partial z^2} - \varepsilon\mu\frac{\partial^2 z_\pm}{\partial t^2}\right)\frac{df}{dz_\pm} + \left[\left(\frac{\partial z_\pm}{\partial z}\right)^2 - \varepsilon\mu\left(\frac{\partial z_\pm}{\partial t}\right)^2\right]\frac{d^2 f}{dz_\pm^2};\end{aligned} \tag{1.45}$$

since $z_\pm$ are linear functions of $z$ and $t$,

$$\frac{\partial^2 z_\pm}{\partial z^2} = \frac{\partial^2 z_\pm}{\partial t^2} = 0, \tag{1.46}$$

and we find that

$$\left(\frac{\partial z_\pm}{\partial z}\right)^2 - \varepsilon\mu\left(\frac{\partial z_\pm}{\partial t}\right)^2 = 1 - \varepsilon\mu\left(\frac{\pm 1}{\sqrt{\varepsilon\mu}}\right)^2 = 0. \tag{1.47}$$

The solutions $\mathbf{H}_\pm(z_\pm)$ correspond to plane wave-packets propagating along the z-axis with the velocity $\pm 1/\sqrt{\varepsilon\mu}$; in the special case of vacuum, we have $\varepsilon = \varepsilon_0$, and $\mu = \mu_0$; the propagation velocity is then $1/\sqrt{\varepsilon_0\mu_0} = c$.

The solution to the propagation equation 1.43 can also be sought in Fourier space, in which case, the field is represented as

$$\mathbf{H}(\mathbf{x}, t) = \frac{1}{(\sqrt{2\pi})^4}\iiiint \tilde{\mathbf{H}}(\mathbf{k}, \omega)\exp[i(\omega t - \mathbf{k}\cdot\mathbf{x})]\, d\omega\, d^3\mathbf{k}; \tag{1.48}$$

with this, the d'Alembertian operator can be identified as follows:

$$(\Delta - \varepsilon\mu\partial_t^2) \equiv \varepsilon\mu\omega^2 - \mathbf{k}^2. \tag{1.49}$$

It is then easily seen that any plane wave satisfying the dispersion relation

$$\mathbf{k}^2 = \varepsilon\mu\omega^2, \tag{1.50}$$

is a solution to the wave equation 1.43; in addition, since this particular problem is linear, any superposition of such plane waves is also a solution. Furthermore, as will be discussed in Chapter 5, from the dispersion relation given in Equation 1.50, we can also derive the group velocity of the wave,

$$\frac{\partial\omega}{\partial k} = \pm\frac{1}{\sqrt{\varepsilon\mu}}. \tag{1.51}$$

Although we have derived a solution to the wave equation 1.43, we have omitted considering a second, physically important condition on electromagnetic waves propagating in a linear, isotropic medium, the so-called transversality condition. To properly include this constraint in our description, we simply consider the divergence of the electric induction, as described by Equation 1.10; in the present case, where external sources are absent, we have

$$\nabla\cdot\mathbf{D} = \nabla\cdot(\varepsilon\mathbf{E}) = 0, \tag{1.52}$$

which implies that $\nabla\cdot\mathbf{E} = 0$, since the permittivity, $\varepsilon$, is a constant. In Fourier space, where

$$\mathbf{E}(\mathbf{x}, t) = \frac{1}{(\sqrt{2\pi})^4}\iiiint \tilde{\mathbf{E}}(\mathbf{k}, \omega)\exp[i(\omega t - \mathbf{k}\cdot\mathbf{x})]\, d\omega\, d^3\mathbf{k}; \tag{1.53}$$

this translates into the simple condition

$$\mathbf{k} \cdot \tilde{\mathbf{E}} = 0, \tag{1.54}$$

which physically corresponds to the fact that the electromagnetic waves have only transverse components: there is no component of **E** or **H** along the direction of propagation.

Finally, we also note that one could be tempted to generalize the notion of plane waves to a spherical solution to the wave equation using a variable of the form $r \pm t/\sqrt{\varepsilon\mu}$; however, the condition of transversality, which is related to the gauge condition, would not be satisfied in that case. Indeed, the simplest three-dimensional solution of the wave equation in vacuum corresponds to a dipole radiation pattern; more complex solutions appear as so-called multipole expansions and are discussed in Chapter 5.

## 1.8 Notes to Chapter 1

Our first task is to show that the divergence of a curl is zero. This can first be seen intuitively by treating the operator $\nabla$ as a regular vector: we know that $\mathbf{a} \cdot (\mathbf{a} \times \mathbf{b}) = 0$, so $\nabla \cdot (\nabla \times \mathbf{b}) = 0$ seems like a reasonable statement. A rigorous proof can be given easily in Cartesian coordinates: we start from Equation 1.6, which states that

$$\nabla \times \mathbf{V} = (\partial_y V_z - \partial_z V_y)\hat{x} + (\partial_z V_x - \partial_x V_z)\hat{y} + (\partial_x V_y - \partial_y V_x)\hat{z}\,; \tag{1.55}$$

we then apply the divergence operator, as defined in Equation 1.5, where

$$\nabla \cdot \mathbf{V} = \partial_x V_x + \partial_y V_y + \partial_z V_z; \tag{1.56}$$

we then have

$$\begin{aligned} \nabla \cdot (\nabla \times \mathbf{V}) &= \partial_x(\partial_y V_z - \partial_z V_y) + \partial_y(\partial_z V_x - \partial_x V_z) + \partial_z(\partial_x V_y - \partial_y V_x) \\ &= [\partial_x\partial_y - \partial_y\partial_x]V_z + [\partial_y\partial_z - \partial_z\partial_y]V_x + [\partial_z\partial_x - \partial_x\partial_z]V_y, \end{aligned} \tag{1.57}$$

which is obtained by grouping terms. Because the partial derivative operators commute, this last result is identically equal to zero. Commutation is ensured by the fact that the spatial coordinates lie on orthogonal axes.

We now demonstrate that Equation 1.19 is correct: working in Cartesian coordinates and using the definitions given above, we first have

$$\nabla \times (\nabla \times \mathbf{A}) = \nabla \times [(\partial_y A_z - \partial_z A_y)\hat{x} + (\partial_z A_x - \partial_x A_z)\hat{y} + (\partial_x A_y - \partial_y A_x)\hat{z}]. \tag{1.58}$$

Let us now consider the x-component of $\nabla \times (\nabla \times \mathbf{A})$:

$$
\begin{aligned}
[\nabla \times (\nabla \times \mathbf{A})]_x &= \partial_y[\nabla \times \mathbf{A}]_z - \partial_z[\nabla \times \mathbf{A}]_y \\
&= \partial_y(\partial_x A_y - \partial_y A_x) - \partial_z(\partial_z A_x - \partial_x A_z);
\end{aligned}
\tag{1.59}
$$

we first add and subtract $\partial_x\partial_x A_x = \partial_x^2 A_x$, to obtain

$$
[\nabla \times (\nabla \times \mathbf{A})]_x = \partial_y(\partial_x A_y - \partial_y A_x) - \partial_z(\partial_z A_x - \partial_x A_z) + \partial_x^2 A_x - \partial_x^2 A_x; \tag{1.60}
$$

we then group terms,

$$
[\nabla \times (\nabla \times \mathbf{A})]_x = \partial_y\partial_x A_y + \partial_z\partial_x A_z + \partial_x^2 A_x - [\partial_x^2 + \partial_y^2 + \partial_z^2]A_x; \tag{1.61}
$$

rearranging the first three terms on the right-hand side and using the fact that the partial derivative operators commute, we finally have

$$
[\nabla \times (\nabla \times \mathbf{A})]_x = \partial_x(\partial_x A_x + \partial_y A_y + \partial_z A_z) - [\partial_x^2 + \partial_y^2 + \partial_z^2]A_x. \tag{1.61}
$$

We now clearly recognize the divergence of **A** and the Laplacian operator; we can then identify the x-component of $\nabla \times (\nabla \times \mathbf{A})$ as follows:

$$
[\nabla \times (\nabla \times \mathbf{A})]_x = \partial_x(\nabla \cdot \mathbf{A}) - \Delta A_x. \tag{1.62}
$$

It is clear that the result derived for the x-component also holds for the y- and z-components; therefore, we finally have

$$
\nabla \times (\nabla \times \mathbf{A}) = \nabla(\nabla \cdot \mathbf{A}) - \Delta \mathbf{A}. \tag{1.63}
$$

For completeness, we also give the expression of the gradient, curl, and Laplacian operators in cylindrical coordinates, where position is given by $\mathbf{x} = (r, \theta, z)$. The gradient operator is

$$
\nabla \equiv \left(\frac{1}{r}\partial_r r, \frac{1}{r}\partial_\theta, \partial_z\right); \tag{1.64}
$$

here, the explicit result of applying the divergence operator to a vector field $\mathbf{A}(r, \theta, z)$, is

$$
\nabla \cdot \mathbf{A} = \frac{1}{r}\partial_r(rA_r) + \frac{1}{r}\partial_\theta A_\theta + \partial_z A_z; \tag{1.65}
$$

and the curl of **V** reads

$$\nabla \times \mathbf{A} = \left(\frac{1}{r}\partial_\theta A_z - \partial_z A_\theta\right)\hat{r} + (\partial_z A_r - \partial_r A_z)\hat{\theta} + \frac{1}{r}[\partial_r(rA_\theta) - \partial_\theta A_r]\hat{z}; \quad (1.66)$$

Finally, the Laplacian of a scalar field $\varphi$ is explicitly given by

$$\Delta\varphi = \frac{1}{r}\partial_r(r\partial_r\varphi) + \frac{1}{r^2}\partial_\theta^2\varphi + \partial_z^2\varphi. \quad (1.67)$$

In spherical coordinates, where position is parameterized by $\mathbf{x} = (r, \theta, \phi)$, the gradient takes the form

$$\nabla \equiv \left(\frac{1}{r^2}\partial_r r^2, \frac{1}{r\sin\theta}\partial_\theta \sin\theta, \frac{1}{r\sin\theta}\partial_\phi\right). \quad (1.68)$$

The divergence of a scalar field **A** is

$$\nabla \cdot \mathbf{A} = \frac{1}{r^2}\partial_r(r^2 A_r) + \frac{1}{r\sin\theta}\partial_\theta(\sin\theta A_\theta) + \frac{1}{r\sin\theta}\partial_\phi A_\phi, \quad (1.69)$$

while its curl reads

$$r(\nabla \times \mathbf{A}) = \frac{1}{\sin\theta}[\partial_\theta(\sin\theta A_\phi) - \partial_\phi A_\theta]\hat{r} + \left[\frac{1}{\sin\theta}\partial_\phi A_r - \partial_r(rA_\phi)\right]\hat{\theta}$$
$$+ [\partial_r(rA_\theta) - \partial_\theta A_r]\hat{\phi}. \quad (1.70)$$

The Laplacian of the scalar field $\varphi$, expressed in spherical coordinates, is

$$r^2(\Delta\varphi) = \partial_r[r^2(\partial_r\varphi)] + \frac{1}{\sin\theta}\partial_\theta[\sin\theta(\partial_\theta\varphi)] + \frac{1}{\sin^2\theta}\partial_\phi^2\varphi. \quad (1.71)$$

We now turn our attention to the demonstration that the Gaussian charge distribution used to model a three-dimensional delta function with spherical symmetry in the limit where its radial scale tends to zero is appropriate. We have already shown that the integrated charge is constant and equal to $-e$, as shown in Equations 1.31 and 1.32.

We first consider the one-dimensional case. The Dirac delta-function is defined by the two following properties:

$$\int_{-\infty}^{+\infty} \delta(x - x_0)dx = 1, \quad (1.72)$$

and

$$\int_{-\infty}^{+\infty} \delta(x - x_0) f(x) dx = f(x_0). \tag{1.73}$$

Let us now consider a Gaussian with variable scale $\Delta x$:

$$G(x, x_0, \Delta x) = \frac{1}{\sqrt{\pi}\Delta x} \exp\left[-\left(\frac{x - x_0}{\Delta x}\right)^2\right]; \tag{1.74}$$

we want to show that

$$\lim_{\Delta x \to 0} [G(x, x_0, \Delta x)] = \delta(x - x_0). \tag{1.75}$$

It is clear that $G(x, x_0, \Delta x)$ is normalized; in other words,

$$\int_{-\infty}^{+\infty} G(x, x_0, \Delta x) dx = \frac{1}{\sqrt{\pi}\Delta x} \int_{-\infty}^{+\infty} \exp\left[-\left(\frac{x - x_0}{\Delta x}\right)^2\right] dx = 1. \tag{1.76}$$

Since Equation 1.76 is true independently from the particular value of the scale $\Delta x$, it will hold in the limit where the scale goes to zero. We now focus on the second property of delta functions, and consider the integral

$$\int_{-\infty}^{+\infty} G(x, x_0, \Delta x) f(x) dx = \frac{1}{\sqrt{\pi}\Delta x} \int_{-\infty}^{+\infty} \exp\left[-\left(\frac{x - x_0}{\Delta x}\right)^2\right] f(x) dx. \tag{1.77}$$

To evaluate Equation 1.77 in the limit where $\Delta x \to 0$, we first Taylor-expand the arbitrary function $f(x)$ around $x = x_0$:

$$f(x) = \sum_{n=0}^{\infty} \frac{1}{n!} (x - x_0)^n \frac{d^n f}{dx^n}(x_0); \tag{1.78}$$

using this result into Equation 1.77, we have

$$\int_{-\infty}^{+\infty} G(x, x_0, \Delta x) f(x) dx = \sum_{n=0}^{\infty} \frac{d^n f}{dx^n}(x_0) \int_{-\infty}^{+\infty} \exp\left[-\left(\frac{x - x_0}{\Delta x}\right)^2\right] (x - x_0)^n \frac{dx}{\sqrt{\pi}\Delta x}. \tag{1.79}$$

The first term in the series can be isolated:

$$\int_{-\infty}^{+\infty} G(x, x_0, \Delta x) f(x) dx = f(x_0) + \sum_{n=1}^{\infty} \frac{d^n f}{dx^n}(x_0) \int_{-\infty}^{+\infty} \exp\left[-\left(\frac{x - x_0}{\Delta x}\right)^2\right] (x - x_0)^n \frac{dx}{\sqrt{\pi}\Delta x}; \tag{1.80}$$

the other terms can be evaluated by performing a simple change of variable, where $z = x - x_0$, which is a translation. We then have

$$\int_{-\infty}^{+\infty} G(x, x_0, \Delta x) f(x) dx = f(x_0) + \sum_{n=1}^{\infty} \frac{d^n f}{dx^n}(x_0) \int_{-\infty}^{+\infty} \exp\left[-\left(\frac{z}{\Delta x}\right)^2\right] z^n \frac{dz}{\sqrt{\pi}\Delta x}. \tag{1.81}$$

It is clear that for odd values of $n$, the integral is antisymmetrical and vanishes, because the Gaussian itself is an even function of $z$; for even values of $n$, one can use integration by parts:

$$\begin{aligned}\int_{-\infty}^{+\infty} \exp\left[-\left(\frac{z}{\Delta x}\right)^2\right] z^{2m} dz &= \left[-\frac{\Delta x^2}{2} \exp\left[-\left(\frac{z}{\Delta x}\right)^2\right] z^{2m-1}\right]_{-\infty}^{+\infty} \\ &\quad + \frac{2m-1}{2} \Delta x^2 \int_{-\infty}^{+\infty} \exp\left[-\left(\frac{z}{\Delta x}\right)^2\right] z^{2m-2} dz,\end{aligned} \tag{1.82}$$

and we see that the first term on the right-hand side of Equation 1.82 vanishes because the exponential tends to zero faster that the power diverges at infinity. The remaining integral now includes a lower-degree even polynomial in $z$, and we can repeat the integration by part. It is clear that this procedure can be applied to all the terms in the series to yield the following result:

$$\begin{aligned}&\sum_{m=1}^{\infty} \frac{d^{2m} f}{dx^{2m}}(x_0) \int_{-\infty}^{+\infty} \exp\left[-\left(\frac{z}{\Delta x}\right)^2\right] z^{2m} \frac{dz}{\sqrt{\pi}\Delta x} \\ &= \frac{1}{\sqrt{\pi}\Delta x} \sum_{m=1}^{\infty} \frac{d^{2m} f}{dx^{2m}}(x_0) \Delta x^{2m} \left(\frac{2m-1}{2}\right) \times \left(\frac{2m-3}{2}\right) \\ &\quad \times \cdots \times \frac{3}{2} \times \frac{1}{2} \int_{-\infty}^{+\infty} \exp\left[-\left(\frac{z}{\Delta x}\right)^2\right] dz.\end{aligned} \tag{1.83}$$

Since the Gaussian is normalized, we end up with

$$\begin{aligned}\int_{-\infty}^{+\infty} G(x, x_0, \Delta x) f(x) dx &= f(x_0) + \sum_{m=1}^{\infty} \frac{d^{2m} f}{dx^{2m}}(x_0) \Delta x^{2m} \left(\frac{2m-1}{2}\right) \\ &\quad \times \left(\frac{2m-3}{2}\right) \times \cdots \times \frac{3}{2} \times \frac{1}{2},\end{aligned} \tag{1.84}$$

and the limit of Equation 1.84 when $\Delta x \to 0$ yields the sought-after result.

We now consider the three-dimensional case, and the connection between Cartesian coordinates, where the three-dimensional delta-function takes the simple form

$$\delta^3(\mathbf{x}-\mathbf{x}_0) = \delta(x-x_0)\delta(y-y_0)\delta(z-z_0), \tag{1.85}$$

and spherical coordinates.

The relation between Cartesian and spherical coordinates is

$$\begin{aligned} x &= r\sin\theta\,\sin\varphi, \\ y &= r\sin\theta\,\cos\varphi, \\ z &= r\cos\theta, \\ r &= \sqrt{x^2+y^2+z^2}, \\ \theta &= \arccos\left(\frac{z}{r}\right), \\ \varphi &= \arctan\left(\frac{x}{y}\right). \end{aligned} \tag{1.86}$$

Therefore, the three-dimensional delta-function can be expressed in spherical coordinates as

$$\delta^3(\mathbf{x}-\mathbf{x}_0) = \frac{1}{r^2}\delta(r-r_0)\delta(\cos\theta-\cos\theta_0)\delta(\varphi-\varphi_0), \tag{1.87}$$

where $r_0$, $\theta_0$, and $\varphi_0$ are the radius, polar, and azimuthal angle of $\mathbf{x}_0$, respectively. We also note that the factor $r^{-2}$ has the correct dimensionality, since $[\delta(u - u_0)] \equiv [u]^{-1}$, where the brackets denote the unit of the bracketed quantity; this guarantees the normalization of the delta-function. However, the precise mathematical origin of this factor can be traced to the Jacobian of the transformation from Cartesian to spherical coordinates. In the most general case of $n$-dimensional space, we have

$$\begin{aligned} \delta(\mathbf{x}-\mathbf{a}) &= \delta(x_1-a_1)\delta(x_2-a_2)\ldots\delta(x_n-a_n) \\ &= \frac{1}{|J(x_i, x'_j)|}\delta(x'_1 - a'_1)\delta(x'_2 - a'_2)\ldots\delta(x'_n - a'_n), \end{aligned} \tag{1.88}$$

where $x_i$ and $x'_j$ are two different coordinate systems, and where the Jacobian of the transformation relating the coordinate systems is given by the $n \times n$ matrix

$$J(x_i, x'_j) = \left(\frac{\partial x_i}{\partial x'_j}\right). \tag{1.89}$$

The reason behind this is that the relevant quantity is the delta-function, multiplied by the differential volume, $\delta(\mathbf{x} - \mathbf{a})dx_1dx_2\ldots dx_n$, because the properties defining the generalized delta-function are expressed in terms of volume integrals:

$$\iiint \ldots \int \delta(\mathbf{x}-\mathbf{a})dx_1 dx_2 \ldots dx_n = 1, \qquad (1.90)$$
$$\iiint \ldots \int \delta(\mathbf{x}-\mathbf{a})f(\mathbf{x})dx_1 dx_2 \ldots dx_n = f(\mathbf{a}).$$

Therefore, we require that

$$\delta(x_1 - a_1)\ldots\delta(x_n - a_n)\, dx_1 \ldots dx_n = \delta(x_1' - a_1')\ldots\delta(x_n' - a_n')\, dx_1' \ldots dx_n'; \qquad (1.91)$$

since the differential volume elements are related by

$$dx_1 \ldots dx_n = \left|\frac{\partial x_i}{\partial x_j'}\right| dx_1' \ldots dx_n' = |J(x_i, x_j')| dx_1' \ldots dx_n', \qquad (1.92)$$

Equation 1.88 is correct.

Going back to the case of spherical coordinates, we can now use the Jacobian of the transformation,

$$J = \begin{pmatrix} \dfrac{\partial x}{\partial r} & \dfrac{\partial x}{\partial \theta} & \dfrac{\partial x}{\partial \varphi} \\ \dfrac{\partial y}{\partial r} & \dfrac{\partial y}{\partial \theta} & \dfrac{\partial y}{\partial \varphi} \\ \dfrac{\partial z}{\partial r} & \dfrac{\partial z}{\partial \theta} & \dfrac{\partial z}{\partial \varphi} \end{pmatrix} = \begin{pmatrix} \sin\theta\sin\varphi & r\cos\theta\sin\varphi & r\sin\theta\cos\varphi \\ \sin\theta\cos\varphi & r\cos\theta\cos\varphi & -r\sin\theta\sin\varphi \\ \cos\theta & -r\sin\theta & 0 \end{pmatrix}; \qquad (1.93)$$

we then find that $|J| = r^2$, which is the result sought. We also note in passing that for the inverse transformation, one can use the relation $|J^{-1}| = 1/|J|$.

Finally, when $\mathbf{x}_0 = \mathbf{0}$, the three-dimensional delta-function, as expressed in cylindrical coordinates, reduces to

$$\delta^3(\mathbf{x}) = \frac{1}{2\pi r^2}\delta(r), \qquad (1.94)$$

where the factor $1/2\pi$ results from the angular terms. From our previous discussion, it is easily seen that

$$\lim_{r_0 \to 0}\left\{\frac{1}{2\pi^{3/2} r_0 r^2}\exp\left[-\left(\frac{r}{r_0}\right)^2\right]\right\} = \delta(\mathbf{x}). \qquad (1.95)$$

Equation 1.95 supports the Gaussian charge distribution model given in Equation 1.31.

## 1.9 References for Chapter 1

Note: the numbers listed below refer to the main bibliography and reference sections at the end of this book.

1, 4, 8, 9, 11, 27, 30, 31, 32, 33, 35, 36, 44, 47, 48, 55, 56, 64, 69, 70, 71, 72, 73, 74, 76, 94, 96, 99, 100, 101, 102, 116, 120, 125, 144, 149, 152, 158, 162, 165, 194, 195, 209, 210, 213, 220, 221, 225, 250, 286, 298, 321, 324, 344, 345, 375, 416, 421, 455, 456, 458, 459, 460, 461, 469, 471, 524, 556, 557, 588, 793, 794, 795, 796, 797, 798, 799, 800, 801, 802, 803, 804, 805, 807.

# 2

# The Lorentz Transformation

## 2.1 Introduction

There are many different conceptual contexts within which one can introduce special relativity: historical, philosophical, mathematical, group theoretical, or physical, to name a few. Each approach has its own specific merits, and it is difficult to define an optimized, streamlined description of the field. Therefore, in this chapter, we will endeavor to give a broad view of the subject by first following the excellent, classic presentations of Pauli and Barut. The unusual approach taken by Schwinger and co-authors will also be reviewed, as it gives new insight on the deep connection between special relativity and electrodynamics. Finally, here and in Chapter 4, a number of important extensions of special relativity will be injected in the discussion to further broaden our overview of the foundations of electrodynamics. These include descriptions of spinors, dual tensors, and some mathematical tools also useful in general relativity.

The main physical fact underlying the theory of special relativity is the invariance of the speed of light under a change of inertial reference frame; this fact, which is theoretically borne out of Maxwell's equations, was first experimentally verified with precision by the well-known Michelson-Morley experiment. The reference frames used in special relativity are defined so that free particles, in the absence of external fields, move with constant velocities in such inertial or "Galilean," to use Einstein's terminology, reference frames. Near the end of the nineteenth century, whereas the laws of Newtonian mechanics were thought to obey Galilean transforms from one inertial frame to another, it became clear that Maxwell's equations could not be written in invariant form under such transformations, which Barut writes in the following form:

$$\mathbf{x}' = O(\mathbf{x}) + \mathbf{v}t, \quad t' = t. \tag{2.1}$$

Here, $O$ represents an arbitrary orthogonal transformation of the spatial coordinates; that is, a transformation conserving both lengths and angles, while $\mathbf{v}$ is the relative velocity between the two frames.

It is very important to stress that only the spatial coordinates are modified under a Galilean transform; this is directly related to the fact that, in Newtonian mechanics, time is absolute, and forces or interactions propagate instantaneously, in sharp contrast with Maxwell's theory. As Pauli writes: "as long ago as 1887, in a paper still written from the point of view of the elastic-solid theory of light, Voigt mentioned that it was mathematically convenient to introduce a local time $t'$ into a moving reference system. The origin of $t'$ was taken to be a linear function of the space coordinates, while the time scale was assumed to be unchanged. In this way the wave equation,

$$\Delta\phi - \frac{1}{c^2}\frac{\partial^2\phi}{\partial t^2} = 0, \tag{2.2}$$

could be made to remain valid in the moving reference system, too. These remarks, however, remained completely unnoticed, and a similar transformation was not again suggested until 1892 and 1895, when H. A. Lorentz published his fundamental papers on the subject." At that point, Lorentz formally introduced a transformation of the space and time coordinates under which Maxwell's equations remained unchanged. It is important to note that Lorentz's theory was still developed under the assumption that electromagnetic waves correspond to vibrations of an underlying substance permeating all space—the ether. As Pauli points out, Lorentz obtained physical results with a first formulation similar to that of Voigt; in particular, it was recognized that all first order effects in $\mathbf{v}/c$ that were experimentally observed could be explained within this approach, if the motion of electrons in the ether was accounted for. However, the results of the Michelson-Morley experiment, which was designed to probe second-order effects scaling as $(\mathbf{v}/c)^2$, could not be explained without further modifying the transformation by assuming that all bodies change their length along the direction of their translation velocity, $\mathbf{v}$. This new postulate, independently proposed by Lorentz and Fitzgerald, indicated that a contraction by a factor $\sqrt{1-(\mathbf{v}/c)^2}$ would be required to match the observed results consistently. While no physical explanation was given as to the origin of the Lorentz-Fitzgerald contraction, Lorentz included this new hypothesis in a set of coordinate transformation formulas:

$$\begin{aligned} x' &= \frac{x - vt}{\sqrt{1-(v/c)^2}}, \\ y' &= y, \\ z' &= z, \\ t' &= \frac{t-(vx/c^2)}{\sqrt{1-(v/c)^2}}. \end{aligned} \tag{2.3}$$

Here, $v$ is the relative velocity between the inertial frame $L$, with spatial coordinates $x$, $y$, $z$, and time coordinate $t$, and the inertial frame $L'$, with new spatial coordinates $x', y', z'$, and time coordinate $t'$.

The relative velocity between the two frames is along the $x$ and $x'$ axes, which are collinear. Note that when $v = 0$, $L$ and $L'$ completely coincide. The full physical significance of the Lorentz transformation presented in Equation 2.3 was not realized until Einstein formulated the theory of special relativity, completely abandoning the ether model in the process. We also note that the important modification in the transformation of the time coordinate is due to Larmor in 1900; this puts the temporal part of the Lorentz transform on an equal footing with its spatial counterpart and can be considered as a first step toward the unification of space and time into a single new concept. In 1904, using the transformation equations given in Equation 2.3, Lorentz was able to demonstrate that Maxwell's equations, in the absence of sources, were indeed invariant, provided that the strength of the electromagnetic fields was also properly transformed. He also demonstrated that a universal length contraction effect should be expected if one assumes that all masses transform like electromagnetic mass and that all interactions are electromagnetic in origin. Furthermore, this approach explained why optical experiments could not measure the motion of the earth within the postulated ether. However, a fundamental gap remained between Lorentz's formulation and the special theory of relativity, as the interpretation of Equation 2.3 was still linked to matter and interactions, instead of the underlying space–time continuum.

Poincaré further generalized the work of Lorentz by considering the behavior of Maxwell's equations, now including sources, under the coordinate transformation given in Equation 2.3, and proved their complete covariance. Furthermore, Poincaré clearly stated the relativity principle, requiring that the laws of physics be independent of the coordinate system used to express them; in other words, such laws must be valid in any inertial frame. In fact, one should not restrict this principle to Galilean frames; this is the foundation of the program of general relativity.

The deep paradigm shift induced by the theory of relativity was, however, introduced by Einstein in 1905. To quote Pauli: "It was Einstein, finally, who in a way completed the basic formulation of this new discipline. His paper of 1905 was submitted at almost the same time as Poincaré's article and had been written without previous knowledge of Lorentz's paper of 1904. It includes not only all the essential results contained in the two other papers, but shows an entirely novel, and much more profound, understanding of the whole problem." Indeed, Einstein's breakthrough mainly resides in his fundamental physical interpretation of the mathematical facts born out of the work of Lorentz and Poincaré, in relation to the electromagnetic theory of Maxwell. This approach begins with a single underlying axiom: the velocity of light is independent from the state of motion of the light source. From this basic physical fact of nature, Einstein proceeds to construct a complete, coherent, Lorentz-invariant theory unifying electromagnetism and classical mechanics. In modern physics, covariance, the frame-independent character of physical laws, plays a ubiquitous role: it is both a prerequisite condition and a guiding principle in the search for new, unified theories. The most profound aspect of the aforementioned paradigmatic shift is the new role played by space and time. Instead of an absolute background, spatial distances and temporal durations become relative, as they now fundamentally depend on the reference frame in which they are measured; this guarantees the invariance of natural laws under Lorentz transformations. We also note that in general relativity, the relative character of space–time becomes dynamical, mass distributions shaping the geometry of space–time, which, in turn influences the motion matter.

## 2.2 The Special Lorentz Transform

At this point, the derivation of the Lorentz transformation can be performed, starting from the postulate that the speed of light is the same in any inertial frame. We consider two such frames: $L$, with coordinates $x$, $y$, $z$, $t$, and $L'$, with primed coordinates $x'$, $y'$, $z'$, $t'$. The two frames are in uniform relative motion; furthermore, we chose the $x$- and $x'$-axis such that the velocity of $L'$, as measured in $L$, is $\mathbf{v} = v\hat{x}$; finally, the spatial origins and initial times are chosen to coincide. Because both frames are inertial, we require that uniform rectilinear motion observed in one frame must also be uniform and rectilinear in the other frame; this implies that the sought-after transformation be linear in the space–time coordinates. The propagation of a spherical light wave from the origin of reference frame $L$ can be modeled by the equation describing the expansion of a spherical shell at the speed of light:

$$x^2 + y^2 + z^2 = c^2 t^2. \tag{2.4}$$

The constancy of the speed of light requires that the same phenomenon, as viewed in inertial frame $L'$, satisfies the following equation:

$$x'^2 + y'^2 + z'^2 = c^2 t'^2. \tag{2.5}$$

As required by covariance, Equations 2.4 and 2.5 are identical. Equation 2.4 can be written in a more suggestive form by introducing the complex vector

$$x_1 = x, \quad x_2 = y, \quad x_3 = z, \quad x_0 = ict, \tag{2.6}$$

as first proposed by Poincaré. Within this context, Equation 2.4 now reads

$$x_1^2 + x_2^2 + x_3^2 + x_0^2 = x^2 + y^2 + z^2 + (ict)^2 = 0, \tag{2.7}$$

which indicates that the length of the so-called four-vector under consideration is null. By applying the same analogy to Equation 2.5, we see that the Lorentz transform must conserve the length of four-vectors, as defined above. In three-dimensional space, the coordinate transformations conserving length are translations and rotations; in four-dimensional space–time, the same statement is valid. Therefore, within a translation in four-dimensional space–time, that is a simple shift of the spatial origin and initial time, and a rotation in three-dimensional space, the sought-after Lorentz transform reduces to

$$\begin{aligned} x_1' &= x_1 \cos\alpha + x_0 \sin\alpha, \\ x_0' &= -x_1 \sin\alpha + x_0 \cos\alpha. \end{aligned} \tag{2.8}$$

To obtain the relation between the rotation angle $\alpha$ and the relative velocity between the inertial frames, we simply note that for the origin of the x-axis in $L'$, $x_1' = 0$, we must have $x = vt$; therefore,

$$\tan\alpha = i\frac{v}{c} = i\beta. \tag{2.9}$$

Using simple trigonometry, specifically,

$$\tan\alpha = \frac{\sin\alpha}{\cos\alpha} = \frac{\sqrt{1-\cos^2\alpha}}{\cos\alpha}, \tag{2.10}$$

we can rewrite Equation 2.9 as follows:

$$i\beta\cos\alpha = \sqrt{1-\cos^2\alpha}. \tag{2.11}$$

We may then solve Equation 2.11 to obtain

$$\begin{aligned} \cos\alpha &= \frac{1}{\sqrt{1-\beta^2}} = \gamma, \\ \sin\alpha &= \frac{i\beta}{\sqrt{1-\beta^2}} = i\gamma\beta, \end{aligned} \tag{2.12}$$

and we can write the special Lorentz transformation as

$$\begin{aligned} x' &= \gamma(x - \beta ct), \\ y' &= y, \\ z' &= z, \\ t' &= \gamma\left(t - \beta\frac{x}{c}\right). \end{aligned} \tag{2.13}$$

The inverse transformation is obtained by simply replacing $\beta$ by $-\beta$.

Since the Lorentz transform leaves the transverse dimension invariant, it is often useful to project the spatial components along the direction of the relative velocity by defining the following quantities:

$$x_{\parallel} = \frac{\mathbf{x}\cdot\mathbf{v}}{|\mathbf{v}|}, \quad \mathbf{x}_{\parallel} = x_{\parallel}\frac{\mathbf{v}}{|\mathbf{v}|}, \quad \mathbf{x}_{\perp} = \mathbf{x}-\mathbf{x}_{\parallel}; \tag{2.14}$$

the special Lorentz transform then takes the form

$$\begin{aligned} \mathbf{x}'_{\parallel} &= \gamma(\mathbf{x}_{\parallel}-\mathbf{v}t), \\ \mathbf{x}'_{\perp} &= \mathbf{x}_{\perp}, \\ t' &= \gamma\left(t-\frac{\mathbf{v}\cdot\mathbf{x}_{\parallel}}{c^2}\right) = \gamma\left(t-\frac{\mathbf{v}\cdot\mathbf{x}}{c^2}\right). \end{aligned} \tag{2.15}$$

We now proceed with a number of simple deductions from the transformation derived above; a second approach, which is more technical but also more general, is then presented in Section 2.13. This second derivation uses four-vectors extensively and represents a useful introduction to important mathematical tools that will be used widely in the remainder of this book.

## 2.3 Four-Vectors

The first important point to consider is the notation introduced by Minkowski. Here, the four-vector position is real,

$$x_{\mu} \equiv (x_0, \mathbf{x}) \equiv (ct, \mathbf{x}), \tag{2.16}$$

where the Greek subscript $\mu$ varies between 0 and 3. The 0 subscript refers to the temporal or time-like component of the four-vector, while the three other values of the subscript are related to the spatial or space-like component of the four-vector.

The length of this four-vector is now given by

$$x_{\mu}x^{\mu} = \mathbf{x}^2 - x_0^2 = x^2+y^2+z^2-c^2t^2; \tag{2.17}$$

where we have used Einstein's summation rule over repeated indices:

$$x_{\mu}x^{\mu} = x_1^2+x_2^2+x_3^2-x_0^2; \tag{2.18}$$

the minus sign corresponds to the distinction between covariant and contravariant four-vector, as explained below. Finally, in this case, the rotation angle corresponding to the special Lorentz transform is imaginary:

$$\begin{aligned} x_1' &= x_1 \cosh\varphi - x_0 \sinh\varphi, \\ x_0' &= -x_1 \sinh\varphi + x_0 \cosh\varphi, \end{aligned} \tag{2.19}$$

where we now have

$$\tanh\varphi = \beta. \tag{2.20}$$

Therefore, as explained by Pauli, while the Lorentz transform corresponds to a rotation in an imaginary coordinate system, in a real one it refers to the transformation of one pair of conjugate diameters of the invariant hyperbola

$$x_1^2 - x_0^2 = 1, \tag{2.21}$$

into another. This important fact is directly related to the fact that the geometry of space–time in special relativity is flat and hyperbolic. If we use the imaginary notation introduced by Poincaré, the scalar product of a four-vector is simply given by

$$x_\mu x^\mu = \mathbf{x}^2 + x_0^2 = \mathbf{x}^2 + (ict)^2 = \mathbf{x}^2 - c^2t^2; \tag{2.22}$$

by contrast, using the real Minkowski notation, we have

$$x_\mu x^\mu = \mathbf{x}^2 - x_0^2 = \mathbf{x}^2 - c^2t^2. \tag{2.23}$$

This distinction is critical, when understood in the context of covariant and contravariant four-vectors. The relation between covariant and contravariant four-vectors is provided by the metric

$$x_\mu = g_{\mu\nu} x^\nu, \tag{2.24}$$

where we have used Einstein's notation and introduced the diagonal metric tensor

$$g_{\mu\nu} = \frac{\partial x_\mu}{\partial x^\nu} = \begin{pmatrix} -1 & 0 & 0 & 0 \\ 0 & 1 & 0 & 0 \\ 0 & 0 & 1 & 0 \\ 0 & 0 & 0 & 1 \end{pmatrix}. \tag{2.25}$$

This tensor plays a fundamental role in general relativity, as it describes the curvature of space–time. In special relativity, the diagonal nature of the metric indicates that space–time is flat, while the minus sign of the $g_{00}$ component corresponds to the hyperbolic character of space–time. For a given four-vector, $w_\mu$, the relation between the covariant and contravariant time-like components is $w^0 = -w_0$. This, in turn, guarantees the proper form for the scalar product of two four-vectors:

$$a_\mu b^\mu = a_\mu(g^{\mu\nu}b_\nu) = \mathbf{a}\cdot\mathbf{b} - a_0 b_0. \tag{2.26}$$

In particular, the length of a four-vector is

$$w_\mu w^\mu = \mathbf{w}^2 - w_0^2. \tag{2.27}$$

These considerations lead to a simple rule in flat, hyperbolic space–time: for an arbitrary tensor, the raising or lowering of an index results in a change of sign.

Another extremely important property of four-vectors is the fact that they transform from one inertial frame to another according to the Lorentz transform. Furthermore, since the basic characteristic of the Lorentz transform is that it is a rotation in flat, hyperbolic space–time conserving length and angles, the scalar product of two four-vectors is a conserved quantity. Let us demonstrate this essential aspect of the theory. We consider two four-vectors, $a_\mu = (a_0, \mathbf{a})$, and $b_\mu = (b_0, \mathbf{b})$. Applying the special Lorentz transform to both four-vectors, we have

$$\begin{aligned} a'_1 &= \gamma(a_1 - \beta a_0), \\ a'_2 &= a_2, \\ a'_3 &= a_3, \\ a'_0 &= \gamma(a_0 - \beta a_1); \end{aligned} \tag{2.28}$$

and similar formulae for $b'_\mu$. Let us now derive the scalar product of $a'_\mu$ and $b'_\mu$:

$$\begin{aligned} a'_\mu b'^\mu &= \mathbf{a}'\cdot\mathbf{b}' - a'_0 b'_0 = a'_1 b'_1 + a'_2 b'_2 + a'_3 b'_3 - a'_0 b'_0 \\ &= \gamma(a_1 - \beta a_0)\gamma(b_1 - \beta b_0) + a_2 b_2 + a_3 b_3 - \gamma(a_0 - \beta a_1)\gamma(b_0 - \beta b_1) \\ &= \gamma^2(a_1 b_1 - \beta a_1 b_0 - \beta a_0 b_1 + \beta^2 a_0 b_0 - a_0 b_0 + \beta a_0 b_1 + \beta a_1 b_0 - \beta^2 a_1 b_1) \\ &\quad + a_2 b_2 + a_3 b_3 \\ &= \gamma^2(1 - \beta^2)(a_1 b_1 - a_0 b_0) + a_2 b_2 + a_3 b_3 \\ &= a_1 b_1 + a_2 b_2 + a_3 b_3 - a_0 b_0 = a_\mu b^\mu. \end{aligned} \tag{2.29}$$

In Equation 2.27, we have used the fact that $\gamma^2(1-\beta^2)=1$; this, in turn is directly related to the rotation angle in hyperbolic space–time, $\varphi = \text{argtanh}(\beta)$, and the well-known relation $\cosh^2\varphi - \sinh^2\varphi = 1$. The quantity resulting from taking the scalar product of two four-vectors is generally referred to as a scalar, which is an invariant under the Lorentz transform.

Another quantity of great interest is the proper time; that is time as measured in a frame where the system or particle under consideration is at rest. Proper time is very important, as it allows one to derive new four-vectors from the four-position. The simplest derivation of the proper time, $\tau$, can be performed by considering a frame, $L'$, where the object is at rest; furthermore we choose the spatial origin of that frame to coincide with the position of the particle; using the special Lorentz transform, we have

$$\begin{aligned} x' &= 0 = \gamma(x - \beta ct), \\ t' &= \tau = \gamma\left(t - \beta\frac{x}{c}\right), \end{aligned} \tag{2.30}$$

and the proper time is given by

$$\tau = \gamma(t - \beta^2 t) = t\sqrt{1-\beta^2} = \frac{t}{\gamma}. \tag{2.31}$$

Another, more sophisticated, definition of the proper time can be given by considering the world line of the particle; that is, its four-position measured in an inertial rest frame and parameterized by a Lorentz-invariant quantity: $x_\mu(s)$, where the length of the Lorentz-invariant world-line is defined by the differential equation

$$ds^2 = dx_\mu dx^\mu = dx_1^2 + dx_2^2 + dx_3^2 - dx_0^2. \tag{2.32}$$

The fact that the length, $s = \int\sqrt{ds^2}$, is invariant under the Lorentz transformation is manifest, as it corresponds to the contraction of covariant and contravariant differential elements. The physical meaning of $s$ can best be understood by considering a frame where the particle is at rest. In this frame, the spatial differential element is null: $d\mathbf{x} = \mathbf{0}$, and $ds^2$ reduces to

$$ds^2 = -dx_0^2 = -c^2dt^2 = -c^2d\tau^2; \tag{2.33}$$

the last identity resulting from the fact that Equation 2.33 is valid in the rest frame of the object. Equation 2.33 is easily integrated to yield

$$s = ic(t - t_0) = ic(\tau - \tau_0), \tag{2.34}$$

which clearly shows that $s$ is directly related to the proper time.

We can now define the four-velocity and the four-acceleration. If we derive the four-position with respect to the proper time, we obtain the four-velocity:

$$u_\mu = \frac{dx_\mu}{d\tau} = \frac{d}{d\tau}(x_0, \mathbf{x}) = \left(c\frac{dt}{d\tau}, \frac{d\mathbf{x}}{d\tau}\right) = c\frac{dt}{d\tau}\left(1, \frac{d\mathbf{x}}{cdt}\right) = c\frac{dt}{d\tau}(1, \beta). \quad (2.35)$$

The derivative of the regular time with respect to the proper time is obtained by differentiating Equation 2.31:

$$\frac{dt}{d\tau} = \gamma; \quad (2.36)$$

the four-velocity then reads

$$u_\mu = \frac{dx_\mu}{d\tau} = (u_0, \mathbf{u}) = u_0(1, \beta) = c\gamma(1, \beta). \quad (2.37)$$

The length of the four-velocity can now be calculated:

$$u_\mu u^\mu = \mathbf{u}^2 - u_0^2 = (\gamma^2\beta^2 - \gamma^2)c^2 = -c^2, \quad (2.38)$$

and we note that the four-acceleration, which is defined as

$$a_\mu = \frac{du_\mu}{d\tau} = \frac{d^2 x_\mu}{d\tau^2}, \quad (2.39)$$

is orthogonal to the four-acceleration:

$$\frac{d}{d\tau}(u_\mu u^\mu) = 2u_\mu \frac{du^\mu}{d\tau} = 2u_\mu a^\mu = \frac{d}{d\tau}(-c^2) = 0. \quad (2.40)$$

This result corresponds to the fact that the derivative of a vector with constant length is always perpendicular to the vector. From Equations 2.37 and 2.39, we can now determine the explicit expression of the four-velocity:

$$\begin{aligned} a_\mu &= \frac{du_\mu}{d\tau} = \frac{d}{d\tau}[c\gamma(1, \beta)] = c\left[\frac{d\gamma}{d\tau}, \left(\frac{d\gamma}{d\tau}\beta + \gamma\frac{d\beta}{d\tau}\right)\right] \\ &= c\frac{dt}{d\tau}\left[\frac{d\gamma}{dt}, \left(\frac{d\gamma}{dt}\beta + \gamma\frac{d\beta}{dt}\right)\right] = c\gamma[\dot{\gamma}, (\dot{\gamma}\beta + \gamma\dot{\beta})]. \end{aligned} \quad (2.41)$$

Here, we have again changed variable and used Equation 2.36; we have also introduced the three-acceleration normalized to $c$, which is represented by

$\dot{\beta} = d\mathbf{v}/cdt$; finally the derivative of $\gamma$ with respect to time can be evaluated in terms of $\dot{\beta}$ by differentiating Equation 2.38:

$$\dot{\gamma} = \frac{d\gamma}{dt} = \frac{d}{dt}(1-\beta^2)^{-1/2} = -\frac{1}{2}(1-\beta^2)^{-3/2}(-2\beta\cdot\dot{\beta}) = \gamma^3(\beta\cdot\dot{\beta}). \quad (2.42)$$

Using this last result, we can recast the four-acceleration as

$$a_\mu = c\gamma\{\gamma^3(\beta\cdot\dot{\beta}), [\gamma^3\beta(\beta\cdot\dot{\beta}) + \gamma\dot{\beta}]\}. \quad (2.43)$$

Returning to the four-velocity, it is seen easily that it corresponds to the particle four-momentum, normalized to $m_0$, which is the rest mass of the particle. Indeed,

$$m_0 u_\mu = \gamma m_0 c(1, \beta) = \left(\frac{E}{c}, \mathbf{p}\right) = (p_0, \mathbf{p}) = p_\mu. \quad (2.44)$$

Here, the spatial components of the four-momentum are given by $\mathbf{p} = \gamma m_0 c\beta = \gamma m_0 \mathbf{v} = m\mathbf{v}$; while its time-like component corresponds to the energy: $p_0 = \gamma m_0 c = mc^2/c$.

Note that in order to guarantee that all the components of a four-vector have the same dimension or units, it is customary to multiply or divide the time-like component of four-vectors by $c$.

## 2.4 Addition of Velocities

We now consider the transformation of velocities. Having established the relation between the four-velocity and the regular velocity, this becomes easy because we know how to transform four-vectors using the Lorentz transformation. In the inertial frame $L$, we consider the four-velocity, $u_\mu = (u_0, \mathbf{u})$; transforming to another inertial frame $L'$, we have

$$\begin{aligned} u_1' &= \gamma(u_1 - \beta u_0), \\ u_2' &= u_2, \\ u_3' &= u_3, \\ u_0' &= \gamma(u_0 - \beta u_1). \end{aligned} \quad (2.45)$$

Here, $\beta = v/c$ corresponds to the relative velocity of the frame $L'$, as measured in $L$, and $\gamma = (1-\beta^2)^{-1/2}$; this is not to be confused with the four-velocity components of $u_\mu$, as measured in $L$, where we have

$$u_0 = \frac{c}{\sqrt{1-\left(\frac{\mathbf{w}}{c}\right)^2}}, \tag{2.46}$$

for the time-like component of $u_\mu$, and

$$\begin{aligned} u_1 &= u_\parallel = u_0 \frac{w_\parallel}{c} = u_0 \frac{\mathbf{w} \cdot \mathbf{v}}{c|\mathbf{v}|}, \\ \mathbf{u}_\perp &= u_0 \frac{\mathbf{w}_\perp}{c} = u_0 \left( \frac{\mathbf{w}}{c} - \frac{\mathbf{v}}{|\mathbf{v}|} \frac{w_\parallel}{c} \right), \end{aligned} \tag{2.47}$$

for its spatial components. Here, **w** is the regular three-velocity of the particle under consideration, as measured in $L$; again, it should not be confused with the relative velocity between the inertial frames, **v**. To determine the transform of **w**, we can now use the transformation formulae given in Equation 2.45, together with Equations 2.46 and 2.47, and the relation between the four-velocity and the regular velocity,

$$\mathbf{w} = c\frac{\mathbf{u}}{u_0}, \quad \mathbf{w}' = c\frac{\mathbf{u}'}{u_0'}. \tag{2.48}$$

We begin by separating the spatial components of the four-velocity, as measured in $L$, into a parallel and a transverse component:

$$\begin{aligned} u_1 &= u_\parallel = u_0 \frac{w_\parallel}{c} = u_0 \frac{\mathbf{w} \cdot \mathbf{v}}{cv}, \\ \mathbf{u}_\perp &= u_0 \frac{\mathbf{w}_\perp}{c} = u_0 \left( \frac{\mathbf{w}}{c} - \frac{\mathbf{v}}{v} \frac{w_\parallel}{c} \right), \end{aligned} \tag{2.49}$$

where $v = c\beta$ is the modulus of the relative velocity; applying the special Lorentz transform, as given in Equation 2.45, we then have

$$\begin{aligned} u_1' &= \gamma\left( u_0 \frac{\mathbf{w} \cdot \mathbf{v}}{cv} - \beta u_0 \right) = \gamma u_0 \left( \frac{\mathbf{w} \cdot \mathbf{v}}{cv} - \beta \right), \\ u_0' &= \gamma\left( u_0 - \beta u_0 \frac{\mathbf{w} \cdot \mathbf{v}}{cv} \right) = \gamma u_0 \left( 1 - \frac{\mathbf{w} \cdot \mathbf{v}}{c^2} \right). \end{aligned} \tag{2.50}$$

We now use Equation 2.50 to first derive the parallel component of $\mathbf{w}'$,

$$w_1' = w_\parallel' = c\frac{u_1'}{u_0'} = \frac{\left( \frac{\mathbf{w} \cdot \mathbf{v}}{v} - v \right)}{\left( 1 - \frac{\mathbf{w} \cdot \mathbf{v}}{c^2} \right)}; \tag{2.51}$$

on the other hand, the transverse components can be grouped as follows:

$$w'_{2,3} = c\,\frac{u'_{2,3}}{u'_0} = c\,\frac{u_{2,3}}{\gamma u_0\left(1-\frac{\mathbf{w}\cdot\mathbf{v}}{c^2}\right)} = \frac{w_{2,3}}{\gamma\left(1-\frac{\mathbf{w}\cdot\mathbf{v}}{c^2}\right)},$$
$$\mathbf{w}'_\perp = \frac{\mathbf{w}_\perp}{\gamma\left(1-\frac{\mathbf{w}\cdot\mathbf{v}}{c^2}\right)}. \tag{2.52}$$

The results given in Equations 2.52 and 2.53 can be consolidated into a single equation by proceeding as follows. The transform of the three-velocity is

$$\mathbf{w}' = w'_\parallel \frac{\mathbf{v}}{v} + w'_\perp; \tag{2.53}$$

using Equations 2.52 and 2.53, we find

$$\mathbf{w}' = \frac{1}{\left(1-\frac{\mathbf{w}\cdot\mathbf{v}}{c^2}\right)}\left[\frac{\mathbf{v}}{v}\left(\frac{\mathbf{w}\cdot\mathbf{v}}{v}-v\right)+\frac{\mathbf{w}_\perp}{\gamma}\right], \tag{2.54}$$

which yields

$$\mathbf{w}' = \frac{1}{\left(1-\frac{\mathbf{w}\cdot\mathbf{v}}{c^2}\right)}\left[\frac{\mathbf{w}}{\gamma}-\frac{w_\parallel}{\gamma}\frac{\mathbf{v}}{v}+\frac{\mathbf{v}}{v}\left(\frac{\mathbf{w}\cdot\mathbf{v}}{v}-v\right)\right], \tag{2.55}$$

from which we finally obtain

$$\mathbf{w}' = \frac{1}{\left(1-\frac{\mathbf{w}\cdot\mathbf{v}}{c^2}\right)}\left\{\frac{\mathbf{w}}{\gamma}+\mathbf{v}\left[\frac{\mathbf{w}\cdot\mathbf{v}}{v^2}\left(1-\frac{1}{\gamma}\right)-1\right]\right\}. \tag{2.56}$$

A second, more direct approach is given by Einstein's addition theorem for velocities. Let us consider an arbitrary motion in the inertial frame $L$: $\mathbf{x}(t)$; there is then a corresponding motion $\mathbf{x}'(t')$ in another inertial frame $L'$ in uniform relative motion with respect to $L$. The three-velocities in $L$ and $L'$ are defined as

$$\mathbf{w} = \frac{d\mathbf{x}}{dt},\quad \mathbf{w}' = \frac{d\mathbf{x}'}{dt'}; \tag{2.57}$$

from the special Lorentz transform, given in Equation 2.13, we have

$$\begin{aligned} dx' &= \gamma(dx - \beta c dt), \\ dy' &= dy, \\ dz' &= dz, \\ dt' &= \gamma\left(dt - \beta \frac{dx}{c}\right); \end{aligned} \tag{2.58}$$

which can then be used to determine the relation between **w** and **w**′: we first write

$$\mathbf{w}' = \frac{d\mathbf{x}'}{dt'} = \frac{d\mathbf{x}'}{dt}\frac{dt}{dt'}, \tag{2.59}$$

and evaluate

$$\frac{dt}{dt'} = \left(\frac{dt'}{dt}\right)^{-1}, \quad \frac{dt'}{dt} = \gamma\left(1 - \beta\frac{dx}{cdt}\right) = \gamma\left(1 - \beta\frac{w_x}{c}\right) = \gamma\left(1 - \frac{\mathbf{w}\cdot\mathbf{v}}{c^2}\right); \tag{2.60}$$

we now calculate the components of $d\mathbf{x}'/dt$, with the result that

$$\begin{aligned} w'_x &= \frac{dx'}{dt}\frac{dt}{dt'} = \frac{\gamma\left(\frac{dx}{dt} - \beta c\right)}{\gamma\left(1 - \frac{\mathbf{w}\cdot\mathbf{v}}{c^2}\right)} = \frac{w_x - v}{\left(1 - \frac{\mathbf{w}\cdot\mathbf{v}}{c^2}\right)}, \\ w'_y &= \frac{dy'}{dt}\frac{dt}{dt'} = \frac{\frac{dy}{dt}}{\gamma\left(1 - \frac{\mathbf{w}\cdot\mathbf{v}}{c^2}\right)} = \frac{w_y}{\gamma\left(1 - \frac{\mathbf{w}\cdot\mathbf{v}}{c^2}\right)}, \\ w'_z &= \frac{dz'}{dt}\frac{dt}{dt'} = \frac{\frac{dz}{dt}}{\gamma\left(1 - \frac{\mathbf{w}\cdot\mathbf{v}}{c^2}\right)} = \frac{w_z}{\gamma\left(1 - \frac{\mathbf{w}\cdot\mathbf{v}}{c^2}\right)}. \end{aligned} \tag{2.61}$$

Again, this result can be grouped to yield the sought-after result in vector form by using the fact that $\mathbf{w} = \mathbf{w}_\perp + w_x \frac{\mathbf{v}}{v}$:

$$\mathbf{w}' = \frac{\left[(w_x - v)\frac{\mathbf{v}}{v} + \frac{\mathbf{w}_\perp}{\gamma}\right]}{\left(1 - \frac{\mathbf{w}\cdot\mathbf{v}}{c^2}\right)} = \frac{1}{\left(1 - \frac{\mathbf{w}\cdot\mathbf{v}}{c^2}\right)}\left\{\frac{\mathbf{w}}{\gamma} + \mathbf{v}\left[\frac{\mathbf{w}\cdot\mathbf{v}}{v^2}\left(1 - \frac{1}{\gamma}\right) - 1\right]\right\}. \tag{2.62}$$

This last result is identical to that given in Equation 2.56 and is valid for a Lorentz transformation without rotation; that is, when the axes of $L$ and $L'$ are aligned.

If we introduce the angle $\theta$ between $\mathbf{w}$ and the relative velocity, we can use Equation 2.62 to obtain

$$\begin{aligned} w' \cos\theta' &= \frac{w\cos\theta - v}{1 - \frac{uv}{c^2}\cos\theta}, \\ w' \sin\theta' &= \frac{w\sin\theta}{\gamma\left(1 - \frac{uv}{c^2}\cos\theta\right)}. \end{aligned} \tag{2.63}$$

Here, $\theta'$ is the angle between $\mathbf{w}'$ and the relative velocity. The tangent of this angle is

$$\tan\theta' = \frac{\sin\theta}{\gamma\left(\cos\theta - \frac{v}{w}\right)}; \tag{2.64}$$

using simple trigonometry, we have

$$w' = w\frac{\sqrt{\left(\frac{v}{w} - \cos\theta\right)^2 + \left(\frac{\sin\theta}{\gamma}\right)^2}}{1 - \frac{wv}{c^2}\cos\theta}, \tag{2.65}$$

and the inverse relation is obtained by switching the sign of the relative velocity

$$w = w'\frac{\sqrt{\left(\frac{v}{w} + \cos\theta\right)^2 + \left(\frac{\sin\theta}{\gamma}\right)^2}}{1 + \frac{wv}{c^2}\cos\theta}. \tag{2.66}$$

Let us see what happens when $w = c$. Using Equation 2.65, we have

$$\begin{aligned} w' &= c\frac{\sqrt{(\beta - \cos\theta)^2 + \left(\frac{\sin\theta}{\gamma}\right)^2}}{1 - \beta\cos\theta} \\ &= c\frac{\sqrt{\beta^2 - 2\beta\cos\theta + \cos^2\theta + \frac{\sin^2\theta}{\gamma^2}}}{1 - \beta\cos\theta} \\ &= c\frac{\sqrt{\beta^2 - 2\beta\cos\theta + \cos^2\theta + \sin^2\theta(1 - \beta^2)}}{1 - \beta\cos\theta} \\ &= c\frac{\sqrt{\beta^2(1 - \sin^2\theta) - 2\beta\cos\theta + 1}}{1 - \beta\cos\theta} = c, \end{aligned} \tag{2.67}$$

which shows that $w' = c$, independent of $\theta$. Therefore, the speed of light is, indeed, invariant, as required by the principle of relativity: if a velocity is measured to be equal to $c$ in a given inertial frame, then it has this value in any inertial frame.

Finally, when $\mathbf{w}$ is parallel to the relative velocity, we have $\theta = 0$ and $\mathbf{w}'$ is also parallel to $\mathbf{v}$; furthermore, Equations 2.65 and 2.66 take a very simple form:

$$w' = \frac{w - v}{1 - \frac{wv}{c^2}}, \quad w = \frac{w' + v}{1 + \frac{wv}{c^2}}. \tag{2.68}$$

## 2.5 Four-Acceleration and Hyperbolic Motion

As discussed in Section 2.3, the four-acceleration is defined as the second-order derivative of the four-position with respect to the proper time:

$$a_\mu = \frac{d^2 x_\mu}{d\tau^2} = \frac{du_\mu}{d\tau}. \tag{2.69}$$

In relativistic kinematics the question of what constitutes uniform acceleration arises. We will define a motion for which the acceleration is constant, and has the magnitude $a$, in a reference frame $K$ moving with the particle or object under consideration, as uniformly accelerated. Of course, the reference frame is not inertial; however, at any given point in time, there exists an inertial frame, $L$, instantaneously coinciding with $K$. In the Galilean frame $L$, however, the acceleration of such a motion is not constant in time.

Within the framework of four-vectors, this problem can be solved in a very simple, elegant manner. We define a uniformly accelerated motion by requiring that the Lorentz-invariant length of the four-acceleration be a constant of the motion. This translates into the following equation:

$$a_\mu a^\mu = a^2 = \mathbf{a}^2 - a_0^2 = \left(\frac{d\mathbf{u}}{d\tau}\right)^2 - \left(\frac{du_0}{d\tau}\right)^2 = \left(\frac{d\mathbf{u}}{d\tau}\right)^2 - c^2\left(\frac{d\gamma}{d\tau}\right)^2; \tag{2.70}$$

furthermore, we already know that the length of the four-velocity is constant, as

$$u_\mu u^\mu = -c^2. \tag{2.71}$$

If we normalize Equations 2.70 and 2.71 to the speed of light, we can rearrange the problem in a very suggestive form. We seek a four-vector,

$w_\mu(\tau) = u_\mu/c = [\gamma, (\mathbf{u}/c)]$, such that

$$w_0^2 - \mathbf{w}^2 = 1,$$
$$\left(\frac{d\mathbf{w}}{d\tau}\right)^2 - \left(\frac{dw_0}{d\tau}\right)^2 = \left(\frac{a}{c}\right)^2. \tag{2.72}$$

The four-vector $w_\mu$ is called the unitary four-velocity and will be used extensively in this book. For now, it is clear that the solution to Equation 2.72 takes the form

$$w_\mu(\tau) = \left[\cosh\left(\frac{a\tau}{c}\right), \mathbf{n}\sinh\left(\frac{a\tau}{c}\right)\right], \tag{2.73}$$

provided that $\mathbf{n}$ is a constant, arbitrary vector of unit length. Indeed, we first have

$$w_0^2 - \mathbf{w}^2 = \cosh^2\left(\frac{a\tau}{c}\right) - \mathbf{n}^2\sinh^2\left(\frac{a\tau}{c}\right) = 1, \tag{2.74}$$

and we can now derive each component of $w_\mu$ with respect to the proper time, to obtain

$$\frac{dw_0}{d\tau} = \left(\frac{a}{c}\right)\sinh\left(\frac{a\tau}{c}\right),$$
$$\frac{d\mathbf{w}}{d\tau} = \left(\frac{a}{c}\right)\left[\frac{d\mathbf{n}}{d\tau}\sinh\left(\frac{a\tau}{c}\right) + \mathbf{n}\cosh\left(\frac{a\tau}{c}\right)\right] = \mathbf{n}\left(\frac{a}{c}\right)\cosh\left(\frac{a\tau}{c}\right). \tag{2.75}$$

It is then easily verified that Equation 2.72 is satisfied; furthermore, this solution can be boosted in any Galilean frame by using the appropriate Lorentz transform, and any set of initial conditions can be met in that manner. If the initial value of the normalized four-velocity is measured to be $w'_{\mu 0}$ in a given reference frame, it can easily be transformed away to match the initial value corresponding to our solution, namely, $w_0 = (1,\mathbf{0})$, which indicates that our reference frame initially coincides with the instantaneous rest frame of the particle.

Finally, two points are noteworthy. First, the choice of associating the hyperbolic cosine function to the normalized energy, $\gamma = w_0$, and the hyperbolic sine to the normalized velocity, $\mathbf{w} = \beta$, can easily be remembered by considering the fact that the minimum energy is the rest energy, which is obtained for $\gamma = 1$ and $\mathbf{w} = \mathbf{0}$. Second, we can integrate the solution given in Equation 2.73 to obtain the world line of a uniformly accelerated particle.

We start from the definition,

$$x_\mu = x_{\mu 0} + \int_0^\tau \frac{dx_\mu}{d\tau'} d\tau' = x_{\mu 0} + \int_0^\tau u_\mu(\tau')\, d\tau'; \tag{2.76}$$

using the result obtained in Equation 2.73, we find that

$$x_\mu(\tau) = x_{\mu 0} + \frac{c^2}{a}\left\{ \sinh\left(\frac{a\tau}{c}\right), \mathbf{n}\left[\cosh\left(\frac{a\tau}{c}\right) - 1\right]\right\}. \tag{2.77}$$

Closer inspection of Equation 2.77 reveals the origin of the terminology of hyperbolic motion used here. By shifting the origin of space–time by (0, $-\mathbf{n}$), we can recast this result as

$$[x_\mu(\tau) - x_{\mu 0}][x^\mu(\tau) - x_0^\mu] = \left(\frac{c^2}{a}\right)^2 \left[\mathbf{n}^2 \cosh^2\left(\frac{a\tau}{c}\right) - \sinh^2\left(\frac{a\tau}{c}\right)\right] = \frac{c^4}{a^2}, \tag{2.78}$$

or

$$[\mathbf{x}(\tau) - \mathbf{x}_0]^2 - c^2(t - t_0)^2 = \frac{c^4}{a^2}, \tag{2.79}$$

which describes a hyperbola.

To appreciate the conceptual simplicity of this derivation using the normalized or unitary four-velocity, we can compare it to the more traditional presentation sometimes used elsewhere. If we restrict ourselves to rectilinear, uniformly accelerated motion, we can use the relation between the four-acceleration and the regular acceleration given in Equation 2.43 and the definition given above to derive the equation of motion for uniform acceleration in an inertial frame. We choose the x-axis to coincide with the direction of the three-velocity, $\mathbf{w}$, and three-acceleration, $\dot{\mathbf{w}}$.

In the so-called instantaneous rest frame, we have, by definition

$$\mathbf{w}' = \mathbf{0}, \quad \mathbf{a}' = \dot{\mathbf{w}}', \quad a_0' = 0. \tag{2.80}$$

In the other frame, we have

$$\begin{aligned} \mathbf{a} &= \gamma^2 \dot{\mathbf{w}} + \gamma^4 \mathbf{w}\left(\frac{\mathbf{w}\cdot\dot{\mathbf{w}}}{c^2}\right), \\ a_0 &= \gamma^4 \left(\frac{\mathbf{w}\cdot\dot{\mathbf{w}}}{c^2}\right). \end{aligned} \tag{2.81}$$

Furthermore, using the Lorentz transform, we can express the components of the four-acceleration in one frame in terms of their counterparts in the other frame:

$$\begin{aligned} a_x &= \gamma a'_x, \\ a_y &= a'_y, \\ a_z &= a'_z, \\ a_0 &= \gamma\beta a'_x. \end{aligned} \tag{2.82}$$

We also note that in this case, there is an identification between the particle velocity and the relative velocity between the instantaneous rest frame and the Galilean frame used in our calculations. Returning to Equation 2.81, we also have

$$\begin{aligned} a_x &= \gamma^2 \dot{w}_x + a_0 \frac{w_x}{c}, \\ a_0 &= \gamma^4 \left( \frac{w_x \dot{w}_x}{c} \right). \end{aligned} \tag{2.83}$$

Using Equations 2.80, 2.82, and 2.83, we can now relate the components of the three-acceleration in each frame:

$$\begin{aligned} \gamma \dot{w}'_x &= \gamma^2 \dot{w}_x \left( 1 + \gamma^2 \frac{w_x^2}{c^2} \right) = \gamma^4 \dot{w}_x, \\ \dot{w}'_y &= \gamma^2 \dot{w}_y, \\ \dot{w}'_z &= \gamma^2 \dot{w}_z. \end{aligned} \tag{2.84}$$

From Equation 2.84, follow the relations found in Einstein's first paper on special relativity:

$$\begin{aligned} \dot{w}_x &= \gamma^{-3} \dot{w}'_x = \dot{w}'_x \,(1-\beta^2)^{3/2}, \\ \dot{w}_y &= \gamma^{-2} \dot{w}'_y = \dot{w}'_y \,(1-\beta^2), \\ \dot{w}_z &= \gamma^{-2} \dot{w}'_z = \dot{w}'_z \,(1-\beta^2). \end{aligned} \tag{2.85}$$

Following Pauli, one can integrate Equation 2.85 to obtain

$$[x - x(t_0)]^2 - c^2(t-t_0)^2 = \frac{c^4}{a^2}, \tag{2.86}$$

where $a$ is the constant value of the acceleration. Furthermore, if we choose the origin of the trajectory such that

$$t_0 = 0, \quad w_x(t_0) = \dot{x}(t_0) = 0, \quad x(t_0) = \frac{c^2}{a}, \tag{2.87}$$

we end up with the equation for a hyperbola:

$$x^2 - c^2t^2 = \frac{c^4}{a^2}. \tag{2.88}$$

Finally, using the proper time, we can rewrite Equation 2.88 as

$$x = \frac{c^2}{a}\cosh\left(\frac{a\tau}{c}\right), \quad t = \frac{c}{a}\sinh\left(\frac{a\tau}{c}\right). \tag{2.89}$$

We note that more complex cases can be reduced to the one presented here by means of a Lorentz transformation with rotation.

## 2.6 Variation of the Mass with Velocity

In this section, the well-known relation between mass and velocity is derived by considering the expression of the equations of motion within the context of special relativity. There are two main approaches to this problem. On the one hand, one can focus on a purely electrodynamical system and introduce the Minkowski force density; on the other hand, one can carefully consider the conservation of energy and momentum in two different Galilean frames. The latter formalism being more general, we will use this method, originally proposed by Lewis and Toleman. We begin by examining the concept of momentum. For a particle moving slowly compared to the speed of light, we have

$$\mathbf{p} = m\mathbf{w}, \quad K = \frac{1}{2}m\mathbf{w}^2, \tag{2.90}$$

where $\mathbf{p}$ is the momentum of the particle, $m$ is its mass, $\mathbf{w}$ is its velocity, and $K$ is its kinetic energy. Equation 2.90 represents the limiting case for low velocities; it is clear that for relativistic velocities, the generalization of the momentum must take the form

$$\mathbf{p} = m(w)\mathbf{w}, \tag{2.91}$$

because of the isotropy of space.

We now seek expressions of the energy and momentum consistent with the Lorentz transformation and consider the scattering of two point particles of equal masses, as observed in two Galilean frames moving with relative velocity $\mathbf{v} = v\hat{x}$, aligned with the x-axis of each frame. Before the collision (superscript $-$), particle 1 moves with the velocity

$$\begin{aligned} w_{x1}^{-} &= 0, \\ w_{y1}^{-} &= u^{-}, \end{aligned} \tag{2.92}$$

as measured in $L_1$, while the velocity components of particle 2 are

$$\begin{aligned} w_{x2}^{-} &= 0, \\ w_{y2}^{-} &= -u^{-}, \end{aligned} \tag{2.93}$$

in $L_2$. Using the addition theorem for velocities, we find the velocity components of particle 2, as measured in $L_1$:

$$\begin{aligned} w_{x2}^{-\prime} &= v, \\ w_{y2}^{-\prime} &= -u^{-}\sqrt{1-\frac{v^2}{c^2}}, \end{aligned} \tag{2.94}$$

and the velocity components of particle 1 in $L_2$,

$$\begin{aligned} w_{x1}^{-\prime} &= -v, \\ w_{y1}^{-\prime} &= u^{-}\sqrt{1-\frac{v^2}{c^2}}. \end{aligned} \tag{2.95}$$

After the collision, in reference frame $L_1$, the velocities are

$$\begin{aligned} w_{x1}^{+} &= 0, \\ w_{y1}^{+} &= -u^{+}, \\ w_{x2}^{+\prime} &= v, \\ w_{y2}^{+\prime} &= u^{+}\sqrt{1-\frac{v^2}{c^2}}, \end{aligned} \tag{2.96}$$

while in $L_2$, we have

$$\begin{aligned} w_{x1}^{+\prime} &= -v, \\ w_{y1}^{+\prime} &= -u^+\sqrt{1-\frac{v^2}{c^2}}, \\ w_{x2}^{+} &= 0, \\ w_{y2}^{+} &= u^+. \end{aligned} \tag{2.97}$$

The situation has a high degree of symmetry, which simplifies our calculations. We now use the conservation of momentum in the x-direction, which yields.

$$u^- = u^+ = u. \tag{2.98}$$

The conservation of momentum in the y-direction indicates that

$$m\left[\sqrt{v^2+u^2\left(1-\frac{v^2}{c^2}\right)}\right]u\sqrt{1-\frac{v^2}{c^2}} = m(u)u. \tag{2.99}$$

In the limit where $u \to 0$, we find

$$m(v) = \frac{m(0)}{\sqrt{1-\frac{v^2}{c^2}}}; \tag{2.100}$$

it is customary to identify the mass of the particle at zero velocity with its so-called rest mass, $m_0$. We then obtain the well-known relation

$$m(v) = \frac{m_0}{\sqrt{1-\frac{v^2}{c^2}}} = \frac{m_0}{\sqrt{1-\beta^2}} = \gamma m_0. \tag{2.101}$$

## 2.7 The Energy-Momentum Four-Vector

Let us briefly return to the four-vector energy-momentum, first defined in relation with the four-velocity. Using Equation 2.101, the relativistic momentum of the particle is now given by

$$\mathbf{p} = m\mathbf{v} = m\frac{d\mathbf{x}}{dt} = \gamma m_0\mathbf{v} = \gamma m_0\beta c, \tag{2.102}$$

and can be related to the unitary, or normalized, four-velocity by noting that

$$\gamma\beta = \frac{d\mathbf{x}}{cd\tau} = \mathbf{u}; \tag{2.103}$$

we now have

$$\mathbf{p} = m_0c\frac{d\mathbf{x}}{d\tau} = m_0c\mathbf{u}. \tag{2.104}$$

Within this context, the relativistic momentum is identified with the spatial components of the normalized four-velocity, multiplied by the rest mass of the particle and the speed of light; clearly, the time-like component of the corresponding four-vector must play an important role. Indeed, we have

$$p_0 = m_0c\frac{dt}{d\tau} = m_0c\gamma = \frac{E}{c}, \tag{2.105}$$

where $E$ is the particle's energy. Therefore, the relativistic energy and momentum of the particle form a four-vector:

$$p_\mu = m_0cu_\mu = m_0c\frac{dx_\mu}{d\tau} = m_0c(\gamma, \mathbf{u}) = m_0c\gamma(1, \beta) = \left(\frac{E}{c}, \mathbf{p}\right). \tag{2.106}$$

The invariant length of the four-momentum is related to the particle's rest mass:

$$p_\mu p^\mu = \mathbf{p}^2 - \left(\frac{E}{c}\right)^2 = m_0^2c^2u_\mu u^\mu = -m_0^2c^2, \tag{2.107}$$

which yields the relation between energy and momentum:

$$E^2 = \mathbf{p}^2c^2 + m_0^2c^4; \tag{2.108}$$

in particular, we see that there is an energy associated with a particle at rest with zero momentum, the so-called rest energy, $m_0c^2$. Finally, for photons, we obtain the interesting relation

$$p_\mu p^\mu = \hbar^2k_\mu k^\mu = 0, \tag{2.109}$$

which is directly related to the dispersion relation in vacuum,

$$k_\mu k^\mu = \mathbf{k}^2 - \left(\frac{\omega}{c}\right)^2 = 0. \tag{2.110}$$

We now turn our attention to the expression of the four-momentum in the low-velocity limit. The variation of the mass with velocity can be Taylor-expanded around $\mathbf{v} = \mathbf{0}$, using the well-known relation $(1+\varepsilon)^n; 1 + n\varepsilon, \varepsilon \ll 1$:

$$m(v) = \frac{m_0}{\sqrt{1-\frac{\mathbf{v}^2}{c^2}}} \simeq m_0\left(1+\frac{1}{2}\frac{\mathbf{v}^2}{c^2}\right); \tag{2.111}$$

for the momentum, the first correction is cubic, but for the energy we find

$$E \simeq m_0c^2 + \frac{1}{2}m_0\mathbf{v}^2 = m_0c^2 + K, \tag{2.112}$$

where $K$ is the Newtonian kinetic energy.

For relativistic velocities, the kinetic energy deviates from the Newtonian expression:

$$K = E - m_0c^2 = \sqrt{\mathbf{p}^2c^2 + m_0^2c^4} - m_0c^2 = m_0c^2\left[\sqrt{1+\left(\frac{\mathbf{p}}{m_0c}\right)^2} - 1\right]; \tag{2.113}$$

of course, we can normalize the kinetic energy to recover the well-known relation between the spatial and temporal components of the normalized four-velocity,

$$\begin{aligned} \frac{K}{m_0c^2} &= \sqrt{1+\mathbf{u}^2} - 1 = \gamma - 1, \\ \frac{E}{m_0c^2} &= \gamma. \end{aligned} \tag{2.114}$$

## 2.8 Transformation of Forces

Having derived the variation of mass with velocity, the generalization of the concept of force to special relativity can be achieved by considering the equation of motion and using the fact that energy-momentum is a four-vector. The variation of the momentum is governed by the equation.

$$\frac{d\mathbf{p}}{dt} = \mathbf{f}, \tag{2.115}$$

where $\mathbf{f}$ represents all the forces applied to the particle. Because the momentum corresponds to the spatial components of the four-momentum, we would like to add a fourth equation to the system, so that the four-momentum appears in a manifest manner. This is easily achieved by considering the energy of the particle. Since $p_\mu p^\mu$ is a constant, we have

$$\frac{d}{dt}(p_\mu p^\mu) = 0 = \frac{d}{dt}\left(\mathbf{p}^2 - \frac{E^2}{c^2}\right); \tag{2.116}$$

next, the time derivative of the square of the momentum is

$$\mathbf{p}\cdot\frac{d\mathbf{p}}{dt} = \frac{1}{2}\frac{d\mathbf{p}^2}{dt} = \mathbf{p}\cdot\mathbf{f}; \tag{2.117}$$

finally, combining Equations 2.116 and 2.117, we obtain

$$\frac{dE}{dt} = c\frac{\mathbf{p}}{E}\cdot\frac{d\mathbf{p}}{dt} = c\frac{\mathbf{p}\cdot\mathbf{f}}{E}. \tag{2.118}$$

We can now group Equations 2.115 and 2.118, which correspond to the time derivative of the four-momentum:

$$\frac{d\mathbf{p}}{dt} = \mathbf{f}, \quad \frac{dp_0}{dt} = \frac{c\mathbf{p}\cdot\mathbf{f}}{E} = \beta\cdot\mathbf{f}. \tag{2.119}$$

In order to obtain a fully covariant equation of motion, we now need to replace the derivative with respect to time with one with respect to proper time:

$$\begin{aligned}\frac{d\mathbf{p}}{d\tau} &= \frac{d\mathbf{p}}{dt}\frac{dt}{d\tau} = \gamma\mathbf{f},\\ \frac{dp_0}{d\tau} &= \frac{dp_0}{dt}\frac{dt}{d\tau} = \gamma\beta\cdot\mathbf{f}.\end{aligned} \tag{2.120}$$

The terms on the left-hand side of Equation 2.120 are the spatial and temporal components of a four-vector:

$$\frac{dp_\mu}{d\tau} = \left(\frac{dp_0}{d\tau}, \frac{d\mathbf{p}}{d\tau}\right) = \frac{d}{d\tau}(m_0 c u_\mu) = m_0 c\frac{du_\mu}{d\tau} = m_0 c a_\mu, \tag{2.121}$$

where we recognize the four-acceleration; therefore, the right-hand side must also be a four-vector.

We thus have the covariant equation of motion

$$\frac{dp_\mu}{d\tau} = m_0 c a_\mu = F_\mu, \tag{2.122}$$

where we have introduced the new four-vector

$$F_\mu = \gamma(\beta \cdot \mathbf{f}, \mathbf{f}) = (\mathbf{u} \cdot \mathbf{f}, \gamma\mathbf{f}). \tag{2.123}$$

Before embarking on the derivation of the transformation formula for the force, it is important to remember that in our discussion, $\beta = d\mathbf{x}/cdt = \mathbf{v}c$, refers to the velocity of the particle in the reference system $L$; by contrast, the parameters of the Lorentz transformation between the two Galilean frames, $L$ and $L'$, will be represented by $\beta_L = \mathbf{v}_L/c$ and $\gamma_L = 1/\sqrt{1-\beta_L^2}$.

The transformation of the force can now be derived using the fact that $F_\mu$ is a four-vector. We start by applying the special Lorentz transform to $F_\mu$:

$$\begin{aligned} F'_x &= \gamma_L(F_x - \beta_L F_0), \\ F'_y &= F_y, \\ F'_z &= F_z, \\ F'_0 &= \gamma_L(F_0 - \beta_L F_x); \end{aligned} \tag{2.124}$$

next, we use the relation between $\mathbf{f}$ and $F_\mu$ in both frames, which yields the following equations:

$$\begin{aligned} &F_x = \gamma f_x, \quad F_y = \gamma f_y, \quad F_z = \gamma f_z, \quad F_0 = \gamma\beta f_x, \\ &F'_x = \gamma' f'_x, \quad F'_y = \gamma' f'_y, \quad F'_z = \gamma' f'_z, \quad F'_0 = \gamma'\beta' f'_x. \end{aligned} \tag{2.125}$$

The relation between $\beta$ and $\beta'$ is given either by the addition theorem for velocities or by dividing the temporal component of $F'_\mu$ by its axial component:

$$\beta' = \frac{\beta - \beta_L}{1 - \beta\beta_L} = \frac{F'_0}{F'_x}; \tag{2.126}$$

for $\gamma'$, the transformation is even simpler, as it is the time-like component of the normalized four-velocity, $u_\mu = (\gamma, \mathbf{u}) = \gamma(1, \beta)$; thus, we have

$$\gamma' = \gamma_L(\gamma - \beta_L u) = \gamma_L(\gamma - \beta_L \beta\gamma). \tag{2.127}$$

With this, we obtain the sought-after transformation law for the force:

$$\begin{aligned} f'_x &= f_x, \\ f'_y &= \frac{f_y}{\gamma_L(1-\beta_L\beta)}, \\ f'_z &= \frac{f_z}{\gamma_L(1-\beta_L\beta)}. \end{aligned} \tag{2.128}$$

A more general derivation of the transformation of the force can be obtained by considering the equation of motion in two Galilean frames:

$$\frac{d\mathbf{p}}{dt} = \mathbf{f},$$
$$\frac{d\mathbf{p}'}{dt'} = \mathbf{f}' = \frac{d\mathbf{p}'}{dt}\frac{dt}{dt'}; \tag{2.129}$$

the relation between time, as measured in $L$ and $L'$ is given by the Lorentz transform,

$$t' = \gamma_L\left(t - \beta_L \cdot \frac{\mathbf{x}}{c}\right), \tag{2.130}$$

which can be derived to yield

$$\frac{dt'}{dt} = \gamma_L\left(1 - \beta_L \cdot \frac{d\mathbf{x}}{cdt}\right) = \gamma_L(1 - \beta_L \cdot \beta). \tag{2.131}$$

At this point, we need an expression for $d\mathbf{p}'/dt$; this can be obtained by noting that the momentum vector corresponds to the spatial components of the four-momentum and transforms according to

$$\mathbf{p}' = \mathbf{p} + \gamma_L \mathbf{v}_L\left[(1 - \gamma_L^{-1})\frac{\mathbf{p} \cdot \mathbf{v}_L}{\mathbf{v}_L^2} - \frac{E}{c^2}\right]. \tag{2.132}$$

This last equation is the generalization of the Lorentz transform to a case where the direction of the velocity between the two frames is arbitrary; it is formally derived in Section 2.13, but we can see that the operator $\mathbf{v}_L \mathbf{v}_L \cdot / \mathbf{v}_L^2$ projects a given vector onto the relative velocity between the frames under consideration.

Deriving Equation 2.132 with respect to time in $L$ we find that

$$\frac{d\mathbf{p}'}{dt} = \frac{d\mathbf{p}}{dt} + \gamma_L \mathbf{v}_L\left[(1 - \gamma_L^{-1})\frac{d\mathbf{p}}{dt} \cdot \frac{\mathbf{v}_L}{\mathbf{v}_L^2} - \frac{dE}{c^2 dt}\right]$$
$$= \mathbf{f} + \gamma_L \mathbf{v}_L\left[(1 - \gamma_L^{-1})\frac{\mathbf{f} \cdot \mathbf{v}_L}{\mathbf{v}_L^2} - \frac{\mathbf{f} \cdot \mathbf{v}}{c^2}\right]. \tag{2.133}$$

Here, we have used the fact that the time derivative of the energy is equal to the variation of the work of the applied

$$\mathbf{f}' = \frac{1}{\gamma_L(1 - \beta_L \cdot \beta)}\left\{\mathbf{f} + \gamma_L \beta_L\left[(1 - \gamma_L^{-1})\frac{\mathbf{f} \cdot \beta_L}{\beta_L^2} - \mathbf{f} \cdot \beta\right]\right\}. \tag{2.134}$$

In the specific case where $\mathbf{v}_L = v_L\hat{x}$, and $\beta = \beta\hat{x}$, we can verify that we recover the result presented in Equation 2.128. We have $\mathbf{f}\cdot\beta = f_x\beta$, $\mathbf{f}\cdot\beta_L/\beta_L^2 = f_x/\beta_L$, and $\beta_L\cdot\beta = \beta_L\beta$; with this, Equation 2.134 yields

$$\begin{aligned} f'_x &= \frac{1}{\gamma_L(1-\beta_L\beta)}\left\{f_x + \gamma_L\beta_L\left[(1-\gamma_L^{-1})\frac{f_x}{\beta_L} - f_x\beta\right]\right\} \\ &= \frac{f_x}{\gamma_L(1-\beta_L\beta)}[1+\gamma_L(1-\gamma_L^{-1}) - \gamma_L\beta_L\beta] = f_x, \\ f'_y &= \frac{f_y}{\gamma_L(1-\beta_L\beta)}, \\ f'_z &= \frac{f_z}{\gamma_L(1-\beta_L\beta)}. \end{aligned} \tag{2.135}$$

In closing, we come back to the covariant equation of motion, in the case where the external forces are electromagnetic. This will help us in establishing the expression of the electromagnetic field tensor, which is discussed in more detail in Chapter 4. For an electron, we have

$$\mathbf{f} = -e(\mathbf{E}+\mathbf{v}\times\mathbf{B}); \tag{2.136}$$

therefore,

$$\begin{aligned} \mathbf{F} &= -e(\gamma\mathbf{E}+\mathbf{u}\times c\mathbf{B}), \\ F_0 &= -e\mathbf{u}\cdot\mathbf{E}, \end{aligned} \tag{2.137}$$

where the spatial components of the normalized four-velocity are given by $\mathbf{u} = \gamma\mathbf{v}/c = \gamma\beta = d\mathbf{x}/cd\tau$.

The corresponding covariant equation of motion is

$$\begin{aligned} \frac{dp_\mu}{d\tau} &= m_0 c a_\mu = F_\mu, \\ \frac{d\mathbf{p}}{d\tau} &= -e(\gamma\mathbf{E}+\mathbf{u}\times c\mathbf{B}), \\ \frac{dp_0}{d\tau} &= -e\mathbf{u}\cdot\mathbf{E}. \end{aligned} \tag{2.138}$$

The time-like component of Equation 2.138 describes energy conservation; indeed, we have $p_\mu = m_0 c u_\mu$, and $p_0 = \gamma m_0 c$; therefore,

$$\frac{dp_0}{d\tau} = -e\mathbf{u}\cdot\mathbf{E} = m_0 c\frac{d\gamma}{d\tau} = m_0 c\frac{d\gamma}{dt}\frac{dt}{d\tau} = \gamma m_0 c\frac{d\gamma}{dt}, \tag{2.139}$$

and we recover the well-known relation

$$\frac{d\gamma}{dt} = -\frac{e\mathbf{v}\cdot\mathbf{E}}{m_0c^2}, \tag{2.140}$$

which indicates that only the electric field produces work, because the magnetic force is always perpendicular to the charge velocity. Also note that in the case when the electric field derives from a scalar potential, where

$$\mathbf{E} = -\nabla\varphi, \tag{2.141}$$

we have

$$\frac{d\gamma}{dt} = \frac{e\mathbf{v}\cdot\nabla\varphi}{m_0c^2} = \frac{e}{m_0c^2}\left(\frac{d\mathbf{x}}{dt}\cdot\nabla\varphi\right) = \frac{e}{m_0c^2}\left(\frac{d\mathbf{x}}{dt}\cdot\frac{d\varphi}{d\mathbf{x}}\right) = \frac{d}{dt}\left(\frac{e\varphi}{m_0c^2}\right), \tag{2.142}$$

which can be formally integrated to yield

$$\gamma = 1 + \frac{e(\varphi-\varphi_0)}{m_0c^2}. \tag{2.143}$$

We now return to the covariant equation of motion for a charged particle in an electromagnetic field: a closer inspection of Equation 2.138 reveals that the four-velocity is involved in the expression of the electromagnetic force, as $\gamma$ and $\mathbf{u} = \gamma\beta$ appear in the space-like and time-like components of the covariant electromagnetic force, $F_\mu$. Therefore, our goal is to establish the covariant expression of the Lorentz force, using the four-velocity. Since $F_\mu$ is not directly proportional to $u_\mu$ ($F_\mu \neq \alpha u_\mu$), the simplest possible relation between the two four-vectors must be of the form

$$F_\mu = T_{\mu\nu}u^\mu, \tag{2.144}$$

where $T_{\mu\nu}$ is a tensor to be determined.

This can be easily accomplished by explicitly rewriting Equation 2.138 in terms of the components of the four-velocity:

$$\begin{aligned}
F_1 &= T_{11}u_1 + T_{12}u_2 + T_{13}u_3 - T_{10}u_0 = -e(\gamma E_1 + u_2cB_3 - u_3cB_2),\\
F_2 &= T_{21}u_1 + T_{22}u_2 + T_{23}u_3 - T_{20}u_0 = -e(\gamma E_2 + u_3cB_1 - u_1cB_3),\\
F_3 &= T_{31}u_1 + T_{32}u_2 + T_{33}u_3 - T_{30}u_0 = -e(\gamma E_3 + u_1cB_2 - u_2cB_1),\\
F_0 &= T_{01}u_1 + T_{02}u_2 + T_{03}u_3 - T_{00}u_0 = -e(u_1E_1 + u_2E_2 + u_3E_3);
\end{aligned} \tag{2.145}$$

we then find that the covariant form of the Lorentz force equation is

$$\frac{dp_\mu}{d\tau} = F_\mu = -T_{\mu\nu}u^\mu = -eF_{\mu\nu}u^\mu, \tag{2.146}$$

where we have introduced the electromagnetic field tensor

$$F_{\mu\nu} = \begin{pmatrix} 0 & cB_3 & -cB_2 & -E_1 \\ -cB_3 & 0 & cB_1 & -E_2 \\ cB_2 & -cB_1 & 0 & -E_3 \\ E_1 & E_2 & E_3 & 0 \end{pmatrix}. \tag{2.147}$$

This is a very important result, which will be examined in much detail in Chapter 4; one of the key properties of $F_{\mu\nu}$ is the fact that it is antisymmetrical: $F_{\nu\mu} = -F_{\mu\nu}$. Note that this derives from the cross-product in the magnetic force and from the minus sign for the time-like component of the four-velocity, $u_0 = \gamma$; this, in turn, is related to the relation between covariant and contravariant four-vectors in the flat, hyperbolic space–time of special relativity, as expressed in Equations 2.24 and 2.25.

Finally, in a general context, it is also important to understand that energy conservation is guaranteed by the fact that the four-velocity and four-acceleration are orthogonal. Indeed, we can contract the covariant equation of motion with the four-velocity, to obtain

$$u^\mu \frac{dp_\mu}{d\tau} = m_0 c(u^\mu a_\mu) = m_0 c\left(u^\mu \frac{du_\mu}{d\tau}\right) = m_0 c \frac{d}{d\tau}\left(\frac{u^\mu u_\mu}{2}\right); \tag{2.148}$$

since the normalized four-velocity has a constant length, namely $u_\mu u^\mu = -1$, we first find that

$$u^\mu a_\mu = 0, \tag{2.149}$$

which means that the derivative of a four-vector with constant length is always perpendicular to that four-vector, just as for regular three-vectors. More importantly, this helps us establish the fact that

$$u^\mu \frac{dp_\mu}{d\tau} = 0. \tag{2.150}$$

Let us now analyze this result in terms of the physics involved: we start with

$$u^{\mu}\frac{dp_{\mu}}{d\tau} = \mathbf{u}\cdot\frac{d\mathbf{p}}{d\tau} - \gamma\frac{dp_0}{d\tau} = 0, \tag{2.151}$$

which shows that

$$\mathbf{u}\cdot\frac{d\mathbf{p}}{dt}\frac{dt}{d\tau} = \gamma\mathbf{u}\cdot\mathbf{f} = \gamma\frac{dp_0}{dt}\frac{dt}{d\tau} = \gamma^2\frac{dE}{cdt}; \tag{2.152}$$

reorganizing the various terms in Equation 2.148, we recover the fact that the energy-transfer equation is driven by the work of the external force on the particle

$$\frac{dE}{dt} = \frac{c\mathbf{u}\cdot\mathbf{f}}{\gamma} = \mathbf{v}\cdot\mathbf{f}. \tag{2.153}$$

Therefore, the complete equivalence between the constant length of the four-velocity and energy conservation has been established.

## 2.9 Transformation of Energy

The energy is the time-like component of the energy-momentum four-vector; therefore, it behaves as

$$\frac{E'}{c} = p_0' = \gamma(p_0 - \beta\cdot\mathbf{p}) = \gamma\left(\frac{E}{c} - \beta\cdot\mathbf{p}\right). \tag{2.154}$$

The same result can be obtained by using the fact that the length of the four-momentum is invariant:

$$p_{\mu}p^{\mu} = \mathbf{p}^2 - \frac{E^2}{c^2} = -m_0^2c^2 = (m_0c)^2u_{\mu}u^{\mu}, \tag{2.155}$$

and the transform of the three-momentum is as expressed in Equation 2.132.

## 2.10 Transformation of Angular Momentum

We now consider the angular momentum

$$\mathbf{m} = \mathbf{r} \times \mathbf{p}, \tag{2.156}$$

where $\mathbf{r}$ is the position of the particle, and $\mathbf{p} = m_0 d\mathbf{r}/dt$ is its momentum.

For a covariant generalization of this concept, we introduce the antisymmetrical tensor

$$M_{\mu\nu} = x_\mu p_\nu - x_\nu p_\mu, \tag{2.157}$$

where $x_\mu = (ct, \mathbf{r})$ is the four-position of the particle and $p_\mu = m_0 c(\gamma, \mathbf{u}) = m_0 c dx_\mu / d\tau$ is its four-momentum.

The components of this tensor are given by

$$M_{\mu\nu} = \begin{pmatrix} 0 & xp_y - yp_x & xp_z - zp_x & x\gamma m_0 c - ctp_x \\ yp_x - xp_y & 0 & yp_z - zp_y & y\gamma m_0 c - ctp_y \\ zp_x - xp_z & zp_y - yp_z & 0 & z\gamma m_0 c - ctp_z \\ ctp_x - x\gamma m_0 c & ctp_y - y\gamma m_0 c & ctp_z - z\gamma m_0 c & 0 \end{pmatrix}. \tag{2.158}$$

If the particle velocity is defined as $\mathbf{v} = d\mathbf{x}/dt$, we note that

$$\begin{aligned} M_{10} &= x\gamma m_0 c - ctp_x = \gamma m_0 c(x - tv_x), \\ M_{20} &= y\gamma m_0 c - ctp_y = \gamma m_0 c(y - tv_y), \\ M_{30} &= z\gamma m_0 c - ctp_z = \gamma m_0 c(z - tv_z). \end{aligned} \tag{2.159}$$

Therefore, if the velocity is constant, we can define the initial positions, $x_0$, $y_0$, and $z_0$, as the components of the position vector at $t = 0$, $\mathbf{x}(t = 0) = (x_0, y_0, z_0)$, and simplify Equation 2.59 to read

$$\begin{aligned} M_{10} &= \gamma m_0 c x_0, \\ M_{20} &= \gamma m_0 c y_0, \\ M_{30} &= \gamma m_0 c z_0. \end{aligned} \tag{2.160}$$

We now consider the special Lorentz transform:

$$x' = \gamma_L(x - \beta_L ct), \quad y' = y, \quad z' = z, \quad ct' = \gamma_L(ct - \beta_L x), \tag{2.161}$$

where $\beta_L = v_L/c$, and $\gamma_L = (1-\beta_L^2)^{-1/2}$ represent the normalized velocity and associated Lorentz factor between the two reference frames under consideration.

For the momentum, similar equations hold:

$$p'_x = \gamma_L(p_x - \beta_L p_0), \quad p'_y = p_y, \; p'_z = p_z, \; p'_0 = \gamma_L(p_0 - \beta_L p_x), \tag{2.162}$$

which yields

$$p'_x = \gamma_L(\gamma m_0 v_x - \beta_L \gamma m_0 c) = \gamma_L \gamma m_0 (v_x - v_L). \tag{2.163}$$

We can now derive the components of the transform of the angular momentum tensor:

$$\begin{aligned} M'_{12} &= x'p'_y - y'p'_x = x'p_y - yp'_x \\ &= \gamma_L(x - v_L t)p_y - y m_0 \gamma_L \gamma (v_x - v_L) \\ &= \gamma_L[xp_y - \beta_L ctp_y - yp_x + y\gamma m_0 c\beta_L] \\ &= \gamma_L[(xp_y - yp_x) + \beta_L(y\gamma m_0 c - ctp_y)] \\ &= \gamma_L[M_{12} - \beta_L M_{02}]. \end{aligned} \tag{2.164}$$

Following the same procedure, we also find

$$\begin{aligned} &M'_{12} = \gamma_L(M_{12} - \beta_L M_{02}), \quad M'_{23} = M_{23}, \quad M'_{31} = \gamma_L(M_{31} - \beta_L M_{30}), \\ &M'_{10} = M_{10}, \quad M'_{20} = \gamma_L(M_{20} - \beta_L M_{21}), \quad M'_{30} = \gamma_L(M_{30} - \beta_L M_{31}). \end{aligned} \tag{2.165}$$

Of course, because $M_{\mu\nu}$ is an anti-symmetrical tensor, a more general approach can be used. Let us consider a generic anti-symmetrical tensor,

$$A_{\mu\nu} = a_\mu b_\nu - a_\nu b_\mu, \tag{2.166}$$

where $a_\mu$ and $b_\nu$ are two four-vectors.

The transformation law for these four-vectors is the now familiar Lorentz transform:

$$a'_\mu = \mathcal{L}_{\mu\nu} a^\nu, \tag{2.167}$$

which yields, in the special case,

$$a'_x = \gamma_L(a_x - \beta_L a_0), \quad a'_y = a_y, \quad a'_z = a_z, \quad a'_0 = \gamma_L(a_0 - \beta_L a_x), \tag{2.168}$$

and similar relations for $b'_\mu$.

The transform of $A_{\mu\nu}$ is, by definition,

$$A'_{\mu\nu} = a'_{\mu}b'_{\nu} - a'_{\nu}b'_{\mu}. \tag{2.169}$$

Using the Lorentz transform, as expressed in Equation 2.168, we first find that

$$\begin{aligned} A'_{xy} &= a'_x b'_y - a'_y b'_x \\ &= \gamma_L(a_x - \beta_L a_0)b_y - a_y\gamma_L(b_x - \beta_L b_0) \\ &= \gamma_L[(a_x b_y - a_y b_x) - \beta_L(a_0 b_y - a_y b_0)] \\ &= \gamma_L(A_{xy} - \beta_L A_{0y}); \end{aligned} \tag{2.170}$$

next, we have

$$\begin{aligned} A'_{yz} &= a'_y b'_z - a'_z b'_y \\ &= a_y b_z - a_z b_y = A_{yz}. \end{aligned} \tag{2.171}$$

Similarly to Equation 2.170, we find that

$$\begin{aligned} A'_{xz} &= a'_x b'_z - a'_z b'_x \\ &= \gamma_L(a_x - \beta_L a_0)b_z - a_z\gamma_L(b_x - \beta_L b_0) \\ &= \gamma_L[(a_x b_z - a_z b_x) - \beta_L(a_0 b_z - a_z b_0)] \\ &= \gamma_L(A_{xz} - \beta_L A_{0z}). \end{aligned} \tag{2.172}$$

We now examine $A'_{x0}$:

$$\begin{aligned} A'_{x0} &= a'_x b'_0 - a'_0 b'_x \\ &= \gamma_L(a_x - \beta_L a_0)\gamma_L(b_0 - \beta_L b_x) - \gamma_L(a_0 - \beta_L a_x)\gamma_L(b_x - \beta_L b_0) \\ &= \gamma_L^2(a_x b_0 - \beta_L a_x b_x - \beta_L a_0 b_0 + \beta_L^2 a_0 b_x - a_0 b_x + \beta_L a_0 b_0 + \beta_L a_x b_x - \beta_L^2 a_x b_0) \\ &= \gamma_L^2(a_x b_0 - a_0 b_x)(1 - \beta_L^2) = A_{x0}. \end{aligned} \tag{2.173}$$

The situation for $A'_{y0}$ is different:

$$\begin{aligned} A'_{y0} &= a'_y b'_0 - a'_0 b'_y \\ &= a_y\gamma_L(b_0 - \beta_L b_x) - \gamma_L(a_0 - \beta_L a_x)b_y \\ &= \gamma_L[(a_y b_0 - a_0 b_y) - \beta_L(a_y b_x - a_x b_y)] \\ &= \gamma_L(A_{y0} - \beta_L A_{yx}); \end{aligned} \tag{2.174}$$

similarly, we have

$$\begin{aligned} A'_{z0} &= a'_z b'_0 - a'_0 b'_z \\ &= a_z \gamma_L (b_0 - \beta_L b_x) - \gamma_L (a_0 - \beta_L a_x) b_z \\ &= \gamma_L [(a_z b_0 - a_0 b_z) - \beta_L (a_z b_x - a_x b_z)] \\ &= \gamma_L (A_{z0} - \beta_L A_{zx}). \end{aligned} \tag{2.175}$$

The other components of $A'_{\mu\nu}$ are obtained using the fact that this tensor is antisymmetrical. For the angular momentum, we recover the results derived in Equation 2.165.

## 2.11 Transformation of Length, Surface, Volume, and Density

We begin by considering the simple and well-known case of length contraction; again we use the special Lorentz transform:

$$\begin{aligned} x' &= \gamma(x - vt), \quad y' = y, \quad z' = z, \quad ct' = \gamma(ct - \beta x), \\ x &= \gamma(x' + vt'), \quad y = y', \quad z = z', \quad ct = \gamma(ct' + \beta x'). \end{aligned} \tag{2.176}$$

Suppose that in the reference frame $L'$, we measure a segment parallel to the x′-axis, with a length $\ell'$; the question is what is the length of this rod, as measured in $L$? From Equation 2.176, we know that $x' = \gamma(x - vt)$, and the measurement is done at a given time, $t$, which is constant. We thus have

$$\begin{aligned} x'_2 - x'_1 &= \gamma(x_2 - x_1), \\ \ell' &= \gamma \ell. \end{aligned} \tag{2.177}$$

In the particular case where the reference frame $L'$ coincides with the rest frame of the system, $\ell' = \ell_0$ is the so-called proper length, and we see that in any other frame, the length is contracted, as $\ell = \ell_0/\gamma$. This contraction of length goes hand-in-hand with the dilatation of time. If we now measure the time duration between two events, we find that

$$\Delta t = \gamma \Delta \tau, \tag{2.178}$$

where we have denoted the proper time difference by $\Delta\tau$.

Of course Equation 2.178 can also be considered as an integral expression of the relation between time measured in a Galilean frame *L*, and the proper time:

$$\Delta t = \int_{\tau_1}^{\tau_2} \frac{dt}{d\tau} d\tau = \int_{\tau_1}^{\tau_2} u_0 \, d\tau = \int_{\tau_1}^{\tau_2} \gamma \, d\tau = \Delta\tau. \tag{2.179}$$

In the special case where the system is in uniform motion, $\gamma(\tau) = \gamma$ is a constant, and we recover the result given in Equation 2.178.

Let us now consider the case of surfaces. It is clear that if the surface is perpendicular to the relative velocity, $\mathbf{v}_L$, between inertial frames, it is a Lorentz invariant because neither of the transverse dimensions is affected by the relativistic transform. On the other hand, a surface element containing $\mathbf{v}_L$ will transform like the length:

$$S' = S/\gamma. \tag{2.180}$$

Before considering the case of volumes and densities, it is instructive to examine briefly the propagation of electromagnetic waves in a waveguide from a relativistic viewpoint. It is well known that the dispersion relation for waves propagating in a system with uniform transverse boundary conditions is given by

$$k_\mu k^\mu = \mathbf{k}^2 - k_0^2 = \mathbf{k}^2 - \left(\frac{\omega}{c}\right)^2 = -\frac{\omega_c^2}{c^2}, \tag{2.181}$$

where $\omega_c$ is the cutoff frequency of the waveguide eigenmode under consideration.

The exact origin of this dispersion relation will be discussed in detail in Chapter 5. For the moment, the important fact is that $\omega_c$ is determined by the transverse geometry of the waveguide. If we multiply Equation 2.181 by the square of Planck's constant, an interesting fact emerges. We can interpret the dispersion relation in terms of photon mass, to obtain

$$\hbar^2 k_\mu k^\mu = p_\mu p^\mu = -m_0^2 c^2 = -\hbar^2 \frac{\omega_c^2}{c^2}, \tag{2.182}$$

which yields $m_0 = \hbar\omega_c/c^2$. The photons propagating in the waveguide seem to acquire a mass related to the existence of transverse boundary conditions. The interpretation of this pseudomass allows us to better understand the concept of rest mass: a particle has rest mass if it has nonzero energy at zero momentum. This is fully embodied by the relation between energy and momentum, $p_\mu p^\mu = -m_0^2 c^2$. In the case of the photons in the waveguide, the existence of transverse boundary conditions allows us to trap photons at cutoff: at zero momentum, for $\mathbf{p} = \hbar\mathbf{k} = \mathbf{0}$, we find that $\omega = \omega_c \neq 0$. This means that electromagnetic energy is stored at cutoff, since each photon carries a quantum of energy $\hbar\omega_c$. To understand how this situation does not contradict the fact that photons always propagate at the speed of light in vacuum, we need to consider the fact that the trapped modes, which are also called standing waves, can always be decomposed into two counter-propagating modes. In the simplest example of two parallel, loss-free mirrors, the electric field of the trapped modes is of the form

$$\mathbf{E}(\mathbf{x},t) = \mathbf{E}_\perp(\mathbf{x}_\perp)[e^{i(\omega t - k_\| z)} - e^{i(\omega t + k_\| z)}] = \mathbf{E}_\perp(\mathbf{x}_\perp)2i\sin(k_\| z)e^{i\omega t}, \qquad (2.183)$$

where the axial wavenumber satisfies the boundary condition on the mirrors,

$$\sin(k_\| L) = 0, \quad \frac{2\pi c}{\lambda} = \frac{n\pi}{L}, \quad n \in \mathbb{Z}, \qquad (2.184)$$

and where the four-wavenumber satisfies the vacuum dispersion relation,

$$\left(\frac{\omega}{c}\right)^2 = \mathbf{k}_\perp^2 + k_\|^2 = \mathbf{k}_\perp^2 + \left(\frac{n\pi}{L}\right)^2. \qquad (2.185)$$

Note that here we have rotated the geometry by 90 degrees, as the mirrors play the role of the transverse boundary in the waveguide. Equation 2.185 can now be rearranged to clearly exhibit the cutoff frequency associated with the boundary condition on the mirrors:

$$\left(\frac{\omega}{c}\right)^2 - \mathbf{k}_\perp^2 = \left(\frac{n\pi}{L}\right)^2. \qquad (2.186)$$

The key fact is that the trapped electromagnetic modes, while propagating in vacuum at the speed of light between the mirrors, store energy with a total momentum equal to zero because the counterpropagating photons have opposite momenta. In this situation, the electromagnetic field acquires a rest mass given by $m_0 = \hbar n\pi/Lc$.

We now ask whether the cutoff frequency is Lorentz-invariant. There are at least three different ways to show that it is. The first argument is mathematical: if we consider Equation 2.181, we have $\omega_c = -ck_\mu k^\mu$. Since the right-hand side of this equation is a scalar, as it corresponds to the contraction of two four-vectors, we know that it is Lorentz-invariant, as shown in Equation 2.29. The second approach relies on the fact that the cutoff frequency is determined by the transverse geometry of the waveguide. As the transverse dimensions are left unchanged by the Lorentz transform, the cutoff frequency is invariant. The third argument is the most fundamental, as it is derived from a clear understanding of the principle of relativity: natural phenomena do not depend upon the reference frame in which they are viewed. Following this principle, we can recast our question regarding the Lorentz-invariance of the cutoff frequency as follows: if a wave is below cutoff in a waveguide, as observed in a given reference frame, can we find another frame where we would see the wave propagate? Clearly, the answer must be no. Changing the reference frame does not change the underlying nature of physical reality; it simply provides a different perspective. This basic understanding of relativity allows us to distinguish between natural phenomena and their covariant description. For example, in Chapters 5, 7, and 9 we will see how judicious choices of reference frames will allow us to obtain valuable physical insight and a clear understanding of Čerenkov radiation and free-electron lasers, while the underlying physic remains unchanged.

We now consider the transformation of volumes and densities. For volumes, the transformation is evident:

$$V' = V/\gamma; \tag{2.187}$$

the contraction of length implies a contraction of volumes. In terms of densities, one must distinguish between mass density and charge density. In the first case, let us consider the proper frame of the system, where matter is at rest or moving at nonrelativistic velocities. We note that this concept cannot always be applied. In such a system, the proper mass density can be defined as

$$\rho_0 = m_0/V_0; \tag{2.188}$$

transforming the mass and volume to a Galilean frame, we find

$$\rho' = \frac{m'}{V'} = \gamma^2 \frac{m_0}{V_0} = \gamma^2 \rho_0. \tag{2.189}$$

We also note that the density can be defined locally, in differential form, as

$$\rho_0 = \frac{d^3 m_0}{dxdydz}. \tag{2.190}$$

Using the Lorentz transform on the coordinates and maintaining the time coordinate fixed, we then have

$$dx = \gamma dx', \quad dy = dy', \quad dz = dz'. \tag{2.191}$$

Using Equations 2.190 and 2.191, we find

$$\rho' = \frac{d^3 m'}{dx'dy'dz'} = \frac{\gamma d^3 m_0}{dxdydz} \frac{dx}{dx'} \frac{dy}{dy'} \frac{dz}{dz'} = \gamma^2 \rho_0. \tag{2.192}$$

On the other hand, in the case of charge density, because the charge is a relativistic invariant, as will be discussed in Chapter 4, the transformation simply reflects the aforementioned volume contraction.

## 2.12 Relativistic Plasma Frequency

The transformation of the charge density can be illustrated by considering the relativistic expression of the plasma frequency; furthermore, this specific problem will provide a beautiful example of how the principle of relativity can offer two complementary views of the same phenomenon, thereby adding physical insight to our analysis.

The concept of plasma frequency can be introduced by considering a charged fluid, for which the dynamics are governed by the divergence equation,

$$\nabla \cdot \mathbf{E} = \frac{\rho}{\varepsilon_0} = -\frac{ne}{\varepsilon_0}, \tag{2.193}$$

the Lorentz force equation for the fluid,

$$n\frac{d\mathbf{p}}{dt} = -ne(\mathbf{E} + \mathbf{v} \times \mathbf{B}), \tag{2.194}$$

and the concomitant energy conservation equation

$$n\frac{d\gamma}{dt} = -\frac{ne\mathbf{v} \cdot \mathbf{E}}{m_0 c^2}, \tag{2.195}$$

as well as the continuity, or charge conservation, equation,

$$\frac{\partial \rho}{\partial t} + \nabla \cdot \mathbf{j} = -e\left[\frac{\partial n}{\partial t} + \nabla \cdot (n\mathbf{v})\right] = 0. \tag{2.196}$$

Here, $n = d^3N/dxdydz$ is the so-called number density, and $\rho = -en$ is the charge density of the fluid. In general, solving these coupled, nonlinear equations 2.193 to 2.196 can prove a formidable task; however, we will simplify the problem considerably by considering a one-dimensional fluid in the linear regime. In this case, we first linearize the equations. We then take the one-dimensional limit. As we will use the linearization procedure for other problems in this book, the main steps are presented here as an introduction to the technique. We start by writing the fluid density, velocity, and energy fields as perturbation expansions:

$$\begin{aligned} n(x_\mu) &= n_0(x_\mu) + \delta n(x_\mu), \quad \left|\frac{\delta n}{n_0}\right| \ll 1, \\ \mathbf{v}(x_\mu) &= \mathbf{v}_0(x_\mu) + \delta \mathbf{v}(x_\mu), \quad \left|\frac{\delta \mathbf{v}}{\mathbf{v}_0}\right| \ll 1, \\ \gamma(x_\mu) &= \gamma_0(x_\mu) + \delta \gamma(x_\mu), \quad \left|\frac{\delta \gamma}{\gamma_0}\right| \ll 1. \end{aligned} \tag{2.197}$$

We proceed in a similar manner with the electric and magnetic fields:

$$\begin{aligned} \mathbf{E}(x_\mu) &= \mathbf{E}_0(x_\mu) + \delta \mathbf{E}(x_\mu), \quad \left|\frac{\delta \mathbf{E}}{\mathbf{E}_0}\right| \ll 1, \\ \mathbf{B}(x_\mu) &= \mathbf{B}_0(x_\mu) + \delta \mathbf{B}(x_\mu), \quad \left|\frac{\delta \mathbf{B}}{\mathbf{B}_0}\right| \ll 1. \end{aligned} \tag{2.198}$$

The relation between the energy and velocity perturbations can be derived as follows. We have, by definition,

$$\gamma = \gamma_0 + \delta\gamma = \left[1 - \frac{(\mathbf{v}_0 + \delta\mathbf{v})^2}{c^2}\right]^{-1/2}; \tag{2.199}$$

expanding the square, and neglecting the second-order term, we find

$$\gamma_0 + \delta\gamma = \left[1 - \frac{1}{c^2}(\mathbf{v}_0^2 + 2\mathbf{v}_0 \cdot \delta\mathbf{v} + \delta\mathbf{v}^2)\right]^{-1/2} \simeq \left[1 - \frac{1}{c^2}(\mathbf{v}_0^2 + 2\mathbf{v}_0 \cdot \delta\mathbf{v})\right]^{-1/2}. \tag{2.200}$$

To linearize Equation 2.200, we use the approximation $(1+\varepsilon)^n \simeq 1 + n\varepsilon$, $\varepsilon \ll 1$. We first write

$$\gamma_0 + \delta\gamma \simeq \gamma_0\left[\gamma_0^2\left(1 - \frac{\mathbf{v}_0^2}{c} - 2\frac{\mathbf{v}_0 \cdot \delta\mathbf{v}}{c^2}\right)\right]^{-1/2} = \gamma_0\left(1 - 2\gamma_0^2\frac{\mathbf{v}_0 \cdot \delta\mathbf{v}}{c^2}\right)^{-1/2}. \tag{2.201}$$

Equation 2.201 is correct because, by definition, $\gamma_0 = 1/\sqrt{1 - (\mathbf{v}_0^2/c^2)}$; applying the binomial theorem for $n = -1/2$, and $\varepsilon = 2\gamma_0^2\mathbf{v}_0 \cdot \delta\mathbf{v}/c^2$, we then find that

$$\delta\gamma \simeq \gamma_0^3\frac{\mathbf{v}_0 \cdot \delta\mathbf{v}}{c^2}. \tag{2.202}$$

Returning to Equations 2.193 to 2.196, the so-called equilibrium fields, $n_0$, $\mathbf{v}_0$, $\gamma_0$, $\mathbf{E}_0$, and $\mathbf{B}_0$, satisfy the unperturbed equations for the plasma, namely,

$$\begin{aligned}
\nabla \cdot \mathbf{E}_0 &= -\frac{en_0}{\varepsilon_0}, \\
\frac{d}{dt}(\gamma_0 m_0 \mathbf{v}_0) &= -e(\mathbf{E}_0 + \mathbf{v}_0 \times \mathbf{B}_0), \\
\frac{d\gamma_0}{dt} &= -\frac{e\mathbf{v}_0 \cdot \mathbf{E}_0}{m_0 c^2}, \\
\frac{\partial n_0}{\partial t} + \nabla \cdot (n_0 \mathbf{v}_0) &= 0.
\end{aligned} \tag{2.203}$$

The linearization procedure consists in expanding Equations 2.193 to 2.196 to first order, and taking Equation 2.203 into account, to obtain

$$\begin{aligned}
\nabla \cdot \delta\mathbf{E} &= -\frac{e}{\varepsilon_0}\delta n, \\
\frac{d}{dt}(\gamma_0 \delta\mathbf{v} + \delta\gamma \mathbf{v}_0) &= -\frac{e}{m_0}(\delta\mathbf{E} + \mathbf{v}_0 \times \delta\mathbf{B} + \delta\mathbf{v} \times \mathbf{B}_0), \\
\frac{d}{dt}\delta\gamma &= -\frac{e}{m_0 c^2}(\delta\mathbf{v} \cdot \mathbf{E}_0 + \mathbf{v}_0 \cdot \delta\mathbf{E}), \\
\frac{\partial}{\partial t}\delta n + \nabla \cdot (\delta n \mathbf{v}_0 + n_0 \delta\mathbf{v}) &= 0.
\end{aligned} \tag{2.204}$$

At this point, it is interesting to consider the consistency of the linearized equations; in particular, it is instructive to carefully inspect the linearized version of the Lorentz force equation, the linearized equation governing the evolution of the energy, together with Equation 2.202. To simplify our task, we will assume that the equilibrium fields are such that $d_t\gamma_0 = 0$, and $d_t\mathbf{v}_0 = \mathbf{0}$. In this case, we can take the dot product of the linearized Lorentz force equation to obtain

$$\gamma_0 \mathbf{v}_0 \cdot \frac{d}{dt}\delta\mathbf{v} + \mathbf{v}_0^2 \frac{d}{dt}\delta\gamma = -\frac{e}{m_0}[\mathbf{v}_0 \cdot \delta\mathbf{E} + \mathbf{v}_0 \cdot (\delta\mathbf{v} \times \mathbf{B}_0)]. \tag{2.205}$$

Next, we derive Equation 2.202 with respect to time,

$$\frac{d}{dt}\delta\gamma = \gamma_0^3 \frac{\mathbf{v}_0}{c^2} \cdot \frac{d}{dt}\delta\mathbf{v}, \tag{2.206}$$

and we use this result into Equation 2.205, to find

$$\frac{d}{dt}\delta\gamma = -\frac{e}{m_0 c^2}[\mathbf{v}_0 \cdot \delta\mathbf{E} + \mathbf{v}_0 \cdot (\delta\mathbf{v} \times \mathbf{B}_0)], \tag{2.207}$$

where we have used the fact that

$$\frac{c^2}{\gamma_0^2} + \mathbf{v}_0^2 = c^2(1-\beta_0^2) + \mathbf{v}_0^2 = c^2. \tag{2.208}$$

We need to verify the consistency of Equation 2.207 with the other expression of the linearized equation governing the evolution of energy, as stated in Equation 2.204. We first have

$$\mathbf{v}_0 \cdot (\delta\mathbf{v} \times \mathbf{B}_0) = \delta\mathbf{v} \cdot (\mathbf{B}_0 \times \mathbf{v}_0) = -\delta\mathbf{v} \cdot (\mathbf{v}_0 \times \mathbf{B}_0). \tag{2.209}$$

Furthermore, because of our assumption that the total time derivative of the equilibrium energy and velocity fields are identically zero, we also have

$$\frac{d}{dt}(\gamma_0 m_0 \mathbf{v}_0) = \mathbf{0} = -e(\mathbf{E}_0 + \mathbf{v}_0 \times \mathbf{B}_0), \tag{2.210}$$

and we find that, indeed,

$$\frac{d}{dt}\delta\gamma = -\frac{e}{m_0 c^2}[\mathbf{v}_0 \cdot \delta\mathbf{E} + \mathbf{v}_0 \cdot (\delta\mathbf{v} \times \mathbf{B}_0)] = -\frac{e}{m_0 c^2}[\mathbf{v}_0 \cdot \delta\mathbf{E} + \delta\mathbf{v} \cdot \mathbf{E}_0]. \tag{2.211}$$

Further simplification results from considering a purely one-dimensional plasma, where the divergence operator reduces to $\nabla \equiv \hat{x}\partial_x$:

$$\partial_x \delta E = -\frac{e}{\varepsilon_0}\delta n, \tag{2.212}$$

$$d_t(\gamma_0 \delta v + \delta\gamma v_0) = -\frac{e}{m_0}\delta E, \tag{2.213}$$

$$d_t \delta\gamma = -\frac{e}{m_0 c^2}(\delta v E_0 + v_0 \delta E), \tag{2.214}$$

and

$$\partial_t \delta n + \partial_x (n_0 \delta v + v_0 \delta n) = 0, \tag{2.215}$$

where we have written $\delta E = \delta E_x$, $\delta v = \delta v_x$, $v_0 = v_{0x}$, and $E_0 = E_{0x}$, for short.

Note that because the problem is now purely one-dimensional, the magnetic force disappears. Furthermore, Equation 2.202 becomes

$$\delta\gamma = \gamma_0^3 \frac{v_0 \delta v}{c^2}. \tag{2.216}$$

To resolve this system of coupled, linear, partial differential equations (PDEs), we seek a solution of the form

$$\delta f(x,t) = \delta f e^{i(\omega t - kx)} = \delta f e^{i\phi(x,t)}, \tag{2.217}$$

where $\delta f(x, t)$ represents a generic field, and where the phase of this plane-wave solution is defined as

$$\phi(x^\mu) = \phi(x,t) = \omega t - kx = -k_\mu x^\mu. \tag{2.218}$$

This technique of using a perturbation expansion and plane-wave solutions is used widely in classical and quantum electrodynamics and will prove extremely useful throughout this book. It leads to dispersion relations and corresponds to solving the problem in momentum space as we Fourier transform the linearized equations into k-space, and $\hbar k_\mu = p_\mu$, for photons. We illustrate this technique here and note that for the type of solution given in Equation 2.217 we can substitute the partial differential operators as follows:

$$\nabla \equiv -i\mathbf{k}, \quad \partial_t \equiv i\omega, \quad \partial_\mu \equiv ik_\mu. \tag{2.219}$$

Because we are dealing with fluid fields, it is important to understand that the total derivative with respect to time must be considered as a so-called convective derivative:

$$d_t \equiv \partial_t + \frac{\partial \mathbf{x}}{\partial t} \cdot \frac{\partial}{\partial \mathbf{x}} \equiv \partial_t + \mathbf{v} \cdot \nabla. \tag{2.220}$$

For a covariant equivalent to Equation 2.220, we note that

$$d_\tau \equiv \partial_\tau + \frac{\partial \mathbf{x}}{\partial \tau} \cdot \frac{\partial}{\partial \mathbf{x}} \equiv \frac{\partial t}{\partial \tau}\frac{\partial}{\partial t} + \frac{\partial \mathbf{x}}{\partial \tau} \cdot \frac{\partial}{\partial \mathbf{x}} \equiv \gamma \partial_t + \mathbf{u} \cdot \nabla \equiv u^\mu \partial_\mu. \tag{2.221}$$

In Equation 2.221, we need to remember that the four-gradient operator is defined as

$$\partial_\mu \equiv (-\partial_t, \nabla), \tag{2.222}$$

so that the contraction with the normalized four-velocity does yield Equation 2.221:

$$u^\mu \partial_\mu = \mathbf{u} \cdot \nabla - u_0(-\partial_t) = \mathbf{u} \cdot \nabla + \gamma \partial_t. \tag{2.223}$$

With these considerations in mind, the dynamics of our one-dimensional plasma reduce to a linear system of four equations with four unknowns:

$$\begin{aligned} -ik\delta E &= -\frac{e}{\varepsilon_0}\delta n, \\ i(\omega - kv_0)\delta\gamma &= -\frac{e}{m_0 c^2}(E_0 \delta v + v_0 \delta E), \\ \delta\gamma &= \gamma_0^3 \frac{v_0 \delta v}{c^2}, \\ i\omega\delta n - ik(n_0 \delta v + v_0 \delta n) &= 0. \end{aligned} \tag{2.224}$$

To obtain a nontrivial solution, we need to take the determinant of the system and make it equal to zero. This yields the dispersion relation

$$(\omega - kv_0)^2 - \frac{n_0 e^2}{\gamma_0^3 m_0 \varepsilon_0} - i(\omega - kv_0)\frac{eE_0}{\gamma_0^3 m_0 v_0} = 0. \tag{2.225}$$

To give the physical interpretation of Equation 2.225, we introduce the relativistic plasma frequency,

$$\omega_p = \sqrt{\frac{n_0 e^2}{\gamma_0 m_0 \varepsilon_0}}, \tag{2.226}$$

and we note that the equilibrium condition implies that $\mathbf{E}_0 + \mathbf{v}_0 \times \mathbf{B}_0 = \mathbf{0}$; therefore, we can also introduce the relativistic cyclotron frequency,

$$\Omega_0 = \frac{eB_0}{\gamma_0 m_0} = \frac{eE_0}{\gamma_0 m_0 v_0}. \tag{2.227}$$

We can now recast the dispersion relation as

$$\gamma_0^2(\omega - kv_0)^2 - \omega_p^2 - i(\omega - kv_0)\Omega_0 = 0. \tag{2.228}$$

In the absence of external or equilibrium fields, Equation 2.228 takes the simpler form

$$\gamma_0^2(\omega - kv_0)^2 = \omega_p^2, \tag{2.229}$$

where the left-hand side of the equation corresponds to the dispersion of a so-called beam mode, while the right-hand side is the relativistic plasma frequency.

To understand the physics underlying the beam mode dispersion, let us consider the beam frame, where $v_0 = 0$. In such a frame, a perturbation of the beam density, leading to a periodic electrostatic field with characteristic wavelength $\lambda$, will be associated with a wavenumber $k = 2\pi c/\lambda$, while its characteristic frequency is $\omega$. If we now Lorentz transform the corresponding four-wavenumber, we have

$$\omega' = \gamma_0(\omega - v_0 k), \tag{2.230}$$

in the lab frame, where the beam velocity is $v_0$. Within this context, we find that the dispersion relation simply corresponds to the statement that

$$\omega'^2 = \omega_p^2, \tag{2.231}$$

where the relativistic plasma frequency is the characteristic eigenfrequency of the charged fluid.

In particular, if we examine Equation 2.229 in the case where $v_0 = 0$, and $\gamma_0 = 1$, or Equation 2.231, we find that the beam mode can drive the system resonantly near the plasma frequency, which is a natural mode of oscillation of the plasma. Such space-charge waves are electrostatic modes, which propagate with the beam and can couple to electromagnetic modes, thus offering a way to extract energy from the beam. This will be explored in great detail in Chapter 9. In addition, we now see that $\omega_p = \sqrt{n_0 e^2 / \gamma_0 m_0 \varepsilon_0}$ is a relativistic invariant because we can interpret it either in terms of density contraction in the proper frame of the beam, where we have the relation

$$n = \frac{n_0}{\gamma_0}, \tag{2.232}$$

$n_0$ being the density measured in the lab frame, while $m_0$ is the rest mass of the electrons in the beam frame, or as the relativistic mass increase, where

$$m = \gamma_0 m_0, \tag{2.233}$$

is the electron mass, as measured in the lab frame, where the density is $n_0$. Therefore, the quantity $n_0/\gamma_0 m_0$ is invariant. Furthermore, the other quantities in the relativistic plasma frequency, namely the electron charge, $-e$, and the permittivity of vacuum, $\varepsilon_0$, are clearly Lorentz invariant.

Finally, we briefly address two ancillary issues. First, the particular fluid equilibrium chosen here, where

$$m_0 \frac{d}{dt}(\gamma_0 \mathbf{v}_0) = -e(\mathbf{E}_0 + \mathbf{v}_0 \times \mathbf{B}_0) = 0. \tag{2.234}$$

Such a particular solution of the Lorentz force equation corresponds to the so-called "$\mathbf{v} \times \mathbf{B}$" drift: in the case where the external electric and magnetic fields are constant and orthogonal, we see that

$$\mathbf{v}_0 = \frac{\mathbf{E}_0 \times \mathbf{B}_0}{\mathbf{B}_0^2}, \tag{2.235}$$

is a solution of Equation 2.234 because

$$\mathbf{E}_0 + \mathbf{v}_0 \times \mathbf{B}_0 = \mathbf{E}_0 - \mathbf{B}_0 \times \frac{(\mathbf{E}_0 \times \mathbf{B}_0)}{\mathbf{B}_0^2} = \mathbf{E}_0 - \frac{(\mathbf{B}_0 \cdot \mathbf{B}_0)\mathbf{E}_0 - (\mathbf{B}_0 \cdot \mathbf{E}_0)\mathbf{B}_0}{\mathbf{B}_0^2} = \mathbf{0}. \tag{2.236}$$

The second question regards the relativistic transformation of the continuity or charge conservation equation. If we consider the total charge contained in a volume $V$, we have

$$q = \iiint_V \rho \, dV, \tag{2.237}$$

where $\rho = -ne$ is the charge density for electrons.

The variation of the charge during a time interval $dt$ is

$$dq = \iiint_V \frac{\partial \rho}{\partial t} dV dt, \tag{2.238}$$

and it is also equal to the opposite of the charge flux through the surface $S$ surrounding the volume under consideration:

$$dq = -\iint_S \rho\mathbf{v} \cdot dS\, dt. \tag{2.239}$$

Using the divergence theorem, also called Ostrogradski's theorem, we have

$$dq = -\iiint_V \nabla \cdot (\rho\mathbf{v})\, dV\, dt. \tag{2.240}$$

Combining Equations 2.238 and 2.240, we have, in differential form,

$$\frac{\partial\rho}{\partial t} + \nabla \cdot (\rho\mathbf{v}) = 0, \tag{2.241}$$

which is the charge conservation or continuity equation. This equation can be recast in a manifestly covariant form by introducing the four-current density:

$$j_\mu = -en\frac{dx_\mu}{dt} = -enc\frac{dx_\mu}{c\,d\tau}\frac{d\tau}{dt} = -ec\frac{n}{\gamma}u_\mu, \tag{2.242}$$

which is, indeed, a four-vector, because the normalized four-velocity is a four-vector and $n/\gamma$ is a relativistic invariant. The components of $j_\mu$ are given by

$$j_\mu = (\rho c, \mathbf{j}) = \rho(c, \mathbf{v}) = \frac{\rho c}{\gamma}(u_0, \mathbf{u}), \tag{2.243}$$

and the charge conservation equation can be written as

$$\partial_\mu j^\mu = \nabla \cdot \mathbf{j} + \partial_0 \rho c = \nabla \cdot \mathbf{j} + \partial_t \rho = 0. \tag{2.244}$$

In Chapter 3, a more detailed introduction of the four-current density will be proposed, within the context of a fully covariant theory of classical electrodynamics.

## 2.13 The General Lorentz Transform

We now present the aforementioned derivation of the general Lorentz transform, $\mathcal{L}_{\mu\nu}$, using four-vectors. We closely follow the presentation given by Poincelot, which is an excellent synthesis of the work of Weyl, Fock, and Tonnelat. Again, we start from the principle of relativity and consider two inertial frames in relative, rectilinear uniform motion. We have

$$\begin{aligned} x_\mu &= a_\mu + u_\mu s, \\ x_{\mu'} &= a_{\mu'} + u_{\mu'} s, \end{aligned} \tag{2.245}$$

where time, as measured in frame $L$, corresponds to the first coordinate of the four-vector $x_\mu$, while the spatial position is given by the last three components:

$$x_\mu \equiv (x_0, x_1, x_2, x_3) \equiv (ct, \mathbf{x}) \equiv (ct, x, y, z); \tag{2.246}$$

the last identity is valid for Cartesian spatial coordinates, but the formalism can be extended to any orthogonal coordinate system to measure the spatial position. Here, the four-vector $a_\mu$ refers to the spatial origin and initial time in frame $L$, while $u_\mu$ will be interpreted as a four-velocity; both $a_\mu$ and $u_\mu$ are constant four-vectors, while $s$ is a variable parameter. The primed quantities correspond to the frame $L'$. The equations given in Equation 2.245 can also be written in terms of spatial and temporal components:

$$\begin{aligned} \mathbf{x} &= \mathbf{a} + \mathbf{u}s = \mathbf{a} + \mathbf{w}(t - t_0), \\ ct &= a_0 + u_0 s = ct_0 + c(t - t_0), \end{aligned} \tag{2.247}$$

and

$$\begin{aligned} \mathbf{x}' &= \mathbf{a}' + \mathbf{u}' s = \mathbf{a}' + \mathbf{w}'(t' - t_0'), \\ ct' &= a_0' + u_0' s = ct_0' + c(t' - t_0'). \end{aligned} \tag{2.248}$$

It is important to distinguish between $t_0$, which is the initial time, and the other quantities indexed 0, as they refer to the time-like components of the corresponding four-vectors. The principle of relativity requires that the speed of light, $c$, has the same value in Equations 2.247 and 2.248. The relative velocity of $L'$, as measured in $L$, is $\mathbf{v} = \beta c$; similarly, $\mathbf{v}' = \beta' c$ refers to the velocity of $L$, as observed in $L'$. Of course, this is to be carefully distinguished from the velocity of the test particle, which is measured as $\mathbf{w}$ in $L$, and $\mathbf{w}'$ in $L'$.

We now consider an infinitesimal displacement in space-time, which is described by

$$dx_{\mu'} = \mathcal{L}_{\mu'}^{\nu}(x_\lambda, \beta)\, dx_\nu. \tag{2.249}$$

Here, we suppose that the $4 \times 4$ matrix $\mathcal{L}_{\mu'}^{\nu}(x_\lambda, \beta)$ has a nonzero determinant:

$$\det[\mathcal{L}_{\mu'}^{\nu}(x_\lambda, \beta)] \neq 0. \tag{2.250}$$

Similarly, we write:

$$dx_\mu = \mathcal{L}_\mu^{\nu'}(x_{\lambda'}, \beta')\, dx_{\nu'}. \tag{2.251}$$

The linearity of relations, Equations 2.249 and 2.251, comes from the fact that we are considering infinitesimal displacements, where first-order terms dominate the local expansion of the transformation. Now focusing on space-like components, which we index by the roman letter $i$, we have

$$\frac{dx_i}{dx_0} = \frac{\mathcal{L}_i^{\nu'}}{\mathcal{L}_0^{\nu'}} \frac{dx_{\nu'}}{dx_{\nu'}}, \tag{2.252}$$

and we note that Equation 2.252 corresponds to the spatial components of the four-velocity: $dx_i/dx_0 = d\mathbf{x}/d\tau$.

Let us now consider a fixed point in $L'$: $dx_{i'} = 0$. Using Equation 2.252, we find that

$$\beta = \beta_i = \frac{\mathcal{L}_i^{0'}}{\mathcal{L}_0^{0'}}; \tag{2.253}$$

applying the same approach, we also have

$$\beta' = \beta_{i'} = \frac{\mathcal{L}_{i'}^{0}}{\mathcal{L}_{0'}^{0}}. \tag{2.254}$$

We can also express the spatial components of the four-velocity in $L'$ as

$$\frac{dx_{i'}}{dx_{0'}} = \frac{\mathcal{L}_{i'}^{\nu}}{\mathcal{L}_{0'}^{\nu}} \frac{dx_\nu}{dx_\nu} = \frac{\mathcal{L}_{i'}^{\nu}}{\mathcal{L}_{0'}^{\nu}} \frac{u_\nu}{u_\nu} = \frac{w_{i'}}{c}, \tag{2.255}$$

which is a constant, as we are considering a test particle in uniform motion. We can take the logarithmic derivative of Equation 2.255, to find that

$$\partial_\rho \frac{\mathcal{L}_{i'}^{\nu}}{\mathcal{L}_{i'}^{\nu}} \frac{u_\nu u^\rho}{u_\nu} = \partial_\rho \frac{\mathcal{L}_{0'}^{\nu}}{\mathcal{L}_{0'}^{\nu}} \frac{u_\nu u^\rho}{u_\nu} = k, \tag{2.256}$$

where $k$ is a constant, and Equation 2.256 is valid for $i = 1$, 2, or 3.

Therefore, we can write

$$\partial_\rho \frac{\mathcal{L}_{\mu'}^{\nu}}{\mathcal{L}_{\mu'}^{\nu}} \frac{u^\rho u_\nu}{u_\nu} = 2\varphi_\rho(x_\lambda, \beta_i) u^\rho, \tag{2.257}$$

which also yields

$$\partial_\rho \mathcal{L}_{\mu'}^{\nu} u^\rho u_\nu = 2\mathcal{L}_{\mu'}^{\nu}\varphi_\rho u^\rho u_\nu = (\mathcal{L}_{\mu'}^{\nu}\varphi_\rho + \mathcal{L}_{\mu'}^{\rho}\varphi_\nu)u^\rho u_\nu. \tag{2.258}$$

Since Equation 2.258 is valid independently from $u^\rho$ and $u_\nu$, we can formally divide each side by these four-vectors, to obtain the four-gradient of the differential transformation matrix:

$$\partial_\rho \mathcal{L}_{\mu'}^{\nu} = \mathcal{L}_{\mu'}^{\nu}\varphi_\rho + \mathcal{L}_{\mu'}^{\rho}\varphi_\nu. \tag{2.259}$$

Let us now turn our attention to the metric:

$$ds^2 = g_{\mu\nu} dx^\mu dx^\nu; \tag{2.260}$$

as the reference frame $L$ is inertial, and because the spatial axes are orthogonal, the metric is symmetrical, and we have

$$g_{\mu\nu} = g_{\nu\mu}, \quad g_{00} = 1, \quad g_{0i} = g_{i0} = 0, \quad g_{ij} = -\delta_{ij}. \tag{2.261}$$

This corresponds to the flat, hyperbolic space-time of special relativity. Of course, the same holds for $L'$:

$$\begin{aligned} ds'^2 &= g_{\mu'\nu'} dx^{\mu'} dx^{\nu'}, \\ g_{\mu'\nu'} &= g_{\nu'\mu'}, \quad g_{0'0'} = 1, \quad g_{0'i'} = g_{i'0'} = 0, \quad g_{i'j'} = -\delta_{i'j'}. \end{aligned} \tag{2.262}$$

At this point in the derivation our next step is to consider the synchronization of clocks in the two inertial frames. This allows us to establish a relation between the differential lengths defined in Equations 2.260 and 2.262:

$$ds'^2 = \Lambda(x_\mu, \beta_i) ds^2. \tag{2.263}$$

Thus, we have

$$g_{\rho'\nu'} \mathcal{L}_{\lambda}^{\rho'} \mathcal{L}_{\mu}^{\nu'} dx^\lambda dx^\mu = \Lambda g_{\lambda\mu} dx^\lambda dx^\mu, \tag{2.264}$$

where we have used Equations 2.249, 2.260, 2.263, and 2.264.

Since Equation 2.264 holds independently of the differential elements $dx^\lambda$ and $dx^\mu$, we find that

$$g_{\rho'\nu'} \mathcal{L}_{\lambda}^{\rho'} \mathcal{L}_{\mu}^{\nu'} = \Lambda g_{\lambda\mu}. \tag{2.265}$$

Deriving Equation 2.265 with respect to $x^\gamma$, we have

$$g_{\rho'\nu'}(\mathcal{L}_\mu^{\nu'}\partial_\gamma\mathcal{L}_\lambda^{\rho'} + \mathcal{L}_\lambda^{\rho'}\partial_\gamma\mathcal{L}_\mu^{\nu'}) = g_{\lambda\mu}\partial_\gamma\Lambda. \tag{2.266}$$

We can now use the relation established in Equation 2.259:

$$g_{\rho'\nu'}[\mathcal{L}_\lambda^{\rho'}(\mathcal{L}_\mu^{\nu'}\varphi_\gamma + \mathcal{L}_\gamma^{\nu'}\varphi_\mu) + \mathcal{L}_\mu^{\nu'}(\mathcal{L}_\lambda^{\rho'}\varphi_\gamma + \mathcal{L}_\gamma^{\rho'}\varphi_\lambda)] = g_{\lambda\mu}\partial_\gamma\Lambda, \tag{2.267}$$

where we can group terms to obtain

$$g_{\rho'\nu'}(\mathcal{L}_\lambda^{\rho'}\mathcal{L}_\gamma^{\nu'}\varphi_\mu + \mathcal{L}_\mu^{\nu'}\mathcal{L}_\gamma^{\rho'}\varphi_\lambda + 2\mathcal{L}_\lambda^{\rho'}\mathcal{L}_\mu^{\nu'}\varphi_\gamma) = g_{\lambda\mu}\partial_\gamma\Lambda. \tag{2.268}$$

Now using Equation 2.265, we have

$$g_{\lambda\gamma}\varphi_\mu + g_{\gamma\mu}\varphi_\lambda + 2g_{\lambda\mu}\varphi_\gamma = g_{\lambda\mu}\frac{\partial_\gamma\Lambda}{\Lambda} = g_{\lambda\mu}\partial_\gamma(\ln\Lambda). \tag{2.269}$$

Let us first examine this result for $\gamma = \lambda = \mu$. Using the properties of the metric, we have

$$4\varphi_\gamma = \partial_\gamma(\ln\Lambda). \tag{2.270}$$

This time, let us choose $\gamma \neq \lambda = \mu$:

$$2\varphi_\gamma = \partial_\gamma(\ln\Lambda). \tag{2.271}$$

Equations 2.270 and 2.271 are only consistent if $\partial_\gamma(\ln\Lambda) = 0$; therefore, $\Lambda$ is only a function of the relative velocity: $\Lambda = \Lambda(\beta_i)$. Furthermore,

$$\varphi_\gamma = \frac{1}{2}\partial_\gamma\Lambda = 0, \tag{2.272}$$

and, after Equation 2.259,

$$\partial_\gamma\mathcal{L}_\nu^{\mu'} = 0, \quad \mathcal{L}_\nu^{\mu'} = \mathcal{L}_\nu^{\mu'}(\beta_i). \tag{2.273}$$

Since the infinitesimal transform does not depend on the local coordinates, the integral will result in a linear transformation:

$$x_{\mu'} = \mathcal{L}_{\mu'}^{\nu}x_\nu + b_{\mu'}. \tag{2.274}$$

In addition, if the spatial origins of the two frames coincide at some common initial time, the constant four-vector will be equal to zero. In other cases, $b_\mu$, corresponds to a simple translation in space-time.

We now need to show that $\Lambda(\beta_i) = 1$. We start by considering the fact that

$$\Lambda(\beta_i) = \Lambda(\beta^2), \tag{2.275}$$

because Equation 2.263 must remain unchanged if we invert the direction of the spatial axes in $L$ and $L'$. Of course, we also have

$$\Lambda(\beta_{i'}) = \Lambda(\beta'^2). \tag{2.276}$$

Furthermore, if we apply the transformation twice, it is clear that

$$\Lambda(\beta^2)\Lambda(\beta'^2) = 1. \tag{2.277}$$

Using the relation between the Lorentz transform matrix elements and $\Lambda$, as expressed in Equation 2.265, we have

$$\begin{aligned} g_{\mu'\nu'}\mathcal{L}_\rho^{\mu'}\mathcal{L}_\lambda^{\nu'} &= g_{\rho\lambda}\Lambda(\beta^2), \\ g_{\mu\nu}\mathcal{L}_{\rho'}^{\mu}\mathcal{L}_{\lambda'}^{\nu} &= g_{\rho'\lambda'}\Lambda(\beta'^2). \end{aligned} \tag{2.278}$$

Taking advantage of the fact that $\mathcal{L}_\lambda^\nu \mathcal{L}_\nu^\lambda = 1$, we then obtain

$$\begin{aligned} g_{\mu'\nu'}\mathcal{L}_\rho^{\mu'} &= g_{\rho\lambda}\mathcal{L}_{\nu'}^{\lambda}\Lambda(\beta^2), \\ g_{\mu\nu}\mathcal{L}_{\rho'}^{\mu} &= g_{\rho'\lambda'}\mathcal{L}_\nu^{\lambda'}\Lambda(\beta'^2). \end{aligned} \tag{2.279}$$

We now inspect specific components of Equation 2.279. For the top equality, we choose $\rho = 0$ and $\nu' = 0$, and for the bottom equation, we take $\rho' = 0$ and $\nu = 0$; we find

$$\begin{aligned} \mathcal{L}_0^{0'} &= \Lambda(\beta^2)\mathcal{L}_{0'}^{0}, \\ \mathcal{L}_{0'}^{0} &= \Lambda(\beta'^2)\mathcal{L}_0^{0'}. \end{aligned} \tag{2.280}$$

Similarly, in Equation 2.278, we choose $\rho = \nu = 0$, which yields

$$g_{\rho'\nu'}\mathcal{L}_0^{\rho'}\mathcal{L}_0^{\nu'} = \Lambda(\beta^2); \tag{2.281}$$

this can be expressed as

$$\left(\mathcal{L}_0^{0'}\right)^2 - \delta_{i'j'}\mathcal{L}_0^{i'}\mathcal{L}_0^{j'} = \Lambda(\beta^2), \tag{2.282}$$

by using the properties of the metric.

Going back to Equation 2.254, we can recast Equation 2.282 as

$$(\mathcal{L}_0^{0'})^2(1 - \delta_{i'j'}\beta^{i'}\beta^{j'}) = (\mathcal{L}_0^{0'})^2(1 - \beta'^2) = \Lambda(\beta^2), \tag{2.283}$$

where the second equality derives from the definition of the scalar product in regular three-space.

Switching between frames, we also have

$$(\mathcal{L}_{0'}^{0})^2(1 - \beta^2) = \Lambda(\beta'^2), \tag{2.284}$$

and because of Equations 2.277 and 2.280, we find

$$(\mathcal{L}_{0'}^{0})^2(1 - \beta'^2) = \Lambda(\beta^2)\Lambda(\beta'^2) = \Lambda(\beta'^2). \tag{2.285}$$

Comparing Equations 2.284 and 2.285, we find

$$\frac{1 - \beta'^2}{1 - \beta^2} = 1, \quad \beta^2 = \beta'^2, \tag{2.286}$$

and

$$\Lambda(\beta^2) = \Lambda(\beta'^2) = 1, \tag{2.287}$$

where the positive sign is chosen so that $\Lambda(\mathbf{0}) = 1$.

The conclusion is that the differential space–time length element is invariant:

$$ds'^2 = ds^2 = c^2dt^2 - dx_i^2. \tag{2.288}$$

The Lorentz transform conserves the differential length element in flat, hyperbolic space–time.

We can now express the matrix elements. Using Equation 2.287, we first have

$$\mathcal{L}_0^{0'} = \mathcal{L}_{0'}^{0} = \frac{1}{\sqrt{1 - \beta^2}} = \gamma. \tag{2.289}$$

Here, the positive sign has been selected, so that for $\beta = 0$, the time-like component of a four-vector transforms into itself.

Having established the fact that $\Lambda(\beta^2) = \Lambda(\beta'^2) = 1$, we now derive the other components of the Lorentz transform matrix. In particular, going back to Equation 2.279, we have

$$\begin{aligned} g_{\alpha'\varepsilon'} \mathcal{L}_{\rho}^{\alpha'} \mathcal{L}_{\sigma}^{\varepsilon'} &= g_{\rho\sigma}, \\ g_{\alpha\varepsilon} \mathcal{L}_{\rho'}^{\alpha} \mathcal{L}_{\sigma'}^{\varepsilon} &= g_{\rho'\sigma'}, \end{aligned} \tag{2.290}$$

from which we derive

$$\begin{aligned} g_{\alpha'\varepsilon'} \mathcal{L}_{\rho}^{\alpha'} &= g_{\rho\sigma} \mathcal{L}_{\varepsilon'}^{\sigma}, \\ g_{\alpha\varepsilon} \mathcal{L}_{\rho'}^{\alpha} &= g_{\rho'\sigma'} \mathcal{L}_{\varepsilon}^{\sigma'}. \end{aligned} \tag{2.291}$$

In Equation 2.291, we have used the equality $\mathcal{L}_{\lambda}^{\nu} \mathcal{L}_{\nu}^{\lambda} = 1$; we now choose the following values for the indexes: $\alpha' = i'$, $\rho = 0$, in the top equation, and $\alpha = i$, $\rho' = 0'$, in the bottom equation. With this, we find that

$$\begin{aligned} \mathcal{L}_{0}^{i'} &= -\mathcal{L}_{i'}^{0}, \\ \mathcal{L}_{0'}^{i} &= -\mathcal{L}_{i}^{0'}. \end{aligned} \tag{2.292}$$

Equations 2.253 and 2.254 can now be used in conjunction with the result established above to obtain

$$\begin{aligned} \mathcal{L}_{0'}^{i} &= -\mathcal{L}_{i}^{0'} = \frac{\beta^{i}}{\sqrt{1-\beta^2}} = \gamma\beta^{i}, \\ \mathcal{L}_{0}^{i'} &= -\mathcal{L}_{i'}^{0} = \frac{\beta^{i'}}{\sqrt{1-\beta^2}} = \gamma\beta^{i'}. \end{aligned} \tag{2.293}$$

Now, consider Equation 2.291 again, and take $\alpha' = i'$, $\rho = j$ in the top equality to obtain

$$\mathcal{L}_{j}^{i'} = \mathcal{L}_{i'}^{j}. \tag{2.294}$$

Furthermore, in Equation 2.290 we take $\rho = i$, $\varepsilon = 0$, to find that

$$\begin{aligned} g_{\alpha'\varepsilon'} \mathcal{L}_{i}^{\alpha'} \mathcal{L}_{0}^{\varepsilon'} &= 0, \\ \mathcal{L}_{i}^{0'} \mathcal{L}_{0}^{0'} - \delta_{k'l'} \mathcal{L}_{i}^{k'} \mathcal{L}_{0}^{l'} &= 0. \end{aligned} \tag{2.295}$$

Equation 2.295 can also be expressed as

$$\delta_{k'l'} \mathcal{L}_{i}^{k'} \beta^{l'} = -\mathcal{L}_{0}^{0'} \delta_{ip} \beta^{p}, \tag{2.296}$$

where we have taken into account Equations 2.253 and 2.254. Similarly, we have

$$\delta_{kl}\,\mathcal{L}^{k}_{i'}\beta^{l} = -\mathcal{L}^{0}_{0'}\delta_{i'p'}\beta^{p'}. \tag{2.297}$$

We now take $\rho = i$, $\sigma = j$, in Equation 2.290; we first find

$$g_{\alpha'\varepsilon'}\mathcal{L}^{\alpha'}_{i}\mathcal{L}^{\varepsilon'}_{j} = -\delta_{ij}. \tag{2.298}$$

Using the properties of the metric, we then have

$$\delta_{k'l'}\mathcal{L}^{k'}_{i}\mathcal{L}^{l'}_{j} = \delta_{ij} + \mathcal{L}^{0'}_{i}\mathcal{L}^{0'}_{j}, \tag{2.299}$$

which can also be expressed as

$$\delta_{k'l'}\mathcal{L}^{k'}_{i}\mathcal{L}^{l'}_{j} = \delta_{ij} + (\mathcal{L}^{0'}_{0})^{2}\delta_{ip}\delta_{jq}\beta^{p}\beta^{q}. \tag{2.300}$$

Switching between frames, we also obtain

$$\delta_{kl}\mathcal{L}^{k}_{i'}\mathcal{L}^{l}_{j'} = \delta_{i'j'} + (\mathcal{L}^{0'}_{0})^{2}\delta_{i'p'}\delta_{j'q'}\beta^{p'}\beta^{q'}. \tag{2.301}$$

In summary, we find that the space-like elements, $\mathcal{L}^{j'}_{j}$, of the Lorentz transform matrix are symmetrical, as expressed in Equation 2.294. In order to determine the nine corresponding unknowns, we use a spatial rotation, which aligns the spatial axis of the two frames at the initial time, when their spatial origins coincide. Thus, the spatial components are given by

$$\mathcal{L}^{i'}_{j} = R^{i'}_{j} + \Gamma^{i'}_{j}, \tag{2.302}$$

where $R^{i'}_{j}$ is the aforementioned rotation matrix, and $\Gamma^{i'}_{j}$ reduces to zero for $\beta \to 0$.

We now use Equation 2.302 into Equation 2.297:

$$\begin{aligned} \beta^{j}(R^{i'}_{j} + \Gamma^{i'}_{j}) &= -\mathcal{L}^{0}_{0'}\beta^{i'}, \\ \beta^{j}R^{i'}_{j} &= -\mathcal{L}^{0}_{0'}\beta^{i'} - \Gamma^{i'}_{j}\beta^{j}. \end{aligned} \tag{2.303}$$

The definition of the rotation matrix yields

$$-\beta^{j}R^{i'}_{j} = \beta^{i'}, \tag{2.304}$$

and we can rewrite Equation 2.303 as

$$\Gamma^{i'}_{j}\beta^{j} = -(\mathcal{L}^{0}_{0'} - 1)\beta^{i'}. \tag{2.305}$$

Since $\Gamma_j^{i'}$ depends only on the relative velocities, we have

$$\Gamma_j^{i'} = \kappa\beta_j\beta^{i'}, \tag{2.306}$$

where $\kappa$ is a scalar.

Furthermore, we know that $\beta^j\beta_j = -\beta^2$, so Equation 2.305 becomes

$$\kappa\beta^2 = \mathcal{L}_{0'}^{0} - 1, \tag{2.307}$$

and yields the value of $\kappa$. We can then determine the matrix $\Gamma_j^{i'}$:

$$\Gamma_j^{i'} = \frac{\mathcal{L}_{0'}^{0} - 1}{\beta^2}\beta^{i'}\beta_j. \tag{2.308}$$

From this result and the relation established between the rotation matrix and $\Gamma_j^{i'}$ in Equation 2.303,

$$\Gamma_j^{i'} = \frac{\mathcal{L}_{0'}^{0} - 1}{\beta^2}R_p^{i'}\beta^p\delta_{jq}\beta^p, \tag{2.309}$$

we finally obtain the spatial components of the Lorentz transform matrix as a function of $\mathcal{L}_{0'}^{0}$:

$$\mathcal{L}_j^{i'} = R_p^{i'}\left(\delta_j^p + \frac{\mathcal{L}_0^{0'} - 1}{\beta^2}\beta^p\delta_{jq}\beta_q\right). \tag{2.310}$$

Therefore, the sought-after general expression for the Lorentz transform is

$$\begin{aligned}
x^{0'} &= \frac{x^0 - \delta_{ij}x^i\beta^j}{\sqrt{1-\beta^2}} = \gamma(x^0 - \delta_{ij}x^i\beta^j),\\
x^{i'} &= R_p^{i'}\left[\left(\delta_j^p + \frac{\mathcal{L}_0^{0'} - 1}{\beta^2}\beta^p\delta_{jq}\beta_q\right)x^i - \mathcal{L}_0^{0'}\beta^p x^0\right]\\
&= R_p^{i'}\left[\left(\delta_j^p + \frac{\gamma - 1}{\beta^2}\beta^p\delta_{jq}\beta_q\right)x^i - \gamma\beta^p x^0\right].
\end{aligned} \tag{2.311}$$

## 2.14 Thomas Precession

An interesting application of the formalism underlying the Lorentz transform is the case where three different reference frames are considered. We will show that the combination of two Lorentz transforms without rotation is equivalent to a single transform, with rotation. This rotation corresponds to Thomas precession in the case of spin. Again, we closely follow the excellent presentation given by Poincelot and examine the transformation between three frames, $L_1$, $L_2$, and $L_3$. The Lorentz transformation between the first two frames is given by

$$\begin{aligned} \mathbf{x}_2 &= \mathbf{x}_1 + \gamma(\mathbf{u})\mathbf{u}\left\{[1-\gamma^{-1}(\mathbf{u})]\frac{\mathbf{x}_1\cdot\mathbf{u}}{u^2} - t_1\right\}, \\ t_2 &= \gamma(\mathbf{u})\left(t_1 - \frac{\mathbf{x}_1\cdot\mathbf{u}}{c^2}\right), \end{aligned} \tag{2.312}$$

where $\mathbf{u}$ is the relative velocity between $L_1$ and $L_2$, $\gamma(\mathbf{u}) = 1/\sqrt{1-(\mathbf{u}/c)^2}$, where we are explicitly using the transformation formula for the spatial and temporal coordinates in each frame.

Of course, the indexes refer to the three different frames, not to the Roman numerals used in the previous section.

In order to simplify our discussion, the second transform is infinitesimal; therefore, we shall keep only the linear terms in the relative velocity between $L_2$ and $L_3$, which we call $\mathbf{v}$, with $|\mathbf{v}/c|^2 \ll 1$. Furthermore, we will assume that $\mathbf{v}$ is aligned with the $x_2$ and $x_3$-axes, which coincide. With this, the transformation between $L_2$ and $L_3$ takes the simple form

$$\begin{aligned} x_3 &= x_2 - vt_2, \\ y_3 &= y_2, \\ z_3 &= z_2, \\ t_3 &= t_2 - \frac{\mathbf{x}_2\cdot\mathbf{v}}{c^2} = t_2 - \frac{vx_2}{c^2}. \end{aligned} \tag{2.313}$$

The first three formulae in Equation 2.313 can be grouped in vector form:

$$\mathbf{x}_3 = \mathbf{x}_2 - \mathbf{v}t_2. \tag{2.314}$$

Using the velocity addition theorem, the relative velocity of $L_3$ in $L_1$ is

$$\mathbf{w}_1 = \frac{\gamma^{-1}(\mathbf{u})\mathbf{v} + \mathbf{u}\left\{[1-\gamma^{-1}(\mathbf{u})]\frac{\mathbf{v}\cdot\mathbf{u}}{u^2} + 1\right\}}{1 + \frac{\mathbf{v}\cdot\mathbf{u}}{c^2}}; \tag{2.315}$$

neglecting the quadratic terms, we find

$$\mathbf{w}_1 \simeq \mathbf{u} + \gamma^{-1}(\mathbf{u})\left\{\mathbf{v} - [1-\gamma^{-1}(\mathbf{u})]\mathbf{u}\frac{\mathbf{u}\cdot\mathbf{v}}{u^2}\right\}. \tag{2.316}$$

We can use the same argument to obtain the velocity $\mathbf{w}_3$ of $L_1$, as measured in $L_3$:

$$\mathbf{w}_3 = \frac{\gamma^{-1}(\mathbf{v})\mathbf{u} + \mathbf{v}\left\{[1-\gamma^{-1}(\mathbf{v})]\frac{\mathbf{u}\cdot\mathbf{v}}{u^2} + 1\right\}}{1 + \frac{\mathbf{u}\cdot\mathbf{v}}{c^2}}, \tag{2.317}$$

where $\gamma(\mathbf{v}) = 1/\sqrt{1-(\mathbf{v}/c)^2}$ is the relativistic factor associated with the infinitesimal velocity between $L_2$ and $L_3$.

Of course, because $\gamma(\mathbf{v})$ is an even function of $\mathbf{v}$, the lowest-order perturbation in $\mathbf{v}$ is quadratic. Therefore, we can approximate it by $\gamma(\mathbf{v}) \simeq 1$. Equation 2.317 becomes

$$\mathbf{w}_3 \simeq -\left(\mathbf{u} + \mathbf{v} - \mathbf{u}\frac{\mathbf{u}\cdot\mathbf{v}}{c^2}\right). \tag{2.318}$$

In general, the transform from $L_1$ to $L_3$ takes the form

$$\mathbf{x}_3 = \mathcal{T}^{-1}(\mathbf{x}_1) - \gamma(\mathbf{w}_1)\mathbf{w}_3\left\{[1-\gamma^{-1}(\mathbf{w}_1)]\frac{\mathbf{x}_1\cdot\mathbf{w}_1}{w_1^2} - t_1\right\}, \tag{2.319}$$

where $\gamma(\mathbf{w}_1) = 1/\sqrt{1-(\mathbf{w}_1/c)^2}$.

Using the results derived previously, we first find that

$$\begin{aligned}\mathcal{T}^{-1}(\mathbf{x}_1) &= \mathbf{x}_3 + \gamma(\mathbf{w}_1)\mathbf{w}_3\left\{[1-\gamma^{-1}(\mathbf{w}_1)]\frac{\mathbf{x}_1\cdot\mathbf{w}_1}{w_1^2} - t_1\right\}\\ &= \mathbf{x}_2 - \mathbf{v}t_2 - \gamma(\mathbf{w}_1)\left(\mathbf{u}+\mathbf{v}-\mathbf{u}\frac{\mathbf{u}\cdot\mathbf{v}}{c^2}\right)\left\{[1-\gamma^{-1}(\mathbf{w}_1)]\frac{\mathbf{x}_1\cdot\mathbf{w}_1}{w_1^2} - t_1\right\};\end{aligned} \tag{2.320}$$

we then use Equation 2.312 to obtain

$$\begin{aligned}\mathcal{T}^{-1}(\mathbf{x}_1) &= \mathbf{x}_1 + \gamma(\mathbf{u})\mathbf{u}\left\{[1-\gamma^{-1}(\mathbf{u})]\frac{\mathbf{x}_1\cdot\mathbf{u}}{u^2} - t_1\right\} - \gamma(\mathbf{u})\mathbf{v}\left(t_1 - \frac{\mathbf{x}_1\cdot\mathbf{u}}{c^2}\right)\\ &\quad -\gamma(\mathbf{w}_1)\left(\mathbf{u}+\mathbf{v}-\mathbf{u}\frac{\mathbf{u}\cdot\mathbf{v}}{c^2}\right)\left\{[1-\gamma^{-1}(\mathbf{w}_1)]\frac{\mathbf{x}_1\cdot\mathbf{w}_1}{w_1^2} - t_1\right\}.\end{aligned} \tag{2.321}$$

Further simplification occurs when the quadratic and higher-order terms in $\mathbf{v}^n$, $n \geq 2$, are neglected. We first have

$$w_1^2 \simeq u^2 + 2\gamma^{-2}(\mathbf{u})\mathbf{u}\cdot\mathbf{v} \tag{2.322}$$

and

$$\gamma^{-2}(\mathbf{w}_1) = 1 - \frac{w_1^2}{c^2} \simeq 1 - \frac{u^2}{c^2} - 2\gamma^{-2}(\mathbf{u})\frac{\mathbf{u}\cdot\mathbf{v}}{c^2} = \gamma^{-2}(\mathbf{u})\left(1 - 2\frac{\mathbf{u}\cdot\mathbf{v}}{c^2}\right). \tag{2.323}$$

From Equation 2.323, we easily obtain

$$\gamma(\mathbf{w}_1) \simeq \gamma(\mathbf{u})\left(1 + \frac{\mathbf{u}\cdot\mathbf{v}}{c^2}\right). \tag{2.324}$$

We can now use these results to evaluate the right-hand side of Equation 2.321. We begin by writing

$$\gamma(\mathbf{w}_1)\left(\mathbf{u} + \mathbf{v} - \mathbf{u}\frac{\mathbf{u}\cdot\mathbf{v}}{c^2}\right) \simeq \gamma(\mathbf{u})\left(1 + \frac{\mathbf{u}\cdot\mathbf{v}}{c^2}\right)\left(\mathbf{u} + \mathbf{v} - \mathbf{u}\frac{\mathbf{u}\cdot\mathbf{v}}{c^2}\right) = \gamma(\mathbf{u})(\mathbf{u}+\mathbf{v}). \tag{2.325}$$

Let us now focus on the term $\mathbf{x}_1 \cdot \mathbf{w}_1/w_1^2$:

$$\begin{aligned}\frac{\mathbf{x}_1\cdot\mathbf{w}_1}{w_1^2} &\simeq \frac{\mathbf{x}_1\cdot\mathbf{u} + \gamma^{-1}(\mathbf{u})\left\{\mathbf{x}_1\cdot\mathbf{v} - [1-\gamma^{-1}(\mathbf{u})]\mathbf{x}_1\cdot\mathbf{u}\dfrac{\mathbf{u}\cdot\mathbf{v}}{u^2}\right\}}{u^2 + 2\gamma^{-2}(\mathbf{u})\mathbf{u}\cdot\mathbf{v}} \\ &= \frac{1}{u^2}\left[\!\left[\mathbf{x}_1\cdot\mathbf{u} + \gamma^{-1}(\mathbf{u})\left\{\mathbf{x}_1\cdot\mathbf{v} - \mathbf{x}_1\cdot\mathbf{u}\frac{\mathbf{u}\cdot\mathbf{v}}{u^2}[1-\gamma^{-1}(\mathbf{u})]\right\}\right]\!\right] \\ &\quad\times\left[1 - 2\gamma^{-2}(\mathbf{u})\frac{\mathbf{u}\cdot\mathbf{v}}{u^2}\right].\end{aligned} \tag{2.326}$$

Limiting the expansion to first order in **v**, we find

$$\frac{\mathbf{x}_1\cdot\mathbf{w}_1}{w_1^2} \simeq \frac{1}{u^2}\left[\!\left[\mathbf{x}_1\cdot\mathbf{u}\left\{1 - \gamma^{-1}(\mathbf{u})\frac{\mathbf{u}\cdot\mathbf{v}}{u^2}[1+\gamma^{-1}(\mathbf{u})]\right\} + \gamma^{-1}(\mathbf{u})\mathbf{x}_1\cdot\mathbf{v}\right]\!\right]. \tag{2.327}$$

We now multiply Equation 2.327 by $[1 - \gamma^{-1}(\mathbf{w}_1)]$ and follow the same procedure to find

$$[1-\gamma^{-1}(\mathbf{w}_1)]\frac{\mathbf{x}_1 \cdot \mathbf{w}_1}{w_1^2} \simeq \frac{1}{u^2}[1+\gamma^{-1}(\mathbf{u})][\gamma^{-1}(\mathbf{u})\mathbf{x}_1 \cdot \mathbf{v} + \mathbf{x}_1 \cdot \mathbf{u}]. \quad (2.328)$$

Using this result and Equation 2.325 into Equation 2.321, we have

$$\begin{aligned} \mathcal{T}^{-1}(\mathbf{x}_1) = {} & \mathbf{x}_1 + \gamma(\mathbf{u})\mathbf{u}\left\{[1-\gamma^{-1}(\mathbf{u})]\frac{\mathbf{x}_1 \cdot \mathbf{u}}{u^2} - t_1\right\} - \gamma(\mathbf{u})\mathbf{v}\left(t_1 - \frac{\mathbf{x}_1 \cdot \mathbf{u}}{c^2}\right) \\ & - \gamma(\mathbf{u})(\mathbf{u}+\mathbf{v})\left\{\frac{1}{u^2}[1+\gamma^{-1}(\mathbf{u})][\gamma^{-1}(\mathbf{u})\mathbf{x}_1 \cdot \mathbf{v} + \mathbf{x}_1 \cdot \mathbf{u}] - t_1\right\}, \end{aligned} \quad (2.329)$$

which further simplifies to

$$\mathcal{T}^{-1}(\mathbf{x}_1) = \mathbf{x}_1 + \frac{1}{u^2}[1-\gamma^{-1}(\mathbf{u})][(\mathbf{x}_1 \cdot \mathbf{u})\mathbf{v} - (\mathbf{x}_1 \cdot \mathbf{v})\mathbf{u}]. \quad (2.330)$$

Let us introduce

$$d\mathbf{v} = \mathbf{w}_1 - \mathbf{u}. \quad (2.331)$$

We shall demonstrate that

$$\begin{aligned} \mathcal{T}^{-1}(\mathbf{x}_1) &= \mathbf{x}_1 - \Omega \times \mathbf{x}_1, \\ \mathcal{T}(\mathbf{x}_1) &= \mathbf{x}_1 + \Omega \times \mathbf{x}_1, \end{aligned} \quad (2.332)$$

where the rotation vector $\Omega$ is defined as

$$\Omega = -\frac{1}{u^2}[\gamma(\mathbf{u}) - 1](\mathbf{u} \times d\mathbf{v}). \quad (2.333)$$

As shown in Chapter 1, the double cross-product rule states that

$$\mathbf{a} \times (\mathbf{b} \times \mathbf{c}) = (\mathbf{a} \cdot \mathbf{c})\mathbf{b} - (\mathbf{a} \cdot \mathbf{b})\mathbf{c}; \quad (2.334)$$

we then have

$$\begin{aligned} (\mathbf{u} \times d\mathbf{v}) \times \mathbf{x}_1 &= -[(\mathbf{x}_1 \cdot d\mathbf{v})\mathbf{u} - (\mathbf{x}_1 \cdot \mathbf{u})d\mathbf{v}] \\ &= (\mathbf{x}_1 \cdot \mathbf{u})(\mathbf{w}_1 - \mathbf{u}) - [\mathbf{x}_1 \cdot (\mathbf{w}_1 - \mathbf{u})]\mathbf{u} \\ &= (\mathbf{x}_1 \cdot \mathbf{u})\mathbf{w}_1 - (\mathbf{x}_1 \cdot \mathbf{w}_1)\mathbf{u}. \end{aligned} \quad (2.335)$$

Let us now consider

$$\begin{aligned}(\mathbf{x}_1\cdot\mathbf{u})\mathbf{w}_1-(\mathbf{x}_1\cdot\mathbf{w}_1)\mathbf{u}\\ =(\mathbf{x}_1\cdot\mathbf{u})\left[\!\!\left[\mathbf{u}+\gamma^{-1}(\mathbf{u})\left\{\mathbf{v}-[1-\gamma^{-1}(\mathbf{u})]\mathbf{u}\frac{\mathbf{u}\cdot\mathbf{v}}{u^2}\right\}\right]\!\!\right]\\ -\left[\!\!\left[\mathbf{x}_1\cdot\mathbf{u}+\gamma^{-1}(\mathbf{u})\left\{\mathbf{x}_1\cdot\mathbf{v}-[1-\gamma^{-1}(\mathbf{u})](\mathbf{x}_1\cdot\mathbf{u})\frac{\mathbf{u}\cdot\mathbf{v}}{u^2}\right\}\right]\!\!\right]\mathbf{u};\end{aligned} \tag{2.336}$$

once again, we only keep the linear terms in $\mathbf{v}$ to obtain

$$(\mathbf{x}_1\cdot\mathbf{u})\mathbf{w}_1-(\mathbf{x}_1\cdot\mathbf{w}_1)\mathbf{u}\simeq\gamma^{-1}(\mathbf{u})[(\mathbf{x}_1\cdot\mathbf{u})\mathbf{v}-(\mathbf{x}_1\cdot\mathbf{v})\mathbf{u}]. \tag{2.337}$$

Comparing Equations 2.330 and 2.332, we have

$$\Omega\times\mathbf{x}_1=-\frac{1}{u^2}[1-\gamma^{-1}(\mathbf{u})][(\mathbf{x}_1\cdot\mathbf{u})\mathbf{v}-(\mathbf{x}_1\cdot\mathbf{v})\mathbf{u}]. \tag{2.338}$$

Using Equation 2.337, this becomes

$$\Omega\times\mathbf{x}_1\simeq-\frac{1}{u^2}[\gamma(\mathbf{u})-1][(\mathbf{x}_1\cdot\mathbf{u})\mathbf{w}_1-(\mathbf{x}_1\cdot\mathbf{w}_1)\mathbf{u}]. \tag{2.339}$$

Finally, using Equation 2.335, this easily leads to the formulae characterizing a Thomas precession, namely

$$\begin{aligned}\mathcal{T}(\mathbf{x}_1)&=\mathbf{x}_1+\Omega\times\mathbf{x}_1,\\ \Omega&=-\frac{1}{u^2}[\gamma(\mathbf{u})-1](\mathbf{u}\times d\mathbf{v}).\end{aligned} \tag{2.340}$$

We have thus shown that the combination of two Lorentz transformations without rotation induces a precession, described by Equation 2.340.

## 2.15 Schwinger's Approach

To conclude this chapter, we discuss briefly a very different approach to special relativity, exposed in great detail by Schwinger and co-authors in their book on classical electrodynamics. One of the merits of this view is to show that one can follow rather different paths toward the same conclusion, thus gaining physical insight in the process.

The starting point of this approach consists in noting that, while the relation between energy and momentum takes the form

$$|\mathbf{p}| = \frac{E}{c}, \quad p_\mu p^\mu = \hbar^2 k_\mu k^\mu = 0 \tag{2.341}$$

for electromagnetic waves and fields in vacuum, it is obviously invalid for particles with a nonzero rest mass. To remedy this situation, one proposes that the connection between momentum and energy be modified to read

$$\mathbf{p} = \frac{E}{c^2}\mathbf{v}, \tag{2.342}$$

so that for a particle moving at the speed of light, one recovers Equation 2.341.

If we also consider the well-known relation

$$\mathbf{v} = \frac{\partial E}{\partial \mathbf{p}}, \tag{2.343}$$

which is easily derived within the Hamiltonian formalism of classical mechanics, we find that

$$c^2 \mathbf{p} \cdot d\mathbf{p} = E dE, \tag{2.344}$$

which is easily integrated to yield

$$\mathbf{p}^2 - \frac{E^2}{c^2} = p_\mu p^\mu = \kappa, \tag{2.345}$$

where $\kappa$ is a constant.

Of course, the four-vector interpretation of this equation indicates that the constant is also Lorentz-invariant; therefore, we consider a frame where the particle is at rest. In this case, the momentum is zero, and we are left with the rest energy of the particle, with

$$p_\mu p^\mu = -m_0^2 c^2. \tag{2.346}$$

We then recover the well-known relation

$$E = \sqrt{p^2 c^2 + m_0^2 c^4}. \tag{2.347}$$

Going back to the initial assumption in Equation 2.342, we obtain the explicit values for the energy and momentum:

$$\begin{aligned}\mathbf{p} &= \frac{m_0\mathbf{v}}{\sqrt{1-(\mathbf{v}/c)^2}} = \gamma m_0\mathbf{v} = m_0 c\gamma\beta = m_0 c\mathbf{u} = m_0\frac{d\mathbf{x}}{d\tau},\\ \frac{E}{c} &= \frac{m_0 c}{\sqrt{1-(\mathbf{v}/c)^2}} = \gamma m_0 c = m_0 c u_0 = m_0\frac{dt}{d\tau}.\end{aligned} \tag{2.348}$$

As discussed previously, Equation 2.348 has two important consequences. First, in the limit where $\beta \ll 1$, we recover the Newtonian expression for the kinetic energy, $m_0 v^2/2$; second, we can express the velocity as

$$\beta = \frac{\mathbf{p}}{\sqrt{p^2 + m_0^2 c^2}} \tag{2.349}$$

and see that for a particle with nonzero rest mass, $\lim_{p,E\to\infty}(\beta) = 1,\ \beta \le 1$, which shows that to asymptotically approach the speed of light, an infinite amount of energy is required.

To finalize this digression, we briefly consider the Lagrangian of a relativistic particle in its simplest form, where the particle is free; further elaborations along this line will be presented in later chapters. In the Newtonian approach and in the absence of external potentials, the Lagrangian takes the form

$$\mathcal{L} = \mathbf{p}\cdot\left(\frac{d\mathbf{x}}{dt} - \mathbf{v}\right) + K = \mathbf{p}\cdot\left(\frac{d\mathbf{x}}{dt} - \mathbf{v}\right) + \frac{1}{2}m_0 v^2. \tag{2.350}$$

The expression of the kinetic energy is clearly nonrelativistic; we must therefore replace $K$ by a new function to be determined:

$$\mathcal{L} = \mathbf{p}\cdot\left(\frac{d\mathbf{x}}{dt} - \mathbf{v}\right) + \mathcal{K}(\mathbf{v}). \tag{2.351}$$

The relation between the energy and momentum is now given by

$$E = \mathbf{p}\cdot\mathbf{v} - \mathcal{K}(\mathbf{v}), \tag{2.352}$$

and we have

$$\mathcal{K}(\mathbf{v}) = \gamma m_0\mathbf{v}\cdot\mathbf{v} - \gamma m_0 c^2 = \gamma m_0 c^2(\beta^2 - 1) = -m_0 c^2\gamma^{-1}. \tag{2.353}$$

In the limit where $\beta \ll 1$, we have

$$\mathcal{K}(\mathbf{v}) = -m_0c^2\sqrt{1-\beta^2} \simeq -m_0c^2\left(1-\frac{1}{2}\beta^2\right) = K - m_0c^2, \tag{2.354}$$

which leads to Equation 2.349, within a constant directly related to the rest energy.

If we use the action principle on our new Lagrangian,

$$\mathcal{L} = \mathbf{p}\cdot\left(\frac{d\mathbf{x}}{dt}-\mathbf{v}\right)+\mathcal{K}(\mathbf{v}) = \mathbf{p}\cdot\left(\frac{d\mathbf{x}}{dt}-\mathbf{v}\right)-m_0c^2\gamma^{-1}, \tag{2.355}$$

we find

$$\mathbf{p} = \frac{\partial \mathcal{L}}{\partial \mathbf{v}} = \gamma m_0 \mathbf{v}, \tag{2.356}$$

because

$$\frac{\partial \gamma^{-1}}{\partial \mathbf{v}} = \frac{\partial}{\partial \mathbf{v}}\sqrt{1-\frac{\mathbf{v}\cdot\mathbf{v}}{c^2}} = -\frac{\mathbf{v}}{c^2}\gamma. \tag{2.357}$$

Finally, we can eliminate the velocity to obtain the corresponding Hamiltonian:

$$\begin{aligned}\mathcal{L} &= \mathbf{p}\cdot\frac{d\mathbf{x}}{dt}-\mathcal{H},\\ \mathcal{H} &= \sqrt{p^2c^2+m_0^2c^4} = E.\end{aligned} \tag{2.358}$$

## 2.16 References for Chapter 2

Note: the numbers listed below refer to the main bibliography and reference sections at the end of this book.

8, 9, 38, 59, 60, 63, 64, 69, 70, 71, 73, 74, 75, 92, 102, 105, 116, 117, 120, 128, 135, 136, 149, 150, 157, 165, 170, 172, 184, 185, 195, 199, 210, 220, 221, 234, 298, 301, 302, 303, 304, 362, 364, 391, 392, 393, 394, 395, 401, 402, 410, 414, 433, 515, 549, 565, 700, 709, 738, 798, 828, 844, 904.

# 3

# *Covariant Electrodynamics*

The main object of this chapter is to discuss the formalism underlying the covariant formulation of electrodynamics from both a mathematical and a physical standpoint. We will first consider four-vectors and tensors within the context of special relativity and introduce a few important examples. The Maxwell-Lorentz equations will then be presented in this framework. More advanced problems, including the covariant definition of the Maxwell stress tensor, the electromagnetic Lagrangian density, spinors, and dual tensors are also explored in this chapter.

## 3.1 Four-Vectors and Tensors

Here and throughout this book, we are using Pauli's and Einstein's notation, where Greek symbols correspond to the four coordinates of a four-vector, and Roman letters are used for the three space-like coordinates. Furthermore, one sums over repeated indices:

$$\begin{aligned} v_\mu &= (v_0, \mathbf{v}) = (v_0, v_i), \\ v_\mu &= \frac{\partial x_\mu}{\partial x^\nu} v^\nu = g_{\mu\nu} v^\nu, \\ a_\mu b^\mu &= \mathbf{a} \cdot \mathbf{b} - a_0 b_0 = a_i b_i - a_0 b_0. \end{aligned} \tag{3.1}$$

As mentioned previously, the negative sign in the scalar product of the four-vectors $a_\mu$ and $b^\mu$ reflects the hyperbolic nature of the flat space–time of special relativity. It should be noted that some other authors use a metric, defined such that $a_\mu b^\mu = a_0 b_0 - \mathbf{a} \cdot \mathbf{b}$. Of course, provided that this is used consistently throughout a given problem, the results are the same. In fact, covariant and contravariant four-vectors only take their full perspective within the broader context of general relativity. Finally, we also remark that an imaginary temporal axis is sometimes used in the literature. Again, provided that consistency is maintained throughout a specific analysis, this approach is equivalent to the one taken here.

The definition of four-vectors and higher-order tensors is given in terms of their properties under the Lorentz transform. Specifically, we consider two Galilean reference frames, where the four-positions are given by

$$\begin{aligned} x_\mu &= (ct, \mathbf{x}), \\ \bar{x}_\nu &= (ct', \mathbf{x}'), \end{aligned} \tag{3.2}$$

and the Lorentz transform that allows us to connect the two sets of coordinates:

$$\bar{x}_\nu = \mathcal{L}_\nu^\mu x_\mu = \frac{\partial \bar{x}_\nu}{\partial x_\mu} x_\mu. \tag{3.3}$$

Note that Equation 3.3 holds because the Lorentz transform is linear. We can now define four-vectors. These are mathematical objects which transform according to Equation 3.3. Therefore, we have

$$\bar{v}_\nu = \frac{\partial \bar{x}_\nu}{\partial x_\mu} v_\mu. \tag{3.4}$$

As demonstrated in Chapter 2, the contraction of two four-vectors results in a scalar, which is Lorentz-invariant:

$$a_\mu b^\mu = \bar{a}_\mu \bar{b}^\mu. \tag{3.5}$$

Higher-order tensors follow similar transformation rules:

$$\begin{aligned} \bar{T}_{\mu\nu} &= \frac{\partial x^\lambda}{\partial \bar{x}^\mu} \frac{\partial x^\sigma}{\partial \bar{x}^\nu} T_{\lambda\sigma}, \\ \bar{T}^{\mu\nu} &= \frac{\partial \bar{x}^\mu}{\partial x^\lambda} \frac{\partial \bar{x}^\nu}{\partial x^\sigma} T^{\lambda\sigma}. \end{aligned} \tag{3.6}$$

As previously discussed, the Lorentz transform can be considered as the combination of a space-time translation with a rotation in three-dimensional space, and a special Lorentz transform. For the translation, which is a simple shift of the spatial origins and initial times, we have

$$\bar{x}_\mu = x_\mu + a_\mu, \tag{3.7}$$

whereas a three-dimensional rotation takes the form

$$\begin{aligned} \bar{x}_i &= R_{ij}x_j, \\ \bar{x}_0 &= x_0, \end{aligned} \tag{3.8}$$

with $R_{ij}$ conserving length; in other words,

$$\bar{x}_i\bar{x}_i = \bar{x}_1^2 + \bar{x}_2^2 + \bar{x}_3^2 = x_ix_i = x_1^2 + x_2^2 + x_3^2. \tag{3.9}$$

Finally, the special Lorentz transform is given by

$$\begin{aligned} \bar{x}_1 &= \gamma(x_1 - \beta x_0), \\ \bar{x}_2 &= x_2, \\ \bar{x}_3 &= x_3, \\ \bar{x}_0 &= \gamma(x_0 - \beta x_1). \end{aligned} \tag{3.10}$$

Here $\beta = v/c$ is the normalized relative velocity between the two frames under consideration, and it is aligned with the $x_1$-axis; $\gamma = 1/\sqrt{1-\beta^2}$ is the relativistic mass factor. As discussed in Chapter 2, the special Lorentz transform can be considered as a rotation in flat, hyperbolic space–time. We can either explicitly use hyperbolic functions and write

$$\begin{aligned} \gamma &= \cosh\alpha, \\ \beta &= \tanh\alpha; \end{aligned} \tag{3.11}$$

in which case the length of the normalized relative four-velocity directly results from the well-known relation

$$\cosh^2\alpha - \sinh^2\alpha = 1, \tag{3.12}$$

as we have

$$u_\mu u^\mu = \mathbf{u}^2 - u_0^2 = (\gamma\beta)^2 - \gamma^2 = \sinh^2\alpha - \cosh^2\alpha = -1, \tag{3.13}$$

and Equation 3.10 reads

$$\begin{aligned} \bar{x}_1 &= \cosh\alpha x_1 - \sinh\alpha x_0, \\ \bar{x}_2 &= x_2, \\ \bar{x}_3 &= x_3, \\ \bar{x}_0 &= \cosh\alpha x_0 - \sinh\alpha x_1; \end{aligned} \tag{3.14}$$

or we can use the aforementioned imaginary time-axis, so that the four-position now reads

$$x_\mu = (ix_0, x_1, x_2, x_3) = (ix_0, x_i) = (ix_0, \mathbf{x}), \tag{3.15}$$

and the special Lorentz transform is given by

$$\begin{aligned} \bar{x}_1 &= x_1 \cos i\delta - x_0 \sin i\delta, \\ \bar{x}_2 &= x_2, \\ \bar{x}_3 &= x_3, \\ \bar{x}_0 &= x_0 \cos i\delta + x_1 \sin i\delta, \end{aligned} \tag{3.16}$$

where the rotation angle satisfies the relation $\alpha = i\delta$, as can be easily seen by considering that

$$\begin{aligned} \cosh\alpha &= \frac{e^{\alpha}+e^{-\alpha}}{2} = \frac{e^{i\delta}+e^{-i\delta}}{2} = \cos i\delta, \\ \sinh\alpha &= \frac{e^{\alpha}-e^{-\alpha}}{2} = \frac{e^{i\delta}-e^{-i\delta}}{2} = i \sin i\delta. \end{aligned} \tag{3.17}$$

Within the context of the hyperbolic functions, we see that the special Lorentz transform can easily be performed by introducing the rotation angle

$$\alpha = \arg\cosh\gamma. \tag{3.18}$$

The relative normalized four-velocity can then be given as

$$\mathbf{u} = \gamma\beta = \hat{n}\sinh\alpha, \tag{3.19}$$

where $\hat{n}$ is an arbitrary spatial unit vector. In this way, the relation

$$u_\mu u^\mu = \mathbf{u}^2 - u_0^2 = \mathbf{u}^2 - \gamma^2 = \hat{n}^2 \sinh^2\alpha - \cosh^2\alpha = -1, \tag{3.20}$$

will hold independent of the direction of the relative velocity with the axes of both frames. At this point, however, it is important to point out that these axes are aligned. In other words, no rotation has been compounded with the transform. With this, we have

$$\begin{aligned} \bar{\mathbf{x}} &= \mathbf{x}_\perp + \gamma \mathbf{x}_\parallel - \mathbf{u}x_0 = \mathbf{x}_\perp + u_0 \mathbf{x}_\parallel - \mathbf{u}x_0, \\ \bar{x}_0 &= \gamma x_0 - \mathbf{u}\cdot\mathbf{x} = u_0 x_0 - \mathbf{u}\cdot\mathbf{x}, \end{aligned} \tag{3.21}$$

which explicitly reads

$$\begin{aligned}\bar{\mathbf{x}} &= \mathbf{x} + \mathbf{u}\frac{\mathbf{u}\cdot\mathbf{x}}{\gamma+1} - \mathbf{u}x_0 = \mathbf{x} + \mathbf{u}\frac{\gamma-1}{u^2}(\mathbf{u}\cdot\mathbf{x}) - \mathbf{u}x_0,\\ \bar{x}_0 &= \gamma x_0 - \mathbf{u}\cdot\mathbf{x}.\end{aligned} \tag{3.22}$$

Here, we have used Equation 3.20: $u^2 = \gamma^2 - 1 = (\gamma - 1)(\gamma + 1)$. Of course, Equations 3.21 and 3.22 correspond to the expression

$$\begin{aligned}\bar{\mathbf{x}} &= \mathbf{x} + \mathbf{v}\left[(\gamma-1)\frac{\mathbf{x}\cdot\mathbf{v}}{v^2} - \gamma t\right],\\ \bar{t} &= \gamma\left(t - \frac{\mathbf{x}\cdot\mathbf{v}}{c^2}\right),\end{aligned} \tag{3.23}$$

introduced in the previous chapter.

## 3.2 The Electromagnetic Field Tensor

The main thrust of this section is to establish basic concepts that will help formulate Maxwell's equations and the Lorentz gauge condition in a manifestly covariant form. In Chapter 2, we have seen that by considering the Lorentz force equation within the context of special relativity, we could write it as

$$\frac{dp_\mu}{d\tau} = m_0 c\frac{du_\mu}{d\tau} = -eF_{\mu\nu}u^\nu, \tag{3.24}$$

where we have introduced the antisymmetrical tensor

$$F_{\mu\nu} = \begin{pmatrix} 0 & cB_3 & -cB_2 & -E_1 \\ -cB_3 & 0 & cB_1 & -E_2 \\ cB_2 & -cB_1 & 0 & -E_3 \\ E_1 & E_2 & E_3 & 0 \end{pmatrix}, \tag{3.25}$$

as expressed in terms of the electric field, $\mathbf{E} = (E_1, E_2, E_3)$, and the magnetic induction, $\mathbf{B} = (B_1, B_2, B_3)$. Equation 3.24 was obtained by introducing a change of variable, whereby we used the proper time instead of the laboratory time to express the equation of motion of a charged particle, and Equation 3.25

was derived by demanding that both sides of the Lorentz force equation have the proper four-vector character required by covariance.

The complementary approach followed in this chapter starts by considering the definition of the electric field and magnetic induction in terms of the scalar and vector potentials. This argument will then enable us to give a consistent and manifestly covariant definition of the electromagnetic field tensor, from which the sought-after Lorentz-invariant form of Maxwell's equations will result.

The source-free equations of the Maxwell set are

$$\nabla \cdot \mathbf{B} = 0, \tag{3.26}$$

which indicates the absence of magnetic monopoles, and

$$\nabla \times \mathbf{E} + \partial_t \mathbf{B} = \mathbf{0}. \tag{3.27}$$

It is well known that a curl is always divergence-free, so we introduce the vector potential, **A**, defined such that

$$\mathbf{B} = \nabla \times \mathbf{A}. \tag{3.28}$$

With this, Equation 3.26 is always satisfied. Let us now examine Equation 3.27, where we replace the magnetic induction according to Equation 3.28. We first have

$$\nabla \times \mathbf{E} + \partial_t (\nabla \times \mathbf{A}) = \mathbf{0}. \tag{3.29}$$

Since the curl and partial time derivative operators commute, we can rewrite Equation 3.29 as

$$\nabla \times (\mathbf{E} + \partial_t \mathbf{A}) = \mathbf{0}. \tag{3.30}$$

Moreover, we know that the curl of a gradient is always null, so we can add such a term to Equation 3.30, in order to obtain the most general result:

$$\nabla \times (\mathbf{E} + \partial_t \mathbf{A} + \nabla \phi) = \mathbf{0}. \tag{3.31}$$

Requiring that Equation 3.31 hold at any spatial position finally yields the relation between the electric field and the potentials:

$$\mathbf{E} = -\nabla \phi - \partial_t \mathbf{A}. \tag{3.32}$$

Therefore, the introduction of the scalar and vector potentials guarantees that the source-free equations of the Maxwell set are always satisfied. We now wish to interpret Equations 3.28 and 3.32 in light of the definition of the electromagnetic field tensor, Equation 3.25. This brings us back to the four-gradient operator,

$$\partial_\mu = \left(-\frac{\partial}{c\partial t}, \nabla\right), \tag{3.33}$$

which was also discussed in Chapter 2.

It is clear that to form an antisymmetric tensor, $T_{\mu\nu} = -T_{\nu\mu}$, we must apply the four-divergence to a four-vector in the following manner:

$$T_{\mu\nu} = \partial_\mu V_\nu - \partial_\nu V_\mu . \tag{3.34}$$

With this in mind, we introduce a new four-vector combining the vector and scalar potentials:

$$A_\mu = \left(\frac{\phi}{c}, \mathbf{A}\right). \tag{3.35}$$

Here, a simple dimensional argument dictates the choice of $\phi/c$ as the time-like component of the four-potential. From the definition of the electric field given in Equation 3.32, we know that

$$[\partial_t \mathbf{A}] \equiv [\nabla \phi], \tag{3.36}$$

where the brackets indicate the units or dimension of the bracketed quantity.

Furthermore, it is obvious that $[\partial_t] \equiv T^{-1}$; in other words the partial derivative with respect to time has the units of inverse time; and $[\nabla] \equiv L^{-1}$, where $L$ stands for length. We then find that

$$\frac{[\phi]}{LT^{-1}} \equiv [\mathbf{A}] \equiv \left[\frac{\phi}{c}\right]. \tag{3.37}$$

Note that, at this point, we have introduced $A_\mu$, but we have not demonstrated that it is a four-vector. To do so, one could directly apply the Lorentz transform and verify that

$$\bar{A}_\mu = \frac{\partial \bar{x}_\mu}{\partial x^\nu} A_\nu, \tag{3.38}$$

or one can use a line of reasoning similar to that used in Chapter 2, where we start from a known equation and show that the terms on one side of the equality do, indeed, form a four-vector, thus demonstrating that the other terms have the same character. We will use this approach here, after showing how we can construct the electromagnetic field tensor from the four-potential.

Following Equation 3.34, let us now consider the antisymmetrical tensor

$$T_{\mu\nu} = \partial_\mu A_\nu - \partial_\nu A_\mu. \tag{3.39}$$

It is clear that, by construction, the required symmetry property is satisfied. In addition, we see that there are only six terms that need to be considered, as we have

$$T_{\mu\nu} = \begin{pmatrix} 0 & T_{12} & T_{13} & T_{10} \\ -T_{12} & 0 & T_{23} & T_{20} \\ -T_{13} & -T_{23} & 0 & T_{30} \\ -T_{10} & -T_{20} & -T_{30} & 0 \end{pmatrix}. \tag{3.40}$$

The fact that $T_{\mu\nu}$ contains only six independent terms indicates that we are on the right track, as the electric field and magnetic induction each have three components.

Let us now explicitly derive the six independent components of $T_{\mu\nu}$. For the sake of illustration we will consider Cartesian coordinates; we then have

$$\begin{aligned} T_{12} &= \partial_1 A_2 - \partial_2 A_1 = \partial_x A_y - \partial_y A_x = B_z = B_3, \\ T_{13} &= \partial_1 A_3 - \partial_3 A_1 = \partial_x A_z - \partial_z A_x = -B_y = -B_2, \\ T_{10} &= \partial_1 A_0 - \partial_0 A_1 = \partial_x \frac{\phi}{c} + \frac{\partial_t}{c} A_x = -\frac{E_x}{c} = -\frac{E_1}{c}, \\ T_{23} &= \partial_2 A_3 - \partial_3 A_2 = \partial_y A_z - \partial_z A_y = B_x = B_1, \\ T_{20} &= \partial_2 A_0 - \partial_0 A_2 = \partial_y \frac{\phi}{c} + \frac{\partial_t}{c} A_y = -\frac{E_y}{c} = -\frac{E_2}{c}, \\ T_{30} &= \partial_3 A_0 - \partial_0 A_3 = \partial_z \frac{\phi}{c} + \frac{\partial_t}{c} A_z = -\frac{E_z}{c} = -\frac{E_3}{c}. \end{aligned} \tag{3.41}$$

Here we have explicitly used the definition of the electric field and magnetic induction, in terms of Cartesian coordinates, as given in Equations 3.28 and 3.32:

$$\begin{aligned} &\mathbf{B} = \nabla \times \mathbf{A}, \quad B_x = \partial_y A_z - \partial_z A_y, \quad B_y = \partial_z A_x - \partial_x A_z, \quad B_z = \partial_x A_y - \partial_y A_x, \\ &\mathbf{E} = -\nabla \phi - \partial_t \mathbf{A}, \quad E_x = -\partial_x \phi - \partial_t A_x, \quad E_y = -\partial_y \phi - \partial_t A_y, \quad E_z = -\partial_z \phi - \partial_t A_z. \end{aligned} \tag{3.42}$$

We can now group our results to compare $T_{\mu\nu}$ and $F_{\mu\nu}$:

$$cT_{\mu\nu} = \begin{pmatrix} 0 & cB_3 & -cB_2 & -E_1 \\ -cB_3 & 0 & cB_1 & -E_2 \\ cB_2 & -cB_1 & 0 & -E_3 \\ E_1 & E_2 & E_3 & 0 \end{pmatrix} = F_{\mu\nu}. \tag{3.43}$$

Thus, we have derived the relation between the electromagnetic field tensor and the four-potential. In addition, by considering the Lorentz force equation in view of this development, it is now clear that the four-potential is, indeed, a four-vector. For an electron, we have

$$m_0 c a_\mu = \frac{dp_\mu}{d\tau} = -eF_{\mu\nu}u^\nu = -ec(\partial_\mu A_\nu - \partial_\nu A_\mu)u^\nu, \tag{3.44}$$

and the equality implies that the four-potential has the correct four-vector character under the Lorentz transformation.

To further demonstrate the usefulness of the tensorial approach, a derivation of the electromagnetic field tensor in cylindrical coordinates is presented in the notes at the end of this chapter.

In closing, we also note that one can explicitly express the electric field and magnetic induction in terms of the electromagnetic field tensor as follows:

$$\begin{aligned} E_i &= F_{i0} = -F_{0i}, \\ cB_i &= F_{jk} = -F_{kj}, \end{aligned} \tag{3.45}$$

where $i, j, k$ represent a permutation of the numbers 1, 2, and 3.

## 3.3 Covariant Form of Maxwell–Lorentz Equations

With the definition of the electromagnetic field tensor at hand, we are now in a position to express Maxwell's equations and the Lorentz force equation in a fully covariant manner. As explained earlier, the source-free equations of the Maxwell set are automatically satisfied if we express the fields in terms of potentials:

$$\mathbf{B} = \nabla \times \mathbf{A}, \quad \mathbf{E} = -\nabla\phi - \partial_t \mathbf{A}; \tag{3.46}$$

the covariant form of this equation is

$$\frac{F_{\mu\nu}}{c} = \partial_\mu A_\nu - \partial_\nu A_\mu . \tag{3.47}$$

For the expression of the Lorentz force equation, one starts by introducing the four-momentum, which can also be related to the normalized four-velocity,

$$p_\mu = \left(\frac{E^{\cdot}}{c}, \mathbf{p}\right) = m_0 \frac{dx_\mu}{d\tau} m_0 c u_\mu = m_0 c(\gamma, \mathbf{u}), \tag{3.48}$$

to obtain Equation 3.44

$$m_0 c a_\mu = \frac{dp_\mu}{d\tau} = -eF_{\mu\nu}u^\nu = -ec(\partial_\mu A_\nu - \partial_\nu A_\mu)u^\nu, \tag{3.49}$$

where $a_\mu = d^2x_\mu/d\tau^2 = du_\mu/d\tau$ is the four-acceleration. By comparison, the three-dimensional form of the Lorentz force equation is

$$\frac{d\mathbf{p}}{dt} = -e(\mathbf{E} + \mathbf{v} \times \mathbf{B}). \tag{3.50}$$

It is well known that by taking the dot product of Equation 3.50 with the momentum, we can derive the equation governing the evolution of the energy. We first have

$$\mathbf{p} \cdot \frac{d\mathbf{p}}{dt} = -e\mathbf{p} \cdot \mathbf{E} = \frac{1}{2}\frac{d}{dt}(\mathbf{p}^2). \tag{3.51}$$

Next, we consider the relation between the energy and momentum,

$$E^2 = \mathbf{p}^2 c^2 + m_0^2 c^4. \tag{3.52}$$

Therefore, we find

$$\frac{d}{dt}(E^2) = 2E\frac{dE}{dt} = c^2\frac{d}{dt}(\mathbf{p}^2) = -2c^2 e\mathbf{p} \cdot \mathbf{E}; \tag{3.53}$$

since $E = \gamma m_0 c^2$, and $\mathbf{p} = \gamma m_0 \mathbf{v}$, this finally yields

$$\frac{dE}{dt} = m_0 c^2 \frac{d\gamma}{dt} = -e\mathbf{v} \cdot \mathbf{E}. \tag{3.54}$$

The covariant analog of this derivation is very simple: the energy is the time-like component of the four-momentum. Therefore, we have

$$\frac{dp_0}{d\tau} = m_0 c\frac{d\gamma}{d\tau} = -eF_{0\nu}u^\nu = -ec(\partial_0 A_\nu - \partial_\nu A_0)u^\nu. \tag{3.55}$$

Now, the components of the four-vector $\partial_0 A_\nu - \partial_\nu A_0$ are given by

$$\partial_0 A_\nu - \partial_\nu A_0 = \left(0, -\frac{1}{c}\partial_t \mathbf{A} - \nabla\frac{\phi}{c}\right) = \left(0, \frac{\mathbf{E}}{c}\right), \tag{3.56}$$

where the fact that the time-like component is equal to zero results directly from the antisymmetrical character of the electromagnetic field tensor. Contraction with the normalized four-velocity immediately yields

$$\frac{dp_0}{d\tau} = -e\mathbf{E}\cdot\mathbf{u}, \quad \frac{d\gamma}{d\tau} = -\frac{e\mathbf{u}\cdot\mathbf{E}}{m_0 c}. \tag{3.57}$$

To translate back to Equation 3.54, we simply change variables:

$$\frac{d\gamma}{dt} = \frac{d\gamma}{d\tau}\frac{d\tau}{dt} = \frac{1}{\gamma}\frac{d\gamma}{d\tau} = -\frac{e}{m_0 c}\frac{\mathbf{u}}{\gamma}\cdot\mathbf{E} = -\frac{e\mathbf{v}\cdot\mathbf{E}}{m_0 c^2}. \tag{3.58}$$

Another important and closely related aspect of the Lorentz force equation is the fact that it must satisfy the condition

$$u^\mu a_\mu = 0 = -\frac{e}{m_0 c}u^\mu F_{\mu\nu}u^\nu, \tag{3.59}$$

because the length of the four-velocity is constant:

$$\frac{d}{d\tau}(u_\mu u^\mu) = 0. \tag{3.60}$$

Indeed, Equation 3.59 holds because of the antisymmetrical character of the field tensor:

$$u^\mu F_{\mu\nu}u^\nu = cu^\mu(\partial_\mu A_\nu - \partial_\nu A_\mu)u^\nu = 0. \tag{3.61}$$

For an explicit demonstration of Equation 3.61, we refer the reader to the notes at the end of this chapter. Of course, the length of the normalized four-velocity is essentially a statement of the relation between energy and momentum:

$$u_\mu u^\mu = \mathbf{u}^2 - \gamma^2 = -1, \quad p_\mu p^\mu = m_0^2 c^2 (u_\mu u^\mu) = \mathbf{p}^2 - \frac{E^2}{c^2} = -m_0^2 c^2. \quad (3.62)$$

Our next task consists of examining the remaining equations of the Maxwell set, the equations with source terms:

$$\nabla \cdot \mathbf{E} = \frac{\rho}{\varepsilon_0}, \quad (3.63)$$

which is Gauss's theorem, and

$$\nabla \times \mathbf{B} - \frac{1}{c^2}\partial_t \mathbf{E} = \mu_0 \mathbf{j}, \quad (3.64)$$

which corresponds to Ampère's law, as modified by Maxwell to take into account the displacement current. Here, $\rho = -ne$, is the charge density, and $\mathbf{j} = \rho\mathbf{v} = -en\mathbf{v}$, is the current density, as expressed in terms of the volume density, $n$.

Since we now know the deep relation between the field tensor and the four-potential, let us recast Equations 3.63 and 3.64 in terms of the scalar and vector potentials. We first have

$$\nabla \cdot (-\nabla\phi - \partial_t \mathbf{A}) = \frac{\rho}{\varepsilon_0}, \quad (3.65)$$

while Ampère's law reads

$$\nabla \times (\nabla \times \mathbf{A}) + \frac{1}{c^2}\partial_t(\nabla\phi + \partial_t \mathbf{A}) = \mu_0 \mathbf{j}. \quad (3.66)$$

Taking advantage of the fact that the gradient operator and partial derivative with respect to time commute, Equation 3.65 can be modified slightly to read

$$\Delta\phi - \frac{1}{c^2}\partial_t^2\phi + \frac{\rho}{\varepsilon_0} = -\partial_t\left(\frac{1}{c^2}\partial_t\phi + \nabla \cdot \mathbf{A}\right). \quad (3.67)$$

Here we have added the quantity $-\frac{1}{c^2}\partial_t^2\phi$ on each side of the equality. For Equation 3.66, we begin by expanding the double curl:

$$\nabla \times (\nabla \times \mathbf{A}) = (\nabla \cdot \mathbf{A})\nabla - \nabla^2 \mathbf{A} = \nabla(\nabla \cdot \mathbf{A}) - \Delta\mathbf{A}, \quad (3.68)$$

where we have treated $\nabla$ as a regular vector and used the well-known identity

$$\mathbf{a} \times (\mathbf{b} \times \mathbf{c}) = (\mathbf{a} \cdot \mathbf{c})\mathbf{b} - (\mathbf{a} \cdot \mathbf{b})\mathbf{c}. \tag{3.69}$$

As a result, we find that

$$\nabla(\nabla \cdot \mathbf{A}) + \Delta\mathbf{A} + \frac{1}{c^2}\partial_t(\nabla\phi + \partial_t\mathbf{A}) = \mu_0\mathbf{j}, \tag{3.70}$$

which can be rearranged to yield

$$\Delta\mathbf{A} - \frac{1}{c^2}\partial_t^2\mathbf{A} + \mu_0\mathbf{j} = \nabla\left(\frac{1}{c^2}\partial_t\phi + \nabla \cdot \mathbf{A}\right). \tag{3.71}$$

Equations 3.67 and 3.71 have identical structures; moreover, because of the definition of the electric field and magnetic induction in terms of the potentials, we see that a transformation of the potentials according to

$$A_\mu \to A_\mu + \partial_\mu\Lambda, \tag{3.72}$$

where $\Lambda(x_\mu)$ is an arbitrary function of space-time, will leave the field tensor unchanged:

$$F_{\mu\nu} = c(\partial_\mu A_\nu - \partial_\nu A_\mu) = c[\partial_\mu(A_\nu + \partial_\nu\Lambda) - \partial_\nu(A_\mu + \partial_\mu\Lambda)], \tag{3.73}$$

because $\partial_\mu$ and $\partial_\nu$ commute, so that $\partial_\mu\partial_\nu\Lambda - \partial_\nu\partial_\mu\Lambda = 0$. Such a transformation is called a gauge transform and will be examined in great detail in Chapter 4. Here, we simply need to note that the quantity appearing on the right-hand side of Equations 3.67 and 3.71, namely,

$$\frac{1}{c^2}\partial_t\phi + \nabla \cdot \mathbf{A}, \tag{3.74}$$

has exactly the structure discussed above. In fact, we can use the four-gradient operator and the four-potential to rewrite Equation 3.74 in a covariant manner:

$$\frac{1}{c^2}\partial_t\phi + \nabla \cdot \mathbf{A} = \nabla \cdot \mathbf{A} - \left(-\frac{1}{c}\partial_t\right)\left(\frac{\phi}{c}\right) = \partial_\mu A^\mu. \tag{3.75}$$

Setting this quantity equal to zero yields the Lorentz gauge condition:

$$\partial_\mu A^\mu = 0. \tag{3.76}$$

However, for our purpose, we note that it is sufficient to define the arbitrary function $\Lambda$ as $\Lambda = \partial_\mu A^\mu$, to remove the right-hand side terms in Equations 3.67 and 3.71 by a simple gauge transform. Indeed, if we proceed with our covariant program and group these equations together, we obtain

$$\Box A_\mu + \mu_0 j_\mu = \partial_\mu \Lambda, \tag{3.77}$$

where we have introduced the d'Alembertian operator,

$$\Box \equiv \Delta - \frac{1}{c^2}\partial_t^2 \equiv \nabla^2 - \left(-\frac{1}{c}\partial_t\right)^2 \equiv \nabla \cdot \nabla - \partial_0\partial_0 = \partial_\nu \partial^\nu, \tag{3.78}$$

and the four-current density,

$$j_\mu = \left(\frac{\rho}{\varepsilon_0 \mu_0 c}, \mathbf{j}\right) = (\rho c, \mathbf{j}) = (\rho c, \rho \mathbf{v}) = -enc(1, \beta) = -ec\frac{n}{\gamma}(\gamma, \mathbf{u}) = -ec\frac{n}{\gamma}u_\mu. \tag{3.79}$$

Using a gauge transform or choosing the Lorentz gauge condition, we finally obtain the covariant form of the Maxwell-Lorentz electrodynamics:

$$\begin{aligned} F_{\mu\nu} &= c(\partial_\mu A_\nu - \partial_\nu A_\mu), \\ \Box A_\mu + \mu_0 j_\mu &= 0, \\ \frac{dp_\mu}{d\tau} &= -eF_{\mu\nu}u^\nu; \end{aligned} \tag{3.80}$$

where the four-current density and the four-momentum are both related to the normalized four-velocity, as

$$\begin{aligned} j_\mu &= -e\frac{n}{\gamma}u_\mu, \\ p_\mu &= m_0 c u_\mu. \end{aligned} \tag{3.81}$$

Additional constraints are described by

$$\begin{aligned} \partial_\mu A^\mu &= 0, \\ u_\mu u^\mu &= -1, \\ \partial_\mu j^\mu &= 0, \end{aligned} \tag{3.82}$$

which represent, respectively, the Lorentz gauge condition; the relation between energy and momentum, which defines the rest mass; and the continuity, or charge conservation equation. The three-dimensional equivalent of this last equality is easily obtained by applying the four-vector contraction rule:

$$\partial_\mu j^\mu = \nabla \cdot \mathbf{j} - \left(-\frac{1}{c}\partial_t\right) j_0 = \nabla \cdot \mathbf{j} + \partial_t \rho = 0. \tag{3.83}$$

The fact that $j_\mu$ is a four-vector can be established in two different ways. First, one can simply consider the driven wave equation, $\Box A_\mu + \mu_0 j_\mu = 0$, and take into account the fact that the d'Alembertian, $\Box \equiv \partial_\mu \partial^\mu$, is a scalar operator, the permeability of vacuum, $\mu_0$, is a scalar, as implied by the fundamental Lorentz invariance of vacuum, and the four-potential is a four-vector. Alternatively, one can examine the definition of the four-current density in terms of the four-velocity. In that case, one must verify that $n/\gamma$ is invariant because the electron charge, $-e$, is invariant. Indeed, we have verified in the previous chapter that this is the case, as exemplified by the relativistic plasma frequency concept introduced in Section 2.12.

Finally, in vacuum, where $j_\mu = 0$, one can solve the wave equation by Fourier transforming the problem into momentum space. We seek a solution of the form

$$A_\mu(x_\nu) = \left(\frac{1}{\sqrt{2\pi}}\right)^4 \iiiint \tilde{A}_\mu(k^\nu) \exp(-ik^\nu x_\nu) d^4 k_\lambda, \tag{3.84}$$

where we are considering an infinite, linear superposition of plane waves, and the differential element is $d^4 k_\lambda = \Pi_{\lambda=0}^{\lambda=3} dk_\lambda = dk_0\, dk_1\, dk_2\, dk_3$.

It is easily seen that, within this context, the four-gradient operator can be replaced as follows:

$$\partial_\mu \rightarrow -ik_\mu, \tag{3.85}$$

while the d'Alembertian operator becomes

$$\Box \rightarrow -k_\mu k^\mu. \tag{3.86}$$

Two important new quantities are introduced here: the four-wavenumber,

$$k_\mu = \left(\frac{\omega}{c}, \mathbf{k}\right), \tag{3.87}$$

which is the Fourier conjugate of the four-position, and the invariant phase of a plane wave,

$$\phi = -k_\mu x^\mu = k_0 x_0 - \mathbf{k} \cdot \mathbf{x} = \omega t - \mathbf{k} \cdot \mathbf{x}, \tag{3.88}$$

not to be confused with the scalar potential. The solution of the wave equation in vacuum results in the dispersion relation,

$$k_\mu k^\mu = 0 = \mathbf{k}^2 - \frac{\omega^2}{c^2}. \quad (3.89)$$

In closing, we also note the close analogy with quantum mechanics, where the momentum is replaced by an operator:

$$p_\mu \to \hbar \partial_\mu; \quad (3.90)$$

in Fourier space, this translates into

$$p_\mu \to \hbar \partial_\mu \to -i\hbar k_\mu, \quad (3.91)$$

and the dispersion relation can now be analyzed in terms of the photon rest mass, as we have

$$p_\mu p^\mu = \hbar^2 k_\mu k^\mu = 0 = -m_\gamma^2 c^2. \quad (3.92)$$

## 3.4 A Few Invariants, Four-Vectors, and Tensors Commonly Used

Before considering the implications of the electromagnetic field tensor in terms of the Lorentz transformation of the electric field and magnetic induction, it is instructive to list and consider a few invariants, four-vectors, and tensors commonly used in classical and quantum electrodynamics. Since this is a catalog of sorts, we will merely indicate the components of the four-vector, and salient relations of interest.

We begin with the four-position,

$$x_\mu = (ct, \mathbf{x}), \quad (3.93)$$

and the differential distance element,

$$ds^2 = dx_\mu dx^\mu = dx^2 + dy^2 + dz^2 - c^2 dt^2. \quad (3.94)$$

Successive derivations of the four-position, with respect to the proper time, yield the normalized four-velocity,

$$u_\mu = \frac{dx_\mu}{c\,d\tau} = (\gamma, \mathbf{u}), \quad (3.95)$$

with length

$$u_\mu u^\mu = -1, \tag{3.96}$$

and the four-acceleration,

$$a_\mu = \frac{du_\mu}{d\tau}, \tag{3.97}$$

with the important relation

$$u_\mu a^\mu = 0. \tag{3.98}$$

The four-wavenumber is the Fourier-conjugate of the four-position,

$$k_\mu = \left(\frac{\omega}{c}, \mathbf{k}\right), \tag{3.99}$$

from which we can derive the invariant phase,

$$\phi = -k_\mu x^\mu, \tag{3.100}$$

and the vacuum dispersion relation,

$$k_\mu k^\mu = 0 = \mathbf{k}^2 - \frac{\omega^2}{c^2}. \tag{3.101}$$

Multiplying Equation 3.101 by Planck's constant yields the mass-shell condition for photons,

$$\hbar^2 k_\mu k^\mu = p_\mu p^\mu = -m_\gamma^2 c^2 = 0, \tag{3.102}$$

while introducing a cutoff frequency yields an equivalent rest mass:

$$k_\mu k^\mu = -\frac{\omega_c^2}{c^2}, \quad m_\gamma = \frac{\hbar\omega_c}{c^2}. \tag{3.103}$$

The four-momentum is

$$p_\mu = m_0 c u_\mu = \left(\frac{E}{c}, \mathbf{p}\right), \tag{3.104}$$

and is related to the rest mass by

$$p_\mu p^\mu = \mathbf{p}^2 - \frac{E^2}{c^2} = m_0^2 c^2 u_\mu u^\mu = -m_0^2 c^2. \tag{3.105}$$

The four-gradient is defined as

$$\partial_\mu = \frac{\partial}{\partial x^\mu} = \left(-\frac{1}{c}\partial_t, \nabla\right). \tag{3.106}$$

The exact origin of the minus sign in front of the time-like component is interesting, and Feynman's approach to this problem is summarized in the notes at the end of this chapter.

Next, we have introduced the four-current density,

$$j_\mu = -e\frac{n}{\gamma}u_\mu = (\rho c, \mathbf{j}), \tag{3.107}$$

which is divergence-free, in accordance with the continuity, or charge conservation equation:

$$\partial_\mu j^\mu = 0. \tag{3.108}$$

In Fourier, or momentum, space Equation 3.108 becomes

$$k_\mu \tilde{j}^\mu = 0, \tag{3.109}$$

where $\tilde{j}^\mu$ is the Fourier transform of the four-current.

The four-potential is

$$A_\mu = \left(\frac{\phi}{c}, \mathbf{A}\right), \tag{3.110}$$

and may be constrained by the Lorentz gauge condition

$$\partial_\mu A^\mu = 0, \tag{3.111}$$

the four-potential also satisfies the wave equation,

$$\Box A_\mu + \mu_0 j_\mu = 0, \tag{3.112}$$

where the d'Alembertian operator is defined as

$$\Box \equiv \partial_\mu \partial^\mu \equiv \Delta - \frac{1}{c^2}\partial_t^2. \tag{3.113}$$

Finally, we have also introduced a few important tensors—the electromagnetic field tensor,

$$F_{\mu\nu} = c(\partial_\mu A_\nu - \partial_\nu A_\mu), \tag{3.114}$$

and the angular momentum,

$$M_{\mu\nu} = x_\mu p_\nu - x_\nu p_\mu, \tag{3.115}$$

which are both antisymmetrical, and the metric,

$$g_{\mu\nu} = \frac{\partial x_\mu}{\partial x^\nu}, \tag{3.116}$$

as well as the Lorentz transform,

$$\mathcal{L}_{\mu\nu} = \frac{\partial \bar{x}_\mu}{\partial x^\nu}, \tag{3.117}$$

which are symmetrical.

## 3.5 Transformation of the Fields

One of the important roles played by the electromagnetic field tensor is in helping to define the transformation law for the electric field and the magnetic induction. This, in turn, will have important applications in the study of phenomena, such as Čerenkov radiation.

As stated in Section 3.1, the electromagnetic field tensor transforms as

$$\bar{F}_{\mu\nu} = \frac{\partial x^\sigma}{\partial \bar{x}^\mu}\frac{\partial x^\lambda}{\partial \bar{x}^\nu}F_{\sigma\lambda}. \tag{3.118}$$

From this equation, we obtain the explicit transformation formulae for the electric field and the magnetic induction:

$$\begin{aligned} \bar{\mathbf{E}} &= \gamma\left[\mathbf{E} - (1-\gamma^{-1})(\mathbf{E}\cdot\mathbf{v})\frac{\mathbf{v}}{v^2} + \mathbf{v}\times\mathbf{B}\right], \\ \bar{\mathbf{B}} &= \gamma\left[\mathbf{B} - (1-\gamma^{-1})(\mathbf{B}\cdot\mathbf{v})\frac{\mathbf{v}}{v^2} - \frac{\mathbf{v}}{c^2}\times\mathbf{E}\right]. \end{aligned} \tag{3.119}$$

In addition, since $j_\mu$ is a four-vector, we also easily obtain the transformation law for the current density and the charge density:

$$\bar{\rho} = \gamma\left(\rho - \frac{\mathbf{v}}{c^2} \cdot \mathbf{j}\right),$$
$$\bar{\mathbf{j}} = \mathbf{j} + \gamma\frac{\mathbf{v}}{v^2}[(1-\gamma^{-1})(\mathbf{j} \cdot \mathbf{v}) - \rho v^2]; \tag{3.120}$$

similarly, the Lorentz transformation of the four-potential yields

$$\bar{\phi} = \gamma(\phi - \mathbf{v} \cdot \mathbf{A}),$$
$$\bar{\mathbf{A}} = \mathbf{A} + \gamma\frac{\mathbf{v}}{v^2}[(1-\gamma^{-1})(\mathbf{A} \cdot \mathbf{v}) - \phi\beta^2]. \tag{3.121}$$

One of the interesting aspects of the transformation of the electric field and magnetic induction is the fact that in certain cases, one can transform away either the magnetic induction or the electric field. If one first chooses a relative velocity

$$\mathbf{v} = \frac{c^2}{E^2}(\mathbf{E} \times \mathbf{B}), \tag{3.122}$$

in the limit where

$$\left|\frac{\mathbf{E} \times c\mathbf{B}}{E^2}\right| < 1, \tag{3.123}$$

we then have

$$\begin{aligned}
\bar{\mathbf{B}} &= \gamma\left[\mathbf{B} - (1-\gamma^{-1})(\mathbf{B} \cdot \mathbf{v})\frac{\mathbf{v}}{v^2} - \frac{\mathbf{v}}{c^2} \times \mathbf{E}\right] \\
&= \gamma\left[\mathbf{B} - \frac{1}{c^2}\frac{c^2}{E^2}(\mathbf{E} \times \mathbf{B}) \times \mathbf{E}\right] \\
&= \gamma\left\{\mathbf{B} + \frac{1}{E^2}[(\mathbf{E} \cdot \mathbf{B})\mathbf{E} - E^2\mathbf{B}]\right\} \\
&= \gamma\frac{(\mathbf{E} \cdot \mathbf{B})}{E^2}\mathbf{E}.
\end{aligned} \tag{3.124}$$

When the electric field and magnetic induction are orthogonal, $\bar{\mathbf{B}}$ vanishes. Of course, this situation is quite reminiscent of the $\mathbf{E} \times \mathbf{B}$ drift considered in Section 2.12 and which can be used here to transform away the electric field. In complete analogy with the preceding derivation, we can also choose

$$\mathbf{v} = \frac{\mathbf{B} \times \mathbf{E}}{B^2}, \quad \left|\frac{\mathbf{v}}{c}\right| < 1; \tag{3.125}$$

then

$$\begin{aligned} \bar{\mathbf{E}} &= \gamma\left[\mathbf{E} - (1-\gamma^{-1})(\mathbf{E}\cdot\mathbf{v})\frac{\mathbf{v}}{v^2} + \mathbf{v}\times\mathbf{B}\right] \\ &= \gamma\left\{\mathbf{E} - \frac{1}{B^2}[(\mathbf{B}\cdot\mathbf{E})\mathbf{B} - B^2\mathbf{E}]\right\} \\ &= -\gamma\frac{(\mathbf{B}\cdot\mathbf{E})}{B^2}\mathbf{B}, \end{aligned} \tag{3.126}$$

which vanishes for orthogonal fields.

Using the transformation of the electric field and magnetic induction, we can also recover the Lorentz force equation in the restrictive case where the external field is purely electrical in the rest frame of the charge. In that special case, we first write

$$\bar{\mathbf{f}} = -e\bar{\mathbf{E}}, \tag{3.127}$$

which can be recast as

$$\bar{\mathbf{f}} = -e\gamma\left[\mathbf{E} - (1-\gamma^{-1})(\mathbf{E}\cdot\mathbf{v})\frac{\mathbf{v}}{v^2} + \mathbf{v}\times\mathbf{B}\right]. \tag{3.128}$$

We also know that the force transforms according to

$$\mathbf{f} = \gamma^{-1}\bar{\mathbf{f}} + (1-\gamma^{-1})(\bar{\mathbf{f}}\cdot\mathbf{v})\frac{\mathbf{v}}{v^2}, \tag{3.129}$$

as shown in Section 2.8, and that the charge is a relativistic invariant; therefore, we have

$$\mathbf{f} = -e(\mathbf{E} + \mathbf{v}\times\mathbf{B}). \tag{3.130}$$

It is important to emphasize that this is not a derivation of the Lorentz force equation, as is sometimes claimed in the literature. The conditions on this derivation are quite restrictive, as described above.

Finally, we note that the transformation of the electric induction and magnetic field will be considered in great detail in the chapter concerned with Čerenkov radiation.

## 3.6 Electron and QED Units

We have already introduced the normalized four-velocity, which is the four-velocity as measured in units of the speed of light, *c*. Such normalized notations have the advantage of compactness, as a number of constants no longer appear explicitly in the equations. To generalize this useful approach in the study of electrodynamical phenomena, we need to introduce four independent units: length, time, mass, and charge. From these fundamental units, we can build any other units required by a given problem. It is clear that once the unit of length is set, one can use the speed of light to define the corresponding unit of time. Furthermore, when dealing with electrons or positrons, the natural unit of mass is $m_0 = 9.1093897(54) \times 10^{-31}$ kg, while charge can be given in units of $e = 1.60217733(49) \times 10^{-19}$ C.

To choose a unit of length, one must analyze carefully the type of problem under consideration. For example, if one is dealing with the interaction of electrons with a laser, the central wavelength, $\lambda_0$, of the laser pulse spectrum could be chosen. However, in classical and quantum electrodynamics, two natural scales are available. On the one hand, for classical problems, the classical electron radius,

$$r_0 = \frac{e^2}{4\pi\varepsilon_0 m_0 c^2} = 2.8179409 \times 10^{-15}\ \text{m}, \tag{3.131}$$

is a very useful quantity to measure lengths. On the other hand, for problems involving quantum effects, the reduced Compton wavelength,

$$\bar{\lambda}_c = \frac{\hbar}{m_0 c} = 3.861593(22) \times 10^{-13}\ \text{m}, \tag{3.132}$$

proves very effective. It is important to note that the ratio between the natural classical and the quantum electrodynamical length scales is the fine structure constant:

$$\frac{\bar{\lambda}_c}{r_0} = \frac{1}{\alpha} = \frac{e^2}{2\varepsilon_0 hc} = 137.0359895(61). \tag{3.133}$$

We also remark in passing that when one includes gravitational interactions, another fundamental unit of length appears: the so-called Planck scale, $\bar{\lambda}_P$, which is obtained by combining the Newtonian constant of gravitation,

$G = 6.67259(85) \times 10^{11}$ $m^3kg^{-1}s^{-2}$, with the speed of light in vacuum, $c = 2.99792458 \times 10^8$ $ms^{-1}$, and Planck's constant, $\hbar = 1.05457266(63) \times 10^{-34}$ Js. A simple dimensional analysis yields the desired result. We start by letting

$$ƛ_P = G^l c^m \hbar^n, \tag{3.134}$$

where the exponents, $l$, $m$, and $n$ are determined by requiring that $ƛ_P$ has the correct dimensionality

$$[ƛ_P] \equiv \text{m} \equiv [G]^l[c]^m[\hbar]^n \equiv (\text{m}^3\text{kg}^{-1}\text{s}^{-2})^l(\text{ms}^{-1})^m(\text{Js})^n. \tag{3.135}$$

Since Joules correspond to energy, we have $\text{J} = \text{kg}(\text{ms}^{-1})^2 = \text{kgxm}^2\text{s}^{-2}$, and we obtain a system of three independent equations by examining the length, mass, and time units, respectively:

$$\begin{aligned} 1 &= 3l + m + 2n, \\ 0 &= -l + n, \\ 0 &= -2l - m - n. \end{aligned} \tag{3.136}$$

We find

$$\begin{aligned} l &= \frac{1}{2}, \quad m = -\frac{3}{2}, \quad n = \frac{1}{2}, \\ ƛ_P &= \sqrt{\frac{G\hbar}{c^3}} = 1.6160496 \times 10^{-24} \text{ m}. \end{aligned} \tag{3.137}$$

Note that this unit of length is extremely small and independent of the elementary charge; it is the fundamental unit underlying superstring theories. By the same process, one can obtain the Planck mass,

$$m_P = \sqrt{\frac{\hbar c}{G}} = 2.1767127 \times 10^{-19} \text{ kg}, \tag{3.138}$$

which yields the Planck energy scale,

$$\frac{m_P c^2}{e} = 1.2210457 \times 10^8 \text{ GeV}. \tag{3.139}$$

As most of this book is devoted to high-field classical electrodynamics, we will now define classical electron units, where lengths are measured in units of $r_0$, time is given in units of $r_0/c$, while masses and charges are measured in units $m_0$, and $e$, respectively. With these basic units, we can build other quantities of interest. For example, velocities will be given in units of $r_0/(r_0/c) = c$; similarly, momentum, energy, and action have the following units:

$$m_0c, \quad m_0c^2, \quad m_0r_0c. \tag{3.140}$$

In these units, the value of the reduced Planck constant, $\hbar$, is

$$\frac{\hbar}{m_0r_0c} = \frac{\hbar}{m_0c}\frac{4\pi\varepsilon_0 m_0c^2}{e^2} = \frac{4\pi\hbar\varepsilon_0 c}{e^2} = \frac{1}{\alpha}. \tag{3.141}$$

There is a certain degree of arbitrariness remaining in our system, as the permeability and permittivity of vacuum must simply satisfy the relation

$$\mu_0\varepsilon_0 = 1, \tag{3.142}$$

since the speed of light is equal, by definition, to one in these units. However, it proves useful to let $\mu_0 = 4\pi$ and $\varepsilon_0 = 1/4\pi$, to simplify some expressions.

We can now consider the units of scalar potential, electric field, and magnetic induction. To express the voltage in more fundamental units, we recall the relation between the charge, $q$, stored in a capacitor with a potential drop, $V$: $q = CV$, where $C$ is the capacitance; therefore, $[V] = \text{V} = [q/C] = \text{CF}^{-1}$. Since the unit of permittivity is $[\varepsilon_0] = \text{Fm}^{-1}$, we can build our voltage unit by first defining the unit of capacitance, $\varepsilon_0 r_0$; we then have

$$\frac{e}{\varepsilon_0 r_0} = \frac{e}{\varepsilon_0}\frac{4\pi\varepsilon_0 m_0c^2}{e^2} = \frac{4\pi m_0c^2}{e}. \tag{3.143}$$

The basic units of electric field and magnetic induction are

$$\frac{e}{\varepsilon_0 r_0^2}, \quad \frac{e}{\varepsilon_0 r_0^2 c}, \tag{3.144}$$

respectively.

Let us briefly consider two examples. First, we shall express the Schwinger critical field in classical electron units. We have

$$\bar{E}^* = \frac{E^*}{\frac{e}{\varepsilon_0 r_0^2}} = \frac{m_0^2c^3}{e\hbar}\frac{\varepsilon_0 r_0^2}{e} = \frac{1}{4\pi\alpha}. \tag{3.145}$$

The second case is that of the Coulomb field of an electron:

$$E(r) = \frac{-e}{4\pi\varepsilon_0 r^2}; \tag{3.146}$$

in our units, we find

$$\bar{E}(\bar{r}) = \frac{-e}{4\pi\varepsilon_0 r^2}\frac{\varepsilon_0 r_0^2}{e} = \frac{-1}{4\pi\bar{r}^2}. \quad (3.147)$$

The distance at which the Coulomb field reaches the critical field is

$$\bar{r}^* = \sqrt{\alpha}, \quad (3.148)$$

which translates into

$$r^* = \bar{r}^* r_0 = \sqrt{\alpha}(\alpha\lambda\!\!\!^{-}_c) = \alpha^{3/2}\lambda\!\!\!^{-}_c. \quad (3.149)$$

Finally, in these units, the Maxwell–Lorentz equations take the form

$$\begin{aligned} F_{\mu\nu} &= (\partial_\mu A_\nu - \partial_\nu A_\mu), \\ \Box A_\mu + 4\pi j_\mu &= 0, \\ \frac{du_\mu}{d\tau} &= -F_{\mu\nu}u^\nu; \end{aligned} \quad (3.150)$$

where the four-current density and the four-momentum are now related to the normalized four-velocity, as

$$\begin{aligned} j_\mu &= -\frac{n}{\gamma}u_\mu, \\ p_\mu &= u_\mu. \end{aligned} \quad (3.151)$$

Additional constraints are described by

$$\partial_\mu A^\mu = 0, \quad u_\mu u^\mu = -1, \quad \partial_\mu j^\mu = 0. \quad (3.152)$$

This system will be used extensively in the remainder of this book.

## 3.7 Covariant Electromagnetic Lagrangian and Hamiltonian

To study the relativistic expressions of the Lagrangian and Hamiltonian of a charged particle, we start with the principle of least action. Therefore, our initial goal is to define the action of the electrodynamic system consisting of the charge, radiation field, and interaction. Our approach will be to introduce expressions for the various part of the system and to show that they lead to the correct equations when the Lagrange equations are used.

A brief reminder of the principle of least action will prove useful. The time integral of the Lagrangian is the action,

$$S = \int \mathcal{L}(q_i, \dot{q}_i, t)dt, \tag{3.153}$$

where $q_i$ represents generalized coordinates, with corresponding generalized momenta, $\dot{q}_i = dq_i/dt$.

Application of the principle of least action, $\delta S = 0$, leads to Lagrange's equations:

$$\frac{d}{dt}\left(\frac{\partial \mathcal{L}}{\partial \dot{q}_i}\right) - \frac{\partial \mathcal{L}}{\partial q_i} = 0. \tag{3.154}$$

In the simple case of a charge subjected to external fields, it is easy to show that the principle of least action leads to the Lorentz force equation. We first use the potentials to express the Lorentz force as

$$\mathbf{f} = -e(\mathbf{E} + \mathbf{v} \times \mathbf{B}) = -e[(-\nabla\phi - \partial_t\mathbf{A}) + \mathbf{v} \times (\nabla \times \mathbf{A})], \tag{3.155}$$

which can also be written in component form:

$$\begin{aligned} f_i &= -e[-\partial_i\phi - \partial_t A_i + v_j(\nabla \times \mathbf{A})_k - v_k(\nabla \times \mathbf{A})_j] \\ &= -e[-\partial_i\phi - \partial_t A_i + v_j(\partial_i A_j - \partial_j A_i) - v_k(\partial_k A_i - \partial_i A_k)_j]. \end{aligned} \tag{3.156}$$

Here, *i*, *j*, *k* represent a permutation of the numbers 1, 2, and 3. At this point, we add and subtract the term $v_i\partial_i A_i$, expand Equation 3.156, and group terms:

$$f_i = -e[-\partial_i\phi - (\partial_t A_i + v_i\partial_i A_i + v_j\partial_j A_i + v_k\partial_k A_i) + (v_i\partial_i A_i + v_j\partial_i A_j + v_k\partial_i A_k)]. \tag{3.157}$$

The first group of terms in parentheses can be identified easily with the convective derivative of the vector potential:

$$\partial_t A_i + v_i\partial_i A_i + v_j\partial_j A_i + v_k\partial_k A_i = (\partial_t + \mathbf{v}\cdot\nabla)A_i. \tag{3.158}$$

In Section 2.12, we have seen that for a fluid field, this corresponds to the total time derivative. We have

$$\frac{d}{dt}A_i(t,\mathbf{x}) = \partial_t A_i + \frac{\partial x}{\partial t}\partial_x A_i + \frac{\partial y}{\partial t}\partial_y A_i + \frac{\partial z}{\partial t}\partial_z A_i = (\partial_t + \mathbf{v}\cdot\nabla)A_i. \tag{3.159}$$

As a brief note, the covariant generalization of this convective derivative operator can be obtained by changing variable and using the proper time to obtain

$$\begin{aligned} \frac{d}{d\tau}A_\mu(t,\mathbf{x}) &= \frac{\partial t}{\partial \tau}\partial_t A_\mu + \frac{\partial x}{\partial \tau}\partial_x A_\mu + \frac{\partial y}{\partial \tau}\partial_y A_\mu + \frac{\partial z}{\partial \tau}\partial_z A_\mu \\ &= (\gamma\partial_t + \mathbf{u}\cdot\nabla)A_\mu \\ &= (u_\nu\partial^\nu)A_\mu. \end{aligned} \tag{3.160}$$

At this point, we have established that

$$f_i = -e[-\partial_i\phi - d_t A_i + (v_i\partial_i A_i + v_j\partial_i A_j + v_k\partial_i A_k)]. \tag{3.161}$$

The group of terms in parentheses contain the same component of the four-gradient, namely, $\partial_i$; therefore, we group these terms together with $\partial_i\phi$. We first have

$$-\partial_i(\phi - v_i A_i - v_j A_j - v_k A_k) = -\partial_i(\phi - \mathbf{v}\cdot\mathbf{A}), \tag{3.162}$$

but this apparently produces extra terms:

$$-\partial_i(\phi - \mathbf{v}\cdot\mathbf{A}) = -\partial_i\phi + v_i\partial_i A_i + v_j\partial_i A_j + v_k\partial_i A_k + A_i\partial_i v_i + A_j\partial_i v_j + A_k\partial_i v_k. \tag{3.163}$$

However, in the Lagrangian formalism, the generalized coordinates and momenta must be considered as independent variables, with

$$\frac{\partial q_i}{\partial q_j} = \delta_{ij}, \quad \frac{\partial \dot{q}_i}{\partial \dot{q}_j} = \delta_{ij}, \quad \frac{\partial \dot{q}_i}{\partial q_j} = \frac{\partial q_i}{\partial \dot{q}_j} = 0; \tag{3.164}$$

therefore, we have

$$f_i = -e[-\partial_i(\phi - \mathbf{v}\cdot\mathbf{A}) - d_t A_i]. \tag{3.165}$$

To obtain the sought-after equation in its final form, we note that we can formally write

$$-d_t A_i = d_t\left[\frac{\partial}{\partial v_i}(\phi - \mathbf{v}\cdot\mathbf{A})\right], \tag{3.166}$$

because the external potentials do not depend on the velocity of the charge and because the velocity components are independent variables:

$$\begin{aligned}\frac{\partial}{\partial v_i}(\phi-\mathbf{v}\cdot\mathbf{A}) &= -\frac{\partial}{\partial v_i}(\mathbf{v}\cdot\mathbf{A})\\ &= -\frac{\partial}{\partial v_i}(v_iA_i+v_jA_j+v_kA_k)\\ &= -A_i-v_i\frac{\partial A_i}{\partial v_i}-v_j\frac{\partial A_j}{\partial v_i}-v_k\frac{\partial A_k}{\partial v_i}\\ &= -A_i,\end{aligned} \tag{3.167}$$

where we have used $\partial v_j/\partial v_i = \delta_{ij}$, according to Equation 3.164.

We can now introduce the generalized potential,

$$U = -e(\phi-\mathbf{v}\cdot\mathbf{A}), \tag{3.168}$$

and write the Lorentz force equation as

$$\begin{aligned}f_i &= -e\left\{-\partial_i(\phi-\mathbf{v}\cdot\mathbf{A})+d_t\left[\frac{\partial}{\partial v_i}(\phi-\mathbf{v}\cdot\mathbf{A})\right]\right\}\\ &= -\frac{\partial U}{\partial x_i}+\frac{d}{dt}\frac{\partial U}{\partial v_i}.\end{aligned} \tag{3.169}$$

Including the kinetic energy term, $K$, the Lagrangian for a charge in external electromagnetic fields then takes the form

$$\mathcal{L} = K-U = K+e(\phi-\mathbf{v}\cdot\mathbf{A}). \tag{3.170}$$

Before proceeding with a more advanced derivation of the covariant electrodynamic Lagrangian, we note that a simple covariant generalization of $U$ can be written as follows:

$$\gamma U = -e(\gamma\phi-\mathbf{u}\cdot\mathbf{A}) = eA_\mu u^\mu; \tag{3.171}$$

moreover, the multiplication by $\gamma$ can be interpreted as a change of variable from the laboratory time to the proper time, as we have

$$\frac{dU}{d\tau} = \frac{dU}{dt}\frac{dt}{d\tau} = \gamma\frac{dU}{dt}. \tag{3.172}$$

As stated in the introductory remarks to this section, the action for a system comprising charged particles interacting with an external electromagnetic field consists of three parts:

$$S = S_1 + S_2 + S_3. \tag{3.173}$$

Here, $S_1$ describes action of the electromagnetic field distribution:

$$S_1 = \iiiint \frac{d\mathcal{L}_1}{dv} d^3\mathbf{x}\, dt = -\frac{1}{16\pi}\iiiint F_{\mu\nu}F^{\mu\nu} d^3\mathbf{x}\, dt = \frac{1}{8\pi}\iiiint (E^2 - H^2) d^3\mathbf{x}\, dt. \tag{3.174}$$

Furthermore, the covariant structure of Equation 3.174 indicates that the Lagrangian density, $d\mathcal{L}_1/dv$, is Lorentz-invariant. Next, we consider the action of the electron,

$$S_2 = \int \mathcal{L}_2\, d\tau = -m_0 c \int \sqrt{-\frac{d\bar{x}_\mu}{d\tau}\frac{d\bar{x}^\mu}{d\tau}}\, d\tau = -m_0 c \int \sqrt{-\bar{u}_\mu \bar{u}^\mu}\, d\tau, \tag{3.175}$$

where we have used $\bar{x}_\mu = (c\bar{t}, \bar{\mathbf{x}})$ to label the position of the charge, in contradistinction with the electromagnetic field tensor space-time running coordinates, $x_\mu$.

We also note that because the energy-momentum of the electromagnetic field is nonlocal, the corresponding action is given by the integral over all space–time of a Lagrangian density. By contrast, the point charge is fully localized and can be modeled as the integral over the particle's proper time of a four-dimensional delta-function. This crucial difference between the space–time coordinates, $x_\mu$, and the world line followed by a particle, $x_\mu(\tau)$, is illustrated in Figure 3.1. We now continue by deriving the covariant expression of the four-current of a point charge. By definition, the four-current is

$$j_\mu(x_\nu) = -e\left[\frac{u_\mu}{\gamma}\right](x_\nu) n(x_\nu), \tag{3.176}$$

and we can formally express Equation 3.176 as an integral over all times if we use the properties of the Dirac delta-distribution,

$$j_\mu(x_\nu) = \int u_\mu(\bar{x}_\nu) n(\bar{x}_\nu) \delta(t - \bar{t}) \frac{d\bar{t}}{d\bar{\gamma}}, \tag{3.177}$$

where $\bar{x}_\nu = x_\nu(\bar{t}) = [c\bar{t}, \mathbf{x}(\bar{t})]$.

In the case of a point charge, $\rho = \delta_3(\mathbf{x} - \bar{\mathbf{x}})$, which finally yields the covariant integral expression of the four-current:

$$j_\mu(x_\nu) = -e\int u_\mu(\bar{x}_\nu)\delta_4(x_\nu - \bar{x}_\nu) d\bar{\tau}. \tag{3.178}$$

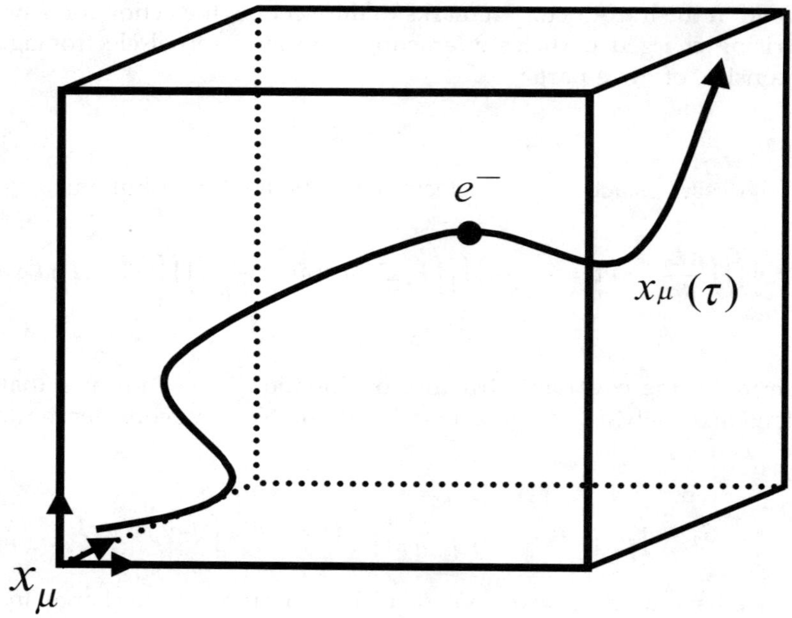


**FIGURE 3.1**
Schematic representation of the world-line, $x_\mu(\tau)$, space–time coordinate, $x_\mu$.

Here, the four-dimensional delta-function is simply given by $\delta_4(x_\nu - \bar{x}_\nu) = \delta_3(\mathbf{x} - \bar{\mathbf{x}})\delta(t - \bar{t})$. This brings us to the third part of the Lagrangian, which describes the interaction between the electron and the external fields:

$$S_3 = e\int\!\!\int\!\!\int\!\!\int\!\!\int \delta_4(x_\nu - \bar{x}_\nu)\bar{u}_\mu(\bar{\tau})A^\mu(x_\nu)d^3\mathbf{x}\,dt\,d\bar{\tau}. \tag{3.179}$$

We now want to show that the Maxwell-Lorentz equations can be derived from this action using the Lagrange equations. We first focus on Maxwell's equations. This part of the derivation can be achieved by considering $S_1$ and $S_3$ only: the corresponding Lagrangian density is

$$\frac{d\mathcal{L}}{dv}(x_\lambda) = \frac{d\mathcal{L}}{d^3\mathbf{x}}(x_\lambda) = -\frac{1}{16\pi}F_{\mu\nu}F^{\mu\nu}(x_\lambda) + e\int \delta_4(x_\lambda - \bar{x}_\lambda)\bar{u}^\mu(\bar{\tau})A_\mu(x_\lambda)d\bar{\tau}. \tag{3.180}$$

Now simply using $\mathcal{L}$ as shorthand for the Lagrangian density, the principle of least action translates into a set of four equations:

$$\frac{\partial}{\partial x^\nu}\frac{\partial \mathcal{L}}{\partial A_{\mu,\nu}} - \frac{\partial \mathcal{L}}{\partial A_\mu} = 0. \tag{3.181}$$

Here, one has to take great care with the mathematical objects involved; in particular, the four-potential plays the role of a generalized coordinate, and we have introduced the corresponding generalized momentum,

$$A_{\mu,\nu} = \frac{\partial A_\mu}{\partial x^\nu} = \partial_\nu A_\mu. \tag{3.182}$$

This can be illustrated by rewriting the electromagnetic field tensor as

$$F_{\mu\nu} = \partial_\mu A_\nu - \partial_\nu A_\mu = A_{\nu,\mu} - A_{\mu,\nu}. \tag{3.183}$$

Since the electromagnetic part of the Lagrangian density depends explicitly on the electromagnetic field tensor in a quadratic-tensorial way, we find that

$$\frac{\partial \mathcal{L}}{\partial A_{\mu,\nu}} = \frac{1}{4\pi} F^{\mu\nu}. \tag{3.184}$$

In fact, this is almost exactly equivalent to the statement that $\partial(x^2)/\partial x = 2x$, where one makes the identification $x \equiv F^{\mu\nu}$. Similarly, because the interaction Lagrangian density is a linear function of the four-potential, we have

$$\frac{\partial \mathcal{L}}{\partial A_\mu} = e\int \delta_4(x_\lambda - \bar{x}_\lambda)\bar{u}^\mu(\bar{\tau})\, d\bar{\tau}, \tag{3.185}$$

and the Lagrange equations become

$$\frac{1}{4\pi}\frac{\partial}{\partial x^\nu} F^{\mu\nu} = e\int \delta_4(x_\lambda - \bar{x}_\lambda)\bar{u}^\mu(\bar{\tau})\, d\bar{\tau}. \tag{3.186}$$

It is seen readily that Equation 3.186 is exactly equivalent to the driven wave equation, written in covariant form:

$$\frac{\partial}{\partial x^\nu} F^{\mu\nu} = \frac{\partial}{\partial x^\nu}(\partial^\mu A^\nu - \partial^\nu A^\mu) = 4\pi e\int \delta_4(x_\lambda - \bar{x}_\lambda)\bar{u}^\mu(\bar{\tau})d\bar{\tau}. \tag{3.187}$$

The drive term is the four-current density of the electron, as given in Equation 3.178, and we have

$$\frac{\partial}{\partial x^\nu}(\partial^\mu A^\nu - \partial^\nu A^\mu) = \partial^\mu(\partial_\nu A^\nu) - (\partial_\nu \partial^\nu) A^\mu = \partial^\mu \Lambda - \Box A^\mu, \tag{3.188}$$

which yields

$$\Box A^\mu + 4\pi j^\mu = \partial^\mu \Lambda, \tag{3.189}$$

where the right-hand side term can be gauge-transformed away, as discussed in Section 3.3, or where one can simply choose the Lorentz gauge condition, $\Lambda = 0$.

The second part of the derivation concerns the Lorentz force equation. In that case the required part of the Lagrangian density is

$$\frac{d\mathcal{L}}{dv}(x_\lambda) = \frac{d\mathcal{L}}{d^3\mathbf{x}}(x_\lambda) = -m_0c\sqrt{-\bar{u}_\mu\bar{u}^\mu} + e\int\delta_4(x_\lambda - \bar{x}_\lambda)\bar{u}^\mu(\bar{\tau})A_\mu(x_\lambda)d^4x_\nu, \quad (3.190)$$

where the differential element $d^4x_\nu = dx_1dx_2dx_3dx_0 = d^3\mathbf{x}dx_0$.

Applying the principle of least action yields the following set of four Lagrange equations:

$$\frac{d}{d\bar{\tau}}\frac{\partial\mathcal{L}}{\partial\bar{u}^\mu} - \frac{\partial\mathcal{L}}{\partial\bar{x}^\mu} = 0. \quad (3.191)$$

Taking the partial derivative of the Lagrangian density with respect to the electron four-velocity, we first find that

$$\frac{\partial\mathcal{L}}{\partial\bar{u}^\mu} = m_0c\bar{u}_\mu(-\bar{u}_\nu\bar{u}^\nu)^{-1/2} + e\int\delta_4(x_\lambda - \bar{x}_\lambda)A_\mu(x_\lambda)d^4x_\nu. \quad (3.192)$$

We now examine the partial derivative of Equation 3.191 with respect to the electron four-position. The kinetic energy term does not depend explicitly on the four-position, and we are left with

$$\frac{\partial\mathcal{L}}{\partial\bar{x}^\mu} = e\int\bar{u}^\nu(\bar{\tau})A_\nu(x_\lambda)\left[\frac{\partial}{\partial\bar{x}^\mu}\delta_4(x_\lambda - \bar{x}_\lambda)\right]d^4x_\nu; \quad (3.193)$$

using the properties of the Dirac delta-function, Equation 3.193 yields

$$\frac{\partial\mathcal{L}}{\partial\bar{x}^\mu} = e\bar{u}^\nu\int\delta_4(x_\lambda - \bar{x}_\lambda)\frac{\partial A_\nu}{\partial x^\mu}d^4x_\nu. \quad (3.194)$$

We now turn our attention to the total derivative of Equation 3.192 with respect to proper time:

$$\frac{d}{d\bar{\tau}}\frac{\partial\mathcal{L}}{\partial\bar{u}^\mu} = \frac{d}{d\bar{\tau}}\left[m_0c\bar{u}_\mu(-\bar{u}_\nu\bar{u}^\nu)^{-1/2} + e\int\delta_4(x_\lambda - \bar{x}_\lambda)A_\mu(x_\lambda)d^4x_\nu\right]. \quad (3.195)$$

We begin with the kinetic energy part:

$$\frac{d}{d\bar{\tau}}[m_0c\bar{u}_\mu(-\bar{u}_\nu\bar{u}^\nu)^{-1/2}] = m_0c\left[\frac{\bar{a}_\mu}{\sqrt{-\bar{u}_\nu\bar{u}^\nu}} + \bar{u}_\mu\frac{\bar{u}_\lambda\bar{a}^\lambda}{\sqrt{-\bar{u}_\nu\bar{u}^\nu}^3}\right], \quad (3.196)$$

where we have introduced the electron four-acceleration, $\bar{a}_\mu = d\bar{u}_\mu/d\bar{t}$. By virtue of the orthogonality of the four-velocity and four-acceleration, Equation 3.196 reduces to

$$\frac{d}{d\bar{\tau}}[m_0 c\bar{u}_\mu(-\bar{u}_\nu\bar{u}^\nu)^{-1/2}] = \frac{m_0 c\bar{a}_\mu}{\sqrt{-\bar{u}_\nu\bar{u}^\nu}}. \tag{3.197}$$

Next, we consider the electromagnetic part of the interaction Lagrangian:

$$\begin{aligned}\frac{d}{d\bar{\tau}}\left[e\int\delta_4(x_\lambda-\bar{x}_\lambda)A_\mu(x_\lambda)d^4x_\nu\right] &= e\int\delta_4(x_\lambda-\bar{x}_\lambda)\frac{d\bar{x}^\nu}{d\bar{\tau}}\frac{\partial A_\mu}{\partial\bar{x}^\nu}(x_\lambda)d^4x_\nu\\ &= e\int\delta_4(x_\lambda-\bar{x}_\lambda)\bar{u}^\nu\frac{\partial A_\mu}{\partial\bar{x}^\nu}(x_\lambda)d^4x_\nu.\end{aligned} \tag{3.198}$$

Grouping terms together, the principle of least action yields

$$\frac{m_0 c\bar{a}_\mu}{\sqrt{-\bar{u}_\nu\bar{u}^\nu}} + e\int\delta_4(x_\lambda-\bar{x}_\lambda)\bar{u}^\nu\frac{\partial A_\mu}{\partial\bar{x}^\nu}d^4x_\nu - e\bar{u}^\nu\int\delta_4(x_\lambda-\bar{x}_\lambda)\frac{\partial A_\nu}{\partial x^\mu}d^4x_\nu; \tag{3.199}$$

using the properties of the delta-function, namely

$$\int\delta(x-\bar{x})f(x)dx = f(\bar{x}), \tag{3.200}$$

Equation 3.199 becomes

$$\frac{m_0 c\bar{a}_\mu}{\sqrt{-\bar{u}_\nu\bar{u}^\nu}} + e\left(\bar{u}^\nu\frac{\partial A_\mu}{\partial\bar{x}^\nu} - \bar{u}^\nu\frac{\partial A_\nu}{\partial\bar{x}^\mu}\right) = 0. \tag{3.201}$$

In Equation 3.201, we recognize the electromagnetic field tensor, as measured along the electron world line:

$$\bar{F}_{\mu\nu} = \frac{\partial A_\nu}{\partial\bar{x}^\mu} - \frac{\partial A_\mu}{\partial\bar{x}^\nu} = F_{\mu\nu}(\bar{x}_\lambda), \tag{3.202}$$

and we can finally recast Equation 3.201 as the covariant form of the Lorentz force equation, as $\bar{u}_\nu\bar{u}^\nu = -1$,

$$m_0 c\bar{a}_\mu = -e\bar{F}_{\mu\nu}\bar{u}^\nu. \tag{3.203}$$

We have thus shown that the covariant Lagrangian densities described in Equations 3.180 and 3.190 lead to Maxwell's equations, Equation 3.189, and to the Lorentz force equation, Equation 3.203.

In closing, we present the three-dimensional equivalents of the Lagrangian density of the electrodynamic system. For the electromagnetic field, we have

$$\frac{d\mathcal{L}}{dv} = \frac{1}{8\pi}(\mathbf{E}^2 - \mathbf{H}^2) + e\delta(\mathbf{x} - \bar{\mathbf{x}})(\mathbf{v} \cdot \mathbf{A} - \phi), \tag{3.204}$$

where we recognize the electromagnetic field energy density, and the generalized potential of the electron, as derived in Equation 3.168. Here, $\bar{\mathbf{x}} = \mathbf{x}(\tau)$ corresponds to the spatial components of the electron world line, and the Dirac distribution provides the density model for the charge distribution. For the electron motion, we have

$$\mathcal{L} = e(\mathbf{v} \cdot \mathbf{A} - \phi) - K = e(\mathbf{v} \cdot \mathbf{A} - \phi) - mc^2 = e(\mathbf{v} \cdot \mathbf{A} - \phi) - \gamma m_0 c^2, \tag{3.205}$$

with $\gamma = 1/\sqrt{1-(\mathbf{v}/c)^2}$, and $\mathbf{v} = d\bar{\mathbf{x}}/dt$.

## 3.8 Field Four-Momentum and Maxwell Stress Tensor

Another important concept in the theory of electrodynamics is that of the energy-momentum of the electromagnetic field. We begin with a brief discussion of Poynting's theorem, which describes the conservation of energy and momentum for a system of charges interacting with an electromagnetic field distribution.

As can be seen by considering the time-like component of the Lorentz force equation, or by taking the dot product of the spatial components of that equation with the momentum, the magnetic induction does not produce any work on charges:

$$\frac{d\gamma}{d\tau} = -\frac{e\mathbf{u} \cdot \mathbf{E}}{m_0 c}. \tag{3.206}$$

In the case where the electromagnetic field is interacting with a charge density distribution, we can examine a volume in space and use the relation between the normalized four-velocity and the four-current, $j_\mu = -encu_\mu/\gamma = \rho c u_\mu/\gamma = (\rho c, \mathbf{j})$, to find that the rate of work produced in that volume is given by

$$\frac{dW}{dt} = \gamma\frac{dW}{d\tau} = \iiint_v \mathbf{j} \cdot \mathbf{E}\, dv. \tag{3.207}$$

This variation of mechanical energy must be balanced by an opposite change of the electromagnetic field energy. We can use Ampère's theorem to replace the current density. We first have

$$\nabla \times \mathbf{H} - \partial_t \mathbf{D} = \mathbf{j}, \tag{3.208}$$

where **H** is the magnetic field and **D** is the electric induction.

With this, Equation 3.207 becomes

$$\frac{dW}{dt} = \iiint_v (\nabla \times \mathbf{H} - \partial_t \mathbf{D}) \cdot \mathbf{E}\, dv. \tag{3.209}$$

Let us now turn our attention to the quantity $\mathbf{E} \cdot (\nabla \times \mathbf{H})$: we want to show that

$$\nabla \cdot (\mathbf{E} \times \mathbf{H}) = \mathbf{H} \cdot (\nabla \times \mathbf{E}) - \mathbf{E} \cdot (\nabla \times \mathbf{H}). \tag{3.210}$$

In Cartesian coordinates, we write

$$\nabla \cdot (\mathbf{E} \times \mathbf{H}) = \partial_x (\mathbf{E} \times \mathbf{H})_x + \partial_y (\mathbf{E} \times \mathbf{H})_y + \partial_z (\mathbf{E} \times \mathbf{H})_z. \tag{3.211}$$

We then expand the cross-product, to obtain

$$\nabla \cdot (\mathbf{E} \times \mathbf{H}) = \partial_x (E_y H_z - E_z H_y) + \partial_y (E_z H_x - E_x H_z) + \partial_z (E_x H_y - E_y H_x). \tag{3.212}$$

Performing the partial derivatives and rearranging terms yields

$$\begin{aligned} \nabla \cdot (\mathbf{E} \times \mathbf{H}) = {} & H_z \partial_x E_y - H_y \partial_x E_z + H_x \partial_y E_z - H_z \partial_y E_x + H_y \partial_z E_x - H_x \partial_z E_y \\ & + E_y \partial_x H_z - E_z \partial_x H_y + E_z \partial_y H_x - E_x \partial_y H_z + E_x \partial_z H_y - E_y \partial_z H_x; \end{aligned} \tag{3.213}$$

further grouping terms, we obtain

$$\begin{aligned} \nabla \cdot (\mathbf{E} \times \mathbf{H}) = {} & H_x (\partial_y E_z - \partial_z E_y) + H_y (\partial_z E_x - \partial_x E_z) + H_z (\partial_x E_y - \partial_y E_x) \\ & + E_x (\partial_z H_y - \partial_y H_z) + E_y (\partial_x H_z - \partial_z H_x) + E_z (\partial_y H_x - \partial_x H_y); \end{aligned} \tag{3.214}$$

identifying the various terms from the curl operator finally yields the sought-after identity,

$$\begin{aligned} \nabla \cdot (\mathbf{E} \times \mathbf{H}) = {} & H_x (\nabla \times \mathbf{E})_x + H_y (\nabla \times \mathbf{E})_y + H_z (\nabla \times \mathbf{E})_z \\ & - E_x (\nabla \times \mathbf{H})_x - E_y (\nabla \times \mathbf{H})_y - E_z (\nabla \times \mathbf{H})_z. \end{aligned} \tag{3.215}$$

We can now use this result with Equation 3.209, to find that

$$\frac{dW}{dt} = \iiint_v [\mathbf{H} \cdot (\nabla \times \mathbf{E}) - \nabla \cdot (\mathbf{E} \times \mathbf{H}) - \mathbf{E} \cdot \partial_t \mathbf{D}] dv. \tag{3.216}$$

The curl of the electric field is related to the magnetic induction through Faraday's law,

$$\nabla \times \mathbf{E} + \partial_t \mathbf{B} = \mathbf{0}, \tag{3.217}$$

and we now have

$$\frac{dW}{dt} = -\iiint_v [\mathbf{H} \cdot \partial_t \mathbf{B} + \nabla \cdot (\mathbf{E} \times \mathbf{H}) + \mathbf{E} \cdot \partial_t \mathbf{D}] dv. \tag{3.218}$$

In Equation 3.218, the terms $\mathbf{H} \cdot \partial_t \mathbf{B} + \mathbf{E} \cdot \partial_t \mathbf{D}$ can be grouped as follows:

$$\mathbf{H} \cdot \partial_t \mathbf{B} + \mathbf{E} \cdot \partial_t \mathbf{D} = \partial_t (\mathbf{H} \cdot \mathbf{B} + \mathbf{E} \cdot \mathbf{D}) - \mathbf{B} \cdot \partial_t \mathbf{H} - \mathbf{D} \cdot \partial_t \mathbf{E}. \tag{3.219}$$

If the medium under consideration is linear and loss-free, the relations between the fields and inductions take the simple form

$$\begin{aligned} \mathbf{D} &= \varepsilon \mathbf{E}, \\ \mathbf{B} &= \mu \mathbf{H}; \end{aligned} \tag{3.220}$$

moreover, if $\partial_t \varepsilon = \partial_t \mu = 0$, Equation 3.219 becomes equivalent to

$$\mathbf{H} \cdot \partial_t \mathbf{B} + \mathbf{E} \cdot \partial_t \mathbf{D} = \frac{1}{2} \partial_t (\mathbf{H} \cdot \mathbf{B} + \mathbf{E} \cdot \mathbf{D}) = \partial_t \left( \frac{\mu \mathbf{H}^2 + \varepsilon \mathbf{E}^2}{2} \right). \tag{3.221}$$

In the case of vacuum, $\varepsilon = \varepsilon_0$, $\mu = \mu_0$, and we can identify

$$\frac{\mu \mathbf{H}^2 + \varepsilon \mathbf{E}^2}{2} = \varepsilon_0 \frac{\mathbf{E}^2}{2} + \frac{\mathbf{B}^2}{2\mu_0}, \tag{3.222}$$

which is the electromagnetic energy density.

We can generalize this concept to the case of a linear, loss-free medium. In this case, Equation 3.218 now reads

$$\frac{dW}{dt} = -\iiint_v [\partial_t \mathcal{E} + \nabla \cdot (\mathbf{E} \times \mathbf{H})] dv, \tag{3.223}$$

where

$$\mathcal{E} = \frac{\mu\mathbf{H}^2 + \varepsilon\mathbf{E}^2}{2}. \tag{3.224}$$

At this point, we emphasize the fact that the assumptions made for the medium under consideration are fairly restrictive; in particular, nonlinear effects, such as those induced by high-intensity lasers, are entirely excluded from this discussion, as well as losses or gains. Such effects will be presented in Part 3, which is concerned with electromagnetic waves in a medium. Returning to our current discussion, we can use the relation between the variation of the electromagnetic energy in the volume and the mechanical work produced by the fields on the charges:

$$\frac{dW}{dt} = \iiint_v \mathbf{j}\cdot\mathbf{E}\,dv = -\iiint_v [\partial_t\mathcal{E} + \nabla\cdot(\mathbf{E}\times\mathbf{H})]dv. \tag{3.225}$$

Furthermore, this can also be expressed in differential form, as Equation 3.225 holds for any volume, and we have

$$\partial_t\mathcal{E} + \nabla\cdot(\mathbf{E}\times\mathbf{H}) = -\mathbf{j}\cdot\mathbf{E}, \tag{3.226}$$

which is often expressed as

$$\partial_t\mathcal{E} + \nabla\cdot\mathbf{S} = -\mathbf{j}\cdot\mathbf{E}, \tag{3.227}$$

where we have introduced the Poynting vector,

$$\mathbf{S} = \mathbf{E}\times\mathbf{H}. \tag{3.228}$$

At this point, a brief return to the question of units is in order. For most of our derivations, we have used MKSA units, where lengths are measured in meters (m), masses in kilograms (kg), time in seconds (s), and current in amperes (A). From these basic units, one can derive secondary units, such as the unit of energy, the joule (J), $\mathrm{J} = \mathrm{kg}\times(\mathrm{m}\times\mathrm{s}^{-1})^2 = \mathrm{kg}\times\mathrm{m}^2\times\mathrm{s}^{-2}$; charge, expressed in coulombs (C), $\mathrm{C} = \mathrm{A}\times\mathrm{s}$; as well as the unit of voltage or potential, the volt (V), $\mathrm{V} = \mathrm{kg}\times\mathrm{m}^2\times\mathrm{s}^{-3}\times\mathrm{A}^{-1}$. For example, it is easily verified that the unit of power, the watt (W), $\mathrm{W} = \mathrm{J}\times\mathrm{s}^{-1} = \mathrm{kg}\times\mathrm{m}^2\times\mathrm{s}^{-3}$, also corresponds to the familiar unit of electrical power: $\mathrm{V}\times\mathrm{A} = (\mathrm{kg}\times\mathrm{m}^2\times\mathrm{s}^{-3}\times\mathrm{A}^{-1})\times\mathrm{A} = \mathrm{W}$. Other important secondary units include the unit of capacitance, the farad(ay) (F), $\mathrm{F} = \mathrm{C}\times\mathrm{V}^{-1}$, and the unit of resistance, the ohm (Ω), $\Omega = \mathrm{V}\times\mathrm{A}^{-1}$.

Furthermore, a few basic equations prove useful for deriving these units and others. To derive the unit of force, one can use $\mathbf{F} = m\mathbf{a} = m(d^2\mathbf{x}/dt^2)$, to find that $[\mathbf{F}] = \text{kg} \times \text{m} \times \text{s}^{-2}$. Similarly useful are the relations $q = CV$, for the charge stored in a capacitor under a potential drop, $V$, and $V = RI$, for the voltage, $V$, across a resistor, $R$, through which a current, $I$, flows. To further illustrate the usefulness of the dimensional approach, we will derive a number of units often employed here. We start with the electric field, for which one can use the equation $\mathbf{E} = -\nabla \phi - \partial_t \mathbf{A}$, to find that $[\mathbf{E}] = \text{V} \times \text{m}^{-1}$, in addition, we also find that the unit of vector potential is $[\mathbf{A}] = \text{V} \times \text{m}^{-1} \times \text{s} = [\phi/c]$. The magnetic induction is then given by $\mathbf{B} = \nabla \times \mathbf{A}$, which yields $[\mathbf{B}] = \text{V} \times \text{m}^{-2} \times \text{s} = [\mathbf{E}/c]$, and we easily see that the cyclotron frequency,

$$[e\mathbf{B}/m_0] = \text{C} \times (\text{V} \times \text{m}^{-2} \times \text{s}) \times \text{kg}^{-1} = \text{J} \times \text{m}^{-2} \times \text{s} \times \text{kg}^{-1} = \text{s}^{-1},$$

has the correct units by using the equation for the work of a charge under a given potential drop, $W = qV$. The unit of permittivity is $[\varepsilon] = \text{F} \times \text{m}^{-1}$, as can be seen by considering a parallel plate capacitor, where $C = \varepsilon S/d$, with $\varepsilon$ the permittivity of the dielectric, $S$ the surface of the plates, and $d$ the distance between the plates. This then yields the unit of electric induction:

$$[\mathbf{D}] = [\varepsilon \mathbf{E}] = (\text{F} \times \text{m}^{-1}) \times (\text{V} \times \text{m}^{-1}) = \text{C} \times \text{m}^{-2},$$

and we can verify that the dimensionality of the equation $\nabla \cdot \mathbf{D} = \rho$ is correct:

$$[\nabla \cdot \mathbf{D}] = \text{m}^{-1} \times (\text{C} \times \text{m}^{-2}) = [\rho] = \text{C} \times \text{m}^{-3}.$$

For permeability, the fundamental vacuum relation, $\varepsilon_0 \mu_0 c^2 = 1$, can be used to yield

$$[\mu_0] = [\mu] = [\varepsilon_0 c^2]^{-1} = \text{F}^{-1} \times \text{m}^{-1} \times \text{s}^2;$$

with this result, we can derive the unit of magnetic field:

$$[\mathbf{H}] = [\mathbf{B}/\mu] = [\mathbf{E}/\mu_0 c] = [\mathbf{E}\varepsilon_0 c] = \text{A} \times \text{m}^{-1}.$$

We can also derive the unit for the Poynting vector:

$$[\mathbf{S}] = [\mathbf{E} \times \mathbf{H}] = (\text{V} \times \text{m}^{-1}) \times (\text{A} \times \text{m}^{-1}) = (\text{V} \times \text{A}) \times \text{m}^{-2} = \text{W} \times \text{m}^{-2},$$

which shows that, indeed, it represents a power flux through a unit surface. Finally, the fact that $\varepsilon \mathbf{E}^2$ and $\mathbf{B}^2/\mu$ have the unit of energy density can be verified;

we start by showing that

$$[\varepsilon \mathbf{E}^2] = [\mathbf{B}^2/\mu]\colon\ [\mathbf{B}^2/\mu] = [(\mathbf{E}/c)^2/\mu] = [\mathbf{E}^2/\mu c^2] = [\varepsilon \mathbf{E}^2];$$

next, we have

$$[\varepsilon \mathbf{E}^2] = (\mathrm{F}\times\mathrm{m}^{-1})\times(\mathrm{V}\times\mathrm{m}^{-1})^2 = (\mathrm{F}\times\mathrm{V})\times\mathrm{m}^{-3} = \mathrm{J}\times\mathrm{m}^{-3}.$$

In closing, we note that this type of analysis does not guarantee that a given derivation is correct; however, it is a powerful tool to "debug" calculations, as a discrepancy in units is a sure sign of trouble.

Further reflecting upon the result given in Equation 3.228, we note that, in vacuum, one could be tempted to express the electromagnetic energy conservation law in a covariant manner by formally introducing the four-vector $S_\mu = (c\mathcal{E}, \mathbf{S})$:

$$\partial_\mu S^\mu = 0. \tag{3.229}$$

For an accelerated point charge, we will see in Chapter 10 that we can calculate the energy-momentum radiated, and write it in a manifestly covariant form; however, the connection between $S_\mu$ and other known four-vectors and tensors is a more delicate question. It is clear that the quadratic scaling of $\mathbf{S}$ and $\mathcal{E}$ with the electric and magnetic fields implies that the electromagnetic tensor, $F_{\mu\nu}$, must also appear with the same scaling. Moreover, in the case where the four-current is not zero, the scalar $\mathbf{j}\cdot\mathbf{E}$ is related to $j_\mu$ and $F_{\mu\nu}$, but to establish an exact relation we need to introduce the Maxwell and Minkowski stress tensors. This is the object of the second part of this section.

Before considering a covariant treatment of the problem, we closely follow the introductory presentations of Greiner and Jackson. Returning to the integral form of Poynting's theorem, as presented in Equation 3.225, we note that one can express the volume integral of the Poynting vector divergence as a surface integral; we then have

$$\frac{d}{dt}\iiint_v (\mathcal{W} + \mathcal{E})dv + \iiint_v \mathbf{j}\cdot\mathbf{E}\,dv + \iint_s \mathbf{S}\cdot\hat{\mathbf{n}}\,ds = 0, \tag{3.230}$$

where we have introduced the mechanical work density, $\mathcal{W}$, defined such that

$$W = \iiint_v \mathcal{W}\,dv. \tag{3.231}$$

The surface $S$ encloses the volume under consideration, and $\hat{\mathbf{n}}$ is the unit vector normal to the surface.

As discussed previously, Equation 3.230 corresponds to the overall conservation of energy for the system comprising the electromagnetic field and the charges present in $v$; we now wish to derive an equivalent equation for the momentum. In this manner, Equation 3.230 will correspond to the time-like component of a conservation law including momentum as it space-like component.

To describe the interaction of an electromagnetic field distribution with a density of charges, we use the Lorentz force density,

$$\frac{d\mathbf{F}}{dv} = \mathbf{f} = \frac{d}{dv}[q(\mathbf{E}+\mathbf{v}\times\mathbf{B})] = \rho(\mathbf{E}+\mathbf{v}\times\mathbf{B}) = \rho\mathbf{E}+\mathbf{j}\times\mathbf{B}. \quad (3.232)$$

The covariant extension of this expression takes the form

$$f_\mu = \frac{d^2 p_\mu}{dv\, d\tau} = \frac{\gamma}{c}F_{\mu\nu}j^\nu = -enF_{\mu\nu}u^\nu, \quad (3.233)$$

and if we use the laboratory time instead of the proper time, this becomes

$$f_\mu = \frac{d^2 p_\mu}{dv\, dt} = \frac{1}{c}F_{\mu\nu}j^\nu = -e\frac{n}{\gamma}F_{\mu\nu}u^\nu. \quad (3.234)$$

Following a procedure analogous to that used in the case of the energy, we can now express the variation of momentum as follows:

$$\frac{d\mathbf{p}}{dt} = \iiint_v \mathbf{f}\, dv = \iiint_v (\rho\mathbf{E}+\mathbf{j}\times\mathbf{B})dv. \quad (3.235)$$

Again, we use Maxwell's equations to express the four-current in terms of the fields and inductions. For the charge density, we have

$$\nabla\cdot\mathbf{D} = \rho, \quad (3.236)$$

and for the current density, we use Ampère's theorem, as expressed in Equation 3.208. With these replacements, Equation 3.235 reads

$$\frac{d\mathbf{p}}{dt} = \iiint_v [(\nabla\cdot\mathbf{D})\mathbf{E}+(\nabla\times\mathbf{H}-\partial_t\mathbf{D})\times\mathbf{B}]dv. \quad (3.237)$$

In order to use the source-free equations of the Maxwell set, we add the null term $\mathbf{H}(\nabla \cdot \mathbf{B})$, and we write

$$\partial_t(\mathbf{D} \times \mathbf{B}) = \partial_t\mathbf{D} \times \mathbf{B} + \mathbf{D} \times \partial_t\mathbf{B}. \tag{3.238}$$

Using these results, we have

$$\frac{d\mathbf{p}}{dt} = \iiint_v [(\nabla \cdot \mathbf{D})\mathbf{E} + (\nabla \cdot \mathbf{B})\mathbf{H} + (\nabla \times \mathbf{H}) \times \mathbf{B} - \partial_t(\mathbf{D} \times \mathbf{B}) + \mathbf{D} \times \partial_t\mathbf{B}]dv, \tag{3.239}$$

and we can replace $\partial_t\mathbf{B}$ by using Faraday's law: $\partial_t\mathbf{B} = -\nabla \times \mathbf{E}$. The momentum transfer equation then reads

$$\frac{d\mathbf{p}}{dt} = \iiint_v [(\nabla \cdot \mathbf{D})\mathbf{E} + (\nabla \cdot \mathbf{B})\mathbf{H} + (\nabla \times \mathbf{H}) \times \mathbf{B} - \partial_t(\mathbf{D} \times \mathbf{B}) - \mathbf{D} \times (\nabla \times \mathbf{E})]dv. \tag{3.240}$$

In vacuum, the term $\mathbf{D} \times \mathbf{B} = \varepsilon_0\mathbf{E} \times \mu_0\mathbf{H} = \mathbf{S}/c^2$ corresponds to the Poynting vector, and we can recast Equation 3.240 in a more suggestive form:

$$\begin{aligned}\frac{d\mathbf{p}}{dt} + \frac{1}{c^2}\iiint_v \partial_t\mathbf{S}\, dv = \iiint_v &[(\nabla \cdot \mathbf{D})\mathbf{E} + (\nabla \cdot \mathbf{B})\mathbf{H} \\ &+ (\nabla \times \mathbf{H}) \times \mathbf{B} - \mathbf{D} \times (\nabla \times \mathbf{E})dv.\end{aligned} \tag{3.241}$$

At this point, we can introduce the momentum of the electromagnetic field,

$$\mathbf{g} = \frac{1}{c^2}\iiint_v \mathbf{S}\, dv, \tag{3.242}$$

and write the momentum transfer equation as

$$\frac{d}{dt}(\mathbf{p} + \mathbf{g}) = \iiint_v \left\{\varepsilon_0[(\nabla \cdot \mathbf{E})\mathbf{E} - \mathbf{E} \times (\nabla \times \mathbf{E})] + \frac{1}{\mu_0}[(\nabla \cdot \mathbf{B})\mathbf{B} - \mathbf{B} \times (\nabla \times \mathbf{B})]\right\}dv, \tag{3.243}$$

where the integrand will be shown to correspond to the divergence of the Maxwell stress tensor.

Using dimensional analysis, we can verify that

$$\begin{aligned}[\mathbf{g}] &= (\mathrm{m} \times \mathrm{s}^{-1})^{-2} \times [\mathbf{S}] \times \mathrm{m}^3 \\ &= \mathrm{m} \times \mathrm{s}^2 \times (\mathrm{W} \times \mathrm{m}^{-2}) \\ &= \mathrm{s}^2 \times \mathrm{m}^{-1} \times (\mathrm{kg} \times \mathrm{m}^2 \times \mathrm{s}^{-3}) \\ &= \mathrm{kg} \times \mathrm{m} \times \mathrm{s}^{-1} \\ &= [\mathbf{p}].\end{aligned} \tag{3.244}$$

We now derive the expression of the Maxwell stress tensor. We first have

$$\begin{aligned}
[(\nabla\cdot\mathbf{E})\mathbf{E}-\mathbf{E}\times(\nabla\times\mathbf{E})]_i &= (\nabla\cdot\mathbf{E})E_i+[(\nabla\times\mathbf{E})\times\mathbf{E}]_i\\
&= (\nabla\cdot\mathbf{E})E_i+(\nabla\times\mathbf{E})_jE_k-(\nabla\times\mathbf{E})_kE_j\\
&= (\nabla\cdot\mathbf{E})E_i+(\partial_kE_i-\partial_iE_k)E_k-(\partial_iE_j-\partial_jE_i)E_j\\
&= (\nabla\cdot\mathbf{E})E_i+(E_i\partial_iE_i+E_j\partial_jE_i+E_k\partial_kE)\\
&\quad-(E_i\partial_iE_i+E_j\partial_iE_j+E_k\partial_iE_k)\\
&= (\nabla\cdot\mathbf{E})E_i+(\mathbf{E}\cdot\nabla)E_i-\partial_i(\mathbf{E}^2/2).
\end{aligned}\tag{3.245}$$

We can use Einstein's convention to sum over repeated indices to recast Equation 3.245 as

$$[(\nabla\cdot\mathbf{E})\mathbf{E}-\mathbf{E}\times(\nabla\times\mathbf{E})]_i = \partial_j\left[E_jE_i-\frac{1}{2}\delta_{ij}(E_kE_k)\right].\tag{3.246}$$

Applying the summation rule, we have

$$\begin{aligned}
\partial_j\left[E_jE_i-\frac{1}{2}\delta_{ij}(E_kE_k)\right] &= E_i(\partial_jE_j)+(E_j\partial_j)E_i-\frac{1}{2}(\delta_{ij}\partial_j)(\mathbf{E}^2)\\
&= E_i(\nabla\cdot\mathbf{E})+(\mathbf{E}\cdot\nabla)E_i-\frac{1}{2}\partial_i(\mathbf{E}^2).
\end{aligned}\tag{3.247}$$

The second term in the integrand of Equation 3.243 is obtained by substituting the magnetic induction and the electric field; therefore, the conservation of momentum takes the form

$$\frac{d}{dt}(\mathbf{p}+\mathbf{g})_i = \iiint_v \partial_j\left\{\varepsilon_0\left[E_jE_i-\frac{1}{2}\delta_{ij}(E_kE_k)\right]+\frac{1}{\mu_0}\left[B_jB_i-\frac{1}{2}\delta_{ij}(B_kB_k)\right]\right\}dv,\tag{3.248}$$

which can be expressed in a more compact fashion as

$$\frac{d}{dt}(\mathbf{p}+\mathbf{g})_i = \iiint_v \partial_jT_{ij}\,dv,\tag{3.249}$$

where we have introduced the Maxwell stress tensor,

$$T_{ij} = \varepsilon_0\left[E_jE_i-\frac{1}{2}\delta_{ij}(E_kE_k)\right]+\frac{1}{\mu_0}\left[B_jB_i-\frac{1}{2}\delta_{ij}(B_kB_k)\right].\tag{3.250}$$

Before considering the properties of this tensor in detail, it is important to understand that, as derived, it is not a tensor in the sense of special relativity. In particular, its behavior under the Lorentz transformation is quite complex. Of course, this is obvious, as this $3 \times 3$ tensor lacks time-like components.

It is clear from Equation 3.250 that the Maxwell stress tensor is symmetric: $T_{ij} = T_{ji}$. The explicit expression of $T_{ij}$ is

$$T_{ij} = \begin{pmatrix} \varepsilon_0 E_x^2 + \mu_0^{-1} B_x^2 - \mathcal{E} & \varepsilon_0 E_x E_y + \mu_0^{-1} B_x B_y & \varepsilon_0 E_x E_z + \mu_0^{-1} B_x B_z \\ \varepsilon_0 E_y E_z + \mu_0^{-1} B_y B_z & \varepsilon_0 E_y^2 + \mu_0^{-1} B_y^2 - \mathcal{E} & \varepsilon_0 E_y E_z + \mu_0^{-1} B_y B_z \\ \varepsilon_0 E_x E_z + \mu_0^{-1} B_x B_z & \varepsilon_0 E_x E_y + \mu_0^{-1} B_x B_y & \varepsilon_0 E_z^2 + \mu_0^{-1} B_x^2 - \mathcal{E} \end{pmatrix}, \tag{3.251}$$

where we have used the electromagnetic energy density, $2\mathcal{E} = \varepsilon \mathbf{E}^2 + \mu_0^{-1} \mathbf{B}^2$.

The trace of the Maxwell stress tensor is then given by

$$\mathrm{Tr}(T_{ij}) = T_{ii} = \varepsilon_0 \mathbf{E}^2 + \mu_0^{-1} \mathbf{B}^2 - 3\mathcal{E} = -\mathcal{E}. \tag{3.252}$$

Given the form of the Maxwell stress tensor, one can replace $\varepsilon_0 \mathbf{E}$ by $\mathbf{D}$, and $\mu_0^{-1} \mathbf{B}$ by $\mathbf{H}$, to generalize this approach to the case a linear medium. The resulting object is called the Minkowski stress tensor and reads

$$S_{ij} = \begin{pmatrix} D_x E_x + B_x H_x - \mathcal{E} & D_x E_y + B_x H_y & D_x E_z + B_x H_z \\ D_y E_z + B_y H_z & D_y E_y + B_y H_y - \mathcal{E} & D_y E_z + B_y H_z \\ D_x E_z + B_x H_z & D_x E_y + B_x H_y & D_z E_z + B_z H_z - \mathcal{E} \end{pmatrix}, \tag{3.253}$$

where the electromagnetic energy density is now given by

$$\mathcal{E} = \frac{\mathbf{E} \cdot \mathbf{D} + \mathbf{H} \cdot \mathbf{B}}{2}. \tag{3.254}$$

Finally, we note that by using the divergence theorem, we can express the conservation of electrodynamical momentum as

$$\frac{d}{dt}(\mathbf{p} + \mathbf{g})_i = \iiint_v \partial_j T_{ij}\, dv = \iint_s T_{ij} n_j\, ds, \tag{3.255}$$

where the surface $S$ encloses the volume under consideration, and where $\hat{\mathbf{n}} = n_x \hat{\mathbf{x}} + n_y \hat{\mathbf{y}} + n_z \hat{\mathbf{z}}$ is the unit vector normal to the surface element.

Within this context, the quantity

$$\frac{dF_i}{ds} = T_{ij} n_j, \tag{3.256}$$

represents the force per unit surface applied by the electromagnetic field.

As we have discussed earlier, the energy and momentum are the time-like and spatial components of a four-vector: $p_\mu = (E/c, \mathbf{p})$. In the case of a particle, this four-vector is related to the normalized four-velocity, and we have $p_\mu = m_0 c u_\mu = m_0(dx_\mu/d\tau)$. Therefore, it is highly desirable to give a covariant expression of the energy-momentum conservation for an electrodynamical system consisting of charges interacting with an electromagnetic field distribution. Such a conservation law would essentially group Equations 3.225 and 3.255 into a single equation of the form

$$\frac{d}{dt}(p_\mu + g_\mu) = \iiint \partial_\nu T^\nu_\mu dv, \tag{3.257}$$

where $g_\mu = (|\mathbf{g}|, \mathbf{g})$ is the four-momentum of the electromagnetic field, which must satisfy the condition $g_\mu g^\mu = 0$, in close analogy with the photon mass-shell condition, $\hbar^2 k_\mu k^\mu = 0$, and where the three-integral is required in our fluid field theory.

Clearly, the covariant form of the Maxwell stress tensor must be closely related to the expression derived previously; however, because of its tensorial character within the context of special relativity, one must be able to express it in terms of other tensors and four-vectors. In particular, as discussed earlier, it is clear that $T_{\mu\nu}$ must scale quadratically with the electromagnetic field tensor, $F_{\mu\nu}$.

To proceed with the derivation, we return to the expression of the Lorentz force density, given in Equation 3.234, $f_\mu = F_{\mu\nu} j^\nu/c$. In close analogy with the calculations presented for the energy and momentum conservation, we now seek to replace the four-current density by its expression provided by Maxwell's equation:

$$\Box A_\mu + \mu_0 j_\mu = 0. \tag{3.258}$$

We can replace the four-potential by the electromagnetic field tensor by using the relation

$$F_{\mu\nu} = c(\partial_\mu A_\nu - \partial_\nu A_\mu), \tag{3.259}$$

supplemented by the requirement that the Lorentz gauge be satisfied, to obtain

$$j^\mu = \varepsilon_0 c \partial_\nu F^{\mu\nu}, \tag{3.260}$$

which can then be used in the Lorentz force density equation:

$$f_\mu = \varepsilon_0 F_{\mu\nu} \partial_\lambda F^{\nu\lambda}. \tag{3.261}$$

The quantity $F_{\mu\nu}\partial_\lambda F^{\nu\lambda}$ can now be expressed as

$$F_{\mu\nu}\partial_\lambda F^{\nu\lambda} = \partial_\lambda(F_{\mu\nu}F^{\nu\lambda}) - F^{\nu\lambda}\partial_\lambda F_{\mu\nu}, \tag{3.262}$$

and the Jacobi identity is used to further simplify the Lorentz force density. We first have

$$\partial_\lambda F_{\mu\nu} + \partial_\mu F_{\nu\lambda} + \partial_\nu F_{\lambda\mu} = 0, \tag{3.263}$$

because $F_{\mu\nu} = c(\partial_\mu A_\nu - \partial_\nu A_\mu)$ is antisymmetric, and the partial derivatives commute:

$$\begin{aligned}\partial_\lambda F_{\mu\nu} + \partial_\mu F_{\nu\lambda} + \partial_\nu F_{\lambda\mu} &= c[\partial_\lambda(\partial_\mu A_\nu - \partial_\nu A_\mu) + \partial_\mu(\partial_\nu A_\lambda - \partial_\lambda A_\nu) + \partial_\nu(\partial_\lambda A_\mu - \partial_\mu A_\lambda)]\\ &= c[(\partial_\nu\partial_\lambda - \partial_\lambda\partial_\nu)A_\mu + (\partial_\lambda\partial_\mu - \partial_\mu\partial_\lambda)A_\nu + (\partial_\mu\partial_\nu - \partial_\nu\partial_\mu)A_\lambda]\\ &= 0.\end{aligned} \tag{3.264}$$

It is important to remark that in the Jacobi identity, no summation is performed and there are no repeated indices in a given term. We also briefly indicate that, following Low, Equation 3.263 can be recast using the completely antisymmetrical Levi-Civita tensor as

$$\partial_\lambda F_{\mu\nu} + \partial_\mu F_{\nu\lambda} + \partial_\nu F_{\lambda\mu} = \partial_\nu \varepsilon^{\mu\nu\lambda\sigma} F_{\lambda\sigma} = 0. \tag{3.265}$$

We also note that this is different from the expression

$$\partial^\lambda F_{\mu\nu} + \partial^\mu F_{\nu\lambda} + \partial^\nu F_{\lambda\mu}; \tag{3.266}$$

to show this, we use the metric, which relates the covariant and contravariant coordinates of four-vectors, as

$$\frac{\partial x_\mu}{\partial x^\nu} = g_{\mu\nu}. \tag{3.267}$$

We then have

$$\begin{aligned}\partial^\lambda F_{\mu\nu} + \partial^\mu F_{\nu\lambda} + \partial^\nu F_{\lambda\mu} &= c[\partial^\lambda(\partial_\mu A_\nu - \partial_\nu A_\mu) + \partial^\mu(\partial_\nu A_\lambda - \partial_\lambda A_\nu)\\ &\quad + \partial^\nu(\partial_\lambda A_\mu - \partial_\mu A_\lambda)]\\ &= c[(\partial^\mu\partial_\nu - \partial^\nu\partial_\mu)A_\lambda + (\partial^\nu\partial_\lambda - \partial^\lambda\partial_\nu)A_\mu\\ &\quad + (\partial^\lambda\partial_\mu - \partial^\mu\partial_\lambda)A_\nu].\end{aligned} \tag{3.268}$$

Using the metric, we now write

$$(\partial^\mu \partial_\nu - \partial^\nu \partial_\mu) A_\lambda = \frac{\partial^2 A_\lambda}{\partial x_\mu \partial x^\nu} - \frac{\partial^2 A_\lambda}{\partial x_\nu \partial x^\mu}$$

$$= \frac{\partial x^\mu}{\partial x_\mu}\left[\frac{\partial}{\partial x^\mu}\left(\frac{\partial A_\lambda}{\partial x^\nu}\right)\right] - \frac{\partial x^\nu}{\partial x_\nu}\left[\frac{\partial}{\partial x^\nu}\left(\frac{\partial A_\lambda}{\partial x^\mu}\right)\right]$$

$$= (g^{\mu\mu} - g^{\nu\nu})\left(\frac{\partial^2 A_\lambda}{\partial x^\mu \partial x^\nu}\right), \qquad (3.269)$$

where the last equality holds because the partial derivatives commute.

However, whether we use a metric with trace 2, or –2, the diagonal coefficients are not all equal:

$$g_{\mu\nu} = \begin{pmatrix} -1 & 0 & 0 & 0 \\ 0 & 1 & 0 & 0 \\ 0 & 0 & 1 & 0 \\ 0 & 0 & 0 & 1 \end{pmatrix}. \qquad (3.270)$$

Returning to the problem at hand, we now use the Jacobi identity to write

$$F^{\nu\lambda}\partial_\lambda F_{\mu\nu} = \frac{1}{2}F^{\nu\lambda}(\partial_\lambda F_{\mu\nu} + \partial_\nu F_{\lambda\mu})$$

$$= -\frac{1}{2}F^{\nu\lambda}\partial_\mu F_{\nu\lambda}$$

$$= -\frac{1}{4}\partial_\mu(F^{\nu\lambda}F_{\nu\lambda})$$

$$= -\frac{1}{4}\delta_\mu^\lambda \partial_\lambda(F^{\alpha\beta}F_{\alpha\beta}). \qquad (3.271)$$

The second equality comes from a direct application of the Jacobi identity, Equation 3.263; the third equality is given by the fact that $\partial_\mu(F^{\nu\lambda}F_{\nu\lambda}) = 2F^{\nu\lambda}\partial_\mu F_{\nu\lambda}$; while the final result is derived from the identity $\delta_\mu^\lambda \partial_\lambda \equiv \partial_\mu$. Therefore, to complete the proof, we must establish that $F^{\nu\lambda}\partial_\lambda F_{\mu\nu} = F^{\nu\lambda}\partial_\nu F_{\lambda\mu}$. This is easily seen, as

$$F^{\nu\lambda}\partial_\lambda F_{\mu\nu} = F^{\lambda\nu}\partial_\nu F_{\mu\lambda}$$

$$= -F^{\nu\lambda}\partial_\nu F_{\mu\lambda}$$

$$= F^{\nu\lambda}\partial_\nu F_{\lambda\mu}. \qquad (3.272)$$

In Equation 3.272, the first identity is obtained by switching the free indices $\lambda$ and $\nu$, while the second and third equalities result from using the antisymmetry of the electromagnetic tensor twice.

The Lorentz force density can now be expressed as the divergence of the covariant Maxwell stress tensor,

$$f_\mu = \varepsilon_0\left[\partial_\lambda(F_{\mu\nu}F^{\nu\lambda}) + \frac{1}{4}\delta_\mu^\lambda\,\partial_\lambda(F^{\alpha\beta}F_{\alpha\beta})\right] = \partial_\lambda T_\mu^\lambda, \tag{3.273}$$

with

$$T_\mu^\lambda = \varepsilon_0\left[(F_{\mu\nu}F^{\nu\lambda}) + \frac{1}{4}\delta_\mu^\lambda(F^{\alpha\beta}F_{\alpha\beta})\right]. \tag{3.274}$$

The aforementioned quadratic scaling of the stress tensor with the electromagnetic field tensor appears clearly, and the stress tensor is symmetrical; in addition, we can express the components of $T_\mu^\lambda$ explicitly and show their relation with the previously derived $3 \times 3$ stress tensor, electromagnetic energy density, and momentum density:

$$T_\mu^\nu = \begin{pmatrix} T_{11} & T_{12} & T_{13} & c\mathcal{g}_1 \\ T_{21} & T_{22} & T_{23} & c\mathcal{g}_2 \\ T_{31} & T_{32} & T_{33} & c\mathcal{g}_3 \\ c\mathcal{g}_1 & c\mathcal{g}_2 & c\mathcal{g}_3 & \mathcal{E} \end{pmatrix}, \tag{3.275}$$

where

$$\mathcal{g} = \frac{\mathbf{S}}{c^2} = \frac{d\mathbf{g}}{dv}, \tag{3.276}$$

is the electromagnetic momentum density. Since $\mathrm{Tr}(T_{ij}) = -\mathcal{E}$, we find that $\mathrm{Tr}(T_\mu^\nu) = 0$. This can also be seen by considering

$$\mathrm{Tr}(T_\mu^\nu) = T_\mu^\mu = \varepsilon_0\left[(F_{\mu\nu}F^{\nu\mu}) + \frac{1}{4}\delta_\mu^\mu(F^{\alpha\beta}F_{\alpha\beta})\right] = 0, \tag{3.277}$$

which holds because $\delta_\mu^\mu = 4$, and $F^{\nu\mu} = -F^{\mu\nu}$. The conservation of energy-momentum for the electrodynamical system then takes the sought-after form:

$$\frac{dp_\mu}{dt} = \iiint f_\mu\,dv = \iiint \partial_\lambda T_\mu^\lambda\,dv. \tag{3.278}$$

In closing, we emphasize the fact that the relation between the Lorentz force density, $f_\mu$, and its covariant counterpart, $f_\mu$, while easily expressed in differential form as

$$f_\mu = \frac{d^2 p_\mu}{dv\,d\tau} = \frac{\gamma}{c} F_{\mu\nu} j^\nu = -enF_{\mu\nu}u^\nu = \gamma f_\mu, \tag{3.279}$$

cannot be recast in a simple way in integral form, because the connection between the proper time and the laboratory time is local, and in this context, $\gamma = dt/d\tau = \gamma(x_\mu)$ is now a fluid field. Finally, the covariant form of the Minkowski stress tensor is

$$S_\mu^\nu = \begin{pmatrix} S_{11} & S_{12} & S_{13} & c\mathcal{g}_1 \\ S_{21} & S_{22} & S_{23} & c\mathcal{g}_2 \\ S_{31} & S_{32} & S_{33} & c\mathcal{g}_3 \\ c\mathcal{g}_1 & c\mathcal{g}_2 & c\mathcal{g}_3 & \mathcal{E} \end{pmatrix}, \tag{3.280}$$

where the electromagnetic energy density has been generalized according to Equation 3.254.

The covariant expression for the Minkowski stress tensor is

$$S_\mu^\lambda = (F_{\mu\nu}H^{\nu\lambda}) + \frac{1}{4}\delta_\mu^\lambda(F_{\alpha\beta}H^{\alpha\beta}), \tag{3.281}$$

where the tensor $H_{\mu\nu}$ is derived by considering Maxwell's equations with sources, in a medium:

$$\begin{aligned} \nabla \times \mathbf{H} - \partial_t \mathbf{D} &= \mathbf{j}, \\ \nabla \cdot \mathbf{D} &= 0. \end{aligned} \tag{3.282}$$

This leads us to write

$$\partial^\mu H_{\mu\nu} = j_\nu, \tag{3.283}$$

and to define

$$H_{\mu\nu} = \begin{pmatrix} 0 & cD_1 & cD_2 & cD_3 \\ -cD_1 & 0 & H_3 & -H_2 \\ -cD_2 & -H_3 & 0 & H_1 \\ -cD_3 & H_2 & -H_1 & 0 \end{pmatrix}. \tag{3.284}$$

In Chapter 7 we will reexamine the covariant formalism, describing the relations between the fields and inductions in great detail.

## 3.9 Metric and Christoffel Symbols

A brief overview of a few basic mathematical tools used in general relativity will prove useful when dealing with the Sagnac effect, as well as Weyl's theory of gauge invariance, which is discussed in Chapter 4, and other topics in this book.

The following is an introductory overview of the geometry of curved space–time; for more detail, we refer the reader to the excellent presentations given by Pauli, Thorne, Wald, Feynman, and Wheeler. Here, we closely follow the classic exposition of Pauli. In Riemannian geometry, one introduces the coordinates $x^1, x^2, \ldots, x^n$ in an $n$-dimensional space or manifold, and the metric can be introduced by considering the length, $s$, of a given world line, $x^\mu(\tau)$, where $\tau$ parameterizes position along the world line. Within this context, the basic axioms of Riemannian geometry are as follows: first, $ds/d\tau$ depends only on the derivatives of the coordinates, $dx^\mu/d\tau$, and not on higher-order derivatives. This is equivalent to using a linear, local Taylor expansion of the differential length element. The second axiom states that

$$ds^2 = g_{\mu\nu} dx^\mu dx^\nu; \tag{3.285}$$

in other words, $ds/d\tau$ is the square root of a quadratic form of the local derivatives. This implies that local orthogonality is reciprocal and that displacement along a closed loop conserves angles. We will see in Chapter 4 that Weyl's theory generalizes Riemannian geometry to allow for a variation of both lengths and angles along a closed loop. We thus see that for an arbitrary point transformation,

$$\bar{x}^\mu = \bar{x}^\mu(x^1, \ldots x^n), \tag{3.286}$$

we have the linear relations

$$\begin{aligned} d\bar{x}^\mu &= \frac{\partial \bar{x}^\mu}{\partial x^\nu} dx^\nu, \\ dx^\nu &= \frac{\partial x^\nu}{\partial \bar{x}^\mu} d\bar{x}^\mu. \end{aligned} \tag{3.287}$$

We now consider the parallel displacement of a vector. We have

$$\frac{dv^\lambda}{d\tau} = -\Gamma^\lambda_{\mu\nu}\frac{dx^\nu}{d\tau}v^\mu, \tag{3.288}$$

where the coefficients $\Gamma^\lambda_{\mu\nu}$ depend only on the coordinates and satisfy the symmetry

$$\Gamma^\lambda_{\mu\nu} = \Gamma^\lambda_{\nu\mu}. \tag{3.289}$$

It is easy to understand the mathematical meaning of Equation 3.288: for an infinitesimal displacement, $dx^\nu/d\tau$, the corresponding variation of the vector $v^\lambda$ is a linear function of both the displacement and the original vector components. We also note that Equation 3.288 implies that $dv^\lambda/d\tau = 0$ in a particular coordinate system, for an infinitesimal transformation. This is the first important definition for parallel displacement; the second one is the conservation of length for the vectors:

$$\frac{d}{d\tau}(v^\mu v_\mu) = \frac{d}{d\tau}(g_{\mu\nu}v^\mu v^\nu) = 0. \tag{3.290}$$

Using Equations 3.288 and 3.290, we have

$$\frac{\partial g_{\lambda\mu}}{\partial x^\nu} = g_{\lambda\sigma}\Gamma^\sigma_{\mu\nu} + g_{\mu\sigma}\Gamma^\sigma_{\lambda\nu} = \Gamma_{\lambda,\mu\nu} + \Gamma_{\mu,\lambda\nu}; \tag{3.291}$$

in addition, following the concepts first introduced by Christoffel and the notations used by Pauli and Lipschitz, we also have

$$\Gamma^\lambda_{\mu\nu} = g^{\lambda\sigma}\Gamma_{\sigma,\mu\nu}. \tag{3.292}$$

From these definitions and Equation 3.291, we find that

$$\Gamma_{\lambda,\mu\nu} = \frac{1}{2}\left(\frac{\partial g_{\lambda\mu}}{\partial x^\nu} + \frac{\partial g_{\lambda\nu}}{\partial x^\mu} - \frac{\partial g_{\mu\nu}}{\partial x^\lambda}\right). \tag{3.293}$$

Using these results and the conservation of length, as given in Equation 3.290, we obtain the important relation

$$\frac{\partial g^{\lambda\mu}}{\partial x^\nu} + g^{\lambda\sigma}\Gamma^\mu_{\sigma\nu} + g^{\mu\sigma}\Gamma^\lambda_{\sigma\nu} = 0. \tag{3.294}$$

In addition, the metric is symmetrical, $g_{\mu\nu} = g_{\nu\mu}$, and if we introduce its determinant,

$$g = \det|g_{\mu\nu}|, \tag{3.295}$$

we have

$$\det|g^{\mu\nu}| = g^{-1}. \tag{3.296}$$

Another important relation is

$$g_{\lambda\nu}g^{\mu\nu} = \delta^{\mu}_{\lambda}. \tag{3.297}$$

Using these equations and differentiating, we find

$$\begin{aligned} dg^{\lambda\mu} &= -g^{\lambda\rho}g^{\mu\sigma}dg_{\rho\sigma}, \\ dg_{\lambda\mu} &= -g_{\lambda\rho}g_{\mu\sigma}dg^{\rho\sigma}, \\ \frac{\partial g^{\lambda\mu}}{\partial x^{\nu}} &= -g^{\lambda\rho}g^{\mu\sigma}\frac{\partial g_{\rho\sigma}}{\partial x^{\nu}}, \\ \frac{\partial g_{\lambda\mu}}{\partial x^{\nu}} &= -g_{\lambda\rho}g_{\mu\sigma}\frac{\partial g^{\rho\sigma}}{\partial x^{\nu}}. \end{aligned} \tag{3.298}$$

In addition, for the determinant, we have

$$\begin{aligned} dg &= gg^{\lambda\mu}dg_{\lambda\mu} = -gg_{\lambda\mu}dg^{\lambda\mu}, \\ \frac{\partial g}{\partial x^{\nu}} &= gg^{\lambda\mu}\frac{\partial g_{\lambda\mu}}{\partial x^{\nu}} = -gg_{\lambda\mu}\frac{\partial g^{\lambda\mu}}{\partial x^{\nu}}. \end{aligned} \tag{3.299}$$

Finally, we can also obtain the following relation for the Christoffel symbols,

$$\Gamma^{\mu}_{\lambda\mu} = g^{\mu\nu}\Gamma_{\mu,\lambda\nu} = \frac{1}{2}g^{\mu\nu}\frac{\partial g_{\mu\nu}}{\partial x^{\lambda}} = \frac{\partial \ln(\sqrt{g})}{\partial x^{\lambda}}, \tag{3.300}$$

since we have established that $\frac{\partial g}{\partial x^{\nu}} = gg^{\lambda\mu}\frac{\partial g_{\lambda\mu}}{\partial x^{\nu}}$, in Equation 3.299. Combining this result with Equations 3.293 and 3.294, we find

$$\frac{1}{\sqrt{g}}\frac{\partial\sqrt{g}g^{\lambda\mu}}{\partial x^{\mu}} + g^{\rho\sigma}\Gamma^{\lambda}_{\rho\sigma} = 0. \tag{3.301}$$

Now considering the world line $x^\mu(\tau)$, we can define the unit tangent vector

$$v^\mu = \frac{dx^\mu}{ds}. \tag{3.302}$$

The fact that the length of this vector is unity can be shown as follows:

$$v^\mu v_\mu = \frac{dx^\mu}{ds} g_{\mu\nu} v^\nu = \frac{dx^\mu}{ds} g_{\mu\nu} \frac{dx^\nu}{ds} = 1, \tag{3.303}$$

where the last equality results from the very definition of length, $ds^2 = g_{\mu\nu}\, dx^\mu dx^\nu$.

The results given in Equations 3.302 and 3.303 are important, as they define a geodesic in the manifold under consideration. Starting at a given point along a world line, a parallel displacement $v^\mu$ of will result in describing the corresponding geodesic. As Pauli puts it, after Weyl, "the geodesic is a curve which always maintain its direction." We can then use the Christoffel symbols to write

$$\frac{dv_\lambda}{ds} = \Gamma_{\mu,\lambda\nu} v^\mu v^\nu = \frac{1}{2}\frac{\partial g_{\mu\nu}}{\partial x^\lambda} v^\mu v^\nu, \tag{3.304}$$

and

$$\frac{dv^\lambda}{ds} = -\Gamma^\lambda_{\mu\nu} v^\mu v^\nu. \tag{3.305}$$

Using the differential definition of the tangent, Equation 3.302, we can recast Equation 3.305 as

$$\frac{d^2 x^\lambda}{ds^2} + \Gamma^\lambda_{\mu\nu} \frac{dx^\mu}{ds}\frac{dx^\nu}{ds} = 0, \tag{3.306}$$

which is the well-known equation for a geodesic. Before proceeding with a few basic considerations regarding space–time curvature, it is interesting to recast Equation 3.206 in terms of a variational principle. Again, following Pauli, we consider

$$\delta \int_{x_1^\lambda}^{x_2^\lambda} ds = \delta \int_{x_1^\lambda}^{x_2^\lambda} \sqrt{g_{\mu\nu} \frac{dx^\mu}{d\sigma}\frac{dx^\nu}{d\sigma}}\, d\sigma = 0, \tag{3.307}$$

where $x_1^\lambda$ and $x_2^\lambda$ represent the initial and end points of the curve, parameterized by $\sigma$.

Without loss of generality, we can in fact identify $\sigma \equiv s$, by requesting that they coincide at the end points of the curve and that they cover the same range. This means that we can use $\sigma = s$ in the differential equations resulting from the variational principle. In complete analogy with Lagrangian formalism, we now define

$$\mathcal{L} = \frac{1}{2} g_{\mu\nu} \frac{dx^\mu}{ds} \frac{dx^\nu}{ds}, \tag{3.308}$$

and Equation 3.307 takes the form

$$\delta \int_{x_1^\lambda}^{x_2^\lambda} ds = \delta \int_{x_1^\lambda}^{x_2^\lambda} \frac{\delta \mathcal{L}}{\sqrt{g_{\mu\nu} \frac{dx^\mu}{ds} \frac{dx^\nu}{ds}}} ds = 0. \tag{3.309}$$

At the end points, the square root is equal to one, and we can simply use

$$\int_{x_1^\lambda}^{x_2^\lambda} \delta \mathcal{L} ds = \delta \int_{x_1^\lambda}^{x_2^\lambda} \mathcal{L} ds, \tag{3.310}$$

for our variational principle. Proceeding as in Section 3.7, the Lagrange equations resulting from Equation 3.310 are

$$\frac{d}{ds} \frac{\partial \mathcal{L}}{\partial \dot{q}^\lambda} - \frac{\partial \mathcal{L}}{\partial q^\lambda} = 0, \tag{3.311}$$

where the generalized coordinates are given by $q^\lambda = x^\lambda$, and the generalized momenta are defined as $\dot{q}^\lambda = dq^\lambda/ds = dx^\lambda/ds$. We now use the definition of the Lagrangian to perform the partial derivative with respect to the generalized momentum:

$$\begin{aligned} \frac{\partial \mathcal{L}}{\partial \dot{q}^\lambda} &= \frac{\partial}{\partial \dot{q}^\lambda} \left( \frac{1}{2} g_{\mu\nu} \frac{dx^\mu}{ds} \frac{dx^\nu}{ds} \right) \\ &= \frac{\partial}{\partial \dot{q}^\lambda} \left( \frac{1}{2} g_{\mu\nu} \dot{q}^\mu \dot{q}^\nu \right). \end{aligned} \tag{3.312}$$

However, when performing the derivative and the summation, some care must be taken regarding Einstein's notation. To illustrate this point, let us consider

$$ds^2 = g_{\mu\nu} dx^\mu dx^\nu, \quad (g_{\mu\nu} = g_{\nu\mu}). \tag{3.313}$$

As we are summing over repeated indices, we see that the combinations $\mu\nu, \mu \neq \nu$ will occur twice, whereas $\mu = \nu$ will occur only once. As a result, when considering a quadratic form, $Q = g_{\mu\nu} a^\mu a^\nu$, we have

$$\frac{\partial Q}{\partial a^\mu} = 2 g_{\mu\nu} a^\nu. \tag{3.314}$$

Applying this result to the Lagrangian, we first find

$$\frac{\partial \mathcal{L}}{\partial \dot{q}^\lambda} = g_{\lambda\rho} \frac{dx^\rho}{ds}, \tag{3.315}$$

and, finally,

$$\frac{d}{ds}\left(g_{\lambda\rho}\frac{dx^\rho}{ds}\right) = \frac{\partial}{\partial x^\lambda}\left(\frac{1}{2} g_{\mu\nu}\dot{q}^\mu \dot{q}^\nu\right) = \frac{1}{2}\frac{\partial g_{\mu\nu}}{\partial x^\lambda}\dot{q}^\mu \dot{q}^\nu = \frac{1}{2}\frac{\partial g_{\mu\nu}}{\partial x^\lambda}\frac{dx^\mu}{ds}\frac{dx^\nu}{ds}, \tag{3.316}$$

which is identical to the geodesic differential equation, Equation 3.306.

Here, we have taken into account the fact that the generalized coordinates and momenta must be treated as independent variables, in particular, $\partial_\lambda \dot{q}^\mu = \partial_\lambda \dot{q}^\nu = 0$.

The concept of curvature was first introduced by Riemann, and considerable work on this question was done by Christoffel, Lipschitz, Weyl, Hessenberg, Ricci, and Levi-Civita. Instead of considering infinitesimal parallel displacement, we now address the question of the variation of a vector undergoing parallel displacement along a closed loop, as illustrated in Figure 3.2. To this end, we consider an arbitrary vector, $v^\lambda$ and we parameterize the loop as follows:

$$x^\mu = x^\mu(p, q). \tag{3.317}$$

The points along the loop are $A(p, q)$, $B(p + \Delta p, q)$, $C(p + \Delta p, q + \Delta q)$, and $D(p, q + \Delta q)$. The difference of the vector, $\Delta v^\lambda$, is proportional to $\Delta p \Delta q$, and the limit,

$$\lim_{\Delta p, \Delta q \to 0}\left(\frac{\Delta v^\lambda}{\Delta p \Delta q}\right) = R^\lambda_{\mu\nu\rho} v^\mu \frac{\partial x^\nu}{\partial p}\frac{\partial x^\rho}{\partial q}, \tag{3.318}$$

defines the Riemann-Christoffel curvature tensor, presented here in mixed form. Equation 3.318 indicates that the variation of the vector under consideration along an infinitesimal closed loop is proportional to the original vector, as well as the partial derivatives of the local coordinates with respect

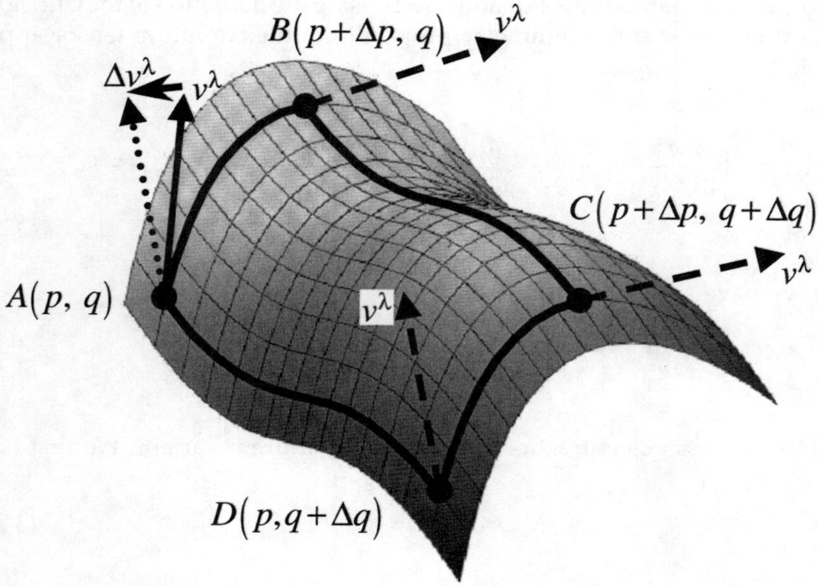


**FIGURE 3.2**
Schematic representation of parallel transport of a vector in curved space–time.

to the variables parameterizing the curve. The curvature tensor can be expressed in terms of Christoffel symbols as follows:

$$R^{\lambda}_{\mu\nu\rho} = \frac{\partial\Gamma^{\lambda}_{\mu\nu}}{\partial x^{\rho}} - \frac{\partial\Gamma^{\lambda}_{\mu\rho}}{\partial x^{\nu}} + \Gamma^{\lambda}_{\rho\sigma}\Gamma^{\sigma}_{\mu\nu} - \Gamma^{\lambda}_{\nu\sigma}\Gamma^{\sigma}_{\mu\rho}. \tag{3.319}$$

For a covariant vector, we have, by a similar procedure

$$\lim_{\Delta p,\Delta q\to 0}\left(\frac{\Delta v_{\lambda}}{\Delta p\Delta q}\right) = R_{\lambda\mu\nu\rho}v^{\mu}\frac{\partial x^{\nu}}{\partial p}\frac{\partial x^{\rho}}{\partial q}, \tag{3.320}$$

with

$$R_{\lambda\mu\nu\rho} = \frac{\partial\Gamma_{\mu,\lambda\rho}}{\partial x^{\nu}} - \frac{\partial\Gamma_{\mu,\lambda\nu}}{\partial x^{\rho}} + g^{\sigma\tau}(\Gamma_{\sigma,\lambda\nu}\Gamma_{\tau,\mu\rho} - \Gamma_{\sigma,\lambda\rho}\Gamma_{\tau,\mu\nu}). \tag{3.321}$$

Just as for covariant and contravariant vectors, we can use the metric to relate $R^{\lambda}_{\mu\nu\rho}$ and $R_{\lambda\mu\nu\rho}$:

$$R_{\lambda\mu\nu\rho} = g_{\lambda\sigma}R^{\sigma}_{\mu\nu\rho}. \tag{3.322}$$

In addition, we can express Equation 3.321 using the definition of the Christoffel derivatives, so that the symmetry properties of the curvature tensor appear explicitly,

$$R_{\lambda\mu\nu\rho} = \frac{1}{2}\left(\frac{\partial^2 g_{\lambda\nu}}{\partial x^\mu \partial x^\rho} + \frac{\partial^2 g_{\mu\rho}}{\partial x^\lambda \partial x^\nu} - \frac{\partial^2 g_{\lambda\rho}}{\partial x^\mu \partial x^\nu} - \frac{\partial^2 g_{\mu\nu}}{\partial x^\lambda \partial x^\rho}\right) + g^{\sigma\tau}(\Gamma_{\sigma,\lambda\nu}\Gamma_{\tau,\mu\rho} - \Gamma_{\sigma,\lambda\rho}\Gamma_{\tau,\mu\nu}), \tag{3.323}$$

and we have

$$\begin{aligned} R_{\lambda\mu\nu\rho} &= -R_{\lambda\mu\rho\nu} = -R_{\mu\lambda\nu\rho} = R_{\nu\rho\lambda\mu}, \\ R_{\lambda\mu\nu\rho} + R_{\lambda\nu\rho\mu} + R_{\lambda\rho\mu\nu} &= 0. \end{aligned} \tag{3.324}$$

Finally, successive contractions yield the curvature invariant. We first have

$$R_{\mu\nu} = R^{\sigma}_{\mu\sigma\nu} = g^{\rho\sigma}R_{\rho\mu\sigma\nu}, \tag{3.325}$$

which can be expressed as

$$R_{\mu\nu} = \frac{\partial \Gamma^{\rho}_{\mu\rho}}{\partial x^\nu} - \frac{\partial \Gamma^{\rho}_{\mu\nu}}{\partial x^\rho} + \Gamma^{\sigma}_{\mu\rho}\Gamma^{\rho}_{\nu\sigma} - \Gamma^{\rho}_{\mu\nu}\Gamma^{\sigma}_{\rho\sigma}, \tag{3.326}$$

and a second contraction yields

$$R = g^{\mu\nu}R_{\mu\nu}. \tag{3.327}$$

As is well known, different solutions of Einstein's equations can be classified according to their curvature: $R > 0$ corresponds to closed universes, while $R = 0$ describes a flat space–time, such as that of special relativity; finally, open universes are characterized by a negative value of $R$. Let us also note that, thus far, we have described the geometry of curved space–time in terms of the metric and geodesics, which indicate how matter moves in such a manifold. The second crucial point of general relativity is the derivation of the equations governing the geometry of space–time, given an energy–mass distribution, which acts as a source term. These are Einstein's equations, which take the form

$$\begin{aligned} G_{\mu\nu} &= R_{\mu\nu} - \frac{1}{2}g_{\mu\nu}R = -\kappa T_{\mu\nu}, \\ R_{\mu\nu} &= -\kappa\left(T_{\mu\nu} - \frac{1}{2}g_{\mu\nu}T\right), \end{aligned} \tag{3.328}$$

where $g_{\mu\nu}$ is the metric, $R_{\mu\nu}$ is the Riemann–Christoffel curvature tensor, $G_{\mu\nu}$ represents the gravitational field tensor, and $T_{\mu\nu}$ is the stress-energy tensor.

The coupling constant has the value

$$\kappa c^2 = \frac{8\pi G}{c^2} = 1.86592(18)\ \mathrm{m} \times \mathrm{kg}^{-1}. \tag{3.329}$$

The fundamental concept at the origin of the theory of general relativity is the equivalence principle, which identifies inertial and gravitational mass, leading to the notion of general covariance of natural laws. Additionally, we note that these equations can also be derived from a variational principle.

Finally, we introduce the notion of covariant derivative: for a given tensor, $F_{\mu\nu}$, we have

$$\begin{aligned}
\nabla_\rho F_{\mu\nu} &= \partial_\rho F_{\mu\nu} - \Gamma^\sigma_{\mu\rho} F_{\sigma\nu} - \Gamma^\sigma_{\nu\rho} F_{\mu\sigma}, \\
\nabla_\mu F_{\nu\rho} &= \partial_\mu F_{\nu\rho} - \Gamma^\sigma_{\nu\mu} F_{\sigma\rho} - \Gamma^\sigma_{\rho\mu} F_{\nu\sigma}, \\
\nabla_\nu F_{\rho\mu} &= \partial_\nu F_{\rho\mu} - \Gamma^\sigma_{\rho\nu} F_{\sigma\mu} - \Gamma^\sigma_{\mu\nu} F_{\rho\sigma}.
\end{aligned} \tag{3.330}$$

In the case of a flat, hyperbolic space–time, only the partial derivatives with respect to the coordinates remain, and we recover the four-gradient.

## 3.10 Solid in Rotation, Sagnac Effect

The description of the geometry of a solid in uniform rotation, within the context of general relativity, leads to the Sagnac effect, which is used in laser ring gyroscopes as a very sensitive accelerometer. In particular, it offers the option of inertial guidance for airplanes and supplement the global positioning system (GPS). Furthermore, this offers a simple application of some of the concepts discussed in the previous section.

To treat the problem under consideration, we use cylindrical coordinates and describe a local metric, corresponding to the tangential flat space–time at a given point moving with the instantaneous velocity, $\mathbf{v} = \omega \times \mathbf{r}$, as measured in the laboratory frame. In the instantaneous tangent frame, we have

$$ds^2 = -d\bar{r}^2 - \bar{r}^2 d\bar{\theta}^2 - d\bar{z}^2 + c^2 d\bar{t}^2; \tag{3.331}$$

the relation with the laboratory frame is

$$\bar{r}\, d\bar{\theta} = \gamma r\, d\theta, \quad d\bar{t} = \frac{dt}{\gamma}, \tag{3.332}$$

with

$$\gamma = \frac{1}{\sqrt{1-(\mathbf{v}/c)^2}} = \frac{1}{\sqrt{1-(\frac{\omega r}{c})^2}}. \tag{3.333}$$

Here, $\omega = d\theta/dt$ is the rotation frequency of the solid, and $r$ is the distance between the contact point and the rotation axis. With this, the metric in the laboratory frame is given by

$$ds^2 = -dr^2 - \frac{r^2 d\theta^2}{1-\frac{\omega^2 r^2}{c^2}} - dz^2 + \left(1 - \frac{\omega^2 r^2}{c^2}\right) dt^2. \tag{3.334}$$

The next step consists in deriving the Christoffel symbols from this metric, which has only diagonal components:

$$g_{\mu\nu} = \begin{pmatrix} -1 & 0 & 0 & 0 \\ 0 & \dfrac{r^2}{1-(\omega r/c)^2} & 0 & 0 \\ 0 & 0 & -1 & 0 \\ 0 & 0 & 0 & 1-(\omega r/c)^2 \end{pmatrix}. \tag{3.335}$$

It is also easy to derive the determinant of the metric,

$$g = \det|g_{\mu\nu}| = -r^2. \tag{3.336}$$

We note that the determinant has the dimensions of a surface; this is due to our using cylindrical coordinates. The differential length produced by an infinitesimal angular displacement is $rd\theta$, thus yielding the result presented in Equation 3.334. The fact that the metric is diagonal has another important consequence: its contravariant components can readily be determined. We have $g^{\mu\mu} = 1/g_{\mu\mu}$, which yields

$$g^{00} = \frac{1}{1-\frac{\omega^2 r^2}{c^2}}, \quad g^{11} = -1, \quad g^{22} = -\left(\frac{1}{r^2} - \frac{\omega^2}{c^2}\right), \quad g^{33} = -1. \tag{3.337}$$

As discussed earlier, the Christoffel symbols are expressed in Equations 3.292 and 3.293 as

$$\Gamma^{\lambda}_{\mu\nu} = g^{\lambda\sigma}\Gamma_{\sigma,\mu\nu}, \quad \Gamma_{\sigma,\mu\nu} = \frac{1}{2}\left(\frac{\partial g_{\sigma\mu}}{\partial x^{\nu}} + \frac{\partial g_{\sigma\nu}}{\partial x^{\mu}} - \frac{\partial g_{\mu\nu}}{\partial x^{\sigma}}\right); \tag{3.338}$$

thus, we have

$$\Gamma^{\lambda}_{\mu\nu} = \frac{1}{2}g^{\lambda\sigma}\left(\frac{\partial g_{\sigma\mu}}{\partial x^{\nu}} + \frac{\partial g_{\sigma\nu}}{\partial x^{\mu}} - \frac{\partial g_{\mu\nu}}{\partial x^{\sigma}}\right) = \frac{1}{2}g^{\lambda\sigma}(\partial_{\mu}g_{\nu\sigma} + \partial_{\nu}g_{\mu\sigma} - \partial_{\sigma}g_{\mu\nu}). \quad (3.339)$$

*A priori*, the Christoffel symbols have $4^3 = 64$ components, which reduce to 40 distinct components because of the symmetry of the metric. This can be seen by considering the different possible cases, and counting the number of distinct component for each situation, $\lambda \neq \mu$, $\lambda \neq \nu$, $\mu \neq \nu$, yields 12 values; $\lambda = \mu \neq \nu$ and $\lambda \neq \mu = \nu$ each yields another 12 values. Finally, $\lambda = \mu = \nu$ yields 4 values, and the total adds up to 40. In the particular case of our diagonal metric, the only distinct, nonzero components are

$$\begin{aligned}
\Gamma^{1}_{00} &= \frac{\omega^2 r}{c^2}, \\
\Gamma^{0}_{01} &= -\frac{\omega^2 r}{c^2 - \omega^2 r^2}, \\
\Gamma^{2}_{21} &= \frac{1}{\left(1 - \frac{\omega^2 r^2}{c^2}\right) r}, \\
\Gamma^{1}_{22} &= -\frac{1}{\left(1 - \frac{\omega^2 r^2}{c^2}\right)^2}.
\end{aligned} \quad (3.340)$$

Following Poincelot, we now consider the simpler problem of the geometry at the surface of the rotating object, $z = 0$, and time is constant. We then have a $2 \times 2$ metric, whereby

$$d\sigma^2 = dr^2 + \frac{r^2 d\theta^2}{1 - \frac{\omega^2 r^2}{c^2}}. \quad (3.341)$$

The corresponding geodesic equation is

$$\frac{d^2 x^{\lambda}}{d\sigma^2} + \Gamma^{\lambda}_{\mu\nu}\frac{dx^{\mu}}{d\sigma}\frac{dx^{\nu}}{d\sigma} = 0, \quad (3.342)$$

where the indices take only two values, corresponding to the radial and azimuthal coordinates.

Let us consider the angular component of Equation 3.342. We have

$$\frac{d^2\theta}{d\sigma^2} + 2\Gamma^{\theta}_{\theta r}\frac{d\theta}{d\sigma}\frac{dr}{d\sigma} = 0, \tag{3.343}$$

where the factor 2 results from the symmetry of the Christoffel symbols. Identifying $\Gamma^{\theta}_{\theta r} = \Gamma^{2}_{21}$, and using Equation 3.340, we find the differential equation

$$\frac{d^2\theta}{d\sigma^2} + \frac{2}{\left(1 - \frac{\omega^2 r^2}{c^2}\right)r}\frac{d\theta}{d\sigma}\frac{dr}{d\sigma} = 0, \tag{3.344}$$

which can also be recast in the form

$$\frac{d}{d\theta}\left[\left(\frac{r^2}{1 - \frac{\omega^2 r^2}{c^2}}\right)\frac{d\theta}{d\sigma}\right] = 0, \tag{3.345}$$

as can be verified easily by performing the differentiation. Equation 3.345 can now be integrated, to yield

$$\left(\frac{r^2}{1 - \frac{\omega^2 r^2}{c^2}}\right)\frac{d\theta}{d\sigma} = k, \tag{3.346}$$

where $k$ is a constant.

Going back to the metric equation, Equation 3.341, we also have

$$\left(\frac{dr}{d\sigma}\right)^2 = 1 - \frac{r^2}{1 - \frac{\omega^2 r^2}{c^2}}\left(\frac{d\theta}{d\sigma}\right)^2. \tag{3.347}$$

Combining Equations 3.346 and 3.347, we can eliminate $d\theta/d\sigma$, to obtain a differential equation in $r$:

$$\frac{dr}{d\sigma} = \pm\left(1 - \frac{r^2}{1 - \frac{\omega^2 r^2}{c^2}}\right)^{1/2} k\frac{\left(1 - \frac{\omega^2 r^2}{c^2}\right)^{1/2}}{r} = \pm\sqrt{1 + k^2\left(\frac{\omega^2}{c^2} - \frac{1}{r^2}\right)}. \tag{3.348}$$

We seek a differential equation relating the radius and angle only; therefore, we eliminate the arc length, $\sigma$, by changing variable,

$$\frac{dr}{d\theta} = \frac{dr}{d\sigma}\frac{d\sigma}{d\theta} = \frac{dr}{d\sigma}\left(\frac{d\theta}{d\sigma}\right)^{-1}. \tag{3.349}$$

Using this procedure and the results derived above, we find

$$\frac{dr}{d\theta} = \pm\frac{1}{k}\frac{r^2}{1-\frac{\omega^2 r^2}{c^2}}\sqrt{1+k^2\left(\frac{\omega^2}{c^2}-\frac{1}{r^2}\right)}. \tag{3.350}$$

Let us now introduce the normalized radius,

$$\rho = \frac{r}{k}\sqrt{1+\frac{k^2\omega^2}{c^2}}, \tag{3.351}$$

and consider the differential equation for the angle,

$$\frac{d\theta}{d\rho} = \left(\frac{dr}{d\theta}\right)^{-1}\frac{r}{\rho} = \pm\left(\frac{1}{\rho^2}-\frac{k^2\omega^2}{c^2+k^2\omega^2}\right)\left(1-\frac{1}{\rho^2}\right)^{-1}. \tag{3.352}$$

The equation for the normalized radius then takes the form

$$\frac{1}{\sqrt{1-\rho^{-2}}}\rho^{-2}\frac{d\rho}{d\theta} = \pm 1 + \left(\frac{k^2\omega^2}{c^2+k^2\omega^2}\right)\frac{1}{\sqrt{1-\rho^{-2}}}\frac{d\rho}{d\theta}, \tag{3.353}$$

and can be integrated to obtain

$$\arg\cos\left(\frac{1}{\rho}\right) = \pm(\theta-\theta_0) + \left(\frac{k^2\omega^2}{c^2+k^2\omega^2}\right)\sqrt{\rho^2-1}. \tag{3.354}$$

We can choose the initial angle such that $\theta_0 = 0$; moreover, we can introduce the constant $a = k/\sqrt{1+(k\omega/c)^2}$ and return to the radius to obtain the sought-after geodesic equation,

$$\theta(r) = \pm\left[\arg\cos\left(\frac{a}{r}\right) - \frac{a\omega^2}{c^2}\sqrt{r^2-a^2}\right]. \tag{3.355}$$

In the case where $k = a = 0$, we can go back to Equation 3.350, to find that $d\theta/dr = 0$, which indicates that the corresponding geodesic lines coincide with the radii. For $k \neq 0$, two families of curves are found, which are symmetrical with respect to $\theta = 0$.

For a detailed description of the geometrical properties of these geodesic lines, we refer the interested reader to the discussion given by Poincelot. Here, we will focus on the question of proper time in this system and its relation to the Sagnac effect. In general relativity, light rays propagate along a geodesic of null length; therefore, we return to Equation 3.331 and consider the particular case where

$$ds^2 = -d\bar{r}^2 - \bar{r}^2 d\bar{\theta}^2 - d\bar{z}^2 + c^2 d\bar{t}^2 = 0. \tag{3.356}$$

The connection between the tangential contact frame and the laboratory frame is

$$r = \bar{r}, \quad z = \bar{z}, \quad \theta = \bar{\theta} \pm \omega\bar{t}. \tag{3.357}$$

For the null geodesic, we then have

$$ds^2 = -dr^2 - r^2(d\theta \pm \omega d\bar{t})^2 - dz^2 + c^2 d\bar{t}^2 = 0. \tag{3.358}$$

On the other hand, the differential length, as measured in the contact frame, is given by

$$d\sigma^2 = dr^2 + \frac{r^2 d\theta^2}{1 - \frac{\omega^2 r^2}{c^2}} + dz^2, \tag{3.359}$$

which yields

$$ds^2 = -d\sigma^2 + r^2\left(1 - \frac{1}{1 - \frac{\omega^2 r^2}{c^2}}\right)d\theta^2 \mp 2\omega r^2 d\theta\, d\bar{t} + c^2\left(1 - \frac{\omega^2 r^2}{c^2}\right)d\bar{t}^2. \tag{3.360}$$

Equation 3.360 can now be rearranged as

$$ds^2 = c^2 d\tau^2 - d\sigma^2 = 0, \tag{3.361}$$

where we have introduced

$$\tau = \sqrt{1 - \frac{\omega^2 r^2}{c^2}}\left[\bar{t} \pm \frac{\omega r^2 \theta}{c^2\left(1 - \frac{\omega^2 r^2}{c^2}\right)}\right]; \tag{3.362}$$

it is then easily seen that Equation 3.361 defines the proper time, $\tau$, as we have

$$\left(\frac{d\sigma}{c\,d\tau}\right)^2 = 1. \tag{3.363}$$

In particular, if we examine two light rays propagating along the same arc, *ds*, of the null geodesic, but in opposite directions, the difference in propagation time will be given by differentiating Equation 3.362 with respect to the angle subtending *ds*:

$$d\tau = \frac{2\omega r^2}{c^2\sqrt{1-\frac{\omega^2 r^2}{c^2}}}\,d\theta. \tag{3.364}$$

Furthermore, the area spanned over the angular displacement is given by

$$dS = \frac{1}{2}r(r\,d\theta), \tag{3.365}$$

so that we finally obtain

$$d\tau = \frac{4\omega}{c^2\sqrt{1-\frac{\omega^2 r^2}{c^2}}}\,dS. \tag{3.366}$$

For a complete rotation of the light rays going in opposite directions, we integrate Equation 3.364 from 0 to $2\pi$, with $r = r(\theta)$; in the case where $|\omega r/c| \ll 1$, Equation 3.366 becomes

$$\frac{d\tau}{dS} \simeq \frac{4\omega}{c^2}, \quad \Delta\tau = \oint \frac{4\omega}{c^2}\,dS = \frac{4\omega S}{c^2}. \tag{3.367}$$

This result corresponds to the Sagnac effect, which can be measured today to a very high degree of accuracy using a laser or fiber ring interferometer. The phase difference between counterpropagating light signals is a direct measurement of the instantaneous rotation frequency of the system, and one can build extremely precise laser gyroscopes based on this principle. Also note that the curve $r(\theta)$ is arbitrary, as our result only depends on the surface enclosed by the loop.

Finally, we briefly present the nonrelativistic calculation of this effect for comparison. In this case, we first consider a circle of radius *r*, and the propagation time of a light ray along the circle is simply

$$\Delta t = \frac{2\pi r}{c}, \tag{3.368}$$

if the path is not rotating. In the case where there is a fixed rotation frequency $\omega$, the equations for the propagation time along (+) and against (−) the direction of rotation are

$$c\Delta t^{\pm} = 2\pi r \pm \omega r \Delta t^{\pm}. \tag{3.369}$$

Solving Equation 3.369 and measuring the time difference between the two rays yield

$$\Delta\tau = \Delta t^{+} - \Delta t^{-} = 2\pi r\left(\frac{1}{c-\omega r} - \frac{1}{c+\omega r}\right) = 4\pi r\frac{\omega r}{c^2 - \omega^2 r^2}. \tag{3.370}$$

The nonrelativistic framework implies that $|\omega r/c| \ll 1$, and we recover

$$\Delta\tau \simeq \frac{4\pi r^2\omega}{c^2} = \frac{4\omega S}{c^2}. \tag{3.371}$$

In terms of laser interferometry, the corresponding phase difference is

$$\Delta\phi = \frac{8\pi S\omega}{\lambda c}, \tag{3.372}$$

where $\lambda$ is the wavelength of the laser used to probe the acceleration.

In closing, a numerical example can give us an idea of the quantities involved: the relative frequency shift associated with $\Delta\phi$ is

$$\frac{\Delta f}{f} = \frac{4S\omega}{Lc}, \tag{3.373}$$

where $L$ is the length of the ring.

For the angular frequency corresponding to an acceleration $a = 0.1g = 0.981\ \text{m} \times \text{s}^{-2}$, at a velocity $v = 98.1\ \text{m} \times \text{s}^{-1} = 220$ mph, we have $\omega = a/v = 0.01\ \text{rad} \times \text{s}^{-1}$. For a ring with $S/L = r/2 = 0.1$ m, we find $\Delta f/f = 1.33 \times 10^{-11}$; at the He-Ne laser wavelength $\lambda = 632.8$ nm, this corresponds to 6.3 kHz; for the rotation velocity of a geo-synchronous satellite, $\omega = 72.7\ \mu\text{rad} \times \text{s}^{-1}$, we find 46 Hz. This is routinely measured by optical heterodyning.

## 3.11 Dual Tensors and Spinors, Dirac Equation

A number of important concepts are briefly described in this section; the concept of dual tensors will be explored in detail in Chapter 10, which deals with the fundamental question of magnetic charges and symmetrized electrodynamics.

We begin with the mathematical definition of dual tensors, as given by Pauli, who considers the surface element

$$\xi^{\mu\nu} = x^{\mu}y^{\nu} - x^{\nu}y^{\mu}, \tag{3.374}$$

and its dual,

$$\tilde{\xi}^{\mu\nu} = \tilde{x}^{\mu}\tilde{y}^{\nu} - \tilde{x}^{\nu}\tilde{y}^{\mu}, \tag{3.375}$$

Here, the key point is that the vectors $\tilde{x}^{\mu}$ and $\tilde{y}^{\mu}$ are orthogonal to $x^{\mu}$ and $y^{\mu}$ :

$$\tilde{x}_{\mu}x^{\mu} = 0, \quad \tilde{x}_{\mu}y^{\mu} = 0, \quad \tilde{y}_{\mu}x^{\mu} = 0, \quad \tilde{y}_{\mu}y^{\mu} = 0. \tag{3.376}$$

In fact, all straight lines in the surface element $\xi^{\mu\nu}$ are perpendicular to all straight lines in $\tilde{\xi}^{\mu\nu}$.

In a four-dimensional manifold, the six independent components of the antisymmetrical dual tensor can then be derived to obtain

$$\begin{aligned} \tilde{\xi}^{14} &= \frac{1}{\sqrt{g}}\xi_{23}, \quad \tilde{\xi}^{24} = \frac{1}{\sqrt{g}}\xi_{31}, \quad \tilde{\xi}^{34} = \frac{1}{\sqrt{g}}\xi_{12}, \\ \tilde{\xi}^{23} &= \frac{1}{\sqrt{g}}\xi_{14}, \quad \tilde{\xi}^{31} = \frac{1}{\sqrt{g}}\xi_{24}, \quad \tilde{\xi}^{12} = \frac{1}{\sqrt{g}}\xi_{34}. \end{aligned} \tag{3.377}$$

The conjugated relations are obtained by exchanging the $\xi^{\mu\nu}$ and $\tilde{\xi}^{\mu\nu}$ and multiplying by $\sqrt{g}$ instead of dividing by that factor:

$$\begin{aligned} \tilde{\xi}_{14} &= \sqrt{g}\,\xi^{23}, \quad \tilde{\xi}_{24} = \sqrt{g}\,\xi^{31}, \quad \tilde{\xi}_{34} = \sqrt{g}\,\xi^{12}, \\ \tilde{\xi}_{23} &= \sqrt{g}\,\xi^{14}, \quad \tilde{\xi}_{31} = \sqrt{g}\,\xi^{24}, \quad \tilde{\xi}_{12} = \sqrt{g}\,\xi^{34}. \end{aligned} \tag{3.378}$$

The relations obtained using Equation 3.374 can be generalized to any second-rank tensor; in addition, the contraction of a tensor by its dual leads to the invariant

$$\frac{1}{2}\xi_{\mu\nu}\tilde{\xi}^{\mu\nu} = \frac{1}{\sqrt{g}}(\xi_{12}\xi_{34} + \xi_{13}\xi_{24} + \xi_{14}\xi_{23}). \tag{3.379}$$

To illustrate the usefulness of the dual tensor concept, we present a cursory introduction to the symmetrized version of Maxwell's equations, which will be discussed in considerable detail in Chapter 10. If magnetic charges are allowed, following Dirac's proposal to explain charge quantization, Maxwell's equations take the form

$$\begin{aligned}\partial_\nu F^{\mu\nu} &= 4\pi j^\mu,\\ \partial_\nu \tilde{F}^{\mu\nu} &= 4\pi g^\mu,\end{aligned} \tag{3.380}$$

where $\tilde{F}^{\mu\nu}$ is the dual to the electromagnetic field tensor, and $g^\mu$ is the magnetic four-current.

In addition we are using electron units, as introduced in Section 3.6. We can also introduce the electric and magnetic four-potentials,

$$A^\mu = (\phi, \mathbf{A}), \quad V^\mu = (\varphi, \mathbf{V}), \tag{3.381}$$

which can then be used to derive the electromagnetic field tensor and its dual:

$$\begin{aligned}F^{\mu\nu} &= \partial^\mu A^\nu - \partial^\nu A^\mu - \varepsilon^{\mu\nu\alpha\beta}\partial_\alpha V_\beta,\\ \tilde{F}^{\mu\nu} &= \varepsilon^{\mu\nu\alpha\beta}\partial_\alpha A_\beta + \partial^\mu V^\nu - \partial^\nu V^\mu.\end{aligned} \tag{3.382}$$

Here, $\varepsilon^{\mu\nu\alpha\beta}$ is the completely antisymmetrical Levi-Civita tensor. Now applying the Lorentz gauge condition to both potentials, we have

$$\partial_\mu A^\mu = \partial_\mu V^\mu = 0, \tag{3.383}$$

and

$$\begin{aligned}\partial_\nu \partial^\nu A_\mu &= \Box A_\mu = -4\pi j_\mu,\\ \partial_\nu \partial^\nu V_\mu &= \Box V_\mu = -4\pi g_\mu.\end{aligned} \tag{3.384}$$

Finally, if the four-currents are rotated as

$$\begin{aligned}j'_\mu &= j_\mu \cos\theta + g_\mu \sin\theta,\\ g'_\mu &= g_\mu \cos\theta - j_\mu \sin\theta,\end{aligned} \tag{3.385}$$

and we also apply this dual transform to the potentials,

$$\begin{aligned}A'_\mu &= A_\mu \cos\theta + V_\mu \sin\theta,\\ V'_\mu &= V_\mu \cos\theta - A_\mu \sin\theta,\end{aligned} \tag{3.386}$$

Equation 3.385 remains invariant. This illustrates the dual invariance of electrodynamics, which is further explored in Chapter 10.

Our final topic is the description of spinors, which are introduced by considering the Dirac equation. The transition from classical mechanics to quantum mechanics can be effected by replacing the energy-momentum four-vector by the four-gradient operator, multiplied by Planck's constant:

$$p_\mu = \left(\frac{E}{c}, \mathbf{p}\right) \rightarrow -i\hbar\partial_\mu = -i\hbar(-\partial_0, \nabla) = i\hbar\left(\frac{1}{c}\partial_t, -\nabla\right). \tag{3.387}$$

Equation 3.387 describes the correspondence principle in quantum mechanics and is manifestly covariant; moreover, the wavefunctions underlying the formalism of quantum mechanics appear implicitly in the fact that in Fourier space the operator $-i\hbar\partial_\mu$ formally becomes $i\hbar k_\mu$, which is the Fourier conjugate of the four-position, $x_\mu$. Therefore, the canonical conjugation between the four-position and the four-momentum is reflected in the Fourier conjugation between the four-position and the four-wavenumber in momentum space.

If we now use the relation between energy and momentum, namely,

$$E = \sqrt{\mathbf{p}^2c^2 + m_0^2c^4}, \tag{3.388}$$

and the correspondence presented in Equation 3.387, we obtain the so-called square root Klein-Gordon equation:

$$i\hbar\frac{\partial}{\partial t}\psi(\mathbf{x}, t) = \sqrt{-c^2\hbar^2\Delta + m_0^2c^4}\,\psi(\mathbf{x}, t). \tag{3.389}$$

We also note that we can use the relation directly

$$p_\mu p^\mu = m_0^2c^2(u_\mu u^\mu) = -m_0^2c^2, \tag{3.390}$$

to find

$$-\hbar^2\partial_\mu\partial^\mu\psi(x_\nu) = -m_0^2c^2\psi(x_\nu), \tag{3.391}$$

or

$$\left[\hbar^2\left(\Delta - \frac{1}{c^2}\partial_t^2\right) + m_0^2c^2\right]\psi(x_\nu) = 0, \tag{3.392}$$

which is the Klein-Gordon equation. In the case of an external electromagnetic field, we can use the relativistic Hamiltonian, as derived in Section 3.7, and replace the momentum and momentum operator as follows:

$$p_\mu \rightarrow p_\mu - eA_\mu \rightarrow -i\hbar\partial_\mu - eA_\mu, \tag{3.393}$$

where $A_\mu$ is the four-potential from which the external field derives.

There are two main problems with Equation 3.392, when applied to the description of electrons: first, the Klein-Gordon equation operates on a scalar wavefunction, which cannot describe spin-half particles; second, it is a second-order equation in time, which is incompatible with an evolution equation for the wavefunction. In particular, it can lead to unphysical negative probabilities. These difficulties provided the basis for Dirac's work and derivation of a new, covariant, linearized equation for the electron wavefunction. Indeed, Dirac's brilliant insight provides an elegant solution to the problem at hand. We write

$$E = c\alpha_i p_i + \beta m_0 c^2 = c\boldsymbol{\alpha}\cdot\mathbf{p} + \beta m_0 c^2, \tag{3.394}$$

where $\alpha = (\alpha_1, \alpha_2, \alpha_3)$ and $\beta$ will be determined by comparing Equation 3.394 with the original equation, Equation 3.388.

If we consider the square of the energy, we have

$$\begin{aligned} E^2 &= c^2\mathbf{p}^2 + m_0^2c^4 \\ &= c^2 p_i p_i + m_0^2 c^4 \\ &= c^2(\alpha_i p_i + m_0 c)(\alpha_k p_k + m_0 c), \end{aligned} \tag{3.395}$$

which implies that the following relations must hold:

$$\begin{aligned} \alpha_i\alpha_j + \alpha_k\alpha_j &= 2\delta_{ij}\mathbb{1}, \\ \alpha_i\beta + \beta\alpha_i &= \mathbb{0}, \\ \beta^2 &= \mathbb{1}, \end{aligned} \tag{3.396}$$

where $\mathbb{0}$ and $\mathbb{1}$ represent the $n \times n$ null and unit matrices, respectively.

Our next task is to determine the dimension of these objects; thus, following Thaller, we examine the traces of the various matrices involved. It is easily seen that

$$\mathrm{Tr}(\alpha_i) = \mathrm{Tr}(\beta^2\alpha_i) = -\mathrm{Tr}(\beta\alpha_i\beta) = -\mathrm{Tr}(\alpha_i\beta^2) = -\mathrm{Tr}(\alpha_i); \tag{3.397}$$

therefore, the traces must be equal to zero. Additionally, quantum mechanics require that these matrices be Hermitian. Finally, the fact that $\alpha_i^2 = 1$ implies that its eigenvalue is ±1. These facts considered together lead to the requirement that $n$ be an even number, with the simplest possibility, $n = 2$. In that case, the Pauli matrices offer the only set of three independent matrices with the required properties:

$$\sigma_1 = \begin{pmatrix} 0 & 1 \\ 1 & 0 \end{pmatrix}, \quad \sigma_2 = \begin{pmatrix} 0 & -i \\ i & 0 \end{pmatrix}, \quad \sigma_3 = \begin{pmatrix} 1 & 0 \\ 0 & -1 \end{pmatrix}. \tag{3.398}$$

However, four matrices are required: $\alpha_i$, $\alpha_j$, $\alpha_k$ and $\beta$; therefore, $n = 4$ must be considered. In that case, Dirac found the proper set of matrices, namely

$$\alpha_i = \begin{pmatrix} 0 & \sigma_i \\ \sigma_i & 0 \end{pmatrix}, \quad \beta = \begin{pmatrix} 1 & 0 \\ 0 & -1 \end{pmatrix}. \tag{3.399}$$

This is Dirac's standard notation. In a more explicit form, we have

$$\alpha_1 = \begin{pmatrix} 0 & 0 & 0 & 1 \\ 0 & 0 & 1 & 0 \\ 0 & 1 & 0 & 0 \\ 1 & 0 & 0 & 0 \end{pmatrix}, \quad \alpha_2 = \begin{pmatrix} 0 & 0 & 0 & -i \\ 0 & 0 & i & 0 \\ 0 & -i & 0 & 0 \\ i & 0 & 0 & 0 \end{pmatrix},$$

$$\alpha_3 = \begin{pmatrix} 0 & 0 & 1 & 0 \\ 0 & 0 & 0 & -1 \\ 1 & 0 & 0 & 0 \\ 0 & -1 & 0 & 0 \end{pmatrix}, \quad \beta = \begin{pmatrix} 1 & 0 & 0 & 0 \\ 0 & 1 & 0 & 0 \\ 0 & 0 & -1 & 0 \\ 0 & 0 & 0 & -1 \end{pmatrix}. \tag{3.400}$$

We now return to the Dirac equation. Replacing the four-momentum by the four-gradient operator, as prescribed in Equation 3.387, we find

$$i\hbar \frac{\partial}{\partial t} \psi_\mu(x_\lambda) = [i\hbar c \boldsymbol{\alpha} \cdot \nabla + \beta m_0 c^2] \psi_\mu(x_\lambda), \tag{3.401}$$

where the operator

$$[-i\hbar c \boldsymbol{\alpha} \cdot \nabla + \beta m_0 c^2] = \begin{pmatrix} m_0 c^2 \mathbb{1} & -i\hbar c \boldsymbol{\sigma} \cdot \nabla \\ -i\hbar c \boldsymbol{\sigma} \cdot \nabla & -m_0 c^2 \mathbb{1} \end{pmatrix}, \tag{3.402}$$

and where the wavefunction has four-components, and $\boldsymbol{\alpha} = (\alpha_1, \alpha_2, \alpha_3)$ and $\boldsymbol{\sigma} = (\sigma_1, \sigma_2, \sigma_3)$ represent a triplet of matrices.

From the wavefunctions satisfying the Dirac equation, a number of important operators can be used to define observable quantities: the position, $\mathbf{x}$, and momentum, $\mathbf{p}$, as well as the orbital angular momentum, $\ell = \mathbf{x} \times \mathbf{p}$, the spin operator, $\mathbf{s} = -\frac{i}{4}\alpha \times \alpha$, and the total angular momentum, $j = \ell + \mathbf{s}$. For the spin, the cross product between the $\alpha$-matrices must be defined with care:

$$\alpha \times \alpha = \varepsilon_{ijk}\alpha_j\alpha_k, \tag{3.403}$$

where $\varepsilon_{ijk}$ is the completely antisymmetric unit tensor.

Following the formalism of quantum mechanics, the action of an operator, $O$, on a given wavefunction is described by

$$\langle \psi | O | \psi \rangle = \int \psi O \psi^* d^3x\,; \tag{3.404}$$

furthermore, the quantity

$$|\psi(\mathbf{x})|^2 = \psi_\mu(\mathbf{x})\psi^*{}_\mu(\mathbf{x}) \tag{3.405}$$

is interpreted as the probability density for the electron.

Within this physical context, we can introduce and briefly discuss spinors. First, we review some important properties of the free Dirac operator, including its energy spectrum, which led to the discovery of antiparticles. This operator corresponds to the Hamiltonian in Equation 3.402:

$$\mathcal{D} = [-i\hbar c\alpha \cdot \nabla + \beta m_0 c^2] = \begin{pmatrix} m_0 c^2 \mathbb{1} & -i\hbar c\sigma \cdot \nabla \\ -i\hbar c\sigma \cdot \nabla & -m_0 c^2 \mathbb{1} \end{pmatrix}, \tag{3.406}$$

and can best be analyzed in momentum space. We Fourier transform the wavefunction as

$$\tilde{\psi}_\mu(\mathbf{p}) = \mathcal{F}\psi_\mu(\mathbf{p}) = \frac{1}{(\sqrt{2\pi})^3}\iiint \psi_\mu(\mathbf{x})\exp(-i\mathbf{p}\cdot\mathbf{x})d^3x, \tag{3.407}$$

which is the three-dimensional equivalent of the covariant plane wave expansion,

$$\tilde{\psi}_\mu(p^\lambda) = \frac{1}{(\sqrt{2\pi})^4}\iiiint \psi_\mu(x_\lambda)\exp(-ip^\lambda x_\lambda)d^4x. \tag{3.408}$$

The inverse Fourier transform is given by

$$\psi_\mu(\mathbf{x}) = \mathcal{F}^{-1}\tilde{\psi}_\mu(\mathbf{x}) = \frac{1}{(\sqrt{2\pi})^3}\iiint \tilde{\psi}_\mu(\mathbf{p})\exp(+i\mathbf{p}\cdot\mathbf{x})d^3x, \tag{3.409}$$

and we can now consider the operator

$$[\mathcal{F}\mathcal{D}\mathcal{F}^{-1}](\mathbf{p}) = \begin{pmatrix} m_0c^2\mathbb{1} & c\boldsymbol{\sigma}\cdot\mathbf{p} \\ c\boldsymbol{\sigma}\cdot\mathbf{p} & -m_0c^2\mathbb{1} \end{pmatrix}. \tag{3.410}$$

This $4 \times 4$ matrix has four eigenvalues,

$$\lambda_1(\mathbf{p}) = \lambda_2(\mathbf{p}) = -\lambda_3(\mathbf{p}) = -\lambda_4(\mathbf{p}) = \sqrt{c^2\mathbf{p}^2 + m_0^2c^4}, \tag{3.411}$$

and can be diagonalized by means of the unitary transform

$$\mathrm{U}(\mathbf{p}) = \frac{[m_0c^2 + \lambda_1(\mathbf{p})]\mathbb{1} + \beta c\boldsymbol{\alpha}\cdot\mathbf{p}}{\sqrt{2\lambda_1(\mathbf{p})[m_0c^2 + \lambda_1(\mathbf{p})]}}. \tag{3.412}$$

The free Dirac operator spectrum includes negative eigenvalues, which correspond to negative energies for the Hamiltonian, and the question of the physical interpretation of such eigenstates arises. A deeper insight can be attained by considering the Dirac equation for a charge $e$ subjected to an external electromagnetic field. In that case, we have

$$\mathcal{D}(e) = c\boldsymbol{\alpha}\cdot[\mathbf{p} - e\mathbf{A}(\mathbf{x}, t)] + \beta m_0c^2 + e\phi(\mathbf{x}, t), \tag{3.413}$$

and it can be shown that if a wavefunction $\psi$ is a solution of the Dirac equation for $\mathcal{D}(e)$, the charge conjugate wavefunction, $\mathrm{C}\psi$, is a solution for $\mathcal{D}(-e)$. Moreover,

$$\mathrm{C}\mathcal{D}(e)\mathrm{C}^{-1} = -\mathcal{D}(-e). \tag{3.414}$$

The physical interpretation of this symmetry is the existence of antiparticles, which appear as negative energy states; more precisely, an electron with a negative eigenvalue for the Hamiltonian is interpreted as a positron with positive energy. Furthermore, it can also be shown that the symmetries underlying the covariant equations of electrodynamics are such that a positron moving forward in time can be interpreted as its charge conjugate, the electron, moving backward in time. This property is clearly exhibited in Feynman diagrams,

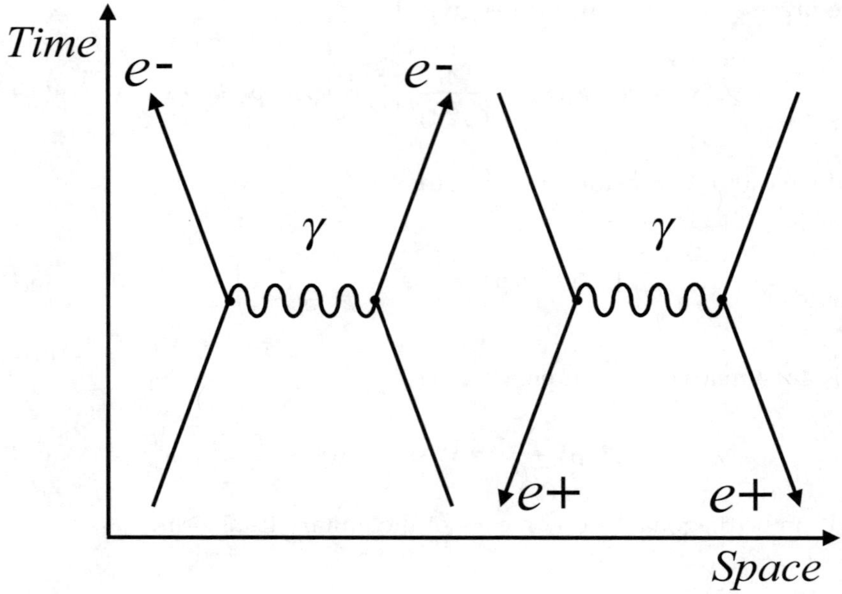


**FIGURE 3.3**
Feynman diagram for the exchange of a photon mediating the Coulomb interaction between an electron and a positron; the diagram is manifestly time-reversible.

as illustrated in Figure 3.3. Therefore, the Dirac equation introduces a profound connection between spin, charge conjugation, and time reversal.

Our final discussion is focused on the mathematical description of spinors, and their relation with the Lorentz transform and the covariant form of electrodynamics. This material is exposed in depth in the monograph by Barut, and the introductory remarks presented here are largely inspired by that work. In our discussion of the Dirac equation, we have introduced the Pauli matrices (see Equation 3.398) and the unity, or identity, $2 \times 2$ matrix, $\mathbb{1}$. There is a well-known relation between these objects and rotations in three dimensions; moreover, as we have seen that the Lorentz transformation reduces to a rotation in hyperbolic space–time, it is natural to seek establishing a connection with the so-called two-dimensional, unimodular, unitary group. We begin by considering a $2 \times 2$ Hermitian matrix, which can always be projected onto the aforementioned Pauli and unity matrices:

$$M = m^0 \mathbb{1} + m^1 \sigma_1 + m^2 \sigma_2 + m^3 \sigma_3, \quad m^\mu \in \mathbb{R}. \tag{3.415}$$

Using Equation 3.398, we find that the explicit form of $M$ is:

$$M = (m_{ij}) = \begin{pmatrix} m^0 + m^3 & m^1 - im^2 \\ m^1 + im^2 & m^0 - m^3 \end{pmatrix}. \tag{3.416}$$

Following Barut, we write

$$M = m^{\mu}\sigma_{\mu}; \tag{3.417}$$

however, it is important to keep in mind that we are dealing with matrices and not four-vectors. The exact connection between these objects will de discussed shortly. At this point, an important concept can be introduced: if we consider an arbitrary 2 × 2 complex matrix, $L$, such that $\det|L| = 1$, and we apply the matrix transformation

$$\overline{M} = LML^{\dagger} = \bar{x}^{\mu}\sigma_{\mu}, \tag{3.418}$$

we have

$$\det|\overline{M}| = (\bar{x}^0)^2 - (\bar{x}^1)^2 - (\bar{x}^2)^2 - (\bar{x}^3)^2 = \det|M| = (x^0)^2 - (x^1)^2 - (x^2)^2 - (x^3)^2. \tag{3.419}$$

If we now interpret the matrix components in terms of four-vector coordinates, we recognize the basic light-cone invariance underlying the Lorentz transform. This fundamental property can be written more explicitly,

$$\mathcal{L}^{\mu}_{\nu}x^{\nu}\sigma_{\mu} = Lx^{\mu}\sigma_{\mu}L^{\dagger} = x^{\mu}L\sigma_{\mu}L^{\dagger}, \tag{3.420}$$

which yields

$$\mathcal{L}^{\mu}_{\nu}\sigma_{\mu} = L\sigma_{\nu}L^{\dagger}. \tag{3.421}$$

Contracting this equation by $\sigma_{\mu}$ and taking the trace of each side of the equality we find

$$\mathcal{L}^{\mu}_{\nu} = \frac{1}{2}\mathrm{Tr}(\sigma_{\mu}L\sigma_{\nu}L^{\dagger}) \equiv \mathcal{L}(L). \tag{3.422}$$

Therefore, we have demonstrated the relation between the matrix $L$ and the Lorentz transform, $\mathcal{L}(L)$ It is important to note that $\mathcal{L}(L) = \mathcal{L}(-L)$; in addition, multiplication is conserved:

$$\mathcal{L}(K)\mathcal{L}(L) = \mathcal{L}(KL). \tag{3.423}$$

The relation is a homomorphism, and the 2 × 2 unimodular matrices form a two-dimensional representation of the Lorentz group. The ambiguity on the sign of the transformation matrix is connected with the fact that for spin

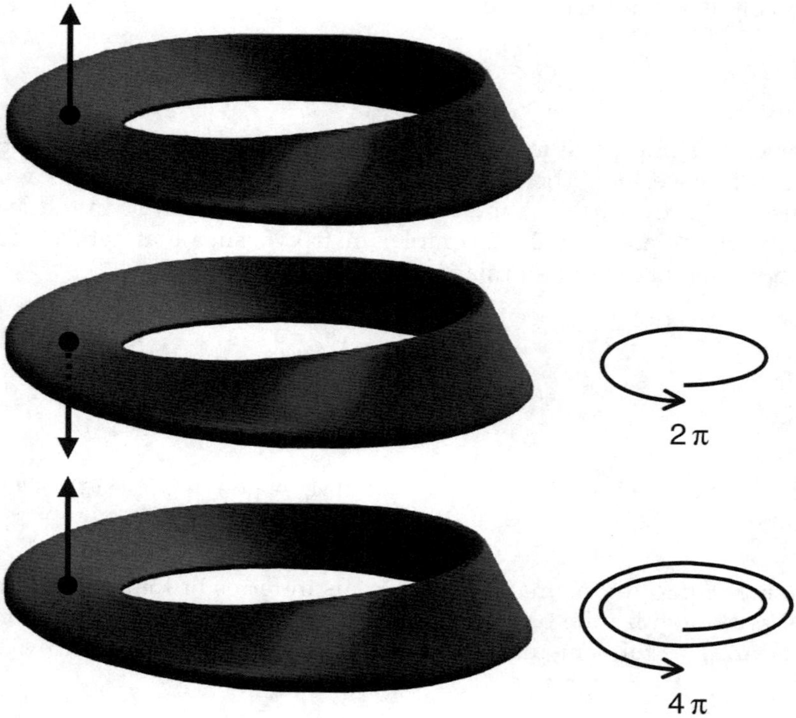


**FIGURE 3.4**
The normal an a Möbius strip behaves like a half-integer spin: both are reversed through a rotation of $2\pi$, and require a rotation of $4\pi$ to be conserved.

half particles, a rotation of $2\pi$ results in a change of sign for the spin, while a rotation of $4\pi$ leaves the spin invariant. This can be modeled by considering the normal along a Möbius strip, as illustrated in Figure 3.4. If one follows the normal over a $2\pi$ loop, this unit vector is flipped; a $4\pi$ loop restores the original vector. This idea was also described by Feynman, after Finkelstein, in his 1986 Dirac memorial lectures with Weinberg.

Within this context, we can now introduce spinors. An object satisfying the definition

$$\xi = \begin{pmatrix} \xi^1 \\ \xi^2 \end{pmatrix}, \quad \xi \in \mathbb{C}^2 \tag{3.424}$$

is called a spinor if its behavior under the Lorentz transform can be described by

$$\bar{\xi} = L(\mathcal{L})\xi, \quad \det|L(\mathcal{L})| = 1. \tag{3.425}$$

Let us now briefly highlight a few basic properties of such objects. As we have introduced spinors in relation with the Lorentz transform, it is interesting to first consider the invariant

$$\xi^{\mu} K_{\mu\nu} \zeta^{\nu}. \tag{3.426}$$

Using the Lorentz transform, as expressed in Equation 3.420, we see that the $2 \times 2$ matrix $K$ must be antisymmetrical:

$$L_{\mu\alpha} K^{\mu\nu} L_{\nu\beta} = -L_{\mu\alpha} K^{\nu\mu} L_{\nu\beta} = -L_{\mu\beta} K^{\mu\nu} L_{\nu\alpha}. \tag{3.427}$$

Of course, for $2 \times 2$ matrices, this means that

$$K = \kappa i \sigma_2 = \kappa \begin{pmatrix} 0 & 1 \\ -1 & 0 \end{pmatrix}, \tag{3.428}$$

where $i\sigma_2 = C$ is called the metric spinor.

With this, in close analogy with the connection between covariant and contravariant four-vector, where $x_{\mu} = g_{\mu\nu} x^{\nu}$, we have

$$\xi_{\mu} = C_{\mu\nu} \xi^{\nu}, \quad \text{and} \quad \xi^{\mu} \zeta_{\mu} = -\xi_{\mu} \zeta^{\mu}. \tag{3.429}$$

We also note that higher-order spinors can be constructed, just as tensors can be generated from four-vectors. For example, we can consider the spinor

$$\xi^{\mu\nu} = \xi^{\mu} \xi^{\nu}, \tag{3.430}$$

which Lorentz transforms according to

$$\bar{\xi}^{\mu\nu} = L^{\mu}_{\alpha} L^{\nu}_{\beta} \xi^{\alpha\beta}. \tag{3.431}$$

To finalize this introduction, we briefly indicate the relation between spinors, four-vectors, and the metric, and discuss the spinor form of Maxwell's equations. Starting from a $2 \times 2$ Hermitian matrix, $M = \sigma_{\mu} m^{\mu}$, and using the Lorentz transformation, $\overline{M} = LML^{\dagger}$, we obtain

$$\overline{M}_{\mu\dot{\mu}} = L^{\nu}_{\mu} L^{*\dot{\nu}}_{\dot{\mu}} M_{\nu\dot{\nu}}, \tag{3.432}$$

which yields the connection between four-vectors and second-order spinors,

$$M_{\mu\dot{\mu}} = \sigma_{\nu\mu\dot{\mu}} m^{\nu}, \tag{3.433}$$

where $\sigma_{\nu\mu\dot{\mu}}$ corresponds to the Pauli matrices, and where the notation $\xi^{\dot{\mu}}$ is defined by the transformation relation

$$\bar{\xi}^{\dot{\mu}} = L^{*\dot{\mu}}_{\dot{\nu}}\xi^{\dot{\nu}}, \tag{3.434}$$

in contradistinction with

$$\bar{\xi}^{\mu} = L^{\mu}_{\nu}\xi^{\nu}, \tag{3.435}$$

which is equivalent to Equation 3.425. With this, the relations between the metric, the spinor metric, and the Pauli matrices are

$$\begin{aligned} \sigma^{\alpha}_{\mu\gamma}\sigma^{\beta\dot{\gamma}}_{\gamma} + \sigma^{\alpha}_{\nu\dot{\gamma}}\sigma^{\beta\dot{\gamma}}_{\mu} &= g_{\mu\nu}C^{\alpha\beta}, \\ \sigma^{\mu}_{\alpha\dot{\alpha}} &= g^{\mu\nu}\sigma^{\gamma\dot{\delta}}_{\nu}C_{\gamma\alpha}C_{\dot{\delta}\dot{\alpha}}, \\ g_{\mu\dot{\mu}\nu\dot{\nu}} &= C_{\mu\nu}C_{\dot{\mu}\dot{\nu}}. \end{aligned} \tag{3.436}$$

Finally, the spinor form of Maxwell's equations can be given as follows: following Barut, we first introduce the quantity

$$F^{\alpha\dot{\alpha}\beta\dot{\beta}} = \sigma^{\alpha\dot{\alpha}}_{\mu}\sigma^{\beta\dot{\beta}}_{\nu}F^{\mu\nu}, \tag{3.437}$$

where $F^{\mu\nu}$ is the electromagnetic field tensor.

The fact that it is antisymmetric allows us to further write

$$F^{\alpha\dot{\alpha}\beta\dot{\beta}} = \frac{1}{2}(\varphi^{\alpha\beta}C^{\dot{\alpha}\dot{\beta}} + C^{\alpha\beta}\varphi^{*\dot{\alpha}\dot{\beta}}), \tag{3.438}$$

where $\varphi^{\alpha\beta}$ is a symmetric, second-order spinor.

We now introduce the spinor equivalent of the four-current density

$$j^{\alpha\dot{\alpha}} = \sigma^{\alpha\dot{\alpha}}_{\mu}j^{\mu}, \tag{3.439}$$

and write

$$\partial^{\alpha\dot{\alpha}}\varphi_{\alpha\beta} = -4\pi j^{\dot{\alpha}}_{\beta}, \tag{3.440}$$

where we have defined the derivative with respect to the spinor indices as

$$\partial^{\alpha\dot{\alpha}} = \sigma^{\alpha\dot{\alpha}}_{\mu}\partial^{\mu}. \tag{3.441}$$

In closing, we remark that with Barut, Maxwell's equations can also be written as

$$\beta_{\nu}\partial^{\nu}G_{\mu} + 4\pi j_{\mu} = 0, \tag{3.442}$$

if the four-vector $G_\mu$ is defined as

$$G_\mu = (0, \mathbf{E} + i\mathbf{B}), \tag{3.443}$$

and $\beta_\mu$ represents $4 \times 4$ Hermitian matrices, such that

$$\begin{aligned} \beta_0\beta_\mu &= \beta_\mu\beta_0, \\ \beta_i\beta_j &= \delta_{ij} - i\varepsilon_{ijk}\beta_k, \\ \beta_\mu^2 &= 1. \end{aligned} \tag{3.444}$$

This yields

$$\beta_0 = \begin{pmatrix} 1 & 0 \\ 0 & 1 \end{pmatrix}, \quad \beta_1 = \begin{pmatrix} -\sigma_2 & 0 \\ 0 & \sigma_2 \end{pmatrix}, \quad \beta_2 = i\begin{pmatrix} 0 & 1 \\ -1 & 0 \end{pmatrix}, \quad \beta_3 = \begin{pmatrix} 0 & \sigma_2 \\ \sigma_2 & 0 \end{pmatrix}. \tag{3.445}$$

Equation 3.442 is similar to the so-called Weyl or neutrino equation, which is essentially the Dirac equation for a massless particle of spin half:

$$i\hbar\partial_t\psi(t) = c\boldsymbol{\sigma}\cdot\mathbf{p}\,\psi(t). \tag{3.446}$$

This last equation has the very interesting property that it is not invariant under space reflections, which describes the famous parity violation in neutrino experiments.

For an in-depth treatment of the Dirac equation, we refer the reader to the excellent monograph on this subject by Thaller, as well as the classic series by Greiner and co-authors; two other books, by Bjorken and Drell, and Peskin and Schroeder, also prove very useful, as well as the classic text from Heitler. In addition, Schweber's historical description of the development of QED provides a superb and authoritative reading experience. For a detailed discussion of spinors and the Lorentz group, the monograph by Barut offers very detailed insights.

## 3.12 Notes to Chapter 3

We consider the electromagnetic field tensor problem in cylindrical coordinates, to show the generality of the tensorial approach. It is also a useful exercise that will prove helpful when considering cylindrical waveguides and Bessel functions. The gradient operator now takes the form

$$\nabla \equiv \hat{r}\partial_r + \hat{\theta}\frac{\partial_\theta}{r} + \hat{z}\partial_z, \tag{3.447}$$

and one must pay close attention to the fact that the derivatives of the unit vectors are nonzero:

$$\begin{aligned} \partial_\theta \hat{r} &= \hat{\theta}, \\ \partial_\theta \hat{\theta} &= -\hat{r}. \end{aligned} \tag{3.448}$$

This can easily be shown by considering Figure 3.5:

$$\begin{aligned} \hat{r} &= \hat{x}\cos\theta + \hat{y}\sin\theta, \\ \hat{\theta} &= -\hat{x}\sin\theta + \hat{y}\cos\theta. \end{aligned} \tag{3.449}$$

Taking the partial derivative with respect to the azimuthal angle, we find

$$\begin{aligned} \partial_\theta \hat{r} &= -\hat{x}\sin\theta + \hat{y}\cos\theta = \hat{\theta}, \\ \partial_\theta \hat{\theta} &= -\hat{x}\cos\theta - \hat{y}\sin\theta = -\hat{r}. \end{aligned} \tag{3.450}$$

Also note that because $\hat{r}$ and $\hat{\theta}$ are unit vectors, we have

$$\begin{aligned} \partial_\theta(\hat{r}^2) &= \partial_\theta(1) = 0 = 2\hat{r}\cdot\partial_\theta\hat{r}, \\ \partial_\theta(\hat{\theta}^2) &= \partial_\theta(1) = 0 = 2\hat{\theta}\cdot\partial_\theta\hat{\theta}. \end{aligned} \tag{3.451}$$

The derivative of a vector with constant length is always orthogonal to that vector; this is also true in space–time, as $u_\mu u^\mu = -1$ implies that $u_\mu a^\mu = -0$. Finally, the partial derivatives of the unit vectors with respect to $r$ are $z$ equal to zero because an infinitesimal change in either direction does not affect these vectors, as is shown in Figure 3.5 and Equation 3.449.

To see how this works, let us first consider the divergence in cylindrical coordinates:

$$\begin{aligned} \nabla\cdot\mathbf{A} &= \left(\hat{r}\partial_r + \hat{\theta}\frac{\partial_\theta}{r} + \hat{z}\partial_z\right)\cdot(\hat{r}A_r + \hat{\theta}A_\theta + \hat{z}A_z) \\ &= \hat{r}\cdot\partial_r(\hat{r}A_r + \hat{\theta}A_\theta + \hat{z}A_z) + \frac{\hat{\theta}}{r}\cdot\partial_\theta(\hat{r}A_r + \hat{\theta}A_\theta + \hat{z}A_z) \\ &\quad + \hat{z}\cdot\partial_z(\hat{r}A_r + \hat{\theta}A_\theta + \hat{z}A_z) \\ &= \hat{r}^2\partial_r A_r + \frac{\hat{\theta}}{r}\cdot[(\partial_\theta\hat{r})A_r + \hat{r}(\partial_\theta A_r) + (\partial_\theta\hat{\theta})A_\theta + \hat{\theta}(\partial_\theta A_\theta) \\ &\quad + (\partial_\theta\hat{z})A_z + \hat{z}(\partial_\theta A_z)] + \hat{z}^2\partial_z A_z \\ &= \partial_r A_r + \frac{\hat{\theta}}{r}\cdot[\hat{\theta}A_r + \hat{r}(\partial_\theta A_r) - \hat{r}A_\theta + \hat{\theta}(\partial_\theta A_\theta) + \hat{z}\partial_\theta A_z] + \partial_z A_z \\ &= \partial_r A_r + \frac{\hat{\theta}^2}{r}(A_r + \partial_\theta A_\theta) + \partial_z A_z = \frac{1}{r}\partial_r(rA_r) + \frac{1}{r}\partial_\theta A_\theta + \partial_z A_z. \end{aligned} \tag{3.452}$$

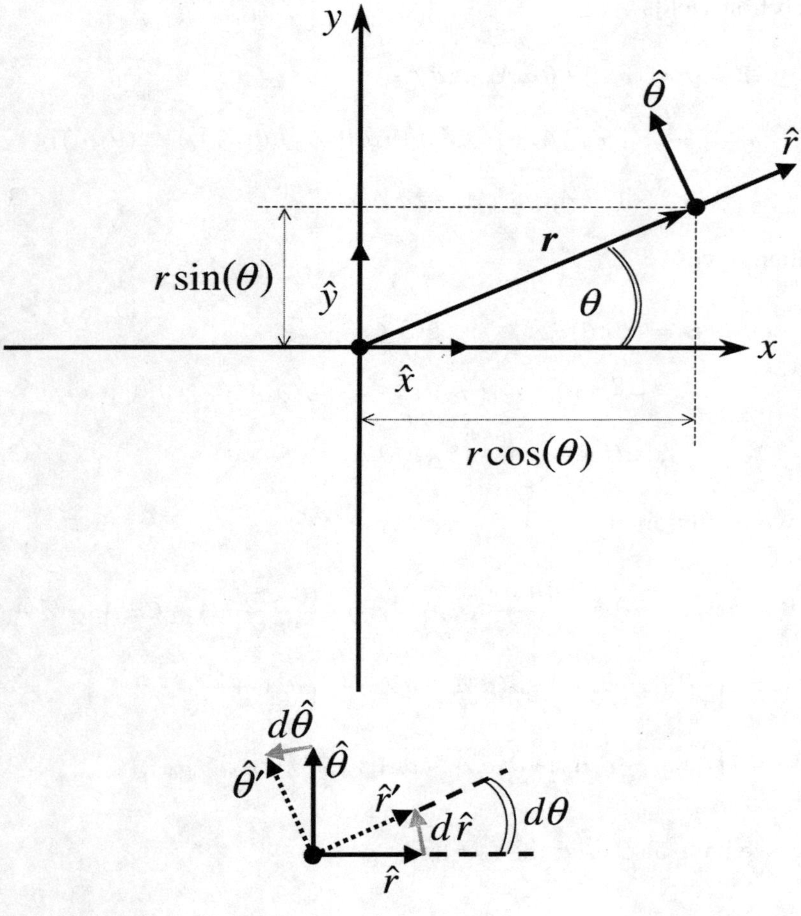


**FIGURE 3.5**
Cylindrical coordinates and the variation of the radial and azimuthal unit vectors under an infinitesimal rotation.

We now derive the explicit expressions for the components of the electric field and magnetic induction, using Equations 3.28, 3.32, and 3.447. We have, by definition,

$$\begin{aligned}\mathbf{B} = \nabla \times \mathbf{A} &= \left(\hat{r}\partial_r + \hat{\theta}\frac{\partial_\theta}{r} + \hat{z}\partial_z\right) \times (\hat{r}A_r + \hat{\theta}A_\theta + \hat{z}A_z) \\ &= \hat{r} \times [\partial_r(\hat{r}A_r + \hat{\theta}A_\theta + \hat{z}A_z)] + \frac{\hat{\theta}}{r} \times [\partial_\theta(\hat{r}A_r + \hat{\theta}A_\theta + \hat{z}A_z)] \\ &\quad + \hat{z} \times [\partial_z(\hat{r}A_r + \hat{\theta}A_\theta + \hat{z}A_z)], \qquad (3.453)\end{aligned}$$

which first yields

$$\begin{aligned}\mathbf{B} &= \hat{r}\times(\hat{r}\partial_r A_r+\hat{\theta}\partial_r A_\theta+\hat{z}\partial_r A_z)\\ &+\frac{\hat{\theta}}{r}\times[(\partial_\theta\hat{r})A_r+\hat{r}(\partial_\theta A_r)+(\partial_\theta\hat{\theta})A_\theta+\hat{\theta}(\partial_\theta A_\theta)+\hat{z}(\partial_\theta A_z)]\\ &+\hat{z}\times(\hat{r}\partial_z A_r+\hat{\theta}\partial_z A_\theta+\hat{z}\partial_z A_z).\end{aligned}\tag{3.454}$$

we then have

$$\begin{aligned}\mathbf{B} &= (\hat{r}\times\hat{\theta})\partial_r A_\theta+(\hat{r}\times\hat{z})\partial_r A_z\\ &+\frac{\hat{\theta}}{r}\times[\hat{\theta}A_r+\hat{r}(\partial_\theta A_r)-\hat{r}A_\theta+\hat{\theta}(\partial_\theta A_\theta)+\hat{z}(\partial_\theta A_z)]\\ &+(\hat{z}\times\hat{r})\partial_z A_r+(\hat{z}\times\hat{\theta})\partial_z A_\theta,\end{aligned}\tag{3.455}$$

and we finally obtain

$$\begin{aligned}\mathbf{B} &= \hat{z}\partial_r A_\theta-\hat{\theta}\partial_r A_z+\frac{(\hat{\theta}\times\hat{r})}{r}(\partial_\theta A_r-A_\theta)+\frac{(\hat{\theta}\times\hat{z})}{r}\partial_\theta A_z+\hat{\theta}\partial_z A_r-\hat{r}\partial_z A_\theta\\ &= \hat{r}\left(\frac{1}{r}\partial_\theta A_z-\partial_z A_\theta\right)+\hat{\theta}(\partial_z A_r-\partial_r A_z)+\hat{z}\left(\partial_r A_\theta+\frac{A_\theta}{r}-\frac{1}{r}\partial_\theta A_r\right)\\ &= \hat{r}\left(\frac{1}{r}\partial_\theta A_z-\partial_z A_\theta\right)+\hat{\theta}(\partial_z A_r-\partial_r A_z)+\frac{\hat{z}}{r}[\partial_r(rA_\theta)-\partial_\theta A_r].\end{aligned}\tag{3.456}$$

Here, we have also used the fact that

$$\hat{r}\times\hat{\theta} = \hat{z},\quad \hat{\theta}\times\hat{z} = \hat{r},\quad \hat{z}\times\hat{r} = \hat{\theta}.\tag{3.457}$$

For the electric field, the situation is considerably simpler:

$$\begin{aligned}\mathbf{E} &= -\nabla\phi-\partial_t\mathbf{A}\\ &= -\left(\hat{r}\partial_r+\hat{\theta}\frac{\partial_\theta}{r}+\hat{z}\partial_z\right)\phi-\partial_t(\hat{r}A_r+\hat{\theta}A_\theta+\hat{z}A_z)\\ &= -\hat{r}(\partial_r\phi+\partial_t A_r)-\hat{\theta}\left(\frac{\partial_\theta\phi}{r}+\partial_t A_\theta\right)-\hat{z}(\partial_z\phi+\partial_t A_z).\end{aligned}\tag{3.458}$$

We can now examine the electromagnetic field tensor; of course, the same care must be taken when dealing with the unit vectors and the partial derivative with respect to the azimuthal angle. For this reason, we write the four-gradient operator as

$$\partial_\mu = \hat{t}\,\frac{\partial_t}{c} + \hat{r}\partial_r + \hat{\theta}\,\frac{\partial_\theta}{r} + \hat{z}\partial_z, \tag{3.459}$$

where $\hat{t}$ is a unit vector along the temporal axis, in complete analogy with three-dimensional space.

The electromagnetic field tensor is then formally defined as

$$\mathbb{F} = c\left(\hat{t}\,\frac{\partial_t}{c} + \hat{r}\partial_r + \hat{\theta}\,\frac{\partial_\theta}{r} + \hat{z}\partial_z\right) \otimes \left(\hat{t}\,\frac{\phi}{c} + \hat{r}A_r + \hat{\theta}A_\theta + \hat{z}A_z\right). \tag{3.460}$$

The antisymmetrical nature of the field tensor is now reflected by the fact that

$$\hat{t} \otimes \hat{t} = \hat{r} \otimes \hat{r} = \hat{\theta} \otimes \hat{\theta} = \hat{z} \otimes \hat{z} = 0, \tag{3.461}$$

in close analogy with the cross-product, and

$$\begin{aligned}
\mathbb{F} &= F_{rt}(\hat{r} \otimes \hat{t}) + F_{\theta t}(\hat{\theta} \otimes \hat{t}) + F_{zt}(\hat{z} \otimes \hat{t}) + F_{r\theta}(\hat{r} \otimes \hat{\theta}) + F_{rz}(\hat{r} \otimes \hat{z}) + F_{\theta z}(\hat{\theta} \otimes \hat{z}) \\
&= -F_{tr}(\hat{r} \otimes \hat{t}) - F_{t\theta}(\hat{\theta} \otimes \hat{t}) - F_{tz}(\hat{z} \otimes \hat{t}) - F_{\theta r}(\hat{r} \otimes \hat{\theta}) \\
&\quad -F_{zr}(\hat{r} \otimes \hat{z}) - F_{z\theta}(\hat{\theta} \otimes \hat{z});
\end{aligned} \tag{3.462}$$

we also have

$$\begin{aligned}
\hat{r} \otimes \hat{\theta} &= -\hat{\theta} \otimes \hat{r}, \\
\hat{r} \otimes \hat{z} &= -\hat{z} \otimes \hat{r}, \\
\hat{\theta} \otimes \hat{z} &= -\hat{z} \otimes \hat{\theta},
\end{aligned} \tag{3.463}$$

again, just like the cross-product, but,

$$\begin{aligned}
\hat{r} \otimes \hat{t} &= \hat{t} \otimes \hat{r}, \\
\hat{\theta} \otimes \hat{t} &= \hat{t} \otimes \hat{\theta}, \\
\hat{z} \otimes \hat{t} &= \hat{t} \otimes \hat{z}.
\end{aligned} \tag{3.464}$$

We can now proceed by expanding the expression in Equation 3.462:

$$\begin{aligned}
\mathbb{F} = \hat{t} \otimes &\left[\partial_t\left(\hat{t}\,\frac{\phi}{c} + \hat{r}A_r + \hat{\theta}A_\theta + \hat{z}A_z\right)\right] \\
&+ c\hat{r} \otimes \left[\partial_r\left(\hat{t}\,\frac{\phi}{c} + \hat{r}A_r + \hat{\theta}A_\theta + \hat{z}A_z\right)\right] \\
&+ \frac{c\hat{\theta}}{r} \otimes \left[\partial_\theta\left(\hat{t}\,\frac{\phi}{c} + \hat{r}A_r + \hat{\theta}A_\theta + \hat{z}A_z\right)\right] \\
&+ c\hat{z} \otimes \left[\partial_z\left(\hat{t}\,\frac{\phi}{c} + \hat{r}A_r + \hat{\theta}A_\theta + \hat{z}A_z\right)\right].
\end{aligned} \tag{3.465}$$

Using the rules established in Equations 3.460 through 3.463, we first find that

$$\begin{aligned}\mathbb{F} = {} & (\hat{t}\otimes\hat{r})\partial_t A_r + (\hat{t}\otimes\hat{\theta})\partial_t A_\theta + (\hat{t}\otimes\hat{z})\partial_t A_z \\ & + (\hat{r}\otimes\hat{t})\partial_r\phi + (\hat{r}\otimes\hat{\theta})c\partial_r A_\theta + (\hat{r}\otimes\hat{z})c\partial_r A_z \\ & + (\hat{\theta}\otimes\hat{t})\frac{1}{r}\partial_\theta\phi + \frac{c\hat{\theta}}{r}\otimes[\partial_\theta(\hat{r}A_r + \hat{\theta}A_\theta)] + (\hat{\theta}\otimes\hat{z})\frac{c}{r}\partial_\theta A_z \\ & + (\hat{z}\otimes\hat{t})c\partial_z\phi + (\hat{z}\otimes\hat{r})c\partial_z A_r + (\hat{z}\otimes\hat{\theta})c\partial_z A_\theta.\end{aligned} \tag{3.466}$$

Moreover, with the help of Equation 3.448, we have

$$\begin{aligned}\frac{c\hat{\theta}}{r}\otimes[\partial_\theta(\hat{r}A_r + \hat{\theta}A_\theta)] &= \frac{c\hat{\theta}}{r}\otimes(\hat{\theta}A_r + \hat{r}\partial_\theta A_r - \hat{r}A_\theta + \hat{\theta}\partial_\theta A_\theta) \\ &= (\hat{\theta}\otimes\hat{r})\frac{c}{r}(\partial_\theta A_r - A_\theta).\end{aligned} \tag{3.467}$$

Finally, grouping terms and using Equation 3.463, this yields each of the six independent components of the field tensor:

$$\begin{aligned} F_{rt} &= -F_{tr} = \partial_r\phi + \partial_t A_r = -E_r, \\ F_{\theta t} &= -F_{t\theta} = \frac{1}{r}\partial_\theta\phi + \partial_t A_\theta = -E_\theta, \\ F_{zt} &= -F_{tz} = \partial_z\phi + \partial_t A_z = -E_z, \\ F_{r\theta} &= -F_{\theta r} = \frac{c}{r}[\partial_r(rA_\theta) - \partial_\theta A_r] = cB_z, \\ F_{rz} &= -F_{zr} = c(\partial_r A_z - \partial_z A_r) = -cB_\theta, \\ F_{\theta z} &= -F_{z\theta} = c\left(\frac{1}{r}\partial_\theta A_z - \partial_z A_\theta\right) = cB_r. \end{aligned} \tag{3.468}$$

This somewhat convoluted method is required to take into account properly the fact that the derivative of the unit vector $\hat{r}$ and $\hat{\theta}$ with respect to the angle is not equal to zero; however, the result also shows the efficiency of the tensorial formalism, which is independent from the particular choice of coordinates used to map four-dimensional space-time.

As an exercise, we now want to explicitly demonstrate that

$$u^\mu(\partial_\mu A_\nu - \partial_\nu A_\mu)u^\nu = 0. \tag{3.469}$$

Let us start by introducing the four-vector

$$h_\mu = (\partial_\mu A_\nu - \partial_\nu A_\mu)u^\nu; \tag{3.470}$$

then, by definition,

$$u^\mu(\partial_\mu A_\nu - \partial_\nu A_\mu)u^\nu = u^\mu h_\mu = \mathbf{u}\cdot\mathbf{h} - u_0 h_0, \tag{3.471}$$

and we need to evaluate the components of $h_\mu$. Proceeding systematically, we first have

$$\begin{aligned} h_0 &= (\partial_0 A_\nu - \partial_\nu A_0)u^\nu, \\ \mathbf{h} &= (\nabla A_\nu - \partial_\nu \mathbf{A})u^\nu. \end{aligned} \tag{3.472}$$

Focusing on the time-like component of $h_\mu$, we obtain

$$\begin{aligned} h_0 &= (\partial_0 A_\nu - \partial_\nu A_0)u^\nu \\ &= \mathbf{u}\cdot(\partial_0\mathbf{A} - \nabla A_0) - u_0(\partial_0 A_0 - \partial_0 A_0) \\ &= \mathbf{u}\cdot\left(-\frac{1}{c}\partial_t\mathbf{A} - \nabla\frac{\phi}{c}\right) \\ &= \mathbf{u}\cdot\frac{\mathbf{E}}{c}. \end{aligned} \tag{3.473}$$

Now considering the spatial components of $h_\mu$, we have

$$\begin{aligned} h_i &= (\partial_i A_\nu - \partial_\nu A_i)u^\nu \\ &= \mathbf{u}\cdot(\partial_i\mathbf{A} - \nabla A_i) - u_0\left(\partial_i A_0 + \frac{1}{c}\partial_t A_i\right) \\ &= \mathbf{u}\cdot(\partial_i\mathbf{A} - \nabla A_i) - \frac{u_0}{c}(\partial_i\phi + \partial_t A_i) \\ &= \mathbf{u}\cdot(\partial_i\mathbf{A} - \nabla A_i) + u_0\frac{E_i}{c}. \end{aligned} \tag{3.474}$$

The contraction of $h_\mu$ with the normalized four-velocity is given by

$$\begin{aligned} u^\mu h_\mu &= \mathbf{u}\cdot\mathbf{h} - u_0 h_0 \\ &= u_i h_i - u_0 h_0 \\ &= u_i[\mathbf{u}\cdot(\partial_i\mathbf{A} - \nabla A_i)] + u_i u_0\frac{E_i}{c} - u_0\mathbf{u}\cdot\frac{\mathbf{E}}{c} \\ &= u_i[\mathbf{u}\cdot(\partial_i\mathbf{A} - \nabla A_i)]. \end{aligned} \tag{3.475}$$

We can expand the remaining terms:

$$\begin{aligned} u_i[\mathbf{u}\cdot(\partial_i\mathbf{A}-\nabla A_i)] = {} & u_x[u_x(\partial_x A_x-\partial_x A_x)+u_y(\partial_x A_y-\partial_y A_x) \\ & +u_z(\partial_x A_z-\partial_z A_x)]+u_y[u_x(\partial_y A_x-\partial_x A_y) \\ & +u_y(\partial_y A_y-\partial_y A_y)+u_z(\partial_y A_z-\partial_z A_y)] \\ & +u_z[u_x(\partial_z A_x-\partial_x A_z)+u_y(\partial_z A_y-\partial_y A_z) \\ & +u_z(\partial_z A_z-\partial_z A_z)], \end{aligned} \tag{3.476}$$

which simplifies to read

$$\begin{aligned} u_i[\mathbf{u}\cdot(\partial_i\mathbf{A}-\nabla A_i)] = {} & u_x(u_yB_z-u_zB_y)+u_y(-u_xB_z+u_zB_x) \\ & +u_z(u_xB_y-u_yB_x). \end{aligned} \tag{3.477}$$

This last expression is, indeed, identically equal to zero, thus yielding the sought-after result.

Finally, we discuss the choice of sign in the four-gradient operator: *a priori*, as noted by Feynman, one would be tempted to write

$$\partial_\mu = \left(\frac{1}{c}\partial_t, \nabla\right). \tag{3.478}$$

However, this operator does not behave like a four-vector under the Lorentz transform. To demonstrate this, we follow the discussion given by Feynman and co-authors and consider a scalar function, $\phi(x, t)$. We first vary time, while keeping $x$ constant. In the laboratory frame, we have

$$\Delta\phi = \partial_t\phi\Delta t. \tag{3.479}$$

However, in another Galilean frame, one measures a variation

$$\Delta\phi = \partial_{\bar{x}}\Delta\bar{x}+\partial_{\bar{t}}\Delta\bar{t}. \tag{3.480}$$

Using the special Lorentz transform between the two frames and requiring that $\Delta x = 0$, since we are keeping $x$ constant, we find

$$\Delta\bar{x} = -\gamma v\Delta t, \quad \text{and} \quad \Delta\bar{t} = \gamma\Delta t. \tag{3.481}$$

We now use these results in Equation 3.480:

$$\Delta\phi = \partial_{\bar{x}}\phi(-\gamma v\Delta t)+\partial_{\bar{t}}\phi(\gamma\Delta t) = [(\partial_{\bar{t}}-v\partial_{\bar{x}})\phi]\gamma\Delta t. \tag{3.482}$$

Now using the result established in Equation 3.482, together with Equation 3.479, we obtain

$$\partial_t \phi = \gamma(\partial_{\bar{t}} - v\partial_{\bar{x}})\phi. \tag{3.483}$$

For the spatial gradient, we switch $\bar{x}$ and $\bar{t}$:

$$\partial_x \phi = \gamma(\partial_{\bar{x}} - v\partial_{\bar{t}})\phi. \tag{3.484}$$

"Now we can see that the gradient is rather strange," to use Feynman's words; indeed, if we use the inverse Lorentz transform, $(v \to -v)$, to express $x$ and $t$ in terms of $\bar{x}$ and $\bar{t}$, we have

$$t = \gamma(\bar{t} + v\bar{x}), \quad x = \gamma(\bar{x} + v\bar{t}), \tag{3.485}$$

which is the proper transformation for a four-vector; comparing Equation 3.485 with Equations 3.483 and 3.484, a clear discrepancy in the signs appears. This problem is resolved by defining the correct four-gradient operator

$$\partial_\mu = (-\partial_0, \nabla) = (-\partial_t, \nabla). \tag{3.486}$$

Within this context, the quantity

$$\partial_\mu \phi = k_\mu, \tag{3.487}$$

where $\phi(x_\lambda)$ is an arbitrary scalar field, transforms as a four-vector, and we can build tensors by applying the four-gradient operator to four-vectors. For example, the antisymmetric electromagnetic field tensor is given by

$$F_{\mu\nu} = \partial_\mu A_\nu - \partial_\nu A_\mu, \tag{3.488}$$

while the symmetrical metric tensor is defined as

$$g_{\mu\nu} = \partial_\mu x_\nu = \partial_\nu x_\mu. \tag{3.489}$$

## 3.13 References for Chapter 3

Note: the numbers listed below refer to the main bibliography and reference sections at the end of this book.

8, 9, 27, 48, 55, 59, 60, 63, 64, 69, 71, 73, 74, 75, 96, 97, 99, 102, 105, 116, 120, 125, 135, 143, 149, 150, 158, 165, 172, 181, 185, 193, 194, 195, 207, 208, 209, 210, 213, 220, 221, 225, 226, 229, 238, 239, 298, 416, 421, 432, 455, 456, 458, 459, 460, 461, 572, 573, 588, 592, 593, 793, 794, 895, 796, 797, 799, 800, 801, 802, 803, 804, 805, 806.

# 4

# *Gauge Condition and Transform*

## 4.1 Introduction

Gauge invariance and Lorentz covariance are the two fundamental physical principles underlying both classical and quantum electrodynamics. While Lorentz covariance embodies the global invariance of electrodynamics under the space–time rotations of special relativity in hyperbolic space–time, gauge invariance is a local symmetry reflecting the fact that the four-potential can be modified according to

$$A_\mu(x^\lambda) \to A_\mu(x^\lambda) + \partial_\mu \Lambda(x^\lambda), \tag{4.1}$$

without changing the electromagnetic field; here, $\Lambda(x^\lambda)$ is an arbitrary function of the four-position. The term *local symmetry* can be explained as follows: while the local character of the gauge transform is obvious, the fact that it is a symmetry may seem surprising at first glance because it looks quite different from usual symmetries, such as a $2\pi/n$ rotation leaving an $n$-polygon unchanged. However, by definition, a symmetry is an operation that transforms an object into itself; for the gauge symmetry, that object is $F_{\mu\nu}$. This idea is illustrated schematically in Figure 4.1. Moreover, as prescribed by Noether's theorem, there is an invariant quantity associated with this local symmetry: the four-current satisfies the conservation equation

$$\partial_\mu j^\mu = 0. \tag{4.2}$$

In this chapter we explore the concept of electrodynamical gauge invariance and its connection with charge conservation. We review the approach first taken by Weyl, who introduced pure infinitesimal geometry to describe gauge invariance, and we introduce a few fundamental ideas, including Noether's theorem and the relativistic field Lagrangian, which prove extremely useful when extended to QED and modern quantum field theories based on gauge invariance.

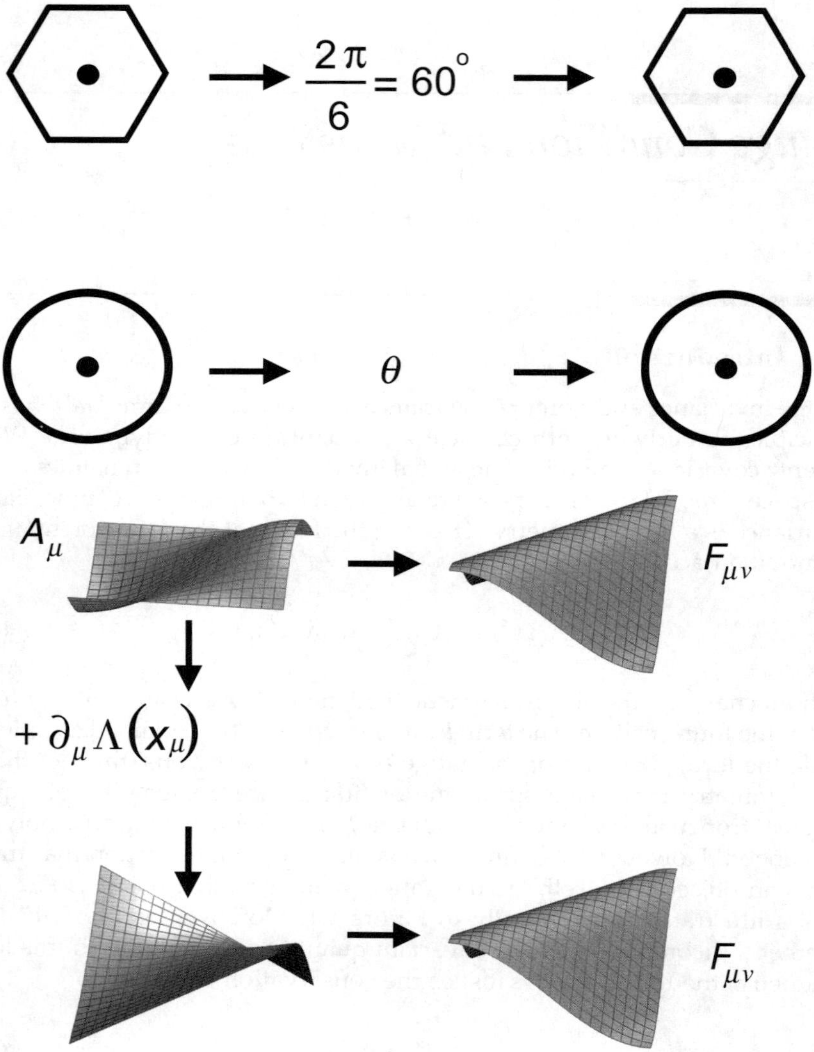


**FIGURE 4.1**
Schematic representation of gauge invariance: on top, the hexagon is conserved through a rotation of $\pi/3$; in the middle, a circle remains invariant for any rotation about its axis; for gauge invariance, adding the (local) quantity $\partial_\mu \Lambda(x_\nu)$ to the 4-potential, $A_\mu$, leaves the electromagnetic field tensor, $F_{\mu\nu}$, unchanged.

Before investigating the details of gauge invariance, let us first summarize briefly the main relevant features of Maxwell's theory. As discussed earlier, the electric field and magnetic induction are generated by sources according to

$$\nabla \cdot \mathbf{E} = \frac{1}{\varepsilon_0}\rho, \tag{4.3}$$

and

$$\nabla \times \mathbf{B} - \frac{1}{c^2}\partial_t \mathbf{E} = \mu_0 \mathbf{j}, \tag{4.4}$$

where the sources satisfy the charge conservation or continuity equation

$$\partial_t \rho + \nabla \cdot \mathbf{j} = 0. \tag{4.5}$$

More importantly within the present context, the source-free equations,

$$\nabla \cdot \mathbf{B} = 0 \tag{4.6}$$

and

$$\nabla \times \mathbf{E} + \partial_t \mathbf{B} = 0, \tag{4.7}$$

suggest the introduction of the vector and scalar potentials, defined such that

$$\mathbf{B} = \nabla \times \mathbf{A}, \tag{4.8}$$

and

$$\mathbf{E} = -\nabla \varphi - \partial_t \mathbf{A}. \tag{4.9}$$

Because the divergence of a curl is zero, Equation 4.6 is now automatically satisfied; similarly, since the curl of a gradient is also identically zero, we see that Equation 4.7 is satisfied as well.

As discussed in Chapter 3, the introduction of the four-potential,

$$A_\mu = \left(\frac{1}{c}\varphi, \mathbf{A}\right), \tag{4.10}$$

four-current,

$$j_\mu = (c\rho, \mathbf{j}), \tag{4.11}$$

and four-gradient operator,

$$\partial_\mu = \left(-\frac{1}{c}\partial_t, \nabla\right), \tag{4.12}$$

allows us to rewrite Maxwell's equations in a manifestly covariant way. The antisymmetrical electromagnetic field tensor is defined in terms of the four-potential as

$$\frac{1}{c}F_{\mu\nu} = \partial_\mu A_\nu - \partial_\nu A_\mu, \tag{4.13}$$

and the equations with source terms now take the compact form

$$\partial_\mu F^{\mu\nu} + \mu_0 c j^\nu = 0, \tag{4.14}$$

while the source-free equations are automatically satisfied. Finally, the charge conservation equation is now expressed by Equation 4.2.

The components of the electromagnetic field tensor correspond to the electric field and magnetic induction:

$$F_{\mu\nu} = \begin{bmatrix} 0 & E_1 & E_2 & E_3 \\ -E_1 & 0 & cB_3 & -cB_2 \\ -E_2 & -cB_3 & 0 & cB_1 \\ -E_3 & cB_2 & -cB_1 & 0 \end{bmatrix}. \tag{4.15}$$

As was discussed in detail in Chapter 3, the equation governing the evolution of the electromagnetic field tensor in vacuum can be derived from the principle of least action by using the following Lagrangian density:

$$\mathcal{L} = -\frac{\varepsilon_0}{4}F_{\mu\nu}F^{\mu\nu} = \frac{\varepsilon_0}{2}(\mathbf{E}^2 - c^2\mathbf{B}^2). \tag{4.16}$$

We then obtain the wave equation in vacuum,

$$\partial_\mu F^{\mu\nu} = 0. \tag{4.17}$$

It is now seen easily that the gauge transformation of Equation 4.1 leaves the electromagnetic field tensor unchanged:

$$F_{\mu\nu} \rightarrow F_{\mu\nu} + \partial_\mu\partial_\nu\Lambda - \partial_\nu\partial_\mu\Lambda = F_{\mu\nu}. \tag{4.18}$$

This important result is very general, as the partial derivative operators commute under most circumstances. The gauge transformation is a local symmetry, as it depends on the four-position through the generating function $\Lambda(x_\lambda)$. In addition, gauge transformations form a group, whose structure corresponds to U(1). The invariance of the electromagnetic field tensor implies that the Lagrangian density for the free field is also invariant. Following Kaku, we note that a direct application of Noether's theorem would yield an energy–momentum tensor of the form

$$\varepsilon_0^{-1} T_{\mu\nu} = -cF_{\mu\lambda}\partial_\nu A^\lambda + \frac{1}{4} g_{\mu\nu} F^{\alpha\beta} F_{\alpha\beta}. \tag{4.19}$$

Unfortunately, this tensor is not symmetrical, which precludes the existence of a conserved angular momentum tensor; neither is it gauge invariant. This problem can be alleviated by using a gauge transformation:

$$T_{\mu\nu} \rightarrow T_{\mu\nu} + \varepsilon_0 c \partial^\lambda (F_{\mu\lambda} A_\nu), \tag{4.20}$$

and we obtain

$$\varepsilon_0^{-1} T_{\mu\nu} = F_{\mu\lambda} F_\nu^\lambda + \frac{1}{4} g_{\mu\nu} F^{\alpha\beta} F_{\alpha\beta}, \tag{4.21}$$

which is both symmetric and gauge invariant.

The Dirac equation was introduced in Section 3.11, and it is interesting to revisit it within the context of the present chapter addressing gauge invariance. The Dirac equation is

$$i\hbar\partial_t \psi(x_\lambda) = [i\hbar c\alpha \cdot \nabla + \beta m_0 c^2]\psi(x_\lambda), \tag{4.22}$$

and we can define new matrices,

$$\gamma^\mu = (\beta, \beta\alpha) = \beta(1, \alpha_i), \tag{4.23}$$

so that the Dirac equation now reads

$$[i\hbar\gamma^\mu \partial_\mu - m_0 c]\psi(x_\lambda) = 0. \tag{4.24}$$

Moreover, Equation 4.24 can be derived by applying the principle of least action to the following Lagrangian density:

$$\mathcal{L} = \psi^*(i\hbar c\gamma^\mu \partial_\mu - m_0 c^2)\psi - \frac{\varepsilon_0}{4} F_{\alpha\beta} F^{\alpha\beta}. \tag{4.25}$$

Here, $\psi^* \psi$ is the electron probability density, which has the correct dimension $[\psi^* \psi] = m^{-3}$. The electron four-current is then given by

$$j_\mu(x_\lambda) = -e\psi^*(x_\lambda)\gamma_\mu \psi(x_\lambda), \tag{4.26}$$

and corresponds to the invariance of the Dirac equation under the wavefunction transformation

$$\psi(x_\lambda) \rightarrow \psi(x_\lambda)e^{i\Lambda(x_\lambda)}. \tag{4.27}$$

We can now present briefly the connection between this invariance of the Dirac equation and gauge invariance. Again, referring to Kaku's discussion, we consider the coupling term

$$A_\mu j^\mu = -eA_\mu \psi^* \gamma^\mu \psi, \tag{4.28}$$

and the concurrent transformations,

$$\begin{aligned} \psi(x_\lambda) &\rightarrow \psi(x_\lambda)e^{i\Lambda(x_\lambda)}, \\ A_\mu(x_\nu) &\rightarrow A_\mu(x_\nu) - \partial_\mu\Lambda(x_\nu). \end{aligned} \tag{4.29}$$

The main difficulty with this approach is that $\partial_\mu\psi$ is not covariant, because it now includes the term $\partial_\mu\Lambda$. This problem is resolved by considering the covariant derivative,

$$D_\mu \equiv \partial_\mu - ieA_\mu. \tag{4.30}$$

At this point, we note that this is analogous to the covariant derivative introduced in Section 3.9, which generalizes the four-gradient to curved space–time; the exact connection will be discussed shortly. If we now apply this operator to the transformed wavefunction, we have

$$\begin{aligned} D_\mu(\psi e^{-ie\Lambda}) &= [\partial_\mu - ieA_\mu](\psi e^{ie\Lambda}) \\ &= e^{ie\Lambda}D_\mu\psi + \psi(ie\partial_\mu\Lambda - ieA_\mu)e^{ie\Lambda} \\ &= e^{ie\Lambda}D_\mu\psi, \end{aligned} \tag{4.31}$$

because the term $\partial_\mu\Lambda - A_\mu$ is transformed away by the new gauge condition.

Finally, in terms of the Lagrangian density from which the Dirac equation derives, we have

$$\mathcal{L} = \psi^*(i\hbar c\gamma^\mu D_\mu - m_0c^2)\psi - \frac{\varepsilon_0}{4}F_{\alpha\beta}F^{\alpha\beta}. \tag{4.32}$$

## 4.2 Lorentz Gauge

The Lorentz gauge condition is prevalent in classical electrodynamics, as it takes a manifestly covariant form and occurs naturally when one derives the wave equation from the potential. We have derived it in vector form in Section 3.3, and we present the covariant derivation here, for completeness. As discussed earlier, the four-potential is defined so that the source-free equations of the Maxwell set are automatically satisfied. In turn, the field tensor is introduced as the antisymmetrical second-rank tensor deriving from the four-potential, as written in Equation 4.13. Moreover, the field tensor remains unchanged if we modify the four-potential according to a gauge transformation, as established in Equation 4.18.

Once the field tensor is introduced, it is easy to verify that Maxwell's equations with sources can be written as $\partial_\mu F^{\mu\nu} + \mu_0 c j^\nu = 0$. Let us start by briefly examining the time-like component of this equation. We have

$$\partial_\mu F^{\mu 0} = -\mu_0 c j^0, \tag{4.33}$$

which yields

$$\partial_i F^{i0} - \partial_0 F^{00} = -\mu_0 \rho c^2 = -\frac{\rho}{\varepsilon_0}. \tag{4.34}$$

Since $E^i = F^{0i} = -F^{i0}$, we find Poisson's equation,

$$\nabla \cdot \mathbf{E} = \frac{\rho}{\varepsilon_0}. \tag{4.35}$$

Now considering the spatial components, we start with

$$\partial_\mu F^{\mu i} + \mu_0 c j^i = 0, \tag{4.36}$$

which translates into

$$\partial_j F^{ji} - \partial_0 F^{0i} + \mu_0 c j^i = 0. \tag{4.37}$$

We now use the indices, $i$, $j$, $k$, as a circular permutation of the numbers 1, 2, and 3:

$$\partial_i F^{ii} + \partial_j F^{ji} + \partial_k F^{ki} - \partial_0 F^{0i} + \mu_0 c j^i = 0. \tag{4.38}$$

The field tensor is antisymmetric, $F^{ii} = 0$, and we identify the remaining components as follows: $F^{ij} = -F^{ji} = cB^k$, $F^{ki} = -F^{ik} = cB^j$, while $F^{0i} = E^i$, and $\partial_0 = -c^{-1}\partial_t$; with this, we obtain

$$-(\nabla \times c\mathbf{B})^i + \frac{1}{c}\partial_t E^i + \mu_0 c j^i = 0. \tag{4.39}$$

Dividing Equation 4.39 by $-c$, we can finally rewrite it in the sought-after vector form:

$$\nabla \times \mathbf{B} - \frac{1}{c^2}\partial_t \mathbf{E} = \mu_0 \mathbf{j}. \tag{4.40}$$

To obtain the covariant driven wave equation, as expressed in terms of the four-potential, we first use Equation 4.14 and replace the field tensor by $F_{\mu\nu} = c(\partial_\mu A_\nu - \partial_\nu A_\mu)$:

$$\partial_\mu(\partial^\mu A^\nu - \partial^\nu A^\mu) + \mu_0 j^\nu = 0. \tag{4.41}$$

This is then recast as

$$(\partial_\mu \partial^\mu) A^\nu + \mu_0 j^\nu = \partial^\nu(\partial_\mu A^\mu), \tag{4.42}$$

where we recognize the d'Alembertian operator acting on the four-potential, and the four-current drive term,

$$\begin{aligned} \Box A^\nu + \mu_0 j^\nu &= \partial^\nu(\partial_\mu A^\mu) = \partial^\nu \Lambda, \\ \Box \mathbf{A} + \mu_0 \mathbf{j} &= \nabla\left(\frac{1}{c^2}\partial_t \varphi + \nabla \cdot \mathbf{A}\right), \\ \Box \frac{\varphi}{c} + \mu_0 \rho c &= -\frac{1}{c}\partial_t\left(\frac{1}{c^2}\partial_t \varphi + \nabla \cdot \mathbf{A}\right). \end{aligned} \tag{4.43}$$

The right-hand side term can be removed by means of a gauge transform, or we can choose to work in the Lorentz gauge, where

$$\partial_\mu A^\mu = \nabla \cdot \mathbf{A} + \frac{1}{c^2}\partial_t \varphi = 0. \tag{4.44}$$

We have thus shown that the Lorentz gauge condition naturally appears in the derivation of the wave equation. Furthermore, it is manifestly covariant. Finally, in Fourier space, we have

$$\tilde{A}^\mu(k_\lambda) = \frac{1}{(\sqrt{2\pi})^4}\iiiint A^\mu(x_\lambda) e^{ik_\lambda x^\lambda} d^4x, \tag{4.45}$$

and the Lorentz condition takes the simple form

$$k_\mu \tilde{A}^\mu = 0. \tag{4.46}$$

## 4.3 Coulomb Gauge and Instantaneous Scalar Potential

Another very important gauge condition is the Coulomb gauge,

$$\nabla \cdot \mathbf{A} = 0, \tag{4.47}$$

which is also referred to as radiation gauge or transverse gauge. The transversality condition is evident if we consider a plane wave:

$$\mathbf{A}(x_\lambda) = \mathbf{A}_0 \exp(-ik^\lambda x_\lambda) = \mathbf{A}_0 \exp[i(\omega t - \mathbf{k} \cdot \mathbf{x})]; \tag{4.48}$$

in this case, we find that

$$\nabla \cdot \mathbf{A} = 0 = -i\mathbf{k} \cdot \mathbf{A}, \tag{4.49}$$

and we see that the vector potential and the wavenumber must be orthogonal.

The Coulomb gauge condition plays a crucial role in electrodynamics because it effectively allows one to evaluate the scalar potential as an *instantaneous* function of the charge distribution. For example, for a number of problems involving the interaction of point charges, adopting the Coulomb gauge alleviates the near-intractable difficulties associated with the finite propagation time of the electrostatic field. This is exemplified by the two-time interaction potential characterizing the rectilinear Dirac-Lorentz equation for two electrons, or an electron and a positron, interacting via their Coulomb fields.

In order to establish this important fact clearly, we return to the general vector form of the wave equation:

$$\left(\Delta - \frac{1}{c^2}\partial_t^2\right)\mathbf{A} + \mu_0 \mathbf{j} = \nabla\left(\frac{1}{c^2}\partial_t \varphi + \nabla \cdot \mathbf{A}\right), \tag{4.50}$$

$$\left(\Delta - \frac{1}{c^2}\partial_t^2\right)\frac{\varphi}{c} + \mu_0 \rho c = -\frac{1}{c}\partial_t\left(\frac{1}{c^2}\partial_t \varphi + \nabla \cdot \mathbf{A}\right), \tag{4.51}$$

where the coupling between the scalar and vector potentials is clear.

If we now choose to work in the Coulomb gauge and set $\nabla \cdot \mathbf{A} = 0$, we see that the time derivative operators in Equation 4.51 cancel, leaving Poisson's equation for an *electrostatic* potential,

$$\Delta\varphi = -\frac{\rho}{\varepsilon_0}; \tag{4.52}$$

the scalar potential is given by the *instantaneous* distribution of charges.

This is not surprising if we remember that we started with Poisson's equation,

$$\nabla \cdot \mathbf{E} = \frac{\rho}{\varepsilon_0} = \nabla \cdot (-\nabla\varphi - \partial_t \mathbf{A}); \tag{4.53}$$

since the partial derivative with respect to time and divergence commute, we see that choosing the Coulomb gauge immediately yields Equation 4.52. Of course, the finite propagation time effects are still present, but they have been moved to the vector potential. To see this, let us now examine Equation 4.50 within the framework of the Coulomb gauge. We first have

$$\left(\Delta - \frac{1}{c^2}\partial_t^2\right)\mathbf{A} + \mu_0 \mathbf{j} = \frac{1}{c^2}\nabla(\partial_t \varphi); \tag{4.54}$$

we can then define the transverse current

$$\mathbf{j}_\perp = \mathbf{j} - \varepsilon_0 \nabla(\partial_t \varphi) = \mathbf{j} - \mathbf{j}_\parallel, \tag{4.55}$$

and recast Equation 4.54 as

$$\left(\Delta - \frac{1}{c^2}\partial_t^2\right)\mathbf{A} + \mu_0 \mathbf{j}_\perp = \mathbf{0}. \tag{4.56}$$

Since the axial component of the current is the gradient of $\varepsilon_0 \partial_t \varphi$, it is clear that

$$\nabla \times \mathbf{j}_\parallel = \mathbf{0}. \tag{4.57}$$

Furthermore, the fact that the transverse component of the current drives Equation 4.56 comes from the transversality condition on the vector potential, as expressed in Equation 4.49, but it can also be demonstrated as follows. Poisson's equation can be formally integrated as

$$\varphi = \iiint \frac{\rho(\bar{\mathbf{x}}, t)}{4\pi\varepsilon_0 |\mathbf{x} - \bar{\mathbf{x}}|} d\bar{x}\, d\bar{y}\, d\bar{z}, \tag{4.58}$$

where we have used the electrostatic Green function, which is presented in greater detail in Chapter 5.

To determine the axial current, we take the gradient of the time derivative of the scalar potential. We have

$$\mathbf{j}_{\|} = \varepsilon_0 \nabla(\partial_t \varphi) = \nabla\left\{\partial_t\left[\iiint \frac{\rho(\bar{\mathbf{x}},t)}{4\pi|\mathbf{x}-\bar{\mathbf{x}}|}d\bar{v}\right]\right\}, \tag{4.59}$$

and the partial derivative with respect to time can be moved over in the volume integral,

$$\mathbf{j}_{\|} = \nabla\left[\iiint \frac{\partial_t \rho(\bar{\mathbf{x}},t)}{4\pi|\mathbf{x}-\bar{\mathbf{x}}|}d\bar{v}\right]. \tag{4.60}$$

The continuity equation can now be used,

$$\partial_t \rho(\bar{\mathbf{x}},t) = -\bar{\nabla}\cdot\mathbf{j}(\bar{\mathbf{x}},t), \tag{4.61}$$

and we find

$$\mathbf{j}_{\|} = -\nabla\left[\iiint \frac{\bar{\nabla}\cdot\mathbf{j}(\bar{\mathbf{x}},t)}{4\pi|\mathbf{x}-\bar{\mathbf{x}}|}d\bar{v}\right]. \tag{4.62}$$

At this point we use Gauss' theorem and add the null term

$$\iiint \bar{\nabla}\cdot\left[\frac{\mathbf{j}(\bar{\mathbf{x}},t)}{4\pi|\mathbf{x}-\bar{\mathbf{x}}|}\right]d\bar{v} = 0, \tag{4.63}$$

to Equation 4.62:

$$\begin{aligned}
\mathbf{j}_{\|} &= \nabla\left\{\iiint \bar{\nabla}\cdot\left[\frac{\mathbf{j}(\bar{\mathbf{x}},t)}{4\pi|\mathbf{x}-\bar{\mathbf{x}}|}\right]d\bar{v} - \iiint \frac{\bar{\nabla}\cdot\mathbf{j}(\bar{\mathbf{x}},t)}{4\pi|\mathbf{x}-\bar{\mathbf{x}}|}d\bar{v}\right\} \\
&= \nabla\left[\iiint \mathbf{j}(\bar{\mathbf{x}},t)\cdot\bar{\nabla}\left(\frac{1}{4\pi|\mathbf{x}-\bar{\mathbf{x}}|}\right)d\bar{v}\right] \\
&= -\nabla\left[\iiint \mathbf{j}(\bar{\mathbf{x}},t)\cdot\nabla\left(\frac{1}{4\pi|\mathbf{x}-\bar{\mathbf{x}}|}\right)d\bar{v}\right] \\
&= -\nabla\left\{\nabla\cdot\left[\iiint \frac{\mathbf{j}(\bar{\mathbf{x}},t)}{4\pi|\mathbf{x}-\bar{\mathbf{x}}|}d\bar{v}\right]\right\}.
\end{aligned} \tag{4.64}$$

The second equality results from the fact that $\nabla \cdot (\alpha \mathbf{v}) = \mathbf{v} \cdot \nabla \alpha + \alpha(\nabla \cdot \mathbf{v})$, while we have switched the variables $\mathbf{x}$ and $\bar{\mathbf{x}}$ to obtain the third equality; finally, the last equality holds because $\nabla$ operates on $\mathbf{x}$ only.

If we now introduce

$$\mathbf{a} = \iiint \frac{\mathbf{j}(\bar{\mathbf{x}}, t)}{4\pi|\mathbf{x} - \bar{\mathbf{x}}|} d\bar{v} \tag{4.65}$$

and use the identity

$$\nabla \times (\nabla \times \mathbf{a}) = \nabla(\nabla \cdot \mathbf{a}) - \Delta \mathbf{a}, \tag{4.66}$$

we obtain

$$\mathbf{j}_{\parallel} = -\nabla \times \left[ \nabla \times \iiint \frac{\mathbf{j}(\bar{\mathbf{x}}, t)}{4\pi|\mathbf{x} - \bar{\mathbf{x}}|} d\bar{v} \right] - \Delta \iiint \frac{\mathbf{j}(\bar{\mathbf{x}}, t)}{4\pi|\mathbf{x} - \bar{\mathbf{x}}|} d\bar{v}. \tag{4.67}$$

At this point, we apply the Laplacian operator to the quantity,

$$\Delta \frac{1}{|\mathbf{x} - \bar{\mathbf{x}}|} = -4\pi\delta(\mathbf{x} - \bar{\mathbf{x}}), \tag{4.68}$$

and we find that

$$\begin{aligned} -\Delta \iiint \frac{\mathbf{j}(\bar{\mathbf{x}}, t)}{4\pi|\mathbf{x} - \bar{\mathbf{x}}|} d\bar{v} &= -\iiint \Delta \left[ \frac{\mathbf{j}(\bar{\mathbf{x}}, t)}{4\pi|\mathbf{x} - \bar{\mathbf{x}}|} \right] d\bar{v} \\ &= -\iiint \left[ \Delta \left( \frac{1}{4\pi|\mathbf{x} - \bar{\mathbf{x}}|} \right) \right] \mathbf{j}(\bar{\mathbf{x}}, t) d\bar{v} \\ &= \iiint \delta(\mathbf{x} - \bar{\mathbf{x}}) \mathbf{j}(\bar{\mathbf{x}}, t) d\bar{v} \\ &= \mathbf{j}(\mathbf{x}, t). \end{aligned} \tag{4.69}$$

Again, this holds because $\Delta \equiv \nabla^2$ applies only on $\mathbf{x}$, not $\bar{\mathbf{x}}$. We can now identify

$$\mathbf{j}_{\parallel} = \mathbf{j} - \nabla \times \left[ \nabla \times \iiint \frac{\mathbf{j}(\bar{\mathbf{x}}, t)}{4\pi|\mathbf{x} - \bar{\mathbf{x}}|} d\bar{v} \right] = \mathbf{j} - \mathbf{j}_{\perp}. \tag{4.70}$$

Since the transverse current component is the curl of a vector, we also have

$$\nabla \cdot \mathbf{j}_{\perp} = 0, \tag{4.71}$$

which completes our demonstration.

There is also an important connection between the Coulomb gauge and the concept of virtual or longitudinally polarized photons. If we consider the Coulomb field surrounding a point charge, as expressed in spherical coordinates, we have

$$\mathbf{E}(r) = \hat{r}\frac{-e}{4\pi\varepsilon_0 r^2} = -\nabla\varphi(r), \tag{4.72}$$

where the scalar potential is given by

$$\varphi(r) = \frac{-e}{4\pi\varepsilon_0 r}. \tag{4.73}$$

In this particular gauge, we clearly have $\nabla \cdot \mathbf{A} = 0$, as the vector potential is identically equal to zero. Furthermore, if we try to express the Coulomb field in the form

$$\mathbf{E}(r) = -\partial_t \mathbf{A}(r,t), \tag{4.74}$$

two potential difficulties appear. First, as the electric field does not depend explicitly on time, we require the vector potential to be of the form

$$\mathbf{A}(r,t) = (t_0 - t)\mathbf{E}(r); \tag{4.75}$$

second, Poisson's equation yields

$$\nabla \cdot \mathbf{E} = -\partial_t(\nabla \cdot \mathbf{A}) = -\frac{e}{\varepsilon_0}\delta_3(\mathbf{x} - \mathbf{x}_0). \tag{4.76}$$

Within this context, a suitable gauge condition is the temporal gauge,

$$A_0 = \frac{\varphi}{c} = 0, \tag{4.77}$$

as the Lorentz and Coulomb gauge conditions read

$$\partial_\mu A^\mu = \nabla \cdot \mathbf{A} = \frac{e(t - t_0)}{\varepsilon_0}\delta_3(\mathbf{x} - \mathbf{x}_0), \tag{4.78}$$

which diverges both at $t \to \pm\infty$, and $\mathbf{x} = \mathbf{x}_0$. These difficulties become particularly problematic in view of the definition of the canonical four-momentum,

$$\pi_\mu = p_\mu - eA_\mu. \tag{4.79}$$

Moreover, this is also related to the fact that the Coulomb field cannot be expressed in terms of plane waves, as it does not satisfy the transversality condition: the electrostatic field is polarized longitudinally. The dependence of the Coulomb field with distance, $r^{-2}$, is also different from that of a radiated wave in free space, which behaves as $r^{-1}$. In terms of photons, a radiative mode can be described as a superposition of transversally polarized plane waves, while an electrostatic mode corresponds to longitudinally polarized virtual photons. This can be explained by considering electromagnetic modes in a system with boundary conditions. In this case, we have seen that trapped modes exist, which are characterized by a cutoff frequency $\omega_0$, and disperse as

$$k_\mu k^\mu = \mathbf{k}^2 - \frac{\omega^2}{c^2} = -\frac{\omega_0^2}{c^2}, \tag{4.80}$$

thus leading to an equivalent rest mass, $\hbar\omega_0/c^2$. Furthermore, because of the existence of the boundary conditions, such modes can have an electric field component in the direction of propagation, in close analogy with the Coulomb field. For example, in a waveguide, electromagnetic modes are categorized as transverse electric (TE) or transverse magnetic (TM), which form two independent subsets of the eigenmodes supported by the structure. Again, in close analogy with the Coulomb field, nonradiating modes are evanescent, as they are not supported in vacuum, far from charge distributions.

The fact that the Coulomb field itself is nonradiative is clear, as it has no explicit time dependence. We can also examine the Poynting vector,

$$\mathbf{S} = \mathbf{E} \times \mathbf{H} = 0, \tag{4.81}$$

as no magnetic field is produced by a static electric charge.

## 4.4 Other Gauge Conditions

In this section, we briefly consider two different items. First, we list a few different gauge conditions often used in classical and quantum electrodynamics; second, we give a cursory presentation of the QED concept of propagator and its form in different gauges.

The most common gauge constraints apply either directly to the four-potential or to an additional term in the action, of the form,

$$\kappa(\partial_\mu A^\mu)^2. \tag{4.82}$$

Within this context, we have

$$
\begin{aligned}
&\text{Lorentz gauge,} && \partial_\mu A^\mu = 0;\\
&\text{Coulomb gauge,} && \nabla \cdot \mathbf{A} = 0;\\
&\text{Temporal gauge,} && \varphi = 0;\\
&\text{Axial gauge,} && A_\parallel = 0;\\
&\text{Feynman gauge,} && \kappa = 1;\\
&\text{Unitary gauge,} && \kappa = 0;\\
&\text{Landau gauge,} && \kappa \to \infty.
\end{aligned}
$$

In quantum electrodynamics, an important concept is that of propagator, which is briefly introduced here. When considering photons, this approach very nearly coincides with the Green function analysis, which is described in Chapter 5. Therefore, we will first focus on this particular case. As we have seen before, the general driven wave equation for the electromagnetic four-potential is

$$\square A^\mu + \mu_0 j^\mu = \partial^\mu(\partial_\nu A^\nu). \tag{4.83}$$

One of the most fundamental properties of the wave equation is the fact that it is linear. If we know the solutions for two different four-current sources, $j_1^\mu$ and $j_2^\mu$,

$$
\begin{aligned}
\square A_1^\mu + \mu_0 j_1^\mu &= \partial^\mu(\partial_\nu A_1^\nu),\\
\square A_2^\mu + \mu_0 j_2^\mu &= \partial^\mu(\partial_\nu A_2^\nu),
\end{aligned}
\tag{4.84}
$$

then, the solution for a linear superposition of sources is also known,

$$\square(\lambda_1 A_1^\mu + \lambda_2 A_2^\mu) + \mu_0(\lambda_1 j_1^\mu + \lambda_2 j_2^\mu) = \partial^\mu[\partial_\nu(\lambda_1 A_1^\mu + \lambda_2 A_2^\mu)]. \tag{4.85}$$

Now, we can express any four-current as a linear superposition of delta-functions:

$$j^\mu(x_\lambda) = \iiiint j^\mu(\bar{x}_\lambda)\delta_4(\bar{x}_\lambda - x_\lambda)\, d^4x. \tag{4.86}$$

Therefore, the solution to the wave equation driven by a delta-function is of particular importance, as we can write, in the Lorentz gauge,

$$\square G(\bar{x}_\lambda - x_\lambda) + \mu_0 \delta_4(\bar{x}_\lambda - x_\lambda) = 0, \tag{4.87}$$

and the general solution to the wave equation, with an arbitrary four-current source is:

$$A^{\mu}(x_{\lambda}) = \iiiint j^{\mu}(\bar{x}_{\lambda})G(\bar{x}_{\lambda} - x_{\lambda})\, d^4x. \tag{4.88}$$

The solution $G(\bar{x}_{\lambda} - x_{\lambda})$ is known as the Green function, or propagator, for the linear wave equation under consideration. For photons, the expression of the propagator in Fourier or momentum space is easily derived. We first have

$$G(\bar{x}_{\lambda} - x_{\lambda}) = \frac{1}{(\sqrt{2\pi})^4}\iiiint \tilde{G}(k_{\lambda})\exp[-ik_{\lambda}(\bar{x}^{\lambda} - x^{\lambda})]\, d^4k; \tag{4.89}$$

furthermore, in momentum space, the propagator only depends on $k^2 = k_{\lambda}k^{\lambda}$, and Equation 4.67 reduces to

$$G(\bar{x}_{\lambda} - x_{\lambda}) = \frac{1}{(\sqrt{2\pi})^4}\iiiint \tilde{G}(k^2)\exp[-ik_{\lambda}(\bar{x}^{\lambda} - x^{\lambda})]\, d^4k. \tag{4.90}$$

The wave equation is then resolved by using the Fourier transform of the delta-function,

$$\delta_4(\bar{x}_{\lambda} - x_{\lambda}) = \frac{1}{(\sqrt{2\pi})^4}\iiiint \exp[-ik^{\lambda}(\bar{x}_{\lambda} - x_{\lambda})]\, d^4k, \tag{4.91}$$

and we obtain

$$\tilde{G}(k^2) = \frac{\mu_0}{k^2}, \tag{4.92}$$

because the d'Alembertian operator takes the very simple form

$$\Box \equiv -k_{\lambda}k^{\lambda} = -k^2, \tag{4.93}$$

in Fourier space.

Of course the difficulty is the presence of poles on the light-cone, or photon mass-shell, where $k^2 = 0$; to avoid this, one adds a small, negative, imaginary mass to the photons, which also guarantees causality:

$$\tilde{G}(k^2) = \frac{\mu_0}{k^2 + i\varepsilon}, \tag{4.94}$$

and the aforementioned mass is given by

$$p_\lambda p^\lambda = \hbar^2 k_\lambda k^\lambda = \hbar^2(k^2 + i\varepsilon) = -m_\gamma^2 c^2, \tag{4.95}$$

in the limit where $k^2 \to 0$.

This calculation can now be repeated for various gauge conditions, which yield different propagators; additionally, a tensor can be formed for the propagator by using the metric. Employing the conventional notation, we have, for the Lorentz gauge,

$$\tilde{G}_{\mu\nu} = -\frac{4\pi}{k^2 + i\varepsilon} g_{\mu\nu}; \tag{4.96}$$

in the Landau gauge,

$$\tilde{G}_{\mu\nu} = -\frac{4\pi}{k^2 + i\varepsilon}\left(g_{\mu\nu} - \frac{k_\mu k_\nu}{k^2 + i\varepsilon}\right); \tag{4.97}$$

finally, in the Coulomb gauge, the propagator is defined as

$$\tilde{G}_{\mu\nu} = 4\pi \begin{bmatrix} \dfrac{1}{k^2} & 0 \\ 0 & \dfrac{\delta_{ij} - (k_i k_j / k^2)}{k^2 + i\varepsilon} \end{bmatrix}. \tag{4.98}$$

For a full presentation of propagators and their application to the Feynman rules in QED, we refer the reader to Greiner and Reinhardt.

## 4.5 Charge Conservation

As stated in the introduction, there is a conserved quantity related to the symmetry of gauge invariance, the electric charge. In this section, we derive this invariant, first in vector form and then in a covariant manner.

To compute the effect of a gauge transformation on a given electrodynamic system, we consider its influence upon the interaction Lagrangian and start by modifying the canonical momentum, along with the scalar and vector potentials:

$$\begin{aligned} \mathbf{A} &\to \mathbf{A} + \nabla\Lambda, \\ \varphi &\to \varphi - \frac{1}{c}\partial_t \Lambda, \\ \mathbf{p} &\to \mathbf{p} + e\nabla\Lambda. \end{aligned} \tag{4.99}$$

The original, nonrelativistic Lagrangian for a distribution of charges, $q_i$, is

$$\mathcal{L} = \varepsilon_0 \iiint \left[ \mathbf{E} \cdot (-\partial_t \mathbf{A} - \nabla \varphi) - c^2 \mathbf{B} \cdot (\nabla \times \mathbf{A}) + \frac{1}{2}(c^2 \mathbf{B}^2 - \mathbf{E}^2) \right] d^3x + \sum_i \mathcal{L}_i, \quad (4.100)$$

where

$$\mathcal{L}_i = \mathbf{p}_i \cdot \left( \frac{d\mathbf{x}_i}{dt} - \mathbf{v}_i \right) + \frac{1}{2} m_0 \mathbf{v}_i^2 - q_i \varphi(\mathbf{x}_i) + q_i \mathbf{v}_i \cdot \mathbf{A}(\mathbf{x}_i). \quad (4.101)$$

This can be established by consulting Section 3.7 and applying the appropriate modifications. The effect of the gauge transform on this Lagrangian is then

$$\mathcal{L} \rightarrow \mathcal{L} + \sum_i q_i \left[ \nabla \Lambda \cdot \left( \frac{d\mathbf{x}_i}{dt} - \mathbf{v}_i \right) + \partial_t \Lambda + \mathbf{v}_i \cdot \nabla \Lambda \right], \quad (4.102)$$

which simplifies to yield

$$\mathcal{L} \rightarrow \mathcal{L} + \sum_i q_i \left[ \partial_t \Lambda + \frac{d\mathbf{x}_i}{dt} \cdot \nabla \Lambda \right]. \quad (4.103)$$

In Equation 4.103, we recognize the convective derivative operator; therefore, we can further simplify the new Lagrangian by introducing a total time derivative:

$$\mathcal{L} \rightarrow \mathcal{L} + \frac{d}{dt} \left[ \sum_i q_i \Lambda(\mathbf{x}_i, t) \right]. \quad (4.104)$$

For a continuous charge distribution, Equation 4.103 takes the form

$$\mathcal{L} \rightarrow \mathcal{L} + \iiint (\rho \partial_t \Lambda + \mathbf{j} \cdot \nabla \Lambda) dv. \quad (4.105)$$

We now vary $\Lambda$ by $\delta\Lambda$:

$$\delta S = -\int dt \iiint \delta\Lambda (\partial_t \rho + \nabla \cdot \mathbf{j}) dv. \quad (4.106)$$

The principle of least action requires that $\delta S = 0$, independently of the variation $\delta\Lambda$, which establishes charge conservation:

$$\partial_t \rho + \nabla \cdot \mathbf{j}. \quad (4.107)$$

For the covariant generalization of this derivation, we rewrite Equation 4.103 as

$$\mathcal{L} \to \mathcal{L} + \sum_i q_i \gamma_i \left[ \partial_t \Lambda + \frac{d\mathbf{x}_i}{dt} \cdot \nabla \Lambda \right] = \mathcal{L} + \sum_i q_i u_i^\mu \partial_\mu \Lambda. \tag{4.108}$$

Taking the limit of a continuous four-current distribution, we make the replacement

$$\sum_i q_i u_i^\mu \partial_\mu \Lambda \to \iiint (q n u^\mu \partial_\mu \Lambda) dv, \tag{4.109}$$

which yields

$$\mathcal{L} \to \mathcal{L} + \iiint (j_\mu \partial^\mu \Lambda) dv. \tag{4.110}$$

Again, we vary the gauge function, to obtain

$$\delta S = -\int d\tau \iiint \delta\Lambda (\partial^\mu j_\mu) dv, \tag{4.111}$$

which must hold independent of $\delta\Lambda$; this finally yields

$$\partial^\mu j_\mu = 0. \tag{4.112}$$

## 4.6 Noether's Theorem

In the previous section, we have seen that for a given symmetry, one can associate an invariant or conserved quantity. This concept is, in fact, very general and fully embodied by Noether's theorem. The presentation given here is somewhat restricted in scope but provides a good understanding of the fundamental ideas behind the connection between symmetry and invariance, a ubiquitous theme in modern field theories.

We will consider both infinitesimal coordinate transformations,

$$x_\mu \to \bar{x}_\mu = x_\mu + \delta x_\mu, \tag{4.113}$$

and the corresponding variations in the field components,

$$F_\mu(x_\lambda) \to \bar{F}_\mu(x_\lambda + \delta x_\lambda) = F_\mu(x_\lambda) + \delta F_\mu(x_\lambda), \tag{4.114}$$

in contradistinction with a field variation at a fixed point,

$$F_\mu(x_\lambda) \to \bar{F}_\mu(x_\lambda) = F_\mu(x_\lambda) + dF_\mu(x_\lambda). \tag{4.115}$$

Such relations characterize continuous transformations, and the object of Noether's theorem is to derive their influence on the Lagrangian, which we take to be a function of the position, field components, and their derivatives with respect to $x_\mu$:

$$\mathcal{L} = \mathcal{L}[x_\lambda, F_\mu(x_\lambda), \partial_\nu F_\mu(x_\lambda)]. \tag{4.116}$$

Under an infinitesimal transformation, the Lagrangian becomes

$$\mathcal{L} \to \bar{\mathcal{L}}[\bar{x}_\lambda, \bar{F}_\mu(\bar{x}_\lambda), \partial_\nu \bar{F}_\mu(\bar{x}_\lambda)]. \tag{4.117}$$

Following Goldstein, one of the key assumptions that is used in this derivation is the functional identity between the transformed and original Lagrangians:

$$\bar{\mathcal{L}}[\bar{x}_\lambda, \bar{F}_\mu(\bar{x}_\lambda), \partial_\nu \bar{F}_\mu(\bar{x}_\lambda)] = \mathcal{L}[\bar{x}_\lambda, \bar{F}_\mu(\bar{x}_\lambda), \partial_\nu \bar{F}_\mu(\bar{x}_\lambda)]. \tag{4.118}$$

Furthermore, we also consider that the action remains unchanged under the coordinate and field transformation:

$$\bar{S} = \int_{\bar{X}} \bar{\mathcal{L}}[\bar{x}_\lambda, \bar{F}_\mu(\bar{x}_\lambda), \partial_\nu \bar{F}_\mu(\bar{x}_\lambda)] d^4\bar{x}_\lambda = S = \int_X \mathcal{L}[x_\lambda, F_\mu(x_\lambda), \partial_\nu F_\mu(x_\lambda)] d^4 x_\lambda. \tag{4.119}$$

This is a reasonable assumption, as we will consider the action near its local minimum, where $\delta S = 0$. The functional identity between the Lagrangians and the requirement that the action be invariant, immediately yield

$$\int_{\bar{X}} \mathcal{L}[\bar{x}_\lambda, \bar{F}_\mu(\bar{x}_\lambda), \partial_\nu \bar{F}_\mu(\bar{x}_\lambda)] d^4\bar{x}_\lambda = \int_X \mathcal{L}[x_\lambda, F_\mu(x_\lambda), \partial_\nu F_\mu(x_\lambda)] d^4 x_\lambda. \tag{4.120}$$

In Equation 4.120, the four-position is simply a running integration parameter indicating where the field and its derivatives must be evaluated, and we can rewrite

$$\int_{\bar{X}} \mathcal{L}[x_\lambda, \bar{F}_\mu(x_\lambda), \partial_\nu \bar{F}_\mu(x_\lambda)] d^4 x_\lambda = \int_X \mathcal{L}[x_\lambda, F_\mu(x_\lambda), \partial_\nu F_\mu(x_\lambda)] d^4 x_\lambda. \tag{4.121}$$

At this point, we need to evaluate the difference between each integral due to the infinitesimal transformation under consideration, which will then be set equal to zero by virtue of Equation 4.121. To see how this works, we simplify the problem by considering a one-dimensional "toy model." In this case, we have

$$\delta S = \int_{a+\delta a}^{b+\delta b} [\mathcal{L}(x) + \delta\mathcal{L}(x)]dx - \int_a^b \mathcal{L}(x)dx, \tag{4.122}$$

which can be split into three parts,

$$\delta S = -\int_a^{a+\delta a} \mathcal{L}(x)dx + \int_b^{b+\delta b} \mathcal{L}(x)dx + \int_{a+\delta a}^{b+\delta b} \delta\mathcal{L}(x)dx. \tag{4.123}$$

This can also be recast as

$$\delta S = \int_a^b \delta\mathcal{L}(x)\,dx - \int_a^{a+\delta a} [\mathcal{L}(x) + \delta\mathcal{L}(x)]\,dx + \int_b^{b+\delta b} [\mathcal{L}(x) + \delta\mathcal{L}(x)]dx. \tag{4.124}$$

To lowest order, we have

$$\int_a^{a+\delta a} [\mathcal{L}(x) + \delta\mathcal{L}(x)]dx \simeq \int_a^{a+\delta a} \mathcal{L}(x)\,dx \simeq \int_a^{a+\delta a} \mathcal{L}(a)\,dx = \mathcal{L}(a)\,\delta a; \tag{4.125}$$

similarly,

$$\int_b^{b+\delta b} [\mathcal{L}(x) + \delta\mathcal{L}(x)]\,dx \simeq \mathcal{L}(b)\,\delta b. \tag{4.126}$$

Thus, we now have

$$\begin{aligned} \delta S &\simeq \int_a^b \delta\mathcal{L}(x)\,dx + \mathcal{L}(b)\,\delta b - \mathcal{L}(a)\,\delta a \\ &= \int_a^b \delta\mathcal{L}(x)\,dx + [\mathcal{L}(x)\,\delta x]_a^b \\ &= \int_a^b \left\{ \delta\mathcal{L}(x) + \frac{d}{dx}[\mathcal{L}(x)\,\delta x] \right\} dx. \end{aligned} \tag{4.127}$$

Returning to the original problem, Equation 4.127 can be generalized to four dimensions:

$$\begin{aligned} \delta S \simeq \int_x \Big( &\mathcal{L}[x_\lambda, \bar{F}_\mu(x_\lambda), \partial_\nu \bar{F}_\mu(x_\lambda)] - \mathcal{L}[x_\lambda, F_\mu(x_\lambda), \partial_\nu F_\mu(x_\lambda)] \\ &+ \frac{d}{dx^\rho}\{ \mathcal{L}[x_\lambda, F_\mu(x_\lambda), \partial_\nu F_\mu(x_\lambda)]\,\delta x^\rho \} \Big) d^4 x_\lambda. \end{aligned} \tag{4.128}$$

The next step consists in evaluating, to lowest order, the difference

$$\mathcal{L}[x_\lambda, \bar{F}_\mu(x_\lambda), \partial_\nu \bar{F}_\mu(x_\lambda)] - \mathcal{L}[x_\lambda, F_\mu(x_\lambda), \partial_\nu F_\mu(x_\lambda)]. \tag{4.129}$$

This can be achieved by using the fact that $\bar{F}_\mu(x_\lambda) = F_\mu(x_\lambda) + dF_\mu(x_\lambda)$, as expressed in Equation 4.115:

$$\mathcal{L}[x_\lambda, \bar{F}_\mu(x_\lambda), \partial_\nu \bar{F}_\mu(x_\lambda)] - \mathcal{L}[x_\lambda, F_\mu(x_\lambda), \partial_\nu F_\mu(x_\lambda)] \simeq \frac{\partial \mathcal{L}}{\partial F_\mu} dF_\mu + \frac{\partial \mathcal{L}}{\partial(\partial_\nu F_\mu)} d(\partial_\nu F_\mu). \tag{4.130}$$

Using the notation introduced in Section 3.7, where $F_{\mu,\nu} = \partial_\nu F_\mu$, we can alternately write

$$\mathcal{L}[x_\lambda, \bar{F}_\mu(x_\lambda), \partial_\nu \bar{F}_\mu(x_\lambda)] - \mathcal{L}[x_\lambda, F_\mu(x_\lambda), \partial_\nu F_\mu(x_\lambda)] \simeq \frac{\partial \mathcal{L}}{\partial F_\mu} dF_\mu + \frac{\partial \mathcal{L}}{\partial F_{\mu,\nu}} dF_{\mu,\nu}. \tag{4.131}$$

Because $dF_\mu$ corresponds to a variation of the field for a fixed position, we have

$$d(\partial_\nu F_\mu) = d_\nu(dF_\mu), \qquad \text{or} \qquad dF_{\mu,\nu} = \frac{d}{dx^\nu}(dF_\mu). \tag{4.132}$$

In other words, the infinitesimal variation and the spatial derivative commute; in addition, the fixed four-position condition translates into a total derivative for the differential element. Moreover the Lagrange equations indicate that

$$\frac{d}{dx^\nu}\left(\frac{\partial \mathcal{L}}{\partial F_{\mu,\nu}}\right) = \frac{\partial \mathcal{L}}{\partial F_\mu}, \tag{4.133}$$

which yield

$$\begin{aligned} \frac{\partial \mathcal{L}}{\partial F_\mu} dF_\mu + \frac{\partial \mathcal{L}}{\partial F_{\mu,\nu}} dF_{\mu,\nu} &= \frac{d}{dx^\nu}\left(\frac{\partial \mathcal{L}}{\partial F_{\mu,\nu}}\right) dF_\mu + \frac{\partial \mathcal{L}}{\partial F_{\mu,\nu}}\left(\frac{d}{dx^\nu} dF_\mu\right) \\ &= \frac{d}{dx^\nu}\left(\frac{\partial \mathcal{L}}{\partial F_{\mu,\nu}} dF_\mu\right). \end{aligned} \tag{4.134}$$

Using this result into Equation 4.128, we find that

$$\delta S = 0 = \int_x \frac{d}{dx^\nu}\left(\frac{\partial \mathcal{L}}{\partial F_{\mu,\nu}} dF_\mu + \mathcal{L}\delta x^\nu\right) d^4 x_\lambda. \tag{4.135}$$

As this result holds independently of the variation and integration domain, the integrand must be zero, and the four-current

$$j^\nu = \frac{\partial \mathcal{L}}{\partial F_{\mu,\nu}} dF_\mu + \mathcal{L}\delta x^\nu, \tag{4.136}$$

is conserved.

To further illustrate the importance of Noether's theorem, we now consider a more explicit form for the continuous transformations introduced in Equations 4.113 and 4.114, as discussed by Goldstein in classical mechanics. Specifically, we examine linear transformations characterized by two local tensors, $\chi_{i\mu}$, and $f_{i\mu}$,defined such that

$$\delta x_\mu = \delta_i \chi_{i\mu}, \qquad \delta F_\mu = \delta_i f_{i\mu}. \tag{4.137}$$

Here, the quantities $\delta_i$ parameterize the infinitesimal variation. We can then use Equation 4.137 to express the field differential at a fixed position defined in Equation 4.115:

$$\delta F_\mu = dF_\mu + \frac{\partial F_\mu}{\partial x^\nu}\delta x^\nu, \tag{4.138}$$

which yields

$$dF_\mu = \delta F_\mu - \frac{\partial F_\mu}{\partial x^\nu}\delta x^\nu = \delta_i\left(f_{i\mu} - \frac{\partial F_\mu}{\partial x^\nu}\chi_i^\nu\right). \tag{4.139}$$

This result can now be used in Equation 4.135 to obtain

$$\delta S = 0 = \int_x \frac{d}{dx^\nu}\left[\frac{\partial \mathcal{L}}{\partial F_{\mu,\nu}}\delta_i(f_{i\mu} - F_{\mu,\rho}\chi_i^\rho) + \mathcal{L}\delta_i\chi_i^\nu\right] d^4 x_\lambda. \tag{4.140}$$

Grouping terms, we finally find

$$\delta S = 0 = \int_x \delta_i \frac{d}{dx^\nu}\left[\left(\mathcal{L}\delta_\rho^\nu - \frac{\partial \mathcal{L}}{\partial F_{\mu,\nu}} F_{\mu,\rho}\right)\chi_i^\rho + \frac{\partial \mathcal{L}}{\partial F_{\mu,\nu}} f_{i\mu}\right] d^4 x_\lambda. \tag{4.141}$$

As this result must hold independent of the variations $\delta_i$, we can write

$$\frac{d}{dx^\nu} j_i^\nu = 0, \tag{4.142}$$

where for each value of $i$, the conserved four-current is defined as

$$j_i^\nu = \left( \mathcal{L}\delta_\rho^\nu - \frac{\partial \mathcal{L}}{\partial F_{\mu,\nu}} F_{\mu,\rho} \right) \chi_i^\rho + \frac{\partial \mathcal{L}}{\partial F_{\mu,\nu}} f_{i\mu}. \tag{4.143}$$

We have thus shown that, in a very general context, symmetries imply conservation laws, as expressed by Noether's theorem. In the specific case of gauge transformations, the invariance of the electromagnetic field translates into the conservation of the four-current density. Other important conservation laws can be inferred from symmetries, including the existence of an energy–stress tensor, discussed in Section 3.8.

## 4.7 Yang–Mills and Non-Abelian Gauge Fields

This section is intended as a series of introductory remarks on the concept of gauge field theories in modern physics; further details can found in the monograph by Kaku on quantum field theory, as well as in Greiner's textbook on quantum chromodynamics and Weinberg's quantum theory of fields.

The basic idea behind the Maxwell and Yang–Mills theory is their invariance under local transformations as opposed to global symmetries. This property, called gauge invariance, yields specific constraints, including the conservation of the four-current in the case of electrodynamics. The extension of these fundamental concepts to the electroweak interaction was one of the great contributions of Yang and Mills, as well as 'tHooft, who showed that such theories could be renormalizable both for massless and massive gauge bosons. Of course, in the massless case, the photon of quantum electrodynamics is the vector of the electromagnetic interaction, connecting the electric charges conserved by gauge invariance.

One of the key points of the extension of gauge theories is the concept of non-Abelian gauge group, in contradistinction with the simple concept of gauge invariance in electrodynamics. In this case, the massless bosons carrying the interaction, namely the photons, do not carry electric charge and the U(1) group underlying QED is an Abelian gauge group. In the case of non-Abelian gauge groups, the vector bosons can be charged, yielding much more complex interactions. Two examples can be considered: on the one hand, the graviton, which mediates the gravitational interaction, carries energy, which is equivalent to mass or charge in this case; on the other hand, the gluons of quantum chromodynamics are, indeed, carrying color.

In order to bring our presentation further, a brief discussion of electrodynamics within the context of gauge theories and local symmetries proves useful. In this case, the free electromagnetic field is described by the Lagrangian density

$$\mathcal{L}_0 = -\frac{1}{4}F_{\mu\nu}F^{\mu\nu}, \tag{4.144}$$

where the antisymmetric electromagnetic field tensor derives from the four-potential:

$$F_{\mu\nu} = \partial_\mu A_\nu - \partial_\nu A_\mu. \tag{4.145}$$

The interaction Lagrangian for an electron is given by

$$\mathcal{L}_1 = \overline{\psi}(p_\mu - eA_\mu)\gamma^\mu \psi, \tag{4.146}$$

where the spinor wavefunction $\psi$ was introduced in Section 3.11.

The fundamental property of the QED Lagrangian densities presented in Equations 4.144 and 4.146 is their invariance under the local transformations:

$$\begin{aligned} A_\mu &\to A_\mu + \partial_\mu \Lambda, \\ \psi &\to e^{-ie\Lambda}\psi. \end{aligned} \tag{4.147}$$

The phase shift of the spinor field is important and can be connected indirectly to the notion of canonical momentum, as exemplified by the Bohm–Aharanov effect described in Section 4.12. It can also be related to the concept of charge quantization proposed by Dirac, where the quantized angular momentum of a system comprising a magnetic and an electric monopole is proportional to the product of their charge. This will be discussed in Chapter 10.

The generalization of gauge symmetry from QED to larger groups can be achieved by considering matrices instead of the four-potential. In this case, and adopting the conventional notation, we have

$$\mathcal{L}_0 = \frac{1}{4}\mathrm{Tr}(F_{\mu\nu}F^{\mu\nu}), \tag{4.148}$$

where the field is now given by

$$F_{\mu\nu} = \partial_\mu A_\nu - \partial_\nu A_\mu + ig[A_\mu, A_\nu], \tag{4.149}$$

and $g$ is a coupling constant.

Equation 4.149 can be recast in terms of the commutator of the covariant derivative introduced in Section 4.1. If we let $D_\mu \equiv \partial_\mu - igA_\mu$, we have

$$F_{\mu\nu} = \frac{i}{g}[D_\mu, D_\nu]. \tag{4.150}$$

The interaction Lagrangian now becomes

$$\mathcal{L}_1 = \bar{\psi}(p_\mu - gA_\mu)\gamma^\mu \psi, \tag{4.151}$$

and the total Lagrangian is invariant under the generalized gauge transformations

$$\begin{aligned} A_\mu &\to e^{ig\Lambda}\left(A_\mu - \frac{i}{g}\partial_\mu\right)e^{-ig\Lambda}, \\ \psi &\to e^{ig\Lambda}\psi. \end{aligned} \tag{4.152}$$

Here, it is important to understand that the gauge angle, $\Lambda(x_\lambda)$, is a local function; furthermore, the object corresponding to the QED vector potential is now more complex:

$$A_\mu = \lambda_i A_\mu^i, \tag{4.153}$$

where the $\lambda^i$ represents a matrix base. Furthermore, the gauge angle is now also given in terms of theses matrices,

$$\Lambda = \lambda_i \Lambda^i. \tag{4.154}$$

This important property can be shown as follows. Let us start with the fermion field, $\psi$, which explicitly transforms as

$$\psi_i(x_\lambda) \to \Omega_{ij}(x_\lambda)\psi_j(x_\lambda), \tag{4.155}$$

where the matrix $\Omega \in \mathrm{SU}(n)$ is defined by the relations

$$\Omega_{ij}(x_\lambda) = [e^{-i\Lambda_\alpha(x_\lambda)t_\alpha}]_{ij}. \tag{4.156}$$

Here, the $t_\alpha$, which is called the isospin matrices for SU(3) satisfies the commutation rule

$$[t_\alpha, t_\beta] = iC^\gamma_{\alpha\beta} t_\gamma, \tag{4.157}$$

where $C^\gamma_{\alpha\beta}$ represents the structure constant of the Lie group under consideration.

From these, we can construct the generalized version of the four-potential,

$$A_\mu(x_\lambda) = A^\alpha_\mu(x_\lambda) t_\alpha, \tag{4.158}$$

and we formally introduce the operator,

$$D_\mu \equiv \partial_\mu - igA_\mu, \tag{4.159}$$

by generalizing the principle of correspondence to the canonical momentum

$$\pi_\mu = p_\mu - eA_\mu \to i\hbar\partial_\mu - eA_\mu = i\hbar\left(\partial_\mu - i\frac{e}{\hbar}A_\mu\right) = i\hbar D_\mu. \tag{4.160}$$

Before providing a geometrical interpretation of this operator, it is crucial to note that it provides a gauge covariant derivative of the fermion field. Using the abbreviated notation

$$\begin{aligned} \psi' &= \Omega\psi, \\ A'_\mu &= \Omega A_\mu \Omega^{-1} - \frac{i}{g}(\partial_\mu \Omega)\Omega^{-1}, \end{aligned} \tag{4.161}$$

we have

$$\begin{aligned} D_\mu \psi' &= (\partial_\mu - igA'_\mu)\psi' \\ &= (\partial_\mu - igA'_\mu)\Omega\psi \\ &= \Omega(\partial_\mu\psi) + (\partial_\mu\Omega)\psi - igA'_\mu\Omega\psi \\ &= \Omega(\partial_\mu\psi) + (\partial_\mu\Omega)\psi - ig\left[\Omega A_\mu\Omega^{-1} - \frac{i}{g}(\partial_\mu\Omega)\Omega^{-1}\right]\Omega\psi \\ &= \Omega(\partial_\mu\psi - igA_\mu\Omega^{-1}\Omega\psi) + (\partial_\mu\Omega)\psi + i^2(\partial_\mu\Omega)\Omega^{-1}\Omega\psi \\ &= \Omega(\partial_\mu - igA_\mu)\psi \\ &= D_\mu\psi. \end{aligned} \tag{4.162}$$

To conclude this section, we emphasize the deep connection between the gauge covariant derivative operator presented here and the notion of covariant derivative used in general relativity. This can be achieved by giving a geometrical interpretation of $D_\mu$in terms of parallel displacement of a four-vector, as discussed in Section 3.9. Moreover, in Sections 4.8 and 4.9, we will further investigate the fundamental relation between gauge invariance, local symmetry, and the geometry of space–time.

We want to compare the local symmetry

$$\psi' = \Omega\psi = e^{-ig\Lambda}\psi, \tag{4.163}$$

and the Lorentz transform

$$u_\mu = \mathcal{L}_\mu^\nu u_\nu. \tag{4.164}$$

As discussed in Section 3.9, in curved space–time, one is led to introduce the covariant derivative, defined as

$$\nabla_\mu F_{\nu\rho} = \partial_\mu F_{\nu\rho} - \Gamma_{\nu\mu}^\sigma F_{\sigma\rho} - \Gamma_{\rho\mu}^\sigma F_{\nu\sigma}, \tag{4.165}$$

where the Christoffel symbols depend upon the local metric:

$$\Gamma_{\mu\nu}^\lambda = g^{\lambda\sigma}\Gamma_{\sigma,\mu\nu}, \qquad \Gamma_{\sigma,\mu\nu} = \frac{1}{2}\left(\frac{\partial g_{\sigma\mu}}{\partial x^\nu} + \frac{\partial g_{\sigma\nu}}{\partial x^\mu} - \frac{\partial g_{\mu\nu}}{\partial x^\sigma}\right). \tag{4.166}$$

In the simpler case of a four-vector, we can introduce the tensor

$$\nabla_\mu u^\nu = u_\mu^\nu = \partial_\mu u^\nu + \Gamma_{\lambda\mu}^\nu u^\lambda; \tag{4.167}$$

while for the spinor field, we have a nearly identical equation,

$$D_\mu\psi = (\partial_\mu + \Gamma_\mu)\psi, \tag{4.168}$$

where $\Gamma_\mu(x_\lambda)$ is a local rotation matrix, with an imaginary angle corresponding to the so-called gauge angle, $\Lambda(x_\lambda)$ .

In this analogy, the Lorentz transformation is also a local rotation in curved space–time, where the rotation angle corresponds to the boost parameter or rapidity: $\text{argsinh}\,\varphi = \gamma\beta$. In curved space–time, Equation 4.167 indicates the difference between the coordinate-independent variation of the four-vector under parallel displacement and the effects due to the local coordinate representation, described by the Christoffel symbols. As the gauge transformation is local, the same type of effect occurs, which is compensated for by the definition of the gauge covariant derivative operator, $D_\mu$. This is best seen by considering the physically observable, namely the electromagnetic field tensor, which can be expressed as the commutator of the gauge covariant derivative: $[D_\mu, D_\nu] = -igF_{\mu\nu}$. To demonstrate this, we consider the operation of the commutator in question upon a trial function, $\varphi$:

$$\begin{aligned}
[D_\mu, D_\nu]\varphi &= [(\partial_\mu - igA_\mu), (\partial_\nu - igA_\nu)]\varphi \\
&= (\partial_\mu - igA_\mu)(\partial_\nu\varphi - igA_\nu\varphi) - (\partial_\nu - igA_\nu)(\partial_\mu\varphi - igA_\mu\varphi) \\
&= \partial_\mu\partial_\nu\varphi - ig\partial_\mu(A_\nu\varphi) - igA_\mu\partial_\nu\varphi - g^2 A_\mu A_\nu\varphi \\
&\quad - \partial_\nu\partial_\mu\varphi + ig\partial_\nu(A_\mu\varphi) + igA_\nu\partial_\mu\varphi + g^2 A_\nu A_\mu\varphi \\
&= -ig\varphi\partial_\mu A_\nu - igA_\nu\partial_\mu\varphi - igA_\mu\partial_\nu\varphi + ig\varphi\partial_\nu A_\mu + igA_\mu\partial_\nu\varphi + igA_\nu\partial_\mu\varphi \\
&= -ig\varphi(\partial_\mu A_\nu - \partial_\nu A_\mu) \\
&= -igF_{\mu\nu}\varphi.
\end{aligned} \tag{4.169}$$

## 4.8 Weyl's Theory

At this point, it is interesting to address the question of gauge invariance within the context of Weyl's theory. We will follow the presentation given by Pauli in his monograph on the theory of relativity. Although Weyl's original goal to unify gravitation and electromagnetism within the framework of a classical field theory no longer presents its original interest because it does not involve quantum principles, it does provide powerful insight into the physics underlying gauge symmetries. In particular, Weyl's theory is developed as a natural extension of Riemannian geometry, called pure infinitesimal geometry, which fully comprises covariance and space–time curvature, thus including general relativity and gravitation, and introduces gauge invariance as a local symmetry modifying the length of a vector upon infinitesimal parallel displacement. In this manner, gauge symmetries are seen to result from a local property of the space–time metric.

Riemannian curved space–time is introduced as a natural extension of Euclidean geometry by considering the parallel displacement of a vector over an infinitesimal loop; curvature manifests itself in the fact that the vector orientation is modified by such a displacement. Weyl's took the next step by assuming a change of length, in addition to the change of direction, induced by curvature. Following Riemann's formalism, the space–time interval is first defined in terms of a metric as

$$ds^2 = g_{\mu\nu}dx^\mu dx^\nu, \tag{4.170}$$

where the metric, $g_{\mu\nu}$, is a function of the four-position: $g_{\mu\nu} = g_{\mu\nu}(x_\lambda)$.

We now consider an infinitesimal displacement, parameterized by $\tau$. The square of the length, $ds^2 = \lambda$, as measured at the initial point $\sigma(\tau)$, and at the final point $\sigma(\tau + d\tau)$, along the world curve $x_\mu(\tau)$, is changed. Following Weyl and Pauli, we shall assume axiomatically that

$$\frac{d\lambda}{d\tau} = -\lambda\frac{d\phi}{d\tau}, \tag{4.171}$$

where $\phi(\tau)$ is independent of the length $\lambda$ and plays the role of a potential.

Additionally, Weyl requires that $d\phi/d\tau$ depends only on the first derivatives of the displacement curve, $dx_\mu/d\tau$ this second axiom can also be viewed as resulting from the Taylor expansion of the local variation of length induced by the potential $\phi$. Equation 4.171 must be independent of the choice of parameterization, which implies that $d\phi/d\tau$ is a homogeneous function of first degree in $dx_\mu/d\tau$. Using Christoffel symbols, the variation of a four-vector resulting from the displacement along the world-curve can be expressed as

$$\frac{dv^\mu}{d\tau} = -\Gamma^\mu_{\nu\lambda}\frac{dx^\nu}{d\tau}v^\lambda. \tag{4.172}$$

On the other hand, the variation of length postulated in Equation 4.171 implies that

$$\frac{d}{d\tau}(g_{\mu\nu}v^\mu v^\nu) = \frac{d}{d\tau}(v_\mu v^\mu) = -g_{\mu\nu}v^\mu v^\nu\frac{d\phi}{d\tau}. \tag{4.173}$$

Considering Equations 4.172 and 4.173 together, we find that $d\phi/d\tau$ must depend linearly on $dx^\mu/d\tau$:

$$d\phi = \phi_\mu dx^\mu. \tag{4.174}$$

The corresponding geodesics can now be determined:

$$\frac{\partial g_{\mu\nu}}{\partial x^\lambda} + g_{\mu\nu}\phi_\lambda = g_{\mu\rho}\Gamma^\rho_{\nu\lambda} + g_{\nu\rho}\Gamma^\rho_{\mu\lambda} = \Gamma_{\mu,\nu\lambda} + \Gamma_{\nu,\mu\lambda}. \tag{4.175}$$

This is to be compared with the geodesic of Riemannian geometry, derived in Section 3.9:

$$\frac{d^2x^\lambda}{ds^2} + \Gamma^\lambda_{\mu\nu}\frac{dx^\mu}{ds}\frac{dx^\nu}{ds} = 0, \tag{4.176}$$

which is seen to constitute the limit of Equation 4.175 when $\phi_\mu = 0$. More explicitly, we can relate the Christoffel symbols of Weyl's theory to those of Riemannian geometry as follows:

$$\Gamma_{\mu,\nu\lambda} - \bar{\Gamma}_{\mu,\nu\lambda} = \frac{1}{2}(g_{\mu\nu}\phi_\lambda + g_{\mu\lambda}\phi_\nu - g_{\nu\lambda}\phi_\mu), \tag{4.177}$$

where $\bar{\Gamma}_{\mu,\nu\lambda}$ represents the Riemannian quantities.

We now approach the central argument behind Weyl's theory and its direct relation to the concept of local symmetries and gauge invariance. We consider a local transformation to the metric, namely,

$$g_{\mu\nu} \to \Lambda(x_\rho) g_{\mu\nu}, \tag{4.178}$$

which simply corresponds to a local regauging of the length element. When this is done in conjunction with satisfying Equation 4.171, we find that the quantities $\phi_\mu$ must be shifted as follows:

$$\phi_\mu \to \phi_\mu - \frac{\partial \ln \Lambda}{\partial x^\mu} = \phi_\mu - \frac{1}{\Lambda}\frac{\partial \Lambda}{\partial x^\mu}. \tag{4.179}$$

To quote Pauli: "Just as we had postulated the invariance of all geometrical relations and physical laws under arbitrary coordinate transformations, so we have now to postulate, in addition, their invariance with respect to … changes in the gauge." This constitutes gauge invariance within the context of Weyl's theory, where this concept was first introduced.

Let us now proceed to see how this requirement allows Weyl to derive Maxwell's equations from this new space–time geometry. Following Pauli, we first formally integrate Equation 4.171 between the points $\sigma_1$ and $\sigma_2$, to find

$$[\ln \lambda]_{\sigma_1}^{\sigma_2} = -\int_{\sigma_1}^{\sigma_2} \phi_\mu dx^\mu, \qquad \lambda(\sigma_2) = \lambda(\sigma_1)\exp\left[-\int_{\sigma_1}^{\sigma_2} \phi_\mu dx^\mu\right]. \tag{4.180}$$

Next, we examine the required conditions under which the length of a vector becomes independent from the path chosen, as in Riemannian geometry: the necessary and sufficient constraint is

$$F_{\mu\nu} = \frac{\partial \phi_\mu}{\partial x^\nu} - \frac{\partial \phi_\nu}{\partial x^\mu} = 0. \tag{4.181}$$

Moreover, when this condition is satisfied, the potential $\phi_\mu$ can be gauge-transformed away, according to Equation 4.179. Returning to the more general case of Weyl's theory, which is an extension of Riemannian geometry, we find that the field tensor can be nonzero; furthermore, the antisymmetrical tensor also satisfies the Jacobi identity

$$\partial_\lambda F_{\mu\nu} + \partial_\mu F_{\nu\lambda} + \partial_\nu F_{\lambda\mu} = 0. \tag{4.182}$$

The definition of the field tensor as a function of the potential, together with Equation 4.182, are identical in form with Maxwell's equations; following Weyl, we can therefore identify the quantities $\phi_\mu$ with the four-potential, $A_\mu$. This approach yields a beautiful connection between geometry and electromagnetism; however, a number of problems have rendered this theory "obsolete," to use the term of Schwinger and co-authors. One such difficulty was pointed out by Einstein: if we use Equation 4.180 in the simple case of a static potential, where $\phi_i = 0$, we find that $\tau = \tau_0 e^{\alpha\phi_0 t}$. This can be interpreted as follows. The time flow is perturbed by the local potential; two clocks located at different spatial positions under different static potentials cannot be synchronized. From an experimental point of view, such effects could be measurable by observing atomic transitions, for example. Moreover, the exponential nature of the time differential means that the value of the proportionality constant, $\alpha$, yields the duration of the experiment over which discrepancies could be measured. For an in-depth discussion of Weyl's theory, both from physical and mathematical standpoints, we refer the reader to Pauli's presentation.

## 4.9 Kaluza–Klein Five-Dimensional Space–Time

Following a similar attempt to unify gravitation and electrodynamics, Kaluza and Klein proposed a revolutionary approach, which is now becoming one of the fundamental concepts underlying superstring theories. Therefore, a brief presentation is appropriate, as it also pertains to the general concept of gauge invariance discussed in the present chapter.

In 1921, Kaluza discovered a geometrical representation of Maxwell's equations, which was extended by Klein in 1926, to form what is now known as higher-dimensional Kaluza–Klein theories. In its original form, the theory considers a five-dimensional manifold, where a cylindrical metric, $\gamma_{\mu\nu}$, is defined:

$$ds^2 = \gamma_{\mu\nu} dx^\mu dx^\nu, \quad \mu, \nu = 0, \dots 4. \tag{4.183}$$

The cylindrical character of the metric simply corresponds to the fact that there exists a coordinate system where

$$\frac{\partial \gamma_{\mu\nu}}{\partial x^4} = 0; \tag{4.184}$$

in this case, the $x^4$-axis is the axis of the hypercylinder. Moreover, Kaluza and Klein chose a space-like fifth dimension, with $\gamma_{44} = 1$.

From this starting point, Kaluza and Klein separate the distance element as follows:

$$ds^2 = (dx^4 + \gamma_{i4} dx^i)^2 + g_{ik} dx^i dx^k, \quad i,k = 0,\ldots 3. \tag{4.185}$$

Thus, the metric $g_{ik}$ corresponds to that of general relativity in a four-dimensional space–time, with three spatial coordinates, $x_1$, $x_2$, $x_3$, and one time coordinate, $x_0 = ct$. Of course, the coordinate transformations of general relativity are still allowed on the $x_i$; in addition, it is clear that transformations of the form

$$\bar{x}^4 = x^4 + f(x^i) \tag{4.186}$$

leave the $4 \times 4$ portion of the metric invariant,

$$\bar{g}_{ik} = g_{ik}. \tag{4.187}$$

The effect of the local transformation given in Equation 4.186 on the other elements of $\gamma_{\mu\nu}$ is described by

$$\bar{\gamma}_{i4} = \gamma_{i4} - \frac{\partial f}{\partial x^i}. \tag{4.188}$$

At this point, the elements of the five-dimensional metric can be determined. First, we can compare Equations 4.183 and 4.185, to find that

$$\gamma_{ik} = g_{ik} + \gamma_{i4}\gamma_{k4}. \tag{4.189}$$

Moreover, if we consider the relation between $\gamma_{\mu\nu}$ and $\gamma^{\mu\nu}$, we can establish that

$$\det|\gamma_{\mu\nu}| = \det|g_{ik}|, \tag{4.190}$$

from which we can obtain

$$\begin{aligned} \gamma^{44} &= 1 + \gamma^{jk}\gamma_{i4}\gamma_{k4}, \\ \gamma^{i4} &= -g^{ik}\gamma_{k5}, \\ \gamma^{jk} &= g^{ik}. \end{aligned} \tag{4.191}$$

This can be summarized by writing

$$\gamma^{\mu\nu} = \begin{pmatrix} \gamma^{ik} & \gamma^{i4} \\ \gamma^{4k} & \gamma^{44} \end{pmatrix} = \begin{pmatrix} g^{ik} & -g^{ik}\gamma_{k4} \\ \gamma^{4k} & 1+g^{ik}\gamma_{i4}\gamma_{k4} \end{pmatrix}. \tag{4.192}$$

Using the results summarized in Equation 4.192, along with the gauge transform in Equation 4.188, Kaluza and Klein were led to suggest the identification of $\gamma_{i4}$ with the four-potential, $A_i$. In fact, the conditions

$$\begin{aligned} \gamma^{44} &= 1+g^{ik}\gamma_{i4}\gamma_{k4} = 1, \\ \gamma^{i4} &= -g^{ik}\gamma_{k4}, \\ \bar{\gamma}_{i4} &= \gamma_{i4} - \partial_i f, \\ \bar{g}_{ik} &= g_{ik} \end{aligned} \tag{4.193}$$

naturally lead one to introduce the antisymmetrical tensor

$$\alpha F_{ik} = \frac{\partial \gamma_{k4}}{\partial x^i} - \frac{\partial \gamma_{i4}}{\partial x^k}, \tag{4.194}$$

which exhibits the fundamental property of invariance under the aforementioned gauge transformation,

$$\bar{F}_{ik} = F_{ik}. \tag{4.195}$$

At this point, we can write

$$\gamma_{\mu\nu} = \begin{pmatrix} g_{ik} & \alpha A_i \\ \alpha A_k & 1 \end{pmatrix}, \tag{4.196}$$

where $\alpha$ is a proportionality constant.

Next, we consider the geodesics for the Kaluza–Klein metric. It is clear that for the usual four-dimensional space–time components, we have

$$g_{ik}\frac{dx^i}{ds}\frac{dx^k}{ds} = -1, \tag{4.197}$$

while Equation 4.184 yields

$$\frac{dx^4}{ds} + \gamma_{i4}\frac{dx^i}{ds} = k, \tag{4.198}$$

where $k$ is a constant.

Equations 4.197 and 4.198 can be combined to give

$$\frac{d}{ds}\left(g_{ik}\frac{dx^k}{ds}\right) - \frac{1}{2}\frac{\partial g_{jk}}{\partial x^i}\frac{dx^j}{ds}\frac{dx^k}{ds} = k\alpha F_{ik}\frac{dx^k}{ds}, \tag{4.199}$$

which is identical to the equation of motion of a charged particle in curved space–time with an applied electromagnetic field, provided that we choose $\alpha k = -e/m_0$ for an electron.

In this formalism, the fifth dimension is expressed physically in the form of electromagnetic phenomena; furthermore, the cylindrical geometry allows one to consider that the fifth dimension is extremely small, to the extent that is not detectable experimentally. This idea, originally put forth by Klein, has been considerably amplified in modern superstring theories. One of the basic ideas underlying such theories is the replacement of point particles interacting with quantized fields by covariant strings, thus alleviating the renormalization problems deriving from the point-like nature of the interaction vertices. The origin of such difficulties is both classical, as discussed in Chapter 10, and quantum, as exemplified by the problem of the zero-point energy of the vacuum, which is briefly outlined in Chapter 6.

## 4.10 Charged Black Holes, Quantum Gravity, and Inflation

Before outlining a few salient features of superstring theories, we briefly consider a few interesting points regarding the nature of space–time within the context of modern physics. In particular, the methods used in quantum cosmology are noteworthy, as they combine elements from both general relativity and quantized field theory.

A useful line of approach to quantum cosmology is the study of quantum black holes. Despite its extraordinary weakness compared to the other forces, gravity is (almost) always attractive and becomes a dominant factor in at least two arenas: large-scale interactions, on the one hand, such as the dynamics of galaxies and clusters, and black holes, or gravitationally collapsed objects, on the other hand. As demonstrated by Chandrasekhar, for stars above roughly one and a half solar masses (the precise value is 1.37 $M_\odot$), the gravitational forces are sufficient to overcome the degeneracy pressure of electrons and neutrons, and the object can collapse beyond its Schwarzschild radius to become a black hole. The Schwarzschild radius is defined by considering the metric of a nonrotating black hole, in units where $G = c = 1$,

$$ds^2 = -\left(1-2\frac{M}{r}\right)dt^2 + \left(1-2\frac{M}{r}\right)^{-1} dr^2 + r^2(d\theta^2 + \sin^2\theta d\varphi^2), \quad (4.200)$$

and is given by $r_0 = 2GM/c^2 \simeq 3(MM_\odot)$ km, where $M$ is the mass of the black hole, and $M_\odot = 1.9891 \times 10^{30}$ kg is one solar mass. Let us also briefly remark that the apparent singularity at the Schwarzschild radius can be removed by an appropriate change of coordinates, while the central singularity, at $r = 0$, is an essential one. One can introduce the new radial coordinate, $\rho = 4M\sqrt{1-2Mr^{-1}}$, to obtain the so-called Euclidean-Schwarzschild metric,

$$ds^2 = \rho^2\left(\frac{d\bar{t}}{4M}\right)^2 + \left(\frac{r^2}{4M^2}\right)^2 d\rho^2 + r^2(d\theta^2 + \sin^2\theta d\varphi^2), \quad (4.201)$$

where $\bar{t} = -it$.

Another extremely important result in general relativity concerns the case where a black hole has initial angular momentum and rotates; this corresponds to the solution derived by Kerr and later generalized to charged, rotating black holes by Newman and co-workers. In this case, the metric reads

$$\begin{aligned} ds^2 = &-\left(\frac{\Omega^2 - \omega^2\sin^2\theta}{r^2 + \omega^2\cos^2\theta}\right)dt^2 - \left[\frac{2\omega\sin^2\theta(r^2+\omega^2-\Omega^2)}{r^2+\omega^2\cos^2\theta}\right]dt\,d\varphi \\ &+ \left[\frac{(r^2+\omega^2)^2 - \Omega^2\omega^2\sin^2\theta}{r^2+\omega^2\cos^2\theta}\right]\sin^2\theta d\varphi^2 + \left(\frac{r^2+\omega^2\cos^2\theta}{\Omega^2}\right)dr^2 \\ &+ (r^2+\omega^2\cos^2\theta)d\theta^2, \end{aligned} \quad (4.202)$$

where

$$\Omega^2 = r^2 + \omega^2 + q^2 - 2Mr. \quad (4.203)$$

This result is quite fundamental because it shows that a black hole is characterized by three parameters only, its mass, $M$, charge, $q$, and rotation frequency, $\omega$. In this respect it closely resembles an elementary particle, and such models have been proposed. A large number of investigations have also been performed regarding questions such as the entropy of a black hole, which was shown to scale as the surface of the event horizon by Hawking.

For a brief overview of this important result and an introduction to the concept of path integrals in curved space–time, we can follow Hawking and examine the metric given in Equation 4.201. In the $\rho$-$\bar{t}$ plane, it looks just like polar coordinates, and the full angular sweep is obtained for a period equal to $8\pi M$, in units of $\bar{t}$. If we now consider the transition between the field configurations $\psi_1(t_1)$ and $\psi_2(t_2)$, the corresponding matrix element is

$$\exp[-i\mathcal{H}(t_2 - t_1)], \tag{4.204}$$

where $\mathcal{H}$ is the Hamiltonian of the interaction in question.

If we further choose the particular case where $t_2 - t_1 = -i8\pi M$ and where $\psi_2 = \psi_1$, we recover the thermodynamic partition function $Z$,

$$Z = \sum_n \langle \psi_n | \exp(-8\pi M\mathcal{H}) | \psi_n \rangle, \tag{4.205}$$

for the temperature $T = 1/8\pi M$. If, on the other hand, we think in terms of path integrals, Equation 4.205 also reads

$$Z = \sum_n \langle \psi_n | \exp(-\beta\mathcal{H}) | \psi_n \rangle = \int \mathcal{D}(\psi) \exp[-iS(\psi)], \tag{4.206}$$

where $S(\psi)$ is the action functional, as defined by Feynman, and where $\beta = 8\pi M$.

This identity underlies the fact that the black hole emits thermal radiation and that its entropy is proportional to the area of its event horizon.

Black hole thermal radiation can be understood as follows. Near the horizon, the quantum fluctuations of the vacuum can result in the creation of virtual pairs; if one of the particles tunnels through the horizon to be absorbed by the black hole, its partner becomes a free particle which participates in the thermal radiation, as illustrated in Figure 4.2. As the black hole effectively radiates gravitational energy away, it evaporates. Within this context, the identification of elementary particles with microscopic, charged, rotating black holes is problematic, as they would evaporate on very short time scales. However, one can consider an interesting situation: suppose that the electron is actually a mini-black hole. The creation of a virtual electron–positron pair near its horizon would result in an asymmetric process, as the positron would be attracted preferentially by the electron, both gravitationally and electrostatically. If the original electron absorbs the positron, we are left with another electron, the virtual electron, which is now a free particle! This mechanism, illustrated in Figure 4.3, quenches the evaporation process and may stabilize the mini-black hole.

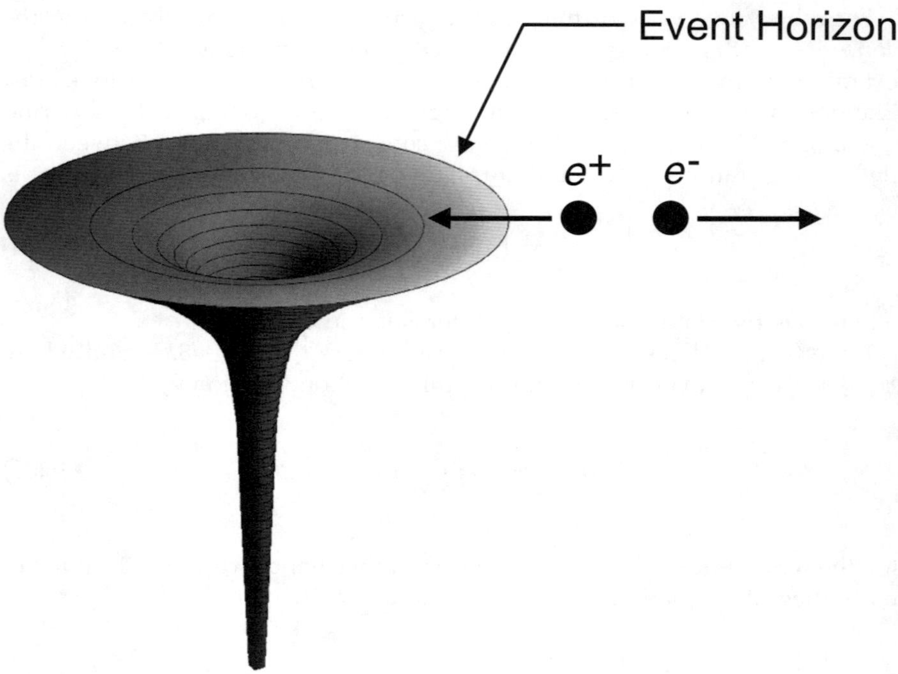


**FIGURE 4.2**
Near the event horizon of a black hole, vacuum fluctuation can lead to the absorption of a virtual particle in the black hole, leaving its counterpart free to escape the black hole in the form of Hawking radiation.

Furthermore, black hole thermal radiation is closely related to the Unruh effect, where a uniformly accelerated detector records the vacuum fluctuations as blackbody radiation, with a temperature scaling linearly with the acceleration, $a$:

$$T = \frac{a}{2\pi}\left(\frac{\hbar}{ck_B}\right), \tag{4.207}$$

where $k_B = 1.380\ 658(12) \times 10^{-23}$ J/°K is the Boltzmann constant.

Returning to Hawking's derivation, the action for the black-hole metric contains two terms, the usual volume integral of the scalar curvature and a surface term, which can be shown to give rise to

$$Z = \exp\left(-\frac{\beta^2}{16\pi}\right), \tag{4.208}$$

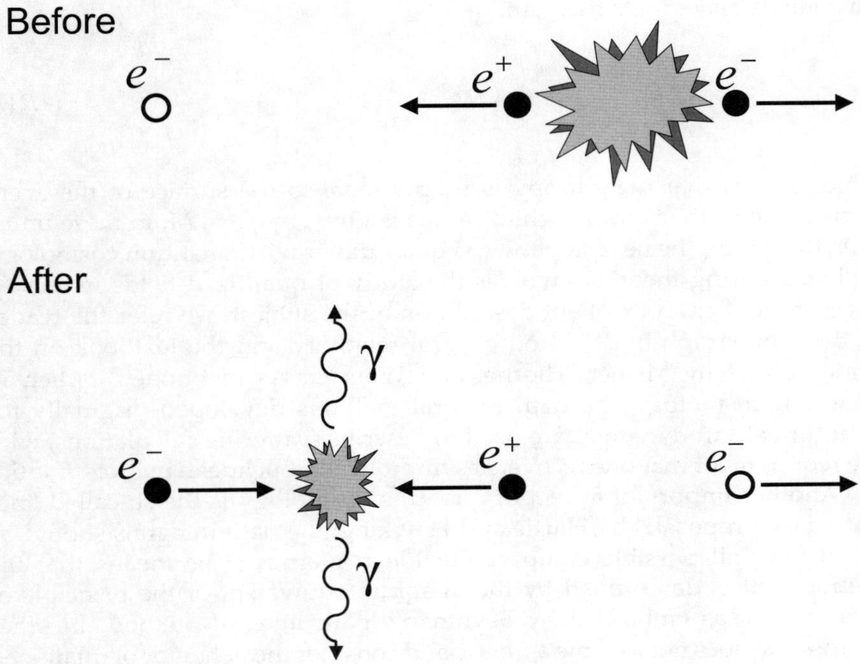


**FIGURE 4.3**
A mini-black hole electron absorbs a virtual positron, leaving the virtual electron free: the evaporation mechanism is quenched.

once the path integral defined in Equation 4.206 has been performed. From the partition function, we can verify the expectation value for the mass:

$$\langle W \rangle = M = \frac{d}{d\beta}\ln(Z) = \frac{\beta}{8\pi}. \tag{4.209}$$

Moreover, it is well known in thermodynamics that the relation between the free energy and the temperature is

$$\ln(Z) = -\frac{F}{T}. \tag{4.210}$$

The derivation is finalized by remembering the definition of the entropy, $E$:

$$F = \langle W \rangle + TE. \tag{4.211}$$

This yields Hawking's theorem,

$$E = 4\pi M^2 = \frac{1}{4}(4\pi r_0^2), \tag{4.212}$$

which shows that the entropy is proportional to the surface of the event horizon, since the Schwarzschild radius is given by $r_0 = 2M$, in these units.

At this point, the next step toward quantum gravity, quantum cosmology, and superstring theories involves the study of quantized fields in curved space–time. For an excellent description of the subject, we refer the reader to the monograph by Wald on general relativity and the textbook on the same subject by Misner, Thorne, and Wheeler. As mentioned earlier, in quantum cosmology, the path integral methods developed originally for quantum electrodynamics are used to describe scattering calculations, with the requirement that one recovers asymptotically Euclidean metrics at infinity. Another important aspect of quantum cosmology is the so-called "no-boundary proposal," by Hartle and Hawking: the path integrals should be taken over all possible compact Euclidean metrics. This means that the metric itself is determined by the quantum equivalent of the principle of least action, as embodied by Feynman's path integral method. In other words, one does not assume a metric and consider the behavior of quantized fields in that particular geometry, as in the study of quantized fields in curved space–time. Rather, the full dynamical interplay of geometry, quantum effects, and the matter–energy fields is subjected to the path integral formulation. In fact, the fundamental question of the geometry of space–time is well illustrated by the approach taken in quantum cosmology, where a number of researchers, including Wheeler and Hawking, have proposed that one should consider an equation governing the behavior of a so-called universal wave function, $\psi$. Within this context, the Wheeler–De Witt equation purports to describe the evolution of the wavefunction of the universe:

$$G_{ijkl}\left[\frac{\partial^2}{\partial h_{ij}\partial h_{kl}} - \sqrt{h}R\right]\psi = 0. \tag{4.213}$$

In this approach, the very geometry of the universe becomes a dynamical variable. Its large-scale behavior closely matches the well-known solutions of general relativity, but at extremely small scales, near the Planck length, the topology reaches the so-called "quantum foam" state, illustrated in Figure 4.4, where vacuum fluctuations are thought to produce rapidly changing bubbles, mini-black holes, wormholes, and other exotic geometries.

Of interest and closely related to the subject of quantum cosmology, the question of the matter–antimatter asymmetry and the horizon–flatness problem

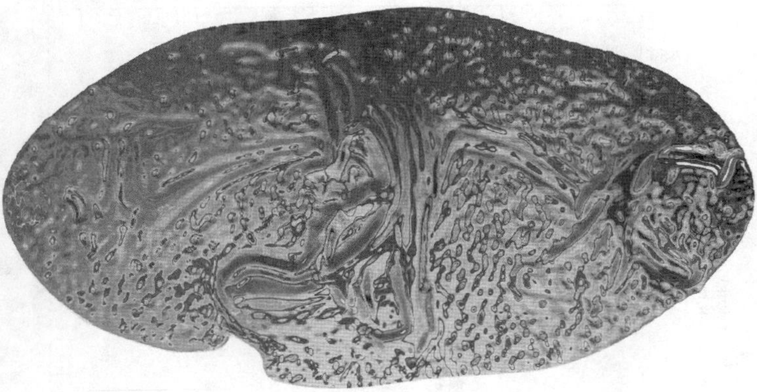

**FIGURE 4.4**
Artist rendering of quantum foam, a possible model for the fabric of space–time at the Planck scale.

have played an important role in establishing grand unified theories (GUTs) and the inflationary universe model. To provide the proper background to address these issues, a few words about the so-called big bang model are in order. General relativity has provided a theoretical framework for the three fundamental pieces of experimental evidence supporting the big bang theory. First, the observed recession of distant galaxies, which appear more redshifted as they are further away, as summarized by Hubble's law, which provides a linear relation between the recession velocity and the distance as shown in Figure 4.5: $v = Hd$, where $H$ is Hubble's constant, measured at $H \simeq 55.2$ km $\times$ s$^{-1}$/megaparsec. A megaparsec corresponds to $3.26 \times 10^6$ light-years, as a parsec is the distance at which the average distance between the earth and the sun subtends a one-second arc. Second, the three-degree background radiation: this is the very nearly isotropic (to within $10^{-5}$, when the dipole component due to the motion of the earth and local cluster, approximately 600 km/s, is subtracted) microwave radiation spectrum bathing the cosmos, that has a spectral distribution in superb agreement with the blackbody radiation spectrum, where the energy density is given by

$$\frac{du}{d\omega} = \frac{\hbar\omega^3}{\pi^2 c^3\left[\exp\left(\frac{\hbar\omega}{k_B T}\right) - 1\right]}, \tag{4.214}$$

for a temperature $T = 2.736$°K as illustrated in Figure 4.6. The microwave cosmic background corresponds to the thermal photons produced by the big

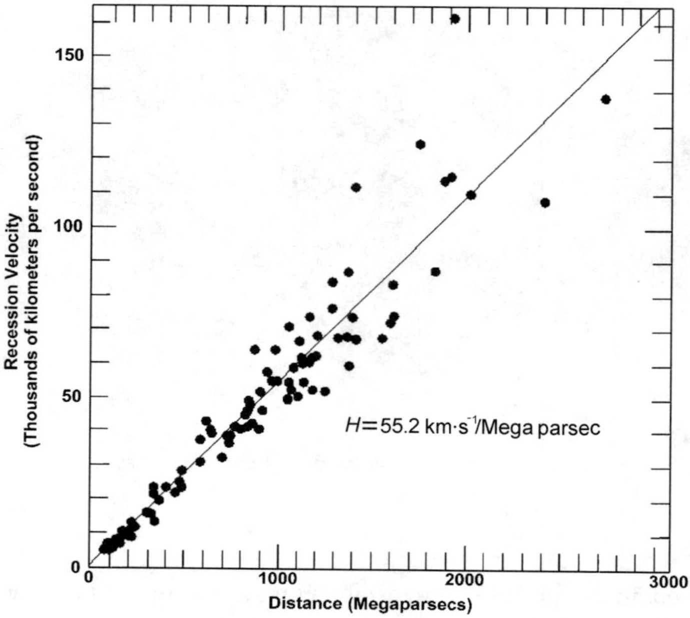


**FIGURE 4.5**
Recent measurement of Hubble's constant (adapted from the data published by J. Kristian, et al.).

bang, which decoupled from the electron–proton plasma approximately 300,000 years after the big bang and were stretched in wavelength, as the universe continued to expand. Finally, the relative abundance of elements, as produced by nucleosynthesis and observed in deep space, very nicely fits models developed on the big bang hypothesis. In particular, such models correctly predict that approximately 25% of all the primordial hydrogen should have fused to produce helium.

The metric typically used for such models is generally of the Robertson–Walker form,

$$ds^2 = dt^2 - \rho^2(t)\left[\frac{dr^2}{1-\alpha^2 r^2} + r^2(d\theta^2 + \sin^2\theta d\varphi^2)\right], \tag{4.215}$$

where $\alpha$ is a constant related to the global curvature, and $\rho(t)$ determines the scale of the universe.

If we consider an isotropic universe characterized by the pressure $p(t)$ and the density $\eta(t)$, the metric yields two coupled equations describing energy

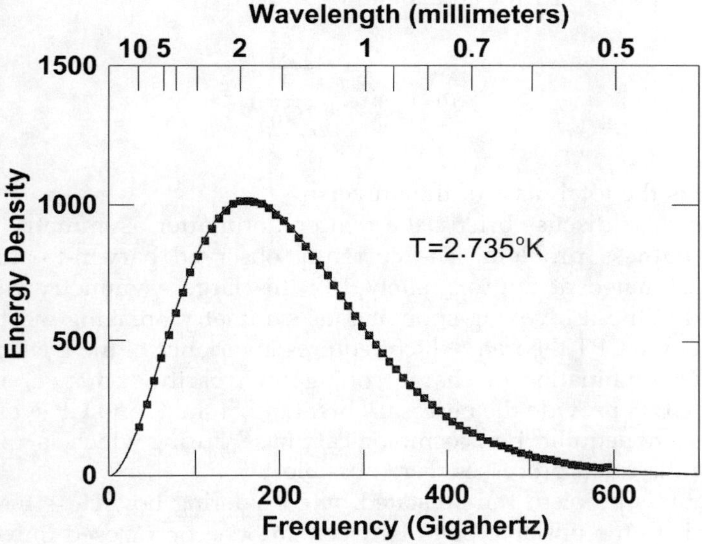


**FIGURE 4.6**
The 2.735°K blackbody radiation resulting from the afterglow of the big bang; the solid line is from theory (data from the COBE team; see bibliography).

conservation:

$$\left(\frac{1}{\rho}\frac{d\rho}{dt}\right)^2 = \frac{8\pi}{3}G\eta - \frac{\alpha^2}{\rho^2},$$
$$\frac{d^2\rho}{dt^2} = -4\pi G\left(p + \frac{\eta}{3}\right) + \frac{\Lambda}{3}. \tag{4.216}$$

Here, $\Lambda$ is the cosmological constant, which we assume to be equal to zero. In this particular case and using a perfect gas model to couple the density and pressure, we obtain as single evolution equation for the scale:

$$2\rho\frac{d^2\rho}{dt^2} + \left(\frac{d\rho}{dt}\right)^2 + \alpha^2 = 0. \tag{4.217}$$

If we further assume that the universe is very nearly flat, that is $\alpha \simeq 0$, this equation can be solved asymptotically for large values of time, to yield

Friedman's power law for an expanding universe,

$$\rho(t) \simeq t^{2/3} \sqrt[3]{\frac{9}{2GM}}, \tag{4.218}$$

where $M$ is the total mass of the universe.

We can now discuss briefly the matter–antimatter asymmetry and the horizon–flatness problem. The currently observed baryon-to-antibaryon ratio is estimated at approximately $10^9$; this large asymmetry cannot be explained without invoking spontaneous symmetry breaking mechanisms, because of the CPT theorem which requires invariance of the laws of nature under the combination of charge conjugation, parity exchange, and time reversal. GUTs provide the necessary breaking of the C and CP symmetries, as well as a nonequilibrium cosmological phase, during which asymmetrical baryon–antibaryon processes were possible.

The flatness problem can be stated by considering how close the current universe is to the unstable point between an open or a closed universe: the observed density of the universe, $\rho$, is quite close to the so-called critical density, $\rho_c$. In fact, it is estimated that the ratio $\Omega = \rho/\rho_c$, is in the range 0.1 to 10. Although this looks like a fairly large difference, once extrapolated back in time, it is easily seen how extraordinarily fine-tuned the universe turns out to be. For this, we briefly return to the Robertson–Walker metric in Equation 4.215, from which one can derive the so-called cosmological equation,

$$H^2 = \left(\frac{1}{\rho}\frac{d\rho}{dt}\right)^2 = \frac{8}{3}\pi G\eta \pm \frac{\alpha^2}{\rho^2} + \frac{\Lambda}{3}, \tag{4.219}$$

where $H$ is Hubble's constant.

As seen earlier, when the density dominates, Equation 4.219 can be approximated by

$$H^2 = \left(\frac{1}{\rho}\frac{d\rho}{dt}\right)^2 \simeq \frac{8}{3}\pi G\eta. \tag{4.220}$$

In addition, we assume that the density evolves as $\eta \propto \rho^{-3}(t)$, and we recover Friedman's expansion law and the relation between Hubble's constant and the age of the universe:

$$\rho(t) = t^{2/3} \sqrt[3]{\frac{9}{2GM}}, \qquad t = \frac{2}{3H} = \frac{1}{\sqrt{6\pi G\eta}}. \tag{4.221}$$

We now introduce the redshift parameter,

$$z = \frac{\lambda_0}{\lambda} - 1 \simeq H_0 \frac{d}{c} + \beta, \tag{4.222}$$

where $\lambda_0$ is the wavelength emitted by a reference atomic transition in a galaxy located at a distance $d$, while $\lambda$ is the measured wavelength, and $\beta$ is the normalized line-of-sight velocity component of the galaxy relative to the average Hubble recession velocity; $H_0$ is the current value of $H$.

With this, we can recast the cosmological equation as

$$H^2 = \left(\frac{1}{\rho}\frac{d\rho}{dt}\right)^2 = \left(\frac{1}{1+z}\frac{dz}{dt}\right)^2 = H_0^2[\Omega(1+z)^3 + \Omega_\alpha(1+z)^2 + \Omega_\Lambda], \tag{4.223}$$

where $\Omega$ is the density parameter, defined in terms of the current density $\eta_0$,

$$\Omega = \frac{8\pi G\eta_0}{3H_0^2}, \tag{4.224}$$

while the curvature parameter is given by

$$\Omega_\alpha = \frac{\alpha^2}{(\rho_0 H_0)^2}. \tag{4.225}$$

$\Omega_\alpha > 0$ corresponds to an open universe, while negative values yield closed solutions. Finally, the cosmological constant yields a third parameter,

$$\Omega_\Lambda = \frac{\Lambda}{3H_0^2}; \tag{4.226}$$

moreover, we have the constraint

$$\Omega + \Omega_\alpha + \Omega_\Lambda = 1. \tag{4.227}$$

Using Equation 4.223 to perform the aforementioned extrapolation backward in time, we find that at the time when the GUT symmetry was broken, for $t \simeq 10^{-36}$s, the density parameter must have been extraordinarily close to $1: \Omega - 1 \approx 10^{-55}$.

The second feature of interest here is the horizon problem. Briefly stated, the cosmic background radiation looks too isotropic to have been produced by causally disconnected regions of space at the time when the hydrogen plasma recombined and photons were decoupled.

The inflationary universe scenario, first proposed by Guth, resolves these problems in a very ingenious way. First, as we have already discussed, the matter–antimatter asymmetry can be explained by GUTs; more importantly, GUTs also lead to a phase transition corresponding to spontaneous symmetry breaking, where the cosmological constant and parameter $\Omega_\Lambda$ become dominant factors in the cosmological equation. In this case, the solution corresponds to an exponential expansion, where the scale now behaves as

$$\rho(t) = \rho_0 \exp\left[\sqrt{\frac{8\pi\eta_b}{3}}\frac{t}{M_b}\right], \tag{4.228}$$

with $\eta_b$ and $M_b$ referring to the density and mass of the expanding "bubble," respectively.

The e-folding time turns out to be very short, and the inflationary expansion can reach very high values, well in excess of $10^{30}$. In that case, the visible universe may simply correspond to an extremely small fraction of the entire universe, and the observed curvature seems small because the overall radius is very large. Furthermore, as the visible universe corresponds to the expansion of a very small "bubble" causally connected at a very early age, the apparent horizon problem is removed: the horizon was simply pushed away by inflation.

For further details, we refer the reader to Guth's introduction on inflation and Kaku's more technical discussion in his textbook on quantum field theories. Peebles' monograph on the principles of physical cosmology should also prove very useful.

## 4.11 Superstrings and Dimensionality

Before embarking on a cursory discussion of superstring theories, a strong word of caution is required: the unification energy scale for quantum gravity and the other forces is extraordinarily high, many orders of magnitude beyond what can be achieved by the most powerful accelerators developed to date. This can be illustrated by asking a simple question: at what scale do quantum fluctuations become so strong that they can self-collapse under the gravitational pull of the corresponding energy? To answer this question, we begin with the uncertainty principle, which indicates that

$$\Delta x \Delta p \geq \hbar, \qquad \Delta t \Delta E \geq \hbar. \tag{4.229}$$

Next, we return to the Schwarzschild radius, which can be derived to within a numerical factor by considering the escape velocity of a spherical mass distribution and equating it to the speed of light. The radial gravitational force is given by

$$\mathbf{F} = \hat{r} G \frac{Mm}{r^2} = -\nabla U(r) = -\hat{r} \frac{dU}{dr}, \tag{4.230}$$

where $M$ is the mass of the sphere, $m$ is a test mass, and where the gravitational potential is

$$U(r) = G \frac{Mm}{r}. \tag{4.231}$$

The difference in potential energy for the test mass leaving the surface of the sphere, of radius $r$, and escaping at infinity is

$$\Delta U = U(\infty) - U(r) = G \frac{Mm}{r}. \tag{4.232}$$

Equating this to the initial kinetic energy of the test particle yields the escape velocity:

$$\begin{aligned} \Delta U &= G \frac{Mm}{r} = \frac{1}{2} m v^2. \\ v &= \sqrt{\frac{2GM}{r}}. \end{aligned} \tag{4.233}$$

The maximum escape velocity is, by definition, $c$, and we find that gravitational collapse will occur if

$$r = \sqrt{\frac{2GM}{c^2}}, \tag{4.234}$$

which is the Schwarzschild radius. Of course, the fact that this classical Newtonian derivation yields the correct answer is fortuitous; however, in terms for order of magnitude calculations, it is adequate. Let us now confine the quantum vacuum within this radius. Using Equation 4.229 and taking

$\Delta t = r/c$, we find that the corresponding energy is $\Delta E \approx \hbar c/r$. Finally, using the well-known relation between mass and energy, we obtain $M \approx \hbar/cr$, which can be used in Equation 4.234 to obtain

$$r \approx \frac{2GM}{c^2} \approx \frac{2G}{c^2}\frac{\hbar}{cr}, \quad r \simeq \sqrt{\frac{2G\hbar}{c^3}}. \tag{4.235}$$

Dropping the numerical factor, we recover the Planck length, given in Section 3.6:

$$\lambda\!\!\!^{-}_P = \sqrt{\frac{G\hbar}{c^3}}. \tag{4.236}$$

At such scales, the energy density of the vacuum fluctuations is sufficient to radically alter the local metric; gravitational effects become comparable to the other forces at the Planck scale. The main problem stemming from these considerations is that fully unified quantum gravity and superstring theories operate at these extremely high energy densities. As a corollary, they cannot be distinguished from more conventional theories in our present, low energy world. Therefore, an experimental validation of such ideas represents a formidable challenge. The mathematics involved is equally challenging, and only partial, so-called zero-mass problems can currently be tackled. Nevertheless, superstring theories represent the first fully renormalizable theories capable of unifying the electroweak, strong, and gravitational interactions; as such, they are generating large quantities of speculative work in parts of the theoretical physics community.

For a detailed introduction to superstring theories, we encourage the reader to consult Kaku's excellent presentation. In this brief overview, we will follow the main lines of that description and use a similar notation. The main concept underlying the string approach is the elimination of divergences due to the point-like nature of vertices in the scattering description of interactions in conventional quantized field theories. The similarity of the string approach to Kaluza–Klein theories is the use of a higher-dimensional space to accommodate the extra degrees of freedom required by the string structure; finally, the powerful mathematical by-product of the string methodology is a wealth of gauge and conformal symmetries capable of encompassing the strong and electroweak gauge groups, as well as quantum gravity.

We now consider a string moving through space–time: instead of a world line, $x^\mu(\tau)$, as is the case for a point particle, the string trajectory describes a so-called world sheet, $\sigma^a(\tau, \upsilon)$. The string action is defined by

$$S = -\frac{1}{4\pi\rho}\int \sqrt{g}g^{ab}\partial_a y_\mu \partial_b y_\nu \eta^{\mu\nu} d^2\sigma, \tag{4.237}$$

where $g^{ab}$ is a metric for the surface, while $\eta^{\mu\nu}$ is a flat metric in $d$-dimensional space–time, and the vector $y_\mu(\tau,\upsilon)$ connects the origin to a point on the world sheet.

For open strings, $\rho = 1/2$, and for closed strings, $\rho = 1/4$. Finally, the Roman indices $a$ and $b$ represent the two dimensions of the world sheet, parameterized by $\tau$ and $\upsilon$, while $\mu = 0,1,\ldots,d-1$.

As explained by Kaku, the string action postulated in Equation 4.237 is invariant under a number of transformations, including general coordinate reparameterization on the world-sheet; furthermore, the two-dimensional metric can be replaced by

$$g_{ab} \sim \partial_a y_\mu \partial_b y^\mu, \tag{4.238}$$

which can be thought of as the dynamical equation for the metric. With this, the string action reduces to

$$S = \frac{1}{2\pi\rho}\int\sqrt{(\partial_\tau y^\mu)^2(\partial_\upsilon y^\mu)^2 - [(\partial_\tau y_\mu)y^\mu]^2}, \tag{4.239}$$

which is simply proportional to the surface of the world sheet.

At this point, the string action can be quantized, to derive the corresponding spectrum of states for the free theory. One can choose to work in a gauge where $g_{ab} = \delta_{ab}$, the so-called conformal gauge; in this case, the Lagrangian density takes a particularly simple form:

$$\mathcal{L} = \frac{1}{4\pi\rho}[(\partial_\tau y_\mu)^2 + (\partial_\upsilon y_\mu)^2]. \tag{4.240}$$

The two parameters on the world sheet can be described by a single complex variable: if we set $z = \tau + i\upsilon$, the Lagrangian density becomes

$$\mathcal{L} = \frac{1}{2\pi}\partial_z y_\mu \partial_{\bar{z}} y^\mu, \tag{4.241}$$

and the principle of least action yields the equation of motion for the string,

$$(\partial_\tau^2 + \partial_\upsilon^2)y_\mu = 0, \tag{4.242}$$

provided that we have set the following boundary conditions: $\partial_\upsilon y_\mu = 0$. This eliminates a surface term and yields Equation 4.242.

The generalized momenta can now be introduced,

$$p_\mu = \frac{\delta\mathcal{L}}{\delta(\partial_\tau y^\mu)}, \tag{4.243}$$

where the variation $\delta$ corresponds to a conformal transformation of the string:

$$z \to f(z), \qquad \delta y_\mu(z, \bar{z}) = (\varepsilon\partial_z + \bar{\varepsilon}\partial_{\bar{z}})y_\mu. \tag{4.244}$$

Quantization is then performed by requiring that the generalized coordinates and their conjugates, the momenta, satisfy the following commutation rule:

$$[p_\mu(\tau), y_\nu(\tau')] = -i\eta_{\mu\nu}\delta(\tau - \tau'). \tag{4.245}$$

Creation and annihilation operators can be introduced by Fourier analysis:

$$\begin{aligned} y_\mu(\tau) &= y_\mu^0 + i\sum_n \frac{1}{\sqrt{n}}(a_\mu^n - a_\mu^{-n})\cos n\tau, \\ \pi p_\mu(\tau) &= p_\mu^0 + \sum_n \sqrt{n}(a_\mu^n - a_\mu^{-n})\cos n\tau, \end{aligned} \tag{4.246}$$

with the commutation rules

$$[a_\mu^n, a_\nu^m] = \delta^{n,-m}\eta_{\mu\nu}. \tag{4.247}$$

Finally, the Hamiltonian yields the particle spectrum,

$$\mathcal{H} = \int_0^\pi [p_\mu(\partial_\tau y^\mu) - \mathcal{L}]d\tau = \sum_n n a_{-n,\mu} a_n^\mu + \rho p_\mu^2, \tag{4.248}$$

and corresponds to a set of independent harmonic oscillators. Particles are created by operating on the vacuum state, labeled $|0\rangle$. The lowest excited state corresponds to a vector meson with zero mass, $a_{-1}^\mu|0\rangle$, which can be identified with the photon of electromagnetism or electroweak theory, if isospin is introduced. Of considerable interest is the next excitation above the vacuum, a massive spin-2 vector, $a_{-1}^\mu a_{-1}^\nu|0\rangle$. This particle has the same properties as the graviton, which carries the gravitational interaction.

In closing, two further points are noteworthy. First, supersymmetry allows the transition from strings to so-called superstrings; second, the dimensionality of space–time must be different from the usual four-dimensional manifold of general relativity in order for the theory to be self-consistent. The first point implies that for every fermion, there exists a supersymmetric partner, which is a boson, and vice-versa. For example, the sleptons are the supersymmetric partners of the leptons. The second point can be introduced by returning to the string theory outlined above and using the so-called light-cone gauge by introducing the light-cone coordinates,

$$y^\pm = \frac{1}{\sqrt{2}}(y^0 \pm y^{d-1}). \tag{4.249}$$

The basic idea is that the momenta defined in Equation 4.243 are not all independent; by using the light-cone gauge approach, one can remove the unwanted momenta and obtain a set of purely transverse harmonic oscillators, provided that the quantity

$$\frac{n}{12}(26-d)+\frac{1}{n}\left(\frac{d-26}{12}+2-2s\right) \tag{4.250}$$

is identically zero, as required by Lorentz invariance. Here, $s$ is the spin of the zero-mass excitation, and the solution to Equation 4.250 is $s=1$, and $d=26$! This means that the dimensionality of space–time required for a fully self-consistent theory must be 26: from the Kaluza–Klein viewpoint, 22 dimensions are compacted down to the Planck scale. For superstrings, one of the leading candidates requires that $d=10$.

This last discussion brings us to close this section with a final speculation regarding the geometry of space–time. We have seen that, in quantum cosmology, the topology becomes a dynamical variable, on par with the matter–energy fields. Furthermore, in Kaluza–Klein and superstring theories, the number of dimensions of space–time was different from that of our intuitive four-dimensional world. In these paragraphs, we propose that in a fully self-consistent theory of space–time, one should insist that the dimensionality of the manifold must be considered as a local dynamical variable itself, governed by quantum field equations, and that its eigenvalues are not necessarily integers.

Although this represents a fairly radical departure from more conventional approaches, it is, nevertheless, but a natural extension of the concept of a dynamical space–time, first introduced by Einstein with general relativity and greatly expanded by recent advances in supergravity, superstring, and M-theory. In addition, this idea is also closely related to that presented by Hartle and Hawking regarding the boundary condition on the universal wavefunction. Finally, it is hoped that this new angle might shed some light on the dynamics of the Calabi–Yau hyperspace and on the question of what physical principles determine the dimensionality of the dynamical space–time of modern quantum field theories, including gravitation.

Within this framework, we introduce the compacity $\zeta$, of a flat, $\eta$-dimensional generalized hyperspace, which is defined as the ratio of the volume of a hypersphere of radius $R$ to the volume of the corresponding hypercube, $R^\eta$; alternatively, this is simply the volume of a hypersphere of unit radius

$$\zeta(\eta)=\frac{\sqrt{\pi^\eta}}{\Gamma\left(\frac{\eta}{2}+1\right)}, \quad \eta\in\mathbb{R}^+. \tag{4.251}$$

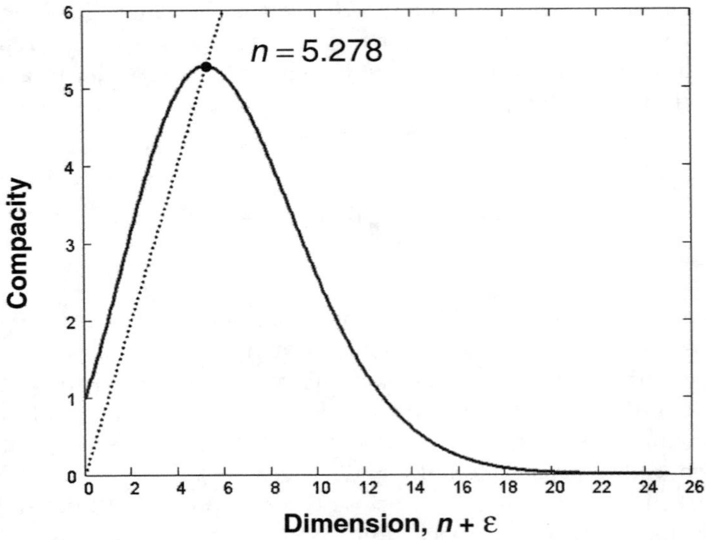


**FIGURE 4.7**
Variation of the volume of a unit hypersphere, as a function of the dimension of space.

Here, it is crucial to note that we allow the dimensionality $x$ to take any positive real value. By analogy with the principle of least action, one of the simplest extremum principles that can be applied to the problem of determining the "natural" dimensionality of hyperspace is to require that its compacity be maximal. Therefore, we seek a solution to the equation $\partial_\eta \zeta = 0$. Before proceeding further, it is important to note that, within the context of a quantum field theory, the dimensional operator will become an action operator: $\hbar \partial_\eta$ has the units of Planck's constant, as the dimension is a pure number.

The solution to the maximum compacity problem is illustrated in Figure 4.7, and it is easily verified that there is a unique value of $\eta$ for which $\zeta(\eta)$ is maximal; in addition, for this value of the dimensionality, the compacity is precisely equal to $x$, as expressed in Equation 4.251 and shown graphically in Figure 4.7. It also shows that for higher dimensions, the compacity decays quasi-exponentially, and that for some intermediate values, the equivalent of $\pi$ can take integer values; transcendence and integrality can be exchanged between dimensionality and geometry.

The value one finds is approximately $\eta^* = \zeta(\eta^*) = 5.278$, close to the original Kaluza–Klein five-dimensional theory and about half of the value required by some modern superstring theories. Of course, one immediate question comes to mind: what is the physical significance of the fractional dimension? As will be shown in the next paragraphs, a possible geometrical interpretation of the generalized hypersphere requires a different definition of distance. In fact, a somewhat unexpected result is that the problem can be reduced as follows: we separate the dimension into its integer part and its irreducible noninteger signature by writing $\eta = n + \varepsilon$, where $n = \mathrm{int}(\eta) \in \mathbb{N}$, and $0 \leq \varepsilon < 1$. In the following, we will refer to a manifold of dimension $\varepsilon$ as a "hypospace." This reduction can be accomplished by comparing the volume of a hypersphere, as derived by using Euler's gamma function, to the result obtained by introducing a trial function of length based on a dimensionality argument. Let us label the position of a point in $\eta$-dimensional flat hyperspace with the coordinates $(x_1, x_2, \ldots, x_n, y)$. It is clear that if each $x_i$ has the dimension of, say, length, $y^{1/\varepsilon}$ also has the dimension of length; therefore, an intuitive trial equation for the unit hypersphere in that space is

$$x_1^2 + x_2^2 + \cdots + x_n^2 + y^{2/\varepsilon} = R^2 = 1. \tag{4.252}$$

For our purpose, we will first study a (2 + $\varepsilon$)-dimensional hyperspace; in that case, our initial guess is $x^2 + y^2 + z^{2/\varepsilon} = R^2$, and the hypervolume is obtained by computing

$$\begin{aligned} V(2+\varepsilon, R) &= \int_{-R^\varepsilon}^{+R^\varepsilon} dz \int_{-\sqrt{R^2 - z^{2/\varepsilon}}}^{+\sqrt{(R^2 - z^{2/\varepsilon})}} dy \int_{-\sqrt{R^2 - y^2 - z^{2/\varepsilon}}}^{+\sqrt{R^2 - y^2 - z^{2/\varepsilon}}} dx \\ &= 2\int_{-R^\varepsilon}^{+R^\varepsilon} dz \int_{-\sqrt{R^2 - z^{2/\varepsilon}}}^{+\sqrt{R^2 - z^{2/\varepsilon}}} \sqrt{R^2 - y^2 - z^{2/\varepsilon}}\, dy. \end{aligned} \tag{4.253}$$

A simple change of variable, where $y' = y/\sqrt{R^2 - z^{2/\varepsilon}}$, allows us to rewrite Equation 4.253 as

$$V(2+\varepsilon, R) = 2\int_{-R^\varepsilon}^{+R^\varepsilon} (R^2 - z^{2/\varepsilon}) dz \int_{-1}^{+1} \sqrt{1 - y'^2}\, dy' = \pi \int_{-R^\varepsilon}^{+R^\varepsilon} (R^2 - z^{2/\varepsilon}) dz; \tag{4.254}$$

normalizing the last variable as $z' = z/R^\varepsilon$, we find

$$V(2+\varepsilon, R) = 2\pi R^{2+\varepsilon} \int_0^1 (1 - z'^{2/\varepsilon})\, dz'. \tag{4.255}$$

The last integral is easily performed to finally obtain

$$V(2+\varepsilon, R) = \frac{4\pi}{2+\varepsilon} R^{2+\varepsilon}, \tag{4.256}$$

which is to be compared to $\sqrt{\pi^{2+\varepsilon}}\,\Gamma^{-1}(2 + \frac{\varepsilon}{2}) R^{2+\varepsilon}$. For $\varepsilon = 1$, which corresponds to the three-dimensional case, the two formulae agree, and we recover $\frac{4}{3}\pi R^3$; however, for $\varepsilon = 0$, the trial equation yields a volume which is twice as large as the correct one: $2\pi R^2$, instead of $\pi R^2$.

Before seeking a correction for this discrepancy, we study the $(1 + \varepsilon)$-dimensional case. The sphere is now defined by the equation $x^2 + y^{2/\varepsilon} = R^2$, and we have

$$V(1+\varepsilon) = \int_{-R^{\varepsilon}}^{+R^{\varepsilon}} dy \int_{-\sqrt{R^2-y^{2/\varepsilon}}}^{+\sqrt{R^2-y^{2/\varepsilon}}} dx = 2\int_{-R^{\varepsilon}}^{+R^{\varepsilon}} \sqrt{R^2 - y^{2/\varepsilon}}\, dy. \tag{4.257}$$

Again, we normalize the variable to the radius to obtain

$$V(1+\varepsilon, R) = 4R^{1+\varepsilon}\int_0^1 \sqrt{1 - y'^{2/2}}\, dy'. \tag{4.258}$$

Although this last integral cannot be performed analytically, it is easily verified that for $\varepsilon = 1$, Equation 4.258 yields the correct two-dimensional volume $\pi R^2$, while in the limit where $\varepsilon \to 0$, $V(1,R) = 4R$, which is, again, twice as large as the correct answer. This last limit is easily obtained by noting that for $0 \le y' < 1$, $\lim_{\varepsilon \to 0}\sqrt{1 - y'^{2/\varepsilon}} = 1$: the function becomes a square between –1 and +1.

The $(3 + \varepsilon)$-dimensional case is also instructive: we start from $x_1^2 + x_2^2 + x_3^2 + x^{2/\varepsilon} = R^2$; the volume is now

$$V(3+\varepsilon, R) = \int_{-R^{\varepsilon}}^{+R^{\varepsilon}} dx \int_{-\sqrt{R^2-x^{2/\varepsilon}}}^{+\sqrt{R^2-x^{2/\varepsilon}}} dx_1 \int_{-\sqrt{R^2-x_1^2-x^{2/\varepsilon}}}^{+\sqrt{R^2-x_1^2-x^{2/\varepsilon}}} dx_2 \int_{-\sqrt{R^2-x_1^2-x_2^2-x^{2/\varepsilon}}}^{+\sqrt{R^2-x_1^2-x_2^2-x^{2/\varepsilon}}} dx_3. \tag{4.259}$$

The integral over $x_3$ is immediately performed; we then introduce the new variable $x_2' = x_2/\sqrt{R^2 - x_1^2 - x^{2/\varepsilon}}$, to obtain

$$\begin{aligned}\int_{-\sqrt{R^2-x_1^2-x^{2/\varepsilon}}}^{+\sqrt{R^2-x_1^2-x^{2/\varepsilon}}} \sqrt{R^2 - x_1^2 - x_2^2 - x^{2/\varepsilon}}\, dx_2 &= (R^2 - x_1^2 - x^{2/\varepsilon})\int_{-1}^{+1}\sqrt{1 - x_2'^2}\, dx_2' \\ &= \frac{\pi}{2}(R^2 - x_1^2 - x^{2/\varepsilon}).\end{aligned} \tag{4.260}$$

Introducing $x_1' = x_1/\sqrt{R^2 - x^{2/\varepsilon}}$, noting that $\int_{-1}^{+1}(1 - x_1'^2)dx_1' = \frac{4}{3}$, and normalizing $x$ to the radius, we finally find that

$$V(3+\varepsilon, R) = \frac{8}{3}\pi R^{3+\varepsilon}\int_0^1 (1 - y^{2/\varepsilon})^{3/2}\, dy. \tag{4.261}$$

An interesting pattern emerges: for $(n + \varepsilon)$-dimensional hyperspace, where $n$ is odd, we are left with an irreducible integral that cannot be evaluated

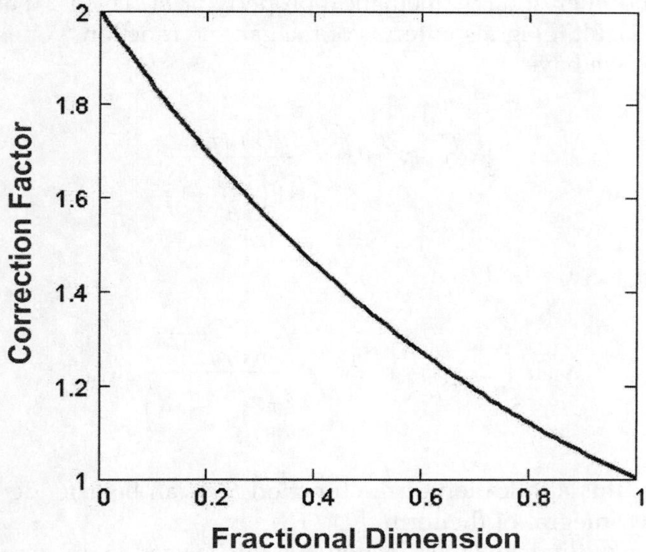


**FIGURE 4.8**
Correction factor for fractional dimension, as described in the text.

analytically, whereas, if $n$ is even, the integral is easily performed. In addition, for the limiting case where $\varepsilon = 1$, the four-dimensional hypervolume is recovered correctly, whereas the factor of 2 previously encountered for $\varepsilon = 0$ is present again. This is illustrated in Figure 4.8, where we plot the ratio $\frac{V(n+\varepsilon,R)}{\sqrt{\pi^{n+\varepsilon}}R^{n+\varepsilon}}\Gamma(\frac{n+\varepsilon}{2}+1)$, as a function of the fractional dimension $\varepsilon$, for $n = 1,2,$ and 3. The discrepancy is exactly the same for all three cases: it has the limiting value of 2 for $\varepsilon \to 0$, while it tends to unity for $\varepsilon \to 1$. This has two important consequences. First and foremost, this reduction of the problem of ($n$ + $\varepsilon$)-dimensional space to the $n$-dimensional case means that we can use the simple case of $\varepsilon$-dimensional hypospace to evaluate the correction factor exactly: $V(\varepsilon, R) = \int_{-R^\varepsilon}^{+R^\varepsilon} dx = 2R^\varepsilon$; therefore, we have

$$\frac{V(n+\varepsilon,R)}{\sqrt{\pi^{n+\varepsilon}}R^{n+\varepsilon}}\Gamma\left(\frac{n+\varepsilon}{2}+1\right) = \frac{2}{\sqrt{\pi^\varepsilon}}\Gamma\left(1+\frac{\varepsilon}{2}\right) = \hbar(\varepsilon). \quad (4.262)$$

This is verified graphically in Figure 4.8, where the correction factor of Equation 4.262 matches exactly the cases where $n = 1, 2,$ and 3. Equation 4.262 is important as it indicates how one must modify the notion of distance in $\varepsilon$-dimensional hypospace, which, in turn, can be extended to any ($n$ + $\varepsilon$)-dimensional hyperspace by virtue of the reduction property demonstrated above.

The second interesting mathematical property of $\hbar(\varepsilon)$ is that it allows us to evaluate difficult integrals in terms of the gamma function. For example, by using $n = 1$, we have

$$\int_0^1 \sqrt{1 - x^{2/\varepsilon}}\,dx = \frac{\hbar(\varepsilon)\sqrt{\pi}^{1+\varepsilon}}{4\Gamma\left(\frac{1+\varepsilon}{2} + 1\right)}, \tag{4.263}$$

while, for $n = 3$, we find that

$$\int_0^1 (1 - x^{2/\varepsilon})^{3/2}\,dx = \frac{\hbar(\varepsilon)\sqrt{\pi}^{3+\varepsilon}}{\frac{8}{3}\pi\Gamma\left(\frac{3+\varepsilon}{2} + 1\right)}. \tag{4.264}$$

In addition, this application of the function $\hbar(\varepsilon)$ can be extended by recurrence to any integral of the form $\int_0^1 (1 - x^{2/\varepsilon})^{2/\varepsilon}\,dx$.

Finally, we outline how the function $\hbar(\varepsilon)$ can be used to redefine the notion of distance in $(n + \varepsilon)$-dimensional hyperspace, to match the gamma function results. We can attempt to regauge the coordinate for the fractional dimension, so that the equation for the hypersphere now reads

$$x_1^2 + x_2^2 + \cdots + x_n^2 + [y\hbar(\varepsilon)]^{2/\varepsilon} = R^2. \tag{4.265}$$

In particular, in the simple case where $n = 0$, we have $y^2\hbar^2(\varepsilon) = R^{2\varepsilon}$, and the volume integral is now

$$\int_{-R^\varepsilon/\hbar(\varepsilon)}^{+R^\varepsilon/\hbar(\varepsilon)} dy = \frac{2R^\varepsilon}{\hbar(\varepsilon)} = \frac{\sqrt{\pi}^\varepsilon}{\Gamma\left(1 + \frac{\varepsilon}{2}\right)} R^\varepsilon, \tag{4.266}$$

which precisely matches the gamma function result.

For $n = 1$, the validity of this approach is also easily verified:

$$\begin{aligned} V(1+\varepsilon, R) &= \int_{-R^\varepsilon/\hbar(\varepsilon)}^{+R^\varepsilon/\hbar(\varepsilon)} dy \int_{-\sqrt{R^2 - [y\hbar(\varepsilon)]^{2/\varepsilon}}}^{+\sqrt{R^2 - [y\hbar(\varepsilon)]^{2/\varepsilon}}} dx \\ &= 2\int_{-R^\varepsilon/\hbar(\varepsilon)}^{+R^\varepsilon/\hbar(\varepsilon)} \sqrt{R^2 - [y\hbar(\varepsilon)]^{2/\varepsilon}}\,dy \\ &= 4\int_0^1 R\sqrt{1 - y'^2}\,\frac{R^\varepsilon}{\hbar(\varepsilon)}\,dy', \end{aligned} \tag{4.267}$$

and we can use Equation 4.263 to obtain the sought-after result,

$$V(1+\varepsilon, R) = \frac{4R^{1+\varepsilon}}{\hbar(\varepsilon)} \frac{\hbar(\varepsilon)\sqrt{\pi}^{1+\varepsilon}}{4\Gamma\left(\frac{1+\varepsilon}{2}+1\right)} = \frac{\sqrt{\pi}^{1+\varepsilon}}{\Gamma\left(\frac{1+\varepsilon}{2}+1\right)}. \tag{4.268}$$

Following the same path, higher-dimensional cases can be verified by recurrence. However, this simple regauging technique is insufficient, because the volume of a hypercube will also be modified,

$$R^{n+\varepsilon} \rightarrow \hbar(\varepsilon) R^{n+\varepsilon}, \tag{4.269}$$

as well as the compacity, which becomes

$$\frac{\sqrt{\pi}^{1+\varepsilon}}{\Gamma\left(\frac{1+\varepsilon}{2}+1\right)} \rightarrow \frac{1}{\hbar(\varepsilon)} \frac{\sqrt{\pi}^{1+\varepsilon}}{\Gamma\left(\frac{1+\varepsilon}{2}+1\right)}. \tag{4.270}$$

Clearly, a more detailed approach must be followed to generalize properly the dimensionality of space–time. Nevertheless, this mathematical exercise shows that spaces with noninteger dimensions could be envisioned by extending fairly simple concepts.

Three remarks are in order to conclude this section. First, a simple physical interpretation of the maximally compact hyperspace can be given by considering the energy density of the vacuum,

$$\frac{d^{n+\varepsilon}W}{dx_1 dx_2 \ldots dx_n dy} = \langle 0|\mathcal{H}|0\rangle, \tag{4.271}$$

for $n + \varepsilon$ spatial dimensions and one time-like coordinate.

Equation 4.271 has the correct units, as $\langle\psi|\psi\rangle$ is a probability density, with units of $L^{-(n+\varepsilon)}$, while the Hamiltonian has the units of energy. The isotropic propagation of a signal at the speed of light is then described by

$$x_1^2 + x_2^2 + \cdots + x_n^2 + y^{2/\varepsilon} - x_0^2 = 0, \tag{4.272}$$

and the volume swept by the signal in a time interval $\Delta x_0$ is

$$\Delta x_0^{n+\varepsilon} \frac{\sqrt{\pi}^{n+\varepsilon}}{\Gamma\left(\frac{n+\varepsilon}{2}+1\right)}. \tag{4.273}$$

The corresponding action is then simply given by

$$S = \int_0^{\Delta x_0} dx_0 \int_{x_1^2+\cdots+x_n^2+y^{2/\varepsilon}\le \Delta x_0^2} \frac{d^{n+\varepsilon}W}{dx_1 dx_2 \ldots dx_n dy} dx_1 \ldots dx_n dy$$

$$= \Delta x_0^{n+1+\varepsilon} \frac{\sqrt{\pi^{n+\varepsilon}}}{\Gamma\left(\frac{n+\varepsilon}{2}+1\right)} \langle 0|\mathcal{H}|0\rangle . \tag{4.274}$$

In order to analyze properly the dimensional behavior of the action, we choose a unit of time and length, and set $\Delta x_0 = 1$; in this gauge, the action is extremal for $\frac{\partial S}{\partial \eta} = 0$, which corresponds to $\eta = n + \varepsilon + 1 = \eta^* + 1$.

The second remark concerns the fact that when $\varepsilon$ goes from 0 to 1, the discrepancy between Equations 4.252 and 4.265 goes from 2 to 1. This suggests a possible analogy with the discussion on spin given in Section 3.11.

Finally, our third remark is that an important element of this approach is the evolution of the notion of distance between zero and one dimension. In zero dimension, for a point space, we are dealing with pure numbers, and there is no connection with physics. In one dimension, we recover the concept of scale and coupling connecting mathematics and physics. The reason one should focus on this problem is that, once it is resolved, the generalization to higher dimensions is quite straightforward.

## 4.12 The Bohm–Aharanov Effect

To conclude this chapter, we return to the question of the indetermination of the potentials. Gauge invariance is based on the notion that one can transform the potentials without modifying the resulting electromagnetic field. The potentials cannot be observed, while the fields are measurable. The Bohm–Aharanov effect brings this concept under sharp examination, as a modification of the canonical momentum of electrons by a static magnetic field results in a measurable phase-shift of the electron wavefunctions, as illustrated in Figure 4.9. This is a quantum effect, as one relies on the interference of the electron wavefunctions produced by the diffraction of the electron wavepacket through a narrow slit.

The differential phase for a wavefunction is

$$d\varphi = k_\mu dx^\mu = \hbar^{-1} p_\mu dx^\mu = \mathbf{k} \cdot d\mathbf{x} - \omega dt; \tag{4.275}$$

for a given trajectory, the invariant phase variation will be of the form

$$e^{i\varphi} = \exp\left(i\int d\varphi\right) = \exp\left(i\hbar^{-1}\int p_\mu dx^\mu\right) = \exp\left[i\int(\mathbf{k}\cdot d\mathbf{x} - \omega dt)\right], \tag{4.276}$$

where the integral is taken along the trajectory.

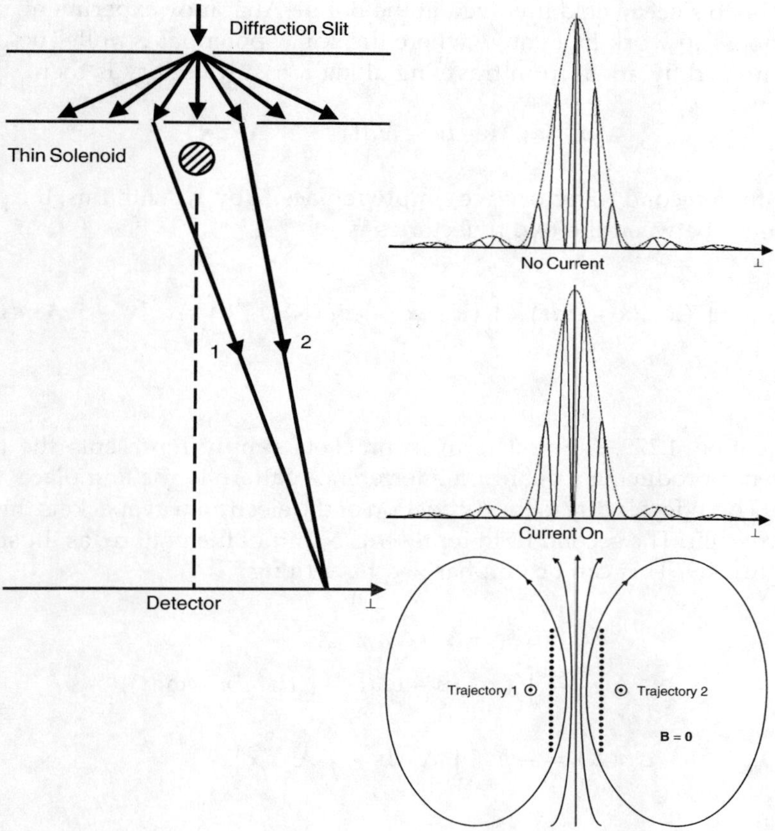


**FIGURE 4.9**
The Bohm–Aharanov effect: electron waves are diffracted and propagated close to a long, thin solenoid. The interference fringes are displaced when the solenoid carries current, although the electrons propagate in a field-free region, as shown in the lower right corner.

In the case of a charge interacting with electromagnetic fields, we must use the canonical momentum to obtain a gauge invariant equation:

$$\begin{aligned} e^{i\varphi} &= \exp\left(i\hbar^{-1}\int \pi_\mu dx^\mu\right) \\ &= \exp\left[i\hbar^{-1}\int (p_\mu - eA_\mu)dx^\mu\right] \\ &= \exp\left\{i\hbar^{-1}\int [(\mathbf{p} - e\mathbf{A})\cdot d\mathbf{x} - (E - e\phi)dt]\right\}. \end{aligned} \tag{4.277}$$

There is no electric field involved in the Bohm–Aharanov experiment, so we can choose to work in a gauge where the scalar potential is null. The phase accumulated by an electron traveling along a first trajectory is then

$$\varphi_1 = \int_1 (\mathbf{k}\cdot d\mathbf{x} - \omega dt) - e\hbar^{-1}\mathbf{A}\cdot d\mathbf{x}, \tag{4.278}$$

while for a second trajectory, we simply replace $\int_1$ by $\int_2$: with this, the phase difference between the two trajectories is

$$\Delta\varphi = \left[\int_1 (\mathbf{k}\cdot d\mathbf{x} - \omega dt) - \int_2 (\mathbf{k}\cdot d\mathbf{x} - \omega dt)\right] - e\hbar^{-1}\left[\int_1 \mathbf{A}\cdot d\mathbf{x} - \int_2 \mathbf{A}\cdot d\mathbf{x}\right]. \tag{4.279}$$

In Equation 4.279, the first term in brackets simply represents the phase difference producing a quantum interference pattern in the first place, when $\mathbf{A} = \mathbf{0}$. This effect is due to the diffraction of the electron wavepackets through a narrow slit. The second term represents a shift of the pattern, as illustrated in Figure 4.9. This can be emphasized by writing

$$\begin{aligned} \Delta\varphi &= \Delta\varphi_{12} + \Delta\varphi(\mathbf{A}), \\ \Delta\varphi_{12} &= \int_1 (\mathbf{k}\cdot d\mathbf{x} - \omega dt) - \int_2 (\mathbf{k}\cdot d\mathbf{x} - \omega dt), \\ \Delta\varphi(\mathbf{A}) &= -e\hbar^{-1}\left(\int_1 \mathbf{A}\cdot d\mathbf{x} - \int_2 \mathbf{A}\cdot d\mathbf{x}\right). \end{aligned} \tag{4.280}$$

In order to avoid any question regarding the possibility of a different $\mathbf{v}\times\mathbf{B}$ Lorentz force acting on the electrons along different trajectories, a very long, thin solenoid is used, where the field is confined inside the coil; outside the field is null. In addition, to obtain a good fringe contrast, the energy and momentum distribution of the electrons must be very narrow, so that the four-wavenumber spectrum of the associated de Broglie waves is nearly monochromatic. In the absence of current in the solenoid, one records an interference pattern, which corresponds to the diffraction of the electron wavefunction through the slit. When the solenoid carries a current, one observes a global shift of the electron interference pattern, which is attributed to the term $-e\hbar^{-1}(\int_1 \mathbf{A}\cdot d\mathbf{x} - \int_2 \mathbf{A}\cdot d\mathbf{x})$, and is equal to

$$-\frac{e}{\hbar}\oint \mathbf{A}\cdot d\mathbf{x}. \tag{4.281}$$

This is simply the flux of the magnetic field inside the solenoid, measured in quantum units.

At first glance, the fact that one can measure a quantity directly related to the potential seems to be in conflict with gauge invariance. Let us, however, continue the derivation by considering the effect of a gauge transform:

$$\mathbf{B} = \nabla \times \mathbf{A}, \quad \mathbf{A} \rightarrow \mathbf{A} + \nabla f. \tag{4.282}$$

But what happens to the phase shift? We have

$$\Delta\varphi(\mathbf{A}) \rightarrow -\frac{e}{\hbar}\left[\int_1 (\mathbf{A} + \nabla f) \cdot d\mathbf{x} - \int_2 (\mathbf{A} + \nabla f) \cdot d\mathbf{x}\right], \tag{4.283}$$

or

$$\Delta\varphi(\mathbf{A}) \rightarrow \Delta\varphi(\mathbf{A}) - \frac{e}{\hbar}\left(\int_1 \nabla f \cdot d\mathbf{x} - \int_2 \nabla f \cdot d\mathbf{x}\right). \tag{4.284}$$

The quantity

$$\nabla f \cdot d\mathbf{x} = \frac{\partial f}{\partial x} dx + \frac{\partial f}{\partial y} dy + \frac{\partial f}{\partial z} dz = df \tag{4.285}$$

is a total differential, and the integrals in Equation 4.279 can be performed to yield

$$\Delta\varphi(\mathbf{A}) \rightarrow \Delta\varphi(\mathbf{A}) - \frac{e}{\hbar}([f]_1 - [f]_2). \tag{4.286}$$

At this point, it is important to note that, as the electrons are detected as particles, a meaningful phase measurement implies that the trajectories under comparison must have the same end points: the electrons following trajectories 1 and 2 originate and are detected at the same points. In turn, this means that

$$[f]_1 = [f]_2, \quad \Delta\varphi(\mathbf{A}) \rightarrow \Delta\varphi(\mathbf{A}): \tag{4.287}$$

the Bohm–Aharanov experiment is gauge invariant and demonstrates the reality of the canonical momentum.

## 4.13 References for Chapter 4

Note: the numbers listed below refer to the main bibliography and reference sections at the end of this book.

8, 9, 11, 27, 28, 29, 40, 43, 59, 69, 71, 73, 74, 75, 92, 95, 96, 98, 99, 102, 105, 109, 111, 112, 120, 125, 126, 149, 150, 156, 158, 172, 175, 176, 181, 209, 210, 213, 217, 221, 227, 234, 235, 237, 238, 239, 240, 241, 242, 297, 309, 310, 419, 420, 432, 450, 469, 471, 472, 473, 474, 475, 477, 513, 540, 542, 599, 612, 613, 614, 615, 616, 617, 618, 664, 665, 666, 695, 702, 708, 732, 734, 737, 738, 759, 795, 799, 851, 852, 853, 854, 855, 859, 860, 861, 864, 865, 866, 867, 868, 869, 870, 871, 872, 873, 874, 875, 878, 879, 880, 881, 884, 885, 896, 897, 898.

# Part II

# Electromagnetic Waves

# 5

# Green and Delta-Functions, Eigenmode Theory of Waveguides

## 5.1 Introduction

As we have seen before, the general driven wave equation for the electromagnetic four-potential is

$$\Box A^{\mu} + \mu_0 j^{\mu} = \partial^{\mu}(\partial_{\nu} A^{\nu}). \tag{5.1}$$

One of the most fundamental properties of the wave equation is the fact that it is linear: if we know the solutions for two different four-current sources, $j_1^{\mu}$ and $j_2^{\mu}$,

$$\begin{aligned} \Box A_1^{\mu} + \mu_0 j_1^{\mu} &= \partial^{\mu}(\partial_{\nu} A_1^{\nu}), \\ \Box A_2^{\mu} + \mu_0 j_2^{\mu} &= \partial^{\mu}(\partial_{\nu} A_2^{\nu}), \end{aligned} \tag{5.2}$$

then, the solution for a linear superposition of sources is also known;

$$\Box(\lambda_1 A_1^{\mu} + \lambda_2 A_2^{\mu}) + \mu_0(\lambda_1 j_1^{\mu} + \lambda_2 j_2^{\mu}) = \partial^{\mu}[\partial_{\nu}(\lambda_1 A_1^{\mu} + \lambda_2 A_2^{\mu})]. \tag{5.3}$$

Now, we can express any four-current as a linear superposition of delta-functions:

$$j^{\mu}(x_{\lambda}) = \iiiint j^{\mu}(\bar{x}_{\lambda}) \delta_4(x_{\lambda} - \bar{x}_{\lambda})\, d^4\bar{x}. \tag{5.4}$$

Therefore, the solution to the wave equation driven by a delta-function is of particular importance, as we can write, in the Lorentz gauge,

$$\Box G(x_{\lambda} - \bar{x}_{\lambda}) + \mu_0 \delta_4(x_{\lambda} - \bar{x}_{\lambda}) = 0, \tag{5.5}$$

and the general solution to the wave equation, with an arbitrary four-current source is

$$A^{\mu}(x_{\lambda}) = \iiiint j^{\mu}(\bar{x}_{\lambda})G(x_{\lambda} - \bar{x}_{\lambda})\, d^4\bar{x}. \tag{5.6}$$

The solution $G(x_{\lambda} - \bar{x}_{\lambda})$ is known as the Green function or propagator for the linear wave equation under consideration.

In order to determine the Green function for the d'Alembertian operator, a few basic properties of the delta-function will be used, as well as momentum space representation. Therefore, these are first reviewed in this chapter.

## 5.2 The Dirac Delta-Function

In mathematical terms, the Dirac delta-function is a distribution, defined by its operation on a given function:

$$\int_{-\infty}^{+\infty} \delta(x - \bar{x}) f(x)\, dx = f(\bar{x}). \tag{5.7}$$

It essentially acts as a localization operator, as it reduces a continuous function to a single value at a given point.

We will first discuss the delta-function in relation to orthonormal functions and Hilbert spaces. We consider a set of functions that are both orthogonal and normalized. Typically, the set is discretely or continuously infinite, and we label the functions as $\psi_n(x)$. The normalization condition is

$$\int_{-\infty}^{+\infty} \psi_n(x)\psi_n^*(x)\, dx = \int_{-\infty}^{+\infty} |\psi_n(x)|^2 dx = 1, \tag{5.8}$$

while the orthogonality is described by

$$\int_{-\infty}^{+\infty} \psi_n(x)\psi_m^*(x) dx = 0, \quad m \neq n. \tag{5.9}$$

Equations 5.8 and 5.9 can be grouped by introducing the Kronecker symbol:

$$\int_{-\infty}^{+\infty} \psi_m^*(x)\psi_n(x)\, dx = \delta_{mn}. \tag{5.10}$$

In Hilbert space, an arbitrary function can be represented by an infinite series of orthonormal functions, as

$$f(x) = \sum_n f_n \psi_n(x), \tag{5.11}$$

where the coefficients can be determined by using the orthonormalization condition given in Equation 5.10. We have

$$\begin{aligned}\int_{-\infty}^{+\infty} \psi_m^*(x) f(x)\, dx &= \int_{-\infty}^{+\infty} \psi_m^*(x) \sum_n f_n \psi_n(x)\, dx \\ &= \sum_n f_n \int_{-\infty}^{+\infty} \psi_m^*(x)\, \psi_n(x)\, dx \\ &= \sum_n f_n\, \delta_{mn} \\ &= f_m. \end{aligned} \tag{5.12}$$

The procedure outlined above is called a projection, where we diagonalize the series by taking advantage of the orthonormality of the basis functions, $\psi_n$.

Having determined the coefficients of the series, the arbitrary function can be formally expressed as

$$\begin{aligned} f(x) &= \sum_n f_n \psi_n(x) \\ &= \sum_n \left[ \int_{-\infty}^{+\infty} \psi_n^*(\bar{x}) f(\bar{x}) d\bar{x} \right] \psi_n(x) \\ &= \sum_n \int_{-\infty}^{+\infty} d\bar{x}\ f(\bar{x})\, \psi_n^*(\bar{x})\, \psi_n(x); \end{aligned} \tag{5.13}$$

however, as the sum $\sum_n \psi_n^*(\bar{x})\, \psi_n(x)$ may diverge for $x = \bar{x}$, one typically introduces auxiliary factors, so that

$$\sum_n \alpha_n \psi_n^*(\bar{x})\, \psi_n(x) \tag{5.14}$$

remains finite. We can the write

$$f(x) = \lim_{\alpha_n \to 1} \int_{-\infty}^{+\infty} d\bar{x}\ f(\bar{x}) \sum_n \alpha_n \psi_n^*(\bar{x})\, \psi_n(x). \tag{5.15}$$

With this, we can define the Dirac delta-function,

$$\delta(x-\bar{x}) = \sum_n \psi_n^*(\bar{x})\,\psi_n(x), \tag{5.16}$$

which has the required property that

$$f(x) = \int_{-\infty}^{+\infty} f(\bar{x})\,\delta(x-\bar{x})\,d\bar{x}. \tag{5.17}$$

In the particular case where the orthonormal functions are the complex exponentials corresponding to Fourier series and transforms, we can write

$$\psi_n(x) = \frac{1}{\sqrt{a}}\exp(ik_n x), \qquad k_n = n\frac{2\pi}{a}, \tag{5.18}$$

where $a$ represents a spatial unit cell.

This representation proves very convenient and is used, for example, to model the quantized modes of the free electromagnetic field, as reviewed in detail in Chapter 6.

The orthogonality and normalization of such eigenfunctions are quite straightforward: we have

$$\begin{aligned}\int \psi_m^*(x)\,\psi_n(x)\,dx &= \int_{-a/2}^{+a/2} \frac{1}{\sqrt{a}}e^{-ik_m x}\,\frac{1}{\sqrt{a}}e^{ik_n x}\,dx \\ &= \frac{1}{a}\int_{-a/2}^{+a/2} \exp\left[i\frac{2\pi}{a}(n-m)x\right]dx \\ &= \frac{1}{a}\left[\frac{a}{2\pi i(n-m)}\exp\left\{i\frac{2\pi}{a}(n-m)x\right\}\right]_{-a/2}^{+a/2} \\ &= \frac{\sin\left[\pi(n-m)\right]}{\pi(n-m)}.\end{aligned} \tag{5.19}$$

For $n = m$, we use the limit of the *sinc* function:

$$\lim_{x\to 0}[\mathrm{sinc}\,(x)] = \lim_{x\to 0}\left(\frac{\sin x}{x}\right) = 1, \tag{5.20}$$

whereas, for $n \neq m$, we can write $n - m = p \in \mathbb{N}$, and it is clear that

$$\sin\left[\pi(n-m)\right] = \sin(p\pi) = 0. \tag{5.21}$$

Therefore, we have shown the orthonormality of the complex exponentials of Fourier series:

$$\int \psi_m^*(x)\,\psi_n(x)\,dx \;=\; \mathrm{sinc}\,[\pi(n-m)] \;=\; \delta_{nm}. \tag{5.22}$$

Within this context, we can use the formalism described in Equation 5.16 to obtain a representation of the delta-function in terms of complex exponentials:

$$\begin{aligned}\delta(x-\bar{x}) &= \sum_n \psi_n^*(\bar{x})\,\psi_n(x)\\ &= \sum_n \frac{1}{\sqrt{a}}\exp(-ik_n\bar{x})\,\frac{1}{\sqrt{a}}\exp(ik_n x)\\ &= \frac{1}{a}\sum_n \exp\left[i\frac{2\pi n}{a}(x-\bar{x})\right].\end{aligned} \tag{5.23}$$

By taking the limit where the period, $a$, goes to infinity, we can obtain the Fourier transform of the delta-function. We first recast Equation 5.23 as

$$\delta(x-\bar{x}) \;=\; \frac{1}{2\pi}\sum_n \exp\left[i\frac{2\pi n}{a}(x-\bar{x})\right]\frac{2\pi}{a}, \tag{5.24}$$

and introduce the wavenumber $k = 2\pi n/a$. In the continuous limit, the corresponding differential element is $dk/dn = 2\pi/a$, and we can write

$$\begin{aligned}\lim_{a\to\infty}\left\{\frac{1}{2\pi}\sum_n \exp\left[i\frac{2\pi n}{a}(x-\bar{x})\right]\frac{2\pi}{a}\right\} &= \lim_{a\to\infty}\left\{\frac{1}{2\pi}\sum_n \exp[ik(x-\bar{x})]\,dk\right\}\\ &= \frac{1}{2\pi}\int_{-\infty}^{+\infty}\exp[ik(x-\bar{x})]\,dx.\end{aligned} \tag{5.25}$$

Of course, one can also use the property defining the Dirac delta-function to obtain the same result. We define the symmetrical Fourier transform of a given function, $f(x)$, and its inverse, as

$$\begin{aligned}\tilde{f}(k) &= \frac{1}{\sqrt{2\pi}}\int_{-\infty}^{+\infty} f(x)e^{-ikx}\,dx,\\ f(x) &= \frac{1}{\sqrt{2\pi}}\int_{-\infty}^{+\infty} \tilde{f}(k)e^{ikx}\,dk.\end{aligned} \tag{5.26}$$

In the special case where $f(x) = \delta(x-\bar{x})$, we have

$$\tilde{f}(k) \;=\; \frac{1}{\sqrt{2\pi}}\int_{-\infty}^{+\infty} \delta(x-\bar{x})\,e^{-ikx}\,dx, \tag{5.27}$$

which is equal, by definition, to

$$\tilde{f}(k) = \frac{1}{\sqrt{2\pi}} \int_{-\infty}^{+\infty} \delta(x-\bar{x})\, e^{-ikx}\, dx = \frac{e^{-ik\bar{x}}}{\sqrt{2\pi}}. \tag{5.28}$$

Inserting that result into the inverse Fourier transform, we find that

$$\delta(x-\bar{x}) = \frac{1}{\sqrt{2\pi}} \int_{-\infty}^{+\infty} \frac{e^{-ik\bar{x}}}{\sqrt{2\pi}} e^{ikx}\, dk = \frac{1}{2\pi} \int_{-\infty}^{+\infty} \exp[ik(x-\bar{x})]\, dk, \tag{5.29}$$

which is identical to the limit derived in Equation 5.25.

At this point, it is important to emphasize that the Dirac delta-function is really a distribution. For example, if we try to evaluate the integral in Equation 5.29 for $(x = \bar{x})$, we see that it diverges. To eliminate this problem, we can use the same technique as for the sum in Equations 5.14 and 5.15 and introduce a "stretching" function; alternatively, we can consider a limit and introduce the function

$$\delta(x-\bar{x}, \alpha) = \frac{1}{2\pi} \int_{-1/\alpha}^{+1/\alpha} \exp[ik(x-\bar{x})]\, dk, \tag{5.30}$$

which is defined such that

$$\lim_{\alpha \to 0} [\delta(x-\bar{x}, \alpha)] = \delta(x-\bar{x}). \tag{5.31}$$

The integral in Equation 5.30 can easily be performed to yield

$$\begin{aligned} \delta(x-\bar{x}, \alpha) &= \frac{1}{2\pi} \left[ \frac{e^{ik(x-\bar{x})}}{i(x-\bar{x})} \right]_{-1/\alpha}^{+1/\alpha} \\ &= \frac{1}{\pi} \left[ \frac{e^{i(x-\bar{x})/\alpha} - e^{-i(x-\bar{x})/\alpha}}{2i(x-\bar{x})} \right] \\ &= \frac{1}{\pi\alpha} \operatorname{sinc}\left( \frac{x-\bar{x}}{\alpha} \right). \end{aligned} \tag{5.32}$$

This function is illustrated in Figure 5.1; for $x = \bar{x}$, we find that $\delta(0, \alpha) = 1/\pi\alpha$. Furthermore, the integral of $\delta(x-\bar{x}, \alpha)$ is normalized:

$$\begin{aligned} \int_{-\infty}^{+\infty} \delta(x-\bar{x}, \alpha) dx &= \frac{1}{\pi\alpha} \int_{-\infty}^{+\infty} \operatorname{sinc}\left( \frac{x-\bar{x}}{\alpha} \right) dx \\ &= \frac{1}{\pi} \int_{-\infty}^{+\infty} \frac{\sin z}{z} dz \\ &= 1. \end{aligned} \tag{5.33}$$

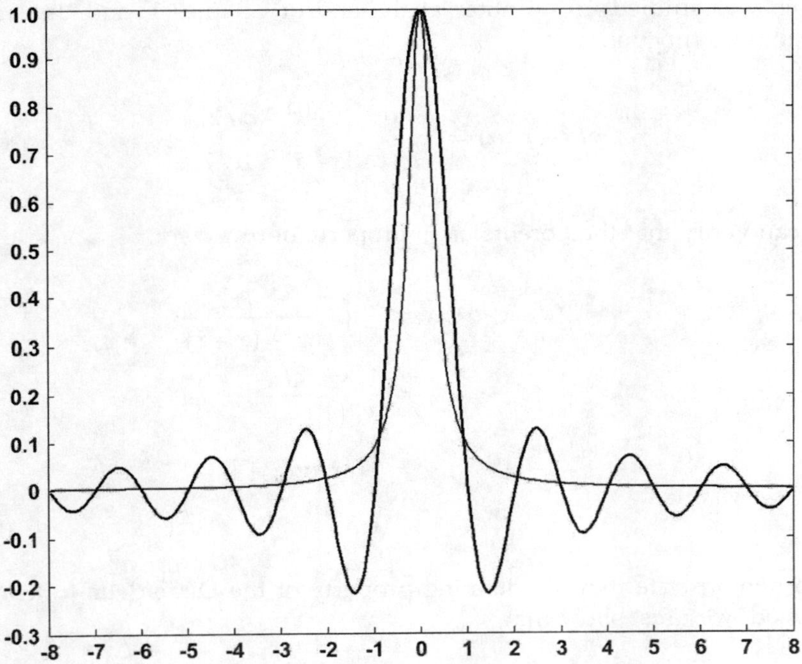


**FIGURE 5.1**
Normalized sinc and Lorentzian functions, both representations of the Dirac delta-function.

With this, we have

$$\delta(x-\bar{x}) = \lim_{\alpha\to 0}\left[\frac{1}{\pi\alpha}\,\text{sinc}\left(\frac{x-\bar{x}}{\alpha}\right)\right]. \tag{5.34}$$

In fact, this notion can be extended to any function defined such that

$$\begin{aligned} &\int_{-\infty}^{+\infty} g(x-\bar{x},\,\alpha)\,dx = 1, \\ &\lim_{\alpha\to 0}\,[g(x-\bar{x},\,\alpha)] = 0, \quad x \neq \bar{x}. \end{aligned} \tag{5.35}$$

Three important examples of such functions are the Lorentzian

$$L(x-\bar{x},\,\alpha) = \frac{\alpha}{\pi}\,\frac{1}{\alpha^2+(x-\bar{x})^2}, \tag{5.36}$$

the Gaussian

$$G(x-\bar{x},\,\alpha) = \frac{1}{\alpha\sqrt{\pi}}\,\exp\left[-\left(\frac{x-\bar{x}}{\alpha}\right)^2\right], \tag{5.37}$$

which was studied in the notes at the end of Chapter 1, and the simple rectangular function,

$$g(x-\bar{x},\alpha) = \begin{cases} 0, & |x-\bar{x}| > \alpha/2, \\ \alpha^{-1}, & |x-\bar{x}| \le \alpha/2. \end{cases} \tag{5.38}$$

We can verify that the Lorentzian is properly normalized:

$$\begin{aligned} \int_{-\infty}^{+\infty} L(x-\bar{x},\alpha)\,dx &= \frac{\alpha}{\pi}\int_{-\infty}^{+\infty}\frac{dx}{\alpha^2+(x-\bar{x})^2} \\ &= \frac{1}{\pi}\int_{-\infty}^{+\infty}\frac{dz}{1+z^2} \\ &= \frac{1}{\pi}[\operatorname{arctg}(z)]_{-\infty}^{+\infty} \\ &= 1. \end{aligned} \tag{5.39}$$

To demonstrate that the defining property of the Dirac delta-function is satisfied, we must show that

$$\lim_{\alpha\to 0}\left[\int_{-\infty}^{+\infty} g(x-\bar{x},\alpha)f(x)\,dx\right] = f(\bar{x}). \tag{5.40}$$

In general, the approach is similar to that demonstrated in the notes at the end of Chapter 1. The test function, $f(x)$, is Taylor expanded around $x = \bar{x}$;

$$f(x) = f(\bar{x}) + \sum_{n=1}^{\infty}\frac{(x-\bar{x})^n}{n!}\frac{d^n f(\bar{x})}{dx^n}; \tag{5.41}$$

the leading term is the constant $f(\bar{x})$, which can be extracted from the integral,

$$\begin{aligned} \int_{-\infty}^{+\infty} g(x-\bar{x},\alpha)f(x)\,dx &= f(\bar{x})\int_{-\infty}^{+\infty} g(x-\bar{x},\alpha)\,dx \\ &\quad + \int_{-\infty}^{+\infty} g(x-\bar{x},\alpha)\sum_{n=1}^{\infty}\frac{(x-\bar{x})^n}{n!}\frac{d^n f(\bar{x})}{dx^n}\,dx. \end{aligned} \tag{5.42}$$

Since, by definition, $g$ is normalized, as shown in Equation 5.35, we have

$$\int_{-\infty}^{+\infty} g(x-\bar{x},\alpha)f(x)\,dx f(\bar{x}) + \int_{-\infty}^{+\infty} g(x-\bar{x},\alpha)\sum_{n=1}^{\infty}\frac{(x-\bar{x})^n}{n!}\frac{d^n f(\bar{x})}{dx^n}\,dx. \tag{5.43}$$

We then need to show that

$$\lim_{\alpha\to 0}\left[\int_{-\infty}^{+\infty} g(x-\bar{x},\,\alpha)\sum_{n=1}^{\infty}\frac{(x-\bar{x})^n}{n!}\frac{d^n f(\bar{x})}{dx^n}\,dx\right] = 0. \tag{5.44}$$

As shown in the notes at the end of Chapter 1, one can change the integration variable by introducing $z = x-\bar{x}$ and integrate by parts to reduce the polynomial order. We first have

$$\int_{-\infty}^{+\infty} g(x-\bar{x},\,\alpha)\sum_{n=1}^{\infty}\frac{(x-\bar{x})^n}{n!}\frac{d^n f(\bar{x})}{dx^n}\,dx = \sum_{n=1}^{\infty}\frac{1}{n!}\frac{d^n f(\bar{x})}{dx^n}\int_{-\infty}^{+\infty} g(z,\,\alpha)z^n\,dz. \tag{5.45}$$

Now considering the simple case of a rectangular function, as expressed in Equation 5.38, we can write a particular term of the series as

$$\int_{-\infty}^{+\infty} g(z,\,\alpha)z^n dz = \frac{1}{\alpha}\int_{-\alpha/2}^{+\alpha/2} z^n dz = \frac{1}{\alpha}\left[\frac{z^{n+1}}{n+1}\right]_{-\alpha/2}^{+\alpha/2}$$

$$= \frac{\alpha^n}{(n+1)2^{n+1}}[1-(-1)^{n+1}]. \tag{5.46}$$

It is then clear that

$$\lim_{\alpha\to 0}\left\{\sum_{n=1}^{\infty}\frac{1}{n!}\frac{d^n f(\bar{x})}{dx^n}\frac{\alpha^n}{(n+1)2^{n+1}}[1-(-1)^{n+1}]\right\} = 0. \tag{5.47}$$

We can also consider a larger class of normalized functions, defined such that

$$\int_{-\infty}^{+\infty} g(x)dx = 1, \tag{5.48}$$

in which case we also have

$$\int_{-\infty}^{+\infty} g(x,\,\alpha)dx = \int_{-\infty}^{+\infty}\frac{1}{\alpha}g\left(\frac{x}{\alpha}\right)dx = 1. \tag{5.49}$$

In addition, we require that these functions are nonzero over a finite interval:

$$g(x)\neq 0,\quad a<x<b. \tag{5.50}$$

With this, we can consider a particular term in the series,

$$\int_{-\infty}^{+\infty} g(z,\alpha)z^n dz = \int_{-\infty}^{+\infty} g\left(\frac{z}{\alpha}\right)z^n dz = \alpha^{n+1}\int_{-\infty}^{+\infty} g(y)y^n dy = \alpha^{n+1}\int_a^b g(y)y^n (dy), \tag{5.51}$$

where we have used the change of variable $y = z/\alpha$.

It is then clear that

$$\lim_{\alpha\to 0}\left[\int_{-\infty}^{+\infty} g(z,\alpha)z^n dz\right] = \lim_{\alpha\to 0}\left[\alpha^{n+1}\int_a^b g(y)y^n dy\right] = 0. \tag{5.52}$$

At this point, we will consider a few fundamental properties of the Dirac delta-function, starting with its integral and derivative. For this, we use a Lorentzian representation of the delta-function:

$$\delta(x-\bar{x}) = \lim_{\alpha\to 0}\left[\frac{1}{\pi}\frac{\alpha}{\alpha^2+(x-\bar{x})^2}\right], \tag{5.53}$$

and we see that

$$\begin{aligned}\int_{-\infty}^{x}\delta(x'-\bar{x})dx' &= \lim_{\alpha\to 0}\left[\int_{-\infty}^{x}\frac{1}{\pi}\frac{\alpha dx'}{\alpha^2+(x'-\bar{x})^2}\right]\\ &= \lim_{\alpha\to 0}\left[\frac{1}{\pi}\int_{-\infty}^{(x-\bar{x})/\alpha}\frac{dz}{1+z^2}\right]\\ &= \lim_{\alpha\to 0}\left\{\left[\frac{1}{\pi}\operatorname{arctg}(z)\right]_{-\infty}^{(x-\bar{x})/\alpha}\right\}\\ &= \lim_{\alpha\to 0}\left[\frac{1}{\pi}\operatorname{arctg}\left(\frac{x-\bar{x}}{\alpha}\right)+\frac{1}{2}\right].\end{aligned} \tag{5.54}$$

Here, we have used the change of variable

$$z = \frac{x'-\bar{x}}{\alpha},\quad dz = \frac{dx'}{\alpha},\quad \lim_{x'\to\infty}(z) = -\infty,\quad z(x'=x) = \frac{x-\bar{x}}{\alpha}. \tag{5.55}$$

This function, shown in Figure 5.2, takes the following values,

$$\int_{-\infty}^{x}\delta(x'-x)\,dx' = H(x-\bar{x}) = \begin{cases}0, & x<\bar{x},\\ \frac{1}{2}, & x=\bar{x},\\ 1, & x>\bar{x},\end{cases} \tag{5.56}$$

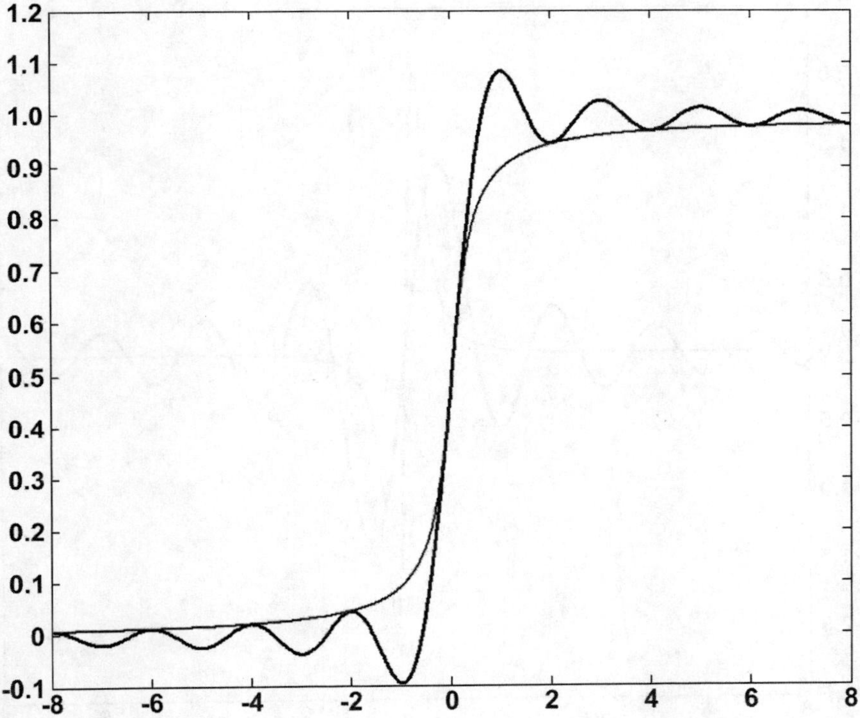


**FIGURE 5.2**
Integrals of a sinc and a Lorentzian.

and is called the Heaviside stepfunction, $H$. Clearly, in the method used here, one must require that if one uses a different representation of the Dirac delta-function, the limit produces the same result. For the derivative, we can proceed in the same manner:

$$\begin{aligned}\partial_x \delta(x-\bar{x}) &= \lim_{\alpha\to 0}\left\{\partial_x\left[\frac{1}{\pi}\frac{\alpha}{\alpha^2+(x-\bar{x})^2}\right]\right\}\\ &= \lim_{\alpha\to 0}\left\{\frac{1}{\pi}\frac{2\alpha(x-\bar{x})}{[\alpha^2+(x-\bar{x})^2]^2}\right\}. \end{aligned} \tag{5.57}$$

This distribution is illustrated in Figure 5.3; its action upon a test function is determined by evaluating the following limit:

$$\int_{-\infty}^{+\infty}\partial_x\delta(x-\bar{x})f(x)\,dx = \lim_{\alpha\to 0}\left\{\int_{-\infty}^{+\infty}\frac{1}{\pi}\frac{2\alpha(x-\bar{x})}{[\alpha^2+(x-\bar{x})^2]^2}f(x)\,dx\right\}. \tag{5.58}$$

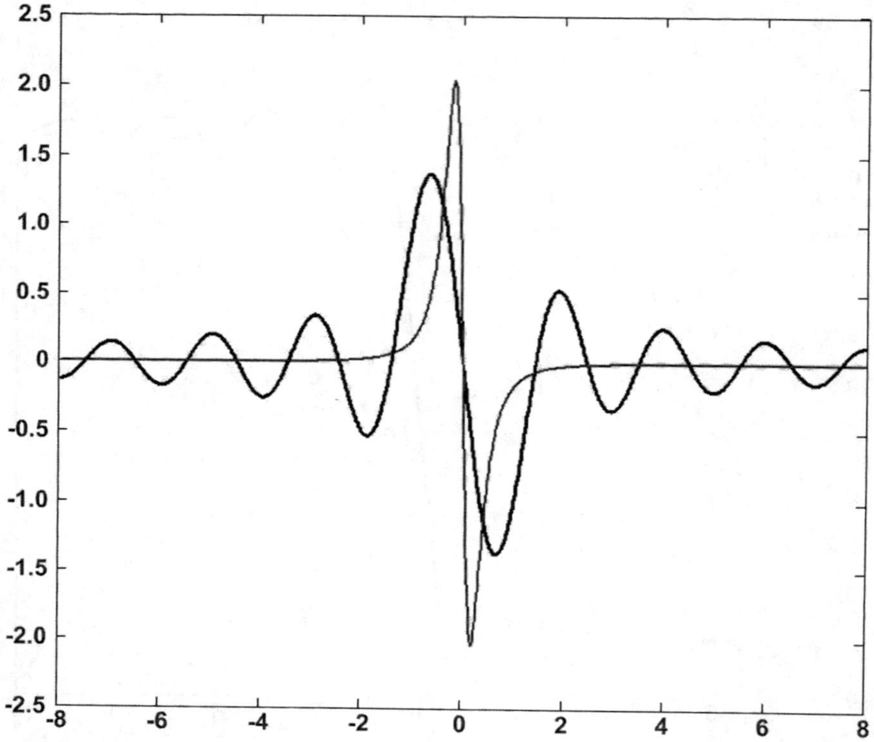


**FIGURE 5.3**
Derivatives of a sinc and a Lorentzian.

To this end, we use integration by parts, which follows from the fact that

$$\begin{aligned} \frac{d}{dx}[u(x)v(x)] &= u'(x)v(x) + u(x)v'(x), \\ \int u'(x)v(x)\,dx &= [u(x)v(x)] - \int u(x)v'(x)\,dx. \end{aligned} \tag{5.59}$$

We then have

$$\begin{aligned} \int_{-\infty}^{+\infty} \frac{1}{\pi} \frac{2\alpha(x-\bar{x})}{[\alpha^2+(x-\bar{x})^2]^2} f(x)dx &= \int_{-\infty}^{+\infty} \partial_x \left[\frac{1}{\pi} \frac{\alpha}{\alpha^2+(x-\bar{x})^2}\right] f(x)dx \\ &= \left[\frac{1}{\pi} \frac{\alpha f(x)}{\alpha^2+(x-\bar{x})^2}\right]_{-\infty}^{+\infty} \\ &\quad - \int_{-\infty}^{+\infty} \frac{1}{\pi} \frac{\alpha}{\alpha^2+(x-\bar{x})^2} \partial_x f(x)(dx). \end{aligned} \tag{5.60}$$

As the Lorentzian function tends to the delta-function in the limit where the scale $\alpha \to 0$, we find that

$$\begin{aligned}\int_{-\infty}^{+\infty} \partial_x \delta(x-\bar{x}) f(x)\,dx &= \lim_{\alpha\to 0}\left[\frac{1}{\pi}\frac{\alpha f(x)}{\alpha^2+(x-\bar{x})^2}\right]_{-\infty}^{+\infty} \\ &\quad -\lim_{\alpha\to 0}\left[\int_{-\infty}^{+\infty}\frac{1}{\pi}\frac{\alpha}{\alpha^2+(x-\bar{x})^2} f'(x)(dx)\right] \\ &= [\delta(x-\bar{x}) f(x)]_{-\infty}^{+\infty} - \int_{-\infty}^{+\infty}\delta(x-\bar{x}) f'(x)\,dx \\ &= f'(\bar{x}). \end{aligned} \tag{5.61}$$

In other words, we can directly apply integration by parts to the delta-function:

$$\begin{aligned}\int_{-\infty}^{+\infty} \delta'(x-\bar{x}) f(x) dx &= [\delta(x-\bar{x}) f(x)]_{-\infty}^{+\infty} - \int_{-\infty}^{+\infty}\delta(x-\bar{x}) f'(x)(dx) \\ &= -f'(\bar{x}). \end{aligned} \tag{5.62}$$

The delta-function is symmetrical, while its derivative is antisymmetrical:

$$\delta(-x) = \delta(x), \qquad \delta'(-x) = -\delta'(x). \tag{5.63}$$

Other important properties of the delta-function can be derived from its action in integral form. We first have

$$\delta(ax) = \frac{1}{|a|}\delta(x), \tag{5.64}$$

which is easily shown by changing the integration variable,

$$\begin{aligned}\int_{-\infty}^{+\infty}\delta(ax) f(x)\,dx &= \frac{1}{|a|}\int_{-\infty}^{+\infty}\delta(y) f(y) dy = \frac{f(0)}{|a|} \\ &= \int_{-\infty}^{+\infty}\frac{\delta(x)}{|a|} f(x)\,dx. \end{aligned} \tag{5.65}$$

In the more complex case where the argument of the delta-function is a function of $x$, we find that

$$\delta[g(x)] = \sum_n \frac{\delta(x-a_n)}{|g'(a_n)|}, \tag{5.66}$$

where the $a_n$ are the roots of $g(x)$: $g(a_n) = 0$.

Equation 5.66 holds provided that, for each root, $g(a_n) \neq 0$. To demonstrate Equation 5.66, we use the integral operation defining the delta-function:

$$\begin{aligned}\int_{-\infty}^{+\infty} \delta[g(x)]f(x)dx &= \sum_n \int_{a_n-\varepsilon}^{a_n+\varepsilon} \delta[g(x)]f(x)dx \\ &= \sum_n \int_{a_n-\varepsilon}^{a_n+\varepsilon} \delta[(x-a_n)g'(a_n)]f(x)dx \\ &= \sum_n \frac{f(a_n)}{|g'(a_n)|} \\ &= \int_{-\infty}^{+\infty} \sum_n \frac{\delta(x-a_n)}{|g'(a_n)|} f(x)dx. \end{aligned} \tag{5.67}$$

The intervals $[a_n - \varepsilon, a_n + \varepsilon]$ surround the roots of $g(x)$, and in the limit where $\varepsilon \to 0$, the Taylor expansion,

$$g(x) \simeq (x-a_n)\frac{dg}{dx}(x = a_n) = (x-a_n)g'(a_n), \tag{5.68}$$

is exact, provided that $g'(a_n) \neq 0$.

Using Equation 5.66, we can also show that

$$\delta(x^2 - a^2) = \frac{\delta(x-a) + \delta(x+a)}{2|a|}. \tag{5.69}$$

This equation is useful to demonstrate the important identity,

$$\delta(x^2) = \frac{\delta(x)}{|x|}. \tag{5.70}$$

For a certain class of suitably defined functions, we wish to demonstrate that

$$\int \delta(x^2) f(x)dx = \int \frac{\delta(x)}{|x|} f(x)dx. \tag{5.71}$$

Starting from Equation 5.69 and defining $g(x) = f(x)/|x|$, we first have

$$\begin{aligned}\int \delta(x^2 - a^2) f(x)dx &= \int \delta(x^2 - a^2)|x| g(x)\, dx \\ &= \frac{|a|[g(a) + g(-a)]}{2|a|} \\ &= \frac{g(a) + g(-a)}{2}. \end{aligned} \tag{5.72}$$

Applying this result to a function $f(x)$, such that

$$\lim_{x\to 0}\left[\frac{f(x)}{|x|}\right] = g(0), \tag{5.73}$$

exists, we can now write

$$\begin{aligned}\lim_{a\to 0}\left[\int \delta(x^2 - a^2) f(x)\, dx\right] &= \lim_{a\to 0}\left[\frac{g(a) + g(-a)}{2}\right] \\ &= g(0) \\ &= \int \delta(x^2) f(x) dx,\end{aligned} \tag{5.74}$$

which is identical to

$$\int \frac{\delta(x)}{|x|} f(x) dx = \int \delta(x) g(x) dx = g(0). \tag{5.75}$$

In this sense, the identity, Equation 5.70, is properly defined.

In the special case where the derivative is zero, one can Taylor-expand $g(x)$ to higher order:

$$g(x) \simeq \frac{(x - a_n)^n}{n!} \frac{d^n g}{dx^n}(a_n), \tag{5.76}$$

where $n$ is the minimum order for which $d^n g/dx^n(a_n) \neq 0$.

In the neighborhood of this particular root, we will need to evaluate

$$\int_{a_n-\varepsilon}^{a_n+\varepsilon} \delta\left[\frac{(x - a_n)^n}{n!} \frac{d^n g}{dx^n}(a_n)\right] f(x) dx. \tag{5.77}$$

For example, for a second-order pole, we have

$$\int_{a_n-\varepsilon}^{a_n+\varepsilon} \delta\left[\frac{(x - a_n)^2}{2} g''(a_n)\right] f(x) dx, \tag{5.78}$$

and we can use Equation 5.70, which states that $\delta(z^2) = \delta(z)/|z|$: we change variable and write

$$z = (x - a_n)\sqrt{\frac{|g''(a_n)|}{2}}, \tag{5.79}$$

so that the integral in Equation 5.78 now reads

$$\int_{-\varepsilon\sqrt{2/|g''|}}^{+\varepsilon\sqrt{2/|g''|}} \delta(z^2) f\left(\sqrt{\frac{2}{|g''(a_n)|}}\, z + a_n\right)\sqrt{\frac{2}{|g''(a_n)|}}\, dz, \tag{5.80}$$

which is equal to

$$\int_{-\varepsilon\sqrt{2/|g''|}}^{+\varepsilon\sqrt{2/|g''|}} \frac{\delta(z)}{|z|} f\left(\sqrt{\frac{2}{|g''(a_n)|}}\, z + a_n\right)\sqrt{\frac{2}{|g''(a_n)|}}\, dz. \tag{5.81}$$

Switching back to $x$ as the variable, we find that

$$\int_{a_n-\varepsilon}^{a_n+\varepsilon} \delta\left[\frac{(x-a_n)^2}{2} g''(a_n)\right] f(x)dx = \int_{a_n-\varepsilon}^{a_n+\varepsilon} \frac{\delta[(x-a_n)\sqrt{|g''(a_n)|/2}]}{(x-a_n)\sqrt{|g''(a_n)|/2}} f(x)dx. \tag{5.82}$$

Using the identity $\delta(ax) = \delta(x)/|a|$ finally yields the sought-after result:

$$\delta\left[\frac{(x-a_n)^2}{2} g''(a_n)\right] \equiv \frac{2\delta(x-a_n)}{|(x-a_n)g''(a_n)|}. \tag{5.83}$$

We also note that in the most general case of $n$-dimensional space, we have

$$\begin{aligned} \delta(\mathbf{x}-\bar{\mathbf{x}}) &= \delta(x_1-\bar{x}_1)\delta(x_2-\bar{x}_2)\ldots\delta(x_n-\bar{x}_n) \\ &= \frac{1}{|J(x_i, x_j')|}\delta(\bar{x}_1'-\bar{x}_1')\delta(\bar{x}_2'-\bar{x}_2')\ldots\delta(\bar{x}_n'-\bar{x}_n'), \end{aligned} \tag{5.84}$$

where $x_i$ and $x_j'$ are two different coordinate systems, and where the Jacobian of the transformation relating the coordinate systems is given by the $n \times n$ matrix

$$J(x_i, x_j') = \left(\frac{\partial x_i}{\partial x_j'}\right). \tag{5.85}$$

The reason behind this is that the relevant quantity is the delta-function, multiplied by the differential volume, $\delta(\mathbf{x}-\bar{\mathbf{x}})\, dx_1 dx_2 \ldots dx_n$, because the properties defining the generalized delta-function are expressed in terms of volume integrals:

$$\begin{aligned} \iiint\cdots\int \delta(\mathbf{x}-\bar{\mathbf{x}})dx_1dx_2\ldots dx_n &= 1, \\ \iiint\cdots\int \delta(\mathbf{x}-\bar{\mathbf{x}})f(\mathbf{x})dx_1dx_2\ldots dx_n &= f(\bar{\mathbf{x}}). \end{aligned} \tag{5.86}$$

Therefore, we request that

$$\delta(x_1-\bar{x}_1)\ldots\delta(x_n-\bar{x}_n)dx_1\ldots dx_n = \delta(x_1'-\bar{x}_1')\ldots\delta(x_n'-\bar{x}_n')dx_1'\ldots dx_n'; \quad (5.87)$$

since the differential volume elements are related by

$$dx_1\ldots dx_n = \left|\frac{\partial x_i}{\partial x_j'}\right| dx_1'\ldots dx_n' = |J(x_i, x_j')|\, dx_1'\ldots dx_n', \quad (5.88)$$

Equation 5.84 is correct. In particular, in cylindrical coordinates, we have

$$\delta(\mathbf{x}-\bar{\mathbf{x}}) = \frac{1}{r}\delta(r-\bar{r})\delta(\theta-\bar{\theta})\delta(z-\bar{z}), \quad (5.89)$$

while in spherical coordinates, we find

$$\delta(\mathbf{x}-\bar{\mathbf{x}}) = \frac{1}{r^2\sin\theta}\delta(r-\bar{r})\delta(\theta-\bar{\theta})\delta(\varphi-\bar{\varphi}). \quad (5.90)$$

Finally, an important expression is that of the delta-function in terms of Bessel functions:

$$\delta(x-\bar{x}) = \frac{x}{2}\int_{-\infty}^{+\infty} \mathrm{J}_n(\alpha x)\mathrm{J}_n(\alpha\bar{x})\alpha\, d\alpha. \quad (5.91)$$

## 5.3 Fourier, Laplace, and Hankel Transforms

This section is intended as a brief review of some of the mathematical properties of Fourier transforms, which are used throughout this book; in particular, the momentum space representation of light and matter waves proves extremely useful in classical and quantum electrodynamics.

To complement our mathematical overview of the Dirac delta-function, a few remarks concerning Fourier and Hankel transforms are in order. We begin with the more general concept of integral transforms, where a function $f(x)$ is associated with its transform,

$$\bar{f}(\alpha) = \int_0^{\infty} f(x)k(\alpha, x)dx. \quad (5.92)$$

The function of two variables $k(\alpha, x)$ is called the kernel of the integral transform; for example, the kernel

$$k(\alpha, x) = e^{-\alpha x}, \tag{5.93}$$

leads to the Laplace transform. The variable $\alpha$ is often called the conjugate of $x$.

The linearity of integral transforms can easily be verified:

$$\int_0^\infty f(x)k(\alpha, x)\,dx + \int_0^\infty g(x)k(\alpha, x)\,dx = \int_0^\infty [f(x) + g(x)]k(\alpha, x)\,dx; \tag{5.94}$$

furthermore,

$$\int_0^\infty cf(x)k(\alpha, x)\,dx = c\int_0^\infty f(x)k(\alpha, x)\,dx. \tag{5.95}$$

A second important kernel leads to the Mellin transform:

$$k(\alpha, x)x^{\alpha-1}, \qquad \hat{f}(\alpha) = \int_0^\infty f(x)x^{\alpha-1}\,dx. \tag{5.96}$$

This particular transform plays an essential role in defining Fourier kernels: for a given kernel, $k(\alpha, x) = k(\alpha x)$, to be a Fourier kernel, it is necessary that

$$\hat{k}(s)\hat{k}(s-1) = 1; \tag{5.97}$$

this condition is related to the existence of an inversion theorem.

Within this context, the cosine Fourier transform is defined as follows. We consider the kernel $k(x) = \kappa \cos x$; the corresponding Mellin transform is

$$\hat{k}(s) = \kappa\int_0^\infty x^{s-1} \cos x\,dx = \frac{\kappa}{2}\left(\int_0^\infty x^{s-1}e^{ix}dx + \int_0^\infty x^{s-1}e^{-ix}\,dx\right); \tag{5.98}$$

moreover, we have a relation with Euler's gamma function,

$$\int_0^\infty e^{-px}x^{s-1}dx = \frac{\Gamma(s)}{p^s}, \quad p > 0. \tag{5.99}$$

Therefore, we can formally recast Equation 5.98 as

$$\int_0^\infty e^{\pm ix}x^{s-1}dx = e^{\pm is\pi/2}\Gamma(s), \tag{5.100}$$

which yields

$$\hat{k}(s) = \kappa\cos\left(\frac{s\pi}{2}\right)\Gamma(s). \tag{5.101}$$

Furthermore, we also have

$$\hat{k}(s-1) = \kappa\cos\left[\frac{(s-1)\pi}{2}\right]\Gamma(s-1) = \kappa\sin\left(\frac{s\pi}{2}\right)\Gamma(s-1), \tag{5.102}$$

and we can use the inversion condition stated in Equation 5.97 to determine the constant:

$$\hat{k}(s)\hat{k}(s-1) = 1 = \kappa^2\sin\left(\frac{s\pi}{2}\right)\cos\left(\frac{s\pi}{2}\right)\Gamma(s)\Gamma(s-1). \tag{5.103}$$

Euler's gamma function has well-defined recurrence properties; in particular,

$$\Gamma(s)\Gamma(s-1) = \frac{\pi}{\sin(s\pi)}. \tag{5.104}$$

Furthermore,

$$\sin(2x) = 2\sin(x)\cos(x), \tag{5.105}$$

and Equation 5.103 reduces to

$$\kappa^2\sin\left(\frac{s\pi}{2}\right)\cos\left(\frac{s\pi}{2}\right)\frac{s\pi}{\sin(s\pi)} = \frac{\pi}{2}\kappa^2 = 1, \quad \kappa = \sqrt{\frac{2}{\pi}}. \tag{5.106}$$

This leads to the cosine Fourier transform and its inverse,

$$\begin{aligned} \tilde{f}_c(\alpha) &= \sqrt{\frac{2}{\pi}}\int_0^\infty f(x)\cos(\alpha x)\,dx, \\ f(x) &= \sqrt{\frac{2}{\pi}}\int_0^\infty \tilde{f}_c(\alpha)\cos(\alpha x)\,d\alpha. \end{aligned} \tag{5.107}$$

The same approach results in the definition of the sine Fourier transform,

$$\begin{aligned} \tilde{f}_s(\alpha) &= \sqrt{\frac{2}{\pi}}\int_0^\infty f(x)\sin(\alpha x)\,dx, \\ f(x) &= \sqrt{\frac{2}{\pi}}\int_0^\infty \tilde{f}_s(\alpha)\sin(\alpha x)\,d\alpha, \end{aligned} \tag{5.108}$$

as well as the complex symmetrical Fourier transform,

$$\tilde{f}(\alpha) = \frac{1}{\sqrt{2\pi}} \int_{-\infty}^{+\infty} f(x) e^{\pm i\alpha x}\, dx,$$
$$f(x) = \frac{1}{\sqrt{2\pi}} \int_{-\infty}^{+\infty} \tilde{f}(\alpha) e^{\pm i\alpha x}\, d\alpha. \tag{5.109}$$

In addition, the kernel

$$k(x) = \sqrt{x}\, \mathrm{J}_\nu(x), \tag{5.110}$$

yields the Mellin transform

$$\hat{k}(s) = \frac{2^s}{\sqrt{2}} \frac{\Gamma\left(\frac{\nu}{2} + \frac{s}{2} + \frac{1}{4}\right)}{\Gamma\left(\frac{\nu}{2} - \frac{s}{2} + \frac{1}{4}\right)}, \tag{5.111}$$

which satisfies the inversion condition, and yields the Hankel transform of order $\nu$:

$$\bar{f}_\nu(\alpha) = \int_0^\infty x f(x) \mathrm{J}_\nu(\alpha x)\, dx,$$
$$f(x) = \int_0^\infty \alpha \bar{f}_\nu(\alpha) \mathrm{J}_\nu(\alpha x)\, d\alpha. \tag{5.112}$$

To complement this brief overview, we mention Fourier's integral theorem:

$$f(x) = \frac{2}{\pi} \int_0^\infty \cos(\alpha x)\, d\alpha \int_0^\infty f(\rho) \cos(\alpha\rho)\, d\rho, \tag{5.113}$$

and its equivalent for Bessel functions, Hankel's integral theorem,

$$f(r) = \int_0^\infty \alpha \mathrm{J}_\nu(\alpha r)\, d\alpha \int_0^\infty \rho f(\rho) \mathrm{J}_\nu(\alpha\rho)\, d\rho. \tag{5.114}$$

Equation 5.114 is also known as a Fourier–Bessel integral and can be used to evaluate Weber's integral by choosing a Gaussian form for $f$:

$$\int_0^\infty e^{-\alpha^2 r^2} \mathrm{J}_\nu(\beta r) r^{\nu+1}\, dr = \frac{\beta^\nu}{(2\alpha^2)^{\nu+1}} \exp\left[-\left(\frac{\beta}{2\alpha}\right)^2\right], \quad \alpha, \beta > 0, \quad \Re(\nu) > -1. \tag{5.115}$$

Another important integral that can be evaluated in the same fashion is

$$\int_0^\infty e^{-\alpha r} \mathrm{J}_0(\beta r)\, dr = \frac{1}{\sqrt{\alpha^2 + \beta^2}}, \quad \alpha, \beta > 0. \tag{5.116}$$

This last integral proves extremely useful when one is concerned with the expression of the field produced by a point charge in rectilinear motion in vacuum, as is discussed in Section 5.7.

The relation between the Fourier transform of a function and its derivatives also proves useful. Under fairly general circumstances, if $\tilde{f}(\omega)$ is the Fourier transform of $f(t)$, the Fourier transform of $d^n f(t)/dt^n$ is $(-i\omega^n)\tilde{f}(\omega)$. For Hankel transforms, an analogous theorem holds. We first recall the definition of the $\nu$-th order Hankel transform of $f(r)$,

$$\bar{f}_\nu(\alpha) = \int_0^\infty r f(r) \mathrm{J}_\nu(\alpha r)\, dr, \tag{5.117}$$

and we introduce the Hankel transform of its derivative,

$$\bar{f}'_\nu(\alpha) = \int_0^\infty r \frac{df}{dr} \mathrm{J}_\nu(\alpha r)\, dr. \tag{5.118}$$

The relation between the two is

$$\bar{f}'_\nu(\alpha) = (\nu - 1)\int_0^\infty f(r) \mathrm{J}_\nu(\alpha r)\, dr - \alpha \bar{f}'_{\nu-1}(\alpha). \tag{5.119}$$

To conclude this section, we present Parseval's theorem for Fourier and Hankel transforms. If we consider two functions, $f$ and $g$, we have the following relation:

$$\begin{aligned}\int_{-\infty}^{+\infty} f(t) g(t)\, dt &= \int_{-\infty}^{+\infty} f(t)\, dt \frac{1}{\sqrt{2\pi}} \int_{-\infty}^{+\infty} \tilde{g}(\omega) e^{i\omega t}\, d\omega \\ &= \int_{-\infty}^{+\infty} \tilde{g}(\omega)\, d\omega \frac{1}{\sqrt{2\pi}} \int_{-\infty}^{+\infty} f(t) e^{i\omega t}\, dt \\ &= \int_{-\infty}^{+\infty} f(-\omega) g(\omega)\, d\omega.\end{aligned} \tag{5.120}$$

In particular, when $f = g$, we find that

$$\begin{aligned}\int_{-\infty}^{+\infty} f^2(t)\, dt &= \int_{-\infty}^{+\infty} \tilde{f}(-\omega) \tilde{f}(-\omega)\, d\omega \\ &= \int_{-\infty}^{+\infty} |f(\omega)|^2\, d\omega.\end{aligned} \tag{5.121}$$

Similarly, for Hankel transforms, we have

$$\begin{aligned}\int_0^\infty \alpha \bar{f}_\nu(\alpha) \bar{g}_\nu(\alpha)\, d\alpha &= \int_0^\infty \alpha \bar{f}_\nu(\alpha)\, d\alpha \int_0^\infty r g(r) \mathrm{J}_\nu(\alpha r)\, dr \\ &= \int_0^\infty r g(r)\, dr \int_0^\infty \alpha \bar{f}_\nu(\alpha) \mathrm{J}_\nu(\alpha r)\, d\alpha \\ &= \int_0^\infty r f(r) g(r)\, dr.\end{aligned} \tag{5.122}$$

## 5.4 Green Functions in Vacuum

At this point, the solution of the wave equation driven by a unit, point excitation in four-dimensional space–time can be derived. This is to be distinguished from the Liénard–Wiechert potentials, which are produced by a point charge in motion. The Green function corresponds to a four-dimensional delta-excitation, while the point charge is described by a three-dimensional delta-function, with an argument explicitly dependent on time or proper time, in general. Mathematically, the Green function is the solution of the wave equation

$$\Box G(x_\lambda - \bar{x}_\lambda) + \mu_0 \delta_4(x_\lambda - \bar{x}_\lambda) = 0, \tag{5.123}$$

while for the point charge, we have

$$\Box A_\mu - \mu_0 e \delta_3[\mathbf{x} - \mathbf{z}(\tau)] \frac{u_\mu(\tau)}{\gamma(\tau)} = 0. \tag{5.124}$$

In Equation 5.124, the position of the charge is described by the world line $z_\mu(\tau) = [z_0, z(\tau)]$, while its four-velocity is $u_\mu(\tau) = dz_\mu/d\tau$, and the relation between the proper time and the laboratory time is $dz_0/d\tau = u_0 = \gamma$. The Liénard–Wiechert potentials will be derived in Section 5.5; here, we focus on the Green function problem.

### 5.4.1 Green Function for Poisson's Equation, Coulomb Potential

In order to simplify our presentation, we start with Poisson's equation and the corresponding three-dimensional Green function; note that in this particular case the Green function coincides with the Coulomb potential produced by a point charge.

In the Coulomb gauge, Poisson's equation reads

$$\Delta\phi = -\frac{\rho}{\varepsilon_0}; \tag{5.125}$$

moreover, in Cartesian coordinates, the Laplacian operator is given by

$$\Delta = \partial_x^2 + \partial_y^2 + \partial_z^2, \tag{5.126}$$

and Poisson's equation takes the form

$$[\partial_x^2 + \partial_y^2 + \partial_z^2]\phi(x, y, z) = -\frac{1}{\varepsilon_0}\rho(x, y, z). \tag{5.127}$$

This equation is more easily resolved by Fourier transforming the potential and the source density into momentum space:

$$\phi(x, y, z) = \frac{1}{(\sqrt{2\pi})^3}\iiint_{\mathbb{R}^3}\tilde{\phi}(k_x, k_y, k_z)\exp[i(k_x x + k_y y + k_z z)]dk_x dk_y dk_z,$$
$$\rho(x, y, z) = \frac{1}{(\sqrt{2\pi})^3}\iiint_{\mathbb{R}^3}\tilde{\rho}(k_x, k_y, k_z)\exp[i(k_x x + k_y y + k_z z)]dk_x dk_y dk_z. \tag{5.128}$$

In this representation, the action of the Laplacian operator is very simple:

$$[\partial_x^2 + \partial_y^2 + \partial_z^2]e^{i(k_x x + k_y y + k_z z)} = -(k_x^2 + k_y^2 + k_z^2)e^{i(k_x x + k_y y + k_z z)}. \tag{5.129}$$

This can be summarized by writing

$$\Delta e^{i\mathbf{k}\cdot\mathbf{x}} = -\mathbf{k}^2 e^{i\mathbf{k}\cdot\mathbf{x}}; \tag{5.130}$$

in other words, the complex exponential function, $e^{i\mathbf{k}\cdot\mathbf{x}}$, is an eigenmode of the Laplacian operator, $\Delta$, with the eigenvalue $-\mathbf{k}^2$; this basic property explains why it is advantageous to work in momentum space.

We can proceed with the problem at hand: the Laplacian and integral operators commute because the Laplacian operates in real space on $\mathbf{x} = (x, y, z)$, while the integral is performed in Fourier space, on the conjugate variable $\mathbf{k} = (k_x, k_y, k_z)$; therefore, we have

$$\begin{aligned}\Delta\left[\iiint_{\mathbb{R}^3}\tilde{\phi}(\mathbf{k})e^{i\mathbf{k}\cdot\mathbf{x}}\,d^3k\right] &= \iiint_{\mathbb{R}^3}\tilde{\phi}(\mathbf{k})\Delta[e^{i\mathbf{k}\cdot\mathbf{x}}]\,d^3k \\ &= \iiint_{\mathbb{R}^3}\tilde{\phi}(\mathbf{k})(-\mathbf{k}^2 e^{i\mathbf{k}\cdot\mathbf{x}})\,d^3k \\ &= -\frac{1}{\varepsilon_0}\iiint_{\mathbb{R}^3}\tilde{\rho}(\mathbf{k})e^{i\mathbf{k}\cdot\mathbf{x}}\,d^3k,\end{aligned} \tag{5.131}$$

where we have introduced the notation $d^3k = dk_x dk_y dk_z$.

Moreover, because the complex exponential eigenmodes are orthogonal, Equation 5.131 can readily be diagonalized by using the equation defining the Dirac delta-function in Fourier space:

$$\int_{-\infty}^{+\infty} e^{i(k-\bar{k})x}\,dx = 2\pi\delta(k - \bar{k}). \tag{5.132}$$

In the present case, we can simply multiply each side of Equation 5.131 by $e^{-i\bar{\mathbf{k}}\cdot\mathbf{x}}$ and perform the triple integral over $x$, $y$, and $z$, to obtain

$$\iiint_{\mathbb{R}^3}\iiint_{\mathbb{R}^3}\tilde{\phi}(\mathbf{k})(-\mathbf{k}^2e^{i\mathbf{k}\cdot\mathbf{x}})e^{-i\bar{\mathbf{k}}\cdot\mathbf{x}}\,d^3k\,d^3x = -\frac{1}{\varepsilon_0}\iiint_{\mathbb{R}^3}\iiint_{\mathbb{R}^3}\tilde{\rho}(\mathbf{k})e^{i\mathbf{k}\cdot\mathbf{x}}e^{-i\bar{\mathbf{k}}\cdot\mathbf{x}}\,d^3k\,d^3x,$$
$$\iiint_{\mathbb{R}^3}\iiint_{\mathbb{R}^3}\tilde{\phi}(\mathbf{k})(-\mathbf{k}^2)e^{i(\mathbf{k}-\bar{\mathbf{k}})\cdot\mathbf{x}}\,d^3k\,d^3x = -\frac{1}{\varepsilon_0}\iiint_{\mathbb{R}^3}\iiint_{\mathbb{R}^3}\tilde{\rho}(\mathbf{k})e^{i(\mathbf{k}-\bar{\mathbf{k}})\cdot\mathbf{x}}\,d^3k\,d^3x,$$
$$\iiint_{\mathbb{R}^3}\delta(\mathbf{k}-\bar{\mathbf{k}})\tilde{\phi}(\mathbf{k})(-\mathbf{k}^2)\,d^3k = -\frac{1}{\varepsilon_0}\iiint_{\mathbb{R}^3}\delta(\mathbf{k}-\bar{\mathbf{k}})\tilde{\rho}(\mathbf{k})\,d^3k, \tag{5.133}$$

where we have divided each side of the last equation by $8\pi^3$.

Integrating both sides of the last equality over the wavenumber and using the defining property of the Dirac delta-function,

$$\int_{-\infty}^{+\infty} f(k)\delta(k-\bar{k})\,dk = f(\bar{k}), \tag{5.134}$$

finally yields the relation between the Fourier transform of the potential and source density:

$$\tilde{\phi}(\mathbf{k}) = \frac{\tilde{\rho}(\mathbf{k})}{\varepsilon_0\mathbf{k}^2}, \tag{5.135}$$

where we have exchanged the dummy variables $\mathbf{k}$ and $\bar{\mathbf{k}}$.

At this point, we note that these mathematical steps are often skipped by stating that, because of the orthogonality of the eigenmodes, the equality between the integral implies the equality for each spectral component:

$$\iiint_{\mathbb{R}^3}-\mathbf{k}^2\tilde{\phi}(\mathbf{k})e^{i\mathbf{k}\cdot\mathbf{x}}\,dx\,dy\,dz = -\iiint_{\mathbb{R}^3}\frac{\tilde{\rho}(\mathbf{k})}{\varepsilon_0}e^{i\mathbf{k}\cdot\mathbf{x}}\,dx\,dy\,dz$$
$$\Rightarrow -\mathbf{k}^2\tilde{\phi}(\mathbf{k}) = -\frac{\tilde{\rho}(\mathbf{k})}{\varepsilon_0}. \tag{5.136}$$

This is also true when one considers series of orthogonal modes and discrete spectra.

Returning to the Green function problem, we must replace the source spectral density, $\tilde{\rho}(\mathbf{k})$, by that corresponding to a delta-function excitation: we start from Poisson's equation,

$$\Delta G = -\frac{1}{\varepsilon_0}\delta(\mathbf{x}-\bar{\mathbf{x}}), \tag{5.137}$$

and we use the Fourier representation of the delta-function to write

$$\Delta G = -\frac{1}{8\pi^3\varepsilon_0}\iiint e^{i\mathbf{k}\cdot(\mathbf{x}-\bar{\mathbf{x}})}\,d^3k. \tag{5.138}$$

In view of the preceding discussion, in Fourier space, this yields

$$-\mathbf{k}^2\frac{\tilde{G}(\mathbf{k})}{\sqrt{(2\pi)^3}} = -\frac{e^{i\mathbf{k}\cdot(\mathbf{x}-\bar{\mathbf{x}})}}{8\pi^3\varepsilon_0}. \tag{5.139}$$

To complete the derivation of the Green function, we need to Fourier-transform back into real space:

$$G(\mathbf{x}-\bar{\mathbf{x}}) = \frac{1}{\sqrt{(2\pi)^3}}\iiint\tilde{G}(\mathbf{k})e^{-i\mathbf{k}\cdot(\mathbf{x}-\bar{\mathbf{x}})}\,d^3k = \frac{1}{8\pi^3\varepsilon_0}\iiint\frac{e^{-i\mathbf{k}\cdot(\mathbf{x}-\bar{\mathbf{x}})}}{\mathbf{k}^2}\,d^3k. \tag{5.140}$$

Because the spectral density of the Green function depends on $\mathbf{k}^2$, and not on the direction of the $k$-vector, spherical coordinates are a natural choice to perform the integral. We have

$$\begin{aligned}
k &= \sqrt{\mathbf{k}^2} = \sqrt{k_x^2+k_y^2+k_z^2},\\
k_x &= k\sin\theta\cos\varphi,\\
k_y &= k\sin\theta\sin\varphi,\\
k_z &= k\cos\theta,\\
d^3k &= dk_x dk_y dk_z = \frac{1}{|J|}dk\,d\theta\,d\varphi = k^2\sin\theta\,dk\,d\theta\,d\varphi.
\end{aligned} \tag{5.141}$$

Moreover, we choose the $k_z$-axis to coincide with the direction of the vector $\Delta\mathbf{x} = \mathbf{x}-\bar{\mathbf{x}}$; with this we have

$$\mathbf{k}\cdot(\mathbf{x}-\bar{\mathbf{x}}) = \mathbf{k}\cdot\Delta\mathbf{x} = k_z\Delta x = k\cos\theta\Delta x, \tag{5.142}$$

and

$$G(\mathbf{x}-\bar{\mathbf{x}}) = \frac{1}{8\pi^3\varepsilon_0}\int_0^\infty dk\int_0^\pi \sin\theta\exp(ik\cos\theta\Delta x)\int_0^{2\pi}d\varphi. \tag{5.143}$$

The integral over $\varphi$ is trivial and yields a factor of $2\pi$. For the integral over $\theta$, we use the following formula,

$$\int_0^\pi e^{a\cos x}\sin x\,dx = \frac{2}{a}\sinh(a); \tag{5.144}$$

switching from $a$ to $ia$, we obtain

$$\int_0^{\pi} e^{ia\cos x}\sin x\,dx = \frac{2}{ia}\sinh(ia) = 2\frac{\sin a}{a} = 2\,\mathrm{sinc}(a). \tag{5.145}$$

Identifying $x \equiv \theta$ and $a \equiv k\Delta x$, we have

$$\begin{aligned} G(\mathbf{x}-\bar{\mathbf{x}}) &= \frac{1}{8\pi^3\varepsilon_0}\int_0^{\infty} dk 2\,\mathrm{sinc}(k\Delta x)2\pi \\ &= \frac{1}{2\pi^2\varepsilon_0\Delta x}\int_0^{\infty}\frac{\sin(k\Delta x)}{k}dk. \end{aligned} \tag{5.146}$$

The last integral, over the modulus of the $k$-vector is well known:

$$\int_0^{\infty}\frac{\sin ax}{x}dx = \frac{\pi}{2}\mathrm{sgn}(a). \tag{5.147}$$

For $a \equiv \Delta x$, this yields

$$G(\mathbf{x}-\bar{\mathbf{x}}) = \frac{1}{4\pi\varepsilon_0|\mathbf{x}-\bar{\mathbf{x}}|}. \tag{5.148}$$

The most general solution to Poisson's equation is therefore

$$\phi(\mathbf{x}) = \iiint\frac{\rho(\bar{\mathbf{x}})}{4\pi\varepsilon_0|\mathbf{x}-\bar{\mathbf{x}}|}d^3\bar{x} + \phi_0(\mathbf{x}), \tag{5.149}$$

where $\phi_0(\mathbf{x})$ is the solution to the homogeneous equation

$$\Delta\phi_0(\mathbf{x}) = 0. \tag{5.150}$$

It is easily shown that if we choose the boundary condition $\lim_{|\mathbf{x}|\to\infty}[\phi_0(\mathbf{x})] = 0$, we must require that $\phi_0(\mathbf{x}) = 0$. Therefore, Equation 5.149 becomes

$$\phi(\mathbf{x}) = \iiint\frac{\rho(\bar{\mathbf{x}})}{4\pi\varepsilon_0|\mathbf{x}-\bar{\mathbf{x}}|}d^3x. \tag{5.151}$$

In the case of an electron, we have

$$\rho(\mathbf{x}) = -e\delta_3(\mathbf{x}), \tag{5.152}$$

and we find that

$$\phi(\mathbf{x}) = \iiint \frac{-e\delta_3(\bar{\mathbf{x}})}{4\pi\varepsilon_0|\mathbf{x}-\bar{\mathbf{x}}|} d^3x = \frac{-e}{4\pi\varepsilon_0|\mathbf{x}|}, \tag{5.153}$$

which is identical to the Coulomb potential.

### 5.4.2 Green Function for the d'Alembertian, Photon Propagator

In the important case of the d'Alembertian operator, the formalism used in the previous subsection must be extended to account properly for the temporal dimension and its different geometrical role in flat, hyperbolic space–time. In particular, the representation of the d'Alembertian operator in momentum space leads to the eigenvalue equation

$$\Box e^{ik_\mu x^\mu} = -k^\nu k_\nu e^{ik_\mu x^\mu} = \left(\frac{\omega^2}{c^2} - \mathbf{k}^2\right) e^{ik_\mu x^\mu}, \tag{5.154}$$

which has two poles on the light-cone: $\omega = \pm c|\mathbf{k}|$.

We start from the wave equation, driven by a four-dimensional delta-function:

$$\Box G(x_\mu, \bar{x}_\mu) + \mu_0 \delta_4(x_\mu - \bar{x}_\mu) = 0. \tag{5.155}$$

Because the excitation depends only on the difference between the four-position of the source and the observation points, the associated Green function will have the same dependence: $G(x_\mu, \bar{x}_\mu) = G(x_\mu - \bar{x}_\mu)$. Furthermore, the technique of Fourier transforming the problem into momentum space will be used here:

$$G(x_\mu - \bar{x}_\mu) = \frac{1}{(\sqrt{2\pi^4})} \iiiint \tilde{G}(k^\mu) e^{ik^\mu x_\mu} d^4k. \tag{5.156}$$

Using Equation 5.156 into the wave equation (5.155), we first have

$$\begin{aligned}
\Box G(x_\mu - \bar{x}_\mu) + \mu_0 \delta_4(x_\mu - \bar{x}_\mu) &= 0 \\
&= \Box\left[\frac{1}{4\pi^2} \iiiint \tilde{G}(k^\mu) e^{ik^\mu x_\mu} d^4k\right] + \mu_0 \delta_4(x_\mu - \bar{x}_\mu) \\
&= \frac{1}{4\pi^2} \iiiint \tilde{G}(k^\mu)[\Box e^{ik^\mu x_\mu}]\, d^4k + \mu_0 \delta_4(x_\mu - \bar{x}_\mu),
\end{aligned} \tag{5.157}$$

where we have used the aforementioned commutation between the differential and integral operators.

As discussed above, the action of the d'Alembertian operator on the complex exponential eigenfunctions yields a simple eigenvalue,

$$\Box e^{ik^\mu x_\mu} = (\partial_\nu \partial^\nu) e^{ik^\mu x_\mu} = \partial_\nu \left( \frac{\partial e^{ik^\mu x_\mu}}{\partial x_\nu} \right) = \partial_\nu (\delta_\mu^\nu i k^\mu e^{ik^\mu x_\mu}) = ik^\nu \frac{\partial e^{ik^\mu x_\mu}}{\partial x_\nu}$$

$$= ik^\nu \frac{\partial x_\lambda}{\partial x^\nu} \frac{\partial e^{ik^\mu x_\mu}}{\partial x_\lambda} = ik^\nu g_{\lambda\nu} \delta_\mu^\lambda i k^\mu = -k^\nu g_{\lambda\nu} k^\lambda = -k^\nu k_\nu e^{ik^\mu x_\mu}, \quad (5.158)$$

and we now have

$$\frac{1}{4\pi^2} \iiiint \tilde{G}(k^\mu) \left( -k_\mu k^\mu e^{ik^\mu x_\mu} \right) d^4k + \mu_0 \delta_4(x_\mu - \bar{x}_\mu) = 0. \quad (5.159)$$

As the source current is a Dirac delta-function, we can easily diagonalize Equation 5.159, by projecting onto another eigenmode, $e^{-i\bar{k}^\mu x_\mu}$:

$$0 = \int_{\mathbb{R}^4} e^{-i\bar{k}^\mu x_\mu} \left[ \frac{1}{4\pi^2} \int_{\mathbb{R}^4} \tilde{G}(k^\mu) \left( -k_\mu k^\mu e^{ik^\mu x_\mu} \right) d^4k + \mu_0 \delta_4(x_\mu - \bar{x}_\mu) \right] d^4x$$

$$= \frac{1}{4\pi^2} \int_{\mathbb{R}^4} \tilde{G}(k^\mu)(-k_\mu k^\mu)\, d^4k \int_{\mathbb{R}^4} e^{i(k^\mu - \bar{k}^\mu)x_\mu} d^4x + \mu_0 \int_{\mathbb{R}^4} e^{-i\bar{k}^\mu x_\mu} \delta_4(x_\mu - \bar{x}_\mu)\, d^4x$$

$$= 4\pi^2 \int_{\mathbb{R}^4} \tilde{G}(k^\mu)(-k_\mu k^\mu)\, \delta_4(k^\mu - \bar{k}^\mu) d^4k + \mu_0 e^{-i\bar{k}^\mu \bar{x}_\mu}$$

$$= -4\pi^2 \bar{k}_\mu \bar{k}^\mu \tilde{G}(\bar{k}^\mu) + \mu_0 e^{-i\bar{k}^\mu \bar{x}_\mu}. \quad (5.160)$$

This yields the sought-after spectral density of the Green function:

$$\tilde{G}(k^\mu) = \frac{\mu_0}{4\pi^2} \frac{e^{-ik^\mu \bar{x}_\mu}}{k_\mu k^\mu}, \quad (5.161)$$

where we have switched the dummy variables, $k^\mu$, and $\bar{k}^\mu$.

In principle, to obtain the Green function, all that is left to do is to perform the integration back into real space:

$$G(x_\mu - \bar{x}_\mu) = \frac{1}{(\sqrt{2\pi})^4} \iiiint \frac{\mu_0}{4\pi^2} \frac{e^{-ik^\mu \bar{x}_\mu}}{k_\mu k^\mu} e^{ik^\mu \bar{x}_\mu} d^4k$$

$$= \frac{\mu_0}{16\pi^4} \int_{\mathbb{R}^4} \frac{e^{ik^\mu (x_\mu - \bar{x}_\mu)}}{k_\mu k^\mu} d^4k, \quad (5.162)$$

which can be recast as

$$G(x_\mu - \bar{x}_\mu) = \frac{1}{16\pi^4 \varepsilon_0 c} \iiint_{\mathbb{R}^3} e^{i\mathbf{k}\cdot\Delta\mathbf{x}} \, d^3k \int_{-\infty}^{+\infty} \frac{e^{-i\omega\Delta t}}{c^2\mathbf{k}^2 - \omega^2} \, d\omega, \tag{5.163}$$

where we have introduced the spatial interval, $\Delta\mathbf{x} = \mathbf{x} - \bar{\mathbf{x}}$, and its temporal counterpart, $c\Delta t = x_0 - \bar{x}_0 = c(t - \bar{t})$, and where we have used the fact that $\varepsilon_0\mu_0 c^2 = 1$.

However, it is clear that the two poles at $\omega/c = k_0 = \pm|\mathbf{k}|$ must be handled properly. To perform the integral over the frequency, namely,

$$\int_{-\infty}^{+\infty} \frac{e^{-i\omega\Delta t}}{c^2\mathbf{k}^2 - \omega^2} \, d\omega, \tag{5.164}$$

we will use Cauchy's residue theorem, which states that for a holomorphic function with $n$ simple poles inside a given integration contour, $C$, we have

$$\oint_C f(z)\, dz = 2\pi i \sum_n \mathrm{Res}[f(z_n)], \tag{5.165}$$

where the residues are calculated by taking the limit

$$\mathrm{Res}[f(z_n)] = \lim_{z \to z_n} [(z - z_n) f(z)]. \tag{5.166}$$

To apply the theorem properly, the poles must either be completely inside or completely outside the integration contour. With this in mind, we shift the poles by introducing the infinitesimal imaginary frequency, $i\varepsilon$. At this point, two different choices are possible: we can shift the poles either in the upper or in the lower complex half-planes. In addition, two separate integration contours can be considered: the real axis and a half-circle of infinite radius in the upper complex half-plane, or the real axis and a half-circle of infinite radius in the lower complex half-plane. This is illustrated in Figure 5.4. The difference between these two integration contours can be understood in physical terms by considering that

$$\begin{aligned} +\oint \frac{e^{-i\omega\Delta t}}{c^2\mathbf{k}^2 - \omega^2} \, d\omega &= -\oint \frac{e^{-i\omega\Delta t}}{c^2\mathbf{k}^2 - \omega^2} \, d\omega \\ &= \lim_{\Omega\to\infty} \left[ \int_{-\Omega}^{+\Omega} \frac{e^{-i\omega\Delta t}}{c^2\mathbf{k}^2 - \omega^2} \, d\omega + \int_0^{\pi} \frac{e^{-i\Delta t \Omega e^{i\theta}}}{c^2\mathbf{k}^2 - (\Omega e^{i\theta})^2} \Omega \, d\theta \right], \end{aligned} \tag{5.167}$$

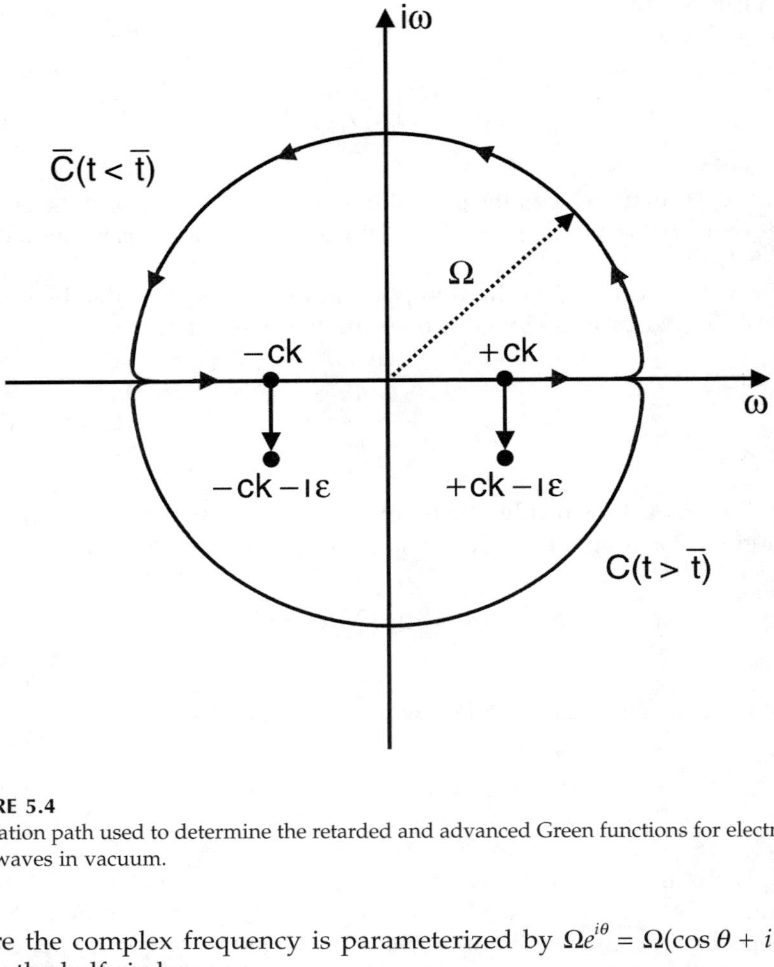


**FIGURE 5.4**
Integration path used to determine the retarded and advanced Green functions for electromagnetic waves in vacuum.

where the complex frequency is parameterized by $\Omega e^{i\theta} = \Omega(\cos\theta + i\sin\theta)$ along the half circles.

The different signs in front of the contour integrals correspond to the fact that they are described in opposite directions, as shown in Figure 5.4; integration along an anticlockwise path resulting in a positive sign. When the limit $\Omega \to \infty$ is taken, under the proper physical conditions, the integrals over the half circles vanish, and we obtain

$$\lim_{\Omega\to\infty}\left[\int_{-\Omega}^{+\Omega} f(\omega)\,d\omega + \int_0^{\pi} f(\Omega e^{i\theta})\Omega\,d\theta\right] = \int_{-\infty}^{+\infty} f(\omega)\,d\omega$$

$$= \begin{cases} 0, \\ \pm 2\pi i\{\mathrm{Res}[f(-c|\mathbf{k}|\pm i\varepsilon)] + \mathrm{Res}[f(+c|\mathbf{k}|\pm i\varepsilon)]\}, \end{cases} \qquad (5.168)$$

depending upon whether the poles lie inside or outside the integration contour. Here the function $f$ corresponds to the integrand in Equation 5.164.

To determine under which circumstances this applies, we now focus our attention on the limit,

$$\lim_{\Omega\to\infty}\left[\int_0^{\pi}\frac{e^{-i\Delta t\Omega e^{i\theta}}}{c^2\mathbf{k}^2-(\Omega e^{i\theta})^2}\Omega\, d\theta\right]. \tag{5.169}$$

An upper limit on the modulus of the integrand can be derived as follows:

$$\begin{aligned}\left|\frac{e^{-i\Delta t\Omega e^{i\theta}}}{c^2\mathbf{k}^2-(\Omega e^{i\theta})^2}\Omega\right| &= \left|\frac{e^{-i\Delta t\Omega(\cos\theta+i\sin\theta)}}{c^2\mathbf{k}^2-\Omega^2 e^{2i\theta}}\Omega\right| \\ &= \left|\frac{e^{\Delta t\Omega\sin\theta}e^{-i\Delta t\Omega\cos\theta}}{c^2\mathbf{k}^2-\Omega^2 e^{2i\theta}}\Omega\right| \\ &= \left|\frac{e^{\Delta t\Omega\sin\theta}}{c^2\mathbf{k}^2-\Omega^2 e^{2i\theta}}\Omega\right|.\end{aligned} \tag{5.170}$$

Taking the limit, we find that

$$\lim_{\Omega\to\infty}\left|\frac{e^{-i\Delta t\Omega e^{i\theta}}}{c^2\mathbf{k}^2-(\Omega e^{i\theta})^2}\Omega\right|, \tag{5.171}$$

behaves like

$$\lim_{\Omega\to\infty}\left|\frac{e^{\Delta t\Omega\sin\theta}}{\Omega}\right|. \tag{5.172}$$

In the upper half-plane, $\sin\theta \geq 0$, so we must require that $\Delta t \leq 0$ to have convergence

$$\lim_{\Omega\to\infty}\left[\int_0^{\pi}\frac{e^{-i\Delta t\Omega e^{i\theta}}}{c^2\mathbf{k}^2-(\Omega e^{i\theta})^2}\Omega\, d\theta\right] \leq \lim_{\Omega\to\infty}\pi\left|\frac{e^{\Delta t\Omega\sin\theta}}{\Omega}\right| = 0, \tag{5.173}$$

because the exponential diverges faster that the linear denominator. In the lower half-plane, $\sin\theta \leq 0$, and the condition becomes $\Delta t \geq 0$. Physically, these conditions correspond to advanced waves and retarded waves, respectfully. The contour in the upper half-plane yields finite answers for waves, such that the observation time, $t$, is smaller than the emission time, $\bar{t}$; this yields a negative value for $\Delta t = t - \bar{t}$; in contrast, causal waves, such that $t \geq \bar{t}$ and $\Delta t \geq 0$, are described by the integration contour in the lower complex half-plane.

Furthermore, the sign of the imaginary frequency shift and the choice of integration contour chosen to apply Cauchy's residue theorem are correlated. If we integrate in the lower complex half-plane and the poles are shifted in the positive imaginary frequency half-plane, the resulting retarded waves will be such that $G = 0$ for $x_0 < \bar{x}_0$, since the residues will be zero. For advanced waves, for which $G = 0$, if $x_0 > \bar{x}_0$, this corresponds to moving the poles in the negative imaginary region. Here, $x_\mu$ is interpreted as the four-position where the field is observed, while $\bar{x}_\mu$ corresponds to the four-location of the source. These possibilities are not entirely surprising, in view of the time-symmetrical character of Maxwell's equations. We also note that, when dealing with radiation reaction and the Dirac–Lorentz equation in Chapter 10, the advanced waves will be used to symmetrize the Green function and renormalize the electromagnetic mass of a point charge.

Having established the physical interpretation of the different configurations used to determine the Green function, we now derive the residue for the poles lying in the lower half of the complex plane for causal waves. The integral is now given by

$$\oint \frac{e^{-i\omega\Delta t}}{c^2\mathbf{k}^2 - (\omega + i\varepsilon)^2} d\omega = \int_{-\infty}^{+\infty} \frac{e^{-i\omega\Delta t}}{c^2\mathbf{k}^2 - (\omega + i\varepsilon)^2} d\omega, \tag{5.174}$$

since we have established that the integral along the half-circle vanishes in this case. The poles lie in the negative imaginary half-plane:

$$\omega^{\pm} = \pm c|\mathbf{k}| - i\varepsilon = \pm ck - i\varepsilon, \tag{5.175}$$

and we need to evaluate

$$\lim_{\omega \to \omega^{\pm}} \left[ (\omega + \omega^{\pm}) \frac{e^{-i\omega\Delta t}}{c^2\mathbf{k}^2 - (\omega + i\varepsilon)^2} \right]. \tag{5.176}$$

The denominator can be expanded as

$$c^2\mathbf{k}^2 - (\omega + i\varepsilon)^2 = [ck - (\omega + i\varepsilon)][ck + (\omega + i\varepsilon)]; \tag{5.177}$$

with this, we have

$$\begin{aligned}(\omega + \omega^{\pm}) \frac{e^{-i\omega\Delta t}}{c^2\mathbf{k}^2 - (\omega + i\varepsilon)^2} &= \frac{[\omega - (\pm ck - i\varepsilon)]e^{-i\omega\Delta t}}{(ck - \omega - i\varepsilon)(ck + \omega + i\varepsilon)} \\ &= \frac{\mp e^{-i\omega\Delta t}}{ck \pm (\omega + i\varepsilon)},\end{aligned} \tag{5.178}$$

and the limit becomes

$$\lim_{\omega\to\omega^{\pm}}\left[\frac{\mp e^{-i\omega\Delta t}}{ck\pm(\omega+i\varepsilon)}\right] = \frac{\mp e^{-i\omega^{\pm}\Delta t}}{ck\pm(\omega^{\pm}+i\varepsilon)}$$
$$= \frac{\mp e^{-i(\pm ck-i\varepsilon)\Delta t}}{ck\pm[(\pm ck - i\varepsilon)+i\varepsilon]}$$
$$= \frac{\mp e^{\pm ick\Delta t}e^{-\varepsilon\Delta t}}{2ck}. \qquad (5.179)$$

When $\varepsilon \to 0$, the sum of the residues yields

$$\lim_{\varepsilon\to 0}\left[\int_{-\infty}^{+\infty}\frac{e^{-i\omega\Delta t}}{c^2\mathbf{k}^2-(\omega+i\varepsilon)^2}d\omega\right] = -2\pi i\left(\frac{e^{ick\Delta t}-e^{-ick\Delta t}}{2ck}\right)$$
$$= 2\pi\frac{\sin(ck\Delta t)}{ck}. \qquad (5.180)$$

In summary, we have shown that for causal waves,

$$\int_{-\infty}^{+\infty}\frac{e^{-i\omega\Delta t}}{c^2\mathbf{k}^2-(\omega+i\varepsilon)^2}d\omega = \begin{cases}0, & \Delta t<0,\\ 2\pi\dfrac{\sin(ck\Delta t)}{ck}, & \Delta t\geq 0.\end{cases} \qquad (5.181)$$

We can now proceed with the remaining steps of the derivation: at this point, for $\Delta t \geq 0$, we need to evaluate the triple integral

$$G^{+}(x_\mu-\bar{x}_\mu) = \frac{1}{16\pi^4\varepsilon_0 c}\iiint_{\mathbb{R}^3} e^{i\mathbf{k}\cdot\Delta\mathbf{x}}\, d^3k\left[2\pi\frac{\sin(ck\Delta t)}{ck}\right]. \qquad (5.182)$$

The choice of notation $G^{+}$ for causal waves and $G^{-}$ for advanced waves will be justified at the end of this section, when we discuss the singular and nonsingular components of the Green functions and use the Heaviside step-function. As in the case of the Green function for Poisson's equation, the use of a cylindrical coordinate system with $k_z$ aligned with the spatial separation $\Delta\mathbf{x}$ allows us to write

$$\begin{aligned} d^3k &= k^2\sin\theta\, dk\, d\theta\, d\varphi,\\ \mathbf{k}\cdot\Delta\mathbf{x} &= k\cos\theta\Delta x, \end{aligned} \qquad (5.183)$$

and the integral of Equation 5.182 is recast as

$$\begin{aligned} G^{+}(x_\mu - \bar{x}_\mu) &= \frac{\mu_0}{8\pi^3}\int_0^{2\pi} d\varphi \int_0^\infty k^2\,dk \int_0^\pi e^{ik\cos\theta\Delta x}\,\frac{\sin(ck\Delta t)}{k}\sin\theta\,d\theta \\ &= \frac{\mu_0}{4\pi^2}\int_0^\infty k\sin(ck\Delta t)\,dk \int_0^\pi e^{ik\cos\theta\Delta x}\sin\theta\,d\theta. \end{aligned} \tag{5.184}$$

The integral over $\theta$ has already been evaluated in Section 5.4.1:

$$\int_0^\pi e^{ik\cos\theta\Delta x}\sin\theta\,d\theta = 2\frac{\sin(k\Delta x)}{k\Delta x}. \tag{5.185}$$

Using the notation $c\Delta t = \Delta x_0$, and $2\sin a \sin b = \cos(a-b) - \cos(a+b)$, we obtain

$$\begin{aligned} G^{+}(x_\mu - \bar{x}_\mu) &= \frac{\mu_0}{2\pi^2 \Delta x}\int_0^\infty \sin(k\Delta x_0)\sin(k\Delta x)\,dk \\ &= \frac{\mu_0}{4\pi^2 \Delta x}\int_0^\infty [\cos k(\Delta x_0 - \Delta x) - \cos k(\Delta x_0 + \Delta x)]\,dk. \end{aligned} \tag{5.186}$$

Finally, we use the Fourier transform of the delta-function to express the integral in Equation 5.186:

$$\begin{aligned} \delta(k - \bar{k}) &= \frac{1}{2\pi}\int_{-\infty}^{+\infty} e^{i(k-\bar{k})x}\,dx \\ &= \frac{1}{2\pi}\int_{-\infty}^{+\infty} [\cos x(k-\bar{k}) + i\sin x(k-\bar{k})]\,dx \\ &= \frac{1}{\pi}\int_{-0}^{+\infty} \cos x(k-\bar{k})\,dx. \end{aligned} \tag{5.187}$$

This yields the expression for the causal Green function,

$$G^{+}(x_\mu - \bar{x}_\mu) = \begin{cases} 0, & \Delta x_0 < 0, \\ \dfrac{\mu_0}{4\pi\Delta x}[\delta(\Delta x_0 - \Delta x) - \delta(\Delta x_0 + \Delta x)], & \Delta x_0 \geq 0, \end{cases} \tag{5.188}$$

which reduces to

$$G^{+}(x_\mu - \bar{x}_\mu) = \frac{\mu_0}{4\pi\Delta x}\delta(\Delta x_0 - \Delta x) = \frac{\mu_0}{4\pi}\frac{\delta[(\Delta x_0 - \bar{x}_0) - |\mathbf{x} - \bar{\mathbf{x}}|]}{|\mathbf{x} - \bar{\mathbf{x}}|}, \tag{5.189}$$

by virtue of the fact that we are restricting the nonzero value of the causal Green function to $\Delta x_0 \geq 0$, in which case, $\delta(\Delta x_0 + \Delta x) = \delta(\Delta x_0 + \Delta|\mathbf{x}|)$ is identically equal to zero.

The Green function for advanced waves can be derived in a similar fashion to obtain

$$G^-(x_\mu - \bar{x}_\mu) = \begin{cases} 0, & \Delta x_0 > 0, \\ \dfrac{\mu_0}{4\pi\Delta x}[\delta(-\Delta x_0 - \Delta x) - \delta(-\Delta x_0 + \Delta x)], & \Delta x_0 \leq 0, \end{cases} \tag{5.190}$$

or

$$G^-(x_\mu - \bar{x}_\mu) = \frac{\mu_0}{4\pi\Delta x}\delta(-\Delta x_0 - \Delta x) = \frac{\mu_0}{4\pi}\frac{\delta[(\Delta\bar{x}_0 - x_0) - |\mathbf{x} - \bar{\mathbf{x}}|]}{|\mathbf{x} - \bar{\mathbf{x}}|}. \tag{5.191}$$

We can also use the Heaviside step-function, introduced in Section 5.2, Equation 5.56:

$$\begin{aligned} G^+(x_\mu - \bar{x}_\mu) &= \frac{\mu_0}{4\pi\Delta x}H(+\Delta x_0)\delta(\Delta x_0 - \Delta x), \\ G^-(x_\mu - \bar{x}_\mu) &= \frac{\mu_0}{4\pi\Delta x}H(-\Delta x_0)\delta(\Delta x_0 + \Delta x). \end{aligned} \tag{5.192}$$

The Green function can then be recast in a manifestly covariant form by using the fact that

$$(x_\mu - \bar{x}_\mu)(x^\mu - \bar{x}^\mu) = (\mathbf{x} - \bar{\mathbf{x}})^2 - (x_0 - \bar{x}_0)^2 = \Delta\mathbf{x}^2 - \Delta x_0^2; \tag{5.193}$$

we can then use the delta-function identity, and its symmetry, to write

$$\delta(x^2 - a^2) = \frac{\delta(x - a) + \delta(x + a)}{|a|} = \delta(a^2 - x^2). \tag{5.194}$$

After identifying $x \equiv \Delta x_0$ and $a \equiv |\Delta\mathbf{x}| = \Delta x$, we find that

$$\delta[(x_\mu - \bar{x}_\mu)(x^\mu - \bar{x}^\mu)] = \frac{\delta(\Delta x_0 - \Delta x) + \delta(\Delta x_0 + \Delta x)}{\Delta x}. \tag{5.195}$$

This last identity can now be used in Equation 5.192 to write

$$\begin{aligned} G^+(x_\mu - \bar{x}_\mu) &= \frac{\mu_0}{4\pi}H(+\Delta x_0)\delta[(x_\mu - \bar{x}_\mu)(x^\mu - \bar{x}^\mu)], \\ G^-(x_\mu - \bar{x}_\mu) &= \frac{\mu_0}{4\pi}H(-\Delta x_0)\delta[(x_\mu - \bar{x}_\mu)(x^\mu - \bar{x}^\mu)]. \end{aligned} \tag{5.196}$$

In electron units, we have $\mu_0 = 4\pi$, and Equation 5.196 can be recast in a very compact form by introducing the shorter notation, $\Delta x_\mu = x_\mu - \bar{x}_\mu$:

$$G^{\pm}(\Delta x_\mu) = H(\pm\Delta x_0)\delta(\Delta x_\mu \Delta x^\mu). \tag{5.197}$$

Finally, we note that the Green function can be separated into its so-called "singular" and "nonsingular" components, $G_1$ and $G_0$, respectively. We have

$$\begin{aligned} 2G_0(\Delta x_\mu) &= \frac{\Delta x_0}{|\Delta x_0|}\delta(\Delta x_\mu \Delta x^\mu), \\ 2G_1(\Delta x_\mu) &= \delta(\Delta x_\mu \Delta x^\mu), \end{aligned} \tag{5.198}$$

and the retarded and advanced Green functions can be expressed as

$$G^{\pm} = G_1 \pm G_0 = \frac{1}{2}\left(1 \pm \frac{\Delta x_0}{|\Delta x_0|}\right)\delta(\Delta x_\mu \Delta x^\mu) = H(\pm\Delta x_0)\delta(\Delta x_\mu \Delta x^\mu). \tag{5.199}$$

Physically, the nonsingular Green function satisfies the homogeneous wave equation,

$$\Box G_0 = 0, \tag{5.200}$$

while the singular Green function is the solution of the driven wave equation,

$$\Box G_1 + 4\pi\delta_4(\Delta x_\mu) = 0. \tag{5.201}$$

## 5.5 Liénard–Wiechert Potentials

Given the Green functions, we can now derive the solution to the wave equation driven by an arbitrary four-current distribution:

$$\Box A_\mu + \mu_0 j_\mu = 0, \qquad A_\mu(x_\nu) = \int_{\mathbb{R}^4} G(x_\nu - \bar{x}_\nu) j_\mu(\bar{x}_\nu)\, d^4\bar{x}_\nu. \tag{5.202}$$

Using the expression for the causal Green function derived in the previous section, we find that

$$\begin{aligned} A_\mu(x_\nu) &= \int_{\mathbb{R}^4} G^{+}(x_\nu - \bar{x}_\nu) j_\mu(\bar{x}_\nu)\, d^4\bar{x} \\ &= \frac{\mu_0}{4\pi}\int_{\mathbb{R}^4} H(x_0 - \bar{x}_0)\delta[(x_\nu - \bar{x}_\nu)(x^\nu - \bar{x}^\nu)] j_\mu(\bar{x}_\nu)\, d^4\bar{x}. \end{aligned} \tag{5.203}$$

More explicitly, the scalar and vector potentials are given by

$$\phi(\mathbf{x}, t) = \frac{1}{4\pi\varepsilon_0}\iiint_{\mathbb{R}^3} d^3\bar{x}\int_{-\infty}^{+\infty}\rho(\bar{\mathbf{x}}, \bar{t})\frac{\delta[c(t-\bar{t}) - |\mathbf{x}-\bar{\mathbf{x}}|]}{|\mathbf{x}-\bar{\mathbf{x}}|}c\,d\bar{t}$$
$$= \frac{1}{4\pi\varepsilon_0}\iiint_{\mathbb{R}^3}\frac{\rho[\bar{\mathbf{x}}, t-(|\mathbf{x}-\bar{\mathbf{x}}|/c)]}{|\mathbf{x}-\bar{\mathbf{x}}|}d^3\bar{x}, \tag{5.204}$$

and

$$\mathbf{A}(\mathbf{x}, t) = \frac{\mu_0}{4\pi}\iiint_{\mathbb{R}^3} d^3\bar{x}\int_{-\infty}^{+\infty}\mathbf{j}(\bar{\mathbf{x}}, \bar{t})\frac{\delta[c(t-\bar{t}) - |\mathbf{x}-\bar{\mathbf{x}}|]}{|\mathbf{x}-\bar{\mathbf{x}}|}c\,d\bar{t}$$
$$= \frac{\mu_0}{4\pi}\iiint_{\mathbb{R}^3}\frac{\mathbf{j}[\bar{\mathbf{x}}, t-(|\mathbf{x}-\bar{\mathbf{x}}|/c)]}{|\mathbf{x}-\bar{\mathbf{x}}|}d^3\bar{x}. \tag{5.205}$$

The retardation condition appears clearly: the potential produced by an excitation located at $\bar{\mathbf{x}}$ and observed at a point $\mathbf{x}$ and time $t$ was produced at a time $\bar{t}$ satisfying the propagation condition

$$t = \bar{t} + \frac{|\mathbf{x}-\bar{\mathbf{x}}|}{c}. \tag{5.206}$$

In Equations 5.204 and 5.205, the integral over the temporal delta-function is easily performed, but the retardation condition makes it difficult, in general, to compute the four-potential. In the case of a point charge, however, the charge density is modeled by a three-dimensional delta-function; as a result, one can perform the triple integral over space to obtain the Liénard–Wiechert potentials. The four-current is described by the charge density,

$$\rho(\mathbf{x}, t) = -e\delta_3[\mathbf{x} - \mathbf{z}(t)], \tag{5.207}$$

and by the current density,

$$\mathbf{j}(\mathbf{x}, t) = \mathbf{v}(t)\rho(\mathbf{x}, t) = -ec\beta(t)\delta[\mathbf{x} - \mathbf{z}(t)], \tag{5.208}$$

so that

$$j_\mu = \rho c(1, \beta) = -e\delta_3[\mathbf{x} - \mathbf{z}(t)][1, \beta(t)]. \tag{5.209}$$

Here $\mathbf{z}(t)$ is the position of the electron, and $\mathbf{v}(t) = c\beta(t) = d\mathbf{z}(t)/dt$ is its velocity. The four-potential is given by

$$A_\mu(x_\nu) = \frac{\mu_0}{4\pi}\int_{\mathbb{R}^4}\frac{\delta(\Delta x_0 - \Delta x)}{\Delta x}j_\mu(\bar{x}_\nu)\,d^4\bar{x} \tag{5.210}$$

for the electron four-current, which yields

$$
\begin{aligned}
A_\mu(x_0, \mathbf{x}) &= \frac{-e\mu_0 c}{4\pi}\int_{\mathbb{R}^4} \frac{\delta(x_0 - \bar{x}_0 - |\mathbf{x} - \bar{\mathbf{x}}|)}{|\mathbf{x} - \bar{\mathbf{x}}|} \delta_3[\bar{\mathbf{x}} - \mathbf{z}(\bar{x}_0)][1, \beta(\bar{x}_0)] d^4\bar{x} \\
&= \frac{-e\mu_0 c}{4\pi}\int_{-\infty}^{+\infty} d\bar{x}_0 \int_{\mathbb{R}^3} \frac{\delta(x_0 - \bar{x}_0 - |\mathbf{x} - \bar{\mathbf{x}}|)}{|\mathbf{x} - \bar{\mathbf{x}}|} \delta_3[\bar{\mathbf{x}} - \mathbf{z}(\bar{x}_0)][1, \beta(\bar{x}_0)] d^3\bar{x} \\
&= \frac{-e\mu_0 c}{4\pi}\int_{-\infty}^{+\infty} \frac{\delta[x_0 - \bar{x}_0 - |\mathbf{x} - \mathbf{z}(\bar{x}_0)|]}{|\mathbf{x} - \mathbf{z}(\bar{x}_0)|}[1, \beta(\bar{x}_0)]\, d\bar{x}_0. \qquad (5.211)
\end{aligned}
$$

To perform the last integral, over the time coordinate, we simply change the integration variable:

$$
s(\bar{x}_0) = x_0 - \bar{x}_0 - |\mathbf{x} - \mathbf{z}(\bar{x}_0)|. \qquad (5.212)
$$

To derive the differential element, we evaluate

$$
\begin{aligned}
\frac{ds}{d\bar{x}_0} &= \frac{d}{d\bar{x}_0}\{x_0 - \bar{x}_0 - \sqrt{[\mathbf{x} - \mathbf{z}(\bar{x}_0)] \cdot [\mathbf{x} - \mathbf{z}(\bar{x}_0)]}\} \\
&= -1 - \frac{1}{2}2\frac{[\mathbf{x} - \mathbf{z}(\bar{x}_0)]}{|\mathbf{x} - \mathbf{z}(\bar{x}_0)|} \cdot \left[-\frac{d\mathbf{z}}{d\bar{x}_0}\right] \\
&= -1 + \frac{[\mathbf{x} - \mathbf{z}(\bar{x}_0)]}{|\mathbf{x} - \mathbf{z}(\bar{x}_0)|} \cdot \beta(\bar{x}_0) \\
&= -1 + \hat{n}(\bar{x}_0) \cdot \beta(\bar{x}_0), \qquad (5.213)
\end{aligned}
$$

where we have introduced the unit vector in the direction of observation, $\hat{n}(\bar{x}_0)$.

The four-potential is then given by

$$
\begin{aligned}
A_\mu(x_0, \mathbf{x}) &= \frac{-e\mu_0 c}{4\pi}\int_{-\infty}^{+\infty} \delta(s)\frac{[1, \beta(\bar{x}_0)]}{|\mathbf{x} - \mathbf{z}(\bar{x}_0)|}\frac{d\bar{x}_0}{ds} ds \\
&= \frac{-e\mu_0 c}{4\pi}\int_{-\infty}^{+\infty} \delta(s)\frac{[1, \beta(s)]}{|\mathbf{x} - \mathbf{z}(s)|}\frac{ds}{1 - \hat{n}(s) \cdot \beta(s)} \\
&= \frac{-e\mu_0 c}{4\pi} \left.\frac{[1, \beta(s)]}{|\mathbf{x} - \mathbf{z}(s)|[1 - \hat{n}(s) \cdot \beta(s)]}\right|_{s=0} \\
&= \frac{-e\mu_0 c}{4\pi} \left.\frac{u_\mu(\bar{x}_0)}{|\mathbf{x} - \mathbf{z}(\bar{x}_0)|[\gamma(\bar{x}_0) - \hat{n}(\bar{x}_0) \cdot \mathbf{u}(\bar{x}_0)]}\right|_{\bar{x}_0 = x_0 - |\mathbf{x} - \mathbf{z}(\bar{x}_0)|}. \qquad (5.214)
\end{aligned}
$$

Here, $u_\mu = dz_\mu/d\tau = \gamma(1, \beta) = (\gamma, \mathbf{u})$ is the normalized four-velocity of the electron.

In addition, we note that in vacuum, the observation vector can be related to the four-wavenumber of the propagating radiation,

$$(1, \hat{n}) = \frac{ck_\mu}{\omega}, \tag{5.215}$$

which allows us to write the four-potential as

$$= \frac{e\mu_0\omega}{4\pi} \frac{u_\mu(\bar{x}_0)}{|\mathbf{x} - \mathbf{z}(\bar{x}_0)|[u_\mu(\bar{x}_0)k^\mu(\bar{x}_0)]}\bigg|_{\bar{x}_0 = x_0 - |\mathbf{x} - \mathbf{z}(\bar{x}_0)|}. \tag{5.216}$$

Finally, we can translate Equation 5.214 into the more familiar form of the Liénard–Wiechert potentials by remembering that $\phi(\mathbf{x}, t) = cA_0(\mathbf{x}, t)$,

$$\begin{aligned} \phi(\mathbf{x}, t) &= \frac{-e}{4\pi\varepsilon_0} \frac{1}{|\mathbf{x} - \mathbf{z}(t^-)|[1 - \hat{n}(t^-) \cdot \beta(t^-)]}, \\ \mathbf{A}(\mathbf{x}, t) &= \frac{-e\mu_0}{4\pi} \frac{\mathbf{v}(t^-)}{|\mathbf{x} - \mathbf{z}(t^-)|[1 - \hat{n}(t^-) \cdot \beta(t^-)]}, \end{aligned} \tag{5.217}$$

where the retarded time satisfies the condition

$$t^- = t - \frac{|\mathbf{x} - \mathbf{z}(t^-)|}{c}. \tag{5.218}$$

In closing, we consider the potential produced by a point charge in uniform motion; we will return to this problem in Section 5.7, using a Lorentz transform of the Coulomb field and a vacuum eigenmode analysis in cylindrical coordinates using Bessel functions. In this simple example, the position of the electron is described by

$$\mathbf{z}(t) = \hat{z}vt = \hat{z}\beta ct, \tag{5.219}$$

as we have aligned the $z$-axis of the reference frame with the constant velocity of the point charge. The retardation equation is easily solved by using cylindrical coordinates. We first write

$$|\mathbf{x} - \mathbf{z}(t^-)| = \sqrt{r^2 + (z - vt^-)^2} = c(t - t^-); \tag{5.220}$$

taking the square of each side of Equation 5.220 leads to a quadratic equation, of the form

$$\begin{aligned} f(ct^-)^2 + 2g(ct^-) + h &= 0, \\ f &= (1 - \beta^2) = \gamma^{-2}, \\ g &= \beta z - ct, \\ h &= c^2t^2 - r^2 - z^2. \end{aligned} \tag{5.221}$$

The solutions are given by

$$ct^{-} = \frac{g \pm \sqrt{\Delta}}{f}, \quad \Delta = g^2 - fh$$
$$= (\beta z - ct)^2 - (1-\beta^2)(c^2t^2 - r^2 - z^2) = \frac{r^2}{\gamma^2} + (z - \beta ct)^2, \qquad (5.222)$$

and we find that

$$ct^{-} = \gamma^2(ct - \beta z) \pm \gamma\sqrt{r^2 + \gamma^2(z - vt)^2}. \qquad (5.223)$$

Temporarily setting $\beta = 0$ and $\gamma = 1$, we see that only the minus sign satisfies the condition $t^{-} \leq t$. We now have $\mathbf{z}(t^{-}) = \hat{z}\beta ct^{-}$, and we can return to the scalar potential, as expressed in Equation 5.217, and use the definition of the observation unit vector to simplify our derivation: as $\hat{n}(t^{-}) = [\mathbf{x} - \mathbf{z}(t^{-})]/|\mathbf{x} - \mathbf{z}(t^{-})|$. Since the velocity is constant, we have

$$\phi(\mathbf{x}, t) = \frac{-e}{4\pi\varepsilon_0}\left\{\frac{1}{|\mathbf{x} - \mathbf{z}(t^{-})| - [\mathbf{x} - \mathbf{z}(t^{-})] \cdot \hat{z}\beta}\right\}; \qquad (5.224)$$

moreover, Equation 5.220 indicates that $|\mathbf{x} - \mathbf{z}(t^{-})| = c(t - t^{-})$, and we can further simplify the expression for the potential, to obtain

$$\phi(\mathbf{x}, t) = \frac{-e}{4\pi\varepsilon_0}\left[\frac{1}{c(t - t^{-}) - \beta z + \beta^2 ct^{-}}\right]$$
$$= \frac{-e}{4\pi\varepsilon_0}\left[\frac{1}{(ct - \beta z) - (1 - \beta^2)ct^{-}}\right]$$
$$= \frac{-e}{4\pi\varepsilon_0}\left[\frac{\gamma^2}{\gamma^2(ct - \beta z) - ct^{-}}\right]. \qquad (5.225)$$

Inserting the result obtained for the retarded time in Equation 5.222, we finally find that

$$\phi(\mathbf{x}, t) = \frac{-e\gamma}{4\pi\varepsilon_0\sqrt{r^2 + \gamma^2(z - vt)^2}}, \qquad \mathbf{A}(\mathbf{x}, t) = \hat{z}\beta\frac{\phi(\mathbf{x}, t)}{c}. \qquad (5.226)$$

## 5.6 Green Function with Boundary Conditions: Cylindrical Waveguide

In this section, our goal is to derive the Green function for a system with cylindrical symmetry and boundary conditions. In this case and for bounded systems in general, the Green function is quite different from its vacuum counterpart, as a discrete spectrum of eigenmodes emerges; furthermore, this can also be considered as an introduction to Chapter 9, which deals with synchrotron radiation in a waveguide. Finally, we also note that there is a very close correspondence between the mathematical tools used here to describe eigenmodes and eigenvalues in a bounded system and the formalism underlying quantum mechanics, including Hilbert spaces and orthonormal functions.

### 5.6.1 Cylindrical Vacuum Eigenmodes

In cylindrical coordinates, the four-potential takes the form

$$A_\mu = \left(\frac{\phi}{c}, \mathbf{A}\right) = \left(\frac{\phi}{c}, \hat{r}A_r + \hat{\theta}A_\theta + \hat{z}A_z\right), \tag{5.227}$$

and the corresponding electric field is

$$\begin{aligned}
\mathbf{E} &= -\nabla\phi - \partial_t\mathbf{A},\\
E_r &= -\partial_r\phi - \partial_t A_r,\\
E_\theta &= -\frac{1}{r}\partial_\theta\phi - \partial_t A_\theta,\\
E_z &= -\partial_z\phi - \partial_t A_z,
\end{aligned} \tag{5.228}$$

while the magnetic induction is given by

$$\begin{aligned}
\mathbf{B} &= \nabla\times\mathbf{A},\\
B_r &= \frac{1}{r}\partial_\theta A_z - \partial_z A_\theta,\\
B_\theta &= \partial_z A_r - \partial_r A_z,\\
B_z &= \frac{1}{r}[\partial_r(rA_\theta) - \partial_\theta A_r].
\end{aligned} \tag{5.229}$$

With this definition of the fields, the source-free equations of the Maxwell set are automatically satisfied,

$$\nabla\cdot\mathbf{B} = 0, \qquad \nabla\times\mathbf{E} + \partial_t\mathbf{B} = \mathbf{0}, \tag{5.230}$$

and we will study the coupling of the four-current produced by a unit point excitation,

$$\delta_4(x_\mu - \bar{x}_\mu) = \frac{1}{r}\delta(r-\bar{r})\delta(\theta-\bar{\theta})\delta(z-\bar{z})\delta(t-\bar{t}), \tag{5.231}$$

or by a point charge, where

$$\begin{aligned} j_\mu &= (\rho c, \mathbf{j}), \\ \rho(x_\lambda) &= -e\delta[r-r(t)]\frac{1}{r}\delta[\theta-\theta(t)]\delta[z-z(t)], \\ \mathbf{j}(x_\lambda) &= \rho\frac{d\mathbf{r}}{dt}, \end{aligned} \tag{5.232}$$

and where the electron trajectory is described by

$$\mathbf{r}(t) = \hat{r}r(t) + \hat{z}z(t). \tag{5.233}$$

In Equations 5.231 and 5.232, the $r^{-1}$ factor in front of the angular delta-function results from the Jacobian of the transformation from Cartesian to cylindrical coordinates:

$$\begin{aligned} x &= r\cos\theta, \quad y = r\sin\theta, \\ dxdy &= Jdrd\theta = \begin{vmatrix} \frac{\partial x}{\partial r} & \frac{\partial x}{\partial \theta} \\ \frac{\partial y}{\partial r} & \frac{\partial y}{\partial \theta} \end{vmatrix} drd\theta \\ &= \begin{vmatrix} \cos\theta & -r\sin\theta \\ \sin\theta & r\cos\theta \end{vmatrix} drd\theta = rdrd\theta. \end{aligned} \tag{5.234}$$

Furthermore, when we work in cylindrical coordinates, the $\hat{r}$ and $\hat{\theta}$ unit vectors depend upon the angle $\theta$:

$$\begin{aligned} \hat{r} &= \hat{x}\cos\theta + \hat{y}\sin\theta, \\ \hat{\theta} &= -\hat{x}\sin\theta + \hat{y}\cos\theta, \\ \frac{\partial\hat{r}}{\partial\theta} &= \hat{\theta}, \quad \frac{\partial\hat{\theta}}{\partial\theta} = -\hat{r}. \end{aligned} \tag{5.235}$$

Thus, the velocity of the point charge is

$$\begin{aligned} \frac{d\mathbf{r}}{dt} &= \frac{d}{dt}(r\hat{r} + z\hat{z}) = \dot{r}\hat{r} + r\frac{d\hat{r}}{dt} + \dot{z}\hat{z} \\ &= \dot{r}\hat{r} + r\frac{d\hat{r}}{d\theta}\frac{d\theta}{dt} + \dot{z}\hat{z} = \dot{r}\hat{r} + r\dot{\theta}\hat{\theta} + \dot{z}\hat{z}, \end{aligned} \tag{5.236}$$

where we have used the notation $\dot{r} = dr/dt$, $\dot{\theta} = d\theta/dt$, $\dot{z} = dz/dt$.

In the Lorentz gauge,

$$\partial_\mu A^\mu = \frac{1}{c^2}\partial_t\phi + \nabla\cdot\mathbf{A} = \frac{1}{c^2}\partial_t\phi + \frac{1}{r}[\partial_r rA_r] + \frac{1}{r}\partial_\theta A_\theta + \partial_z A_z = 0, \tag{5.237}$$

the wave equation is given by

$$\begin{aligned} \Box A_\mu + \mu_0 j_\mu &= 0, \\ \Box\mathbf{A} + \mu_0\mathbf{j} &= \mathbf{0}, \\ \Box\rho + \frac{1}{\varepsilon_0}\phi &= 0. \end{aligned} \tag{5.238}$$

Note that the d'Alembertian operates on the vector potential: this means that the partial derivative with respect to the angle will operate both on the components of the vector potential and on the unit vectors $\hat{r}$ and $\hat{\theta}$. We start the detailed derivation by writing the d'Alembertian in terms of the Laplacian, as expressed in cylindrical coordinates:

$$\Box \equiv \Delta - \frac{1}{c^2}\partial_t^2 \equiv \frac{1}{r}\partial_r(r\partial_r) + \frac{1}{r^2}\partial_\theta^2 + \partial_z^2 - \frac{1}{c^2}\partial_t^2. \tag{5.239}$$

For the axial and temporal components of the four-vector, the unit vectors $\hat{z}$ and $\hat{t}$ are constant; therefore the wave equation directly yields

$$\begin{aligned} \Box A_z + \mu_0 j_z &= \left[\frac{1}{r}\partial_r(r\partial_r) + \frac{1}{r^2}\partial_\theta^2 + \partial_z^2 - \frac{1}{c^2}\partial_t^2\right]A_z + \mu_0 j_z = 0, \\ \Box\phi + \frac{1}{\varepsilon_0}\phi &= \left[\frac{1}{r}\partial_r(r\partial_r) + \frac{1}{r^2}\partial_\theta^2 + \partial_z^2 - \frac{1}{c^2}\partial_t^2\right]\phi + \frac{1}{\varepsilon_0}\phi = 0. \end{aligned} \tag{5.240}$$

For the radial and azimuthal components, matters are a bit more complicated. We first have

$$\begin{aligned}\Box(\hat{r}A_r + \hat{\theta}A_\theta) &= \hat{r}\left[\frac{1}{r}\partial_r(r\partial_r) + \partial_z^2 - \frac{1}{c^2}\partial_t^2\right]A_r + \hat{\theta}\left[\frac{1}{r}\partial_r(r\partial_r) + \partial_z^2 - \frac{1}{c^2}\partial_t^2\right]A_\theta \\ &\quad + \frac{1}{r^2}\partial_\theta^2(\hat{r}A_r + \hat{\theta}A_\theta). \qquad (5.241)\end{aligned}$$

Focusing on the last term, we proceed systematically:

$$\begin{aligned}\partial_\theta^2(\hat{r}A_r + \hat{\theta}A_\theta) &= \partial_\theta[(\partial_\theta\hat{r})A_r + \hat{r}(\partial_\theta A_r) + (\partial_\theta\hat{\theta})A_\theta + \hat{\theta}(\partial_\theta A_\theta)] \\ &= \partial_\theta[\hat{\theta}A_r + \hat{r}(\partial_\theta A_r) - \hat{r}A_\theta + \hat{\theta}(\partial_\theta A_\theta)] \\ &= \partial_\theta[\hat{r}(\partial_\theta A_r - A_\theta) + \hat{\theta}(\partial_\theta A_\theta + A_r)] \\ &= (\partial_\theta\hat{r})(\partial_\theta A_r - A_\theta) + \hat{r}(\partial_\theta^2 A_r - \partial_\theta A_\theta) \\ &\quad + (\partial_\theta\hat{\theta})(\partial_\theta A_\theta + A_r) + \hat{\theta}(\partial_\theta^2 A_\theta + \partial_\theta A_r) \\ &= \hat{r}(\partial_\theta^2 A_r - \partial_\theta A_\theta - \partial_\theta A_\theta - A_r) \\ &\quad + \hat{\theta}(\partial_\theta^2 A_\theta + \partial_\theta A_r + \partial_\theta A_r - A_\theta) \\ &= \hat{r}(\partial_\theta^2 A_r - 2\partial_\theta A_\theta - A_r) + \hat{\theta}(\partial_\theta^2 A_\theta + 2\partial_\theta A_r - A_\theta). \qquad (5.242)\end{aligned}$$

Using this result in Equation 5.241, we find that the radial component of $\Box(\hat{r}A_r + \hat{\theta}A_\theta)$ is

$$\left[\frac{1}{r}\partial_r(r\partial_r) + \partial_z^2 - \frac{1}{c^2}\partial_t^2\right]A_r + \frac{1}{r^2}(\partial_\theta^2 A_r - 2\partial_\theta A_\theta - A_r) = \Box A_r - \frac{1}{r^2}(A_r + 2\partial_\theta A_\theta), \qquad (5.243)$$

while the azimuthal component reads

$$\left[\frac{1}{r}\partial_r(r\partial_r) + \partial_z^2 - \frac{1}{c^2}\partial_t^2\right]A_\theta + \frac{1}{r^2}(\partial_\theta^2 A_\theta + 2\partial_\theta A_r - A_\theta) = \Box A_\theta - \frac{1}{r^2}(A_\theta - 2\partial_\theta A_r). \qquad (5.244)$$

In cylindrical coordinates, the radial and azimuthal components of the wave equation are coupled. Moreover, for waves propagating along the $z$-axis, in the absence of sources, we can define transverse electric (TE) and transverse magnetic (TM) modes, by requiring that $E_z = 0$, or $B_z = 0$, respectively. It is then easily seen, by consulting Equations 5.228 and 5.229, that the two families of modes are generated by $A_r$ and $A_\theta$, for TE modes, and $A_z$ and $\phi$, for TM modes. The presence of four-current source terms or boundary conditions can give rise to hybrid modes, as will be discussed shortly.

We now focus on the so-called vacuum modes, which satisfy the wave equation

$$\begin{aligned}
\Box A_r - \frac{1}{r^2}(A_r + 2\partial_\theta A_\theta) &= 0,\\
\Box A_\theta - \frac{1}{r^2}(A_\theta - 2\partial_\theta A_r) &= 0,\\
\Box A_r &= 0,\\
\Box \phi &= 0.
\end{aligned} \tag{5.245}$$

and the Lorentz gauge condition, Equation 5.237.

The structure of the differential equations indicates that variables can be separated, as no operator cross-coupling the different degrees of freedom is present. Furthermore, we consider the azimuthal symmetry of the system, and Fourier transform the four-potential into momentum space:

$$\begin{aligned}
A_\mu(r, \theta, z, t) &= \sum_m \frac{1}{(\sqrt{2\pi})^2}\int_{-\infty}^{+\infty} d\omega\\
&\quad \times \int_{-\infty}^{+\infty} dk_\parallel \tilde{A}_\mu^m(\omega, k_\parallel) f_\mu^m(r) \exp[i(\omega t - k_\parallel z + m\theta)],
\end{aligned} \tag{5.246}$$

where the azimuthal mode index is an integer, $m \in \mathbb{N}$.

To write things in a more compact manner, we introduce the phase

$$\psi_m(\omega, k_\parallel) = \omega t - k_\parallel z + m\theta; \tag{5.247}$$

we then have

$$\begin{aligned}
\frac{1}{c^2}\partial_t^2[\tilde{A}_\mu^m(\omega, k_\parallel) f_\mu^m(r) e^{i\psi_m}] &= -\frac{\omega^2}{c^2}\tilde{A}_\mu^m(\omega, k_\parallel) f_\mu^m(r) e^{i\psi_m},\\
\partial_z^2[\tilde{A}_\mu^m(\omega, k_\parallel) f_\mu^m(r) e^{i\psi_m}] &= -k_\parallel^2 \tilde{A}_\mu^m(\omega, k_\parallel) f_\mu^m(r) e^{i\psi_m},\\
\frac{1}{r^2}\partial_\theta^2[\tilde{A}_\mu^m(\omega, k_\parallel) f_\mu^m(r) e^{i\psi_m}] &= -\frac{m^2}{r^2}\tilde{A}_\mu^m(\omega, k_\parallel) f_\mu^m(r) e^{i\psi_m},\\
\frac{2}{r^2}\partial_\theta[\tilde{A}_\mu^m(\omega, k_\parallel) f_\mu^m(r) e^{i\psi_m}] &= \frac{2im}{r^2}\tilde{A}_\mu^m(\omega, k_\parallel) f_\mu^m(r) e^{i\psi_m}.
\end{aligned} \tag{5.248}$$

With this, the operation of the d'Alembertian takes a somewhat simpler form, as we have established the following identities:

$$\frac{1}{c^2}\partial_t^2 \equiv -\frac{\omega^2}{c^2}, \quad \frac{1}{r^2}\partial_\theta^2 \equiv -\frac{m^2}{r^2}, \quad \frac{2}{r^2}\partial_\theta \equiv \frac{2im}{r^2}, \quad \partial_z^2 \equiv -k_\parallel^2. \tag{5.249}$$

To further simplify our notation, we recast Equation 5.246 as

$$2\pi A_\mu(x_\lambda) = \sum_m \iint_{\mathbb{R}^2} d\omega dk_\| \tilde{A}_\mu^m f_\mu^m \mathrm{e}^{i\psi_m}. \tag{5.250}$$

Let us first consider the TM mode generators:

$$\begin{aligned}
\Box 2\pi A_z &= \Box\left(\sum_m \iint_{\mathbb{R}^2} d\omega dk_\| \tilde{A}_z^m f_z^m \mathrm{e}^{i\psi_m}\right) \\
&= \sum_m \iint_{\mathbb{R}^2} d\omega dk_\| \Box(\tilde{A}_z^m f_z^m \mathrm{e}^{i\psi_m}) \\
&= \sum_m \iint_{\mathbb{R}^2} d\omega dk_\| \tilde{A}_z^m \Box(f_z^m \mathrm{e}^{i\psi_m}) \\
&= \sum_m \iint_{\mathbb{R}^2} d\omega dk_\| \left\{\frac{1}{r}\partial_r[r(\partial_r f_z^m)]\right. \\
&\quad \left. + \left(\frac{\omega^2}{c^2} - k_\|^2 - \frac{m^2}{r^2}\right) f_z^m \right\} \tilde{A}_z^m \mathrm{e}^{i\psi_m} \\
&= 0.
\end{aligned} \tag{5.251}$$

The fact that the d'Alembertian and series-integral operators commute is clear, as the d'Alembertian operates in real space, while the series-integral operators act in Fourier, or momentum, space. For the scalar potential, we have a similar equation, and we can simply make the substitution $\tilde{A}_z^m f_z^m \to \tilde{A}_0^m f_0^m = \tilde{\phi}^m f_0^m$.

For the radial and azimuthal components of the wave equation, we have to include the extra terms derived in Equations 5.243 and 5.244. We first have

$$\begin{aligned}
2\pi\left[\Box A_r - \frac{1}{r^2}(A_r + 2\partial_\theta A_\theta)\right] &= \left[\Box - \frac{1}{r^2}\right]\left(\sum_m \iint_{\mathbb{R}^2} d\omega dk_\| \tilde{A}_r^m f_r^m \mathrm{e}^{i\psi_m}\right) \\
&\quad - \frac{2}{r^2}\partial_\theta\left(\sum_m \iint_{\mathbb{R}^2} d\omega dk_\| \tilde{A}_\theta^m f_\theta^m \mathrm{e}^{i\psi_m}\right).
\end{aligned} \tag{5.252}$$

Next, we use the fact that the d'Alembertian and series-integral operators commute:

$$\begin{aligned}
2\pi\left[\Box A_r - \frac{1}{r^2}(A_r + 2\partial_\theta A_\theta)\right] &= \sum_m \iint_{\mathbb{R}^2} d\omega dk_\| \tilde{A}_r^m \left[\Box - \frac{1}{r^2}\right](f_r^m \mathrm{e}^{i\psi_m}) \\
&\quad - \sum_m \iint_{\mathbb{R}^2} d\omega dk_\| \tilde{A}_\theta^m \frac{2}{r^2}\partial_\theta(f_\theta^m \mathrm{e}^{i\psi_m}),
\end{aligned} \tag{5.253}$$

and the operational identities established in Equation 5.249, to obtain

$$2\pi\left[\Box A_r - \frac{1}{r^2}(A_r + 2\partial_\theta A_\theta)\right] = \sum_m \iint_{\mathbb{R}^2} d\omega\, dk_\parallel \left\{ \frac{1}{r}\partial_r[r(\partial_r f_z^m)] + \left(\frac{\omega^2}{c^2} - k_\parallel^2 - \frac{m^2+1}{r^2}\right) f_z^m \right\} \tilde{A}_r^m \mathrm{e}^{i\psi_m} - \sum_m \iint_{\mathbb{R}^2} d\omega\, dk_\parallel \left(\frac{2im}{r^2} f_\theta^m\right) \tilde{A}_\theta^m \mathrm{e}^{i\psi_m}. \quad (5.254)$$

For the azimuthal component, a similar derivation yields

$$2\pi\left[\Box A_\theta - \frac{1}{r^2}(A_\theta - 2\partial_\theta A_r)\right] = \sum_m \iint_{\mathbb{R}^2} d\omega\, dk_\parallel \left\{ \frac{1}{r}\partial_r[r(\partial_r f_\theta^m)] + \left(\frac{\omega^2}{c^2} - k_\parallel^2 - \frac{m^2+1}{r^2}\right) f_\theta^m \right\} \tilde{A}_\theta^m \mathrm{e}^{i\psi_m} + \sum_m \iint_{\mathbb{R}^2} d\omega\, dk_\parallel \left(\frac{2im}{r^2} f_r^m\right) \tilde{A}_r^m \mathrm{e}^{i\psi_m}.$$

As the complex exponentials, $e^{i\psi_m}$, are orthogonal functions, we can now consider the radial equation for the spectral density of a given azimuthal mode:

$$\begin{aligned}
&\left\{ \frac{1}{r}d_r[r(d_r f_r^m)] + \left(\frac{\omega^2}{c^2} - k_\parallel^2 - \frac{m^2+1}{r^2}\right) f_r^m \right\} \tilde{A}_r^m - \frac{2im}{r^2} f_\theta^m \tilde{A}_\theta^m = 0, \\
&\left\{ \frac{1}{r}d_r[r(d_r f_\theta^m)] + \left(\frac{\omega^2}{c^2} - k_\parallel^2 - \frac{m^2+1}{r^2}\right) f_\theta^m \right\} \tilde{A}_\theta^m + \frac{2im}{r^2} f_r^m \tilde{A}_r^m = 0, \\
&\left\{ \frac{1}{r}d_r[r(d_r f_z^m)] + \left(\frac{\omega^2}{c^2} - k_\parallel^2 - \frac{m^2}{r^2}\right) f_z^m \right\} \tilde{A}_z^m = 0, \\
&\left\{ \frac{1}{r}d_r[r(d_r f_0^m)] + \left(\frac{\omega^2}{c^2} - k_\parallel^2 - \frac{m^2}{r^2}\right) f_0^m \right\} \tilde{\phi}^m = 0.
\end{aligned} \quad (5.255)$$

More technically, the orthogonality of the complex exponentials allows us to diagonalize the wave equation, which means that to obtain a null result, each *individual* component must be zero. In other words, when we are dealing with a *linear* system of equations involving orthogonal functions, we can group terms for each eigenfunction and separately solve the equation for each term, individually. There is no cross-coupling between the orthogonal functions. This will be reviewed extensively when we solve the Green function problem at hand; we now return to the radial equations and consider the Lorentz gauge condition, which now reads

$$\frac{i\omega}{c^2} f_0^m \tilde{\phi}^m + \frac{1}{r}[d_r(r f_r^m)]\tilde{A}_r^m + \frac{im}{r} f_\theta^m \tilde{A}_\theta^m - ik_\parallel f_z^m \tilde{A}_z^m = 0, \tag{5.256}$$

and couples the various spectral amplitudes.

To solve the radial wave equation for the TM modes, we use the equation

$$\frac{1}{x}\frac{d}{dx}(xy') + \left(\beta^2 - \frac{p^2}{x^2}\right) y = 0, \tag{5.257}$$

which is satisfied by the Bessel functions, $y(x) = Z_p(\beta x)$, where $Z_p = c_1 \mathrm{J}_p + c_2 \mathrm{N}_2$ is a linear combination of Bessel functions of the first and second kind. Indeed, if we make the following identifications,

$$x \equiv r, \qquad y \equiv f_z^m, \qquad \beta^2 \equiv \frac{\omega^2}{c^2} - k_\parallel^2, \qquad p \equiv m, \tag{5.258}$$

we find that

$$f_z^m(r) = Z_m(k_\perp r) \tag{5.259}$$

is a solution of the radial wave equation,

$$\frac{1}{r} d_r[r(d_r f_z^m)] + \left(\frac{\omega^2}{c^2} - k_\parallel^2 - \frac{m^2}{r^2}\right) f_z^m = 0, \tag{5.260}$$

provided that the condition

$$\frac{\omega^2}{c^2} - k_\parallel^2 = k_\perp^2, \tag{5.261}$$

is satisfied. This last equation describes the dispersion of the eigenmode in vacuum. Because Bessel functions of the second kind diverge on axis, we further restrict our solution to

$$f_z^m(r) = \mathrm{J}_m(k_\perp r), \tag{5.262}$$

where we have absorbed the proportionality constant in the spectral mode amplitude, $\tilde{A}_z^m$. It is clear that we have an identical solution for the scalar potential:

$$f_0^m(r) = \mathrm{J}_m(k_\perp r). \tag{5.263}$$

So far, the spectral mode amplitudes, $\tilde{A}_z^m$ and $\tilde{\phi}^m$, can take arbitrary values; however, we will see that the Lorentz gauge condition yields a relation between these amplitudes. As a result, the amplitude of a given TM mode will be set by a single parameter.

For the TE modes, it is clear that the radial dependence of the eigenmodes is different from that derived for TM modes, as the radial and azimuthal components of the wave equation have a different structure than their axial and temporal counterparts. In particular, the equations are obviously coupled. To determine the TE eigenmodes, we consider the equation

$$y'' + \frac{1-2\alpha}{x}y' + \left(\beta^2 + \frac{\alpha^2 - p^2}{x^2}\right)y = 0, \tag{5.264}$$

and its solution,

$$y(x) = x^\alpha Z_p(\beta x). \tag{5.265}$$

For $\alpha = -1$, we have

$$y'' + \frac{3}{x}y' + \left(\beta^2 + \frac{1-p^2}{x^2}\right)y = 0, \qquad y(x) = \frac{1}{x}Z_p(\beta x). \tag{5.266}$$

Using the notation $\beta = k_\perp$ and $p = m$, we can recast Equation 5.266 as

$$y'' + \frac{1}{x}y' + \left(k_\perp^2 - \frac{m^2+1}{x^2}\right)y + \frac{2}{x}\left(y' + \frac{1}{x}y\right) = 0. \tag{5.267}$$

Continuing with our identification procedure, we write $x = r$, and $f_r^m(r) = 1/(k_\perp r)$ $\mathrm{j}_m(k_\perp r)$. Multiplying Equation 5.267 by $k_\perp^{-1}$, we find that

$$f_r^{m''} + \frac{1}{r}f_r^{m'} + \left(k_\perp^2 - \frac{m^2+1}{r^2}\right)f_r^m + \frac{2}{r}\left(f_r^{m'} + \frac{1}{r}f_r^m\right) = 0. \tag{5.268}$$

Finally, we note that

$$\begin{aligned} f_r^{m'} &= \frac{d}{dr}\left[\frac{1}{k_\perp r}\mathrm{J}_m(k_\perp r)\right] \\ &= \frac{1}{k_\perp}\left[-\frac{1}{r^2}\mathrm{J}_m(k_\perp r) + \frac{1}{r}k_\perp \mathrm{J}_m'(k_\perp r)\right] \\ &= \frac{1}{r}\left[\mathrm{J}_m'(k_\perp r) - \frac{1}{r}f_r^m\right]. \end{aligned} \tag{5.269}$$

Using this last result in Equation 5.268, we obtain

$$f_r^{m''} + \frac{1}{r} f_r^{m'} + \left(k_\perp^2 - \frac{m^2+1}{r^2}\right) f_r^m + \frac{2}{r}\left\{\frac{1}{r}\left[\mathrm{J}_m(k_\perp r) - \frac{1}{r} f_r^m\right] + \frac{1}{r} f_r^m\right\}$$

$$= f_r^{m''} + \frac{1}{r} f_r^{m'} + \left(k_\perp^2 - \frac{m^2+1}{r^2}\right) f_r^m + \frac{2}{r}\mathrm{J}'_m(k_\perp r) = 0. \tag{5.270}$$

Again, the reason we choose the Bessel function of the first kind in our solution is because it is bounded on axis. We now compare this equation with the radial component of the wave equation,

$$\left\{\frac{1}{r} d_r[r(d_r f_r^m)] + \left(\frac{\omega^2}{c^2} - k_\parallel^2 - \frac{m^2+1}{r^2}\right) f_r^m\right\} \tilde{A}_r^m - \frac{2im}{r^2} f_\theta^m \tilde{A}_\theta^m = 0. \tag{5.271}$$

It is easily seen that

$$\frac{1}{r} d_r[r(d_r f_r^m)] = \frac{1}{r}(d_r f_r^m) + d_r^2 f_r^m = \frac{1}{r} f_r^{m'} + f_r^{m''}; \tag{5.272}$$

furthermore, if the spectral mode amplitudes satisfy the relation

$$\tilde{A}_\theta^m = \frac{i}{m} \tilde{A}_r^m, \tag{5.273}$$

we can rewrite Equation 5.271 as

$$\left\{f_r^{m''} + \frac{1}{r} f_r^{m'} + \left(\frac{\omega^2}{c^2} - k_\parallel^2 - \frac{m^2+1}{r^2}\right) f_r^m - \frac{2im}{r^2} f_\theta^m \frac{i}{m}\right\} \tilde{A}_r^m = 0. \tag{5.274}$$

Finally, if we identify

$$\frac{\omega^2}{c^2} - k_\parallel^2 = k_\perp^2, \qquad f_\theta^m = \mathrm{J}'_m(k_\perp r), \tag{5.275}$$

the radial wave equation is satisfied, independent of the spectral mode amplitude.

To complete the derivation, we must show the consistency of our solution with the two other equations constraining the radial and azimuthal components of the four-vector. We will first address this question for the Lorentz gauge condition. With the solutions derived for the TM and TE modes, we have, so far,

$$f_z^m(r) = f_0^m(r) = \mathrm{J}_m(k_\perp r), \quad f_r^m(r) = \frac{1}{k_\perp r}\mathrm{J}_m(k_\perp r), \quad f_\theta^m(r)\mathrm{J}'_m(k_\perp r), \tag{5.276}$$

where the spectral amplitudes $\tilde{A}_z^m$, $\tilde{\phi}^m$, and $\tilde{A}_r^m$ are still arbitrary, but the azimuthal spectral amplitude satisfies the constraint $m\tilde{A}_\theta^m = \tilde{A}_r^m$.

Finally the argument of the Bessel functions satisfies the vacuum dispersion relation, $\omega^2 = (k_\parallel^2 + k_\perp^2)c^2$.

Using these results in the Lorentz gauge condition, we find that

$$\begin{aligned} &\left(\frac{i\omega}{c^2}\tilde{\phi}^m - ik_\parallel\tilde{A}_z^m\right)\mathrm{J}_m(k_\perp r) + \frac{1}{r}\left\{d_r\left[r\frac{1}{k_\perp r}\mathrm{J}_m(k_\perp r)\right]\right\}\tilde{A}_r^m + \frac{im}{r}\mathrm{J}'_m(k_\perp r)\tilde{A}_\theta^m \\ &= i\left(\frac{\omega}{c^2}\tilde{\phi}^m - k_\parallel\tilde{A}_z^m\right)\mathrm{J}_m(k_\perp r) + \frac{1}{r}\left\{d_r\left[\frac{1}{k_\perp}\mathrm{J}_m(k_\perp r)\right] - \mathrm{J}'_m(k_\perp r)\right\}\tilde{A}_r^m = 0. \end{aligned} \tag{5.277}$$

The term multiplying the radial component of the spectral is identically equal to zero, so we are left with a constraint on the axial and temporal amplitudes:

$$\frac{\omega}{c^2}\tilde{\phi}^m = k_\parallel\tilde{A}_z^m. \tag{5.278}$$

Thus, we see that for a given TE or TM mode, we can define the spectral amplitudes in terms of two generators, $\tilde{A}_{TE}^m(\omega, k_\parallel)$ and $\tilde{A}_{TM}^m(\omega, k_\parallel)$, as follows:

$$\begin{aligned} \tilde{A}_r^m(\omega, k_\parallel) &= \tilde{A}_{TE}^m(\omega, k_\parallel), \quad \tilde{A}_\theta^m(\omega, k_\parallel) = \frac{i}{m}\tilde{A}_{TE}^m(\omega, k_\parallel), \\ \tilde{A}_z^m(\omega, k_\parallel) &= \tilde{A}_{TM}^m(\omega, k_\parallel), \quad \tilde{\phi}^m(\omega, k_\parallel) = \frac{k_\parallel c^2}{\omega}\tilde{A}_{TM}^m(\omega, k_\parallel). \end{aligned} \tag{5.279}$$

Moreover, the fact that the Lorentz gauge condition is automatically decoupled in terms of TE and TM modes reinforces this basic decomposition of guided waves into two independent families. The last equation that must be checked for consistency is the azimuthal component of the wave equation. To verify that it is, indeed, satisfied by our solution, it proves useful to express the derivative of the Bessel function as follows:

$$\frac{dZ_p}{dx} = \frac{p}{x}Z_p - Z_{p+1}. \tag{5.280}$$

With this, we proceed with the identification used earlier, to write

$$f_\theta^m(r) = J_m'(k_\perp r) = \frac{m}{k_\perp r} J_m(k_\perp r) - J_{m+1}(k_\perp r) = m f_r^m(r) - J_{m+1}(k_\perp r). \quad (5.281)$$

In addition, with the relations between the amplitudes derived previously, the azimuthal component of the wave equation reads

$$\left\{ \frac{1}{r} d_r [r(d_r f_\theta^m)] + \left( k_\perp^2 - \frac{m^2+1}{r^2} \right) f_\theta^m + \frac{2m^2}{r^2} f_r^m \right\} \tilde{A}_\theta^m = 0, \quad (5.282)$$

or

$$f_\theta^{m''} + \frac{1}{r} f_\theta^{m'} + \left( k_\perp^2 - \frac{m^2+1}{r^2} \right) f_\theta^m + \frac{2m^2}{r^2} f_r^m = 0. \quad (5.283)$$

Using Equation 5.281, we have

$$\begin{aligned} f_\theta^{m''} &= m f_r^m - J_{m+1}, \\ f_\theta^{m'} &= m f_r^{m'} - J_{m+1}', \\ f_\theta^{m''} &= m f_r^{m''} - J_{m+1}'', \end{aligned} \quad (5.284)$$

which yields

$$\begin{aligned} & m\left[ f_r^{m''} + \frac{1}{r} f_r^{m'} + \left( k_\perp^2 - \frac{m^2+1}{r^2} \right) f_r^m \right] + \frac{2m^2}{r^2} f_r^m \\ & - \left[ J_{m+1}'' + \frac{1}{r} J_{m+1}' + \left( k_\perp^2 - \frac{m^2+1}{r^2} \right) J_{m+1} \right] = 0. \end{aligned} \quad (5.285)$$

Since $f_r^m(r) = 1/(k_\perp r) J_m(k_\perp r)$, it obeys the equation

$$f_r^{m''} + \frac{3}{r} f_r^{m'} + \left( k_\perp^2 - \frac{m^2-1}{r^2} \right) f_r^m = 0, \quad (5.286)$$

while for $J_{m+1}(k_\perp r)$, we have

$$J_{m+1}'' + \frac{1}{r} J_{m+1}' + \left[ k_\perp^2 - \frac{(m+1)}{r^2} \right] J_{m+1} = 0. \quad (5.287)$$

Therefore, we can write

$$f_r^{m''} + \frac{1}{r} f_r^{m'} + \left(k_\perp^2 - \frac{m^2+1}{r^2}\right) f_r^m = -\frac{2}{r} f_r^{m'} - \frac{2}{r^2} f_r^m,$$
$$\mathrm{J}''_{m+1} + \frac{1}{r}\mathrm{J}'_{m+1} + \left(k_\perp^2 - \frac{m^2+1}{r^2}\right)\mathrm{J}_{m+1} = \frac{2m}{r^2}\mathrm{J}_{m+1}, \tag{5.288}$$

as $(m+1)^2 = (m^2+1) + 2m$.

These results can now be used in Equation 5.285 to yield

$$m\left(-\frac{2}{r} f_r^{m'} - \frac{2}{r^2} f_r^m\right) + \frac{2m^2}{r^2} f_r^m - \frac{2m}{r^2}\mathrm{J}_{m+1} = \frac{2m}{r}\left(-f_r^{m'} + \frac{m-1}{r} f_r^m - \frac{1}{r}\mathrm{J}_{m+1}\right) = 0. \tag{5.289}$$

Remembering that $f_r^{m'} = (1/r)[\mathrm{J}'_m(k_\perp r) - (1/r) f_r^m]$, as derived in Equation 5.269, and using Equation 5.281 to write $\mathrm{J}_{m+1}(k_\perp r) = m/(k_\perp r)\mathrm{J}_m(k_\perp r) - \mathrm{J}'_m(k_\perp r)$, we can verify that Equation 5.289 is correctly satisfied:

$$-\frac{1}{r}\left[\mathrm{J}'_m(k_\perp r) - \frac{1}{r} f_r^m\right] + \frac{m-1}{r} f_r^m - \frac{1}{r}\left[\frac{m}{k_\perp r}\mathrm{J}_m(k_\perp r) - \mathrm{J}'_m(k_\perp r)\right] = 0. \tag{5.290}$$

We have thus shown that the solution given in Equations 5.276 and 5.279 is fully self-consistent. The delicate balance of terms in the Bessel function solution appears clearly, and the vacuum dispersion relation corresponds to the eigenvalues associated with these eigenmodes. Also note that the vacuum spectrum is continuous.

### 5.6.2 Cylindrical Waveguide Eigenmodes

The presence of boundaries modifies the situation, as a discrete eigenvalue spectrum emerges. Let us consider a perfectly conducting, cylindrical surface, described by the equation $r = a$. In general, the boundary conditions take the form

$$\hat{n} \times \mathbf{E} = \mathbf{0}, \qquad \hat{n} \cdot \mathbf{B} = 0, \tag{5.291}$$

where $\hat{n}$ is the normal to the surface.

The tangential electric field is zero, as the image currents it drives cancel the field at the surface; the normal magnetic induction is also null because of the induced currents. In the case of a cylindrical waveguide, $\hat{n} = -\hat{r}$, and

Equation 5.291 reduces to

$$E_\theta(r = a) = 0, \qquad E_z(r = a) = 0, \qquad B_r(r = a) = 0. \tag{5.292}$$

For TE modes, we have $E_z = 0$, and using the definition of the fields and inductions, we see that

$$E_\theta = -\partial_t A_\theta \propto \mathrm{J}'_m(k_\perp r), \qquad B_r = -\partial_z A_\theta \propto \mathrm{J}'_m(k_\perp r). \tag{5.293}$$

The time-independent boundary condition is satisfied if

$$\mathrm{J}'_m(k_\perp a) = 0, \qquad k_\perp = \frac{\chi'_{mn}}{a}, \tag{5.294}$$

where $\chi'_{mn}$ is the $n$-th zero of $\mathrm{J}'_m$. The dispersion relation is now

$$\frac{\omega^2}{c^2} = k_\parallel^2 + \frac{\chi'^2_{mn}}{a^2}, \tag{5.295}$$

and the aforementioned discrete eigenvalue spectrum appears clearly, as exemplified by the cutoff frequencies, $\chi'_{mn}c/a$. If we compare this dispersion relation with that derived for the unbounded geometry, it is important to note that in a waveguide, the transverse wavenumber has a nonzero minimum value given by the lowest cutoff frequency; in vacuum, by contrast, the transverse wavenumber component for a nondiffracting plane wave simply corresponds to the off-axis propagation of the wave and can be eliminated by a simple rotation.

Considering the TM modes, we now have

$$\begin{aligned} E_\theta &= -\frac{1}{r}\partial_\theta \phi \propto \mathrm{J}_m(k_\perp r), \\ E_z &= -\partial_z \phi - \partial_t A_z \propto \mathrm{J}_m(k_\perp r), \\ B_r &= \frac{1}{r}\partial_\theta A_z \propto \mathrm{J}_m(k_\perp r). \end{aligned} \tag{5.296}$$

and the boundary condition is

$$\mathrm{J}_m(k_\perp a) = 0, \qquad k_\perp = \frac{\chi_{mn}}{a}, \tag{5.297}$$

where $\chi_{mn}$ is the $n$-th zero of $\mathrm{J}_m$. Again, a discrete eigenvalue spectrum results from the boundary condition; it is, however, different from the TE mode spectrum.

At this level, a few important remarks are in order. First, we note that the cylindrical boundary condition, which has the correct symmetry for these eigenmodes, is a natural constraint. In vacuum, the modes have the same structure, but there is no radial scale to the problem; therefore, a continuous spectrum emerges. The radial boundary condition simply sets the radial scale, and its Fourier-conjugate, the transverse wavenumber, with a discrete mode spectrum. This explains why the eigenwaves propagating in the cylindrical waveguide retain a flat phase front. These modes are also vacuum eigenmodes; in other words, they are plane waves. The second point is that one can take advantage of the vacuum eigenmode structure to build natural boundary conditions for a given system. For example, by beating vacuum eigenmodes with the same frequency but different axial and transverse wavenumbers, it is possible, in certain cases, to generate a complex surface where the boundary conditions given in Equation 5.291 are satisfied in a time-independent manner. This will be studied in the next section, which is concerned with the eigenmodes of a cylindrical corrugated waveguide and their excitation by a point charge traveling at constant velocity along the waveguide axis.

We now summarize our results. The bounded TE modes are described by

$$\begin{aligned}
A_r(x_\lambda) &= \frac{1}{2\pi}\sum_m\sum_n\iint_{\mathbb{R}^2} d\omega\, dk_\parallel \tilde{A}^m_{TE}(\omega, k_\parallel) \\
&\quad\times\frac{a}{\chi'_{mn}r}\mathrm{J}_m\left(\chi'_{mn}\frac{r}{a}\right)\exp[i(\omega t - k_\parallel z + m\theta)], \\
A_\theta(x_\lambda) &= \frac{1}{2\pi}\sum_m\sum_n\iint_{\mathbb{R}^2} d\omega\, dk_\parallel \frac{i}{m}\tilde{A}^{mn}_{TE}(\omega, k_\parallel) \\
&\quad\times\mathrm{J}'_m\left(\chi'_{mn}\frac{r}{a}\right)\exp[i(\omega t - k_\parallel z + m\theta)], \\
\frac{\omega^2}{c^2} &= k_\parallel^2 + \frac{\chi'^2_{mn}}{a^2},
\end{aligned} \tag{5.298}$$

while the TM modes are given by

$$\begin{aligned}
A_z(x_\lambda) &= \frac{1}{2\pi}\sum_m\sum_n\iint_{\mathbb{R}^2} d\omega\, dk_\parallel \tilde{A}^{mn}_{TM}(\omega, k_\parallel) \\
&\quad\times\mathrm{J}_m\left(\chi_{mn}\frac{r}{a}\right)\exp[i(\omega t - k_\parallel z + m\theta)], \\
\phi(x_\lambda) &= \frac{1}{2\pi}\sum_m\sum_n\iint_{\mathbb{R}^2} d\omega\, dk_\parallel \frac{k_\parallel c^2}{\omega}\tilde{A}^{mn}_{TM}(\omega, k_\parallel) \\
&\quad\times\mathrm{J}_m\left(\chi_{mn}\frac{r}{a}\right)\exp[i(\omega t - k_\parallel z + m\theta)], \\
\frac{\omega^2}{c^2} &= k_\parallel^2 + \frac{\chi^2_{mn}}{a^2}.
\end{aligned} \tag{5.299}$$

The fields and inductions can be determined by using Equations 5.228 and 5.229: for the TE modes, the electric field is

$$\begin{aligned}
E_r &= \frac{-1}{2\pi}\sum_m\sum_n\iint_{\mathbb{R}^2} d\omega\, dk_\parallel \frac{i\omega a}{\chi'_{mn} r}\tilde{A}_{TE}^{mn}(\omega, k_\parallel)\mathrm{J}_m\left(\chi'_{mn}\frac{r}{a}\right)e^{i\psi_m(\omega,k_\parallel)},\\
E_\theta &= \frac{1}{2\pi}\sum_m\sum_n\iint_{\mathbb{R}^2} d\omega\, dk_\parallel \frac{\omega}{m}\tilde{A}_{TE}^{mn}(\omega, k_\parallel)\mathrm{J}'_m\left(\chi'_{mn}\frac{r}{a}\right)e^{i\psi_m(\omega,k_\parallel)},\\
E_z &= 0,
\end{aligned} \tag{5.300}$$

and the magnetic induction is

$$\begin{aligned}
B_r &= \frac{-1}{2\pi}\sum_m\sum_n\iint_{\mathbb{R}^2} d\omega\, dk_\parallel \frac{k_\parallel}{m}\tilde{A}_{TE}^{mn}(\omega, k_\parallel)\mathrm{J}'_m\left(\chi'_{mn}\frac{r}{a}\right)e^{i\psi_m(\omega,k_\parallel)},\\
B_\theta &= \frac{-1}{2\pi}\sum_m\sum_n\iint_{\mathbb{R}^2} d\omega\, dk_\parallel \frac{ik_\parallel a}{\chi'_{mn} r}\tilde{A}_{TE}^{mn}(\omega, k_\parallel)\mathrm{J}_m\left(\chi'_{mn}\frac{r}{a}\right)e^{i\psi_m(\omega,k_\parallel)},\\
B_z &= \frac{-1}{2\pi}\sum_m\sum_n\iint_{\mathbb{R}^2} d\omega\, dk_\parallel \frac{i\chi'_{mn}}{ma}\tilde{A}_{TE}^{mn}(\omega, k_\parallel)\mathrm{J}_m\left(\chi'_{mn}\frac{r}{a}\right)e^{i\psi_m(\omega,k_\parallel)}.
\end{aligned} \tag{5.301}$$

Here, we have used the notation

$$\psi_m(\omega, k_\parallel) = \omega t - k_\parallel z + m\theta, \tag{5.302}$$

for the phase. The exact functional dependence of $B_z$ is obtained as follows. We first have

$$B_z = \frac{1}{r}[\partial_r(rA_\theta) - \partial_\theta A_r] = \frac{1}{r}[A_\theta + r(\partial_r A_\theta) - \partial_\theta A_r]; \tag{5.303}$$

considering a particular $\mathrm{TE}_{mn}$ mode, this yields

$$B_{z,mn} \propto \frac{1}{r}\left(\frac{i}{m}\mathrm{J}'_m + r\frac{i}{m}k_\perp \mathrm{J}''_m - im\frac{1}{k_\perp r}\mathrm{J}_m\right), \tag{5.304}$$

where we have used the notation $k_\perp = \chi'_{mn}/a$.

We can factorize Equation 5.304, to find

$$B_{z,mn} \propto i\frac{k_\perp}{m}\left[\mathrm{J}''_m + \frac{1}{k_\perp r}\mathrm{J}'_m - \left(\frac{m}{k_\perp r}\right)^2 \mathrm{J}_m\right]. \tag{5.305}$$

Furthermore, we know that $J_m$ satisfies the differential equation

$$J''_m + \frac{1}{x}J'_m + \left(1 - \frac{m^2}{x^2}\right)J_m = 0, \tag{5.306}$$

for $x = k_\perp r$. Therefore, we have

$$J''_m + \frac{1}{x}J'_m - \frac{m^2}{x^2}J_m = J''_m + \frac{1}{k_\perp r}J'_m - \left(\frac{m}{k_\perp r}\right)^2 J_m = -J_m, \tag{5.307}$$

which yields the sought-after result.

In the case of TM modes, the electric field is

$$\begin{aligned}
E_r &= \frac{-1}{2\pi}\sum_m\sum_n\iint_{\mathbb{R}^2} d\omega\, dk_\parallel \frac{\chi_{mn}k_\parallel c^2}{a\omega}\tilde{A}^{mn}_{TM}(\omega, k_\parallel)J'_m\left(\chi_{mn}\frac{r}{a}\right)e^{i\psi_m(\omega,k_\parallel)},\\
E_\theta &= \frac{-1}{2\pi}\sum_m\sum_n\iint_{\mathbb{R}^2} d\omega\, dk_\parallel \frac{imk_\parallel c^2}{r\omega}\tilde{A}^{mn}_{TM}(\omega, k_\parallel)J_m\left(\chi_{mn}\frac{r}{a}\right)e^{i\psi_m(\omega,k_\parallel)},\\
E_z &= \frac{-1}{2\pi}\sum_m\sum_n\iint_{\mathbb{R}^2} d\omega\, dk_\parallel \frac{i\chi^2_{mn}c^2}{a^2\omega}\tilde{A}^{mn}_{TM}(\omega, k_\parallel)J_m\left(\chi_{mn}\frac{r}{a}\right)e^{i\psi_m(\omega,k_\parallel)},
\end{aligned} \tag{5.308}$$

and for the magnetic induction, we have

$$\begin{aligned}
B_r &= \frac{1}{2\pi}\sum_m\sum_n\iint_{\mathbb{R}^2} d\omega\, dk_\parallel \frac{im}{r}\tilde{A}^{mn}_{TM}(\omega, k_\parallel)J_m\left(\chi_{mn}\frac{r}{a}\right)e^{i\psi_m(\omega,k_\parallel)},\\
B_\theta &= \frac{-1}{2\pi}\sum_m\sum_n\iint_{\mathbb{R}^2} d\omega\, dk_\parallel \frac{\chi_{mn}}{a}\tilde{A}^{mn}_{TM}(\omega, k_\parallel)J'_m\left(\chi_{mn}\frac{r}{a}\right)e^{i\psi_m(\omega,k_\parallel)},\\
B_z &= 0.
\end{aligned} \tag{5.309}$$

For the derivation of the axial electric field component, one considers a given $TM_{mn}$ mode and uses the dispersion relation:

$$E_{z,mn} \propto \left(ik_\parallel\frac{k_\parallel c^2}{\omega} - i\omega\right)J_m = \frac{i}{\omega}(k^2_\parallel c^2 - \omega^2)J_m = -\frac{ik^2_\perp c^2}{\omega}J_m = -\frac{i\chi^2_{mn}c^2}{a^2\omega}J_m. \tag{5.310}$$

To solve the Green function problem, we will use the vacuum eigenmodes derived here and the fact that they form a complete set of orthogonal functions, capable of describing any excitation in the waveguide. At this point, we also note that the transverse components of the four-current will drive TE modes, while the charge density and axial current excite TM modes. The other method consists of finding the new eigenmodes of the system, including the four-current source; this approach will be studied in detail in Chapter 9.

### 5.6.3 Orthogonality of the Cylindrical Waveguide Eigenmodes

Continuing with our program to derive the Green function in a cylindrical waveguide, we now examine the orthogonality properties of the vacuum eigenmodes. A few mathematical identities will prove useful in our endeavor to diagonalize the driven wave equation. We start with the azimuthal integral

$$\frac{1}{2\pi}\int_0^{2\pi} e^{i(m-p)\theta}\,d\theta = \delta_{mp}, \tag{5.311}$$

which is easily verified by first considering the case $m = p$, which clearly yields $\delta_{mm} = \delta_{pp} = 1$; for $m \neq p$, we have

$$\frac{1}{2\pi}\int_0^{2\pi} e^{i(m-p)\theta}\,d\theta = \frac{1}{2\pi}\left[\frac{e^{i(m-p)\theta}}{i(m-p)}\right]_0^{2\pi} = 0, \tag{5.312}$$

since $m - p \in \mathbb{N}$. Next, we consider the Fourier projection,

$$\frac{1}{2\pi}\int_{-\infty}^{+\infty} \exp[i(\omega-\omega')t]dt = \delta(\omega-\omega'), \tag{5.313}$$

which we prove by considering the limit

$$\frac{1}{2\pi}\int_{-\infty}^{+\infty} \exp[i(\omega-\omega')t]\,dt = \lim_{a\to\infty}\left\{\frac{1}{2\pi}\int_{-a}^{+a} \exp[i(\omega-\omega')t]\,dt\right\}. \tag{5.314}$$

We then have

$$\begin{aligned}\frac{1}{2\pi}\int_{-a}^{+a} \exp[i(\omega-\omega')t]\,dt &= \frac{1}{2\pi}\left[\frac{e^{i(\omega-\omega')t}}{i(\omega-\omega')}\right]_{-a}^{+a} \\ &= \frac{1}{2\pi}\left[\frac{e^{i(\omega-\omega')a}+e^{i(\omega-\omega')a}}{i(\omega-\omega')}\right] \\ &= \frac{\sin[(\omega-\omega')a]}{\pi(\omega-\omega')}.\end{aligned} \tag{5.315}$$

This function is plotted in Figure 5.1; the maximum is obtained for $\omega = \omega'$, with the value $a/\pi$; moreover, its integral is normalized. If we define $x = \omega - \omega'$, we have

$$\int_{-\infty}^{+\infty} \frac{\sin ax}{\pi x}\, dx = 1. \tag{5.316}$$

Therefore, as discussed in Section 5.2,

$$\lim_{a\to\infty}\left\{\frac{1}{2\pi}\int_{-a}^{+a} \exp[i(\omega-\omega')t]\, dt\right\} = \delta(\omega-\omega'). \tag{5.317}$$

For the Bessel function equivalent of these orthogonality conditions, we must distinguish between the TE and TM modes. In the latter case, we will use the well-known integral

$$\begin{aligned}\int_0^1 x\mathrm{J}_p(\chi_{pq})\mathrm{J}_p(\chi_{pn})\, dx &= \frac{1}{2}\mathrm{J}_{p+1}(\chi_{pq})\mathrm{J}_{p+1}(\chi_{pn})\delta_{qn}\\ &= \frac{1}{2}\mathrm{J}^2_{p+1}(\chi_{pq})\delta_{qn}.\end{aligned} \tag{5.318}$$

For the TE modes, we start from the indefinite integral

$$\int xZ_p(\alpha x)B_p(\beta x)\, dx = \frac{\beta xZ_p(\alpha x)B_{p-1}(\beta x) - \alpha xZ_{p-1}(\alpha x)B_p(\beta x)}{\alpha^2-\beta^2}, \tag{5.319}$$

where $\alpha \neq \beta$, and $Z_p$ and $B_p$ are arbitrary Bessel functions of the same order.
On the other hand, in the case where $\alpha = \beta$, we have

$$\int xZ_p^2(\alpha x)\, dx = \frac{x^2}{2}[Z_p^2(\alpha x) - Z_{p-1}(\alpha x)Z_{p+1}(\alpha x)]. \tag{5.320}$$

Returning to Equation 5.319, we consider the special case where

$$\begin{aligned}&\int_0^1 x\mathrm{J}_p(\chi'_{pq}x)\mathrm{J}_p(\chi'_{pn}x)\, dx\\ &\quad= \frac{1}{\chi'^2_{pq}-\chi'^2_{pn}}[\chi'_{pn}x\mathrm{J}_p(\chi'_{pq}x)\mathrm{J}_{p-1}(\chi'_{pn}x) - \chi'_{pq}x\mathrm{J}_{p-1}(\chi'_{pq}x)\mathrm{J}_p(\chi'_{pn}x)]_0^1\\ &\quad= \frac{1}{\chi'^2_{pq}-\chi'^2_{pn}}[\chi'_{pn}\mathrm{J}_p(\chi'_{pq})\mathrm{J}_{p-1}(\chi'_{pn}) - \chi'_{pn}\mathrm{J}_{p-1}(\chi'_{pq})\mathrm{J}_p(\chi'_{pn})].\end{aligned} \tag{5.321}$$

We now use the relation between Bessel functions and their derivatives:

$$Z'_p(x) = Z_{p-1}(x) - \frac{p}{x} Z_p(x); \tag{5.322}$$

in the special case when $Z_p = J_p$ and $x = \chi'_{pq}$, we find that

$$J'_p(\chi'_{pq}) = 0 = J_{p-1}(\chi'_{pq}) - \frac{p}{\chi'_{pq}} J_p(\chi'_{pq}). \tag{5.323}$$

Of course, the same holds for $x = \chi'_{pn}$, and we can use these results in Equation 5.321:

$$\int_0^1 x J_p(\chi'_{pq}x) J_p(\chi'_{pn}x)\, dx = \frac{1}{\chi'^2_{pq} - \chi'^2_{pn}} \left[ \chi'_{pn} J_p(\chi'_{pq}) \frac{p}{\chi'_{pn}} J_p(\chi'_{pq}) - \chi'_{pq} \frac{p}{\chi'_{pq}} J_p(\chi'_{pq}) J_p(\chi'_{pn}) \right] = 0. \tag{5.324}$$

Finally, for $q = n$, Equation 5.320 yields

$$\begin{aligned}\int_0^1 x J_p^2(\chi'_{pq}x)\, dx &= \left[ \frac{x^2}{2} \left\{ J_p^2(\chi'_{pq}x) - J_{p-1}(\chi'_{pq}x) J_{p+1}(\chi'_{pn}x) \right\} \right]_0^1 \\ &= \frac{1}{2} [J_p^2(\chi'_{pq}) - J_{p-1}(\chi'_{pq}) J_{p+1}(\chi'_{pq})] \\ &= \frac{1}{2} J_p^2(\chi'_{pq}) \left( 1 - \frac{p^2}{\chi'^2_{pq}} \right),\end{aligned} \tag{5.325}$$

where we have used Equations 5.322 and 5.323, and their counterparts,

$$\begin{aligned} Z'_p(x) &= \frac{p}{x} Z_p(x) - Z_{p+1}(x), \\ J'_p(\chi'_{pq}) &= 0 = \frac{p}{\chi'_{pq}} J_p(\chi'_{pq}) - J_{p+1}(\chi'_{pq}). \end{aligned} \tag{5.326}$$

Therefore, we can recast the orthonormalization condition for the TE modes as

$$\int_0^1 x J_p(\chi'_{pq}x) J_p(\chi'_{pq}x)\, dx = \delta_{qn} \frac{1}{2} J_p^2(\chi'_{pq}) \left( 1 - \frac{p^2}{\chi'^2_{pq}} \right). \tag{5.327}$$

### 5.6.4 Eigenmode Decomposition of the Four-Current

Having established the necessary mathematical identities, our next goal is to solve the axial and temporal components of the driven wave equation:

$$\square A_z + \mu_0 j_z = 0, \qquad \square \phi + \frac{1}{\varepsilon_0}\phi = 0. \tag{5.328}$$

The fundamental idea is that by expanding the solution of the driven wave equation into orthogonal vacuum eigenmodes, we automatically satisfy the boundary conditions, and that the operation of the d'Alembertian upon the eigenfunctions takes a particularly simple form. For example, for a given $\mathrm{TM}_{mn}$ mode, we have

$$\begin{aligned} \square\left[\mathrm{J}_m\left(\chi_{mn}\frac{r}{a}\right)e^{i\psi_{mn}(\omega,k_\parallel)}\right] &= D_{mn}(\omega,k_\parallel)\left[\mathrm{J}_m\left(\chi_{mn}\frac{r}{a}\right)e^{i\psi_\mu(\omega,k\|)}\right], \\ D_{mn}(\omega,k_\parallel) &= \left(\frac{\omega^2}{c^2} - k_\parallel^2 - \frac{\chi_{mn}^2}{a^2}\right); \end{aligned} \tag{5.329}$$

therefore,

$$\begin{aligned} \square A_z &= \frac{1}{2\pi}\sum_m\sum_n\iint_{\mathbb{R}^2} d\omega\, dk_\parallel \square\left[\tilde{A}_{TM}^{mn}(\omega,k_\parallel)\mathrm{J}_m\left(\chi_{mn}\frac{r}{a}\right)\mathrm{e}^{i\psi_{mn}(\omega,k_\parallel)}\right] \\ &= \frac{1}{2\pi}\sum_m\sum_n\iint_{\mathbb{R}^2} d\omega\, dk_\parallel D_{mn}(\omega,\, k_\parallel)\tilde{A}_{TM}^{mn}(\omega,k_\parallel)\mathrm{J}_m\left(\chi_{mn}\frac{r}{a}\right)\mathrm{e}^{i\psi_m(\omega,k_\parallel)}. \end{aligned}$$

In the terminology of classical electrodynamics, the eigenvalue $D_{mn}(\omega, k_\parallel)$ is called the dispersion relation, whereas in QED, it is analogous to the propagator.

Proceeding with our derivation, we now add the axial drive current: Equation 5.328 becomes

$$\frac{1}{2\pi}\sum_m\sum_n\iint_{\mathbb{R}^2} d\omega\, dk_\parallel D_{mn}(\omega,\, k_\parallel)\tilde{A}_{TM}^{mn}(\omega,\, k_\parallel)\mathrm{J}_m\left(\chi_{mn}\frac{r}{a}\right)\mathrm{e}^{i\psi_m(\omega,k_\parallel)} = -\mu_0 j_z. \tag{5.330}$$

The initial problem has been replaced by an infinite set of coupled equations, and our objective is now to determine the mode amplitudes, $\tilde{A}_{TM}^{mn}(\omega, k_\parallel)$. Knowledge of these spectral densities is equivalent to a complete solution of the problem because the normal TE and TM modes form a complete set of Hilbert functions.

To solve Equation 5.330, we need to diagonalize the left-hand side by using the orthogonality of the eigenmodes; this operation will simultaneously result into a projection of the drive current on the eigenmodes. We will know how each vacuum normal mode is excited by the source.

We first use the azimuthal integral of Equation 5.311 by applying the integral operator,

$$\int_0^{2\pi} e^{-ip\theta}\,d\theta, \tag{5.331}$$

to Equation 5.330. By virtue of Equation 5.311, the sum over the azimuthal mode index vanishes, as we are diagonalizing the azimuthal eigenfunctions and projecting the source term:

$$\begin{aligned}
&\frac{1}{2\pi}\int_0^{2\pi} e^{-ip\theta}\,d\theta \sum_m \sum_n \iint_{\mathbb{R}^2} d\omega\, dk_\parallel D_{mn}(\omega, k_\parallel)\tilde{A}_{TM}^{mn}(\omega, k_\parallel)\\
&\quad\times \mathrm{J}_m\left(\chi_{mn}\frac{r}{a}\right)\exp[i(\omega t - k_\parallel z + m\theta)]\\
&\quad= \sum_m \sum_n \iint_{\mathbb{R}^2} d\omega\, dk_\parallel D_{mn}(\omega, k_\parallel)\tilde{A}_{TM}^{mn}(\omega, k_\parallel)\\
&\qquad\times \mathrm{J}_m\left(\chi_{mn}\frac{r}{a}\right)\exp[i(\omega t - k_\parallel z)]\frac{1}{2\pi}\int_0^{2\pi} e^{i(m-p)\theta}\,d\theta\\
&\quad= \sum_m \sum_n \iint_{\mathbb{R}^2} d\omega\, dk_\parallel D_{mn}(\omega, k_\parallel)\tilde{A}_{TM}^{mn}(\omega, k_\parallel)\mathrm{J}_m\left(\chi_{mn}\frac{r}{a}\right)\exp[i(\omega t - k_\parallel z)]\delta_{mp}\\
&\quad= \sum_n \iint_{\mathbb{R}^2} d\omega\, dk_\parallel D_{pn}(\omega, k_\parallel)\tilde{A}_{TM}^{pn}(\omega, k_\parallel)\mathrm{J}_p\left(\chi_{pn}\frac{r}{a}\right)\exp[i(\omega t - k_\parallel z)]\\
&\quad= -\mu_0\int_0^{2\pi} e^{-ip\theta} j_z(r, \theta, z, t)\,d\theta.
\end{aligned} \tag{5.332}$$

The next step consists in diagonalizing the complex exponentials, which is equivalent to Fourier transforming the source current:

$$\begin{aligned}
&\frac{1}{2\pi}\int_{-\infty}^{+\infty} e^{-i\omega' t}\,d\theta \sum_n \iint_{\mathbb{R}^2} d\omega\, dk_\parallel D_{pn}(\omega, k_\parallel)\tilde{A}_{TM}^{pn}(\omega, k_\parallel)\mathrm{J}_p\left(\chi_{pn}\frac{r}{a}\right)\exp[i(\omega t - k_\parallel z)]\\
&\quad= \sum_n \iint_{\mathbb{R}^2} d\omega\, dk_\parallel D_{pn}(\omega, k_\parallel)\tilde{A}_{TM}^{pn}(\omega, k_\parallel)\mathrm{J}_p\left(\chi_{pn}\frac{r}{a}\right)e^{-ik_\parallel z}\frac{1}{2\pi}\int_{-\infty}^{+\infty} e^{i(\omega-\omega')t}\,dt\\
&\quad= \sum_n \iint_{\mathbb{R}^2} d\omega\, dk_\parallel D_{pn}(\omega, k_\parallel)\tilde{A}_{TM}^{pn}(\omega, k_\parallel)\mathrm{J}_p\left(\chi_{pn}\frac{r}{a}\right)e^{-ik_\parallel z}\delta(\omega-\omega')\\
&\quad= \sum_n \int_{\mathbb{R}^2} dk_\parallel D_{pn}(\omega', k_\parallel)\tilde{A}_{TM}^{pn}(\omega', k_\parallel)\mathrm{J}_p\left(\chi_{pn}\frac{r}{a}\right)e^{-ik_\parallel z}\\
&\quad= -\frac{\mu_0}{2\pi}\int_{-\infty}^{+\infty} e^{-i\omega' t}\,dt\int_0^{2\pi} e^{-ip\theta} j_z(r, \theta, z, t)\,d\theta.
\end{aligned} \tag{5.333}$$

A similar calculation, using the Fourier operator,

$$\frac{1}{2\pi}\int_{-\infty}^{+\infty} e^{-ik'_{\parallel}z}\, dz, \tag{5.334}$$

leads to

$$\sum_n D_{pn}(\omega', k'_{\parallel})\tilde{A}_{TM}^{pn}(\omega', k'_{\parallel})J_p\left(\chi_{pn}\frac{r}{a}\right)$$
$$= -\frac{\mu_0}{4\pi^2}\int_{\mathbb{R}} e^{ik'_{\parallel}z}\, dz\int_{\mathbb{R}} e^{-i\omega' t}\, dt\int_0^{2\pi} e^{-ip\theta} j_z(r,\, \theta,\, z,\, t)\, d\theta. \tag{5.335}$$

Finally, we need to diagonalize the Bessel functions; we use Equation 5.318 by applying the integral operator,

$$\int_0^1 xJ_p(\chi_{pn}x)\, dx = \int_0^1 \frac{r}{a}J_p\left(\chi_{pq}\frac{r}{a}\right)\frac{dr}{a}, \tag{5.336}$$

to Equation 5.335. This last operation yields the sought-after result,

$$\int_0^1 \frac{r}{a}J_p\left(\chi_{pq}\frac{r}{a}\right)\frac{dr}{a}\sum_n D_{pn}(\omega', k'_{\parallel})\tilde{A}_{TM}^{pn}(\omega', k'_{\parallel})J_p\left(\chi_{pn}\frac{r}{a}\right)$$
$$= \sum_n D_{pn}(\omega', k'_{\parallel})\tilde{A}_{TM}^{pn}(\omega', k'_{\parallel})\int_0^1 \frac{r}{a}J_p\left(\chi_{pq}\frac{r}{a}\right)J_p\left(\chi_{pn}\frac{r}{a}\right)\frac{dr}{a}$$
$$= \sum_n D_{pn}(\omega', k'_{\parallel})\tilde{A}_{TM}^{pn}(\omega', k'_{\parallel})\frac{1}{2}J_{p+1}^2(\chi_{pq})\delta_{nq}$$
$$= \frac{1}{2}J_{p+1}^2(\chi_{pq})D_{pq}(\omega', k'_{\parallel})\tilde{A}_{TM}^{pq}(\omega', k'_{\parallel})$$
$$= -\frac{\mu_0}{4\pi^2}\int_0^1 \frac{r}{a}J_p\left(\chi_{pq}\frac{r}{a}\right)\frac{dr}{a}\int_{\mathbb{R}} e^{ik'_{\parallel}z}\, dz\int_{\mathbb{R}} e^{i\omega' t}\, dt \times \int_0^{2\pi} e^{-ip\theta} j_z(r,\, \theta,\, z,\, t)\, d\theta, \tag{5.337}$$

from which we obtain

$$\tilde{A}_{TM}^{pq}(\omega, k_{\parallel}) = \frac{-\mu_0 \tilde{j}_z^{pq}(\omega, k_{\parallel})}{2\pi^2 J_{p+1}^2(\chi_{pq})\left(\frac{\omega^2}{c^2} - k_{\parallel}^2 - \frac{\chi_{pq}^2}{a^2}\right)}, \tag{5.338}$$

where we have defined the modal current spectral density

$$\tilde{j}_z^{pq}(\omega, k_{\parallel}) = \int_0^1 \frac{r}{a}J_p\left(\chi_{pq}\frac{r}{a}\right)\frac{dr}{a}\int_{\mathbb{R}} e^{ik_{\parallel}z}\, dz\int_{\mathbb{R}} e^{-i\omega t}\, dt\int_0^{2\pi} e^{-ip\theta} j_z(r,\, \theta,\, z,\, t)\, d\theta, \tag{5.339}$$

and replaced the dummy variables, $\omega'$ and $k'_{\parallel}$, by $\omega$ and $k_{\parallel}$.

Before addressing the Green function problem, we note that a similar derivation yields the spectral amplitude of the TE eigenmodes. We start from the radial component of the driven wave equation,

$$\square A_r - \frac{1}{r^2}(A_r + 2\partial_\theta A_\theta) + \mu_0 j_r = 0, \tag{5.340}$$

and consider a particular $TE_{mn}$ eigenmode, which satisfies the eigenvalue equation

$$\left(\square - \frac{1}{r^2}\right)\left[\frac{a}{\chi'_{mn} r} J_m\left(\chi'_{mn}\frac{r}{a}\right) e^{i\psi_m(\omega,k_\parallel)}\right] + 2\partial_\theta\left[\frac{i}{m} J'_m\left(\chi'_{mn}\frac{r}{a}\right) e^{i\psi_m(\omega,k_\parallel)}\right]$$
$$= D_{mn}(\omega, k_\parallel)\left[\frac{a}{\chi'_{mn} r} J_m\left(\chi'_{mn}\frac{r}{a}\right) e^{i\psi_m(\omega,k_\parallel)}\right], \tag{5.341}$$

with the dispersion relation

$$D_{mn}(\omega, k_\parallel) = \frac{\omega^2}{c^2} - k_\parallel^2 - \frac{\chi'^2_{mn}}{a^2}. \tag{5.342}$$

With this, we now have

$$\frac{1}{2\pi}\sum_m \sum_n \iint_{\mathbb{R}^2} d\omega dk_\parallel D_{mn}(\omega, k_\parallel)\tilde{A}^{mn}_{TE}(\omega, k_\parallel) \times \frac{a}{\chi'_{mn} r} J_m\left(\chi'_{mn}\frac{r}{a}\right) e^{i\psi_m(\omega,k_\parallel)} = -\mu_0 j_r. \tag{5.343}$$

The procedure is virtually identical to that followed for the TM modes, except for the diagonalization of the Bessel functions. For the TE modes, the projection operator is

$$\int_0^1 x^2 J_p(\chi'_{pq}\, x)\, dx = \int_0^1 \left(\frac{r}{a}\right)^2 J_p\left(\chi'_{pq}\frac{r}{a}\right)\frac{dr}{a}, \tag{5.344}$$

so that we can use the identity established in Equation 5.327. The diagonalization leads to

$$\sum_n D_{pn}(\omega', k'_\parallel)\tilde{A}^{pn}_{TE}(\omega', k'_\parallel)\frac{a}{\chi'_{pn} r} J_p\left(\chi'_{pn}\frac{r}{a}\right)$$
$$= -\frac{\mu_0}{4\pi^2}\int_{\mathbb{R}} e^{ik'_\parallel z}\, dz \int_{\mathbb{R}} e^{-i\omega' t}\, dt \int_0^{2\pi} e^{-ip\theta} j_r(r, \theta, z, t)\, d\theta. \tag{5.345}$$

We now apply the operator defined in Equation 5.344, to obtain

$$\begin{aligned}
&\int_0^1 \left(\frac{r}{a}\right)^2 \mathrm{J}_p\left(\chi'_{pq}\frac{r}{a}\right)\frac{dr}{a}\sum_n D_{pn}(\omega', k'_\parallel)\tilde{A}^{pn}_{TE}(\omega', k'_\parallel)\frac{a}{\chi'_{pn}r}\mathrm{J}_p\left(\chi'_{pn}\frac{r}{a}\right)\\
&= \sum_n D_{pn}(\omega', k'_\parallel)\tilde{A}^{pn}_{TE}(\omega', k'_\parallel)\int_0^1 \left(\frac{r}{a}\right)^2 \mathrm{J}_p\left(\chi'_{pq}\frac{r}{a}\right)\frac{a}{\chi'_{pn}r}\mathrm{J}_p\left(\chi'_{pn}\frac{r}{a}\right)\frac{dr}{a}\\
&= \sum_n D_{pn}(\omega', k'_\parallel)\tilde{A}^{pn}_{TE}(\omega', k'_\parallel)\frac{1}{\chi'_{pn}}\int_0^1 \left(\frac{r}{a}\right)\mathrm{J}_p\left(\chi'_{pq}\frac{r}{a}\right)\mathrm{J}_p\left(\chi'_{pn}\frac{r}{a}\right)\frac{dr}{a}\\
&= \sum_n D_{pn}(\omega', k'_\parallel)\tilde{A}^{pn}_{TE}(\omega', k'_\parallel)\frac{1}{\chi'_{pn}}\,\delta_{pq}\frac{1}{2}\mathrm{J}^2_p(\chi'_{pq})\left(1-\frac{p^2}{\chi'^2_{pq}}\right)\\
&= D_{pq}(\omega', k'_\parallel)\tilde{A}^{pq}_{TE}(\omega', k'_\parallel)\frac{\mathrm{J}^2_p(\chi'_{pq})}{2\chi'_{pq}}\left(1-\frac{p^2}{\chi'^2_{pq}}\right).
\end{aligned} \tag{5.346}$$

This is also proportional to the projected current,

$$\begin{aligned}
\tilde{j}^{pq}_r(\omega, k_\parallel) = &\int_0^1\left(\frac{r}{a}\right)^2 \mathrm{J}_p\left(\chi'_{pq}\frac{r}{a}\right)\frac{dr}{a}\int_{\mathbb{R}} e^{ik_\parallel z}\,dz\\
&\times\int_{\mathbb{R}} e^{-i\omega t}dt\int_0^{2\pi} e^{-ip\theta} j_r(r,\theta,z,t)\,d\theta,
\end{aligned} \tag{5.347}$$

and we finally find that

$$\tilde{A}^{pq}_{TE}(\omega, k_\parallel) = \frac{-\mu_0\chi'_{pq}\,\tilde{j}^{pq}_r(\omega, k_\parallel)}{2\pi^2\mathrm{J}^2_p(\chi'_{pq})\left(1-\frac{p^2}{\chi'^2_{pq}}\right)\left(\frac{\omega^2}{c^2}-k^2_\parallel-\frac{\chi'^2_{pq}}{a^2}\right)}. \tag{5.348}$$

### 5.6.5 Gauge Condition and Continuity Equation

At this point, one might remark that we have used only one equation for each family of modes, namely the axial component of the wave equation for the TM modes, and the radial component for the TE modes, to determine the spectral mode amplitudes, $\tilde{A}^{mn}_{TM}(\omega, k_\parallel)$ and $\tilde{A}^{mn}_{TE}(\omega, k_\parallel)$. However, we have two other equations, one for the scalar potential and one for the azimuthal component of the vector potential. Therefore, it appears at first glance that there is a potential inconsistency in the formalism. To elucidate this difficulty, we note that, as required by gauge invariance, the charge conservation must be taken into account:

$$\partial_\mu j^\mu = \partial_t\rho + \nabla\cdot\mathbf{j} = 0. \tag{5.349}$$

If we consider a given eigenmode and the corresponding modal excitation, $\tilde{j}_\mu^{pq}(\omega, k_\parallel)$, we have, by definition, a proportionality relation, as given in Equation 5.338 for TM modes, and Equation 5.348 for TE modes. As a result, the gauge condition and the charge conservation equation are proportional. The projection of the four-current density source term implicitly yields constraints between $\tilde{j}_r^{pq}(\omega, k_\parallel)$ and $\tilde{j}_\theta^{pq}(\omega, k_\parallel)$ for TE modes, and $\tilde{j}_z^{pq}(\omega, k_\parallel)$ and $\tilde{j}_0^{pq}(\omega, k_\parallel) = c\tilde{\rho}^{pq}(\omega, k_\parallel)$. More explicitly, by virtue of the diagonalization procedure described in the previous subsection, we have

$$\begin{aligned} j_r(x_\lambda) &= \frac{1}{2\pi}\sum_m\sum_n\iint_{\mathbb{R}^2} d\omega dk_\parallel \tilde{j}_r^{mn}(\omega, k_\parallel)\frac{a}{\chi'_{mn}r}\mathrm{J}_m\left(\chi'_{mn}\frac{r}{a}\right) \\ &\quad\times \exp[i(\omega t - k_\parallel z + m\theta)], \\ j_\theta(x_\lambda) &= \frac{1}{2\pi}\sum_m\sum_n\iint_{\mathbb{R}^2} d\omega dk_\parallel \tilde{j}_\theta^{mn}(\omega, k_\parallel)\mathrm{J}'_m\left(\chi'_{mn}\frac{r}{a}\right) \\ &\quad\times \exp[i(\omega t - k_\parallel z + m\theta)], \end{aligned} \tag{5.350}$$

with the constraint

$$\tilde{j}_\theta^{mn}(\omega, k_\parallel) = \frac{i}{m}\tilde{j}_r^{mn}(\omega, k_\parallel), \tag{5.351}$$

for TE modes; for TM modes,

$$\begin{aligned} j_z(x_\lambda) &= \frac{1}{2\pi}\sum_m\sum_n\iint_{\mathbb{R}^2} d\omega dk_\parallel \tilde{j}_z^{mn}(\omega, k_\parallel)\mathrm{J}_m\left(\chi_{mn}\frac{r}{a}\right)\exp[i(\omega t - k_\parallel z + m\theta)], \\ \rho(x_\lambda) &= \frac{1}{2\pi}\sum_m\sum_n\iint_{\mathbb{R}^2} d\omega dk_\parallel \tilde{\rho}^{mn}(\omega, k_\parallel)\mathrm{J}_m\left(\chi_{mn}\frac{r}{a}\right)\exp[i(\omega t - k_\parallel z + m\theta)], \end{aligned} \tag{5.352}$$

with the constraint

$$\omega\tilde{\phi}^{mn}(\omega, k_\parallel) = k_\parallel \tilde{j}_z^{mn}(\omega, k_\parallel). \tag{3.353}$$

With this, we have

$$\begin{aligned} \partial_t\left[\tilde{\rho}^{mn}(\omega, k_\parallel)\mathrm{J}_m\left(\chi_{mn}\frac{r}{a}\right)e^{i\psi_m(\omega,k_\parallel)}\right] &= i\omega\left[\tilde{\rho}^{mn}(\omega, k_\parallel)\mathrm{J}_m\left(\chi_{mn}\frac{r}{a}\right)e^{i\psi_m(\omega,k_\parallel)}\right], \\ \partial_z\left[\tilde{j}_z^{mn}(\omega, k_\parallel)\mathrm{J}_m\left(\chi_{mn}\frac{r}{a}\right)e^{i\psi_m(\omega,k_\parallel)}\right] &= -ik_\parallel\left[\tilde{j}_z^{mn}(\omega, k_\parallel)\mathrm{J}_m\left(\chi_{mn}\frac{r}{a}\right)e^{i\psi_m(\omega,k_\parallel)}\right], \\ \partial_r\left[r\tilde{j}_r^{mn}(\omega, k_\parallel)\frac{a}{\chi'_{mn}r}\mathrm{J}_m\left(\chi'_{mn}\frac{r}{a}\right)e^{i\psi_m(\omega,k_\parallel)}\right] &= \left[\tilde{j}_r^{mn}(\omega, k_\parallel)\mathrm{J}'_m\left(\chi'_{mn}\frac{r}{a}\right)e^{i\psi_m(\omega,k_\parallel)}\right], \\ \partial_\theta\left[\tilde{j}_\theta^{mn}(\omega, k_\parallel)\mathrm{J}'_m\left(\chi'_{mn}\frac{r}{a}\right)e^{i\psi_m(\omega,k_\parallel)}\right] &= im\left[\tilde{j}_\theta^{mn}(\omega, k_\parallel)\mathrm{J}'_m\left(\chi'_{mn}\frac{r}{a}\right)e^{i\psi_m(\omega,k_\parallel)}\right]. \end{aligned} \tag{5.354}$$

Applying the constraints given Equations 5.351 and 5.353, we see that

$$\partial_t\left[\tilde{\rho}^{mn}(\omega, k_{\|})\mathrm{J}_m\left(\chi_{mn}\frac{r}{a}\right)e^{i\psi_m(\omega,k_{\|})}\right] + \partial_z\left[\tilde{j}_z^{mn}(\omega, k_{\|})\mathrm{J}_m\left(\chi_{mn}\frac{r}{a}\right)e^{i\psi_m(\omega,k_{\|})}\right] = 0, \quad (5.355)$$

and

$$\frac{1}{r}\left\{\partial_r\left[r\tilde{j}_r^{mn}(\omega, k_{\|})\frac{a}{\chi'_{mn}r}\mathrm{J}_m\left(\chi'_{mn}\frac{r}{a}\right)e^{i\psi_m(\omega,k_{\|})}\right]\right.$$
$$\left. + \partial_\theta\left[\tilde{j}_\theta^{mn}(\omega, k_{\|})\mathrm{J}'_m\left(\chi'_{mn}\frac{r}{a}\right)e^{i\psi_m(\omega,k_{\|})}\right]\right\} = 0. \quad (5.356)$$

The charge conservation equation is satisfied separately for each TE and TM modal excitation due to the four-current source term. In other words, for eigenmodes, the four-divergence and d'Alembertian operators commute. If we take the four-divergence of the driven wave equation, we have

$$\partial^\mu\left(\Box A_\mu^m + \mu_0 j_\mu^m\right) = \partial^\mu \Box A_\mu^m + \mu_0 \partial^\mu j_\mu^m = 0. \quad (5.357)$$

Using the commutation property, we find that

$$\partial^\mu \Box A_\mu^m + \mu_0 \partial^\mu j_\mu^m = \Box \partial^\mu A_\mu^m + \mu_0 \partial^\mu j_\mu^m = 0; \quad (5.358)$$

as a result, the Lorentz gauge condition, $\partial^\mu A_\mu^m$, implies charge conservation,

$$\partial^\mu j_\mu^m = 0. \quad (5.359)$$

We emphasize that the commutation property is valid for eigenmodes because we have, by definition,

$$\Box A_\mu^m = D_m A_\mu^m, \quad \partial^\mu \Box A_\mu^m = \partial^\mu D_m A_\mu^m = D_m \partial^\mu A_\mu^m. \quad (5.360)$$

### 5.6.6 Green Function in a Cylindrical Waveguide

We now consider a unit excitation in the cylindrical waveguide,

$$\delta_4(x_\mu - \bar{x}_\mu) = \frac{1}{r}\delta(r-\bar{r})\delta(\theta-\bar{\theta})\delta(z-\bar{z})\delta[c(t-\bar{t})]. \quad (5.361)$$

For TM modes, a charge density perturbation, or an axial current, will provide the required excitation, while for TE modes we need a transverse current. Furthermore, to obtain the proper units, we will multiply Equation 5.361 by $-ec$.

Using Equation 5.338, we find that the Green function spectral density for the TM modes is given by

$$\tilde{G}_{TM}^{pq}(\omega, k_\parallel, \bar{x}_\mu) = \frac{-\mu_0 \tilde{g}_z^{pq}(\omega, k_\parallel, \bar{x}_\mu)}{2\pi^2 J_{p+1}^2(\chi_{pq})\left(\frac{\omega^2}{c^2} - k_\parallel^2 - \frac{\chi_{pq}^2}{a^2}\right)}, \tag{5.362}$$

where the spectral current density is obtained by applying Equation 5.339 to the delta-function defined above:

$$\tilde{g}_z^{pq}(\omega, k_\parallel, \bar{x}_\mu) = -ec\int_0^1 \frac{r}{a} J_p\left(\chi_{pq}\frac{r}{a}\right)\frac{dr}{a}\int_{\mathbb{R}} e^{ik_\parallel z}dz\int_{\mathbb{R}} e^{-i\omega t}dt\int_0^{2\pi} e^{-ip\theta}\delta_4(x_\mu - \bar{x}_\mu)d\theta. \tag{5.363}$$

The calculation is easy, as we are dealing with delta-functions:

$$\tilde{g}_z^{pq}(\omega, k_\parallel, \bar{x}_\mu) = -\frac{ec}{a^2} J_p\left(\chi_{pq}\frac{\bar{r}}{a}\right)e^{-i(\omega\bar{t} - k_\parallel \bar{z} + p\bar{\theta})}. \tag{5.364}$$

The final result for the TM mode Green spectral amplitude is

$$\tilde{G}_{TM}^{pq}(\omega, k_\parallel, \bar{x}_\mu) = \frac{\mu_0 ec J_p\left(\chi_{pq}\frac{\bar{r}}{a}\right)e^{-i(\omega\bar{t} - k_\parallel \bar{z} + p\bar{\theta})}}{a^2 2\pi^2 J_{p+1}^2(\chi_{pq})\left(\frac{\omega^2}{c^2} - k_\parallel^2 - \frac{\chi_{pq}^2}{a^2}\right)}. \tag{5.365}$$

In particular, we see that if the excitation occurs at a radial position where $J_p(\chi_{pq}\bar{r}/a) = 0$, the corresponding mode will have zero amplitude. This type of selection rule is often used in microwave tube amplifiers and oscillators, such as gyrotrons and free-electron masers.

For the TE modes, a similar approach is followed:

$$\tilde{G}_{TE}^{pq}(\omega, k_\parallel, \bar{x}_\mu) = \frac{-\mu_0 \chi'_{pq} \tilde{g}_r^{pq}(\omega, k_\parallel, \bar{x}_\mu)}{2\pi^2 J_p^2(\chi'_{pq})\left(1 - \frac{p^2}{\chi'^2_{pq}}\right)\left(\frac{\omega^2}{c^2} - k_\parallel^2 - \frac{\chi'^2_{pq}}{a^2}\right)}, \tag{5.366}$$

and

$$\tilde{g}_r^{pq}(\omega, k_\parallel, \bar{x}_\mu) = -ec\int_0^1 \left(\frac{r}{a}\right)^2 J_p\left(\chi'_{pq}\frac{r}{a}\right)\frac{dr}{a}\int_{\mathbb{R}} e^{ik_\parallel z}dz\int_{\mathbb{R}} e^{-i\omega t}dt\int_0^{2\pi} e^{-ip\theta}\delta_4(x_\mu - \bar{x}_\mu)d\theta. \tag{5.367}$$

Performing the integrals over the delta-functions, we first find that

$$\tilde{g}_r^{pq}(\omega, k_\parallel, \bar{x}_\mu) = -\frac{ec}{a^2}\left(\frac{\bar{r}}{a}\right)\mathrm{J}_p\left(\chi'_{pq}\frac{\bar{r}}{a}\right)e^{-i(\omega\bar{t}-k_\parallel\bar{z}+p\bar{\theta})}; \tag{5.368}$$

finally, the spectral density of the Green function is

$$\tilde{G}_{TE}^{pq}(\omega, k_\parallel, \bar{x}_\mu) = \frac{\mu_0 ec\chi'_{pq}\left(\frac{\bar{r}}{a}\right)\mathrm{J}_p\left(\chi'_{pq}\frac{\bar{r}}{a}\right)e^{-i(\omega\bar{t}-k_\parallel\bar{z}+p\bar{\theta})}}{a^2 2\pi^2\mathrm{J}_p^2(\chi'_{pq})\left(1-\frac{p^2}{\chi'^2_{pq}}\right)\left(\frac{\omega^2}{c^2}-k_\parallel^2-\frac{\chi'^2_{pq}}{a^2}\right)}. \tag{5.369}$$

### 5.6.7 Fast-Wave Excitation in a Cylindrical Waveguide

In this subsection, we study the coupling between a point charge in helical motion and TE modes in a cylindrical waveguide; this is intended as an overview, and a considerably more detailed treatment of the problem is given in Chapter 9. To simplify the presentation, we consider an electron describing a helical trajectory with a well-defined radius of gyration, $r_w$; thus, the four-current can be described by

$$j_\mu = -ec\frac{1}{r}\delta(r-r_w)\delta(z-v_\parallel t)[1,\ \beta_\perp(\hat{x}\cos k_w z) - \hat{y}\sin k_w z)\ \beta_\parallel\hat{z}], \tag{5.370}$$

where $k_w = 2\pi/\ell_w$ is the wiggler wavenumber, of period $\ell_w$, and the transverse and axial normalized velocities are related to the electron energy, $\gamma$:

$$\frac{1}{\gamma^2} = 1-\beta_\perp^2-\beta_\parallel^2. \tag{5.371}$$

Furthermore, we will show in Chapter 9 that the transverse electron momentum is given by the strength of the magnetic wiggler induction:

$$\gamma\beta_\perp = \frac{eB_w}{m_0 k_w c}. \tag{5.372}$$

The radius of the helix is obtained by requiring that in the time it takes to travel one wiggler period, the electron has completed a full gyration:

$$\left(\frac{\ell_w}{\beta_\parallel c}\right)\beta_\perp c = 2\pi r_w \Rightarrow k_w r_w = \frac{\beta_\perp}{\beta_\parallel}. \tag{5.373}$$

The axial and time-like components of the four-current density excite TM modes, while the fast-wave excitation of TE modes is driven by the transverse components of $j_\mu$. In cylindrical coordinates, the radial and azimuthal components of the four-current can be derived as follows:

$$\begin{aligned} j_r &= j_x \cos\theta + j_y \sin\theta \\ &= -ec\beta_\perp \frac{1}{r}\delta(r-r_w)\delta(z-v_\parallel t)(\cos\theta\cos k_w z - \sin\theta \sin k_w z) \\ &= -ec\beta_\perp \frac{1}{r}\delta(r-r_w)\delta(z-v_\parallel t)\cos(k_w z+\theta), \end{aligned} \tag{5.374}$$

and

$$\begin{aligned} j_\theta &= j_x \sin\theta - j_y \cos\theta \\ &= -ec\beta_\perp \frac{1}{r}\delta(r-r_w)\delta(z-v_\parallel t)(\sin\theta\cos k_w z + \cos\theta \sin k_w z) \\ &= -ec\beta_\perp \frac{1}{r}\delta(r-r_w)\delta(z-v_\parallel t)\sin(k_w z+\theta). \end{aligned} \tag{5.375}$$

To determine the coupling to the TE modes, we can use Equations 5.347 and 5.348. We first project the helical current, as derived above, using

$$\tilde{j}_r^{pq}(\omega, k_\parallel) = \int_0^1 \left(\frac{r}{a}\right)^2 \mathrm{J}_p\left(\chi'_{pq}\frac{r}{a}\right)\frac{dr}{a}\int_{\mathbb{R}} e^{ik_\parallel z}dz\int_{\mathbb{R}} e^{-i\omega t}dt\int_0^{2\pi} e^{-ip\theta} j_r(r,\theta,z,t)\,d\theta. \tag{5.376}$$

Thus, the integrals to perform are

$$\int_0^1 \frac{1}{r}\delta(r-r_w)\left(\frac{r}{a}\right)^2 \mathrm{J}_p\left(\chi'_{pq}\frac{r}{a}\right)\frac{dr}{a}\int_{\mathbb{R}} e^{ik_\parallel z}\,dz\int_{\mathbb{R}} \delta(z-v_\parallel t)e^{-i\omega t}\,dt\int_0^{2\pi} \cos(k_w z+\theta)e^{-ip\theta}\,d\theta. \tag{5.377}$$

The radial integral is essentially an overlap integral between the source current and the radial mode under consideration:

$$\int_0^1 \frac{1}{r}\delta(r-r_w)\left(\frac{r}{a}\right)^2 \mathrm{J}_p\left(\chi'_{pq}\frac{r}{a}\right)\frac{dr}{a} = \frac{r_w}{a^3}\mathrm{J}_p\left(\chi'_{pq}\frac{r_w}{a}\right); \tag{5.378}$$

note that for a zero radius of gyration, no coupling will occur.

The integral over the azimuthal angle yields the well-known free-electron laser helicity selection rule:

$$\begin{aligned} \int_0^{2\pi} \cos(k_w z+\theta)e^{-ip\theta}d\theta &= \frac{1}{2}\int_0^{2\pi}[e^{-i(k_w z+\theta)} + e^{-i(k_w z+\theta)}]e^{-ip\theta}\,d\theta \\ &= \frac{1}{2}\int_0^{2\pi}[e^{ik_w z}e^{-i(p-1)\theta} + e^{-ik_w z}e^{-i(p+1)\theta}]\,d\theta \\ &= \pi(e^{ik_w z}\delta_{1p} + e^{-ik_w z}\delta_{-1p}). \end{aligned} \tag{5.379}$$

This shows that the helical symmetry of the wiggler-induced selects the $p = \pm 1$ modes.

Finally, the integrals over time and $z$ are simple,

$$\int_{\mathbb{R}} e^{ik_{\parallel}z} e^{\pm ik_w z} dz \int_{\mathbb{R}} \delta(z - v_{\parallel}t) e^{-i\omega t} dt = \int_{-\infty}^{+\infty} \exp\left[ iz\left( k_{\parallel} \pm k_w - \frac{\omega}{v_{\parallel}} \right) \right] dz$$
$$= 2\pi\delta\left( k_{\parallel} \pm k_w - \frac{\omega}{v_{\parallel}} \right) \tag{5.380}$$

and give rise to the so-called wiggler-shifted beam mode, which disperses as

$$\omega = (k_{\parallel} + k_w)v_{\parallel}. \tag{5.381}$$

Grouping our results together, we first find that

$$\begin{aligned} \tilde{j}_r^{pq}(\omega, k_{\parallel}) &= -ec\beta_{\perp} 2\pi^2 \delta_{\pm 1p} \delta\left( k_{\parallel} \pm k_w - \frac{\omega}{v_{\parallel}} \right) \frac{r_w}{a^3} \mathrm{J}_p\left( \chi'_{pq} \frac{r_w}{a} \right), \\ \tilde{j}_r^{\pm 1q}(\omega, k_{\parallel}) &= -ec\beta_{\perp} 2\pi^2 \delta\left( k_{\parallel} \pm k_w - \frac{\omega}{v_{\parallel}} \right) \frac{r_w}{a^3} \mathrm{J}_{\pm 1}\left( \chi'_{\pm 1q} \frac{r_w}{a} \right). \end{aligned} \tag{5.382}$$

Finally, the amplitude of the free-electron laser mode excited by the helical electron beam is

$$\tilde{A}_{TE}^{\pm 1q}(\omega, k_{\parallel}) = \frac{\mu_0 ec\beta_{\perp}\chi'_{\pm 1q} \frac{r_w}{a^3} \mathrm{J}_{\pm 1}(\chi'_{\pm 1q} \frac{r_w}{a}) \delta(k_{\parallel} \pm k_w - \frac{\omega}{v_{\parallel}})}{\mathrm{J}_{\pm 1}^2(\chi'_{\pm 1q}) \left( 1 - \frac{1}{\chi'^2_{\pm 1q}} \right) \left( \frac{\omega^2}{c^2} - k_{\parallel}^2 - \frac{\chi'^2_{\pm 1q}}{a^2} \right)}. \tag{5.383}$$

Resonant interaction is achieved when both the beam mode and the waveguide mode dispersion condition are satisfied simultaneously. In other words, we request that

$$\begin{aligned} k_{\parallel} \pm k_w - \frac{\omega}{v_{\parallel}} &= 0, \\ \frac{\omega^2}{c^2} - k_{\parallel}^2 - \frac{\chi'^2_{\pm 1q}}{a^2} &= 0. \end{aligned} \tag{5.384}$$

Eliminating the axial wavenumber from the dispersion equations, we obtain the Doppler upshifted and downshifted free-electron laser frequencies. We first have

$$k_{\parallel}^2 = \left( \frac{\omega}{v_{\parallel}} - k_w \right)^2 = \frac{\omega^2}{c^2} - \frac{\chi'^2_{1q}}{a^2}, \tag{5.385}$$

where we have chosen the $p = +1$ azimuthal modes.

We then obtain a quadratic equation for the interaction frequencies:

$$\omega^2\left(\frac{1}{\beta_\parallel^2}-1\right)-2\omega\frac{ck_w}{\beta_\parallel}+c^2\left(k_w^2-\frac{\chi_{1q}'^2}{a^2}\right)=0. \tag{5.386}$$

Introducing the parameters

$$\gamma_\parallel=\frac{1}{\sqrt{1-\beta_\parallel^2}},\quad f=\frac{1}{\gamma_\parallel^2\beta_\parallel^2},\quad g'=-\frac{k_w c}{\beta_\parallel},\quad h=c^2\left(k_w^2+\frac{\chi_{1q}'}{a^2}\right), \tag{5.387}$$

the solutions are given by

$$\omega^\pm=\frac{-g'\pm\sqrt{\Delta}}{f},\quad \Delta=g'^2-fh=k_w^2c^2\left[1-\left(\frac{\chi_{1q}'}{\beta_\parallel\gamma_\parallel k_w a}\right)^2\right], \tag{5.388}$$

which finally yields

$$\omega^\pm=\gamma_\parallel^2\beta_\parallel ck_w\left[1\pm\beta_\parallel\sqrt{1-\left(\frac{\chi_{1q}'}{\beta_\parallel\gamma_\parallel k_w a}\right)^2}\right]. \tag{5.389}$$

This is illustrated in Figure 5.5; the salient physical features of this solution are the following. First, in the free-space limit, where $a\to\infty$, we recover the usual Doppler shift, with

$$\omega^\pm=\gamma_\parallel^2\beta_\parallel ck_w(1\pm\beta_\parallel); \tag{5.390}$$

in the ultrarelativistic limit, where $\beta_\parallel\to\infty$, the Doppler upshifted frequency scales as

$$\omega^+\simeq 2k_w c\gamma_\parallel^2. \tag{5.391}$$

Thus, short wavelength radiation can be obtained by combining a macroscopic wiggler period and a high-energy electron beam. This is one advantage of fast-wave devices, where the relativistic Doppler upshift is used to produce high-energy photons.

Another interesting feature of the free-electron laser frequency in a waveguide is the existence of a so-called "grazing" interaction, when the condition

$$\frac{\chi_{1q}'c}{a}=\omega_c=\beta_\parallel\gamma_\parallel k_w c, \tag{5.392}$$

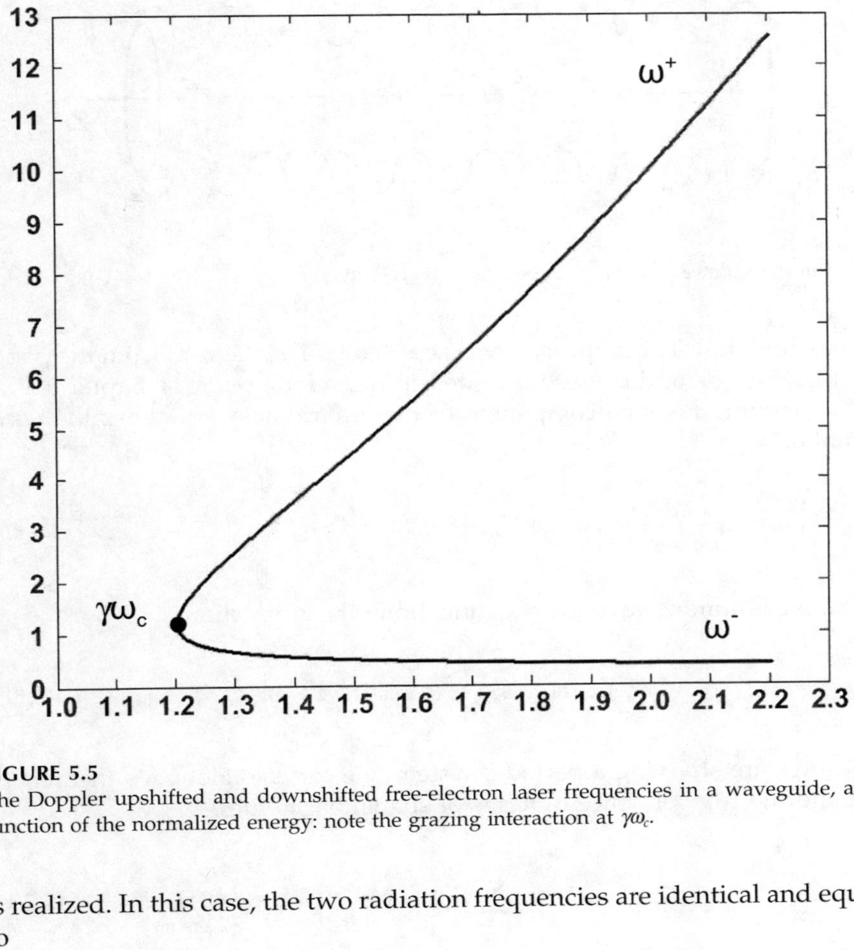


**FIGURE 5.5**
The Doppler upshifted and downshifted free-electron laser frequencies in a waveguide, as a function of the normalized energy: note the grazing interaction at $\gamma\omega_c$.

is realized. In this case, the two radiation frequencies are identical and equal to

$$\omega^+ = \omega^- = \gamma_\|^2 \beta_\| k_w c = \gamma_\| \omega_c. \tag{5.393}$$

### 5.6.8 Slow-Wave Excitation in a Corrugated Waveguide

In this section, we first discuss the effects of a cylindrically symmetric, corrugated, periodic boundary condition on the cylindrical eigenmodes and their dispersion characteristics, as illustrated in Figure 5.6. We then outline the derivation of the mode excitation of a point charge moving with constant velocity along the waveguide axis.

We are considering azimuthally symmetric modes, where $\partial_\theta \equiv 0$, and consequently, we will use the $J_0$ Bessel function to model the radial dependence of the waves propagating in the waveguide. Furthermore, to illustrate a slightly different method, we will use the axial electric field instead

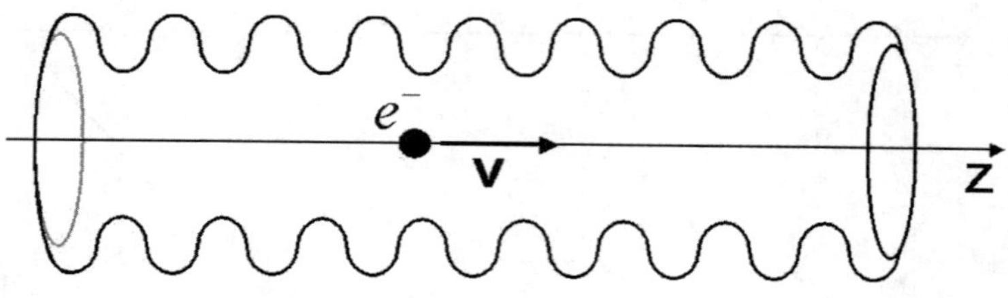


**FIGURE 5.6**
Electron in rectilinear motion in a corrugated waveguide.

of the potential. This approach was adapted by D. J. Gibson from the paper by J. A. Swegle and co-authors, cited in the references for Chapter 5.

In vacuum, the axial component of the electric field, $E_z$, obeys the wave equation

$$\nabla^2 E_z - \frac{1}{c^2}\partial_t^2 E_z = 0. \tag{5.394}$$

We then assume a wave propagating along the $z$-direction:

$$E_z(\mathbf{x}, t) = E_z(r, \theta)\exp[i(kz - \omega t)]. \tag{5.395}$$

Since we are studying a periodic system, we can use Floquet's theorem to expand the $z$-dependence in terms of spatial harmonics,

$$E_z(r, \theta, z, t) = \sum_{n=-\infty}^{\infty} E_{zn}(r, \theta)\exp[i(k_n z - \omega t)], \tag{5.396}$$

where the axial wavenumber is decomposed as follows:

$$k_n = q + nh_0, \tag{5.397}$$

with $h_0 = 2\pi/z_0$, and $z_0$ is the period of the corrugation in the waveguide.

Furthermore, the "free" part of the wavenumber, $q$, must lie between $-k_0/2$ and $+k_0/2$. Combining the last three equations and using cylindrical coordinates, we have

$$\sum_{n=-\infty}^{\infty}\left\{\frac{1}{r}\partial_r[r\partial_r E_{nz}(r, \theta)] + \frac{1}{r^2}\partial_\theta^2 E_{nz}(r, \theta) - k_n^2 E_{nz}(r, \theta) + \frac{\omega^2}{c^2}E_{nz}(r, \theta)\right\}$$
$$\times \exp[i(k_n z - \omega t)] = 0, \tag{5.398}$$

Since the problem is azimuthally symmetric, there is no explicit dependence on $\theta$; moreover, each term of the sum must be equal to 0 independently because of the orthogonality of complex exponentials. Therefore, Equation 5.398 simplifies to read

$$\frac{1}{r}\partial_r[r\partial_r E_{nz}(r)] + \left(\frac{\omega^2}{c^2} - k_n^2\right)E_{nz}(r) = 0, \tag{5.399}$$

This is a standard Bessel equation, and the general solution is

$$E_{nz}(r) = A_n \mathrm{J}_0(\Gamma_n r) + B_n \mathrm{N}_0(\Gamma_n r), \tag{5.400}$$

where the transverse wavenumber is related to the axial wavenumber and frequency by the vacuum dispersion relation,

$$\Gamma_n^2 = \frac{\omega^2}{c^2} - k_n^2. \tag{5.401}$$

Boundedness on-axis requires that the $\mathrm{N}_0$ part of the solution be discarded. Note, however, that when an intense annular beam is present, the modified Bessel function must be taken into account to model properly the excited modes in the waveguide, as is discussed in great detail in the aforementioned paper by J. A. Swegle and co-authors.

The Floquet expansion for $E_z$ now takes the form

$$E_z(\mathbf{x}, t) = \sum_{n=-\infty}^{\infty} A_n \mathrm{J}_0(\Gamma_n r)\exp[i(k_n z - \omega t)]. \tag{5.402}$$

The next step of the process is to apply boundary conditions to produce the sought-after eigenmodes. Assuming a perfectly conducting wall, the tangential electric field at the wall must be zero:

$$E_{\tan} = \left(E_r \frac{dr_w}{dz} + E_z\right)_{r=r_w(z)} = 0. \tag{5.403}$$

As we have been working with the electric field, the radial component, $E_r$ can be found from $E_z$ by using the divergence equation:

$$\nabla \cdot \mathbf{E} = 0 = \frac{1}{r}\partial_r(rE_r) + \partial_z E_z. \tag{5.404}$$

Note that the azimuthal component of the electric does not contribute, as $\partial_\theta \equiv 0$. In fact, since we are working with TM modes, the four-potential has only two components, $A_z$ and $\phi$; the electric field is therefore given by

$$\begin{aligned} E_r &= -\partial_r \phi, \\ E_\theta &= -\frac{1}{r}\partial_\theta \phi = 0, \\ E_z &= -\partial_z \phi - \partial_t A_z. \end{aligned} \tag{5.405}$$

If, in addition, we use the Lorentz gauge condition, we have an additional constraint between the two remaining components of the four-potential:

$$\frac{1}{c^2}\partial_t \phi + \partial_z A_z = 0. \tag{5.406}$$

It is then easy to see that, with the functional dependence in Equation 5.401, we have, for the spectral amplitudes,

$$i\left(-\frac{\omega}{c^2}\phi_n + k_n A_{zn}\right) = 0, \tag{5.407}$$

from the Lorentz gauge condition, and

$$E_{zn} = -ik_n\phi_n + i\omega A_{zn} = \frac{i}{k_n}\left(k_n^2 - \frac{\omega^2}{c^2}\right)\phi_n = -\frac{i\Gamma_n^2}{k_n c^2}\phi_n. \tag{5.408}$$

This leads to

$$\begin{aligned} \phi(\mathbf{x}, t) &= \sum_{n=-\infty}^{\infty} \frac{ik_n c^2}{\Gamma_n^2} A_n \mathrm{J}_0(\Gamma_n r)\exp[i(k_n z - \omega t)], \\ A_z(\mathbf{x}, t) &= \sum_{n=-\infty}^{\infty} \frac{i\omega}{\Gamma_n^2} A_n \mathrm{J}_0(\Gamma_n r)\exp[i(k_n z - \omega t)], \\ E_r(\mathbf{x}, t) &= \sum_{n=-\infty}^{\infty} \frac{ik_n}{\Gamma_n^2} A_n [\partial_r \mathrm{J}_0(\Gamma_n r)]\exp[i(k_n z - \omega t)]. \end{aligned} \tag{5.409}$$

Returning to the boundary condition, we first have

$$\sum_{n=-\infty}^{\infty} \left(\frac{ik_n}{\Gamma_n^2}\partial_r E_{zn}\frac{dr_w}{dz} + E_{zn}\right)_{r=r_w(z)} \exp[i(k_n z - \omega t)] = 0. \tag{5.410}$$

We can remove the part of the complex exponential that does not depend on the Floquet harmonic number from the sum, to obtain

$$e^{i(qz-\omega t)}\sum_{n=-\infty}^{\infty}\left\{\frac{ik_n dE_{zn}[r_w(z)]}{\Gamma_n^2}\frac{dr_w}{dr_w}\frac{dr_w}{dz}+E_{zn}[r_w(z)]\right\}e^{ink_0 z}=0. \tag{5.411}$$

Moreover, the chain rule yields

$$\frac{d}{dr}E_{zn}[r_w(z)]=\frac{dr_w}{dz}\frac{d}{dz}E_{zn}[r_w(z)], \tag{5.412}$$

so we finally have

$$\sum_{n=-\infty}^{\infty}e^{ink_0 z}\left(1+\frac{ik_n}{\Gamma_n^2}\frac{d}{dz}\right)E_{zn}[r_w(z)]=0. \tag{5.413}$$

We now eliminate the $z$-dependence in Equation 5.413 by multiplying both sides by $\frac{1}{z_0}e^{-imh_0 z}$, and integrating over one period. We start with

$$\begin{aligned}&\sum_{n=-\infty}^{\infty}\left\{\frac{1}{z_0}\int_{-z_0/2}^{+z_0/2}\mathrm{e}^{-imh_0 z}\mathrm{e}^{inh_0 z}\left(1+\frac{ik_n}{\Gamma_n^2}\frac{d}{dz}\right)E_{zn}[r_w(z)]dz\right\}=0,\\&\sum_{n=-\infty}^{\infty}\left\{\frac{1}{z_0}\int_{-z_0/2}^{+z_0/2}\left[\mathrm{e}^{i(n-\mathrm{m})h_0 z}E_{zn}(r_w)+\mathrm{e}^{i(n-\mathrm{m})h_0 z}\frac{ik_n}{\Gamma_n^2}\frac{d}{dz}E_{zn}(r_w)\right]dz\right\}=0.\end{aligned} \tag{5.414}$$

Integrating by parts then yields

$$\begin{aligned}&\sum_{n=-\infty}^{\infty}\left\{\frac{1}{z_0}\int_{+z_0/2}^{+z_0/2}\mathrm{e}^{i(n-m)h_0 z}E_{zn}(r_w)+\frac{d}{dz}\left[\mathrm{e}^{i(n-m)h_0 z}\frac{ik_n}{\Gamma_n^2}E_{zn}(r_w)\right]\right.\\&\left.-\frac{ik_n}{\Gamma_n^2}E_{zn}(r_w)+\frac{d}{dz}\mathrm{e}^{i(n-m)h_0 z}dz\right\}=0,\\&\sum_{n=-\infty}^{\infty}\left\{\frac{1}{z_0}\int_{-z_0/2}^{+z_0/2}\mathrm{e}^{i(n-m)h_0 z}E_{zn}(r_w)-\frac{ik_n}{\Gamma_n^2}E_{zn}(r_w)\frac{d}{dz}\mathrm{e}^{i(n-m)h_0 z}dz\right.\\&\left.+\left[\mathrm{e}^{i(n-m)h_0 z}\frac{ik_n}{\Gamma_n^2}E_{zn}(r_w)\right]_{-z_0/2}^{+z_0/2}\right\}=0.\end{aligned} \tag{5.415}$$

We now assume that the corrugation of the wall is a harmonic function,

$$r_w(z) = r_0(1+\varepsilon \sin h_0 z); \tag{5.416}$$

as a result, the term outside the integral is identically equal to zero because the corrugation is symmetrical about the origin, and the other functions involved are also periodic in $z$, with the period $z_0$. Furthermore, we can differentiate the exponent in the second term inside the integral, giving as our boundary condition

$$\sum_{n=-\infty}^{\infty}\left\{\left[1+\frac{(n-m)h_0k_n}{\Gamma_n^2}\right]\frac{1}{z_0}\int_{-z_0/2}^{+z_0/2} e^{i(n-m)h_0 z} E_{zn}[r_0(1+\varepsilon\sin h_0 z)]dz\right\} = 0. \tag{5.417}$$

The explicit radial dependence of the Floquet harmonics of the axial electric field can be used to recast Equation 5.417 as

$$\sum_{n=-\infty}^{\infty}\left\{A_n\left[1+\frac{(n-m)h_0k_n}{\Gamma_n^2}\right]\frac{1}{z_0}\int_{-z_0/2}^{+z_0/2} e^{i(n-m)h_0 z} \mathrm{J}_0[\Gamma_n r_0(1+\varepsilon\sin h_0 z)]dz\right\} = 0. \tag{5.418}$$

Moreover, if we define the matrix elements

$$C_{mn} = \frac{1}{z_0}\left[1+\frac{(n-m)h_0k_n}{\Gamma_n^2}\right]\int_{-z_0/2}^{+z_0/2} e^{i(n-m)h_0 z} \mathrm{J}_0[\Gamma_n r_0(1+\varepsilon\sin h_0 z)]dz, \tag{5.419}$$

the boundary condition takes the simple form

$$\sum_{n=-\infty}^{\infty} C_{mn}A_n = 0, \tag{5.420}$$

or, $\mathbb{C}\cdot\mathbf{A} = 0$, where $\mathbf{A}$ is a vector composed of the coefficients $A_n$. To ensure that this infinite system of linear equations has nontrivial solutions, we must require that the determinant of the matrix be zero:

$$\det\mathbb{C} = 0. \tag{5.421}$$

This represents the sought-after eigenvalue equation for the dispersion relations in the corrugated waveguide. Physically, we see that the corrugation $r_w(z)$ couples the radial and axial degrees of freedom of the system. The separation of variables used effectively for the cylindrical eigenmodes in an uncorrugated waveguide is no longer valid, and each new eigenmode is represented by an infinite series of cylindrical waveguide modes. From Equation 5.421 we obtain a set of allowed values for $\Gamma_n$, called $\Gamma_{mn}$, and we can use the relation

$$\Gamma_{mn}^2 = \frac{\omega^2}{c^2} - k_{mn}^2 = \frac{\omega^2}{c^2} - (q + nh_0)^2, \tag{5.422}$$

to find the dispersion relations,

$$q(\omega) = \pm\sqrt{\frac{\omega^2}{c^2} - \Gamma_{mn}^2} - nh_0, \tag{5.423}$$

or each hybrid $TM_m$ mode.

We can simplify the expression of the matrix $\mathbb{C}$ with a change of variables: letting $v = \sin h_0 z$ and

$$dv = h_0 \cos h_0 z = h_0\sqrt{1 - v^2}\, dz, \tag{5.424}$$

we find that

$$\begin{aligned} &\frac{1}{z_0}\int(\cos h_0 z + i\sin h_0 z)^{n-m} \frac{1}{h_0\sqrt{1-v^2}} J_0[\Gamma_n r_0(1+\varepsilon v)]dv \\ &= \frac{h_0}{2\pi}\int\left[\sqrt{1-v^2} + iv\right]^{n-m} \frac{1}{h_0\sqrt{1-v^2}} J_0[\Gamma_n r_0(1+\varepsilon v)]dv. \end{aligned} \tag{5.425}$$

When the first term is raised to any power, *n-m*, it produces two types of terms, those that include a $\sqrt{1-v^2}$ term, and those that do not. Because the argument of the Bessel function varies as sin $h_0 z$, over the intervals $[-\frac{z_0}{2}, 0]$ and $[0, \frac{z_0}{2}]$, the function will be symmetric about $-z_0/4$ and $z_0/4$, respectively. Similarly, any power of sin $h_0 z$ and the even powers of cos $h_0 z$ are symmetric about the same points. This means that we can integrate from $-z_0/4$ to $z_0/4$. Moreover, odd powers of cos $h_0 z$ are antisymmetric about $-z_0/4$ and $z_0/4$, and those terms yield zero when integrated with the Bessel function. The terms with a $\sqrt{1-v^2}$ component above correspond to terms with an odd power of the cosine and give zero when integrated. Therefore, the final integral is

$$\frac{1}{\pi}\int_{-1}^{1} \frac{P_{mn}(v)}{\sqrt{1-v^2}} J_0[\Gamma_n r_0(1+\varepsilon v)]dv, \tag{5.426}$$

where the $P_{mn}(v)$ are terms that result from the expansion to the ($n$-$m$)th power without the $\sqrt{1-v^2}$ component:

$$\begin{aligned} P_{mn}(v) &= 1 && n-m=0, \\ &= \pm iv && n-m=\pm 1, \\ &= 1-2v^2 && n-m=\pm 2, \\ &= \pm i(3v-4v^3) && n-m=\pm 3. \end{aligned} \tag{5.427}$$

We also note that

$$\begin{aligned} 1+\frac{(n-m)h_0k_n}{\Gamma_n^2} &= \frac{\frac{\omega^2}{c^2}-k_n^2+(n-m)h_0k_n}{\Gamma_n^2} \\ &= \frac{\frac{\omega^2}{c^2}-k_n[h_0+nh_0-(n-m)h_0]}{\Gamma_n^2} \\ &= \frac{\omega^2-k_nk_mc^2}{\Gamma_n^2c^2}, \end{aligned} \tag{5.428}$$

so the final definition for $\mathbb{C}$ is

$$C_{mn} = \left(\frac{\omega^2-k_nk_mc^2}{\Gamma_n^2c^2}\right)\frac{1}{\pi}\int_{-1}^{1}\frac{P_{mn}(v)}{\sqrt{1-v^2}}\mathrm{J}_0[\Gamma_n r_0(1+\varepsilon v)]dv. \tag{5.429}$$

We now briefly outline how one can use these eigenmodes of a corrugated waveguide to study the excitation produced by a point charge in rectilinear motion along the waveguide axis. This will help us have a better physical understanding of the nature of the new eigenmodes of the periodic system, as represented by an infinite series of coupled vacuum cylindrical eigenmodes. Using the relation between the axial electric field and the scalar established in Equation 5.409, we first write

$$\phi(r,z,t) = \frac{1}{\sqrt{2\pi}}\int_{-\infty}^{+\infty}d\omega\int_{-h_0/2}^{+h_0/2}dq\tilde{\phi}(\omega,q)\left[\sum_n F_n\mathrm{J}_0(\Gamma_n r)e^{i(\omega t-k_nz)}\right]. \tag{5.430}$$

The key point here is that the coefficients,

$$F_n(\omega,q) = \frac{ik_nc^2}{\Gamma_n^2}A_n(\omega,q), \tag{5.431}$$

are not arbitrary; they are determined by solving the infinite system of coupled linear equations, $\Sigma_{n=-\infty}^{\infty}C_{mn}A_n = 0$, as described in the previous paragraphs. The associated dispersion is illustrated in Figure 5.7. Within this

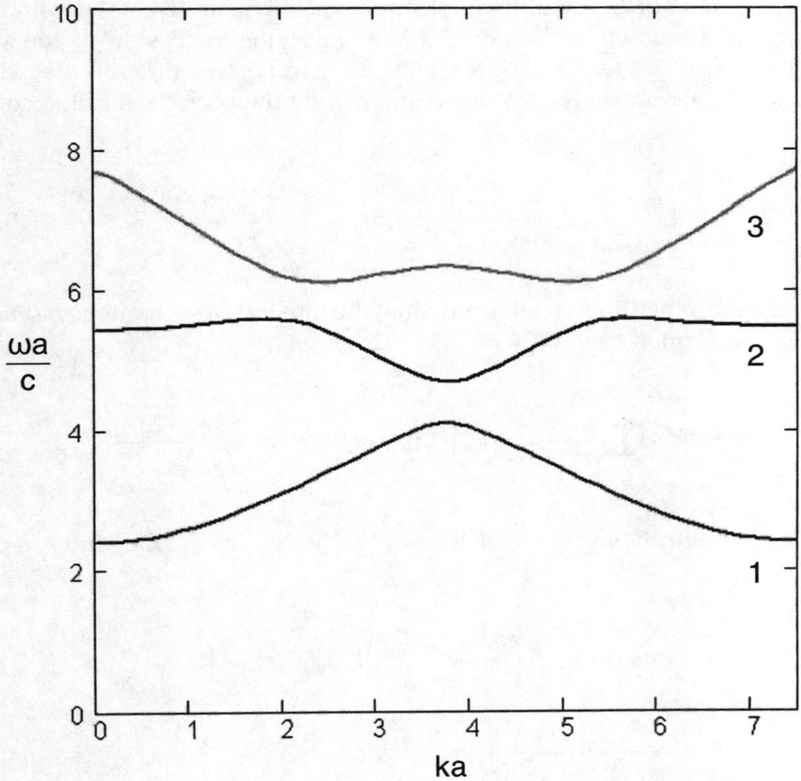


**FIGURE 5.7**
The first three modes in a corrugated waveguide, as obtained by solving the determinant of the coupling matrix; $a$ is the waveguide radius. The corrugation depth is $\varepsilon = 0.007$.

context, the quantity in the brackets in Equation 5.430 is the equivalent of a regular eigenmode, albeit expressed as an infinite series. The operation of the d'Alembertian yields the well-known eigenvalue equations,

$$\square\left[\sum_n F_n \mathrm{J}_0(\Gamma_n r) e^{i(\omega t - k_n z)}\right] = \sum_n \left(\frac{\omega^2}{c^2} - k_n^2 - \Gamma_n^2\right) F_n \mathrm{J}_0(\Gamma_n r) e^{i(\omega t - k_n z)}, \quad (5.432)$$

and the wave equation, driven by the point charge, reads

$$\begin{aligned} &\square \phi - \frac{e}{2\pi\varepsilon_0}\frac{\delta(r)}{r}\delta(z - vt) = 0, \\ &\frac{1}{\sqrt{2\pi}}\int_{-\infty}^{+\infty} d\omega \int_{-h_0/2}^{+h_0/2} dq\, \tilde{\phi}(\omega, q) \sum_n \left(\frac{\omega^2}{c^2} - k_n^2 - \Gamma_n^2\right) F_n \mathrm{J}_0(\Gamma_n r) e^{i(\omega t - k_n z)} \\ &\quad = \frac{e}{2\pi\varepsilon_0}\frac{\delta(r)}{r}\delta(z - vt). \end{aligned} \quad (5.433)$$

Here, the factor of $2\pi$ dividing the electron current results from the cylindrical correction discussed in Section 5.7.2. Multiplying each side of the wave equation by $e^{-i\bar{\omega}t}/\sqrt{2\pi}$, and integrating over time, we find a Dirac delta-function, $\delta(\omega-\bar{\omega})$, on the left-hand side, while the projection of the current reads

$$\int_{-\infty}^{+\infty}\frac{e^{-i\bar{\omega}t}}{\sqrt{2\pi}}dt\frac{e}{2\pi\varepsilon_0}\frac{\delta(r)}{r}\delta(z-vt) = \frac{e}{\sqrt{2\pi}^3\varepsilon_0}\frac{\delta(r)}{r}e^{-i\frac{\bar{\omega}}{v}z}. \tag{5.434}$$

With the aforementioned delta-function, the integral over frequency is immediately performed to yield

$$\int_{-h_0/2}^{+h_0/2}dq\tilde{\phi}(\bar{\omega},q)\sum_n\left(\frac{\bar{\omega}^2}{c^2}-k_n^2-\Gamma_n^2\right)F_n\mathrm{J}_0(\Gamma_n r)e^{-ik_nz} = \frac{e}{\sqrt{2\pi}^3\varepsilon_0}\frac{\delta(r)}{r}e^{-i\frac{\bar{\omega}}{v}z}. \tag{5.435}$$

Next, we multiply each side of the wave equation by $e^{i\bar{k}z}/2\pi$, and integrate over $z$:

$$\int_{-h_0/2}^{+h_0/2}dq\tilde{\phi}(\bar{\omega},q)\sum_n\left(\frac{\bar{\omega}^2}{c^2}-k_n^2-\Gamma_n^2\right)F_n\mathrm{J}_0(\Gamma_n r)\frac{1}{2\pi}\int_{-\infty}^{+\infty}e^{i(\bar{k}-k_n)z}dz$$
$$= \frac{e}{\sqrt{2\pi}^3\varepsilon_0}\frac{\delta(r)}{r}\frac{1}{2\pi}\int_{-\infty}^{+\infty}e^{i\left(\bar{k}-\frac{\bar{\omega}}{v}\right)z}dz. \tag{5.436}$$

This operation results in delta-functions on both sides of the equation:

$$\int_{-h_0/2}^{+h_0/2}dq\tilde{\phi}(\bar{\omega},q)\sum_n\left(\frac{\bar{\omega}^2}{c^2}-k_n^2-\Gamma_n^2\right)F_n\mathrm{J}_0(\Gamma_n r)\delta(\bar{k}-k_n) = \frac{e}{\sqrt{2\pi}^3\varepsilon_0}\frac{\delta(r)}{r}\delta\left(\bar{k}-\frac{\bar{\omega}}{v}\right). \tag{5.437}$$

Some care must be taken when dealing with the integral over the "free" part of the wavenumber. If the quantity $\bar{k}-k_n=\bar{k}-nh_0-q$ has a zero within the integration interval, we obtain the value of the function at the pole; otherwise, we get a zero. The net result is that we obtain a selection rule. For a given value of $\bar{k}$, there is a unique integer $\bar{n}$, such that $\bar{q} = \bar{k}-\bar{n}h_0$ lies in the interval $\left[-\frac{h_0}{2},\frac{h_0}{2}\right]$. Moreover, since we are dealing with a series, we can always find a value of $\bar{n}$ that will yield a pole in the integration interval. We obtain a so-called "Dirac comb" and find that

$$\tilde{\phi}(\bar{\omega},\bar{k}-\bar{n}h_0)\left(\frac{\bar{\omega}^2}{c^2}-\bar{k}^2-\Gamma_{\bar{n}}^2\right)F_{\bar{n}}\mathrm{J}_0(\Gamma_{\bar{n}}r) = \frac{e}{\sqrt{2\pi}^3\varepsilon_0}\frac{\delta(r)}{r}\delta\left(\bar{k}-\frac{\bar{\omega}}{v}\right). \tag{5.438}$$

The series is removed by virtue of the selection rule discussed above. This particular feature of the solution is a direct consequence from the use of a Floquet expansion. Finally, the radial projection can be performed by multiplying each side of Equation 5.438 by $r\,J_0(\chi_{0\bar{n}}r)$, where $\chi_{0\bar{n}}$ is the $\bar{n}$-th zero of $J_0$; with this, we have

$$\int_0^1 r J_0(\chi_{0\bar{n}}r)\frac{e}{\sqrt{2\pi}^3\varepsilon_0}\frac{\delta(r)}{r}\delta\left(\bar{k}-\frac{\overline{\omega}}{v}\right) = \frac{e}{\sqrt{2\pi}^3\varepsilon_0}J_0(0)\delta\left(\bar{k}-\frac{\overline{\omega}}{v}\right)$$
$$= \frac{e}{\sqrt{2\pi}^3\varepsilon_0}\delta\left(\bar{k}-\frac{\overline{\omega}}{v}\right), \quad (5.439)$$

for the current. For the series, we can use Equation 5.319,

$$\int xZ_p(\alpha x)B_p(\beta x)dx = \frac{\beta x Z_p(\alpha x)B_{p-1}(\beta x) - \alpha x Z_{p-1}(\alpha x)B_p(\beta x)}{\alpha^2-\beta^2}, \quad (5.440)$$

to evaluate

$$\int_0^1 rJ_0(\chi_{0\bar{n}}r)J_0(\Gamma_{\bar{n}}r)dr = \frac{\chi_{0\bar{n}}J_{-1}(\chi_{0\bar{n}})J_0(\Gamma_{\bar{n}})}{\chi_{0\bar{n}}^2-\Gamma_{\bar{n}}^2}. \quad (5.441)$$

Switching back the dummy variables, the final result is

$$\tilde{\phi}(\omega, k-nh_0) = \frac{e}{\sqrt{2\pi}^3\varepsilon_0}\frac{\delta(k-\frac{\omega}{v})}{\frac{ikc^2}{\Gamma_n^2}A_n(\omega,k)\left(\frac{\omega^2}{c^2}-k^2-\Gamma_n^2\right)\left[\frac{\chi_{0n}J_{-1}(\chi_{0n})J_0(\Gamma_n)}{\chi_{0n}^2-\Gamma_{\bar{n}}^2}\right]}. \quad (5.442)$$

Note that, as we do not have an explicit diagonalization procedure for the series, the radial integral is somewhat arbitrary. Here we chose to integrate up to a zero of the Bessel function to obtain a shorter expression. In summary, the coefficients $F_n$ and the Bessel function arguments $\Gamma_n$ are first derived from the corrugated waveguide boundary condition the excitation of each mode can then be derived using Equation 5.442. From a practical standpoint, one generally limits the number of coupled modes to obtain a small, manageable matrix, $\mathbb{C}$; the other calculations can then be performed easily.

## 5.7 Point Charge in Rectilinear Motion in Vacuum

The equivalence between the Lorentz transformation of the Coulomb field and a cylindrical, unbounded vacuum eigenmode analysis is demonstrated here. This approach also provides the basis for the study of Čerenkov radiation, which is presented in Chapter 7.

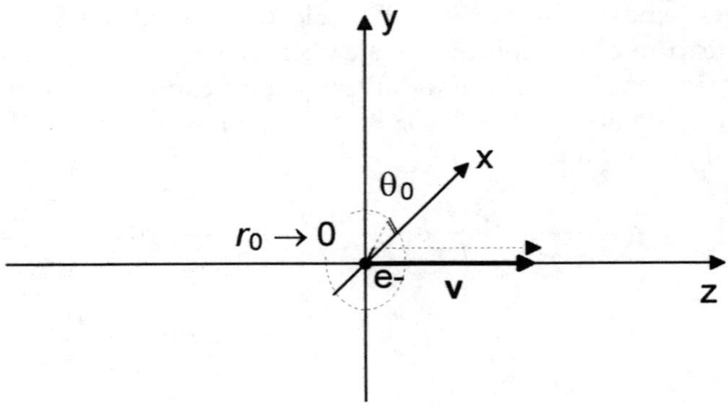


**FIGURE 5.8**
Geometry for an electron in rectilinear motion, including the infinitesimal radial offset, $r_0$, used to define the angle, $\theta_0$, as described in the text.

### 5.7.1 Coulomb Field and Lorentz Transform

The system under consideration is illustrated in Figure 5.8. We consider an electron in rectilinear motion along the $z$-axis, with constant velocity $v\hat{z}$. In a frame co-moving with the electron, the four-potential is simply the Coulomb scalar potential:

$$\bar{A}_\mu = \left(\frac{\bar{\phi}}{c}, \bar{\mathbf{A}}\right) = \left(\frac{-e}{4\pi\varepsilon_0 c\sqrt{\bar{r}^2+\bar{z}^2}}, \mathbf{0}\right); \tag{5.443}$$

to derive the field in the laboratory frame, we use the Lorentz transform, which yields

$$\begin{aligned} \mathbf{A}_\perp &= \bar{\mathbf{A}}_\perp = \mathbf{0}, \\ A_z &= \gamma(\bar{A}_z + \beta\bar{A}_0) = \gamma\beta\frac{-e}{4\pi\varepsilon_0 c\sqrt{\bar{r}^2+\bar{z}^2}}, \\ \phi &= \gamma(\bar{\phi} + v\bar{A}_z) = \gamma\frac{-e}{4\pi\varepsilon_0\sqrt{\bar{r}^2+\bar{z}^2}}. \end{aligned} \tag{5.444}$$

To complete the derivation, we need to relate the space–time coordinates in both frames:

$$\begin{aligned} r &= \bar{r}, \\ z &= \gamma(\bar{z} + v\bar{t}), \quad \bar{z} = \gamma(z - vt), \\ t &= \gamma\left(\bar{t} + \beta\frac{\bar{z}}{c}\right), \quad \bar{t} = \gamma\left(t - \beta\frac{z}{c}\right). \end{aligned} \tag{5.445}$$

We then find the time-dependent field produced in unbounded vacuum by an electron in rectilinear motion, which is identical to that derived in Section 5.5, Equation 5.226:

$$\phi(r, z, t) = \frac{-\gamma e}{4\pi\varepsilon_0\sqrt{r^2 + \gamma^2(z - vt)^2}}, \qquad A_z = \beta\frac{\phi}{c}. \tag{5.446}$$

### 5.7.2 Bessel Vacuum Eigenmode Excitation

We now resolve the same problem using the vacuum eigenmodes for cylindrical geometry. At first glance, it seems that one can choose directly the electron velocity to define the axis of symmetry of the system so that there is no explicit angular dependence, and $\partial_\theta \equiv 0$. Consequently, we are tempted to write the scalar potential as

$$\phi(r, z, t) = \frac{1}{(\sqrt{2\pi})^2}\int_{-\infty}^{+\infty} d\omega \int_{-\infty}^{+\infty} dk_\parallel \int_0^\infty k_\perp dk_\perp \tilde{\phi}(\omega, k_\parallel, k_\perp) e^{i(\omega t - k_\parallel z)} J_0(k_\perp r), \tag{5.447}$$

and the drive charge density as

$$\rho(r, z, t) = -e\frac{1}{r}\delta(r)\delta(z - vt). \tag{5.448}$$

Unfortunately, this approach is flawed. The correct derivation makes use of the complete TM vacuum eigenmode set and will enable us to pinpoint the exact origin of the difficulty. Therefore, instead of Equation 5.447, we write

$$\phi(r, \theta, z, t) = \frac{1}{(\sqrt{2\pi})^2}\sum_m \int_{-\infty}^{+\infty} d\omega \int_{-\infty}^{+\infty} dk_\parallel \int_0^\infty k_\perp dk_\perp \tilde{\phi}_m(\omega, k_\parallel, k_\perp) e^{i(\omega t - k_\parallel z + m\theta)} J_m(k_\perp r), \tag{5.449}$$

whereas the charge density is modeled by

$$\rho(r, \theta, z, t) = -e\frac{1}{r}\delta(r - r_0)\delta(\theta - \theta_0)\delta(z - vt). \tag{5.450}$$

The case where the axis of cylindrical symmetry coincides with the electron velocity is obtained by taking the limit: $r_0 \to 0$. Here, the variables $\omega$ and $t$, $k_\parallel$ and $z$, $m$ and $\theta$, and $k_\perp$ and $r$, respectively, are conjugate. Our first task is to determine the spectral density, $\tilde{\phi}_m(\omega, k_\parallel, k_\perp)$, which satisfies the wave equation driven by the electron four-current,

$$\begin{aligned}
&\Box\phi - \frac{e}{\varepsilon_0}\frac{1}{r}\delta(r-r_0)\delta(\theta-\theta_0)\delta(z-vt) = 0,\\
&\Box A_z - \mu_0 ev\frac{1}{r}\delta(r-r_0)\delta(\theta-\theta_0)\delta(z-vt) = 0.
\end{aligned} \tag{5.451}$$

It is clear that the two equations are proportional, which implies that

$$\frac{A_z}{\phi} = \mu_0\varepsilon_0 v = \frac{\beta}{c}. \tag{5.452}$$

Therefore, we will focus on the scalar potential. The driven wave equation can be written as

$$\begin{aligned}
&\Box\left[\frac{1}{2\pi}\sum_m\int_{-\infty}^{+\infty}d\omega\int_{-\infty}^{+\infty}dk_\parallel\int_0^\infty k_\perp dk_\perp\tilde{\phi}_m(\omega, k_\parallel, k_\perp)e^{i(\omega t-k_\parallel z+m\theta)}J_m(k_\perp r)\right]\\
&= \frac{e}{\varepsilon_0}\frac{1}{r}\delta(r-r_0)\delta(\theta-\theta_0)\delta(z-vt).
\end{aligned} \tag{5.453}$$

For the vacuum eigenmodes, the action of the d'Alembertian yields the simple dispersion eigenvalue:

$$\Box[e^{i(\omega t-k_\parallel z+m\theta)}J_m(k_\perp r)] = \left(\frac{\omega^2}{c^2}-k_\parallel^2-k_\perp^2\right)e^{i(\omega t-k_\parallel z+m\theta)}J_m(k_\perp r). \tag{5.454}$$

The wave equation now reads

$$\begin{aligned}
&\frac{1}{2\pi}\sum_m\int_{-\infty}^{+\infty}d\omega\int_{-\infty}^{+\infty}dk_\parallel\int_0^\infty k_\perp dk_\perp\tilde{\phi}_m(\omega, k_\parallel, k_\perp)\left(\frac{\omega^2}{c^2}-k_\parallel^2-k_\perp^2\right)e^{i(\omega t-k_\parallel z+m\theta)}J_m(k_\perp r)\\
&= \frac{e}{\varepsilon_0}\frac{1}{r}\delta(r-r_0)\delta(\theta-\theta_0)\delta(z-vt).
\end{aligned} \tag{5.455}$$

At this point, we need to project the current and diagonalize the potential. We first multiply each side of Equation 5.455 by $e^{ik'_\parallel z}$ and integrate from $-\infty$ to $+\infty$:

$$\begin{aligned}
&\sum_m\int_{-\infty}^{+\infty}d\omega\int_{-\infty}^{+\infty}dk_\parallel\int_0^\infty k_\perp dk_\perp\tilde{\phi}_m(\omega, k_\parallel, k_\perp)\left(\frac{\omega^2}{c^2}-k_\parallel^2-k_\perp^2\right)e^{i(\omega t+m\theta)}\\
&\times J_m(k_\perp r)\frac{1}{2\pi}\int_{-\infty}^{+\infty}e^{i(k'_\parallel-k_\parallel)z}dz\\
&= \int_{-\infty}^{+\infty}\frac{e}{\varepsilon_0}\frac{1}{r}\delta(r-r_0)\delta(\theta-\theta_0)\delta(z-vt)e^{ik'_\parallel z}dz.
\end{aligned} \tag{5.456}$$

This operation yields a delta-function on the left-hand side of the equation and a complex exponential harmonic oscillation of the current:

$$\sum_m \int_{-\infty}^{+\infty} d\omega \int_{-\infty}^{+\infty} dk_\parallel \int_0^\infty k_\perp dk_\perp \tilde{\phi}_m(\omega, k_\parallel, k_\perp)\left(\frac{\omega^2}{c^2} - k_\parallel^2 - k_\perp^2\right) e^{i(\omega t + m\theta)}$$
$$\times J_m(k_\perp r)\delta(k_\parallel' - k_\parallel) = \frac{e}{\varepsilon_0}\frac{1}{r}\delta(r - r_0)\delta(\theta - \theta_0)e^{ik_\parallel' vt}. \tag{5.457}$$

The integral over $k_\parallel$ can now be readily performed, to obtain

$$\sum_m \int_{-\infty}^{+\infty} d\omega \int_0^\infty k_\perp dk_\perp \tilde{\phi}_m(\omega, k_\parallel', k_\perp)\left(\frac{\omega^2}{c^2} - k_\parallel'^2 - k_\perp^2\right) e^{i(\omega t + m\theta)} J_m(k_\perp r)$$
$$= \frac{e}{\varepsilon_0}\frac{1}{r}\delta(r - r_0)\delta(\theta - \theta_0)e^{ik_\parallel' vt}. \tag{5.458}$$

We repeat the same steps for the frequency and time:

$$\sum_m \int_{-\infty}^{+\infty} d\omega \int_0^\infty k_\perp dk_\perp \tilde{\phi}_m(\omega, k_\parallel', k_\perp)\left(\frac{\omega^2}{c^2} - k_\parallel'^2 - k_\perp^2\right) e^{im\theta} J_m(k_\perp r) \int_{-\infty}^{+\infty} e^{i(\omega - \omega')t} dt$$
$$= \int_{-\infty}^{+\infty} \frac{e}{\varepsilon_0}\frac{1}{r}\delta(r - r_0)\delta(\theta - \theta_0)e^{i\left(k_\parallel' v - \omega'\right)t} dt; \tag{5.459}$$

$$\sum_m \int_{-\infty}^{+\infty} d\omega \int_0^\infty k_\perp dk_\perp \tilde{\phi}_m(\omega, k_\parallel', k_\perp)\left(\frac{\omega^2}{c^2} - k_\parallel'^2 - k_\perp^2\right) e^{im\theta} J_m(k_\perp r) 2\pi\delta(\omega - \omega')$$
$$= \frac{e}{\varepsilon_0}\frac{1}{r}\delta(r - r_0)\delta(\theta - \theta_0)2\pi\delta(k_\parallel' v - \omega'); \tag{5.460}$$

$$\sum_m \int_0^\infty k_\perp dk_\perp \tilde{\phi}_m(\omega', k_\parallel', k_\perp)\left(\frac{\omega'^2}{c^2} - k_\parallel'^2 - k_\perp^2\right) e^{im\theta} J_m(k_\perp r)$$
$$= \frac{e}{\varepsilon_0}\frac{1}{r}\delta(r - r_0)\delta(\theta - \theta_0)\delta(k_\parallel' v - \omega'). \tag{5.461}$$

At this point, we need to diagonalize the azimuthal mode series. We multiply each side of Equation 5.461 by $\frac{e^{-im'\theta}}{2\pi}$, and integrate from 0 to $2\pi$:

$$\sum_m \int_0^\infty k_\perp dk_\perp \tilde{\phi}_m(\omega', k_\parallel', k_\perp)\left(\frac{\omega'^2}{c^2} - k_\parallel'^2 - k_\perp^2\right) J_m(k_\perp r)\frac{1}{2\pi}\int_0^{2\pi} e^{i(m - m')\theta} d\theta$$
$$= \frac{1}{2\pi}\int_0^{2\pi} e^{-im'\theta} d\theta \frac{e}{\varepsilon_0}\frac{1}{r}\delta(r - r_0)\delta(\theta - \theta_0)\delta(k_\parallel' v - \omega'), \tag{5.462}$$

which yields

$$\sum_m \int_0^\infty k_\perp dk_\perp \tilde{\phi}_m(\omega', k'_\parallel, k_\perp)\left(\frac{\omega'^2}{c^2} - k_\parallel'^2 - k_\perp^2\right) J_m(k_\perp r)\delta_{mm'}$$
$$= \int_0^\infty k_\perp dk_\perp \tilde{\phi}_{m'}(\omega', k'_\parallel, k_\perp)\left(\frac{\omega'^2}{c^2} - k_\parallel'^2 - k_\perp^2\right) J_{m'}(k_\perp r)$$
$$= \frac{e}{2\pi\varepsilon_0}\frac{1}{r}\delta(r - r_0)e^{-im'\theta_0}\delta(k'_\parallel v - \omega'). \tag{5.463}$$

Finally, for the radius, we use the expression of the delta-function in terms of Bessel functions given in Equation 5.91. We multiply Equation 5.463 by $rJ_{m'}(k'_\perp r)$, and integrate from 0 to infinity, to obtain

$$\int_0^\infty dk_\perp \tilde{\phi}_{m'}(\omega', k'_\parallel, k_\perp)\left(\frac{\omega'^2}{c^2} - k_\parallel'^2 - k_\perp^2\right) k_\perp \int_0^\infty rJ_{m'}(k'_\perp r)J_{m'}(k_\perp r)dr$$
$$= \int_0^\infty dk_\perp \tilde{\phi}_{m'}(\omega', k'_\parallel, k_\perp)\left(\frac{\omega'^2}{c^2} - k_\parallel'^2 - k_\perp^2\right)\delta(k_\perp - k'_\perp)$$
$$= \tilde{\phi}_{m'}(\omega', k'_\parallel, k'_\perp)\left(\frac{\omega'^2}{c^2} - k_\parallel'^2 - k_\perp'^2\right), \tag{5.464}$$

which is also equal to

$$\int_0^\infty \frac{e}{2\pi\varepsilon_0}\frac{1}{r}\delta(r - r_0)e^{-im'\theta_0}\delta(k'_\parallel v - \omega')rJ_{m'}(k'_\perp r)dr$$
$$= \frac{e}{2\pi\varepsilon_0}\delta(k'_\parallel v - \omega')e^{-im'\theta_0}J_{m'}(k'_\perp r_0). \tag{5.465}$$

We now take the limit when $r_0 \to 0$: the drive term is given by

$$\lim_{r_0\to 0}\left[\frac{e}{2\pi\varepsilon_0}\delta(k'_\parallel v - \omega')e^{-im'\theta_0}J_{m'}(k'_\perp r_0)\right] = \frac{e}{2\pi\varepsilon_0}\delta(k'_\parallel v - \omega')e^{-im'\theta_0}J_{m'}(0), \tag{5.466}$$

and we see that only the $m' = 0$ mode survives because $J_m(0) = \delta_{m0}$. As a result, the spectral amplitude of the scalar potential is given by

$$\tilde{\phi}_0(\omega, k_\parallel, k_\perp) = \frac{e\delta(k_\parallel v - \omega)}{2\pi\varepsilon_0\left(\frac{\omega^2}{c^2} - k_\parallel^2 - k_\perp^2\right)}. \tag{5.467}$$

By comparison, using the restricted eigenmode set of Equation 5.447 and the on-axis charge density described in Equation 5.448 and diagonalizing would result in an error by a factor of $2\pi$. This is because the azimuthal angle cannot be defined for zero radius; instead, one needs to use an arbitrary radius, $r_0$, and the associated, well-defined angle $\theta_0$, and to take the limit $r_0 \to 0$, to properly treat the problem.

To obtain the solution in real space, we have to transform back from momentum space:

$$\phi(r, z, t) = \frac{1}{(\sqrt{2\pi})^2}\int_{-\infty}^{+\infty} d\omega \int_{-\infty}^{+\infty} dk_\parallel \int_0^\infty k_\perp dk_\perp \frac{e}{2\pi\varepsilon_0} \frac{\delta(k_\parallel v - \omega)}{\left(\frac{\omega^2}{c^2} - k_\parallel^2 - k_\perp^2\right)} e^{i(\omega t - k_\parallel z)} J_0(k_\perp r). \tag{5.468}$$

The integral over frequency is performed easily because of the delta-function describing the beam mode:

$$\begin{aligned}\phi(r, z, t) &= \frac{1}{(\sqrt{2\pi})^2}\int_{-\infty}^{+\infty} dk_\parallel \int_0^\infty k_\perp dk_\perp \frac{e}{2\pi\varepsilon_0} \frac{e^{i(k_\parallel vt - k_\parallel z)}}{\left[\frac{(k_\parallel v)^2}{c^2} - k_\parallel^2 - k_\perp^2\right]} J_0(k_\perp r) \\ &= \frac{-e}{4\pi^2\varepsilon_0}\int_{-\infty}^{+\infty} dk_\parallel \int_0^\infty k_\perp dk_\perp \frac{e^{ik_\parallel(vt - z)}}{\left(\frac{k_\parallel^2}{\gamma^2} + k_\perp^2\right)} J_0(k_\perp r). \end{aligned} \tag{5.469}$$

Here, we have used the fact that $\gamma^{-2} = 1 - \beta^2$. To integrate over the axial wavenumber, we recast Equation 5.469 as

$$\phi(r, z, t) = \frac{-e\gamma^2}{2\pi^2\varepsilon_0}\int_0^\infty k_\perp J_0(k_\perp r) dk_\perp \int_0^\infty dk_\parallel \frac{\cos[k_\parallel(vt - z)]}{(k_\parallel^2 + \gamma^2 k_\perp^2)}, \tag{5.470}$$

where we have used the fact that

$$\int_{-\infty}^{+\infty} \frac{e^{iax}}{b^2 + x^2} dx = 2\int_0^\infty \frac{\cos(ax)}{b^2 + x^2} dx, \tag{5.471}$$

because the sine component is antisymmetrical and integrates out. We now use the formula

$$\int_0^\infty \frac{\cos(ax)}{b^2 + x^2} dx = \frac{\pi}{2b} e^{-ab}, \tag{5.472}$$

with this, we make the following identifications:

$$x \equiv k_{\parallel}, \quad a \equiv vt - z, \quad b \equiv \gamma k_{\perp}, \tag{5.473}$$

and obtain

$$\begin{aligned}\phi(r,z,t) &= \frac{-e\gamma^2}{2\pi^2\varepsilon_0}\int_0^{\infty} k_{\perp}\mathrm{J}_0(k_{\perp}r)dk_{\perp}\frac{\pi}{2\gamma k_{\perp}}e^{-k_{\perp}\gamma(vt-z)} \\ &= \frac{-e\gamma}{4\pi\varepsilon_0}\int_0^{\infty}\mathrm{J}_0(k_{\perp}r)dk_{\perp}e^{k_{\perp}\tilde{z}}.\end{aligned} \tag{5.474}$$

Finally, for the integral over the transverse wavenumber, we use the following equation:

$$\int_0^{\infty} e^{-ax}\mathrm{J}_{\nu}(bx)dx = \frac{b^{-\nu}\left[\sqrt{a^2+b^2}-a\right]^{\nu}}{\sqrt{a^2+b^2}}, \quad \mathcal{R}(\nu) > -1, \quad \mathcal{R}(a \pm ib) > 0. \tag{5.475}$$

We identify

$$x \equiv k_{\perp}, \quad a \equiv \gamma(z - vt), \quad \nu \equiv 0, \quad b \equiv r, \tag{5.476}$$

and find that the potential is identical to that derived in Sections 5.5 and 5.7.1:

$$\phi(r,z,t) = \frac{-e\gamma}{4\pi\varepsilon_0\sqrt{r^2+\gamma^2(z-vt)^2}}. \tag{5.477}$$

## 5.8 Multipoles, Spherical Harmonics, and the Hydrogen Atom

A theme that has been developed extensively in this chapter is the utilization of eigenmodes to resolve wave equations with boundary conditions. So far, we have considered electromagnetic waves propagating in systems with a given symmetry and shown how the d'Alembertian operator could reduce to a simple eigenvalue equation, yielding the well-known dispersion relation.

To offer a complementary perspective on these mathematical techniques, we will first introduce the notion of decomposition of the electromagnetic field in multipoles, which leads naturally to spherical harmonics and Legendre polynomials. Finally, we will show how these mathematical objects can be used to solve the wave equation in a different case: the Schrödinger equation for the hydrogen atom. In this manner we will show how these various areas of physics are fundamentally interrelated at the conceptual level. Whereas the wave equation of electrodynamics governs the electromagnetic field and photons, the Schrödinger, Klein–Gordon, and Dirac equations model the evolution of the probability density of a given particle, as first postulated by Born.

Multipole expansions and spherical harmonics originate from the Green function for the Laplacian operator, $|\mathbf{x}-\bar{\mathbf{x}}|^{-1}$, as studied in Section 5.4.1. If we define the angle between $\mathbf{x}$ and $\bar{\mathbf{x}}$ by $\theta$, we first have

$$\frac{1}{|\mathbf{x}-\bar{\mathbf{x}}|} = \frac{1}{\sqrt{x^2+\bar{x}^2-2x\bar{x}\cos\theta}}. \tag{5.478}$$

This expression can be expanded into a power series by using the ratio,

$$\varepsilon = \begin{cases} x/\bar{x}, & \bar{x} > x. \\ \bar{x}/x, & x > \bar{x}; \end{cases} \tag{5.479}$$

we then write

$$\frac{1}{|\mathbf{x}-\bar{\mathbf{x}}|} = \frac{1}{x^+\sqrt{1-2\varepsilon\cos\theta+\varepsilon^2}} = \frac{1}{x^+}\sum_n \varepsilon^n P_n(\cos\theta), \tag{5.480}$$

where we have defined $x^+ = \max(x, \bar{\mathbf{x}})$, and where the polynomials, $P_n$, are the Legendre polynomials:

$$\begin{aligned} P_0(x) &= 1, \\ P_1(x) &= x, \\ P_2(x) &= \frac{1}{2}(3x^2-1), \\ P_3(x) &= \frac{1}{2}(5x^3-3x), \\ P_n(x) &= \frac{1}{2^n n!}\frac{d^n}{dx^n}(x^2-1)^n. \end{aligned} \tag{5.481}$$

Moreover, the orthogonality condition for the Legendre polynomials is

$$\int_{-1}^{+1} P_m(x)P_n(x)dx = -\int_{-\pi}^{+\pi} P_m(\cos\theta)P_n\cos(\theta)\sin\theta\, d\theta = \frac{2\delta_{mn}}{2n+1}. \quad (5.482)$$

The full significance of this expansion becomes clear when associated with the spherical harmonics, which form a complete set of orthogonal functions for the angular spherical coordinates, $\theta$ and $\varphi$. We have

$$Y_l^m(\theta, \varphi) = \sqrt{\frac{2l+1}{4\pi}\frac{(l-m)!}{(l+m)!}} P_l^m(\cos\theta)e^{im\varphi}, \quad |m| \le l, \quad (5.483)$$

where the $P_l^m$ are the associated Legendre polynomials, with

$$\begin{aligned} P_l^m(x) &= (-\sqrt{1-x^2})^m \frac{d^m}{dx^m} P_l(x), \quad m > 0, \\ P_l^{-m}(x) &= (-1)^m \frac{(l-m)!}{(l+m)!} P_l^m(x), \quad m < 0. \end{aligned} \quad (5.484)$$

The orthogonality of these polynomials is expressed as

$$\int_{-1}^{+1} P_l^m(x)P_n^m(x)dx = \frac{2}{2l+1}\frac{(l-m)!}{(l+m)!}\delta_{ln}, \quad (5.485)$$

while the orthogonality of spherical harmonics is described by

$$\int_0^{2\pi} d\varphi \int_0^{\theta} Y_l^{m*}(\theta, \varphi)Y_n^p(\theta, \varphi)\sin\theta\, d\theta = \delta_{ln}\delta_{mp}. \quad (5.486)$$

Moreover, the spherical harmonics form a complete set of orthogonal functions:

$$\sum_{l=0}^{\infty}\sum_{m=-l}^{+l} Y_l^{m*}(\theta, \varphi)Y_l^m(\bar{\theta}, \bar{\varphi}) = \delta(\cos\theta - \cos\bar{\theta})\delta(\varphi - \bar{\varphi}). \quad (5.487)$$

As will be described shortly, these functions play an important role both for multipole expansions in electrodynamics and to describe the hydrogen atom in quantum mechanics. The reason behind this resides in the fact that the spherical harmonics are the eigenmodes of the Laplacian operator in spherical coordinates. Finally, the so-called addition theorem yields a relation between spherical harmonics and the Legendre polynomials:

$$P_l[\sin\theta\sin\bar{\theta}\cos(\varphi - \bar{\varphi}) + \cos\theta\cos\bar{\theta}] = \frac{4\pi}{2l+1}\sum_{m=-l}^{l} Y_l^{m*}(\bar{\theta}, \bar{\varphi})Y_l^m(\theta, \varphi). \quad (5.488)$$

We now return to the Green function for the Laplacian. For an arbitrary charge distribution, the scalar potential is given by

$$\phi(\mathbf{x}) = \int \frac{\rho(\bar{\mathbf{x}}) d^3\bar{x}}{4\pi\varepsilon_0 |\mathbf{x} - \bar{\mathbf{x}}|}, \tag{5.489}$$

and we can use the addition theorem to expand the potential into multipoles, to obtain

$$\begin{aligned} \varepsilon_0 \phi(\mathbf{x}) &= \sum_{l=0}^{\infty} \sum_{m=-l}^{+l} \frac{1}{2l+1} \frac{Y_l^m(\theta, \varphi)}{|x|^{l+1}} \int \rho(\bar{\mathbf{x}}) |\bar{\mathbf{x}}|^l Y_l^{m^*}(\bar{\theta}, \bar{\varphi}) d^3\bar{x} \\ &= \sum_{l=0}^{\infty} \sum_{m=-l}^{+l} \frac{1}{2l+1} \frac{Y_l^m(\theta, \varphi)}{|x|^{l+1}} \rho_{lm}. \end{aligned} \tag{5.490}$$

As the spherical harmonics form a complete set of orthogonal functions, the multipole moments, $\rho_{lm}$, of the charge distribution fully characterize its electrostatic properties; the most often used are the monopole, dipole, quadrupole, and octupole moments.

Using Equation 5.483, we can express the first spherical harmonics, with which we can derive the monopole, dipole, and quadrupole moments. We have

$$\begin{aligned} Y_0^0 &= \frac{1}{\sqrt{4\pi}}, \\ Y_1^0 &= \sqrt{\frac{3}{4\pi}} \cos\theta, \quad Y_1^1 = -\sqrt{\frac{3}{8\pi}} \sin\theta e^{i\varphi}, \\ Y_2^0 &= \sqrt{\frac{5}{16\pi}} (3\cos^2\theta - 1), \quad Y_2^1 = -\sqrt{\frac{15}{8\pi}} \sin\theta \cos\theta\, e^{i\varphi}, \quad Y_2^2 = \sqrt{\frac{15}{32\pi}} \sin^2\theta e^{i2\varphi}. \end{aligned} \tag{5.491}$$

With this, the monopole moment is given by

$$\rho_{00} = \int \rho(\bar{\mathbf{x}}) Y_0^{0^*}(\bar{\theta}, \bar{\varphi}) d^3\bar{x} = \frac{1}{\sqrt{4\pi}} \int \rho(\bar{\mathbf{x}}) d^3\bar{x} = \frac{q}{\sqrt{4\pi}}; \tag{5.492}$$

for the dipole, we have

$$\begin{aligned} \rho_{10} = \int \rho(\bar{\mathbf{x}}) |\bar{\mathbf{x}}| Y_1^{0^*}(\bar{\theta}, \bar{\varphi}) d^3\bar{x} &= \sqrt{\frac{3}{4\pi}} \int \rho(\bar{\mathbf{x}}) |\bar{\mathbf{x}}| \cos\theta d^3\bar{x} \\ &= \sqrt{\frac{3}{4\pi}} \int \bar{z} \rho(\bar{\mathbf{x}}) d^3\bar{x}, \end{aligned} \tag{5.493}$$

and

$$\begin{aligned}\rho_{11} &= \int \rho(\bar{\mathbf{x}})|\bar{\mathbf{x}}|Y_1^{1^*}(\bar{\theta},\ \bar{\varphi})d^3\bar{x}\\ &= -\sqrt{\frac{3}{8\pi}}\int \rho(\bar{\mathbf{x}})|\bar{\mathbf{x}}|\sin\bar{\theta}e^{-i\bar{\varphi}}d^3\bar{x}\\ &= -\sqrt{\frac{3}{8\pi}}\int \rho(\bar{\mathbf{x}})|\bar{\mathbf{x}}|\sin\bar{\theta}(\cos\bar{\varphi}-i\sin\bar{\varphi})d^3\bar{x}\\ &= -\sqrt{\frac{3}{8\pi}}\int \rho(\bar{\mathbf{x}})(\bar{x}-i\bar{y}\sin\bar{\theta})d^3\bar{x}.\end{aligned} \tag{5.494}$$

Finally, we derive the components of the quadrupole moment. We first have

$$\begin{aligned}\rho_{20} &= \int \rho(\bar{\mathbf{x}})|\bar{\mathbf{x}}|^2Y_2^{0^*}(\bar{\theta},\ \bar{\varphi})d^3\bar{x}\\ &= \sqrt{\frac{5}{16\pi}}\int \rho(\bar{\mathbf{x}})|\bar{\mathbf{x}}|^2(3\cos^2\theta-1)d^3\bar{x}\\ &= \sqrt{\frac{5}{16\pi}}\int \rho(\bar{\mathbf{x}})(3\bar{z}^2-|\bar{\mathbf{x}}|^2)d^3\bar{x};\end{aligned} \tag{5.495}$$

next, we consider $\rho_{21}$,

$$\begin{aligned}\rho_{21} &= \int \rho(\bar{\mathbf{x}})|\bar{\mathbf{x}}|^2Y_2^{1^*}(\bar{\theta},\bar{\varphi})d^3\bar{x}\\ &= -\sqrt{\frac{15}{8\pi}}\int \rho(\bar{\mathbf{x}})|\bar{\mathbf{x}}|^2\sin\bar{\theta}\cos\bar{\theta}e^{-i\bar{\varphi}}d^3\bar{x}\\ &= -\sqrt{\frac{15}{8\pi}}\int \rho(\bar{\mathbf{x}})\bar{z}(\bar{x}-i\bar{y})d^3\bar{x}.\end{aligned} \tag{5.496}$$

Finally, for $\rho_{22}$, we have

$$\begin{aligned}\rho_{22} &= \int \rho(\bar{\mathbf{x}})|\bar{\mathbf{x}}|^2Y_2^{2^*}(\bar{\theta},\bar{\varphi})d^3\bar{x}\\ &= \sqrt{\frac{15}{8\pi}}\int \rho(\bar{\mathbf{x}})|\bar{\mathbf{x}}|^2\sin^2\bar{\theta}e^{-i2\bar{\varphi}}d^3\bar{x}\\ &= \sqrt{\frac{15}{32\pi}}\int \rho(\bar{\mathbf{x}})[|\bar{\mathbf{x}}|\sin\bar{\theta}(\cos\bar{\theta}-i\sin\bar{\theta})]^2d^3\bar{x}\\ &= \sqrt{\frac{15}{32\pi}}\int \rho(\bar{\mathbf{x}})(\bar{x}-i\bar{y})^2d^3\bar{x}.\end{aligned} \tag{5.497}$$

These results can be grouped together by defining the quadrupole tensor,

$$q_{ij} = \int \rho(\bar{\mathbf{x}})(3\bar{x}_i\bar{x}_j-|\bar{\mathbf{x}}|\delta_{ij})d^3\bar{x}. \tag{5.498}$$

We can then make the following identifications:

$$\begin{aligned}\rho_{20} &= \sqrt{\frac{5}{16\pi}} q_{zz}, \\ \rho_{21} &= -\sqrt{\frac{15}{72\pi}}(q_{xz} - iq_{yz}), \\ \rho_{22} &= \sqrt{\frac{15}{288\pi}}(q_{xx} - 2iq_{xy} - q_{yy}).\end{aligned} \tag{5.499}$$

The last identity is obtained by considering that

$$\begin{aligned}\overline{(\bar{x} - i\bar{y})^2} &= \overline{\bar{x}^2 - \bar{y}^2 - 2i\bar{x}\bar{y}} \\ &= \frac{1}{3}[(3\bar{x}^2 - |\bar{\mathbf{x}}|^2) - (3\bar{y}^2 - |\bar{\mathbf{x}}|^2) - 6i\overline{\bar{x}\bar{y}}].\end{aligned} \tag{5.500}$$

For our next example illustrating the use of spherical harmonics, we will consider the nonrelativistic derivation of the energy levels of the hydrogen atom. An excellent, very detailed presentation can be found in Cohen-Tannoudji's textbook, listed in the references to this chapter. Here, we give a short overview of the salient features of the theory. We also note that the Schrödinger equation, in its nonlinear form, can be used to describe the propagation of solitons in optical fibers, as discussed in Section 5.9. Therefore, the synergy between the wave equation and eigenmode formalism developed within the context of electrodynamics and their quantum mechanical counterparts should be considered as a fundamental aspect of modern physics.

The hydrogen atom problem can be reduced to a central potential problem by using the center-of-mass reference frame. In this frame, the problem is reduced to that of a point particle, with mass

$$\mu = \frac{m_p m_0}{m_p + m_0}, \tag{5.501}$$

where $m_p = 1.672\,6231(10) \times 10^{-27}$ kg is the proton rest mass, while $m_0$ is the electron rest mass. Moreover, in this case, the Schrödinger equation reads

$$\left\{-\frac{\hbar^2}{2\mu}\left[\frac{1}{r}\partial_r^2 r + \frac{1}{r^2}\left(\partial_\theta^2 + \frac{1}{\tan\theta}\partial_\theta + \frac{1}{\sin^2\theta}\partial_\varphi^2\right)\right] + U(r)\right\}\psi(\mathbf{x}) = E\psi(\mathbf{x}), \tag{5.502}$$

where $U(r)$ is the central potential due to the Coulomb force between the electron and proton, while $\psi(\mathbf{x})$ is the stationary wavefunction for the problem, and $E$ is the corresponding eigenvalue.

Using spherical coordinates and separating variables, by writing, for a particular angular eigenfunction,

$$\psi_l(\mathbf{x}) = \psi_l(r, \theta, \varphi) = R_l(r) Y_l^m(r, \theta), \tag{5.503}$$

we have

$$\left[-\frac{\hbar^2}{2\mu}\frac{1}{r}\frac{d^2}{dr^2}r + \frac{l(l+1)\hbar^2}{2\mu r^2} + U(r)\right] R_l(r) Y_l^m(\theta, \varphi) = E R_l(r) Y_l^m(\theta, \varphi). \tag{5.504}$$

Here we have used the aforementioned property of the spherical harmonics, which are eigenfunctions of the angular degrees of freedom of the Laplacian operator, as expressed in spherical coordinates:

$$\left(\partial_\theta^2 + \frac{1}{\tan\theta}\partial_\theta + \frac{1}{\sin^2\theta}\partial_\varphi^2\right) Y_l^m(\theta, \varphi) = l(l+1) Y_l^m(\theta, \varphi). \tag{5.505}$$

One can further simplify the radial wave equation by changing variables. We let $P_l(r) = rR_l(r)$, and find that the radial equation now reads

$$\left[\frac{\hbar^2}{2\mu}\frac{d^2}{dr^2} + \frac{l(l+1)\hbar^2}{2\mu r^2} + U(r)\right] \mathrm{P}_l(r) = E_l \mathrm{P}_l(r). \tag{5.506}$$

In the case of the hydrogen atom, the potential is simply given by $U(r) = -e^2/r$, and we have

$$\left[\frac{\hbar^2}{2\mu}\frac{d^2}{dr^2} + \frac{l(l+1)\hbar^2}{2\mu r^2} - \frac{e^2}{r}\right] \mathrm{P}_l(r) = E_l \mathrm{P}_l(r). \tag{5.507}$$

Equation 5.507 can be normalized by introducing the Bohr radius, $a_0 = \hbar^2/m_0 e^2$, which has the numerical value $a_0 = 0.529\,177\,249(24) \times 10^{-10}$ m, and the ionization energy, $E^* = m_0 e^4/2\hbar^2 = 13.605\,698\,1(40)$ eV. Since we are using the reduced mass, $\mu$, we use the normalized radius, $\rho = \mu r/a_0 m_0$, and the quantity $\lambda_l^2 = -m_0 E_l/\mu E^*$; the ratio $m_0/\mu = 1 + (m_0/m_p) = 1 + 5.446\,170\,13(11) \times 10^{-4}$. With this, we obtain

$$\left[\frac{d^2}{d\rho^2} - \frac{l(l+1)}{\rho^2} + \frac{2}{\rho} - \lambda_l^2\right] \mathrm{P}_l(\rho) = 0. \tag{5.508}$$

For large radii, the equation takes the asymptotic form

$$\left[\frac{d^2}{d\rho^2} - \lambda_l^2\right] \mathrm{P}_l(\rho) = 0, \tag{5.509}$$

which has a simple exponential solution, $\mathrm{P}_l(\rho) \sim e^{-\rho\lambda_l}$; therefore, we make a new change of variable: $\mathrm{P}_l(\rho) = e^{-\rho\lambda_l}\mathcal{R}_l(\rho)$. This leads to the differential equation

$$\left\{\frac{d^2}{d\rho^2} - 2\lambda_l\frac{d}{d\rho} + \left[\frac{2}{\rho} - \frac{l(l+1)}{\rho^2}\right]\right\}\mathcal{R}_l(\rho) = 0. \tag{5.510}$$

In addition, the behavior of the radial function at the origin must be bounded, which can be shown to imply $\lim_{\rho\to 0}\mathcal{R}_l(\rho) = 0$. At this point, the standard method of derivation is to use a power series expansion, so that

$$\mathcal{R}_l(\rho) = \rho^s\sum_n c_n\rho^n, \tag{5.511}$$

where $s > 0$ will satisfy the aforementioned boundary condition at zero radius. The first and second derivative of the radial function are easily calculated:

$$\begin{aligned}\frac{d\mathcal{R}_l}{d\rho} &= \sum_n (s+n)c_n\rho^{s+n-1},\\ \frac{d^2\mathcal{R}_l}{d\rho^2} &= \sum_n (s+n)(s+n-1)c_n\rho^{s+n-2},\end{aligned} \tag{5.512}$$

and can be inserted in Equation 5.510, to obtain

$$\sum_n\{[(s+n+2)(s+n+1) - l(l+1)]c_{n+2} + 2[1-\lambda_l(s+n+1)]c_{n+1}\}\rho^{s+n} = 0. \tag{5.513}$$

The lowest-order coefficient in the series is $c_0$, for which we must have

$$s(s-1) = l(l+1), \tag{5.514}$$

to obtain a nonzero value for $c_0$. Once the first coefficient is determined, Equation 5.513 provides a recurrence relation:

$$c_{n+1} = \frac{2[\lambda_l(s+n)-1]}{[(s+n+1)(s+n) - l(l+1)]}c_n. \tag{5.515}$$

The two roots of Equation 5.514 are

$$s = -l, \quad s = l+1, \tag{5.516}$$

but we have required that $s$ be positive; therefore, we can use $s = l + 1$ in the recurrence relation, to find

$$c_n = \frac{2[\lambda_l(n+l)-1]}{[(n+l+1)(n+l)-l(l+1)]}c_{n-1} = \frac{2[\lambda_l(n+l)-1]}{n(n+2l+1)}c_{n-1}. \quad (5.517)$$

For large value of $n$, the series converge, as $c_n/c_{n-1} \sim n^{-1}$. To finalize the derivation and quantize the energy, we seek to truncate the series by requiring that, for some value of $n = k$, we have $\lambda_l\,(k + l) - 1 = 0$. In this manner, beyond this value of $n$, the recurrence yields a series of null values. Using the energy normalization, we find that

$$E_{kl} = \frac{-E^*}{(k+l)^2}\frac{\mu}{m_0}. \quad (5.518)$$

If we introduce the principal quantum number, $m = k + l$, we can write the energy difference between two levels as

$$\Delta E_{mn} = E_m - E_n = h\nu_{mn} = \frac{\mu e^4}{8h^2\varepsilon_0^2}\left(\frac{1}{n^2}-\frac{1}{m_2}\right). \quad (5.519)$$

If $n = 1$, we recover the Lyman series, while $n = 2$ corresponds to the Balmer series; $n = 3$ and $n = 4$ are called the Paschen and Brackett series, respectively. The constant is called the Rydberg constant, $R = E^*$.

Inserting $\lambda_l = (k + l)^{-1}$ in the recurrence formula, we have

$$c_n = \frac{2\left[\frac{(n+l)}{(k+l)}-1\right]}{n(n+2l+1)}c_{n-1} = \frac{2[n-k]}{n(n+2l+1)(k+l)}c_{n-1}, \quad (5.520)$$

and it is straightforward to obtain an explicit expression for the coefficients in the truncated series:

$$\frac{c_n}{c_0} = \left(\frac{-2}{k-l}\right)^n \frac{(k-1)!(2l+1)!}{(k-n-1)!n!(n+2l+1)!}. \quad (5.521)$$

Finally, the coefficient $c_0$ can be determined by normalizing the wave function, which means that the probability density of a given stationary eigenfunction, integrated over all space, must be equal to one:

$$\iiint |\psi_{klm}(r,\theta,\varphi)|^2 r^2 d\Omega = 1. \quad (5.522)$$

With the definitions given previously, this translates into

$$\int_0^\infty |R_{kl}(r)|^2 r^2 dr \int |Y_l^m(\theta,\varphi)| d\Omega, = 1 \quad (5.523)$$

which finally yields

$$R_{kl}(r) = \frac{c_0^{kl}}{r}\exp\left[-\frac{r}{\bar{a}_0(k+l)}\right]\left(\frac{r}{\bar{a}_0}\right)^{l+1}$$
$$\times \sum_{n=0}^{n=k-1}\left(\frac{r}{\bar{a}_0}\right)^n\left(\frac{-2}{k+l}\right)^n \frac{(k-1)!(2l+1)!}{(k-n-1)!n!(n+2l+1)!}, \qquad (5.524)$$

where $\bar{a}_0 = a_0 m_0/\mu$.

Note that the normalization constant, $c_0^{kl}$, must be derived for each eigenwavefunction using Equation 5.523; furthermore, it is defined to within a constant phase factor, $\vartheta \in \mathbb{R}$, as $|c_0| = \left|c_0 e^{i\vartheta}\right|$. For the first few lower energy levels, we have

$$R_{1,0}(r) = \frac{1}{\rho}e^{-\rho}\rho c_0^{1,0} = \frac{2}{\sqrt{\bar{a}_0^3}}\exp\left[-\frac{r}{\bar{a}_0}\right],$$
$$R_{2,0}(r) = \frac{1}{\rho}e^{-\rho/2}\rho(c_0^{2,0} + c_1^{2,0}\rho)$$
$$= e^{-\rho/2}c_0^{2,0}\left(1-\frac{\rho}{2}\right) = \frac{1}{\sqrt{2\bar{a}_0^3}}\left(1-\frac{r}{2\bar{a}_0}\right)\exp\left[-\frac{r}{2\bar{a}_0}\right], \qquad (5.525)$$
$$R_{1,1}(r) = \frac{1}{\rho}e^{-\rho/2}\rho^2 c_0^{1,1} = \frac{1}{\sqrt{24\bar{a}_0^3}}\frac{r}{\bar{a}_0}\exp\left[-\frac{r}{2\bar{a}_0}\right].$$

For a detailed discussion of the properties of the hydrogen atom, we refer the reader to the textbook by Cohen-Tannoudji and co-authors. We also note that the formalism applied to discuss the question of a particle in a central potential has widespread use in quantum mechanics, including the problem of positronium, where an electron and a positron form a transient leptonic atom, which can decay via the reactions

$$e^+ + e^- \rightarrow 2\gamma,$$
$$e^+ + e^- \rightarrow 3\gamma, \qquad (5.526)$$

with half-lives, $\tau_{\|} \simeq 1.25\times 10^{-10}$ s, for para-positronium, where the spins are antiparallel, and $\tau_{\perp} \simeq 1.42\times 10^{-7}$ s, for ortho-positronium, respectively. The three-photon decay of ortho-positronium is required to conserve angular momentum properly; as it relies on a higher-order multipole component, the probability of decay is considerably smaller, leading to a much longer metastable state. Of course, to treat this problem fully, spin and relativistic effects must be taken into account.

Another important class of problems can be treated using the Klein–Gordon equation for a spinless particle. In this case, we can use the principle of correspondence, relating the four-momentum to the four-gradient operator:

$$p_\mu \rightarrow i\hbar\partial_\mu, \tag{5.527}$$

and the relativistic Hamiltonian,

$$\mathcal{H} = \sqrt{(c\mathbf{p} - e\mathbf{A})^2 + m_0^2c^4} - e\phi, \tag{5.528}$$

to formally recast the Schrödinger equation, $\mathcal{H}\psi = E\psi$, as

$$\left[\sqrt{(i\hbar\nabla - e\mathbf{A})^2 + m_0^2c^4} - e\phi\right]\psi = i\hbar\partial_t\psi. \tag{5.529}$$

In the particular case of a central potential, the formalism used for the hydrogen atom can be used to a large extent, because of the similar structures of the Schrödinger and Klein–Gordon equations.

Thus, we have seen that the mathematical formalism and techniques used to solve the wave equation with boundary conditions are virtually identical for electromagnetic and matter waves. In particular, for a given symmetry, one can use eigenfunctions, which form complete sets of orthogonal functions and can be properly normalized. For cylindrical symmetry, Bessel functions represent eigenmodes, whereas for spherical symmetry, spherical harmonics and associated Legendre polynomials form a basis for the corresponding Hilbert space. Furthermore, the eigenmode coupling analysis discussed for a corrugated cylindrical waveguide in Section 5.6.8 is entirely analogous to the derivation of transition probabilities between two eigenstates induced by an external potential in quantum mechanics. Finally, the diagonalization and projection techniques used to derive Green functions in a waveguide are also extremely useful in quantum mechanics and are used extensively to derive transition probabilities or the expectation value of a given operator.

In summary, in both quantum mechanics and electrodynamics, it is particularly advantageous to consider eigenfunctions, $\psi_n$, of a given operator, $O$. These eigenmodes are the solutions of the eigenvalue equation

$$O\psi_n = \lambda_n\psi_n, \tag{5.530}$$

and they can be normalized to form a complete set of orthonormal functions in Hilbert space. Mathematically, the orthonormality of the eigenfunction is expressed by

$$\int \psi_m^*(x)\psi_n(x)dx = \delta_{mn}, \tag{5.531}$$

while the completeness of the set is described by the relation

$$\sum \psi_n^*(x)\psi_n(\bar{x}) = \delta(x-\bar{x}). \tag{5.532}$$

Here, the vector $x$ and the differential element $dx$ are to be considered in an $n$-dimensional space. Also note that the eigenfunctions can form either a discrete or a continuous set; therefore the sign $\Sigma$ is to be considered here as either a series or an integral.

Symmetries and boundary conditions play an important role in defining the eigenfunctions. For example, for the Laplacian and d'Alembertian operators, Bessel functions are the eigenmodes in cylindrical coordinates, while spherical harmonics and associated Legendre polynomials are the eigenfunctions in spherical coordinates. In Cartesian coordinates, harmonic functions, represented by complex exponentials, are the eigenmodes of these operators, as discussed in Chapter 6. Typically, boundary conditions discretize the continuous eigenvalue spectrum by introducing additional constraints on the eigenmodes.

In the case of a wave equation driven by a current, we have

$$\mathcal{O}f = j, \tag{5.533}$$

where $f$ is the sought-after electromagnetic field distribution produced by the current $j$.

The method of resolution consists in expanding $f$ into eigenmodes, by writing

$$f = \sum \alpha_n \psi_n; \tag{5.534}$$

to determine the unknown coefficients, $\alpha_n$, the source current is projected, while the eigenmode expansion is diagonalized. The action of the operator on $f$ is first derived, with the result that

$$\mathcal{O}f = \mathcal{O}\sum \alpha_n \psi_n = \sum \mathcal{O}(\alpha_n \psi_n) = \sum \alpha_n \mathcal{O}\psi_n = \sum \alpha_n \lambda_n \psi_n. \tag{5.535}$$

Here, we have used the fact that $\mathcal{O}$ and $\Sigma$ commute. At this point, the wave equation reads

$$\sum \alpha_n \lambda_n \psi_n = j. \tag{5.536}$$

We now multiply each side of Equation 5.536 by $\psi_m^*$, which simply yields

$$\psi_m^* \sum \alpha_n \lambda_n \psi_n = \sum \alpha_n \lambda_n \psi_n \psi_m^* = \psi_m^* j. \tag{5.537}$$

Next, we integrate each side,

$$\int \sum \alpha_n \lambda_n \psi_n \psi_m^* dx = \sum \alpha_n \lambda_n \int \psi_n \psi_m^* dx = \int \psi_m^* j dx, \tag{5.538}$$

and use the orthonormality condition, Equation 5.531, to obtain

$$\sum \alpha_n \lambda_n \delta_{mn} = \alpha_m \lambda_m = \int \psi_m^* j dx. \tag{5.539}$$

The diagonalization of the left-hand side allows us to reduce the series to a single term, while the source term has been projected onto the corresponding eigenmode; we then obtain the sought-after result:

$$\alpha_n = \frac{\int \psi_m^* j dx}{\lambda_n}. \tag{5.540}$$

Of particular interest is the projection of a delta-function source current, as it defines the Green function for the operator under consideration: for $f = G$ and $j = \delta(x - \bar{x})$, we find that

$$G = \sum \gamma_n \psi_n, \quad \gamma_n = \frac{\int \psi_n^* \delta(x - \bar{x}) dx}{\lambda_n} = \frac{\psi_n^*(\bar{x})}{\lambda_n}. \tag{5.541}$$

This result can also be derived using the completeness relation, Equation 5.532.

Finally, in quantum mechanics, very similar concepts arise, including the transition probability between two eigenstates and the expectation value of an operator, $\mathcal{V}$:

$$\begin{aligned} \langle \psi_m | \mathcal{V} | \psi_n \rangle &= \int \psi_m^* \mathcal{V} \psi_n dx, \\ \langle \mathcal{V} \rangle_n &= \langle \psi_n | \mathcal{V} | \psi_n \rangle. \end{aligned} \tag{5.542}$$

## 5.9 Group Velocity Dispersion, Higher-Order Effects, and Solitons

Having derived the dispersion relation for waves propagating in structures with different symmetries and boundary conditions, the physical interpretation of this important concept is developed by first considering the notion of group velocity and group velocity dispersion (GVD), followed by an introduction to higher-order effects which play a crucial role in a variety of modern applications including chirped-pulse amplification, self-phase modulation, active and passive pulse and gain compression, and solitons.

We start with a brief review of the properties of the dispersion relation, followed by a more detailed discussion of an example, gain compression in a free-electron laser. This example will help illustrate some of the concepts introduced in this chapter.

We consider a guiding structure with boundary conditions and a monochromatic plane wave propagating along the structure, which we will take to be the $z$-axis: $e^{i\phi} = \exp[i(\omega t - kz)]$. The wave equation is

$$[\Delta - \varepsilon\mu\partial_t^2]A = 0, \tag{5.543}$$

where we have included the possibility that the structure contains a dielectric or magnetic material, for example, an optical fiber.

In the wave equation, we can separate the Laplacian operator into a transverse and an axial differential operator; moreover, the action of $\partial_z^2$ on the plane wave simply yields the eigenvalue $-k^2$, while the second-order time derivative gives the identity $-\partial_t^2 \equiv \omega^2$. Therefore, we have

$$[\partial_\perp^2 + \varepsilon\mu\omega^2 - k^2]A = 0. \tag{5.544}$$

One then uses transverse eigenmodes satisfying the boundary conditions, such that

$$\partial_\perp^2 F_{mn} = -k_{\perp mn}^2(\omega)F_{mn}, \tag{5.545}$$

to obtain the dispersion relation,

$$\varepsilon\lambda\omega^2 - k^2 - k_{\perp mn}^2(\omega) = 0, \quad D(\omega, k) = 0. \tag{5.546}$$

Here, we have used two indices to catalog the transverse eigenmodes; this is justified in view of the fact that the two degrees of freedom mapping the transverse space are typically independent. The dispersion relation is illustrated in Figure 5.9; it is symmetrical about the frequency-axis, reflecting the quadratic nature of Equation 5.546, which is correlated to causality. In the passive structure, waves propagating in the positive or negative $z$-direction have the same dispersion characteristics.

There are two distinct regions in the dispersion graph: $v_\phi = \frac{\omega}{k} > c$, the so-called "fast-wave" region, and $v_\phi < c$, the so-called "slow-wave" region. The quantity $v_\phi = \omega/k$ is referred to as the phase velocity; its interpretation is understood readily by recasting the aforementioned plane wave as follows:

$$e^{i\phi} = \exp[i(\omega t - kz)] = \exp[ik(v_\phi t - z)], \tag{5.547}$$

and the phase of the wave becomes $\phi = k(v_\phi t - z)$.

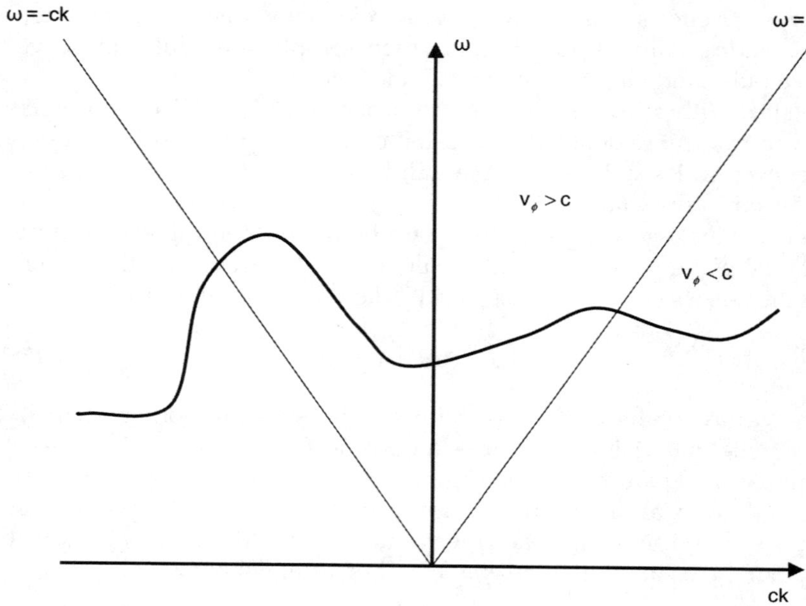


**FIGURE 5.9**
Generic dispersion relation, showing the light-lines, $\omega = \pm ck$, and the fast- and slow-wave regions. The dispersion represented here is unrealistic: it is not symmetrical with respect to the wavenumber, which does not satisfy the principle of reversibility for a passive structure (the fact that waves propagating in either direction disperse in the same manner), and its group velocity, represented by the tangent, $d\omega/dk$, is locally steeper than the light-line.

For example, in the case of an empty rectangular metallic waveguide, the $F_{mn}$ are of the form

$$F_{mn} = \cos\left(m\pi\frac{x}{a}\right)\cos\left(n\pi\frac{y}{b}\right), \tag{5.548}$$

where $a$ is the width of the waveguide along the $x$-axis, and $b$ is its counterpart in the $y$-direction.

In this case, the eigenvalues $k_{\perp mn}$ are independent of the frequency, and the dispersion relation takes the familiar parabolic form

$$\omega^2 = \omega_c^2 + k^2c^2, \quad \omega_c = k_{\perp mn}c = \frac{c}{a}\sqrt{m^2+n^2}, \quad \varepsilon\mu = \varepsilon_0\mu_0 = \frac{1}{c^2}. \tag{5.549}$$

The fact that this relation is called the dispersion relation can be best understood by considering a wavepacket,

$$A(z=0, t) = \int \tilde{A}(\omega)e^{j\omega t}d\omega, \tag{5.550}$$

as measured at the reference plane, $z = 0$. The evolution of the wavepacket as it propagates through the guiding structure is governed by

$$A(z, t) = \int \tilde{A}(\omega)\exp[i(\omega t - kz)]d\omega; \tag{5.551}$$

in other words, we consider the phase shift accumulated by each spectral component of the wavepacket as it propagates in the waveguide. At this point, a word of caution is required. This approach is valid only if the medium under consideration is linear, because we consider that the spectral density, $\tilde{A}(\omega)$, is invariant.

For a narrow band pulse, we can Taylor-expand the dispersion relation around the center frequency $\omega_0$:

$$k(\omega_0 + \Delta\omega) \simeq k(\omega_0) + \Delta\omega\frac{dk}{d\omega}(\omega_0) + \frac{\Delta\omega^2}{2}\frac{d^2k}{d\omega^2}(\omega_0) + \cdots. \tag{5.552}$$

We can identify the following terms:

$$k(\omega_0) = k_0 + i\Gamma_0 \tag{5.553}$$

is the complex phase shift of the center frequency and the gain or loss at that frequency;

$$\frac{dk}{d\omega} = \left(\frac{d\omega}{dk}\right)^{-1} = \frac{1}{v_g(\omega_0)} \tag{5.554}$$

corresponds to the group velocity of the wavepacket; finally,

$$\frac{d^2k}{d\omega^2} = \frac{d}{dk}\left[\frac{1}{v_g(\omega)}\right] = \frac{v_g'}{v_g^2} \tag{5.555}$$

is related to the group velocity dispersion, or GVD, $v_g'$. Note that the higher order derivatives also contain imaginary parts, corresponding to gain bandwidth, for example.

In the case of a Gaussian pulse, of width $\Delta t$,

$$\tilde{A}(\omega) \propto \exp[-(\omega\Delta t)^2]; \tag{5.556}$$

propagating in a passive dispersive medium, the output pulse is given by

$$A(z, t) = \frac{1}{\eta(z)}\exp\left[-\frac{\tau^2}{\Delta t^2(z)}\right]\exp\{i[\phi(z) + \omega_0\tau]\}, \tag{5.557}$$

if we limit ourselves to a second-order expansion of the dispersion relation. Here, the well-known dispersive pulse broadening effect is described by

$$\Delta t(z) = \Delta t\sqrt{1+\left[\frac{2v_g'z}{(v_g\Delta t)^2}\right]^2}; \tag{5.558}$$

note that when $z = 0$, we recover the input pulse width, $\Delta t$. The normalization parameter, $\eta(z)$ conserves the pulse energy. The parameter $\tau$ reflects the propagation time, or delay, of the center frequency,

$$\tau = t - v_g^{-1}z, \tag{5.559}$$

which explains the term "group velocity". Finally, the phase shift is given by

$$\phi(z) = \frac{1}{2}\arctan\left[\frac{2v_g'z}{(v_g\Delta t)^2}\right] - \left(k_0 + \frac{\omega_0}{v_g}\right)z. \tag{5.560}$$

The dispersive effects and their close correspondence to the dispersion relation are evident. With this method, one can study the dispersion of short pulses and chirped pulses in active or passive structures, characterized by gain, gain bandwidth, or losses and dispersive losses.

In the case of a parabolic dispersion relation, of the form

$$\frac{\omega^2}{c^2} = k^2 + \omega_c^2, \quad -k_\mu k^\mu = \omega_c^2, \tag{5.561}$$

one can derive an important relation between the phase and group velocities,

$$\frac{d}{dk}\left(\frac{\omega^2}{c^2}\right) = \frac{2\omega}{c^2}\frac{d\omega}{dk} = 2k, \tag{5.562}$$

which yields

$$\frac{\omega}{k}\frac{d\omega}{dk} = v_\phi(\omega)v_g(\omega) = c^2. \tag{5.563}$$

Therefore, such waves are always fast waves, since the group velocity is bounded by $c$: $v_g \le c \Rightarrow v_\phi \ge c$.

We now briefly consider beam modes, which are electrostatic waves propagating with a charged particle beam. In the case of an electron beam propagating along a structure with the velocity $v_b$, electrostatic beam perturbations, such as density modulations, of spatial scale length $\lambda$, propagate with the beam, as they are nonradiative and their field decays as $1/r^2$ like a Coulomb field. Their dispersion is simply given by

$$\omega = kv_b, \qquad k = \frac{2\pi}{\lambda}, \tag{5.564}$$

and their phase velocity is

$$v_\phi = \frac{\omega}{k} = v_b < c. \tag{5.565}$$

Evidently, such beam modes can only couple to slow waves.

For smooth metallic guiding structures, waves propagate according to a parabolic, fast-wave dispersion relation. The dispersion diagram on Figure 5.10 clearly shows the impossibility of directly coupling a beam mode to these fast waves. Before showing how this problem is solved in fast-wave devices, such as free-electron lasers, gyrotrons, and cyclotron autoresonance masers (CARMs), we complement our brief sketch of the fundamental properties of beam modes by considering the fast and slow, or "negative energy", space–charge waves.

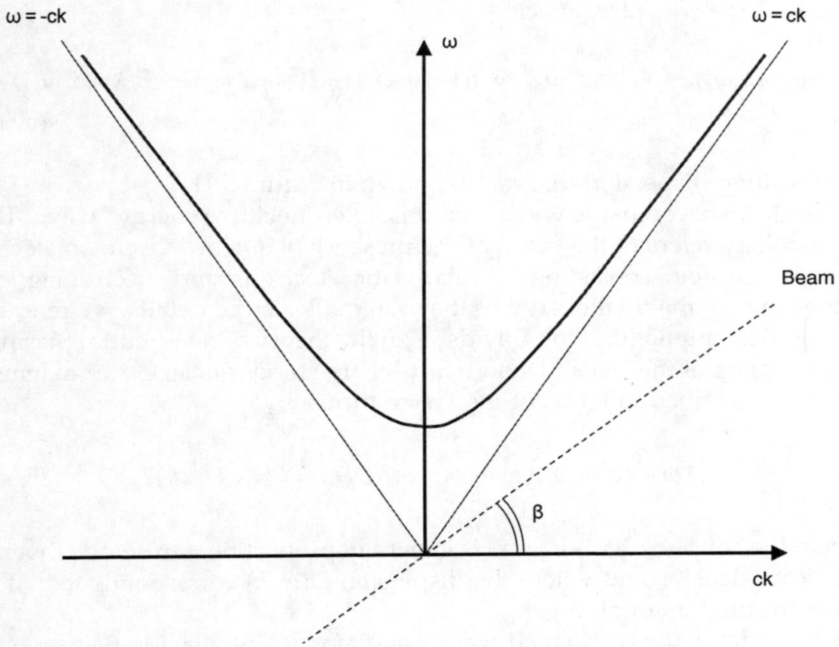


**FIGURE 5.10**
Parabolic dispersion, typical of waves propagating in a waveguide, with a cutoff frequency reached for $k = 0$; the beam mode is represented by a dashed line, while $\beta < 1$ represents normalized velocity. There is no coupling between the electromagnetic mode and the beam mode, as synchronism ($\omega_1 = \omega_2$, $k_1 = k_2$) cannot be achieved.

As discussed in Section 2.12, the electron beam responds to electrostatic perturbations, such as density or velocity modulations, like an harmonic oscillator with eigenfrequency $\omega_p$: it supports electrostatic plasma waves, also called space–charge waves.

The dispersion of these waves can be evaluated by considering the non-linear coupling provided by the Lorentz force. For example, the electric field component oscillating at the plasma frequency, $\mathbf{E}e^{\pm i\omega_p t}$, can drive a velocity component at the same frequency, $\mathbf{v}e^{\pm i\omega_p t}$, which couples to the magnetic field of the beam mode, $\mathbf{B}e^{ikv_b t}$; the Lorentz force then contains components at the beat frequencies $kv_b \pm \omega_p$. We summarize this with the notation

$$\mathbf{E}(\omega_p) \to \mathbf{v}(\omega_p) \times \mathbf{B}(kv_b) = \mathbf{F}(kv_b \pm \omega_p), \tag{5.566}$$

which results from the fact that

$$\exp(ikv_b t) \times \exp(\pm i\omega_p t) = \exp[i(kv_b \pm \omega_p)t]. \tag{5.567}$$

This force, in turn, drives a velocity and density modulation, which gives rise to a source term resulting in a field oscillating at the fast or slow space–charge wave frequencies,

$$\mathbf{F}(kv_b \pm \omega_p) \to \mathbf{v}(kv_b \pm \omega_p) \to n(kv_b \pm \omega_p) \to \mathbf{j}(kv_b \pm \omega_p) \to \mathbf{E}(kv_b \pm \omega_p); \tag{5.568}$$

the resulting dispersion diagram is shown in Figure 5.11.

The slow space–charge wave is an unstable, "negative energy" wave. This terminology refers to the fact that the presence of such waves in an electrodynamic system consisting of relativistic electrons and electromagnetic waves lowers the total energy of the system. For more details, we refer the reader to the monograph by Davidson on the theory of non-neutral plasmas. The coupling of the fast and slow space–charge modes to an electromagnetic mode is described in terms of the Pierce theory by

$$(\omega - kv_b + \omega_p)(\omega - kv_b - \omega_p)D(\omega, k) = C(\omega, k), \tag{5.569}$$

where the coupling term $C(\omega, k)$ is generally proportional to some power of the beam density, and where the dispersion $D(\omega, k)$ corresponds to a slow wave for proper coupling.

For example, the corrugated waveguide modes described in Section 5.6.8 can be excited directly by an electron beam, as discussed by Swegle and co-authors. Another slow-wave device, which plays an important role for radar and communication applications, is the traveling-wave tube (TWT), which typically utilizes a helix to support slow waves that are coupled to the electron beam. Such devices are highly efficient, with more than 50% of the

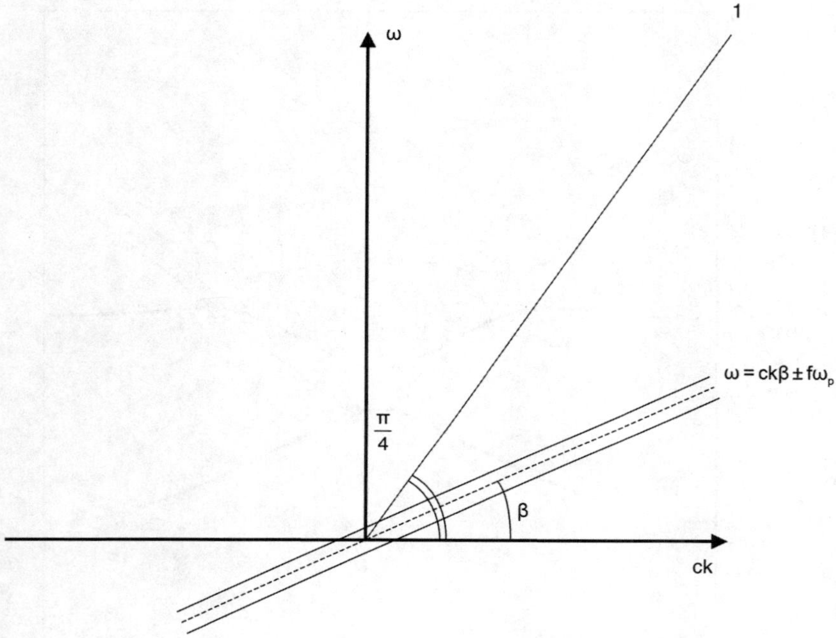


**FIGURE 5.11**
Schematic representation of the two-beam mode sidebands introduced by the beam plasma frequency; $f$ represents a generic filling factor, including the effects of boundary conditions on the space–charge waves.

electron beam kinetic energy transferred into the amplified electromagnetic mode, and are characterized by a very wide interaction bandwidth, which can exceed one octave.

Close to the interaction frequency, illustrated in Figure 5.12, the dispersion relation given in Equation 5.569 can be approximated by a cubic equation, with three roots, corresponding to exponential growth, exponential decay, and interference. This physical property of the coupling results in the practical problem of launching losses, as the signal injected for amplification typically excites all three modes. This regime of operation, where the fast and slow space–charge waves are well separated is called the Raman regime. It implies that the electron beam is "cold" or monokinetic; if the beam has some energy spread, the two space–charge waves coalesce into the so-called single-particle, or Compton, regime, as pictured in Figure 5.13.

We now return to the question of the coupling of beam modes to the fast wave region. In this case, instead of modulating the guiding structure boundary conditions, so that it supports slow waves, we modulate the beam itself with an external electromagnetic field. In free-electron lasers, this modulation is provided by a wiggler magnetic field, $\mathbf{B}_w = B_w[\hat{x}\cos(k_w z) + \sigma\hat{y}\sin(k_w z)]$, of period $\ell_w = 2\pi/k_w$, and circular ($\sigma = 1$) or linear ($\sigma = 0$) polarization;

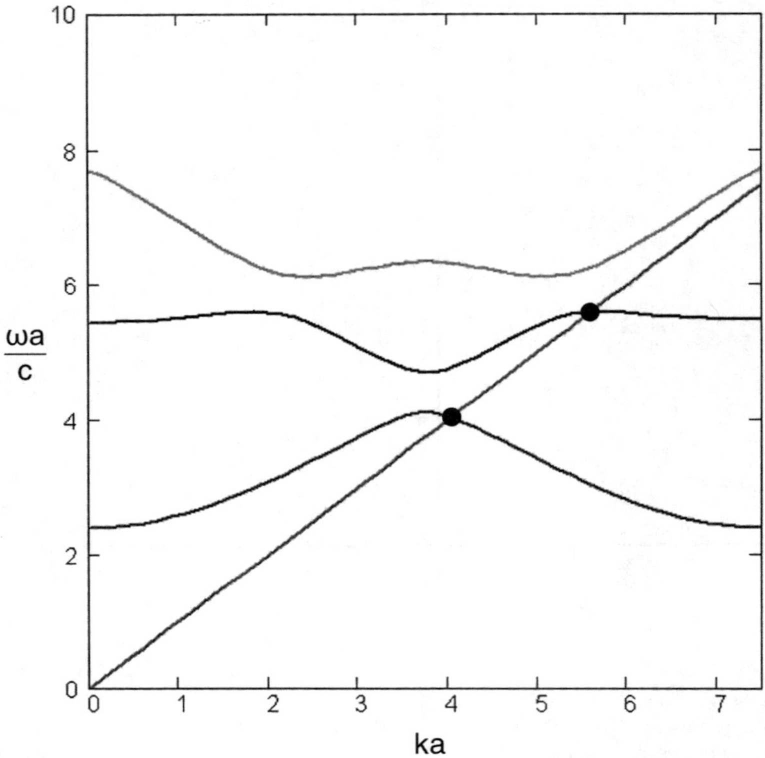


**FIGURE 5.12**
Beam mode coupling with two corrugated waveguide modes; the coupling is possible because all modes here are at least partially in the slow-wave region of the dispersion diagram.

for gyrotrons and CARMs, the modulation is provided by a constant magnetic field, which forces the electrons to gyrate at the relativistic gyrofrequency $\Omega = eB/\gamma m_0$; finally, for Compton scattering, the modulation is due to a laser pulse.

In the case of a wiggler, the nonlinear Lorentz force gives rise to two sidebands,

$$\mathbf{E}(kv_b) \to \mathbf{v}(kv_b) \times \mathbf{B}_w(\pm k_w v_b) = \mathbf{F}[(k \pm k_w)v_b], \tag{5.570}$$

and the resulting dispersion diagram is illustrated in Figure 5.14. As discussed in Section 5.6.7, there are now two interaction frequencies, the Doppler up- and downshifted frequencies. A similar scheme arises in the case of gyrotrons and CARMs:

$$\mathbf{v}(\pm\Omega) \times \mathbf{B}(kv_b) = \mathbf{F}(kv_b \pm \Omega). \tag{5.571}$$

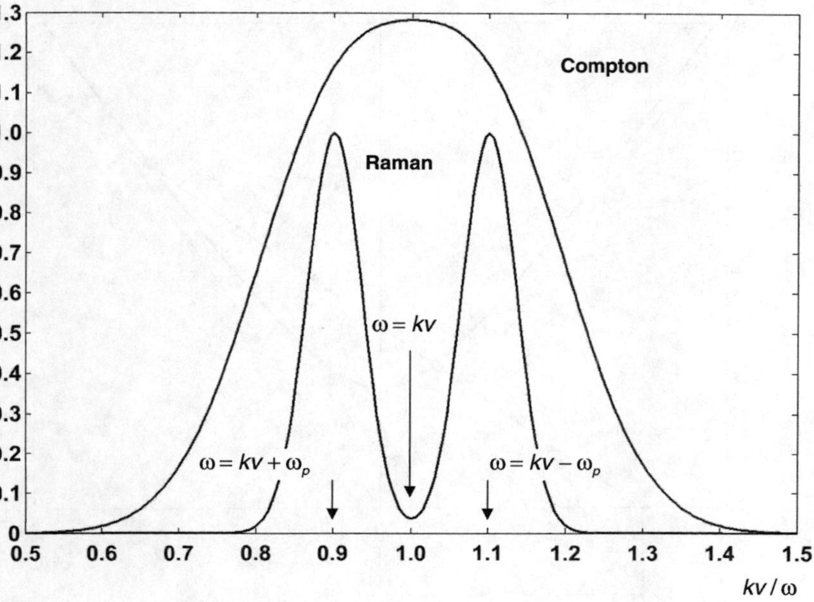


**FIGURE 5.13**
Schematic illustration of the Compton and Raman regime: in the first case, the distribution function of the beam electron is too broad, and the plasma sidebands are mixed; in the second case, the beam is colder, and the two space–charge waves are no longer coalesced.

The interaction frequencies are obtained by simultaneously solving the dispersion relations of the beam and electromagnetic modes:

$$\omega^2 = \omega_c^2 + k^2c^2,$$
$$\omega = kv_b + \begin{cases} k_w v_b, & \text{(FEL)}, \\ \Omega, & \text{(CARM)}. \end{cases} \tag{5.572}$$

Because the relativistic electron beam carries the modulation, these devices have the following basic properties: they are voltage tunable and can scale to high frequencies. However, because the interaction frequency depends strongly on $v_b$, they require high quality electron beams, with $\Delta\gamma/\gamma \ll 1$, unlike the more robust slow wave devices.

We close this brief introduction by considering the covariance of the waveguide FEL dispersion relation. In vacuum, the waves propagate at the speed of light, and we have

$$|\mathbf{k}| = \frac{\omega}{c}, \quad k_\mu k^\mu = 0, \quad p_\mu = \hbar k_\mu, \quad p_\mu p^\mu = 0 = m_\gamma^2 c^2; \tag{5.573}$$

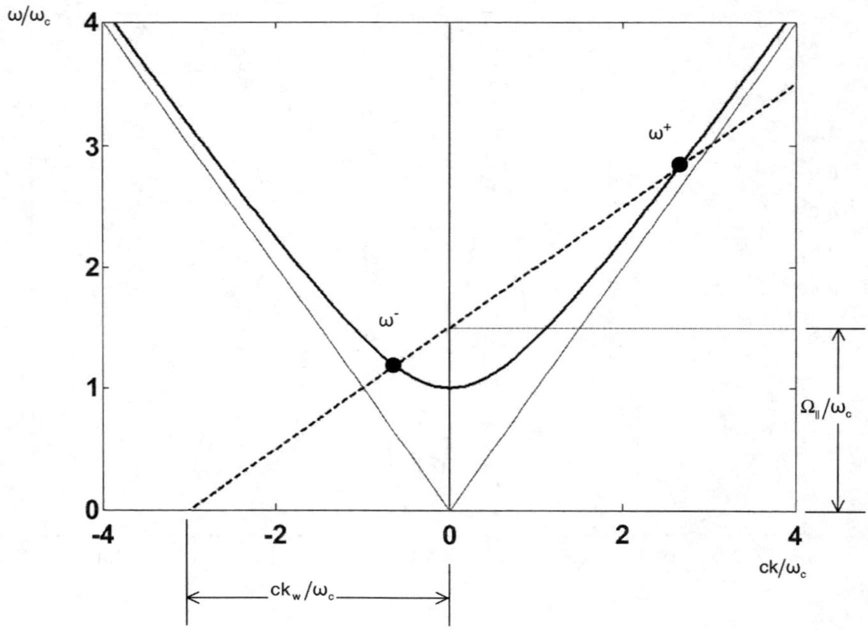


**FIGURE 5.14**
Fast-wave coupling in a waveguide free-electron laser or CARM. In the FEL case, the shift of the beam mode is due to the wiggler, whereas the cyclotron frequency provides the upshift for the CARM. In the case of a gyrotron, the beam mode would graze the electromagnetic mode, near cutoff, as the beam velocity is low.

the photons are real, as opposed to virtual, and massless. In addition, the dispersion relation is manifestly covariant, as it corresponds to the light-cone: the waves propagate at the speed of light, independent of the reference frame. This is precisely the principle of relativity and is illustrated in Figure 5.15, where the FEL interaction in free-space is viewed in the laboratory frame and in the electron frame.

In a waveguide, the situation is somewhat different since there is a cutoff frequency. The mode coupling equations are

$$\begin{aligned} \omega^2 &= k^2c^2 + \omega_c^2, \\ \omega &= (k + k_w)\beta_\| c. \end{aligned} \tag{5.574}$$

Note that we have used the axial velocity in the beam mode. Indeed, if the wiggler strength is high, we have

$$\frac{1}{\gamma^2} = 1 - \beta_\|^2 - \beta_\perp^2, \qquad 1 = \gamma^2 - (\gamma\beta_\|)^2 - A_w^2, \tag{5.575}$$

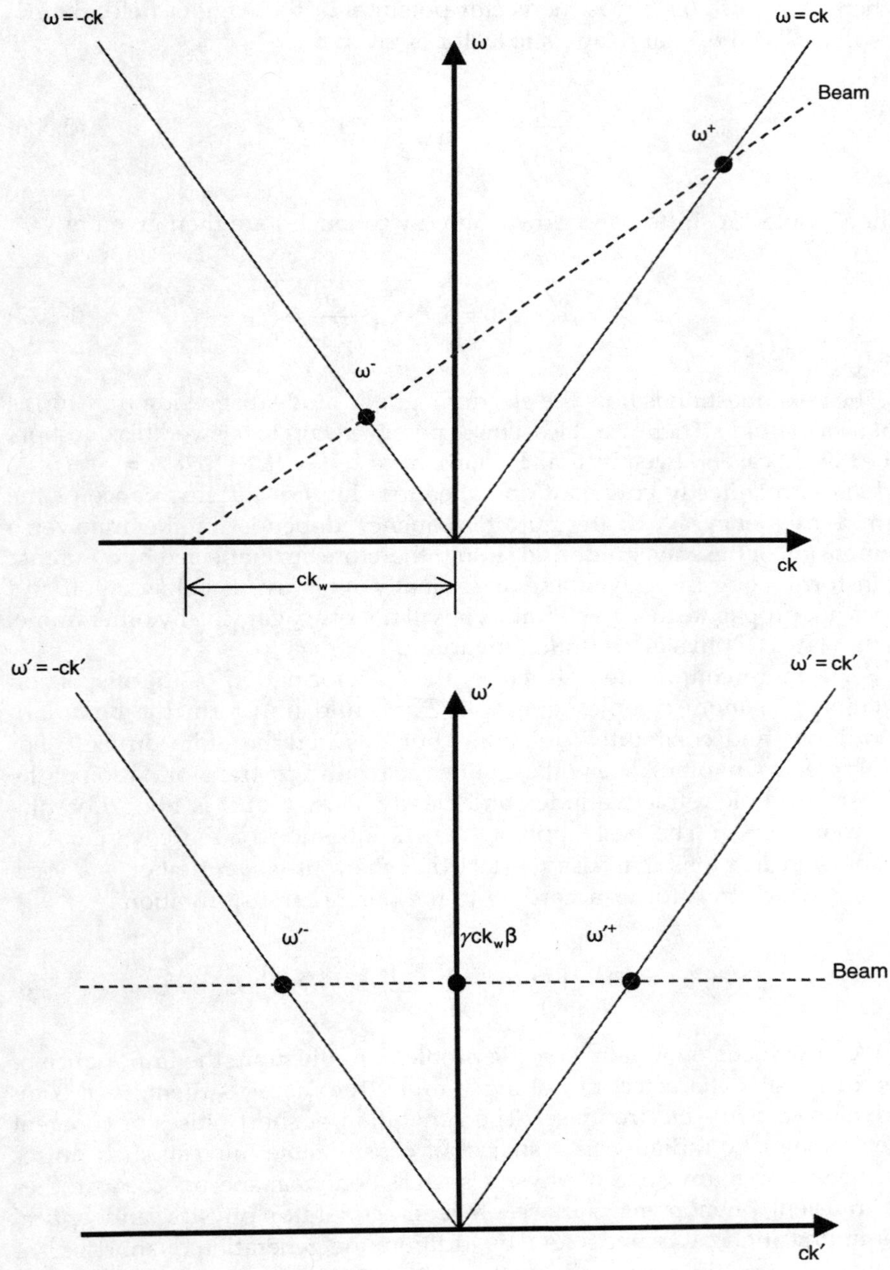


**FIGURE 5.15**
Top: FEL interaction, as seen in the lab frame, where the Doppler upshifted and downshifted frequencies are different. Bottom: the same interaction, as observed in the electron frame; here, the two radiated frequencies are equal, but their wavenumbers are opposite.

where $\mathbf{A}_w = e\mathbf{B}_w/m_0k_wc$ is the vector potential of the wiggler field; the net result is that the axial relativistic factor is given by

$$\gamma_{\|} = \frac{1}{\sqrt{1-\beta_{\|}^2}}, \quad \left(\frac{\gamma}{\gamma_{\|}}\right)^2 = 1 + A_w^2. \tag{5.576}$$

The Doppler upshifted and downshifted frequencies are then given by

$$\omega^{\pm} = \gamma_{\|}^2\beta_{\|}ck_w\left[1 \pm \beta_{\|}\sqrt{1-\left(\frac{\omega_c}{\gamma_{\|}\beta_{\|}ck_w}\right)^2}\right]. \tag{5.577}$$

The next question is how the electromagnetic mode dispersion transforms for a waveguide. There are at least three possible ways to answer this question, one theoretical and two physical. First, we have $k_\mu k^\mu = \mathbf{k}^2 - (\omega/c)^2 = -(\omega_c/c)^2$, which is manifestly covariant and therefore implies the invariance of the cutoff frequency. Next, the cutoff frequency depends on the transverse dimension of the waveguide and should therefore be unaffected by a Lorentz transform along the waveguide axis. Finally, if a wave is below cutoff and does not propagate in a given frame, it will not propagate in any other frame, as the laws of physics are frame-invariant.

For a more complicated structure, the transformation of the dispersion relation is a more complex question. One could transform the boundary conditions and constitutive relations, but this may be quite difficult. For example, in Chapter 7, we will examine the relativistic transformation of the linear, isotropic refractive index and clearly show that it is not a Lorentz-invariant scalar. The best approach to the question of the covariance of dispersion relations is to use the fact that the four-wavenumber is a four-vector, which transforms according to the Lorentz transformation:

$$D(\omega, \mathbf{k}) = D(k_\mu) = 0, \quad \bar{k}_\mu = \mathcal{L}_\mu^\nu k_\nu, \quad D(\mathcal{L}_\mu^{\nu^{-1}}\bar{k}_\nu) = 0 \tag{5.578}$$

We now focus on a more specific problem, to illustrate the importance of the dispersive characteristics of a given electrodynamic system, gain compression in a free-electron laser. The generation of short pulses of coherent electromagnetic radiation is a subject of considerable interest: such pulses are needed in many areas of physics, such as nonlinear spectroscopy, studies of transient phenomena, surface physics, accelerator physics, and optical communications, to name a few. In addition, the generation of short pulses in an active medium can lead to extremely high instantaneous peak powers, as illustrated by recent advances in chirped-pulse amplification (CPA) for solid-state lasers.

In the microwave wavelength range, pulse compression can be achieved by propagating a frequency-chirped pulse in a passive medium, with group velocity dispersion. This effect, which is the direct counterpart of dispersive pulse broadening, is referred to as phase compression. Colliding pulses in a microwave interferometer also result in the generation of short pulses of rf radiation. In the optical part of the electromagnetic spectrum, the generation of short pulses commenced with active mode-locking, was followed by passive mode-locking, and culminated in the colliding pulses mode-locked systems. More recently, soliton formation was observed in optical fibers and led to another compression scheme, the soliton laser. Finally, chirped-pulse amplification has yielded the shortest optical pulses to date, with four-femtosecond pulses representing the current record.

The pulse compression scheme that will be considered here is the following: a frequency-chirped pulse is injected into a FEL amplifier; because of the high gain and narrow bandwidth of the FEL interaction, only the resonant frequency band of the pulse is actively amplified. This results in a high-power, short pulse of coherent electromagnetic radiation at the output of the laser. Furthermore, this scheme is considerably more efficient that the aforementioned passive compression scheme relying on GVD in a waveguide.

We now describe the evolution of a frequency-chirped pulse in a FEL, under the influence of both phase and gain dispersive effects induced by the FEL instability. We make use of a Raman-type FEL dispersion relation, and we also include the effects caused by the waveguide cutoff frequency. In the case of a Gaussian input pulse with a linear frequency chirp, we obtain a simple analytical expression for the output pulse shape after Taylor-expanding the dispersion relation to second order. We first briefly outline the general method used to describe the evolution of the pulse during the FEL interaction. Consider a pulse at the input position, $z = 0$, $E(z = 0, t)$, where $E$ represents the electric field of the pulse. We can Fourier transform this pulse into its spectral frequency components:

$$\tilde{E}(\omega) = \frac{1}{\sqrt{2\pi}} \int_{-\infty}^{+\infty} E(z = 0, t) e^{-i\omega t} dt. \tag{5.579}$$

Under the assumption that the $\omega$-Fourier spectrum remains unchanged during the interaction, by requiring that the interaction be linear for the frequency spectrum, we can obtain the pulse shape at the output of the FEL, at $z = z_0$, by Fourier transforming back again after taking into account the different phase shifts accumulated by the various frequency components of the pulse:

$$E(z = z_0, t) = \frac{1}{\sqrt{2\pi}} \int_{-\infty}^{+\infty} E(\omega) \exp\{i[\omega t - k(\omega) z_0]\} d\omega, \tag{5.580}$$

where $\omega$ and $k$ satisfy the complex dispersion relation, $D(\omega, k) = 0$, associated with the FEL interaction.

In the case of a purely dispersive medium, $k(\omega)$ is real, and only phase shaping effects will affect the pulse. On the other hand, for an active medium such as the FEL, $k(\omega)$ is complex, and we obtain both phase and gain dispersive effects that change the shape of the pulse.

We now take the special case of a Gaussian input pulse, with a linear frequency chirp:

$$E(z=0,\,t) \;=\; E_0 \exp\left(-\frac{t^2}{\Delta t^2}\right)\exp[i(\omega_0+\alpha t)t]. \tag{5.581}$$

Here, $\Delta t$ is the initial pulse width, $\omega_0$ is its center frequency, and $\alpha$ represents the frequency chirp.

The corresponding input spectrum is determined by Fourier transforming Equation 5.581:

$$\tilde{E}(\omega) \;=\; \frac{E_0}{\sqrt{2\pi}}\int_{-\infty}^{+\infty}\exp\left(-\frac{t^2}{\Delta t^2}\right)\exp\{i[(\omega_0-\omega)t+\alpha t^2]\}dt. \tag{5.582}$$

This integral is readily calculated, and we obtain the familiar result:

$$E(\omega) \;=\; \frac{E_0}{\sqrt[4]{4\left(\frac{1}{\Delta t^4}+\alpha^2\right)}}\exp\left[-\frac{(\omega_0-\omega)^2}{\Delta t^2\left(\frac{1}{\Delta t^4}+\alpha^2\right)}\right]\exp\left\{i\left[\theta-\frac{\alpha(\omega_0-\omega)^2}{4\left(\frac{1}{\Delta t^4}+\alpha^2\right)}\right]\right\}, \tag{5.583}$$

where

$$\theta \;=\; \frac{1}{2}\arctan(\alpha\Delta t^2) \tag{5.584}$$

is a constant phase term. This result is illustrated in Figure 5.16.

At this point, we need to make use of the dispersion relation to evaluate the different phase shifts accumulated by the various frequency components in the interaction region. We use a Taylor expansion around $\omega = \omega_0$:

$$k(\omega) \;\simeq\; k(\omega_0)+(\omega-\omega_0)\left.\frac{\partial\omega}{\partial k}\right|_{\omega=\omega_0}+\frac{1}{2}(\omega-\omega_0)^2\left.\frac{\partial^2\omega}{\partial k^2}\right|_{\omega=\omega_0}. \tag{5.585}$$

This equation can be recast as follows:

$$k(\omega) \;\simeq\; k_0+i\chi+(\omega-\omega_0)v_g^{-1}+\frac{1}{2}(\omega-\omega_0)^2(\gamma+i\delta), \tag{5.586}$$

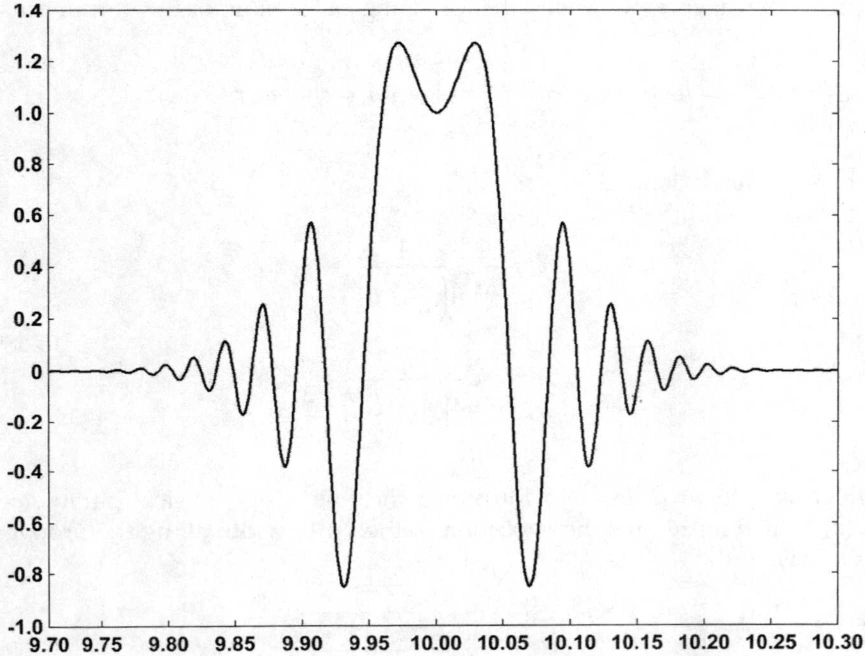


**FIGURE 5.16**
Typical Fourier transform of a Gaussian pulse with a linear frequency chirp.

where we recognize the following terms: $k_0$ is the real phase shift accumulated by the center frequency per unit distance, $\chi$ represents the growth rate of the center frequency, $v_g$ is the complex group velocity at $\omega = \omega_0$, $\gamma$ is the group velocity dispersion (phase dispersive effects), and $\delta$ represents the gain dispersion (bandwidth), corresponding to gain dispersive effects. These parameters are directly determined from the dispersion relation, $D(\omega, k) = 0$.

Making use of the Taylor expansion of the wavenumber and the input spectrum, we can now derive the output pulse shape by evaluating the following integral:

$$\begin{aligned} E(z,t) &= \frac{E_0}{2\sqrt{\pi}\left(\frac{1}{\Delta t^4} + \alpha^2\right)^{1/4}} \\ &\quad \times \int_{-\infty}^{+\infty} \exp[\chi z - (\omega - \omega_0)^2 a(z)] \\ &\quad \times \exp\{i[\theta + \omega t - k_0 z - (\omega - \omega_0) v_g^{-1} z + (\omega - \omega_0)^2 p(z)]\} d\omega. \quad (5.587) \end{aligned}$$

After some algebra, we obtain the following expression for the output pulse:

$$E(z,t) = \frac{E_0}{f(z)}\exp(\chi z)\exp\left[-\frac{\tau^2}{\Delta t^2(z)}\right]\exp\{i[\phi(z)+\omega_0\tau+\alpha(z)\tau^2]\}, \qquad (5.588)$$

where we have defined

$$\begin{aligned} a(z) &= \frac{1}{\Delta t^2}\frac{1}{4\left(\frac{1}{\Delta t^4}+\alpha^2\right)}-\frac{\delta}{2}z, \\ p(z) &= -\alpha\frac{1}{4\left(\frac{1}{\Delta t^4}+\alpha^2\right)}-\frac{\gamma}{2}z, \end{aligned} \qquad (5.589)$$

which are, in turn, used to introduce the following physical parameters: $\Delta t(z)$, which represents the evolution of the pulse width along the propagation axis, with

$$\Delta t(z) = 2\frac{\sqrt{a^2(z)+p^2(z)}}{\sqrt{a(z)}}; \qquad (5.590)$$

the chirp function $\alpha(z)$, defined as

$$\alpha(z) = \frac{-p(z)}{4\sqrt{a^2(z)+p^2(z)}}; \qquad (5.591)$$

the normalization factor, $f(z)$, defined such that, in the absence of gain the energy of the pulse remains constant,

$$\frac{f(z)}{2} = \sqrt[4]{\left(\frac{1}{\Delta t^4}+\alpha^2\right)[a^2(z)+p^2(z)]}; \qquad (5.592)$$

and the phase shift, $\phi(z)$, defined by

$$\phi(z) = \frac{1}{2}\arctan\left[\frac{p(z)}{a(z)}\right]+\theta-(k_0+\omega_0 v_g^{-1})z. \qquad (5.593)$$

Finally,

$$\tau = t - \frac{z}{v_g} \qquad (5.594)$$

reflects the propagation time of the center frequency of the pulse in the interaction region, and the term $e^{\chi z}$ corresponds to the gain induced by the FEL interaction.

It can be verified easily that for $z \to 0$, we recover the input pulse. In the special case of the propagation of an unchirped pulse ($\alpha = 0$) in a passive dispersive medium ($\chi = 0$, $\delta = 0$), we find the well-known dispersive pulse broadening effect,

$$\Delta t(z) = \Delta t \sqrt{1 + \left(\frac{2\gamma z}{\Delta t^2}\right)^2}. \tag{5.595}$$

We now study the evolution of the pulse width, $\Delta t(z)$. We determine the minimum width by taking the derivative of Equation 5.590 with respect to $z$, such that

$$\partial_z \Delta t(z) = 0. \tag{5.596}$$

Taking into account the fact that $\partial_z a(z) = -\delta/2$, and $\partial_z p(z) = -\gamma/2$, we can reduce Equation 5.596 to the following equation:

$$p^2(z) - 2\frac{\gamma}{\delta} a(z) p(z) - a^2(z) = 0. \tag{5.597}$$

Solving Equation 5.597 for $z = z^*$ yields the distance at which the maximum pulse compression is achieved. The corresponding minimum pulse width is then given by

$$\Delta t^* = \Delta t(z^*). \tag{5.598}$$

For example, in the case of a passive dispersive medium ($\chi = \delta = 0$), we find that the maximum compression is obtained at

$$z^* = -\frac{\alpha}{\gamma} \frac{1}{\left(\frac{1}{\Delta t^4} + \alpha^2\right)}, \tag{5.599}$$

where the pulse width is reduced to

$$\Delta t^* = \frac{\Delta t}{\sqrt{1 + \alpha^2 \Delta t^4}}. \tag{5.600}$$

Note that in the case of a passive dispersive medium, the minimum pulse width depends strongly on the width of the input pulse, and that compression is achieved only for $\alpha/\gamma < 0$. We shall henceforth refer to this phenomenon as "phase compression".

We now consider the Raman FEL dispersion relation, from which we can derive the values of the parameters defined above. The two interacting modes, illustrated in Figure 5.14, are the forward propagating waveguide mode,

$$k = \frac{1}{c}\sqrt{\omega^2 - \omega_c^2}, \tag{5.601}$$

where $\omega_c$ is the cutoff frequency of the waveguide in the interaction region, and the beam mode,

$$k = \frac{\omega}{v} - k_w + \frac{\omega_p}{v\gamma_0^{3/2}}, \tag{5.602}$$

corresponding to the unstable slow space–charge wave. Here, $v = c\beta_0$ is the electron beam velocity, $k_w$ is the wiggler wavenumber, $\omega_p$ is the plasma frequency, and $\gamma_0 = (1 - \beta_0^2)^{-1/2}$ is the relativistic factor.

In the limit of small $\omega_p$ and in the vicinity of the upshifted unstable (growing) root, the dispersion relation can be approximated as follows:

$$\left[k - \frac{1}{c}\sqrt{\omega^2 - \omega_c^2}\right]\left[k + k_w - \frac{\omega}{v}\right] = -\chi^2, \tag{5.603}$$

where $\chi$ is a small coupling term which determines the peak growth rate of the FEL instability. Here, we neglect the implicit slow frequency dependence of $\chi$. Its magnitude depends on the wiggler strength, electron beam current, frequency, electromagnetic wave filling factor, etc. We can solve Equation 5.603 for $k(\omega)$, with the result that

$$k(\omega) = \frac{1}{2}\left[\frac{1}{c}\sqrt{\omega^2 - \omega_c^2} + \frac{\omega}{v} - k_w + \sqrt{\left(\frac{1}{c}\sqrt{\omega^2 - \omega_c^2} - \frac{\omega}{v} + k_w\right)^2 - 4\chi^2}\right]. \tag{5.604}$$

We introduce the FEL upshifted frequency,

$$\omega_0 = \omega^+ = \gamma_0^2\beta_0 k_w c\left[1 + \beta_0\sqrt{1 - \left(\frac{\omega_c}{\gamma_0\beta_0 k_w c}\right)^2}\right], \tag{5.605}$$

and wavenumber,

$$k_0 = \frac{1}{c}\sqrt{\omega_0^2 - \omega_c^2}, \tag{5.606}$$

obtained by solving Equations 5.601 and 5.602 simultaneously; we also introduce the small parameter, $\delta\omega = \omega - \omega_0$. Expanding $k(\omega)$ to second order yields the following approximation:

$$k(\omega) \simeq k_0 + i\chi + \delta\omega\frac{1}{2}\left(\frac{1}{v} + \frac{\omega_0}{k_0 c^2}\right) + \delta\omega^2 \frac{1}{4k_0 c^2}\left(1 - \frac{\omega_0^2}{k_0^2 c^2}\right) - i\chi\left(\frac{\delta\omega}{\delta\omega_i}\right)^2. \quad (5.607)$$

This expansion is valid only for $\chi \neq 0$.

Here, we have made use of a parabolic approximation of the gain term,

$$\sqrt{\delta\omega^2 - \delta\omega_i^2} \simeq i\delta\omega_i\left[1 - \left(\frac{\delta\omega}{\delta\omega_i}\right)^2\right], \quad (5.608)$$

with $\delta\omega_i$ as the gain bandwidth of the FEL interaction:

$$\delta\omega_i = \frac{2\chi}{\left(\frac{\omega_0}{k_0 c^2}\right) - \left(\frac{1}{v}\right)}. \quad (5.609)$$

Comparing Equations 5.586 and 5.607, we obtain the sought-after expressions for the expansion parameters. The group velocity, $v_g$, is given by

$$v_g^{-1} = \frac{1}{2v}\left(\frac{1}{v} + \frac{\omega_0}{k_0 c^2}\right). \quad (5.610)$$

The group velocity dispersion, $\gamma$, which vanishes for $\omega_c = 0$, is given by

$$\gamma = \frac{1}{2k_0 c^2}\left[1 - \left(\frac{\omega_0}{k_0 c}\right)^2\right]. \quad (5.611)$$

Finally, the gain dispersive term of the FEL interaction is

$$\delta = -\frac{2\chi}{\delta\omega_i^2}. \quad (5.612)$$

At this point, it is possible to evaluate the distance of maximum "gain compression" induced by the FEL interaction. For the purpose of illustration, we first limit ourselves to the nondispersive case (cutoff frequency, $\omega_c = 0$, $\gamma = 0$). We simplify matters further by considering a frequency-chirped input pulse with constant amplitude ($\Delta t \to \infty$). In this situation, the maximum pulse compression induced by the FEL interaction is obtained after a propagation distance

$$z^* = \frac{\delta\omega_i^2}{\alpha\chi}. \tag{5.613}$$

The corresponding minimum pulse width is readily calculated:

$$\Delta t^* = 2\sqrt{-\delta z^*} = \frac{\delta\omega_i}{\alpha}\sqrt{\frac{8}{\chi z^*}}. \tag{5.614}$$

We obtain the expected $\delta\omega_i/\alpha$ scaling. For a high gain and narrow amplification bandwidth and a strongly chirped pulse, very high compression ratios can be achieved within short interaction distances; conversely, broadband amplification of a nearly monochromatic pulse will result in very small compression. We also note that the minimum pulse width is basically independent of the initial pulse width, as opposed to the usual phase compression. A curve comparing the pulse width as function of distance, for both an active and a passive medium, is given in Figure 5.17. The parameters correspond to an experiment performed at MIT with a central signal frequency $\omega_0/2\pi = 10.0$ GHz, and a chirp rate $\alpha/2\pi = 3.5$ MHz/ns. For experimental details and results, we refer the reader to the paper, "Generation of Short Pulses of Coherent Electromagnetic Radiation in a Free-Electron Laser Amplifier", by Hartemann, Xu, and Bekefi. For an in-depth review of free-electron lasers, the review article by Roberson and Sprangle offers an extensive list of references, as well as good discussions of numerous FEL experiments. The excellent textbook by Freund and Antonsen is quite advanced and very detailed, while the monograph by Marshall is very useful as a more introductory text.

Higher-order dispersive and nonlinear effects also play an important role in the propagation of electromagnetic waves. For example, pulse compression can be achieved in optical fibers by using the self-phase modulation induced by the nonlinear refractive index and group velocity dispersion. Another important example is chirped-pulse amplification; in this case, the basic idea is to avoid nonlinear effects by stretching the pulse temporally. Before the advent of CPA, peak fluxes were reduced by using larger optics, with severe limitations due to the difficulty of producing large, high-quality crystals. In CPA, one takes advantage of the fact that sub-picosecond laser pulses have a large optical bandwidth, typically in the few percent range; these pulses can be stretched by using diffraction gratings to produce wavelength-dependent optical paths. As a result, one obtains a chirped pulse with the same spectral content as the original pulse but with a much longer duration. This pulse can be amplified safely by small optics and recompressed using the conjugate optical system. This is typically achieved by using a –1 magnification on the

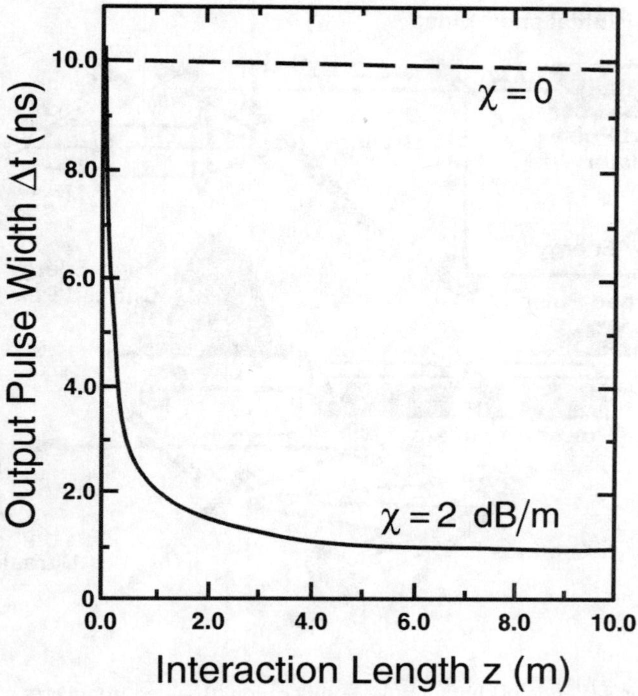


**FIGURE 5.17**
Pulse compression, with gain (solid line), and with GVD only (dashed line). These data correspond to a FEL pulse compression experiment by Hartemann and co-authors.

stretcher, and a +1 magnification on the compressor. Therefore, CPA makes use of the fact that the optics, while restricted in spatial dimensions, are temporally "large". A schematic illustration of this concept is given in Figure 5.18. High-order dispersive effects play a major role in the implementation of CPA because the recompression implies that the spectral content of the amplified pulse is as close to that of the original pulse as possible. Therefore, spectral clipping due to the limited amplification bandwidth of the laser medium and gain shaping effects must be compensated for, as well as B-integral effects, which are related to the nonlinear refractive index in the various optical components of the system. For example, to obtain a gain curve that is nearly frequency-independent, one can introduce frequency-dependent losses in the system, to reduce the gain near its peak and equalize the gain curve. With such techniques, recompression of 20 fs Fourier-limited pulses down to 25 fs have been achieved; furthermore, tabletop, kHz systems, producing 1 TW of peak power, are now available. In such high-average-power laser systems, gain guiding and thermal eigenmodes are used to further improve the quality of the recompressed pulse. For details on CPA and its applications, we have listed a series of review articles in the references to Chapter 5 by Barty, Mourou, Perry, and Umstadter.

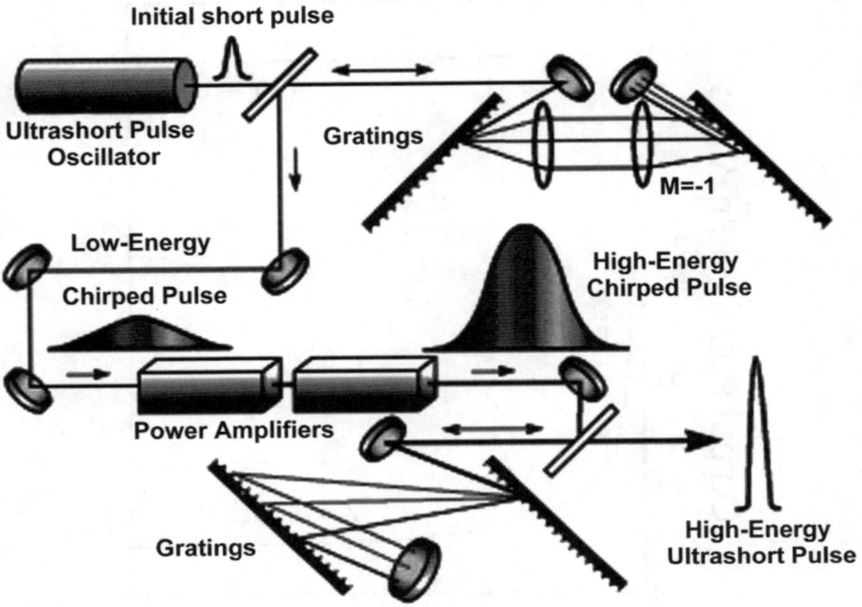


**FIGURE 5.18**
Schematic representation of chirped-pulse amplification (Adapted from Perry et al.

Finally, optical solitons, described by the nonlinear Schrödinger equation, are now routinely produced and provide another good example of nonlinear and dispersive effects in electromagnetic pulse propagation. We give a short review of this very interesting subject here and refer the reader to the considerably more detailed descriptions presented by Haus and co-authors and cited in the references to Chapter 5. The approach and notation used in our presentation are adapted from the invited paper published by Haus and Islam in 1985.

The nonlinear Schrödinger equation can be used to describe the propagation of nondispersive solitary waves, or solitons, in an optical fiber. The method consists in first expanding the axial wavenumber, or propagation constant, around the laser operating frequency $\omega_0$, to include both the GVD and the nonlinear phase shift induced by the soliton:

$$\beta(\omega) \simeq \beta(\omega_0) + (\omega - \omega_0)\frac{d\beta}{d\omega} + \frac{1}{2}(\omega - \omega_0)^2 \frac{d^2\beta}{d\omega^2} + \delta\beta, \qquad (5.615)$$

where $d^2\beta/d\omega^2$ represents the GVD, and $\delta\beta$ is the nonlinear phase shift.

In the next step, we use the slow-varying envelope approximation: we assume that

$$\left|\frac{\partial \tilde{A}}{\partial z}\right| \ll |\beta \tilde{A}|, \tag{5.616}$$

where $\tilde{A}$ is the envelope of the electromagnetic pulse, which is of the form

$$\tilde{A}(\omega - \omega_0, z)\exp[i(\beta z - \omega_0 t)], \tag{5.617}$$

in Fourier space. Multiplying the envelope by the propagation constant, and Fourier transforming back into real space, one obtains an equation governing the evolution of the pulse envelope:

$$i\frac{\partial A}{\partial z} - \frac{1}{2}\frac{\partial^2 \beta}{\partial \omega^2}\frac{\partial^2 A}{\partial \tau^2} + \delta\beta A = 0. \tag{5.618}$$

The parameter $\tau = t - v_g^{-1} z$ corresponds to the propagation of the central frequency component, at the group velocity $v_g = d\omega/d\beta|_{\omega=\omega_0}$.

The variation of the nonlinear refractive index due to the high-intensity electromagnetic wave propagating in the optical fiber can be derived from perturbation theory:

$$\delta\beta = \frac{\omega_0^2}{c^2 \beta(\omega_0)} \frac{\iint n_0 n_2 |U|^4 ds}{\iint |U|^2 ds} |A|^2, \tag{5.619}$$

where the integrals are performed over the optical fiber cross-section, and where the function $U$ represents the transverse mode profile excited in the fiber; $n_2$ is typically measured in units of m$^2$ W$^{-1}$, and characterizes the nonlinear susceptibility of the fiber, near the frequency of interest, $\omega_0$.

It is customary to normalize the fiber eigenmode, so that

$$\iint |U|^2 ds = 1, \tag{5.620}$$

which defines the effective cross-section of the optical waveguide: $|U|^2 = \sigma^{-1}$. For a more detailed introduction to nonlinear susceptibilities, polarizations, and refractive indices, we refer the reader to Chapter 7 and references therein. Following Haus and Islam, one can introduce the parameter $\kappa$, so that the nonlinearity takes the simple form

$$\delta\beta = \kappa |A|^2; \tag{5.621}$$

in this case, we have

$$\kappa = \frac{\omega_0^2 \mu_0 \varepsilon_0 n_0 n_2}{\beta(\omega_0)\sigma} = \left(\frac{2\pi}{\lambda_0}\right)^2 \frac{1}{\beta(\omega_0)} \frac{n_0 n_2}{\sigma}. \tag{5.622}$$

Using the parameter $\kappa$, and normalizing the eigenmode power of the first soliton to $P_1$, a normalized length, $z_1$, can be defined: $\kappa P_1 z_1 = 1$. Equation 5.618 can then be recast as

$$-i\frac{\partial a}{\partial \theta} = \frac{1}{2}\frac{\partial^2 a}{\partial \zeta^2} + |a|^2 a, \tag{5.623}$$

where the normalized wave amplitude, $a = A/\sqrt{P_1}$, the normalized dispersion length is now given by

$$\zeta = \frac{\theta}{\sqrt{z_1 \left|\frac{d^2\beta}{d\omega^2}\right|}}, \tag{5.624}$$

and the normalized fiber propagation length is defined as $\theta = z/z_1$.

Equation 5.623 corresponds to a nonlinear Schrödinger equation, which can be solved by the inverse scattering method, as demonstrated by Zakharov and Shabat. The lowest-order solution is the first soliton,

$$a(\theta, \zeta) = 2\eta_1 \frac{e^{i2\eta_1^2\theta}}{\cosh 2\eta_1\zeta}, \tag{5.625}$$

where the quantity $\eta_1$ corresponds to the energy of the lowest-order solitary wave:

$$P_1 \int_{-\infty}^{+\infty} |a|^2 d\theta = \frac{\sqrt{|d^2\beta/d\omega^2|}}{\kappa\sqrt{z_1}} \int_{-\infty}^{+\infty} |a|^2 d\zeta = 4\eta_1 \frac{\sqrt{|d^2\beta/d\omega^2|}}{\kappa\sqrt{z_1}}. \tag{5.626}$$

In closing, we encourage the reader to consult the papers by Haus and co-authors on this subject, given in the references for Chapter 5, as well as the references listed therein. A few elements of soliton theory are described here, for the sake of completeness. The inverse scattering method builds upon the important concept of reflection-free potential wells in the nonlinear Schrödinger equation, such that incident coupled wavepackets of arbitrary wavelength, $\lambda$, are left unscattered. As well-summarized by Haus and Islam, this means that the coupled scattering equations,

$$
\begin{aligned}
i\frac{dw_1}{d\zeta} + a(\zeta)w_2 &= \lambda w_1, \\
i\frac{dw_2}{d\zeta} + a^*(\zeta)w_1 &= -\lambda w_2,
\end{aligned}
\tag{5.627}
$$

are characterized by complex poles in the wavelength plane. Here, $a(\zeta)$ corresponds to the aforementioned normalized solitary wave amplitude, while $w_{1,2}$ represents the incident, coupled waves. The first-order soliton results from the well-known hyperbolic secant well, sech ($\zeta$); higher-order solitary waves correspond to displaced poles and their residues. The residues can be defined by considering the complex parameter

$$\Lambda_j = \sqrt{\rho_j}\, e^{i\lambda_j\zeta}, \tag{5.628}$$

for $\theta = 0$. With this, the potential wells are uniquely defined,

$$a(\zeta, \theta = 0) = -2\sum_j \Lambda_j^* \psi_{2j}^* \tag{5.629}$$

where the wavefunctions, $\psi$, are the solutions of the coupled equations

$$
\begin{aligned}
\psi_{1,j} + \sum_k \frac{\Lambda_j \Lambda_k^*}{\lambda_j - \lambda_k^*}\psi_{2,k}^* &= 0, \\
\sum_k \frac{\Lambda_k \Lambda_j^*}{\lambda_k - \lambda_j^*}\psi_{1,k}^* + \psi_{2,j}^* &= \Lambda_j^*.
\end{aligned}
\tag{5.630}
$$

Redefining the parameter $\Lambda_j$ as

$$\Lambda_j = \sqrt{\rho_j}\exp(\llbracket -\eta_j[\zeta + 2\mathcal{R}(\lambda_j)\theta] + i\{\mathcal{R}(\lambda_j)\zeta + [\mathcal{R}^2(\lambda_j) - \eta_j^2]\theta\}\rrbracket,) \tag{5.631}$$

accounts for nonzero values of $\theta$. Finally, the second-order soliton can be explicitly expressed as

$$a(\zeta, \theta) = \frac{\dfrac{4\eta_1(\eta_1+\eta_2)}{|\eta_1 - \eta_2|}e^{i2\eta_1^1\theta}\left[\cosh(2\eta_2\zeta) + \dfrac{\eta_2}{\eta_1}\cosh(2\eta_1\zeta)e^{i2(\eta_2^2-\eta_1^2)\theta}\right]}{\cosh[2\zeta(\eta_1+\eta_2)] + \dfrac{4\eta_1\eta_2}{(\eta_1-\eta_2)^2}\cos[2\theta(\eta_2^2 - \eta_1^2)] + \left(\dfrac{\eta_1+\eta_2}{\eta_1-\eta_2}\right)^2\cosh[2\zeta(\eta_1-\eta_2)]} \tag{5.632}$$

## 5.10 References for Chapter 5

Note: the numbers listed below refer to the main bibliography and reference sections at the end of this book.

1, 4, 8, 9, 11, 13, 15, 19, 31, 32, 33, 35, 36, 37, 38, 44, 49, 55, 56, 64, 71, 72, 74, 86, 94, 97, 99, 100, 101, 102, 105, 113, 116, 120, 121, 125, 135, 137, 146, 149, 152, 153, 154, 155, 164, 165, 169, 170, 171, 172, 181, 183, 185, 193, 194, 195, 197, 207, 209, 210, 213, 214, 215, 218, 219, 220, 221, 222, 225, 232, 243, 247, 250, 251, 252, 254, 256, 257, 262, 268, 270, 271, 272, 273, 274, 275, 284, 287, 288, 295, 296, 300, 303, 304, 306, 307, 312, 313, 314, 317, 330, 358, 359, 365, 372, 373, 377, 378, 379, 382, 383, 385, 386, 396, 397, 398, 399, 400, 403, 404, 405, 411, 413, 430, 447, 448, 449, 462, 463, 464, 468, 478, 481, 482, 483, 484, 485, 486, 487, 488, 489, 490, 491, 492, 510, 511, 521, 522, 523, 527, 528, 531, 536, 537, 538, 541, 543, 558, 560, 561, 562, 563, 564, 567, 568, 583, 584, 585, 587, 595, 596, 598, 606, 607, 608, 611, 621, 622, 630, 631, 633, 640, 641, 646, 657, 659, 660, 661, 673, 674, 675, 687, 689, 692, 693, 694, 696, 697, 698, 699, 701, 704, 705, 727, 728, 729, 731, 736, 739, 740, 744, 750, 753, 767, 784, 785, 789, 809, 818, 828, 829, 848, 850, 871, 887, 888, 889, 891, 894, 895, 902, 910, 911, 912.

# 6

# Plane Waves and Photons

## 6.1 Introduction

One of the most important solutions to the propagation equation in vacuum corresponds to plane waves, which are only functions of time and one spatial coordinate. In general, the phase of the plane is defined by the relativistically invariant quantity $\phi = k_\mu x^\mu = \mathbf{k} \cdot \mathbf{x} - \omega t$. Such plane waves represent classical solutions to the electromagnetic wave equation, but they assume a particularly important role in electrodynamics because they offer a natural introduction to the quantum electrodynamical concept of the photon. In addition, plane wave solutions form the basis of the Fourier analysis of wave propagation, which is a powerful mathematical tool to study linear problems where the principle of superposition applies. In this respect, plane waves are closely linked to the Green functions presented in Chapter 5 and represent their natural mathematical complement.

In this chapter, the vacuum propagation equation is first considered, and its plane wave solutions are described. We then consider the same type of waves but now confined in a box by a set of boundary conditions. The connection with the quantization of the free electromagnetic field and photons is then made, and a detailed analysis of the properties of the quantized electromagnetic field is given, including its statistical characteristics. Finally, virtual photons and the Coulomb field are discussed. The goal of this chapter is to provide a useful introduction to some aspects of QED and to offer a background for the question of vacuum fluctuations; this approach is also essential to the understanding of the Casimir effect and Hawking–Unruh radiation.

As discussed in Chapter 2, the propagation of electromagnetic waves in vacuum is described by the wave equation for the free electromagnetic field,

$$\partial_\mu \partial^\mu A_\nu = \Box A_\nu = 0, \tag{6.1}$$

where the four-potential satisfies the Lorentz gauge condition,

$$\partial_\mu A^\mu = 0, \tag{6.2}$$

and where $\partial_\mu = (-\partial_0, \nabla)$ is the four-gradient operator; $\Box = \partial_\mu \partial^\mu = \Delta - \partial_0^2$ is the d'Alembertian operator, also called the electromagnetic wave propagator.

We can seek a general solution of Equation 6.1 in terms of a superposition of plane waves, where the four-potential is described by a Fourier transform:

$$A_\mu(x^\lambda) = \left(\frac{1}{\sqrt{2\pi}}\right)^4 \iiiint \tilde{A}_\mu(k_\lambda) \exp(-ik_\lambda x^\lambda)\, d^4k_\lambda. \tag{6.3}$$

Here, $k_\mu = (\omega/c, \mathbf{k})$ is the four-wavenumber, which is the Fourier conjugate of the four-position, $x_\mu = (ct, \mathbf{x})$.

The terminology often used to describe the Fourier transform in Equation 6.3 refers to the four-wavenumber space as momentum space, since for quantum states, there is a direct correspondence between the four-momentum and four-wavenumber:

$$p_\mu = \hbar k_\mu. \tag{6.4}$$

The obvious advantage of the Fourier transform is that the d'Alembertian operator yields a very simple result when applied to the complex exponential in the Fourier integral: we now have, for the wave equation,

$$\Box A_\mu(x^\lambda) = \left(\frac{1}{\sqrt{2\pi}}\right)^4 \iiiint (-k_\mu k^\mu) \tilde{A}(k_\lambda) \exp(-ik_\lambda x^\lambda)\, d^4k_\lambda = 0. \tag{6.5}$$

At this point, it is important to note that the various Fourier modes are orthogonal; in other words,

$$\begin{aligned} \iiiint \exp(-ik_\lambda x^\lambda) \exp(ik'_\lambda x^\lambda)\, d^4x^\lambda &= \iiiint \exp[-i(k_\lambda - k'_\lambda)x^\lambda]\, d^4x^\lambda \\ &= (2\pi)^4 \delta^4(k_\lambda - k'_\lambda). \end{aligned} \tag{6.6}$$

As a result, the nontrivial solution to Equation 6.5 implies that the following condition be satisfied for all values of the four-wavenumber:

$$-k_\mu k^\mu = \frac{\omega^2}{c^2} - \mathbf{k}^2 = 0. \tag{6.7}$$

This is the well-known dispersion relation for electromagnetic waves propagating in vacuum. Within the context of the aforementioned momentum space terminology, we see that the vacuum dispersion relation corresponds to the mass-shell condition for photons, namely,

$$p_\mu p^\mu = \hbar^2 k_\mu k^\mu = 0 = \mathbf{p}^2 - \frac{E^2}{c^2}, \tag{6.8}$$

which implies that photons are massless, since all their energy corresponds to momentum, as opposed to a particle with rest mass $m_0$, for which $p_\mu p^\mu = -m_0^2 c^2$. In the previous chapter, concerned with the propagation of electromagnetic waves in a vacuum enclosed by boundary conditions, we have seen that, in general, the presence of boundaries discretizes the free modes of the vacuum and introduces a cutoff frequency $\omega_0$, such that the dispersion relation is modified to read $c^2 k_\mu k^\mu + \omega_0^2 = 0$; within this context, the photons are seen to acquire an effective rest mass, $m_0 = \hbar\omega_0/c^2$. Indeed, the photons trapped at cutoff by the boundary conditions have energy but no momentum, which is exactly the definition of rest mass. However, a more detailed analysis of the situation shows that the trapped modes correspond to standing waves, which can be understood in terms of counterpropagating waves; thus, the zero momentum of the trapped wave stems from the fact that photon pairs with finite energy but opposite momenta form the cutoff mode.

Provided that the condition $k_\mu k^\mu = 0$ is satisfied, any superposition of plane waves described by Equation 6.5 is a solution of the wave equation in vacuum. A single free electromagnetic mode, described by the Fourier amplitude $\tilde{A}_\mu(k'_\lambda) = \tilde{A}_{\mu 0}\delta^4(k_\lambda - k'_\lambda)$, corresponds to a plane wave, with four-wavenumber $k_\lambda$, and the scalar $\phi = -k_\lambda x^\lambda$ is the relativistically invariant phase of that wave. In Chapter 8, where the dynamics of an electron in a plane wave is studied, we will generalize plane waves to include free electromagnetic modes where the four-potential is a function of the phase and satisfies the gauge condition, which now corresponds to the transversality condition. We have

$$A_\mu(x_\lambda) = A_\mu(\phi) = A_\mu(k^\lambda x_\lambda), \tag{6.9}$$

and the Lorentz gauge condition now requires that

$$\partial_\mu A^\mu(\phi) = \frac{\partial\phi}{\partial x^\mu}\frac{dA^\mu}{d\phi} = k_\mu \frac{dA^\mu}{d\phi} = 0. \tag{6.10}$$

In particular, if we choose the spatial $z$-axis to coincide with the direction of propagation of the wave, we see that a purely transverse four-potential, where $A_\mu(x_\lambda) = [0, \mathbf{A}_\perp(\phi), 0]$, will automatically satisfy the gauge condition.

## 6.2 Quantization of the Free Electromagnetic Field

We now turn our attention to the question of the quantization of the free electromagnetic field. So far, we have considered the electromagnetic field as a classical field, describable in terms of continuous functions of the four-position, or the four-wavenumber, when we work in momentum space. This model of the field is extremely useful when applied to a large number of phenomena, including wave propagation, diffraction, and interference. In particular, a large variety of optical phenomena can be described within this framework, which forms the basis of geometrical and classical optics, including coherent radiation generation in microwave devices, Fourier optics, and dispersion theory. However, under certain circumstances, this description proves inadequate, and the concept of the photon, or quantum of the electromagnetic field, must be introduced. Such special situations include the physics of vacuum fluctuations, the Casimir effect, and Hawking–Unruh radiation, which require a detailed knowledge of the statistical properties of the free electromagnetic field; the study of the coherence and space–time correlation characteristics of light, as exemplified by squeezed states; and the physics of radiative corrections which play a major role in Compton scattering, particularly in the case of multiphoton interactions at relativistic field intensities; finally, QED concepts, such as the Schwinger critical field and the description of the Coulomb field in terms of virtual quanta rely intrinsically on the concept of the photon.

We start from Maxwell's equations, as expressed in the absence of sources:

$$\nabla \times \mathbf{E} + \partial_t \mathbf{B} = \mathbf{0}, \quad \text{and} \quad \nabla \cdot \mathbf{A} = 0, \tag{6.11}$$

for the source-free equations, and

$$\nabla \times \mathbf{B} - \frac{1}{c^2}\partial_t \mathbf{E} = \mathbf{0}, \quad \text{and} \quad \nabla \cdot \mathbf{E} = 0, \tag{6.12}$$

in vacuum. In this case, we introduce the vector potential only, as we shall work in the Coulomb gauge. We simply have

$$\mathbf{E} = -\partial_t \mathbf{A}, \quad \text{and} \quad \mathbf{B} = \nabla \times \mathbf{A}, \tag{6.13}$$

and the transversality condition, which is equivalent to the fact that the vector potential is divergence-free,

$$\nabla \cdot \mathbf{A} = 0. \tag{6.14}$$

We note that such a divergence-free vector potential will be used in Chapter 8, where the focusing and diffraction of electromagnetic waves will be studied in detail. In that case it will prove useful to introduce a generating vector field $\mathbf{G}$, defined such that $\mathbf{A} = \nabla \times \mathbf{G}$, thus automatically satisfying the Coulomb gauge condition described by Equation 6.14. Finally, using the definition of the fields in terms of the vector potential, we obtain the propagation equation,

$$\left(\Delta - \frac{1}{c^2}\partial_t^2\right)\mathbf{A} = \mathbf{0}. \tag{6.15}$$

To quantize the free electromagnetic field, we need to derive the corresponding Hamiltonian; it will prove useful to start by Fourier transforming the vector potential into three-momentum space, which is the conjugate of the usual three-dimensional space. In addition, we will consider that the field is subjected to spatial boundary conditions. Instead of a continuous spectrum, a discrete spectrum results from this procedure, and the vector potential can be described by a Fourier series. Finally, a cubic box is used, which further simplifies the mathematical expression of the potential. With this, we have

$$\mathbf{A}(x_\mu) = \frac{1}{\sqrt{\varepsilon_0 a^3}} \sum_\ell \sum_m \sum_n A_{\ell mn}(t) \exp\left[i(k_\ell x + k_m y + k_n z)\right], \tag{6.16}$$

where $a$ is the size of the box, and where the wavenumber spectrum satisfies the periodic boundary conditions at the edge of the box:

$$k_\ell = \ell\frac{2\pi}{a}, \quad k_m = m\frac{2\pi}{a}, \quad k_n = n\frac{2\pi}{a}, \quad \ell, m, n \in \mathbb{Z}. \tag{6.17}$$

Finally, it is customary to introduce the normalization factor $1/\sqrt{\varepsilon_0 a^3}$ for convenience. Following Mandel and Wolf, in the remainder of the derivation, we will use a more compact notation, where Equation 6.16 now reads

$$\mathbf{A}(x_\mu) = \frac{1}{\sqrt{\varepsilon_0 a^3}} \sum \mathbf{A}_\mathbf{k}(t) e^{i\mathbf{k}\cdot\mathbf{x}}, \tag{6.18}$$

and the summation is performed over the three spatial indices labeling the normal modes of the box; the wavenumber is defined as $\mathbf{k}_{\ell mn} = \mathbf{k} = \mathbf{k}_\ell \hat{x} + \mathbf{k}_m \hat{y} + \mathbf{k}_n \hat{z}$.

Within this context, the Coulomb gauge condition implies that

$$\frac{1}{\sqrt{\varepsilon_0 a^3}} \sum_\mathbf{k} i\mathbf{k} \cdot \tilde{\mathbf{A}}_\mathbf{k}(t) e^{i\mathbf{k}\cdot\mathbf{x}} = 0; \tag{6.19}$$

because the complex exponential functions are orthogonal, we must also have

$$\mathbf{k} \cdot \tilde{\mathbf{A}}_\mathbf{k}(t) = 0. \tag{6.20}$$

We can now use Equation 6.18 into the wave equation 6.15; the Laplacian operating on complex exponentials takes a very simple form, and we find that

$$\frac{1}{\sqrt{\varepsilon_0 a^3}} \sum_\mathbf{k} \left(-\mathbf{k}^2 - \frac{1}{c^2}\partial_t^2\right) \tilde{\mathbf{A}}_\mathbf{k}(t) e^{i\mathbf{k}\cdot\mathbf{x}} = 0. \tag{6.21}$$

Again, the orthogonality of complex exponentials implies that each term in the sum is identically equal to zero: each mode of the box satisfies the propagation equation

$$\left(\mathbf{k}^2 + \frac{1}{c^2}\partial_t^2\right)\tilde{\mathbf{A}}_{\mathbf{k}}(t) = 0. \tag{6.22}$$

Equation 6.22 corresponds to a harmonic oscillator of frequency $\frac{\omega^2}{c^2} = \mathbf{k}^2$ and can be solved to obtain

$$\tilde{\mathbf{A}}_{\mathbf{k}}(t) = a_{\mathbf{k}}e^{-i\omega t} + a^*_{-\mathbf{k}}e^{i\omega t}, \tag{6.23}$$

where the complex conjugate quantity guarantees that the vector potential is a real vector field.

The transversality condition described in Equation 6.20 must be satisfied by the solution given in Equation 6.23; it is customary to introduce two polarization vectors that are mutually orthogonal and perpendicular to the direction of propagation of the mode under consideration: $e^{\mathbf{k}}_{\perp 1}$ and $e^{\mathbf{k}}_{\perp 2}$ are two unit vectors defined such that $\mathbf{k}\cdot e^{\mathbf{k}}_{\perp 1} = \mathbf{k}\cdot e^{\mathbf{k}}_{\perp 2} = 0$; in addition, $e^{\mathbf{k}}_{\perp 1}\cdot e^{\mathbf{k}}_{\perp 2} = 0$; finally, $e^{\mathbf{k}}_{\perp 1}\times e^{\mathbf{k}}_{\perp 2} = \mathbf{k}/|\mathbf{k}| = e^{\mathbf{k}}_{\parallel}$. The vectors $(e^{\mathbf{k}}_{\perp 1}, e^{\mathbf{k}}_{\perp 2}, e^{\mathbf{k}}_{\parallel})$ form a right-handed, orthonormal basis, and the Fourier coefficients of the vector potential can be projected on this basis:

$$a_{\perp} = a_{\mathbf{k}1}e^{\mathbf{k}}_{\perp 1} + a_{\mathbf{k}2}e^{\mathbf{k}}_{\perp 2}. \tag{6.24}$$

We note that the polarization vectors can be rotated arbitrarily in the plane perpendicular to the direction of propagation of the mode under consideration; furthermore, the basis obviously depends on the mode, as it is defined with respect to $e^{\mathbf{k}}_{\parallel} = \mathbf{k}/|\mathbf{k}|$. Within this context, the indices 1 and 2 represent the two possible polarization states of the mode indexed by $\mathbf{k}$; different combinations of the two vectors will correspond to linear, elliptical, and circular polarization states. For a detailed presentation of the polarization states and their properties under spatial rotations, we refer the reader to Mandel and Wolf.

The vector potential for an arbitrary combination of eigenmodes of the box can now be expressed as

$$\mathbf{A}(x^\mu) = \frac{1}{\sqrt{\varepsilon_0 a^3}}\sum_{\mathbf{k}}\sum_{\sigma}(a_{\mathbf{k}\sigma}e^{\mathbf{k}}_{\perp\sigma}e^{-i\omega t} + a_{-\mathbf{k}\sigma}e^{-\mathbf{k}}_{\perp\sigma}e^{i\omega t})e^{i\mathbf{k}\cdot\mathbf{x}}, \tag{6.25}$$

where $\sigma = 1, 2$ represents the polarization.

Although the presentation given here is not covariant, Equation 6.25 can be recast as

$$\mathbf{A}(x^\mu) = \frac{1}{\sqrt{\varepsilon_0 a^3}}\sum_{\mathbf{k}}\sum_{\sigma}\left(a_{\mathbf{k}\sigma}e^{\mathbf{k}}_{\perp\sigma}e^{ik_\mu x^\mu} + a_{-\mathbf{k}\sigma}e^{-\mathbf{k}}_{\perp\sigma}e^{-ik_\mu x^\mu}\right), \tag{6.26}$$

where $k_\mu = (\omega/c, \mathbf{k})$, and satisfies the vacuum dispersion relation $k_\mu k^\mu = 0$ for each mode of the box.

Again, we emphasize that, strictly speaking, each mode of the box corresponds to an eigenwavenumber, $k_\mu^{\ell mn} = (\omega_{\ell mn}/c, \mathbf{k}_{\ell mn})$, satisfying the dispersion relation

$$\frac{\omega_{\ell mn}^2}{c^2} = \mathbf{k}_{\ell mn}^2 = (\ell^2 + m^2 + n^2)\left(\frac{2\pi}{a}\right)^2. \tag{6.27}$$

As mentioned earlier, for conciseness, the three indices corresponding to the triple set of boundary conditions are implicitly included in the summation $\Sigma_{\mathbf{k}}$. We also note that the oscillation frequency of a given eigenmode is independent of its polarization state; thus there is a degeneracy of the modes in terms of polarization.

To expand the vector potential in terms of the spatial eigenmodes of the box, we write $a_{\mathbf{k}\sigma}(t) = a_{\mathbf{k}\sigma}e^{-i\omega t}$, in which case Equation 6.25 takes the form

$$\mathbf{A}(x_\mu) = \frac{1}{\sqrt{\varepsilon_0 a^3}}\sum_{\mathbf{k}}\sum_{\sigma}[a_{\mathbf{k}\sigma}(t)e_{\perp\sigma}^{\mathbf{k}}e^{i\mathbf{k}\cdot\mathbf{x}} + a_{\mathbf{k}\sigma}^*(t)e_{\perp\sigma}^{-\mathbf{k}}e^{i\mathbf{k}\cdot\mathbf{x}}]. \tag{6.28}$$

The fields can now be evaluated by using Equation 6.13; the partial time derivative and the curl operators take very simple forms, and we find

$$\mathbf{E}(x_\mu) = \frac{i}{\sqrt{\varepsilon_0 a^3}}\sum_{\mathbf{k}}\sum_{\sigma}\omega[a_{\mathbf{k}\sigma}(t)e_{\perp\sigma}^{\mathbf{k}}e^{i\mathbf{k}\cdot\mathbf{x}} - a_{\mathbf{k}\sigma}^*(t)e_{\perp\sigma}^{-\mathbf{k}}e^{-i\mathbf{k}\cdot\mathbf{x}}], \tag{6.29}$$

and

$$\mathbf{B}(x_\mu) = \frac{i}{\sqrt{\varepsilon_0 a^3}}\sum_{\mathbf{k}}\sum_{\sigma}\omega[a_{\mathbf{k}\sigma}(t)(\mathbf{k}\times e_{\perp\sigma}^{\mathbf{k}})e^{i\mathbf{k}\cdot\mathbf{x}} - a_{\mathbf{k}\sigma}^*(t)(\mathbf{k}\times e_{\perp\sigma}^{-\mathbf{k}})e^{-i\mathbf{k}\cdot\mathbf{x}}]. \tag{6.30}$$

To proceed with the quantization of the free electromagnetic field, we now need to derive the Hamiltonian of the system, which corresponds to the field energy of the eigenmodes derived above; we start with the electromagnetic energy density, which is given by $\frac{\varepsilon_0\mathbf{E}^2}{2} + \frac{\mathbf{B}^2}{2\mu_0}$. The total field energy in the box is thus

$$\mathcal{H} = \frac{1}{2}\iiint_{a^3}\left[\varepsilon_0\mathbf{E}^2(x_\mu) + \frac{1}{\mu_0}\mathbf{B}^2(x_\mu)\right]d^3\mathbf{x}, \tag{6.31}$$

where the electric field and magnetic induction are given by Equations 6.29 and 6.30. The orthogonality of the complex exponentials considerably simplifies the energy integral: the eigenmodes of the box do not interfere; in other words, we have

$$\iiint_{a^3} \exp[i(\mathbf{k}-\mathbf{k}')\cdot\mathbf{x}]\,d^3\mathbf{x} = a^3\delta^3_{\mathbf{kk}'}. \tag{6.32}$$

For the magnetic field energy, we are led to evaluate the product

$$(\mathbf{k}\times e^{\mathbf{k}}_{\perp\sigma})\cdot(\mathbf{k}\times e^{-\mathbf{k}}_{\perp\sigma'}) = \mathbf{k}^2(e^{\mathbf{k}}_{\perp\sigma}\cdot e^{-\mathbf{k}}_{\perp\sigma'}) = \mathbf{k}^2\delta_{\sigma\sigma'}. \tag{6.33}$$

The aforementioned orthogonality of the box eigenmodes results in a diagonalization of the summations over modes and polarizations, and we find that the electromagnetic energy contained within the boundaries is

$$\mathcal{H} = 2\sum_{\mathbf{k}}\sum_{\sigma}\omega^2|a_{\mathbf{k}\sigma}(t)|^2; \tag{6.34}$$

as expected, the modes do not interfere. The total field energy is given by the sum of the energy of each of the vacuum eigenmodes excited within the box.

At this point, we need to describe the electromagnetic field within the context of Hamiltonian formalism; the commutation of the corresponding canonical variables will then enable us to quantize the field in terms of photons.

A generalized position, $q_{\mathbf{k}\sigma}$, and a generalized momentum, $p_{\mathbf{k}\sigma}$, must be associated to each eigenmode and polarization state of the system. Furthermore, these canonical variables must be such that the dynamics of the system obey Hamilton's equations,

$$\frac{\partial\mathcal{H}}{\partial p_{\mathbf{k}\sigma}} = \frac{\partial q_{\mathbf{k}\sigma}}{\partial t}, \tag{6.35}$$

and

$$\frac{\partial\mathcal{H}}{\partial q_{\mathbf{k}\sigma}} = \frac{\partial p_{\mathbf{k}\sigma}}{\partial t}. \tag{6.36}$$

The following choice of generalized coordinates yields the desired equations of motion:

$$q_{\mathbf{k}\sigma}(t) = [a_{\mathbf{k}\sigma}(t) + a^*_{\mathbf{k}\sigma}(t)], \tag{6.37}$$

$$p_{\mathbf{k}\sigma}(t) = -i\omega[a_{\mathbf{k}\sigma}(t) - a^*_{\mathbf{k}\sigma}(t)]. \tag{6.38}$$

Using the definition $a_{\mathbf{k}\sigma}(t) = a_{\mathbf{k}\sigma} e^{-i\omega t}$, it is easily seen that

$$\frac{\partial}{\partial t}[q_{\mathbf{k}\sigma}(t)] = p_{\mathbf{k}\sigma}(t). \tag{6.39}$$

In addition, taking the partial derivative of the generalized momentum with respect to time yields the following result:

$$\frac{\partial}{\partial t}[p_{\mathbf{k}\sigma}(t)] = -\omega^2 q_{\mathbf{k}\sigma}(t). \tag{6.40}$$

It is now easy to verify that the Hamiltonian can be expressed as

$$\mathcal{H}(p, q) = \frac{1}{2}\sum_{\mathbf{k}}\sum_{\sigma}[\omega^2 q^2_{\mathbf{k}\sigma}(t) + p^2_{\mathbf{k}\sigma}(t)], \tag{6.41}$$

and that Hamilton's equations 6.35 and 6.36 are satisfied by the choice of generalized coordinates used here. The fact that the eigenmodes are orthogonal is reflected in the fact that each mode contributes energy to the Hamiltonian independently of the other modes: there is no interference between modes. Furthermore, Equation 6.41 indicates that each mode corresponds to an harmonic oscillator, with a frequency satisfying the vacuum dispersion relation, Equation 6.26. In the quantization of the field, each oscillation mode will be identified with a quantum of radiation, thus introducing the concept of the photon.

Using Equations 6.36 and 6.38, we can define the amplitude of the box eigenmodes in terms of the canonical variables,

$$a_{\mathbf{k}\sigma}(t) = \frac{1}{2}\left[q_{\mathbf{k}\sigma}(t) + \frac{i}{\omega}p_{\mathbf{k}\sigma}(t)\right], \tag{6.42}$$

and the potential and fields can then be expressed in terms of the canonical variables:

$$\begin{aligned}\mathbf{A}(x_\mu) = \frac{1}{2\sqrt{\varepsilon_0 a^3}}\sum_{\mathbf{k}}\sum_{\sigma}\Bigg\{&\left[q_{\mathbf{k}\sigma}(t) + \frac{i}{\omega}p_{\mathbf{k}\sigma}(t)\right]e^{\mathbf{k}}_{\perp\sigma}e^{i\mathbf{k}\cdot\mathbf{x}} \\ &+\left[q^*_{\mathbf{k}\sigma}(t) - \frac{i}{\omega}p^*_{\mathbf{k}\sigma}(t)\right]e^{-\mathbf{k}}_{\perp\sigma}e^{-i\mathbf{k}\cdot\mathbf{x}}\Bigg\},\end{aligned} \tag{6.43}$$

$$\begin{aligned}\mathbf{E}(x_\mu) = \frac{i}{2\sqrt{\varepsilon_0 a^3}}\sum_{\mathbf{k}}\sum_{\sigma}\{&[\omega q_{\mathbf{k}\sigma}(t) + ip_{\mathbf{k}\sigma}(t)]e^{\mathbf{k}}_{\perp\sigma}e^{i\mathbf{k}\cdot\mathbf{x}} \\ &-[\omega q^*_{\mathbf{k}\sigma}(t) - ip^*_{\mathbf{k}\sigma}(t)]e^{-\mathbf{k}}_{\perp\sigma}e^{-i\mathbf{k}\cdot\mathbf{x}}\},\end{aligned} \tag{6.44}$$

and

$$\mathbf{B}(x_\mu) = \frac{i}{2\sqrt{\varepsilon_0 a^3}} \sum_{\mathbf{k}} \sum_{\sigma} \left\{ \left[ q_{\mathbf{k}\sigma}(t) + \frac{i}{\omega} p_{\mathbf{k}\sigma}(t) \right] (\mathbf{k} \times e_{\perp\sigma}^{\mathbf{k}}) e^{i\mathbf{k}\cdot\mathbf{x}} - \left[ q_{\mathbf{k}\sigma}^*(t) - \frac{i}{\omega} p_{\mathbf{k}\sigma}^*(t) \right] (\mathbf{k} \times e_{\perp\sigma}^{-\mathbf{k}}) e^{-i\mathbf{k}\cdot\mathbf{x}} \right\}. \tag{6.45}$$

The quantum mechanical description of the electromagnetic field can be achieved by now considering the canonical variables as operators and identifying the corresponding Poisson bracket with the commutator:

$$[q_{\mathbf{k}\sigma}(t), p_{\mathbf{k}'\sigma'}(t)] = i\hbar \delta_{\mathbf{k}\mathbf{k}'}^3 \delta_{\sigma\sigma'}. \tag{6.46}$$

Within the quantum context, the Hamiltonian of Equation 6.41 must now be considered as an operator acting on the state vector of the electromagnetic field; the result of this operation is the corresponding electromagnetic energy level. Because we are considering a system enclosed by a set of boundary conditions, the resulting energy spectrum is discrete. This is a very general result, and quantum numbers can be associated with each set of boundaries. In this case, each mode is indexed by three numbers corresponding to the triple set of spatial boundaries imposed on the electromagnetic field. For unbounded systems, the discrete spectrum is replaced by a continuum; for example, above the ionization threshold, the electron wavefunction can extend to infinity, and the energy spectrum becomes continuous. Additional quantum numbers can reflect internal symmetries, or, in the case of the electromagnetic field, indicate the state of polarization of the corresponding quantum state. The two different possible values for $\sigma$ simply correspond to the fact that the photon is a spin-1 particle. The aforementioned degeneracy of the electromagnetic energy level in terms of the polarization illustrates the fact that the photon spin does not contribute to its energy.

## 6.3 Creation and Annihilation Operators

We now turn our attention to the well-known photon creation and annihilation operators; these can be defined in terms of the generalized position and momentum, $q_{\mathbf{k}\sigma}$, and $p_{\mathbf{k}\sigma}$, and are closely related to the eigenmode amplitudes, $a_{\mathbf{k}\sigma}$:

$$a_{\mathbf{k}\sigma}(t) = \frac{i}{\sqrt{2\hbar\omega}} [\omega q_{\mathbf{k}\sigma}(t) + i p_{\mathbf{k}\sigma}(t)] = \sqrt{\frac{2\omega}{\hbar}}\, a_{\mathbf{k}\sigma}(t), \tag{6.47}$$

and

$$a^{\dagger}_{\mathbf{k}\sigma}(t) = \frac{1}{\sqrt{2\hbar\omega}}[\omega q_{\mathbf{k}\sigma}(t) - ip_{\mathbf{k}\sigma}(t)] = \sqrt{\frac{2\omega}{\hbar}}\, a^{\dagger}_{\mathbf{k}\sigma}(t). \tag{6.48}$$

Here, the notation $a^{\dagger}_{\mathbf{k}\sigma}(t)$ refers to the Hermitian conjugate of the operator $a_{\mathbf{k}\sigma}(t)$; in terms of matrix properties, if $M = (M_{ij})$ is the original matrix, its Hermitian conjugate is given by $M^{\dagger} = (M^{*}_{ji})$. A matrix is said to be Hermitian if $M_{ij} = M^{*}_{ji}$; it is anti-Hermitian if $M_{ij} = -M^{*}_{ji}$; finally, the important property of unitarity is satisfied if $(M^{\dagger}M)_{ij} = \delta_{ij}$. Matrices and operators are closely related in quantum mechanics; in fact, the matrix element of a given operator corresponds to its projection in terms of quantum states: $M_{ij} = \langle i|M|j\rangle$, where we have used Dirac's notation.

Because the new operators introduced in Equations 6.46 and 6.48 are normalized, the commutation relation takes the simple form

$$[a_{\mathbf{k}\sigma}(t), a^{\dagger}_{\mathbf{k}'\sigma'}(t)] = \delta^{3}_{\mathbf{k}\mathbf{k}'}\delta_{\sigma\sigma'}; \tag{6.49}$$

in addition, it is easily verified that each operator commutes with itself.

Next, we need to express the Hamiltonian operator, which yields the energy spectrum of the quantized electromagnetic field in terms of the creation and annihilation operators. We start from Equation 6.41, and replace $q_{\mathbf{k}\sigma}(t)$ and $p_{\mathbf{k}\sigma}(t)$ by their expressions in terms of $a_{\mathbf{k}\sigma}(t)$ and $a^{\dagger}_{\mathbf{k}\sigma}(t)$. Adding Equations 6.46 and 6.48, we first have

$$\sqrt{\frac{2\omega}{\hbar}}\, q_{\mathbf{k}\sigma}(t) = [a_{\mathbf{k}\sigma}(t) + a^{\dagger}_{\mathbf{k}\sigma}(t)]. \tag{6.50}$$

Subtracting Equation 6.48 from Equation 6.46, we also find that

$$i\sqrt{\frac{2\omega}{\hbar}}\, p_{\mathbf{k}\sigma}(t) = [a_{\mathbf{k}\sigma}(t) - a^{\dagger}_{\mathbf{k}\sigma}(t)]. \tag{6.51}$$

Substituting in the definition of the Hamiltonian, given in Equation 6.41, we first have

$$\mathcal{H} = \frac{1}{2}\sum_{\mathbf{k}}\sum_{\sigma}\left\{\frac{\hbar\omega}{2}[a_{\mathbf{k}\sigma}(t) + a^{\dagger}_{\mathbf{k}\sigma}(t)]^{2} - \frac{\hbar\omega}{2}[a_{\mathbf{k}\sigma}(t) - a^{\dagger}_{\mathbf{k}\sigma}(t)]^{2}\right\}. \tag{6.52}$$

Special care must now be taken in expanding the operator products in the brackets; in particular, their order must be preserved, and we can use the fact that each operator commutes with itself:

$$[a_{\mathbf{k}\sigma}(t) + a^{\dagger}_{\mathbf{k}\sigma}(t)]^{2} = 2a_{\mathbf{k}\sigma}(t)a^{\dagger}_{\mathbf{k}\sigma}(t), \tag{6.53}$$

and

$$[a_{\mathbf{k}\sigma}(t) - a^{\dagger}_{\mathbf{k}\sigma}(t)]^2 = -2a^{\dagger}_{\mathbf{k}\sigma}(t)a_{\mathbf{k}\sigma}(t). \tag{6.54}$$

Substituting into the expression for the Hamiltonian, we find

$$\mathcal{H} = \frac{1}{2}\sum_{\mathbf{k}}\sum_{\sigma}\hbar\omega[a_{\mathbf{k}\sigma}(t)a^{\dagger}_{\mathbf{k}\sigma}(t) + a^{\dagger}_{\mathbf{k}\sigma}(t)a_{\mathbf{k}\sigma}(t)]. \tag{6.55}$$

Note that the operators are normalized so that the Hamiltonian has the correct dimension of energy, like the photon energy $\hbar\omega$.

To begin addressing the question of vacuum fluctuations, the Hamiltonian operator can be recast in a more suggestive form. We use the commutation rule given in Equation 6.49 and write it down explicitly:

$$[a_{\mathbf{k}\sigma}(t), a^{\dagger}_{\mathbf{k}'\sigma'}(t)] = \delta^3_{\mathbf{k}\mathbf{k}'}\delta_{\sigma\sigma'} = a_{\mathbf{k}\sigma}(t)a^{\dagger}_{\mathbf{k}'\sigma'}(t) - a^{\dagger}_{\mathbf{k}\sigma}(t)a_{\mathbf{k}'\sigma'}(t). \tag{6.56}$$

We then have

$$a_{\mathbf{k}\sigma}(t)\,a^{\dagger}_{\mathbf{k}'\sigma'}(t) = a^{\dagger}_{\mathbf{k}\sigma}(t)a_{\mathbf{k}'\sigma'}(t) + \delta^3_{\mathbf{k}\mathbf{k}'}\delta_{\sigma\sigma'}, \tag{6.57}$$

and for $\mathbf{k}' = \mathbf{k}$, and $\sigma' = \sigma$, Equation 6.56 yields

$$a_{\mathbf{k}\sigma}(t)\,a^{\dagger}_{\mathbf{k}\sigma}(t) = a^{\dagger}_{\mathbf{k}\sigma}(t)a_{\mathbf{k}\sigma}(t) + 1. \tag{6.58}$$

Finally, the Hamiltonian takes the form

$$\mathcal{H} = \sum_{\mathbf{k}}\sum_{\sigma}\hbar\omega\left[a^{\dagger}_{\mathbf{k}\sigma}(t)a_{\mathbf{k}\sigma}(t) + \frac{1}{2}\right], \tag{6.59}$$

and the zero point fluctuations of the vacuum state appear clearly. The lowest energy level of each harmonic oscillator is $\frac{1}{2}\hbar\omega$, where the frequency is defined as the cutoff frequency of the mode in question: $\omega = \omega_{\ell mn} = \sqrt{\ell^2 + m^2 + n^2}(2\pi/a)$.

## 6.4 Energy and Number Spectra

The energy spectrum generated by the Hamiltonian operator can now be established. In this section, we will continue using Dirac's notation for quantum states and follow the discussion of Messiah, as presented by Mandel and Wolf. Before deriving the energy spectrum, we recall the definition of an operator eigenstate and the associated eigenvalue: for a given operator,

$\Omega$, the quantum state $|\psi_n\rangle$ is an eigenstate with eigenvalue $\lambda_n \in \mathbb{C}$, if the following condition is satisfied

$$\Omega|\psi_n\rangle = \lambda_n|\psi_n\rangle. \tag{6.60}$$

In other words, the action of an operator on one of its eigenstates is very simple; the resulting state is just the original state multiplied by the eigenvalue. This concept is directly related to the diagonalization of the operator. If we consider orthonormalized eigenstates, the matrix elements of the operator take a diagonal form, with

$$\Omega_{mn} = \langle\psi_m|\Omega|\psi_n\rangle = \langle\psi_m|\lambda_n|\psi_n\rangle = \lambda_n\langle\psi_m|\psi_n\rangle = \lambda_n\delta_{mn}. \tag{6.61}$$

For an in-depth presentation of Dirac's notation, as well as Hilbert spaces and their application to quantum mechanics, we refer the reader to Messiah and to Cohen–Tannoudji, Diu, and Laloe.

In the Hamiltonian, the spectrum is governed by the operator $a^\dagger_{\mathbf{k}\sigma}(t)a_{\mathbf{k}\sigma}(t) = N_{\mathbf{k}\sigma}(t)$; the factor $\frac{1}{2}\hbar\omega$ simply corresponds to the vacuum level and appears as a shift of the entire spectrum. Thus, our task is to determine the eigenvalue spectrum of the photon number operator, $N_{\mathbf{k}\sigma}(t)$. The mathematical approach we will follow here was originally presented by Dirac and starts by considering an eigenstate $|\eta_{\mathbf{k}\sigma}\rangle$ of the operator $N_{\mathbf{k}\sigma}(t)$, whereby

$$N_{\mathbf{k}\sigma}(t)|\eta_{\mathbf{k}\sigma}\rangle = n_{\mathbf{k}\sigma}|\eta_{\mathbf{k}\sigma}\rangle, \tag{6.62}$$

and where $n_{\mathbf{k}\sigma}$ is the corresponding eigenvalue.

We now focus on the action of the creation operator on this eigenstate by evaluating the new quantum state $a^\dagger_{\mathbf{k}\sigma}|\eta_{\mathbf{k}\sigma}\rangle$.

At this point, we need to derive the commutation relation between the photon number operator and the creation and annihilation operators:

$$\begin{aligned}[a_{\mathbf{k}\sigma}(t), N_{\mathbf{k}'\sigma'}(t)] &= a_{\mathbf{k}\sigma}(t)N_{\mathbf{k}'\sigma'}(t) - N_{\mathbf{k}'\sigma'}(t)a_{\mathbf{k}\sigma}(t)\\ &= a_{\mathbf{k}\sigma}(t)a^\dagger_{\mathbf{k}'\sigma'}(t)a_{\mathbf{k}'\sigma'}(t) - a^\dagger_{\mathbf{k}'\sigma'}(t)a_{\mathbf{k}'\sigma'}(t)a_{\mathbf{k}\sigma}(t),\end{aligned} \tag{6.63}$$

where we recognize the commutator described in Equation 6.56.

With this, we can rewrite Equation 6.63 as

$$[a_{\mathbf{k}\sigma}(t), N_{\mathbf{k}'\sigma'}(t)] = [a_{\mathbf{k}\sigma}(t), a^\dagger_{\mathbf{k}'\sigma'}(t)]a_{\mathbf{k}'\sigma'}(t) = \delta^3_{\mathbf{k}\mathbf{k}'}\delta_{\sigma\sigma'}a_{\mathbf{k}'\sigma'}(t). \tag{6.64}$$

We can proceed in the same way to evaluate

$$\begin{aligned}[a^\dagger_{\mathbf{k}\sigma}(t), N_{\mathbf{k}'\sigma'}(t)] &= a^\dagger_{\mathbf{k}\sigma}(t)N_{\mathbf{k}'\sigma'}(t) - N_{\mathbf{k}'\sigma'}(t)a^\dagger_{\mathbf{k}\sigma}(t)\\ &= a^\dagger_{\mathbf{k}\sigma}(t)a^\dagger_{\mathbf{k}'\sigma'}(t)a_{\mathbf{k}'\sigma'}(t) - a^\dagger_{\mathbf{k}'\sigma'}(t)a_{\mathbf{k}'\sigma'}(t)a^\dagger_{\mathbf{k}\sigma}(t).\end{aligned} \tag{6.65}$$

Again, we use the commutation relation between $a_{\mathbf{k}\sigma}(t)$ and $a^{\dagger}_{\mathbf{k}\sigma}(t)$ to obtain

$$[a^{\dagger}_{\mathbf{k}\sigma}(t), N_{\mathbf{k}'\sigma'}(t)] = [a^{\dagger}_{\mathbf{k}\sigma}(t), a_{\mathbf{k}'\sigma'}(t)]a^{\dagger}_{\mathbf{k}'\sigma'}(t) = -\delta^{3}_{\mathbf{k}\mathbf{k}'}\delta_{\sigma\sigma'}a^{\dagger}_{\mathbf{k}'\sigma'}(t). \quad (6.66)$$

Equation 6.66 can now be used to evaluate the quantum state $a^{\dagger}_{\mathbf{k}\sigma}|\eta_{\mathbf{k}\sigma}\rangle$: we first consider the action of the photon number operator on this new state,

$$N_{\mathbf{k}\sigma}a^{\dagger}_{\mathbf{k}\sigma}|\eta_{\mathbf{k}\sigma}\rangle = \{a^{\dagger}_{\mathbf{k}\sigma}(t)N_{\mathbf{k}\sigma}(t) - [a^{\dagger}_{\mathbf{k}\sigma}(t), N_{\mathbf{k}\sigma}(t)]\}|\eta_{\mathbf{k}\sigma}\rangle, \quad (6.67)$$

and use Equation 6.66 for $\mathbf{k}' = \mathbf{k}$, and $\sigma' = \sigma$, which yields

$$N_{\mathbf{k}\sigma}a^{\dagger}_{\mathbf{k}\sigma}|\eta_{\mathbf{k}\sigma}\rangle = [a^{\dagger}_{\mathbf{k}\sigma}(t)N_{\mathbf{k}\sigma}(t) + a^{\dagger}_{\mathbf{k}\sigma}(t)]|\eta_{\mathbf{k}\sigma}\rangle = a^{\dagger}_{\mathbf{k}\sigma}(t)[N_{\mathbf{k}\sigma}(t) + 1]|\eta_{\mathbf{k}\sigma}\rangle. \quad (6.68)$$

Since $|\eta_{\mathbf{k}\sigma}\rangle$ is an eigenstate of the photon number operator, Equation 6.68 reduces to

$$N_{\mathbf{k}\sigma}a^{\dagger}_{\mathbf{k}\sigma}|\eta_{\mathbf{k}\sigma}\rangle = (n_{\mathbf{k}\sigma} + 1)a^{\dagger}_{\mathbf{k}\sigma}(t)|\eta_{\mathbf{k}\sigma}\rangle, \quad (6.69)$$

which clearly shows that the quantum state $a^{\dagger}_{\mathbf{k}\sigma}(t)|\eta_{\mathbf{k}\sigma}\rangle$ is also an eigenstate of the photon number operator, with eigenvalue $n_{\mathbf{k}\sigma} + 1$; this can be formally stated as

$$a^{\dagger}_{\mathbf{k}\sigma}(t)|\eta_{\mathbf{k}\sigma}\rangle = \chi_{\mathbf{k}\sigma}|\eta_{\mathbf{k}\sigma} + 1\rangle, \quad (6.70)$$

where $\chi_{\mathbf{k}\sigma}$ is a constant to be determined. Taking the norm of Equation 6.60, we find that

$$\langle\eta_{\mathbf{k}\sigma}|a_{\mathbf{k}\sigma}a^{\dagger}_{\mathbf{k}\sigma}|\eta_{\mathbf{k}\sigma}\rangle = |\chi_{\mathbf{k}\sigma}|^2\langle\eta_{\mathbf{k}\sigma} + 1|\eta_{\mathbf{k}\sigma} + 1\rangle = |\chi_{\mathbf{k}\sigma}|^2. \quad (6.71)$$

The quantum state $a_{\mathbf{k}\sigma}a^{\dagger}_{\mathbf{k}\sigma}|\eta_{\mathbf{k}\sigma}\rangle$ can be evaluated by using the commutator

$$[a_{\mathbf{k}\sigma}, a^{\dagger}_{\mathbf{k}\sigma}] = a_{\mathbf{k}\sigma}a^{\dagger}_{\mathbf{k}\sigma} - a^{\dagger}_{\mathbf{k}\sigma}a_{\mathbf{k}\sigma} = 1; \quad (6.72)$$

we then see that we can rewrite the operator $a_{\mathbf{k}\sigma}a^{\dagger}_{\mathbf{k}\sigma}$ as follows:

$$a_{\mathbf{k}\sigma}a^{\dagger}_{\mathbf{k}\sigma} = [a_{\mathbf{k}\sigma}, a^{\dagger}_{\mathbf{k}\sigma}] + a^{\dagger}_{\mathbf{k}\sigma}a_{\mathbf{k}\sigma} = 1 + a^{\dagger}_{\mathbf{k}\sigma}a_{\mathbf{k}\sigma} = 1 + N_{\mathbf{k}\sigma}, \quad (6.73)$$

and the sought-after quantum states are given by

$$a_{\mathbf{k}\sigma}a^{\dagger}_{\mathbf{k}\sigma}|\eta_{\mathbf{k}\sigma}\rangle = (1 + N_{\mathbf{k}\sigma})|\eta_{\mathbf{k}\sigma}\rangle. \quad (6.74)$$

With this, we can now evaluate the constant $\chi_{\mathbf{k}\sigma}$:

$$|\chi_{\mathbf{k}\sigma}|^2 = \langle\eta_{\mathbf{k}\sigma}|a_{\mathbf{k}\sigma}a^{\dagger}_{\mathbf{k}\sigma}|\eta_{\mathbf{k}\sigma}\rangle = \langle\eta_{\mathbf{k}\sigma}|(1 + N_{\mathbf{k}\sigma})|\eta_{\mathbf{k}\sigma}\rangle, \quad (6.75)$$

and we know that $|\eta_{\mathbf{k}\sigma}\rangle$ is an eigenstate of $N_{\mathbf{k}\sigma}$ as presented in Equation 6.62; therefore, we have

$$|\chi_{\mathbf{k}\sigma}|^2 = (1+n_{\mathbf{k}\sigma})\langle\eta_{\mathbf{k}\sigma}|\eta_{\mathbf{k}\sigma}\rangle = (1+n_{\mathbf{k}\sigma}), \tag{6.76}$$

which yields the important result

$$\chi_{\mathbf{k}\sigma} = \sqrt{1+n_{\mathbf{k}\sigma}}\,e^{i\theta_{\mathbf{k}\sigma}}, \tag{6.77}$$

where $\theta_{\mathbf{k}\sigma}$ is a phase factor.

It is now easy to prove by recurrence that the repeated application of the creation operator increases the photon number by one unit each time. Ignoring the phase factors, which do not contribute to the amplitude of the eigenvalue spectrum, we have

$$[a^{\dagger}_{\mathbf{k}\sigma}(t)]^m|\eta_{\mathbf{k}\sigma}\rangle = \sqrt{(1+n_{\mathbf{k}\sigma})(2+n_{\mathbf{k}\sigma})\cdots(m+n_{\mathbf{k}\sigma})}\,|\eta_{\mathbf{k}\sigma}+m\rangle, \quad m\in\mathbb{N}. \tag{6.78}$$

We now proceed in exactly the same manner with the annihilation operator acting on the eigenstate $|\eta_{\mathbf{k}\sigma}\rangle$, and consider

$$N_{\mathbf{k}\sigma}(t)a_{\mathbf{k}\sigma}(t)|\eta_{\mathbf{k}\sigma}\rangle = \{a_{\mathbf{k}\sigma}(t)N_{\mathbf{k}\sigma}(t) - [a_{\mathbf{k}\sigma}(t), N_{\mathbf{k}\sigma}(t)]\}|\eta_{\mathbf{k}\sigma}\rangle. \tag{6.79}$$

We can now use the commutation rule derived in Equation 6.64, for $\mathbf{k}' = \mathbf{k}$, and $\sigma' = \sigma$, to obtain

$$[a_{\mathbf{k}\sigma}(t), N_{\mathbf{k}\sigma}(t)] = a_{\mathbf{k}\sigma}(t), \tag{6.80}$$

which, in turn, yields

$$\begin{aligned} N_{\mathbf{k}\sigma}(t)a_{\mathbf{k}\sigma}(t)|\eta_{\mathbf{k}\sigma}\rangle &= [a_{\mathbf{k}\sigma}(t)N_{\mathbf{k}\sigma}(t) - a_{\mathbf{k}\sigma}(t)]|\eta_{\mathbf{k}\sigma}\rangle \\ &= a_{\mathbf{k}\sigma}(t)[N_{\mathbf{k}\sigma}(t)-1]|\eta_{\mathbf{k}\sigma}\rangle \\ &= (n_{\mathbf{k}\sigma}-1)a_{\mathbf{k}\sigma}(t)|\eta_{\mathbf{k}\sigma}\rangle. \end{aligned} \tag{6.81}$$

This demonstrates that $a_{\mathbf{k}\sigma}(t)|\eta_{\mathbf{k}\sigma}\rangle$ is an eigenstate of the number operator, with eigenvalue $\eta_{\mathbf{k}\sigma}-1$; thus, we have

$$a_{\mathbf{k}\sigma}(t)|\eta_{\mathbf{k}\sigma}\rangle = \kappa_{\mathbf{k}\sigma}|\eta_{\mathbf{k}\sigma}-1\rangle. \tag{6.82}$$

The normalization constant is evaluated as follows:

$$\begin{aligned} \langle\eta_{\mathbf{k}\sigma}|a^{\dagger}_{\mathbf{k}\sigma}(t)a_{\mathbf{k}\sigma}(t)|\eta_{\mathbf{k}\sigma}\rangle &= |\kappa_{\mathbf{k}\sigma}|^2\langle\eta_{\mathbf{k}\sigma}-1|\eta_{\mathbf{k}\sigma}-1\rangle = |\kappa_{\mathbf{k}\sigma}|^2 \\ &= \langle\eta_{\mathbf{k}\sigma}|N_{\mathbf{k}\sigma}(t)|\eta_{\mathbf{k}\sigma}\rangle = n_{\mathbf{k}\sigma}, \end{aligned} \tag{6.83}$$

and

$$\kappa_{\mathbf{k}\sigma} = \sqrt{n_{\mathbf{k}\sigma}} e^{i\vartheta_{\mathbf{k}\sigma}}. \tag{6.84}$$

Furthermore, the annihilation operator can be applied repeatedly to the eigenstate $|\eta_{\mathbf{k}\sigma}\rangle$, reducing the photon number by one unit each time; this can be summarized as

$$[a_{\mathbf{k}\sigma}(t)]^m |\eta_{\mathbf{k}\sigma}\rangle = \sqrt{n_{\mathbf{k}\sigma}(n_{\mathbf{k}\sigma}-1)\cdots(n_{\mathbf{k}\sigma}-m+1)}\,|\eta_{\mathbf{k}\sigma}-m\rangle, \quad m \in \mathbb{N}. \tag{6.85}$$

Whereas the spectrum generated by the creation operator is unbounded, as shown in Equation 6.68, there is a lower limit to the annihilation spectrum. This is easily understood by considering the fact that the norm of an eigenstate must be definite positive. By examining Equation 6.83, we see that $n_{\mathbf{k}\sigma} = \langle \eta_{\mathbf{k}\sigma}| a^{\dagger}_{\mathbf{k}\sigma}(t) a_{\mathbf{k}\sigma}(t) |\eta_{\mathbf{k}\sigma}\rangle \geq 0$. The only way to guarantee a lower bound to the photon number spectrum is to define the norm of the lowest state as equal to zero:

$$\langle 0| a^{\dagger}_{\mathbf{k}\sigma} a_{\mathbf{k}\sigma} |0\rangle = 0. \tag{6.86}$$

This has an interesting consequence for the spectrum of the Hamiltonian: defining the energy operator for each electromagnetic mode, we have

$$\mathcal{H} = \sum_{\mathbf{k}}\sum_{\sigma} \hbar\omega \left[ N_{\mathbf{k}\sigma}(t) + \frac{1}{2} \right] = \sum_{\mathbf{k}}\sum_{\sigma} H_{\mathbf{k}\sigma}. \tag{6.87}$$

The lowest energy level for a given mode is thus

$$\langle 0| H_{\mathbf{k}\sigma} |0\rangle = \frac{1}{2}\hbar\omega, \tag{6.88}$$

which corresponds to the vacuum energy of the mode under consideration.

## 6.5 Momentum of the Quantized Field

In Section 3.8, we have introduced the four-momentum of the electromagnetic fields and its relation to the Poynting vector. In this section, the same approach is modified to fit within the quantum formalism developed in this chapter.

We begin by considering the form of Maxwell's equation within the photon formalism established in the preceding sections. The temporal evolution of an operator, $O$, is governed by the Heisenberg equation, which states that

$$i\hbar\frac{dO}{dt} = [O,\mathcal{H}], \tag{6.89}$$

where $[O,\mathcal{H}]$ is the commutator between the operator in question and the Hamiltonian.

This is not surprising if we remember the principle of correspondence, which associates the four-gradient operator to the four-momentum: $p_\mu \to i\hbar\partial_\mu$; the time-like component of this equation shows the close relation between the energy, or the Hamiltonian, and the time-derivative operator.

The first result that can be obtained from Equation 6.89 is that the photon number operators are time-independent, since they commute with the Hamiltonian. Furthermore, the evolution of the creation and annihilation operator is governed by

$$i\frac{da_{\mathbf{k}\sigma}}{dt} = \omega a_{\mathbf{k}\sigma}, \quad i\frac{da^\dagger_{\mathbf{k}\sigma}}{dt} = -\omega a^\dagger_{\mathbf{k}\sigma}. \tag{6.90}$$

Using the definition of the electric field and magnetic induction in terms of the generalized coordinates and momenta, as expressed in Equations 6.44 and 6.45, and the relations between the creation and annihilation operators and the conjugate positions and momenta, given in Equations 6.47 and 6.48, we can derive the evolution of the electromagnetic field in the Heisenberg picture. We first have

$$\mathbb{E}(x_\mu) = \frac{1}{\sqrt{a^3}}\sum_{\mathbf{k}}\sum_{\sigma}\sqrt{\frac{\hbar\omega}{2\varepsilon_0}}\left[ia_{\mathbf{k}\sigma}(0)e^{\mathbf{k}}_{\perp\sigma}e^{ik^\mu x_\mu} - ia^\dagger_{\mathbf{k}\sigma}(0)e^{-\mathbf{k}}_{\perp\sigma}e^{-ik^\mu x_\mu}\right], \tag{6.91}$$

for the electric field operator, $\mathbb{E}$, and

$$\mathbb{B}(x_\mu) = \frac{1}{\sqrt{a^3}}\sum_{\mathbf{k}}\sum_{\sigma}\sqrt{\frac{\hbar}{2\omega\varepsilon_0}}\left[ia_{\mathbf{k}\sigma}(0)(\mathbf{k}\times e^{k}_{\perp\sigma})e^{ik^\mu x_\mu} - ia^\dagger_{\mathbf{k}\sigma}(0)(\mathbf{k}\times e^{-\mathbf{k}}_{\perp\sigma})e^{-ik^\mu x_\mu}\right], \tag{6.92}$$

for the magnetic induction operator, $\mathbb{B}$. Using the Heisenberg equation, we then find that

$$\begin{aligned} \nabla\times\mathbb{E}(x_\mu) + \partial_t\mathbb{B}(x_\mu) &= \mathbf{0}, \\ \nabla\times\mathbb{B}(x_\mu) - \frac{1}{c^2}\partial_t\mathbb{E}(x_\mu) &= \mathbf{0}, \end{aligned} \tag{6.93}$$

which are identical in form to Maxwell's equations in vacuum.

The Poynting vector, $\mathbf{S} = \mathbf{E} \times \mathbf{H}$, is closely related to the momentum of the electromagnetic field, as it essentially represents the electromagnetic momentum density; moreover, in vacuum, we have the simple relation, $\mathbf{B} = \mu_0 \mathbf{H}$, between the magnetic induction and field. Therefore, we have

$$\mathbf{G} = \int_{\mathcal{V}} \frac{\mathbf{S}}{c^2} dv = \varepsilon_0 \int_{\mathcal{V}} \mathbf{E} \times \mathbf{B}\, dv = \varepsilon_0 \iiint_{a^3} \mathbf{E} \times \mathbf{B}\, dx\, dy\, dz. \tag{6.94}$$

Here, we have used the fact that for the quantization of the electromagnetic field, a cubic cell of side $a$ is used.

A direct generalization of $\mathbf{G}$ to a Hermitian operator, $\mathbb{G}$, using the substitution of the electric field and magnetic induction by their operator counterparts, is not possible because $\mathbb{E}$ and $\mathbb{B}$ do not commute. Therefore, a slight modification of expression 6.94 is required, and we symmetrize it by writing

$$\mathbb{G} = \frac{\varepsilon_0}{2} \iiint_{a^3} (\mathbb{E} \times \mathbb{B} - \mathbb{B} \times \mathbb{E})\, dx\, dy\, dz. \tag{6.95}$$

The electromagnetic field operators can now be introduced, as expressed in Equations 6.91 and 6.92, and we have

$$\begin{aligned}\mathbb{G} = {} & \frac{\varepsilon_0}{2} \left( \frac{1}{\sqrt{a^3}} \right)^2 \sum_{\mathbf{k}} \sum_{\sigma} \sum_{\bar{\mathbf{k}}} \sum_{\bar{\sigma}} \sqrt{\frac{\hbar \omega}{2\varepsilon_0}} \sqrt{\frac{\hbar}{2\bar{\omega}\varepsilon_0}} \\ & \times \iiint_{a^3} d^3x \left\{ \left[ i a_{\mathbf{k}\sigma} e_{\perp\sigma}^{\mathbf{k}} e^{ik^\mu x_\mu} - i a_{\mathbf{k}\sigma}^\dagger e_{\perp\sigma}^{-\mathbf{k}} e^{-ik^\mu x_\mu} \right] \right. \\ & \left. \times \left[ i a_{\bar{\mathbf{k}}\bar{\sigma}} (\bar{\mathbf{k}} \times e_{\perp\bar{\sigma}}^{\bar{\mathbf{k}}}) e^{i\bar{k}^\mu x_\mu} - i a_{\bar{\mathbf{k}}\bar{\sigma}}^\dagger (\bar{\mathbf{k}} \times e_{\perp\bar{\sigma}}^{-\bar{\mathbf{k}}}) e^{-i\bar{k}^\mu x_\mu} \right] + \dagger \right\}.\end{aligned} \tag{6.96}$$

In Equation 6.96, the symbols $+\dagger$ indicate that the Hermitian conjugate must be added to the original expression. Because of the orthogonality of the photon modes, the volume integrals yield $a^3 \delta^3_{\pm\mathbf{k}\bar{\mathbf{k}}}$, and the sum over the wavenumber is diagonalized. We now have

$$\mathbb{G} = \frac{\hbar}{4} \sum_{\mathbf{k}} \sum_{\sigma} \sum_{\bar{\sigma}} \{ [i a_{\mathbf{k}\sigma} e_{\perp\sigma}^{\mathbf{k}} - i a_{\mathbf{k}\sigma}^\dagger e_{\perp\sigma}^{-\mathbf{k}}] \times [i a_{\mathbf{k}\bar{\sigma}} (\mathbf{k} \times e_{\perp\bar{\sigma}}^{\mathbf{k}}) - i a_{\mathbf{k}\bar{\sigma}}^\dagger (\mathbf{k} \times e_{\perp\bar{\sigma}}^{-\mathbf{k}})] + \dagger \}. \tag{6.97}$$

The double cross-products are

$$\begin{aligned} & e_{\perp\sigma}^{\mathbf{k}} \times (\mathbf{k} \times e_{\perp\bar{\sigma}}^{\mathbf{k}}), \quad e_{\perp\sigma}^{\mathbf{k}} \times (\mathbf{k} \times e_{\perp\bar{\sigma}}^{-\mathbf{k}}) \\ & e_{\perp\sigma}^{-\mathbf{k}} \times (\mathbf{k} \times e_{\perp\bar{\sigma}}^{\mathbf{k}}), \quad e_{\perp\sigma}^{-\mathbf{k}} \times (\mathbf{k} \times e_{\perp\bar{\sigma}}^{-\mathbf{k}}). \end{aligned} \tag{6.98}$$

Using the formula, $\mathbf{a} \times (\mathbf{b} \times \mathbf{c}) = (\mathbf{a} \cdot \mathbf{c})\mathbf{b} - (\mathbf{a} \cdot \mathbf{b})\mathbf{c}$, we have

$$e_{\perp\sigma}^{\mathbf{k}} \times (\mathbf{k} \times e_{\perp\bar{\sigma}}^{\mathbf{k}}) = (e_{\perp\sigma}^{\mathbf{k}} \cdot e_{\perp\bar{\sigma}}^{\mathbf{k}})\mathbf{k} - (e_{\perp\sigma}^{\mathbf{k}} \cdot \mathbf{k})e_{\perp\bar{\sigma}}^{\mathbf{k}} = \delta_{\sigma\bar{\sigma}}\mathbf{k}, \tag{6.99}$$

where the last equality holds because the polarization vector and wavenumber are orthogonal for real photons: $e_{\perp\sigma}^{\mathbf{k}} \cdot \mathbf{k} = 0$. The Kronecker symbol simply reflects the fact that the polarization vectors are orthogonal and of unit length. The double sum over polarization states is now reduced to a single sum, and we finally obtain

$$\mathbb{G} = \frac{\hbar}{2}\sum_{\mathbf{k}}\sum_{\sigma}(a_{\mathbf{k}\sigma}^{\dagger}a_{\mathbf{k}\sigma} + a_{\mathbf{k}\sigma}a_{\mathbf{k}\sigma}^{\dagger})\mathbf{k}, \tag{6.100}$$

where we have used the fact that the terms $a_{\mathbf{k}\sigma}a_{-\mathbf{k}\sigma}$ and $a_{\mathbf{k}\sigma}^{\dagger}a_{-\mathbf{k}\sigma}^{\dagger}$ are antisymmetrical.

This important result can be compared with Equation 6.55 for the energy, and we see that we can group the energy and momentum of the quantized field in a single four-operator:

$$\mathbb{G}_{\mu} = \frac{1}{2}\sum_{\mathbf{k}}\sum_{\sigma}(a_{\mathbf{k}\sigma}^{\dagger}a_{\mathbf{k}\sigma} + a_{\mathbf{k}\sigma}a_{\mathbf{k}\sigma}^{\dagger})\hbar k_{\mu}. \tag{6.101}$$

The transformation properties of this operator under the Lorentz group are identical to that of the four-wavenumber, as the photon number and polarization state must be independent of the observation frame; in other words, photons cannot be created or annihilated, nor can their polarization be flipped by switching reference frame.

## 6.6 Angular Momentum of the Quantized Field

Proceeding in the same manner as for the momentum, we begin with classical theory and define the angular momentum of the electromagnetic field as follows:

$$\mathbf{J}(\mathbf{r}) = \int_{\mathcal{V}}(\mathbf{x} - \mathbf{r}) \times \frac{\mathbf{S}}{c^2}d^3x = \varepsilon_0\int_{\mathcal{V}}(\mathbf{x} - \mathbf{r}) \times [\mathbf{E}(x_{\mu}) \times \mathbf{B}(x_{\mu})]\, d^3x. \tag{6.102}$$

It is immediately seen that we can decompose the angular momentum into two components,

$$\mathbf{J}(\mathbf{r}) = \mathbf{J}(\mathbf{0}) - \mathbf{r} \times \mathbf{G}, \tag{6.103}$$

where **G** is the electromagnetic field momentum discussed in the previous section.

We will return on the classical intrinsic angular momentum of the field, **J**(**0**), in the next section. Here, we follow the procedure outlined in the preceding section; in particular, we symmetrize the operator to guarantee that it is Hermitian:

$$\mathbb{J}(\mathbf{r}) = \frac{\varepsilon_0}{2}\iiint_{a^3}(\mathbf{x}-\mathbf{r})\times(\mathbb{E}\times\mathbb{B}-\mathbb{B}\times\mathbb{E})d^3x = \mathbb{J}(\mathbf{0})-\mathbf{r}\times\mathbb{G}. \tag{6.104}$$

Thus far, we have not shown that this operator does not explicitly depend on time. The fact that for the four-momentum operator we have $\partial_t\mathbb{G}_\mu = 0$ is borne out by the commutation of this operator with the Hamiltonian; therefore, the question reduces to that of the time-independence of the intrinsic angular momentum of the electromagnetic field.

To perform this demonstration, we return to the Heisenberg equations for the field operators, as expressed in Equation 6.93. We then have

$$\begin{aligned}\frac{d\mathbb{J}(\mathbf{0})}{dt} &= \frac{\varepsilon_0}{2}\iiint_{a^3}\mathbf{x}\times\left[\frac{\partial}{\partial t}(\mathbb{E}\times\mathbb{B}-\mathbb{B}\times\mathbb{E})\right]d^3x \\ &= \frac{\varepsilon_0}{2}\iiint_{a^3}\mathbf{x}\times\left[\frac{\partial\mathbb{E}}{\partial t}\times\mathbb{B}+\mathbb{E}\times\frac{\partial\mathbb{B}}{\partial t}-\frac{\partial\mathbb{B}}{\partial t}\times\mathbb{E}-\mathbb{B}\times\frac{\partial\mathbb{E}}{\partial t}\right]d^3x \\ &= \frac{\varepsilon_0}{2}\iiint_{a^3}\mathbf{x}\times[c^2(\nabla\times\mathbb{B})\times\mathbb{B}-\mathbb{E}\times(\nabla\times\mathbb{E}) \\ &\quad +(\nabla\times\mathbb{E})\times\mathbb{E}-\mathbb{B}\times c^2(\nabla\times\mathbb{B})]\,d^3x \\ &= \iiint_{a^3}\mathbf{x}\times[\mu_0^{-1}(\nabla\times\mathbb{B})\times\mathbb{B}+\varepsilon_0(\nabla\times\mathbb{E})\times\mathbb{E}]\,d^3x,\end{aligned} \tag{6.105}$$

where we have used the fact that $\varepsilon_0\mu_0c^2 = 1$, the antisymmetrical nature of the cross-product operator, with $\mathbf{a}\times\mathbf{b} = -\mathbf{b}\times\mathbf{a}$, and the fact that the field operators commute with each other at a fixed time: $[\mathbb{F}_i(\mathbf{x},t),\mathbb{F}_j(\bar{\mathbf{x}},t)] = 0$, where $\mathbb{F}$ represents either $\mathbb{E}$ or $\mathbb{B}$.

Next, we have

$$\begin{aligned}\nabla(\mathbf{a}\cdot\mathbf{b}) &= (\mathbf{a}\cdot\nabla)\mathbf{b}+(\mathbf{b}\cdot\nabla)\mathbf{a}+\mathbf{a}\times(\nabla\times\mathbf{b})+\mathbf{b}\times(\nabla\times\mathbf{a}), \\ \nabla(\mathbf{a}^2) &= 2[(\mathbf{a}\cdot\nabla)\mathbf{a}+\mathbf{a}\times(\nabla\times\mathbf{a})], \\ \mathbf{a}\times(\nabla\times\mathbf{a}) &= \frac{1}{2}\nabla(\mathbf{a}^2)-(\mathbf{a}\cdot\nabla)\mathbf{a}.\end{aligned} \tag{6.106}$$

Identifying $\mathbb{B}\equiv\mathbf{a}$, we find that

$$(\nabla\times\mathbb{B})\times\mathbb{B} = (\mathbb{B}\cdot\nabla)\mathbb{B}-\frac{1}{2}\nabla(\mathbb{B}\cdot\mathbb{B}). \tag{6.107}$$

Taking the cross-product of Equation 6.107 with the position vector, $\mathbf{x}$, we then have

$$\mathbf{x}\times[(\nabla\times\mathbb{B})\times\mathbb{B}] = \mathbf{x}\times\left[(\mathbb{B}\cdot\nabla)\mathbb{B}-\frac{1}{2}\nabla(\mathbb{B}\cdot\mathbb{B})\right]. \tag{6.108}$$

To simplify this expression, we first show that

$$-\frac{1}{2}\mathbf{x}\times[\nabla(\mathbb{B}\cdot\mathbb{B})] = \frac{1}{2}\nabla\times[\mathbf{x}(\mathbb{B}\cdot\mathbb{B})]. \tag{6.109}$$

We start with

$$\nabla\times(f\mathbf{u}) = f(\nabla\times\mathbf{u})-\mathbf{u}\times\nabla f, \tag{6.110}$$

as can be seen by considering a given component of $\nabla\times(f\mathbf{u})$:

$$\begin{aligned}[\nabla\times(f\mathbf{u})]_i &= \partial_j(fu_k)-\partial_k(fu_j)\\ &= u_k\partial_j f + f\partial_j u_k - u_j\partial_k f - f\partial_k u_j\\ &= u_k\partial_j f - u_j\partial_k f + f(\partial_j u_k - \partial_k u_j)\\ &= -(\mathbf{u}\times\nabla f)_i + f(\nabla\times\mathbf{u})_i.\end{aligned} \tag{6.111}$$

We can then use Equation 6.110 to write

$$\begin{aligned}\nabla\times[\mathbf{x}(\mathbb{B}\cdot\mathbb{B})] &= \nabla\times(\mathbb{B}^2\mathbf{x})\\ &= \mathbb{B}^2(\nabla\times\mathbf{x})-\mathbf{x}\times\nabla\mathbb{B}^2\\ &= -\mathbf{x}\times\nabla\mathbb{B}^2,\end{aligned} \tag{6.112}$$

because $\nabla\times\mathbf{x} = \mathbf{0}$.

Multiplying Equation 6.112 by 1/2 yields the desired result.

Next, we consider the term $\mathbf{x}\times[(\mathbb{B}\cdot\nabla)\mathbb{B}]$. In tensorial form, we will show that

$$\mathbf{x}\times[(\mathbb{B}\cdot\nabla)\mathbb{B}] = \nabla\cdot[\mathbb{B}\otimes(\mathbf{x}\times\mathbb{B})], \tag{6.113}$$

where the symbol $\otimes$ denotes the tensorial product:

$$\mathbf{u}\otimes\mathbf{v} = \mathbb{T},\quad T_{ij} = u_i v_j. \tag{6.114}$$

Using Einstein's convention, we can express the three-divergence of the tensorial product as follows:

$$\partial_i T_{ij} = \partial_i(u_i v_j) = v_j(\partial_i u_i)+(u_i\partial_i)v_j, \tag{6.115}$$

or

$$\nabla \cdot (\mathbf{u} \otimes \mathbf{v}) = \mathbf{v}(\nabla \cdot \mathbf{u}) + (\mathbf{u} \cdot \nabla)\mathbf{v}. \tag{6.116}$$

Identifying $\mathbf{u} \equiv \mathbb{B}$, and $\mathbf{v} \equiv \mathbf{x} \times \mathbb{B}$, we then find that

$$\nabla \cdot [\mathbb{B} \otimes (\mathbf{x} \times \mathbb{B})] = (\mathbf{x} \times \mathbb{B})(\nabla \cdot \mathbb{B}) + (\mathbb{B} \cdot \nabla)(\mathbf{x} \times \mathbb{B}). \tag{6.117}$$

As the divergence of the magnetic induction is always equal to zero, we have to show that

$$\nabla \cdot [\mathbb{B} \otimes (\mathbf{x} \times \mathbb{B})] = (\mathbb{B} \cdot \nabla)(\mathbf{x} \times \mathbb{B}) = \mathbf{x} \times [(\mathbb{B} \cdot \nabla)\mathbb{B}]. \tag{6.118}$$

This last equality is easily demonstrated in coordinate form:

$$\begin{aligned}
[(\mathbb{B} \cdot \nabla)(\mathbf{x} \times \mathbb{B})]_i &= (\mathbb{B} \cdot \nabla)[\mathbf{x} \times \mathbb{B}]_i \\
&= (\mathbb{B} \cdot \nabla)(x_j \mathbb{B}_k - x_k \mathbb{B}_j) \\
&= [x_j(\mathbb{B} \cdot \nabla)\mathbb{B}_k - x_k(\mathbb{B} \cdot \nabla)\mathbb{B}_j] + \mathbb{B}_k(\mathbb{B} \cdot \nabla)x_j - \mathbb{B}_j(\mathbb{B} \cdot \nabla)x_k \\
&= \{\mathbf{x} \times [(\mathbb{B} \cdot \nabla)\mathbb{B}]\}_i + \mathbb{B}_k(\mathbb{B} \cdot \nabla)x_j - \mathbb{B}_j(\mathbb{B} \cdot \nabla)x_k \\
&= \{\mathbf{x} \times [(\mathbb{B} \cdot \nabla)\mathbb{B}]\}_i + \mathbb{B}_k(\mathbb{B}_i \partial_i)x_j - \mathbb{B}_j(\mathbb{B}_i \partial_i)x_k \\
&= \{\mathbf{x} \times [(\mathbb{B} \cdot \nabla)\mathbb{B}]\}_i + \mathbb{B}_k \mathbb{B}_i \delta_{ij} - \mathbb{B}_j \mathbb{B}_i \delta_{ik} \\
&= \{\mathbf{x} \times [(\mathbb{B} \cdot \nabla)\mathbb{B}]\}_i + \mathbb{B}_k \mathbb{B}_j - \mathbb{B}_j \mathbb{B}_k \\
&= \{\mathbf{x} \times [(\mathbb{B} \cdot \nabla)\mathbb{B}]\}_i.
\end{aligned} \tag{6.119}$$

Here, we have used the relation $\partial_i x_j = \delta_{ij}$.

Therefore, we have shown that

$$\mathbf{x} \times [(\nabla \times \mathbb{B}) \times \mathbb{B}] = -\frac{1}{2}\nabla \times [\mathbf{x}(\mathbb{B} \cdot \mathbb{B})] - \nabla \cdot [\mathbb{B} \otimes (\mathbf{x} \times \mathbb{B})]. \tag{6.120}$$

A similar relation can be derived for the electric field operator:

$$\mathbf{x} \times [(\nabla \times \mathbb{E}) \times \mathbb{E}] = -\frac{1}{2}\nabla \times [\mathbf{x}(\mathbb{E} \cdot \mathbb{E})] - \nabla \cdot [\mathbb{E} \otimes (\mathbf{x} \times \mathbb{E})]. \tag{6.121}$$

Note, however, that for the electric field, the equivalent to Equation 6.118 works because the absence of charges in vacuum yields $\nabla \cdot \mathbb{E} = 0$.

Having expressed the integrand in Equation 6.105 in terms of a curl and a divergence, we can transform the volume integral into a surface integral by applying the divergence, or Gauss, theorem:

$$\int_{\mathcal{V}} (\nabla \cdot \mathbf{A})\, d^3x = \int_S (\mathbf{n} \cdot \mathbf{A})\, ds, \qquad \int_{\mathcal{V}} (\nabla \times \mathbf{A})\, d^3x = \int_S (\mathbf{n} \times \mathbf{A})\, ds. \tag{6.122}$$

In our case, we find

$$\frac{d\mathbb{J}(\mathbf{0})}{dt} = \frac{1}{2}\int_S [\mathbf{n} \times \mathbf{x}(\mu_0^{-1}\mathbb{B}^2 + \varepsilon_0 \mathbb{E}^2)]\, ds + \int_S \{\mathbf{n} \cdot [\mu_0^{-1}\mathbb{B} \otimes (\mathbf{x} \times \mathbb{B}) + \varepsilon_0 \mathbb{E} \otimes (\mathbf{x} \times \mathbb{E})]\}\, ds, \tag{6.123}$$

where the surface corresponds to that of the $a \times a \times a$ cube used to quantize the free electromagnetic field.

The periodic boundary conditions on each surface of the cube imply that the fields have the same values on opposite surfaces, whereas the vectors $\mathbf{n} \times \mathbf{x}$ are opposite; thus, the first integral vanishes, and Equation 6.123 reduces to

$$\frac{d\mathbb{J}(\mathbf{0})}{dt} = \int_S \{\mathbf{n} \cdot [\mu_0^{-1}\mathbb{B} \otimes (\mathbf{x} \times \mathbb{B}) + \varepsilon_0 \mathbb{E} \otimes (\mathbf{x} \times \mathbb{E})]\}\, ds = \int_S [\mu_0^{-1}(\mathbf{n} \cdot \mathbb{B})(\mathbf{x} \times \mathbb{B}) + \varepsilon_0 (\mathbf{n} \cdot \mathbb{E})(\mathbf{x} \times \mathbb{E})]\, ds. \tag{6.124}$$

The second equality in Equation 6.124 derives from the fact that

$$n_i T_{ij} = n_i(u_i v_j) = (n_i u_i) v_j, \tag{6.125}$$

or

$$\mathbf{n} \cdot \mathbb{T} = \mathbf{n} \cdot (\mathbf{u} \otimes \mathbf{v}) = (\mathbf{n} \cdot \mathbf{u})\mathbf{v}. \tag{6.126}$$

Equation 6.124 can also be written in component form:

$$\frac{d\mathbb{J}_i(\mathbf{0})}{dt} = \int_S [\mu_0^{-1}(n_l \mathbb{B}_l)(x_j \mathbb{B}_k - x_k \mathbb{B}_j) + \varepsilon_0 (n_l \mathbb{E}_l)(x_j \mathbb{E}_k - x_k \mathbb{E}_j)]\, ds = \varepsilon_{ijk} \int_S x_j n_l (\mu_0^{-1} \mathbb{B}_k \mathbb{B}_l + \varepsilon_0 \mathbb{E}_k \mathbb{E}_l)\, ds, \tag{6.127}$$

where $\varepsilon_{ijk}$ is the completely antisymmetrical Levi–Civita tensor.

Using Equations 6.91 and 6.92, the components of the electric field and magnetic induction operators can be expressed in terms of creation and annihilation operators. For the electric field, we have

$$\mathbb{E}_i(x_\mu) = \frac{1}{\sqrt{a^3}} \sum_{\mathbf{k}} \sum_{\sigma} \sqrt{\frac{\hbar\omega}{2\varepsilon_0}} \left[ i a_{\mathbf{k}\sigma}(0) e_{\perp\sigma}^{\mathbf{k}} e^{ik^\mu x_\mu} - i a_{\mathbf{k}\sigma}^{\dagger}(0) e_{\perp\sigma}^{-\mathbf{k}} e^{-ik^\mu x_\mu} \right]_i = \mathbb{E}_i^{+}(x_\mu) + \mathbb{E}_i^{-}(x_\mu), \tag{6.128}$$

where we have introduced

$$\mathbb{E}_i^+(x_\mu) = \frac{1}{\sqrt{a^3}}\sum_{\mathbf{k}}\sum_{\sigma}\sqrt{\frac{\hbar\omega}{2\varepsilon_0}}\left[ia_{\mathbf{k}\sigma}(0)e_{\perp\sigma}^{\mathbf{k}}e^{ik^\mu x_\mu}\right]_i,$$
$$\mathbb{E}_i^-(x_\mu) = \frac{1}{\sqrt{a^3}}\sum_{\mathbf{k}}\sum_{\sigma}\sqrt{\frac{\hbar\omega}{2\varepsilon_0}}\left[-ia_{\mathbf{k}\sigma}^\dagger(0)e_{\perp\sigma}^{-\mathbf{k}}e^{-ik^\mu x_\mu}\right]_i. \tag{6.129}$$

Similar expressions can be defined for the magnetic induction operator.

We can now write

$$\begin{aligned}\mathbb{E}_i(x_\mu)\mathbb{E}_j(x_\mu) &= [\mathbb{E}_i^+(x_\mu)+\mathbb{E}_i^-(x_\mu)][\mathbb{E}_j^+(x_\mu)+\mathbb{E}_j^-(x_\mu)]\\ &= \mathbb{E}_i^+(x_\mu)\mathbb{E}_j^+(x_\mu)+\mathbb{E}_i^+(x_\mu)\mathbb{E}_j^-(x_\mu)\\ &\quad+\mathbb{E}_i^-(x_\mu)\mathbb{E}_j^+(x_\mu)+\mathbb{E}_i^-(x_\mu)\mathbb{E}_j^-(x_\mu).\end{aligned} \tag{6.130}$$

Before considering the order of the operators in Equation 6.130, which is important, a few words about the terminology used here will be useful. The plus and minus signs refer to positive and negative frequencies, respectively; however, the operator labeled with a plus sign corresponds to annihilation operators only, while its counterpart is composed entirely of creation operators. We also note that in the so-called normal or Weyl ordering of operators, creation operators must appear to the left of annihilation operators. In other words, photons must first be created, before annihilation occurs. We see that in Equation 6.130, the only term that is not normally ordered is $\mathbb{E}_i^+\mathbb{E}_j^-$; however, because $\mathbb{E}_i^+$ and $\mathbb{E}_j^-$ commute, the order can be reversed so that Equation 6.130 is now entirely in normal order:

$$\begin{aligned}\mathbb{E}_i(x_\mu)\mathbb{E}_j(x_\mu) &= \mathbb{E}_i^+(x_\mu)\mathbb{E}_j^+(x_\mu)+\mathbb{E}_j^-(x_\mu)\mathbb{E}_i^+(x_\mu)\\ &\quad+\mathbb{E}_i^-(x_\mu)\mathbb{E}_j^+(x_\mu)+\mathbb{E}_i^-(x_\mu)\mathbb{E}_j^-(x_\mu).\end{aligned} \tag{6.131}$$

At this point, a physical argument can be used to reduce the integral remaining in Equation 6.127: one must consider the limit where the cubic volume used to quantize the free electromagnetic field goes to infinity as $a^3$. In this case, the normally ordered operators in Equation 6.131 have a vanishingly small expectation value for photons localized away from the boundary at infinity, and the angular momentum is conserved.

## 6.7 Classical Spin of the Electromagnetic Field

We now return to the intrinsic angular momentum of the electromagnetic field in vacuum, which is related to the spin of photons. In the classical approach to this problem, the magnetic induction is expressed in terms of the scalar potential, so that

$$\mathbf{E}\times\mathbf{B} = \mathbf{E}\times(\nabla\times\mathbf{A}). \tag{6.132}$$

Furthermore, the gauge condition is chosen so that $\phi = 0$ and $\nabla \cdot \mathbf{A} = 0$; we then have

$$\mathbf{E} = -\partial_t\mathbf{A}, \quad \mathbf{B} = \nabla\times\mathbf{A}, \tag{6.133}$$

or, in terms of components,

$$E_i = -\partial_t A_i, \quad B_i = \partial_j A_k - \partial_k A_j = \varepsilon_{ijk}\partial_j A_k. \tag{6.134}$$

The intrinsic angular momentum is then given by

$$\mathbf{J} = \int_{\mathcal{V}} \frac{\mathbf{x}\times\mathbf{S}}{c^2}\, d^3x = \varepsilon_0\int_{\mathcal{V}} \mathbf{x}\times(\mathbf{E}\times\mathbf{B})\, d^3x, \tag{6.135}$$

which translates to

$$\begin{aligned}
J_i &= \varepsilon_0\varepsilon_{ijk}\int_{\mathcal{V}} x_j(\mathbf{E}\times\mathbf{B})_k\, d^3x \\
&= \varepsilon_0\varepsilon_{ijk}\int_{\mathcal{V}} x_j\varepsilon_{klm}E_l B_m\, d^3x \\
&= \varepsilon_0\varepsilon_{ijk}\varepsilon_{klm}\int_{\mathcal{V}} x_j(-\partial_t A_l)B_m\, d^3x \\
&= -\varepsilon_0\varepsilon_{ijk}\varepsilon_{klm}\int_{\mathcal{V}} x_j(\partial_t A_l)\varepsilon_{mnp}(\partial_n A_p)\, d^3x \\
&= -\varepsilon_0\varepsilon_{ijk}\varepsilon_{klm}\varepsilon_{mnp}\int_{\mathcal{V}} x_j(\partial_t A_l)(\partial_n A_p)\, d^3x,
\end{aligned} \tag{6.136}$$

as expressed in terms of coordinates. In order to apply the divergence theorem, as stated in Equation 6.122, the integrand is recast as follows:

$$x_j(\partial_t A_l)(\partial_n A_p) = \partial_n[x_j(\partial_t A_l)A_p] - (\partial_n x_j)(\partial_t A_l)A_p - x_j A_p[\partial_n(\partial_t A_l)]. \tag{6.137}$$

The quantity $\partial_n x_j = \delta_{nj}$, and the surface integral vanishes, so we are left with

$$J_i = \varepsilon_0\varepsilon_{ijk}\varepsilon_{klm}\varepsilon_{mnp}\int_{\mathcal{V}} \{\delta_{nj}(\partial_t A_l)A_p + x_j A_p[\partial_n(\partial_t A_l)]\}\, d^3x. \tag{6.138}$$

We now make use of the contraction formula,

$$\varepsilon_{klm}\varepsilon_{mnp} = \delta_{kn}\delta_{lp} - \delta_{kp}\delta_{ln}, \tag{6.139}$$

to reduce Equation 6.138 to

$$\begin{aligned} J_i &= \varepsilon_0\varepsilon_{ijk}(\delta_{kn}\delta_{lp} - \delta_{kp}\delta_{ln})\int_{\mathcal{V}} \{\delta_{nj}(\partial_t A_l)A_p + x_j A_p[\partial_n(\partial_t A_l)]\}\, d^3x \\ &= \varepsilon_0\varepsilon_{ijk}\int_{\mathcal{V}} (\delta_{kn}\delta_{lp} - \delta_{kp}\delta_{ln})\{\delta_{nj}(\partial_t A_l)A_p + x_j A_p[\partial_n(\partial_t A_l)]\}\, d^3x \\ &= \varepsilon_0\varepsilon_{ijk}\int_{\mathcal{V}} [\delta_{jk}(\partial_t A_l)A_l - (\partial_t A_j)A_k + x_j A_l\partial_k(\partial_t A_l) - x_j A_k\partial_l(\partial_t A_l)]\, d^3x. \end{aligned} \tag{6.140}$$

The term $\partial_l(\partial_t A_l)$ is equal to zero:

$$\partial_l(\partial_t A_l) = \frac{\partial}{\partial x_l}\frac{\partial A_l}{\partial t} = \frac{\partial}{\partial t}\frac{\partial A_l}{\partial x_l} = \partial_t(\nabla\cdot\mathbf{A}) = 0. \tag{6.141}$$

Moreover, we have

$$\varepsilon_{ijk}\delta_{jk} = \varepsilon_{ijj} = 0. \tag{6.142}$$

As a result, Equation 6.140 further simplifies to read

$$J_i = \varepsilon_0\varepsilon_{ijk}\int_{\mathcal{V}} [x_j A_l\partial_k(\partial_t A_l) - (\partial_t A_j)A_k]\, d^3x. \tag{6.143}$$

The first term can be related to the angular momentum operator,

$$\mathbb{L}_i = i\hbar\varepsilon_{ijk}x_j\partial_k, \tag{6.144}$$

and depends on the reference frame because of the term $x_j$. On the other hand, the term

$$\begin{aligned} S_i &= -\varepsilon_0\varepsilon_{ijk}\int_{\mathcal{V}} (\partial_t A_j)A_k\, d^3x, \\ \mathbf{S} &= \varepsilon_0\int_{\mathcal{V}}\left(\mathbf{A}\times\frac{\partial\mathbf{A}}{\partial t}\right)d^3x = \varepsilon_0\int_{\mathcal{V}}(\mathbf{E}\times\mathbf{A})\, d^3x, \end{aligned} \tag{6.145}$$

is frame-independent and corresponds to the spin of the free electromagnetic field. For a circularly polarized plane wave, we have

$$\mathbf{A} = A_0[\hat{x}\cos(k_\mu x^\mu) \pm \hat{y}\sin(k_\mu x^\mu)], \tag{6.146}$$

and

$$\frac{\partial\mathbf{A}}{\partial t} = c\frac{\partial\mathbf{A}}{\partial x_0} = A_0(-ck_0)[-\hat{x}\sin(k_\mu x^\mu) \pm \hat{y}\cos(k_\mu x^\mu)]. \tag{6.147}$$

The corresponding spin density is

$$\begin{aligned}\frac{d\mathbf{S}}{d\mathcal{V}} &= \varepsilon_0\left(\mathbf{A}\times\frac{\partial\mathbf{A}}{\partial t}\right)\\ &= ck_0\varepsilon_0 A_0^2[\hat{x}\cos(k_\mu x^\mu) \pm \hat{y}\sin(k_\mu x^\mu)]\times[\hat{x}\sin(k_\mu x^\mu)\mp\hat{y}\cos(k_\mu x^\mu)]\\ &= ck_0\varepsilon_0 A_0^2\{(\hat{x}\times\hat{y})[\mp\cos^2(k_\mu x^\mu)] + (\hat{y}\times\hat{x})[\pm\sin^2(k_\mu x^\mu)]\}\\ &= \mp\omega\varepsilon_0 A_0^2\hat{z}. \end{aligned} \tag{6.148}$$

## 6.8 Photon Spin

Using Equation 6.145 and the symmetrization technique described for the field momentum operator, we have

$$\mathbb{S} = \frac{\varepsilon_0}{2}\iiint_{a^3}(\mathbb{E}\times\mathbb{A} - \mathbb{A}\times\mathbb{E})\,d^3x, \tag{6.149}$$

and we can replace the electric field and vector potential operators by their expansions, as expressed in Equation 6.91 and

$$\mathbb{A}(x_\mu) = \frac{1}{\sqrt{a^3}}\sum_{\mathbf{k}}\sum_{\sigma}\sqrt{\frac{\hbar}{2\varepsilon_0\omega}}\left[a_{\mathbf{k}\sigma}e_{\perp\sigma}^{\mathbf{k}}e^{ik^\mu x_\mu} + a_{\mathbf{k}\sigma}^{\dagger}e_{\perp\sigma}^{-\mathbf{k}}e^{-ik^\mu x_\mu}\right]. \tag{6.150}$$

With this, we have

$$\begin{aligned}\mathbb{S} = \frac{\hbar}{4a^3}\sum_{\mathbf{k},\sigma}\sum_{\bar{\mathbf{k}},\bar{\sigma}}\sqrt{\frac{\omega}{\bar{\omega}}}\iiint_{a^3}\Big\{&\left[ia_{\mathbf{k}\sigma}e_{\perp\sigma}^{\mathbf{k}}e^{ik^\mu x_\mu} - ia_{\mathbf{k}\sigma}^{\dagger}e_{\perp\sigma}^{-\mathbf{k}}e^{-ik^\mu x_\mu}\right]\\ &\times\left[a_{\bar{\mathbf{k}}\bar{\sigma}}e_{\perp\bar{\sigma}}^{\bar{\mathbf{k}}}e^{i\bar{k}^\mu x_\mu} + a_{\bar{\mathbf{k}}\bar{\sigma}}^{\dagger}e_{\perp\bar{\sigma}}^{-\bar{\mathbf{k}}}e^{-i\bar{k}^\mu x_\mu}\right] + \dagger\Big\}d^3x. \end{aligned} \tag{6.151}$$

The reduction of the volume integral is identical to that used for the field momentum operator, and the commutation relations for the creation and annihilation operators, as expressed in Equation 6.49, can be used to obtain

$$\mathbb{S} = i\hbar\sum_{\mathbf{k},\sigma}\sum_{\bar{\sigma}}\left(a_{\mathbf{k}\bar{\sigma}}^{\dagger}a_{\mathbf{k}\sigma} + \frac{1}{2}\delta_{\sigma\bar{\sigma}}\right)(e_{\perp\sigma}^{\mathbf{k}}\times e_{\perp\bar{\sigma}}^{-\mathbf{k}}). \tag{6.152}$$

Further simplification can be achieved by an appropriate projection basis for the polarization states. In particular, for linear polarization, we have

$$e_{\perp\sigma}^{\mathbf{k}}\times e_{\perp\bar{\sigma}}^{-\mathbf{k}} = \pm\frac{\mathbf{k}}{|\mathbf{k}|}(1-\delta_{\sigma\bar{\sigma}}), \tag{6.153}$$

and we find that

$$\mathbb{S} = i\hbar \sum_{\mathbf{k},\sigma} \sum_{\bar{\sigma}} \left( a_{\mathbf{k}\bar{\sigma}}^{\dagger} a_{\mathbf{k}\sigma} + \frac{1}{2}\delta_{\sigma\bar{\sigma}} \right) \left[ \pm \frac{\mathbf{k}}{|\mathbf{k}|}(1 - \delta_{\sigma\bar{\sigma}}) \right]$$

$$= i \sum_{\mathbf{k}} \frac{\hbar \mathbf{k}}{|\mathbf{k}|} \left( a_{\mathbf{k}\sigma 2}^{\dagger} a_{\mathbf{k}1} - a_{\mathbf{k}\sigma 1}^{\dagger} a_{\mathbf{k}2} \right). \tag{6.154}$$

We recover the classical result derived in Section 6.7: the spin is in the direction of propagation of the wave.

## 6.9 Vacuum Fluctuations

In this section, we give a short overview of the well-known question of vacuum fluctuations of the free quantized electromagnetic field. For considerably more detailed descriptions, we refer the reader to the textbooks by Mandel and Wolf, Loudon, Dirac, and Pauli, which are listed in the references to this chapter.

The key idea is that the lowest energy level for photons, corresponding to the vacuum state, has both a nonzero energy eigenvalue and nonzero fluctuations. In turn, this physical fact is at the origin of some of the divergence problems encountered in QED before the renormalization program was completed by Feynman, Schwinger, Tomonaga, and Dyson.

In the following discussion, the vacuum state will be labeled by the bra and kets $\langle 0|$ and $|0\rangle$, respectively. As discussed in Section 6.4, the expectation value of the creation and annihilation operators is zero for the vacuum state:

$$\langle 0| a_{\mathbf{k}\sigma}^{\dagger} = 0 = a_{\mathbf{k}\sigma}|0\rangle; \tag{6.155}$$

on the other hand, the energy eigenvalue of the vacuum state is

$$\langle 0|H_{\mathbf{k}\sigma}|0\rangle = \frac{1}{2}\hbar\omega, \tag{6.156}$$

as shown in Section 6.4.

We now turn our attention to the expectation value of the field operator in the vacuum state. As discussed earlier, any given field operator can be written in terms of creation and annihilation operators, with

$$\mathbb{F}(x_\mu) = \frac{1}{\sqrt{a^3}} \sum_{\mathbf{k}} \sum_{\sigma} \left[ \mathcal{F}(k_\mu) a_{\mathbf{k}\sigma} e_{\perp\sigma}^{\mathbf{k}} e^{ik^\mu x_\mu} + \dagger \right], \tag{6.157}$$

which reduces to

$$\mathbb{F}(x_\mu) = \frac{1}{\sqrt{a^3}}\sum_{\mathbf{k}}\sum_{\sigma}\left[\mathcal{F}(\omega)a_{\mathbf{k}\sigma}e^{\mathbf{k}}_{\perp\sigma}e^{ik^\mu x_\mu} + \dagger\right], \tag{6.158}$$

because of the mass-shell constraint $k_\mu k^\mu = \mathbf{k}^2 - (\omega/c)^2 = 0$ for photons.

We can now use Equations 6.155 and 6.158 to show that the vacuum expectation value for any field operator $\mathbb{F}$ is zero:

$$\langle 0|\mathbb{F}(x_\mu)|0\rangle = 0 \tag{6.159}$$

The vacuum fluctuations for the field operator $\mathbb{F}$ are defined as

$$\Delta\langle\mathbb{F}\rangle = \sqrt{|\langle 0|\mathbb{F}^2(x_\mu)|0\rangle - [\langle 0|\mathbb{F}(x_\mu)|0\rangle]^2|} = \sqrt{\langle 0|\mathbb{F}^2(x_\mu)|0\rangle}. \tag{6.160}$$

Therefore, we need to evaluate the vacuum expectation for the square of the field operator, namely, $\langle 0|\mathbb{F}^2(x_\mu)|0\rangle$. This quantity will involve four different expectations values:

$$\begin{aligned}\langle 0|a_{\mathbf{k}\sigma}a_{\bar{\mathbf{k}}\bar{\sigma}}|0\rangle &= 0,\\ \langle 0|a_{\mathbf{k}\sigma}a^\dagger_{\bar{\mathbf{k}}\bar{\sigma}}|0\rangle &\neq 0,\\ \langle 0|a^\dagger_{\mathbf{k}\sigma}a_{\bar{\mathbf{k}}\bar{\sigma}}|0\rangle &= 0,\\ \langle 0|a^\dagger_{\mathbf{k}\sigma}a^\dagger_{\bar{\mathbf{k}}\bar{\sigma}}|0\rangle &= 0.\end{aligned} \tag{6.161}$$

Using the nonzero expectation value, we can formally write

$$\langle 0|\mathbb{F}^2(x_\mu)|0\rangle = \frac{1}{a^3}\sum_{\mathbf{k},\sigma}\sum_{\bar{\mathbf{k}},\bar{\sigma}}\mathcal{F}(\omega)\mathcal{F}^*(\bar{\omega})\langle 0|a_{\mathbf{k}\sigma}a^\dagger_{\bar{\mathbf{k}}\bar{\sigma}}|0\rangle(e^{\mathbf{k}}_{\perp\sigma}\cdot e^{\bar{\mathbf{k}}}_{\perp\bar{\sigma}})e^{i(k^\mu-\bar{k}^\mu)x_\mu}. \tag{6.162}$$

The only nonzero value can be derived explicitly by using the commutation relation in Equation 6.57:

$$a_{\mathbf{k}\sigma}a^\dagger_{\bar{\mathbf{k}}\bar{\sigma}} = a^\dagger_{\mathbf{k}\sigma}a_{\bar{\mathbf{k}}\bar{\sigma}} + \delta^3_{\mathbf{k}\bar{\mathbf{k}}}\delta_{\sigma\bar{\sigma}}. \tag{6.163}$$

We then find that

$$\begin{aligned}\langle 0|a_{\mathbf{k}\sigma}a^\dagger_{\bar{\mathbf{k}}\bar{\sigma}}|0\rangle &= \langle 0|a^\dagger_{\mathbf{k}\sigma}a_{\bar{\mathbf{k}}\bar{\sigma}} + \delta^3_{\mathbf{k}\bar{\mathbf{k}}}\delta_{\sigma\bar{\sigma}}|0\rangle\\ &= \langle 0|a^\dagger_{\mathbf{k}\sigma}a_{\bar{\mathbf{k}}\bar{\sigma}}|0\rangle + \langle 0|\delta^3_{\mathbf{k}\bar{\mathbf{k}}}\delta_{\sigma\bar{\sigma}}|0\rangle\\ &= \delta^3_{\mathbf{k}\bar{\mathbf{k}}}\delta_{\sigma\bar{\sigma}}\langle 0|0\rangle\\ &= \delta^3_{\mathbf{k}\bar{\mathbf{k}}}\delta_{\sigma\bar{\sigma}}.\end{aligned} \tag{6.164}$$

Using this last result into Equation 6.162, we have

$$\begin{aligned}\langle 0|\mathbb{F}^2(x_\mu)|0\rangle &= \frac{1}{a^3}\sum_{\mathbf{k},\sigma}\sum_{\bar{\mathbf{k}},\bar{\sigma}} \mathcal{F}(\omega)\,\mathcal{F}^*(\bar{\omega})\,\delta^3_{\mathbf{k}\bar{\mathbf{k}}}\,\delta_{\sigma\bar{\sigma}}\,(e^{\mathbf{k}}_{\perp\sigma}\cdot e^{-\bar{\mathbf{k}}}_{\perp\bar{\sigma}})\,e^{i(k^\mu-\bar{k}^\mu)x_\mu}\\ &= \frac{1}{a^3}\sum_{\mathbf{k},\sigma}\mathcal{F}(\omega)\,\mathcal{F}^*(\omega)\\ &= \frac{1}{a^3}\sum_{\mathbf{k},\sigma}|\mathcal{F}(\omega)|^2\\ &= \frac{2}{a^3}\sum_{\mathbf{k}}|\mathcal{F}(\omega)|^2. \end{aligned}\tag{6.165}$$

Thus, we obtain the sought-after value for the vacuum fluctuations:

$$\Delta\langle\mathbb{F}\rangle = \sqrt{\frac{2}{a^3}\sum_{\mathbf{k}}|\mathcal{F}(\omega)|^2}. \tag{6.166}$$

If we use an infinite series of modes, the series diverges; this is the purely quantum divergence of the vacuum encountered in QED. Introducing a high cutoff frequency allows one to effectively truncate the series, thus yielding a finite result. The other QED divergences include the vacuum polarization problem and the classical Coulomb divergence of the field energy for a point charge; the latter will be discussed in Chapter 10.

## 6.10 The Einstein–Podolsky–Rosen Paradox

The Einstein–Podolsky–Rosen, or EPR, paradox is related to the question of locality in quantum mechanics, as described mathematically by Bell's inequalities. An excellent presentation of this problem is given by Mandel and Wolf, as referenced in the bibliography, and we will restrict our discussion to a basic outline of the ideas underlying the EPR paradox.

The basic idea behind the EPR paradox can be described as follows: for initially correlated, or entangled, two-particle states, such as photons produced in a cascade with $\Delta J = 0$, the measurement of a variable on the first particle completely predetermines the result of the measurement of the corresponding variable on the second particle, independent of the space–time distance at which the particles are located at the time of measurement. Quantum mechanically, the entangled state can be represented by

$$|\zeta\rangle = \frac{1}{\sqrt{2}}(|\xi\rangle_1|\psi\rangle_2 - |\psi\rangle_1|\xi\rangle_2), \tag{6.167}$$

where the numbers 1 and 2 refer to each particle, and where all wavefunctions are normalized.

The difficulty with the EPR paradox arises because of the nonlocal character of the entangled state, which allows the observer to know the state of the second particle without a direct measurement or even the time for the measurement performed on the first particle to perturb the second particle.

A classical analog can be constructed by considering an initial correlated state. For example, one could slice a coin along its middle plane, so that the head and tail are separated. Each half of the original coin can be placed in a box, which can then be transported over a great distance, say one to New York and one to Paris. When one experimentalist opens one of the boxes and looks at the side of the coin that is enclosed, she also immediately knows the "state" of the coin on the other side of the Atlantic. The initial correlation remains, allowing for an instantaneous correlation. Note, however, that no information is transported faster than light in the process. Furthermore, we should strongly emphasize the fact that there are fundamental differences between the classical and quantum cases. In particular, in the quantum experiment, the polarization can be measured against an arbitrary reference axis, provided that this axis is contained in a plane perpendicular to the direction of propagation of the photon.

The EPR paradox can be illustrated by considering an experiment with correlated photons produced in a cascade decay. This example is useful because it closely approximates experiments performed by Aspect and his group, and Mandel and Ou, which clearly demonstrated a violation of Bell's inequality, thus ruling out any hidden-variable interpretation of the EPR paradox. In this analysis, we closely follow the presentation of Mandel and Wolf and strongly encourage the reader to consult their book on optical coherence and quantum optics, referenced in the bibliography, for an in-depth discussion of the subject. Two photons polarized orthogonally are considered:

$$|\zeta\rangle = \frac{1}{\sqrt{2}}(|1_{1x}, 0_{1y}, 0_{2x}, 1_{2y}\rangle - |0_{1x}, 1_{1y}, 1_{2x}, 0_{2y}\rangle). \tag{6.168}$$

Here, the $|1_{1x}, 0_{1y}, 0_{2x}, 1_{2y}\rangle$ state corresponds to the first photon polarized along the $x$-axis, while the second photon is necessarily polarized along the perpendicular axis, the $y$-axis, while the $|0_{1x}, 1_{1y}, 1_{2x}, 0_{2y}\rangle$ state describes the first photon being polarized along the $y$-axis, and the second one, necessarily, parallel to the $x$-axis. The $z$-axis corresponds to the direction of propagation of photon 1, while photon 2 propagates in the opposite direction. In the entangled state $|\zeta\rangle$, the direction of polarization of an individual photon is unknown, but their polarization states are 100% coupled.

A linear polarizer is inserted along the path of each photon, characterized by the angle $\theta_{1,2}$ with respect to the $x$-axis, and a detector is positioned after each polarizer; the quantum efficiency of each detector is $\eta_{1,2}$. We now compute the probabilities $P_i(\theta_i)$ of detection of each photon, when the respective polarizers are set at the angles $\theta_i$:

$$
\begin{aligned}
P_1(\theta_1) &= \eta_1 \langle \zeta | \hat{a}_1^{\dagger} \hat{a}_1 | \zeta \rangle = \frac{\eta_1}{2}, \\
P_2(\theta_2) &= \eta_2 \langle \zeta | \hat{a}_2^{\dagger} \hat{a}_2 | \zeta \rangle = \frac{\eta_2}{2}.
\end{aligned}
\tag{6.169}
$$

Here, we have introduced the field dynamical variables,

$$
\hat{a}_i = \hat{a}_{ix} \cos\theta_i + \hat{a}_{iy} \sin\theta_i, \tag{6.170}
$$

which simply reflect the effect of the polarizers on the photons. The results obtained in Equation 6.169 are readily understood: the probability for each randomly polarized photon to pass through the corresponding polarizer is $\frac{1}{2}$, and the quantum efficiency of the detectors reduces the probability of detection, as reflected in Equation 6.169; these probabilities are independent of the polarizers' settings. Let us now consider the more interesting joint detection probability,

$$
\begin{aligned}
P_{12}(\theta_1,\theta_2) &= \eta_1 \eta_2 \langle \zeta | \hat{a}_1^{\dagger} \hat{a}_2^{\dagger} \hat{a}_2 \hat{a}_1 | \zeta \rangle \\
&= \frac{1}{2} \eta_1 \eta_2 (\sin^2\theta_1 \cos^2\theta_2 + \cos^2\theta_1 \sin^2\theta_2 - 2\sin\theta_1 \cos\theta_2 \cos\theta_1 \sin\theta_2) \\
&= \frac{1}{2} \eta_1 \eta_2 \sin^2(\theta_1 - \theta_2),
\end{aligned}
\tag{6.171}
$$

which clearly introduces a correlation between the polarizers.

Using random variable analysis, the conditional detection probability of the second photon can be expressed in terms of the probability detection of the first photon:

$$
\frac{P_{12}(\theta_1,\theta_2)}{P_1(\theta_1)} = \eta_2 \sin^2(\theta_1 - \theta_2). \tag{6.172}
$$

If the quantum efficiency of the detector for photon 2 is close to 100% and the polarizers are set orthogonally, so that $\theta_1 - \theta_2 = \pm\pi/2$, Equation 6.172 shows that the conditional detection probability of the second photon approaches 100%: the photons are clearly polarized orthogonally. Furthermore, the polarization state of the second photon can be known by determining that of photon 1, without a measurement on photon 2, and instantaneously, independent of the separation between the two photons at the time of the measurement on the first photon. However, as shown by Mandel and Wolf, causality is preserved, as the polarization axis of the first photon is not set by the direction of polarizer 1; it merely serves as a reference axis to measure a random variable.

In closing, we outline the derivation of Bell's inequality, where two observables, $A$ and $B$, parameterized by the variables $\alpha$ and $\beta$, are considered. Moreover, measurements of $A$ and $B$ can only yield two possible values, say 0 or 1. For example, in the case of polarizers the photons can either be transmitted or absorbed, and the parameter is the angle of the polarizer, $\theta$.

Bell considers the average correlation between the observables:

$$C(\alpha,\beta) = \langle A(\alpha)B(\beta)\rangle. \tag{6.173}$$

The key idea behind Bell's derivation is to test the validity of so-called "hidden variable" theories; therefore, the correlation in Equation 6.173 is explicitly expressed as

$$C(\alpha,\beta) = \int A(\alpha,\eta)B(\alpha,\eta)\rho(\eta)\,d\eta, \tag{6.174}$$

where $\eta$ represents the hidden variable, while $\rho(\eta)$ is its normalized probability density:

$$\int \rho(\eta)\,d\eta = 1. \tag{6.175}$$

Locality is implicit in Equation 6.174, in the sense that $A$ does not depend on $\beta$, while $B$ does not depend on $\alpha$.

We now examine the quantities $|C(\alpha, \beta) - C(\alpha, \beta')|$ and $|C(\alpha', \beta) + C(\alpha', \beta')|$. We first have

$$\begin{aligned} |C(\alpha, \beta) - C(\alpha, \beta')| &\leq \int |A(\alpha, \eta)[B(\beta, \eta) - B(\beta', \eta)]|\rho(\eta)\,d\eta, \\ &\leq \int |B(\beta, \eta) - B(\beta', \eta)|\rho(\eta)\,d\eta, \end{aligned} \tag{6.176}$$

since, by definition, $|A(\alpha, \eta)| = 1$. For the same reason, namely $|A(\alpha', \eta)| = 1$, we also have

$$\begin{aligned} |C(\alpha', \beta) + C(\alpha', \beta')| &\leq \int |A(\alpha', \eta)[B(\beta, \eta) + B(\beta', \eta)]|\rho(\eta)d\eta, \\ &\leq \int |B(\beta, \eta) + B(\beta', \eta)|\rho(\eta)\,d\eta. \end{aligned} \tag{6.177}$$

Adding Equations 6.176 and 6.177 together, we can thus write

$$\begin{aligned} &|C(\alpha, \beta) - C(\alpha, \beta')| + |C(\alpha', \beta) + C(\alpha', \beta')| \\ &\quad \leq \int [|B(\beta, \eta) - B(\beta', \eta)| + |B(\beta, \eta) + B(\beta', \eta)|\,]\rho(\eta)\,d\eta. \end{aligned} \tag{6.178}$$

Furthermore, since $|B(\beta, \eta)| = |B(\beta', \eta)| = 1$, we have

$$|B(\beta, \eta) - B(\beta', \eta)| + |B(\beta, \eta) + B(\beta', \eta)| = 2. \tag{6.179}$$

Using the result given in Equation 6.179, together with the normalization of the hidden variable probability density, described in Equation 6.175, we obtain Bell's inequality:

$$|C(\alpha,\beta)-C(\alpha,\beta')|+|C(\alpha',\beta)+C(\alpha',\beta')|\leq 2. \tag{6.180}$$

Note that the hidden variable has now disappeared, by virtue of the integration. Bell's inequality provides a test of the correlation between dichotomic observables, such as those discussed in the case of correlated photons being analyzed by linear polarizers. Using the transmission probability through a polarizer, which can be obtained by setting the quantum efficiencies equal to unity ($\eta_1=\eta_2=1$), we have

$$\begin{aligned} C(\theta_1,\theta_2) &= \frac{1}{2}[\sin^2(\theta_1-\theta_2)+\sin^2(\theta_1-\theta_2)-\cos^2(\theta_1-\theta_2)-\cos^2(\theta_1-\theta_2)] \\ &= \sin^2(\theta_1-\theta_2)-\cos^2(\theta_1-\theta_2) \\ &= -\cos[2(\theta_1-\theta_2)]. \end{aligned} \tag{6.181}$$

It is clear that for a careful choice of polarizing angles, we can set up a violation of Bell's inequality; in other words, we can find a set of angles, $\theta_1$, $\theta_2$, $\theta_1'$, $\theta_2'$, such that

$$\begin{aligned} &|C(\theta_1,\theta_2)-C(\theta_1,\theta_2')|+|C(\theta_1',\theta_2)+C(\theta_1',\theta_2')| \\ &\quad=|-\cos[2(\theta_1-\theta_2)]+\cos[2(\theta_1-\theta_2')]|+|-\cos[2(\theta_1'-\theta_2)]-\cos[2(\theta_1'-\theta_2')]| \\ &\quad=|-\cos[2(\theta_1-\theta_2)]+\cos[2(\theta_1-\theta_2')]|+|\cos[2(\theta_1'-\theta_2)]+\cos[2(\theta_1'-\theta_2')]| \\ &\quad>2. \end{aligned} \tag{6.182}$$

For example, the angles $\theta_1=0$, $\theta_2=\pi/8$, $\theta_2'=3\pi/8$, and $\theta_1'=\pi/4$, yield $2\sqrt{2}$, in clear violation of Bell's inequality. Such experiments have been performed, ruling out the possibility of local, hidden-variable theories for quantum mechanics.

## 6.11 Squeezed States

This interesting example of quantum effects in nonlinear optics is closely related to other subjects, including quantum nondemolition (QND) measurements, optical phase conjugation and degenerate four-wave mixing, as well as parametric down-conversion and the production of entangled quantum states. In the same general field of modern physics, we find topics such as cavity QED, the slowing down of light to extremely low speeds and its storage as a coherent spin state in atoms, as well as Bose–Einstein condensation. A number of recent publications and review articles on these very interesting developments are referenced in the bibliography.

The basic idea behind squeezed states is that conjugate variables for the vacuum state satisfy Heisenberg's uncertainty principle,

$$\Delta p \Delta q \geq \hbar, \tag{6.183}$$

in a very specific way. For properly normalized variables, we have $\Delta p = \Delta q = \sqrt{\hbar}$. Within this context, a squeezed state will have one of its conjugate variables below the vacuum level, while the other variable will be above the vacuum level, to properly satisfy the uncertainty principle. In phase space, this corresponds to a circle of surface $\hbar$ for the vacuum state, and to an ellipse, with the same surface, for the squeezed state.

More precisely, if we consider the normalized operators

$$\begin{aligned} \hat{q} &= \hat{a}^{\dagger} + \hat{a}, \\ \hat{p} &= i(\hat{a}^{\dagger} - \hat{a}), \end{aligned} \tag{6.184}$$

which are defined in terms of the creation and annihilation operators and correspond to generalized coordinates and momenta, we find that their commutator is

$$[\hat{q}, \hat{p}] = 2i, \tag{6.185}$$

while the corresponding Heisenberg uncertainty relation takes the form

$$\sqrt{\langle \Delta\hat{q}^2 \rangle \langle \Delta\hat{p}^2 \rangle} \geq 1. \tag{6.186}$$

With these definitions, a squeezed state can be constructed mathematically by introducing a phase angle, $\theta$, and introducing the new operators

$$\begin{aligned} \widehat{q} &= \hat{a}^{\dagger} e^{i\theta} + \hat{a} e^{-i\theta}, \\ \widehat{p} &= i(\hat{a}^{\dagger} e^{i\theta} - \hat{a} e^{-i\theta}). \end{aligned} \tag{6.187}$$

It is then easily seen that

$$\begin{aligned} \Delta\widehat{q} &= \Delta\hat{q}\cos\theta + \Delta\hat{p}\sin\theta, \\ \Delta\widehat{p} &= -\Delta\hat{q}\sin\theta + \Delta\hat{p}\cos\theta, \end{aligned} \tag{6.188}$$

which also lead to Heisenberg's uncertainty relation,

$$\sqrt{\langle \Delta \hat{q}^2 \rangle \langle \Delta \hat{p}^2 \rangle} \geq 1; \tag{6.189}$$

however, it is clear that for some values of $\theta$, we can have $\langle \Delta \hat{q}^2 \rangle < 1$, $\langle \Delta \hat{p}^2 \rangle > 1$, or $\langle \Delta \hat{q}^2 \rangle > 1$, $\langle \Delta \hat{p}^2 \rangle < 1$, thus constituting a squeezed state.

Degenerate four-wave mixing gives rise to squeezed states, as do other nonlinear interactions. This particular interaction results from the mixing of two input or pump waves in a $\chi^{(3)}$ medium, producing two output signals: a signal and an idler wave. The pump waves can typically be treated classically, while the output signals exhibit quantum mechanical features, including squeezing and optical phase conjugation.

Closely related and of considerable interest is the concept of quantum nondemolition (QND) measurements, where a particular variable, $\hat{\Theta}$, is created experimentally, which obeys the commutation relation

$$[\hat{\Theta}(t_1), \hat{\Theta}(t_2)] = 0, \quad \forall t_1, t_2. \tag{6.190}$$

This means that measurements of this particular variable at different times will yield the same result: the variable is not influenced by the measurement process. A good example of an experimental situation producing a QND variable is the Kerr effect, which also involves a $\chi^{(3)}$ nonlinearity.

For detailed discussions of these concepts, we refer the reader to Mandel and Wolf, as well as the articles listed in the bibliography.

## 6.12 Casimir Effect

The quantum vacuum fluctuations described in Section 6.9 give rise to an interesting phenomenon, the Casimir effect. In the presence of boundary conditions, the mode structure of the vacuum excitations is modified, as a discrete spectrum emerges, with a cutoff frequency, instead of the continuum of free space; in turn, this produces a differential radiation pressure, which is manifested as a force on the boundary surface.

The simple case of two parallel conducting plates is considered here, for the sake of illustration. If the surface of the plates is much larger than their separation ($\sqrt{S} \gg \Delta z$), we can consider that the minimum axial wavenumber will be given by

$$k\Delta z = \pi. \tag{6.191}$$

We can then compute the vacuum energy in the cavity:

$$\begin{aligned} W &= \sum_{|\mathbf{k}|\geq\pi/\Delta z,\sigma} \frac{1}{2}\hbar\omega \\ &\simeq \Delta z S \int_{\pi/\Delta z}^{k*} (\hbar c k) k^2 dk \\ &= \hbar c \Delta z S \left[\frac{k^4}{4}\right]_{\pi/\Delta z}^{k*} \\ &= \hbar c \Delta z S \frac{k^{*4}}{4} - \hbar c S \frac{\pi^4}{4} \frac{1}{\Delta z^3} \\ &= W^* - W_0. \end{aligned} \tag{6.192}$$

Here, we have approximated the sum by an integral and introduced a high wavenumber cutoff, $k^*$, to avoid divergences. The Casimir pressure, $P$, is given by deriving the work of the force on the plates required to balance the variation of the energy between the plates. We have

$$PSd\Delta z + dW_0 = Fd\Delta z + dW_0 = 0, \tag{6.193}$$

which yields

$$P = -\frac{1}{S}\frac{dW_0}{d\Delta z} = \frac{\hbar c \pi^4}{\Delta z^4}. \tag{6.194}$$

Although the numerical factor is wrong, the scaling of the force with the plate separation is correct and has been measured experimentally. We also note that, depending on the type of boundary condition, for example conductor or dielectric, the pressure can be positive or negative. Furthermore, the exact scaling of the Casimir force is related to the dimensionality of space–time, as probed by the quantum vacuum modes. Finally, it has been speculated that this type of effect can give rise to so-called "false vacuum" states, with negative energy densities giving rise to a cosmological constant. It has also been proposed by Thorne and co-authors that stable wormholes and time machines could be built from such false vacua.

## 6.13 Reflection of Plane Waves in Rindler Space

Most of the text and derivations in this section were produced by J. R. Van Meter.

### 6.13.1 Background

In recent years, much attention has been given to the interaction of uniformly accelerating systems with quantum fields, particularly with regards to the thermal Fulling–Davies–Unruh radiation. In contrast, relatively little attention has been given to the interaction of uniformly accelerating systems with classical fields. However, the latter domain seems deserving of study for several reasons. First, this subject represents physics fundamental to both classical electrodynamics and general relativity. For example, it is instructive to study a uniformly accelerating charged particle in a classical context to understand how a radiation field in inertial coordinates can appear as a static field in accelerated coordinates, as well as to explore the approximate behavior of the Coulomb field in Schwarzschild space–time. More generally, because of the mathematical similarity between Rindler and Kruskal coordinates, any result obtained for a uniformly accelerated system may be extendable, at least qualitatively, to a corresponding system in the vicinity of a black hole horizon (as already demonstrated by the deep parallels between Fulling–Davies–Unruh radiation and Hawking radiation).

Another motivation for studying the interaction of uniformly accelerated systems with classical fields is that such analyses might shed some light on corresponding problems in quantum field theory. Boyer's program of approximating quantum electrodynamics with the semiclassical model of stochastic electrodynamics (SED) is noteworthy in this context. In the methodology of SED, the quantum electrodynamical vacuum is approximated by an infinite sum over momenta of plane waves, each with a random phase and an infinitesimal amplitude calculated so as to give a total energy per plane wave of $\frac{1}{2}\hbar\omega$. This model has proven very interesting, as one can match quantum electrodynamical results when calculating the Casimir effect for various boundary configurations. It appears that this model may also be used to derive the thermal effects on a system accelerating uniformly through vacuum, in agreement with quantum field theory.

Of particular interest here is the question of whether Fulling–Davies–Unruh radiation can be backscattered into an inertial laboratory frame, and the possibility of addressing this issue within the framework of semiclassical vacuum fluctuations in Rindler space–time. Various proposals have been put forth for laboratory measurements of backscattered Fulling–Davies–Unruh radiation, including that of Tajima and Chen utilizing an ultrahigh intensity laser to strongly accelerate electrons. Despite some controversy and slight confusion in the literature, the emerging consensus amongst quantum field theorists seems to be that a uniformly accelerating system will not measurably reradiate Fulling–Davies–Unruh radiation into an inertial frame. However, these studies only considered scalar vacuum fields; whether this null radiation result holds for the case of the electromagnetic tensor field has yet to be demonstrated theoretically or experimentally. Whether a uniformly accelerating system will reradiate Fulling–Davies–Unruh radiation into the inertial lab frame thus remains an open question of modern physics. The problem explored in this section might prove germane to the issue.

The present discussion considers the interaction of a uniformly accelerating, perfectly conducting plane mirror with a plane wave at normal incidence. In this regard, the present study is self-contained and represents an original contribution to fundamental classical electrodynamics, particularly by providing physical insight into the relationship between Rindler and Lorentz transformations. This work is especially motivated by its potential relevance to the case of a uniformly accelerating mirror interacting with a quantum field in the vacuum state.

The pertinence of this analysis to the problem of an accelerating mirror interacting with the quantum vacuum may be understood within the stochastic electrodynamical framework as follows. In this model, each virtual photon plane wave incident on the mirror will give rise to a reflected wave that might or might not interfere significantly with the original incident wave. However, each pair of incident/reflected waves will not interfere significantly with any other wave, because of the relative randomization of phases that characterizes the stochastic electrodynamical approach. Thus, in computing the total spectrum, the waves will add incoherently, with the possible exception of each incident wave with its corresponding reflected wave. For the purpose of predicting the qualitative character of the spectrum, it should therefore suffice to consider only an individual incident wave and its reflected wave. The simplest case of normal incidence is the most natural starting point for such an inquiry.

The incident and reflected fields are first transformed to Rindler coordinates and the boundary condition imposed by the mirror, now fixed at a stationary position in Rindler space, is found to determine the reflected wave function. The reflected wave is then expressed in Minkowski coordinates, where its physical meaning is more readily interpreted. To further explicate the physics involved, an alternative means of solving for the reflected wave is presented, which utilizes the Lorentz transform as well as a simple strategy for handling retardation that exploits the unique geometric properties of this problem. Both the case where the mirror accelerates uniformly for all time and the case where the mirror is initially at rest and starts accelerating at $t = 0$ are considered in this section. Finally, some implications of these results are discussed.

### 6.13.2 Derivation of the Reflected Wave Using the Rindler Transform

The problem outlined in the introduction can be summarized more precisely as follows. A mirror moves with uniform proper acceleration such that

$$a_\nu a^\nu = \frac{du_\nu}{d\tau}\frac{du^\nu}{d\tau} = \left(\frac{d^2\mathbf{x}}{d\tau^2}\right)^2 - \left(\frac{d^2t}{d\tau^2}\right)^2 \equiv a^2, \tag{6.195}$$

where $a$ is a constant, and where $\tau$ is the proper time along the mirror's world line $x_\nu(\tau)$, $u_\nu = dx_\nu/d\tau$ is the mirror four-velocity, and $a_\nu$ is its four-acceleration; note that units are normalized so that the speed of light is equal to 1.

We are considering a one-dimensional problem, since the incident and reflected electromagnetic radiation are plane waves at normal incidence; thus Equation 6.195 reduces to

$$a_\nu a^\nu = \left(\frac{\partial^2 z}{\partial \tau^2}\right)^2 - \left(\frac{\partial^2 t}{\partial \tau^2}\right)^2 \equiv a^2. \tag{6.196}$$

To simplify later results, we set $z = a^{-1}$ at $t = 0$. Note that for $t < 0$, $dz/dt < 0$, while for $t > 0$, $dz/dt > 0$. The more realistic case where $dz/dt = 0$ for $t < 0$ will be explored in Section 6.13.3.

A plane wave with wave vector $\mathbf{k} = -k\hat{z}$ is incident on the mirror. Given the geometry of this problem, the electromagnetic field tensor reduces to

$$F^{\mu\nu} = \begin{bmatrix} 0 & E_x & 0 & 0 \\ -E_x & 0 & 0 & -B_y \\ 0 & 0 & 0 & 0 \\ 0 & B_y & 0 & 0 \end{bmatrix}. \tag{6.197}$$

The incident wave is then given by

$$E_x^I = -B_y^I = E_0 \cos(-kz - \omega t) = E_0 \cos[k(z+t)], \tag{6.198}$$

while the reflected wave can be assumed to be of the form

$$E_x^R = B_y^R = -E_0 f(z-t). \tag{6.199}$$

It is easily seen that Equations 6.198 and 6.199 satisfy Maxwell's equations.

We now consider the Rindler transform, which allows us to study the incident and reflected waves in an accelerated frame where the mirror is at rest at all times. Rindler coordinates are related to Minkowski coordinates by

$$\bar{z} = \sqrt{z^2 - t^2}, \quad \bar{t} = \frac{1}{2a}\ln\left(\frac{z+t}{z-t}\right), \tag{6.200}$$

and

$$z = \bar{z}\cosh(a\bar{t}), \quad t = \bar{z}\sinh(a\bar{t}), \tag{6.201}$$

where the coordinate transform has been scaled according to the mirror's acceleration, for convenience.

The Minkowski metric may be transformed to the Rindler metric:

$$ds^2 = -a^2\bar{z}^2 d\bar{t}^2 + dx^2 + dy^2 + d\bar{z}^2. \tag{6.202}$$

The electromagnetic field tensor may be transformed by the well-known formula

$$\bar{F}^{\mu\nu} = \frac{\partial \bar{x}^{\mu}}{\partial x^{\alpha}} \frac{\partial \bar{x}^{\nu}}{\partial x^{\beta}} F^{\alpha\beta}, \tag{6.203}$$

which yields

$$\begin{aligned}
\bar{F}^{01} &= -\bar{F}^{10} = \frac{\partial \bar{t}}{\partial t} F^{01} + \frac{\partial \bar{t}}{\partial z} F^{31} = \frac{\cosh(a\bar{t})}{a\bar{z}} E_x - \frac{\sinh(a\bar{t})}{a\bar{z}} B_y, \\
\bar{F}^{31} &= -\bar{F}^{13} = \frac{\partial \bar{z}}{\partial t} F^{01} + \frac{\partial \bar{z}}{\partial z} F^{31} = -\sinh(a\bar{t}) E_x + \cosh(a\bar{t}) B_y, \\
\bar{F}^{\mu\nu} &= 0 \ \text{otherwise.}
\end{aligned} \tag{6.204}$$

With this, the incident and reflected phase variables become

$$z + t = -\bar{z}\cosh(a\bar{t}) - \bar{z}\sinh(a\bar{t}) = \bar{z}\exp(a\bar{t}), \tag{6.205}$$

and

$$z - t = \bar{z}\cosh(a\bar{t}) - \bar{z}\sinh(a\bar{t}) = \bar{z}\exp(-a\bar{t}). \tag{6.206}$$

We thus obtain

$$\begin{aligned}
\bar{E}_x^I &= E_0 \frac{\exp(a\bar{t})}{a\bar{z}} \cos[k\bar{z}\exp(a\bar{t})], \\
\bar{B}_y^I &= -E_0 \exp(a\bar{t}) \cos[k\bar{z}\exp(a\bar{t})],
\end{aligned} \tag{6.207}$$

and

$$\begin{aligned}
\bar{E}_x^R &= -E_0 \frac{\exp(-a\bar{t})}{a\bar{z}} f[\bar{z}\exp(-a\bar{t})], \\
\bar{B}_y^R &= -E_0 \exp(-a\bar{t}) f[\bar{z}\exp(-a\bar{t})].
\end{aligned} \tag{6.208}$$

It is easy to check whether these expressions satisfy the generally covariant form of Maxwell's equations in vacuum, $\partial_\nu(\sqrt{-g}F^{\mu\nu}) = 0$, and $F_{\mu\nu,\rho} + F_{\rho\mu,\nu} + F_{\nu\rho,\mu} = 0$, which, in the one-dimensional geometry of this problem, reduce to

$$\frac{\partial}{\partial \bar{t}}(a\bar{z}\bar{E}_x) + \frac{\partial}{\partial \bar{z}}(a\bar{z}\bar{B}_y) = 0, \tag{6.209}$$

and

$$\frac{\partial}{\partial \bar{t}}\bar{B}_y + \frac{\partial}{\partial \bar{z}}(a^2\bar{z}^2\bar{E}_x) = 0. \tag{6.210}$$

At this point, we note that the boundary condition for a perfect conductor mandates that there be no transverse electromagnetic forces on the electrons within the mirror. Mathematically, this condition is expressed as

$$e\bar{F}^{1\nu}\bar{u}_\nu\Big|_{\bar{z}=\frac{1}{a}} = e\bar{F}^{2\nu}\bar{u}_\nu\Big|_{\bar{z}=\frac{1}{a}} = 0, \tag{6.211}$$

where $\bar{u}_\nu = d\bar{x}_\nu/d\tau$.

Since the field is transverse and $\bar{u}_3 = 0$, we have

$$\bar{E}_x\Big|_{\bar{z}=\frac{1}{a}} = (\bar{E}_x^I + \bar{E}_x^R)\Big|_{\bar{z}=\frac{1}{a}} = 0. \tag{6.212}$$

In order to solve for the unknown function $f$ in the expression for the reflected wave, the incident and reflected electric fields in Equations 6.207 and 6.208 can now be used in Equation 6.212 to yield

$$E_0\exp(a\bar{t})\cos\left[\frac{k}{a}\exp(a\bar{t})\right] - E_0\exp(-a\bar{t})f[\bar{z}\exp(-a\bar{t})]\Big|_{\bar{z}=\frac{1}{a}} = 0. \tag{6.213}$$

A little algebra reveals that the only value for $f$, which satisfies Equation 6.213 while maintaining its space–time dependence exclusively on $\bar{z}\exp(-a\bar{t})$, in order to satisfy Maxwell's equations, is

$$f[\bar{z}\exp(-a\bar{t})] = \frac{\exp(2a\bar{t})}{a^2\bar{z}^2}\cos\left[\frac{k\exp(a\bar{t})}{a^2\bar{z}}\right]. \tag{6.214}$$

The reflected wave in Rindler coordinates thus becomes

$$\bar{E}_x^R = -E_0\frac{\exp(a\bar{t})}{a^3\bar{z}^3}\cos\left[\frac{k\exp(a\bar{t})}{a^2\bar{z}}\right], \tag{6.215}$$

and

$$\bar{B}_y^R = -E_0\frac{\exp(a\bar{t})}{a^2\bar{z}^2}\cos\left[\frac{k\exp(a\bar{t})}{a^2\bar{z}}\right]. \tag{6.216}$$

Using Equation 6.206, the function $f$ can be expressed in terms of Minkowski coordinates as

$$f(z-t) = \frac{\cos\left[\frac{k}{a^2(z-t)}\right]}{a^2(z-t)^2}, \tag{6.217}$$

and the reflected wave becomes

$$E_x^R = B_x^R = -E_0\frac{\cos\left[\frac{k}{a^2(z-t)}\right]}{a^2(z-t)^2}. \tag{6.218}$$

The physical reason for the unusual dependence on $z - t$ will be made clear in the next section. Here, we only point out that the apparent singularity in the field at $z = t$ does not pose any difficulty. For any finite time $t$, the position of the mirror in our coordinates is greater than $t$: $z_m(t) > t$; thus the point $z = t$ always lies behind the mirror, opposite to the side on which the plane wave is incident, and therefore outside the region for which Equation 6.218 is valid.

### 6.13.3 Derivation of the Reflected Wave Using the Lorentz Transform

We first consider a plane wave at normal incidence to a mirror with constant velocity. In order to solve for the reflected wave, the problem is treated most easily in the frame of the mirror, which requires a Lorentz transform of the original expression for the incident wave:

$$E_x'^I = \gamma(E_x^I - \beta B_y^I) = \gamma(1-\beta)E_x^I, \tag{6.219}$$

and

$$B_y'^I = \gamma(B_y^I - \beta E_x^I) = \gamma(1-\beta)B_y^I, \tag{6.220}$$

where $\beta$ is the relative velocity between the instantaneous rest frame of the mirror and the reference frame.

Since we have

$$z + t = \gamma(z' + \beta t') + \gamma(t' + \beta z') = \gamma(1+\beta)(z'+t'), \tag{6.221}$$

the incident wave can be expressed as

$$E_x'^I(z',t') = -B_y'^I(z',t') = \gamma(1+\beta)E_0\cos[\gamma(1+\beta)k(z'+t')]. \tag{6.222}$$

The boundary condition now reads

$$\mathbf{n}\times\mathbf{E}'\big|_{z'=z'_m} = \hat{y}(E'^{I}_{x} + E'^{R}_{x})\big|_{z'=0} = 0, \tag{6.223}$$

which implies,

$$\mathbf{E}'^{R}\big|_{z'=0} = -\hat{x}\gamma(1+\beta)E_0\cos[\gamma(1+\beta)k(-t')]. \tag{6.224}$$

To satisfy both the boundary condition and Maxwell's equations, the reflected wave can only take the form

$$E'^{R}_{x}(z',t') = B'^{R}_{y}(z',t') = -\gamma(1+\beta)E_0\cos[\gamma(1+\beta)k(z'-t')]. \tag{6.225}$$

Lorentz transforming the reflected wave back to the original lab frame and noting that

$$z'-t' = \gamma(z-\beta t)-\gamma(t-\beta z) = \gamma(1+\beta)(z-t), \tag{6.226}$$

the reflected wave is found to be

$$E^{R}_{x}(z,t) = B^{R}_{y}(z,t) = -\gamma^2(1+\beta)^2E_0\cos[\gamma^2(1+\beta)^2k(z-t)]. \tag{6.227}$$

To extend this result to the case of an accelerating mirror, we observe that a ray of light reflected from an accelerating mirror at some time $t_r$ and position $z_r$, where it has velocity $\beta$, is indistinguishable from a ray of light reflected from an identical mirror at the same time $t_r$ and the same position $z_r$, but with a constant velocity $\beta_0$ which happens to equal $\beta$ at that instant.

The retarded position $z_r$ and retarded time $t_r$ can be expressed in terms of the retarded proper time $\tau_r$ of the mirror as follows:

$$z_r = \frac{1}{a}\cosh(a\tau_r), \quad t_r = \frac{1}{a}\sinh(a\tau_r). \tag{6.228}$$

The retarded Lorentz boost parameters $\gamma$ and $\beta$ thus satisfy the following relations:

$$\gamma\beta = \frac{dz_r}{d\tau_r} = \sinh(a\tau_r), \quad \gamma = \frac{dt_r}{d\tau_r} = \cosh(a\tau_r). \tag{6.229}$$

Hence, using Equations 6.228 and 6.229, and recalling the identity for hyperbolic functions, $\cosh^2 s - \sinh^2 s = 1$, we find

$$(\gamma+\gamma\beta)^2 = \frac{1}{a^2(z_r-t_r)^2}. \tag{6.230}$$

Invoking the light-cone condition, $z_r - t_r = z - t$, the following identity is finally obtained:

$$(\gamma + \gamma\beta)^2 = \frac{1}{a^2(z-t)^2}. \tag{6.231}$$

The reflected wave is therefore described by

$$E_x^R = B_y^R = -E_0 \frac{\cos\left[\frac{k}{a^2(z-t)}\right]}{a^2(z-t)^2}, \tag{6.232}$$

which confirms the result obtained by the Rindler method.

With this approach, several curious features of the reflected wave are understood easily in terms of the Doppler shift. For example, the amplitude of the reflected wave goes to zero as $z$ goes to infinity because of an infinite redshift. At larger $z$, the observed reflected wave originates from a point farther back in the past on the mirror's world line, when the mirror had larger acceleration away from the observer. The resulting Doppler redshift of the reflected wave therefore increases with $z$. Another interesting effect to note is that as time increases, the velocity of the mirror asymptotically approaches the speed of light and, correspondingly, its position asymptotically approaches the reflected wave singularity at $t = z$. Thus, the amplitude of the field near the mirror increases with time, which physically is due, of course, to Doppler blue-shifting.

By the reasoning above, the more realistic case in which the mirror is at rest until uniform proper acceleration is initiated at some finite time can be examined readily. Following the previous light-ray argument and considering the retarded quantities, it is clear that if the mirror is at rest for $t < 0$ and begins to accelerate uniformly at $t = 0$, the reflected wave must be

$$E_x^R = B_y^R = \begin{cases} -E_0 \dfrac{\cos\left[\dfrac{k}{a^2(z-t)}\right]}{a^2(z-t)^2}, & z - t \le \dfrac{1}{a}, \\ -E_0 \cos[k(z-t)], & z - t \ge \dfrac{1}{a}. \end{cases} \tag{6.233}$$

As $z$ increases, the Doppler redshift will decrease the amplitude and frequency of the reflected wave only until $z = t + 1/a$; beyond this point the reflected wave appears as a monochromatic plane wave because its retarded "source" is now stationary.

In conclusion, the reflected wave from a uniformly accelerating mirror has been derived using the Rindler transform and, alternatively, using the Lorentz transform. The physics of the result obtained by the Rindler method have been elucidated by the Lorentz transform approach, and the expected Doppler effects have been plainly demonstrated. Further, both the case where the mirror is always accelerating and the case where the mirror begins acceleration at some finite time have been examined.

We now attempt to interpret these results within the context of SED, wherein the incident wave is taken to represent a virtual photon, so that it has an energy density equal to $\frac{1}{2}\hbar\omega d^3k$. Admittedly, in summing over the infinite momenta of the vacuum, it is not completely clear how to compare meaningfully the total infinite spectrum obtained from the incident and reflected fields with the infinite spectrum of an unbounded vacuum. However, for the purposes of making a qualitative prediction, we note that the amplitude of the reflected field becomes arbitrarily large for small $z - t$. Thus, it seems reasonable to assert that, within the framework of this model, a detector stationed at some fixed position $z$ sufficiently larger than $1/a$ will detect a pulse of radiation that is significantly larger in amplitude than the vacuum noise as soon as the mirror approaches sufficiently close. This semiclassical result seems to be in conflict with the quantum treatment of the same problem; this might indicate that the stochastic electrodynamical model breaks down in this situation. However, the full SED and QED calculations must be performed before definitive statements can be made in this regard; perhaps relevant experiments will also be performed in the not-too-distant future.

### 6.13.4 Mathematical Appendix

It can be shown that the correct plane wave solution is recovered for the reflected wave in the zero-acceleration limit. Taking this limit is not trivial, however, because in the previous expressions it was assumed that the mirror is located at $z = 1/a$ when $t = 0$. To obtain meaningful results in the zero-acceleration limit, it is therefore necessary to shift the $z$ coordinate:

$$z' = z - \frac{1}{a}. \tag{6.234}$$

For this purpose it will simplify matters considerably to use complex fields, such that

$$E_x^I = \mathrm{Re}(\tilde{E}_x^I) = \mathrm{Re}\{E_0 \exp[-ik(z+t)]\}, \tag{6.235}$$

and

$$E_x^R = \mathrm{Re}(\tilde{E}_x^R) = \mathrm{Re}\left\{-E_0 \frac{\exp\left[\frac{-ik}{a^2(z-t)}\right]}{a^2(z-t)^2}\right\}. \tag{6.236}$$

Expressing the incident wave in the new coordinate system,

$$\tilde{E}_x^I = -\tilde{B}_y^I = E_0 \exp\left\{-ik\left[\left(z' + \frac{1}{a}\right) + t\right]\right\} = E_0 \exp\left(-i\frac{k}{a}\right)\exp[-ik(z'+t)], \tag{6.237}$$

the zero-acceleration limit can be taken to yield

$$\lim_{a\to 0}\tilde{E}_x^I = \tilde{E}_0'\exp[-ik(z'+t)], \tag{6.238}$$

where $\tilde{E}_0' \equiv \lim_{a\to 0} E_0 \exp(-i\frac{k}{a})$.

Now expressing the reflected wave in the new coordinate system

$$\tilde{E}_x^R = \tilde{B}_y^R = -E_0\frac{\exp\left\{\frac{-ik}{a[1+a(z'-t)]}\right\}}{[1+a(z'-t)]^2}, \tag{6.239}$$

the zero-acceleration limit for the reflected wave can be taken as follows:

$$\begin{aligned}\lim_{a\to 0}\tilde{E}_x^R &= -\lim_{a\to 0} E_0\exp\left\{\frac{-ik}{a[1+a(z'-t)]}\right\}\\ &= -\lim_{a\to 0} E_0\exp\left\{-i\frac{k}{a}[1-a(z'-t)]\right\}\\ &= -\lim_{a\to 0} E_0\exp\left(-i\frac{k}{a}\right)\exp[ik(z'-t)]\\ &= -\tilde{E}_0'\exp[ik(z'-t)].\end{aligned} \tag{6.240}$$

This result is exactly the reflected wave corresponding to the incident plane wave in Equation 6.239 for a stationary mirror at $z' = 0$.

## 6.14 References for Chapter 6

Note: the numbers listed below refer to the main bibliography and reference sections at the end of this book.

1, 4, 17, 32, 35, 36, 55, 56, 69, 73, 83, 132, 148, 151, 157, 166, 181, 214, 215, 219, 247, 263, 264, 265, 266, 267, 269, 271, 310, 315, 318, 319, 320, 321, 322, 328, 343, 344, 356, 357, 375, 389, 390, 406, 407, 408, 415, 514, 515, 524, 532, 533, 534, 535, 551, 552, 553, 554, 555, 556, 557, 586, 589, 590, 591, 594, 623, 624, 633, 634, 639, 642, 643, 656, 664, 702, 703, 707, 713, 760, 761, 853, 854, 855, 856, 857, 858, 859, 860, 862, 863, 864, 866, 869, 873, 876, 877, 905, 906, 907.

# 7

# *Relativistic Transform of the Refractive Index: Čerenkov Radiation*

## 7.1 Introduction

Čerenkov radiation was discovered experimentally by Vavilov and Čerenkov, and the first theoretical description of this new phenomenon was given by Tamm and Frank. In 1934, Čerenkov noticed that pure sulfuric acid, in a platinum container close to a radium source, emitted a weak, blue radiation. Čerenkov conducted a series of experiments to investigate this new phenomenon, in terms of radiation spectrum, polarization, and influence of both the radioactive source and material involved. In particular, great care was taken to distinguish this phenomenon from ordinary luminescence, and the different time-scales involved provided an important discriminating factor. The theory of Tamm and Frank unambiguously demonstrated that Čerenkov radiation is generated by a charged particle when it is moving through a medium with a constant velocity exceeding that of light in this material; in this sense, it can be considered as an electromagnetic shock wave. The discovery of Čerenkov radiation and its theoretical investigation earned the 1958 Nobel Prize for Čerenkov, Tamm, and Frank. Čerenkov radiation is currently used in high-energy physics to detect charged particles and to provide an estimate of their energy; it is also extremely useful, together with transition radiation, to study the temporal characteristics of electron beams. Picosecond resolution can be achieved routinely in the determination of short electron bunch duration by using Čerenkov or transition radiation and streak cameras. In addition, well-designed experimental setups can provide accurate information on the electron beam energy and momentum spread, as is extensively discussed in the literature.

In this chapter, we first describe Čerenkov radiation in terms of the Tamm–Frank theory, closely following the detailed exposition given by Zrelov in his book on Čerenkov radiation in high-energy physics. We then introduce the relativistic transform of the electromagnetic inductions and perform a detailed analysis of the Čerenkov radiation condition based on the relativistic transform of the refractive index. This approach provides a powerful example of the application of the principle of relativity and illustrates in a clear manner the crucial difference between scalars, as defined within the context of special relativity, and pseudoscalars, such as the refractive index, which behave in a very complex manner under Lorentz transform. It also provides a different insight into the physics of Čerenkov radiation and clearly exemplifies the importance and usefulness of complementary approaches to the description of a given physical phenomenon. Finally, Čerenkov radiation, just like transition radiation, is an interesting form of production of light by a charged particle because it first seems that a particle moving with constant velocity can actually radiate, apparently violating the principle of relativity. The detailed analyses presented in this chapter resolve this apparent contradiction by showing that the particle is in fact subjected to an image current which gives it a small transverse acceleration. This electromagnetic image is induced by boundary conditions in the case of transition radiation or by the polarization of the medium traversed by the charge in the case of Čerenkov radiation.

## 7.2 Classical Theory of Čerenkov Radiation

We start the theoretical description of Čerenkov radiation by considering a charged particle moving with constant velocity in a linear, isotropic medium, characterized by a dielectric constant differing from that of vacuum: $\varepsilon \neq \varepsilon_0$. The constant velocity approximation holds as long as the particle does not experience scattering in the dielectric medium, and for weak radiation, where the energy loss associated with electromagnetic radiation remains small compared to the kinetic energy of the radiating charge. These conditions are satisfied in a wide variety of experimental situations. It should also be noted that, in a symmetrical way, magnetically induced Čerenkov radiation can be excited in a magnetic medium, where $\mu \neq \mu_0$.

The macroscopic description of the problem, namely characterizing the response of the medium by a dielectric constant differing from $\varepsilon_0$, is justified because the Čerenkov radiation wavelength is much longer than the interatomic distances in the medium. In addition, we consider an infinite medium, without boundary conditions. In this case, Maxwell's equations are driven by the four-current of the moving charge, and we have

$$\nabla \times \mathbf{H} - \partial_t \mathbf{D} = \mathbf{j} \tag{7.1}$$

and

$$\nabla \cdot \mathbf{D} = \rho, \tag{7.2}$$

where $\rho$ and $\mathbf{j}$ represent the charge density and the current density of the moving electron, $\mathbf{H}$ is the magnetic field, and $\mathbf{D} = \varepsilon \mathbf{E}$ is the electric induction, as expressed in terms of the electric field $\mathbf{E}$.

The reaction of the dielectric medium to the electric and magnetic field of the moving charge is characterized by $\varepsilon$. The source-free equations are

$$\nabla \cdot \mathbf{B} = 0 \tag{7.3}$$

and

$$\nabla \times \mathbf{E} + \partial_t \mathbf{B} = \mathbf{0}. \tag{7.4}$$

Because we are considering a medium with no magnetic properties, the magnetic induction and field are related by $\mathbf{B} = \mu_0 \mathbf{H}$.

As is customary, we introduce the four-potential, so that the source-free equations are automatically satisfied:

$$\mathbf{E} = -\nabla \phi - \partial_t \mathbf{A}, \tag{7.5}$$

and

$$\mathbf{B} = \nabla \times \mathbf{A}. \tag{7.6}$$

In addition, the four-potential is chosen to satisfy the following gauge condition:

$$\varepsilon \mu_0 \partial_t \phi + \nabla \cdot \mathbf{A} = 0. \tag{7.7}$$

In the vacuum limit, where $\varepsilon \rightarrow \varepsilon_0$, this gauge condition coincides with the Lorentz gauge condition, $\partial_\mu A^\mu = 0$.

Using Equations 7.5 and 7.6 into the equations with source terms, we obtain the driven wave equations for the potentials:

$$\Delta \mathbf{A} - \varepsilon \mu_0 \partial_t^2 \mathbf{A} + \mu_0 \mathbf{j} = \mathbf{0}, \tag{7.8}$$

and

$$\Delta \phi - \varepsilon \mu_0 \partial_t^2 \phi + \frac{1}{\varepsilon} \rho = 0. \tag{7.9}$$

Before proceeding further, a short digression regarding this particular gauge condition and wave equations is useful to illustrate the concepts of gauge invariance and covariance. At first glance, Equations 7.7 to 7.9 seem to lack the symmetry required to allow a simple translation into covariant notation; in particular, the gauge condition resembles the Lorentz condition used in vacuum, but includes a modification of its time-like component. In fact, it is easily seen that if we modify the definition of the four-gradient and four-potential as follows,

$$\bar{\partial}_\mu \equiv (-\sqrt{\varepsilon \mu_0}^{\,a} \partial_t, \nabla), \tag{7.10}$$

and

$$\bar{A}_\mu \equiv (\sqrt{\varepsilon\mu_0}^{\,b}\,\phi,\ \mathbf{A}), \tag{7.11}$$

the gauge condition takes the form $\bar{\partial}_\mu \bar{A}^\mu = 0$, provided that $a + b = 2$. Now, in the driven wave equations, the d'Alembertian operator is also modified. We see that to identify it with $\bar{\partial}_\mu \bar{\partial}^\mu$, the exponent $a$ must be equal to one, in which case, $\bar{\partial}_\mu \bar{\partial}^\mu \equiv \Delta - \varepsilon\mu_0 \partial_t^2$. With $a = b = 1$, the spatial components of the wave equation indicate that no modification of the current density is required. However, the time-like component (Equation 7.10) now involves the modified scalar potential, $\sqrt{\varepsilon\mu_0}\phi$, and the four-current must read $\bar{j}_\mu \equiv (\frac{\rho}{\sqrt{\varepsilon\mu_0}}, \mathbf{j})$, so that the propagation equation takes the apparently covariant form,

$$[\bar{\partial}_\nu \bar{\partial}^\nu]\,\bar{A}_\mu + \mu_0\,\bar{j}_\mu = 0. \tag{7.12}$$

It is also easily verified that the new definitions of the four-gradient, four-potential, and four-current are dimensionally correct. However, one should bear in mind the fact that, as will be shown in detail in this chapter, only the vacuum permeability and permittivity are true scalars in the covariant sense.

At this point, it is important to note that, in general, the dielectric constant is a function of the frequency of the electromagnetic radiation propagating in the medium under consideration. In addition, following the definition first introduced in Chapter 1, the refractive index can be given in terms of the relative dielectric constant: $\mathbf{D}(\omega) = \varepsilon(\omega)\mathbf{E}(\omega) = n^2(\omega)\mathbf{E}(\omega)$. Therefore, a Fourier analysis of Čerenkov radiation is appropriate. We introduce the Fourier-conjugate fields and potentials,

$$F(\mathbf{x}, t) = \frac{1}{\sqrt{2\pi}} \int_{-\infty}^{+\infty} \tilde{F}(\mathbf{x}, \omega) e^{i\omega t} d\omega, \tag{7.13}$$

where $F(\mathbf{x}, t)$ represents any of the variables used in this problem; in frequency space, the partial derivative with respect to time can be replaced as follows: $\partial_t \to i\omega$.

With this, the driven wave equation (7.8) for the vector potential takes the form

$$\Delta\tilde{\mathbf{A}}(\mathbf{x}, \omega) + \varepsilon(\omega)\mu_0\omega^2\tilde{\mathbf{A}}(\mathbf{x}, \omega) + \mu_0\tilde{\mathbf{j}}(\mathbf{x}, \omega) = \mathbf{0}, \tag{7.14}$$

where the Fourier transform of the electron current density now drives the equation.

In addition, it is easily seen that because $\mathbf{j} = \rho\beta c$, we do not need to solve the equation for the scalar potential, provided that we take $\mathbf{A} = \phi\beta/c$.

Here, $\beta = \mathbf{v}/c$ is the normalized velocity of the charge moving through the Čerenkov medium. In frequency space, the gauge condition now reads

$$\varepsilon(\omega)\mu_0 i\omega\tilde{\phi}(\mathbf{x}, \omega) + \nabla \cdot \tilde{\mathbf{A}}(\mathbf{x}, \omega) = i\frac{\varepsilon(\omega)}{\varepsilon_0}\frac{\omega}{\beta^2 c}\beta \cdot \tilde{\mathbf{A}}(\mathbf{x}, \omega) + \nabla \cdot \tilde{\mathbf{A}}(\mathbf{x}, \omega) = 0. \tag{7.15}$$

Finally, for completeness, we note that the definition of the electric and magnetic fields can now be given in terms of the vector potential only and reduce to

$$\tilde{\mathbf{E}}(\omega) = -i\left\{\frac{1}{\varepsilon(\omega)\mu_0\omega}\nabla[\nabla \cdot \tilde{\mathbf{A}}(\omega)] + \omega\tilde{\mathbf{A}}(\omega)\right\}, \tag{7.16}$$

and

$$\tilde{\mathbf{H}}(\omega) = \frac{1}{\mu_0}\nabla \times \tilde{\mathbf{A}}(\omega). \tag{7.17}$$

We now need to derive the Fourier transform of the current density associated with the moving charge. Choosing the z-axis to coincide with the electron velocity, the current density is given by

$$\mathbf{j}(x_\mu) = \hat{z}j_z(x_\mu) = \hat{z}\beta c\rho(x_\mu), \tag{7.18}$$

where the charge density of the point electron is

$$\rho(x_\mu) = -e\delta(x)\delta(y)\delta(z - \beta ct). \tag{7.19}$$

We now need to Fourier transform the current density; we have, by definition,

$$\tilde{j}_z(x, y, z, \omega) = \frac{1}{\sqrt{2\pi}}\int_{-\infty}^{+\infty} j_z(x, y, z, t)e^{-i\omega t}dt. \tag{7.20}$$

Replacing the current density in the integral by the function defined above, we have

$$\tilde{j}_z(\mathbf{x}, \omega) = -\frac{e\beta c}{\sqrt{2\pi}}\delta(x)\delta(y)\int_{-\infty}^{+\infty} \delta(z - \beta ct)e^{-i\omega t}dt. \tag{7.21}$$

At this point, we can proceed in two different ways: we can introduce the new variable $z' = z - \beta ct$, and use the fact that $\int_{-\infty}^{+\infty}\delta(x)f(x)dx = f(0)$, or we can use the property of the Dirac delta-distribution, $\int_{-\infty}^{+\infty} f(x)\delta(x_0 - \alpha x)\,dx = \frac{1}{\alpha} f\left(\frac{x_0}{\alpha}\right)$. Using either approach, we find that

$$\tilde{j}_z(\mathbf{x}, \omega) = -\frac{e}{\sqrt{2\pi}}\delta(x)\delta(y)\exp\left(-i\frac{\omega z}{\beta c}\right). \tag{7.22}$$

The problem is axially symmetrical; therefore, we use cylindrical coordinates $(r, \varphi, z)$. We need to transform the result given in Equation 7.22 to cylindrical coordinates: the transverse delta-functions must be re-expressed in terms of the radius. To this end, we first assume that

$$\delta(x)\delta(y) = \eta(r)\delta(r); \tag{7.23}$$

the normalization function $\eta(r)$ is then determined as follows. We have

$$\int_{-\infty}^{+\infty}\delta(x)\,dx\int_{-\infty}^{+\infty}\delta(y)\,dy = 1 = \int_0^{2\pi} d\varphi\int_0^{\infty}\eta(r)\delta(r)r\,dr = 2\pi\int_0^{\infty}\eta(r)\delta(r)r\,dr; \tag{7.24}$$

it is easily seen that $\eta(r) = \frac{1}{\pi r}$ satisfies Equation 7.24, since

$$\int_0^{\infty}\eta(r)\delta(r)r\,dr = \frac{1}{\pi}\left[\frac{1}{2}\int_{-\infty}^{+\infty}\delta(r)\,dr\right] = \frac{1}{2\pi}; \tag{7.25}$$

therefore, $\delta(x)\delta(y) = \frac{1}{\pi r}\delta(r)$. We note that one can also use the Jacobian of the transform between Cartesian and cylindrical coordinates to obtain the same result.

The current density of the electron, which is moving with constant velocity along the $z$-axis, is thus given in Fourier frequency-space by

$$\tilde{j}_z(\omega) = -\frac{e}{\sqrt{2}\pi^{3/2}r}\delta(r)\exp\left(-i\frac{\omega z}{\beta c}\right). \tag{7.26}$$

It is clear that the radial and azimuthal components of the vector potential are not driven by the axial current of the charge; therefore, we have $A_r = A_\varphi = 0$, and the axial component of the wave equation is:

$$[\Delta + \varepsilon(\omega)\mu_0\omega^2]\tilde{A}_z - \mu_0\frac{e}{\sqrt{2}\pi^{3/2}r}\delta(r)\exp\left(-i\frac{\omega z}{\beta c}\right) = 0. \tag{7.27}$$

Using the expression of the Laplacian operator in cylindrical coordinates, the wave equation governing the evolution of the axial component of the three-potential now reads

$$\begin{aligned}\left[\frac{1}{r}\partial_r(r\partial_r) + \partial_z^2 + \varepsilon(\omega)\frac{\omega^2}{c^2}\right]\tilde{A}_z(r, z, \omega) &= -\mu_0\tilde{j}_z(r, z, \omega)\\ &= \frac{e\mu_0}{\sqrt{2}\pi^{3/2}r}\delta(r)\exp\left(-i\frac{\omega z}{\beta c}\right).\end{aligned} \tag{7.28}$$

Here, the axial symmetry of the problem dictates that $\partial_\varphi \equiv 0$; the vector potential does not depend on the azimuthal angle, $\varphi$.

Following Tamm and Frank, we assume that we can separate variables and that the solution to Equation 7.28 takes the form

$$\tilde{A}_z(r, z\omega) = f(r)\exp\left(-i\frac{\omega z}{\beta c}\right). \tag{7.29}$$

Note that the exponential dependence is chosen simply to match that of the driving current. Using Equation 7.29 into the wave equation (7.28), we see that the function of the radius must satisfy the following equation:

$$\frac{1}{r}\frac{d}{dr}\left(r\frac{df}{dr}\right)+\left[-\frac{\omega^2}{\beta^2 c^2}+\frac{\varepsilon(\omega)}{\varepsilon_0}\frac{\omega^2}{c^2}\right]f = \frac{e\mu_0}{\sqrt{2}\pi^{3/2}}\frac{\delta(r)}{r}, \tag{7.30}$$

which can be recast to read

$$\frac{1}{r}\frac{d}{dr}(rf')+\left[\frac{\varepsilon(\omega)}{\varepsilon_0}-\frac{1}{\beta^2}\right]\frac{\omega^2}{c^2}f = \frac{e\mu_0}{\sqrt{2}\pi^{3/2}}\frac{\delta(r)}{r}, \tag{7.31}$$

or

$$f''+\frac{1}{r}f'+\frac{\omega^2}{v^2}[\beta^2 n^2(\omega)-1]f = \frac{e\mu_0}{\sqrt{2}\pi^{3/2}}\frac{\delta(r)}{r}, \tag{7.32}$$

where we have introduced the refractive index, $n(\omega) = c/v_g = cdk/d\omega = \sqrt{\varepsilon(\omega)/\varepsilon_0}$.

The last equality is valid in the case where the medium under consideration is a linear, isotropic, dielectric, and where $\mu = \mu_0$. For all values of the radius, except $r = 0$, the delta-function drive current is zero, and Equation 7.32 coincides with the well-known Bessel differential equation,

$$y''+\frac{y'}{\rho}+\left(1-\frac{\nu^2}{\rho^2}\right)y = 0, \quad \rho \in \mathbb{C}, \tag{7.33}$$

where $y(\rho) = \mathrm{Z}_\nu(\rho)$. Here, $\mathrm{Z}_\nu$ represents Bessel functions of the first, second, or third kind, and of order $\nu$. As previously discussed, the Bessel functions of the first kind are generally represented by $\mathrm{J}_\nu$, while Bessel functions of the second kind, or Weber functions, correspond to $\mathrm{Y}_\nu$, and are related to $\mathrm{J}_\nu$ as follows:

$$\mathrm{Y}_\nu(\rho) = \frac{\mathrm{J}_\nu(\rho)\cos(\nu\pi)-\mathrm{J}_{-\nu}(\rho)}{\sin(\nu\pi)}; \tag{7.34}$$

finally, for Bessel functions of the third kind, which are also called Hankel functions, we have

$$\begin{aligned} \mathrm{H}_\nu^1(\rho) &= \mathrm{J}_\nu(\rho) + i\mathrm{Y}_\nu(\rho), \\ \mathrm{H}_\nu^2(\rho) &= \mathrm{J}_\nu(\rho) - i\mathrm{Y}_\nu(\rho). \end{aligned} \tag{7.35}$$

We also note that the Bessel functions of the second and third kind are not bounded on-axis; furthermore, $\mathrm{H}_\nu^1$ and $\mathrm{H}_\nu^2$ are linearly independent for any order, while $\mathrm{J}_\nu$ and $\mathrm{J}_{-\nu}$ are independent if $\nu \notin \mathbb{Z}$.

In the case of the classical Tamm–Frank theory of Čerenkov radiation, we make the following identifications. We first recast Equation 7.32 as

$$f'' + \frac{1}{r}f' + s^2 f = 0, \quad r \neq 0, \quad s^2 = \frac{\omega^2}{v^2}[\beta^2 n^2(\omega) - 1]; \tag{7.36}$$

it is then seen easily that by writing $\rho = sr$, we have $df/d\rho = s^{-1}f'$, and $d^2f/d\rho^2 = s^{-2}f''$, which leads to

$$s^2\frac{d^2f}{d\rho^2} + \frac{s}{r}\frac{df}{d\rho} + s^2 f = 0, \tag{7.37}$$

or

$$\frac{d^2f}{d\rho^2} + \frac{1}{sr}\frac{df}{d\rho} + f = \frac{d^2f}{d\rho^2} + \frac{1}{\rho}\frac{df}{d\rho} + f = 0, \tag{7.38}$$

and the Hankel functions of order $\nu = 0$ zero are solutions of Equation 7.38. Thus, we have

$$f(r, \omega) = a_1\mathrm{H}_0^1[s(\omega)r] + a_2\mathrm{H}_0^2[s(\omega)r], \tag{7.39}$$

where the constants $a_1$ and $a_2$ remain undetermined at this point.

The boundary condition of the problem is given by the fact that, for $r \to 0$, the Dirac delta-function drive current must be taken into account; to this end, we use the identity established in Chapter 5, namely,

$$\delta(sr) = \frac{\delta(r)}{|s|}.$$

With this, and using the notations that we have introduced, we can rewrite Equation 7.32 as

$$s^2\left(\frac{d^2f}{d\rho^2} + \frac{1}{\rho}\frac{df}{d\rho} + f\right) = \frac{e\mu_0}{\sqrt{2}\pi^{3/2}}|s|\frac{\delta(sr)}{r}, \tag{7.40}$$

or

$$\frac{1}{\rho}\frac{\partial}{\partial\rho}\left(\rho\frac{\partial \mathbf{f}}{\partial\rho}\right) + \mathbf{f} = \frac{e\mu_0}{\sqrt{2}\pi^{3/2}}\frac{\delta(\rho)}{\rho}. \tag{7.41}$$

To complete the derivation, we need to determine the appropriate boundary condition, taking into account the delta function. Following Tamm and Frank, we integrate Eq. (7.41) over the area of a circle of arbitrary radius $\Delta\rho$:

$$\int_0^{\Delta\rho} 2\pi\rho d\rho \frac{1}{\rho}\frac{\partial}{\partial\rho}\left(\rho\frac{\partial \mathbf{f}}{\partial\rho}\right) + \int_0^{\Delta\rho} 2\pi\rho d\rho \mathbf{f} = \frac{e\mu_0}{\sqrt{2}\pi^{3/2}}\int_0^{\Delta\rho} 2\pi\rho d\rho \frac{\delta(\rho)}{\rho}; \tag{7.42}$$

the first integral can be performed exactly:

$$\int_0^{\Delta\rho} 2\pi\rho d\rho \frac{1}{\rho}\frac{\partial}{\partial\rho}\left(\rho\frac{\partial \mathbf{f}}{\partial\rho}\right) = 2\pi\int_0^{\Delta\rho}\frac{\partial}{\partial\rho}\left(\rho\frac{\partial \mathbf{f}}{\partial\rho}\right)d\rho = 2\pi\left[\rho\frac{\partial \mathbf{f}}{\partial\rho}\right]_0^{\Delta\rho}, \tag{7.43}$$

while we can use a Taylor expansion for the second integral, to obtain

$$\int_0^{\Delta\rho} 2\pi\rho d\rho \mathbf{f} = 2\pi\int_0^{\Delta\rho}\rho d\rho\sum_0^{\infty}\rho^n\frac{1}{n!}\frac{\partial^n \mathbf{f}(0)}{\partial\rho^n} = 2\pi\sum_0^{\infty}\frac{1}{n!}\frac{\partial^n \mathbf{f}(0)}{\partial\rho^n}\int_0^{\Delta\rho}\rho^{n+1}d\rho, \tag{7.44}$$

and

$$\int_0^{\Delta\rho} 2\pi\rho d\rho \mathbf{f} = 2\pi\Delta\rho^2\sum_0^{\infty}\frac{1}{n!}\frac{\partial^n \mathbf{f}(0)}{\partial\rho^n}\frac{\Delta\rho^n}{n+2}. \tag{7.45}$$

Finally, in view of the symmetry of the delta function, the third integral, on the right-hand side of Eq. (7.42) yields:

$$\int_0^{\Delta\rho} 2\pi\rho d\rho\frac{\delta(\rho)}{\rho} = 2\pi\int_0^{\Delta\rho}\delta(\rho)d\rho = 2\pi\frac{1}{2}\int_{-\Delta\rho}^{\Delta\rho}\delta(\rho)d\rho = \pi. \tag{7.46}$$

Therefore, taking the limit where $\Delta\rho \to 0$, the boundary condition reduces to

$$\lim_{\rho\to 0}\left(\rho\frac{\partial \mathbf{f}}{\partial\rho}\right) = \frac{e\mu_0}{(2\pi)^{3/2}}. \tag{7.47}$$

We now return to Equation 7.36: it is clear that two different cases should be considered, as the quantity $s^2 = \frac{\omega^2}{v^2}[\beta^2 n^2(\omega) - 1]$ can lead to a pure imaginary value of the parameter $s$, obtained when the electron velocity is such that $\beta < n^{-1}(\omega)$, whereas for higher velocities, $s \in \mathbb{R}$. This translates into two very distinct behaviors for the corresponding solutions of the wave equation, as can be seen by considering the asymptotic expressions for the Hankel functions. For large values of $z$, we have

$$\mathrm{H}_0^1(\rho) \approx \sqrt{\frac{2}{\pi\rho}}\exp\left[i\left(\rho - \frac{\pi}{4}\right)\right], \tag{7.48}$$

and

$$\mathrm{H}_0^2(\rho) \approx \sqrt{\frac{2}{\pi\rho}}\exp\left[-i\left(\rho - \frac{\pi}{4}\right)\right]. \tag{7.49}$$

Therefore, for low velocities where $s$ is purely imaginary, and large radii, the parameter $\rho = rs$ is large and pure imaginary, and the function $H_0^2(\rho)$ diverges. As a result, we must choose the constant $a_2 = 0$ to ensure that the behavior of the field at large distances is physical. At this point, we can use the boundary condition given in Equation 7.47 to determine the value of the other constant, $a_1$:

$$\lim_{r\to 0}\left(r\frac{\partial f}{\partial r}\right) = \lim_{r\to 0}\left\{a_1 r\frac{\partial H_0^1[s(\omega)r]}{\partial r}\right\} = \frac{2i}{\pi}a_1 = -\frac{e\mu_0}{(2\pi)^{3/2}} \tag{7.50}$$

This yields the sought-after solution for low velocities:

$$f(r,\omega) = a_1 H_0^1[s(\omega)r] = i\frac{e\mu_0}{4\sqrt{2\pi}}H_0^1[s(\omega)r] \approx i\frac{e\mu_0}{4\pi}\sqrt{\frac{1}{s(\omega)r}}\exp\left\{i\left[s(\omega)r - \frac{\pi}{4}\right]\right\}. \tag{7.51}$$

Returning to the asymptotic behavior of Hankel functions, we see that for low velocities, where $\beta n < 1$, the field decreases exponentially for large values of $r$: the electron does not radiate.

The second case corresponds to the condition $\beta n > 1$, which leads to a real value for the parameter $s$. In this case, both Hankel functions are valid solutions, and we must address the problem within the context of causal Green functions. Since we are considering the solution in frequency space, this implies that we must distinguish between positive and negative frequencies. Causality implies that

$$\begin{aligned} f(r,\omega) &= -i\frac{e\mu_0}{4\sqrt{2\pi}}H_0^2[s(\omega)r], \quad \omega > 0, \quad s \in \mathbb{R}, \\ f(r,\omega) &= +i\frac{e\mu_0}{4\sqrt{2\pi}}H_0^1[s(\omega)r], \quad \omega < 0, \quad s \in \mathbb{R}. \end{aligned} \tag{7.52}$$

Grouping these terms and Fourier transforming back into the time domain, we have

$$A_z(r,z,t) = \frac{e\mu_0}{8\pi}\left[\left[\int_0^{+\infty} e^{i\omega\left(t-\frac{z}{v}\right)}\{-iH_0^2[s(\omega)r]\}d\omega + \int_{-\infty}^{0} e^{i\omega\left(t-\frac{z}{v}\right)}\{iH_0^1[s(\omega)r]\}d\omega\right]\right]. \tag{7.53}$$

Although Equation 7.53 cannot be integrated analytically, one can use the asymptotic expansions discussed above to obtain the expression of the radiated potential at large distances:

$$A_z(r,z,t) = \frac{e\mu_0}{4\pi}\int_0^{+\infty}\exp\left\{-i\left[\omega\left(t-\frac{z}{v}\right)+s(\omega)r-\frac{\pi}{4}\right]\right\}\frac{d\omega}{\sqrt{\pi s(\omega)r}}, \tag{7.54}$$

which can also be expressed as

$$A_z(r, z, t) = -i\frac{e\mu_0}{4\pi}\int_{-\infty}^{+\infty} \frac{1}{\sqrt{rs}} \cos \chi \, d\omega, \tag{7.55}$$

where we have defined the Čerenkov angle $\theta = \arccos(1/\beta n)$ and introduced the parameter

$$\chi = \omega\left(t - \frac{z\cos\theta + r\sin\theta}{v\cos\theta}\right) + \frac{\pi}{4}. \tag{7.56}$$

Within this context, Equation 7.55 clearly establishes the geometry of the radiation process, as a sum of conical waves propagating at an angle $\theta$ with respect to the axis of symmetry defined by the velocity of the charge. Moreover, the Čerenkov angle, which is a function of the frequency, indicates the geometrical aperture of the cone for a given radiated wavelength:

$$\cos\theta(\omega) = \frac{1}{n(\omega)\beta}. \tag{7.57}$$

The electromagnetic field radiated is then obtained by using Equations 7.5 and 7.6, while the power radiated is derived from the Poynting vector, $\mathbf{S} = \mathbf{E}\times\mathbf{H} = \mathbf{E}\times\mathbf{B}/\mu_0$. Given the cylindrical symmetry of the radiation process, the electric field has two components, radial and axial. We have already established the relation between the scalar potential and the axial component of the vector potential: $A_z = \beta\phi/c$. Therefore, we simply have

$$\begin{aligned} E_r &= -\partial_r\phi = -v\partial_r A_z, \\ E_z &= -\partial_z\phi - \partial_t A_z = -[v\partial_z + \partial_t]A_z. \end{aligned} \tag{7.58}$$

For the magnetic field, the relation $\mathbf{B} = \mu_0\mathbf{H} = \nabla\times\mathbf{A}$ and the symmetry of the problem yield only one component: the radial component vanishes because $A_\varphi = 0$ and $\partial_\varphi \equiv 0$, while the axial component is identically zero since $A_r = 0$ and $A_\varphi = 0$. We are left with the azimuthal component,

$$H_\varphi = -\frac{1}{\mu_0}\partial_r A_z. \tag{7.59}$$

For the corresponding Fourier components of the electric and magnetic fields, we return to Equations 7.16 and 7.17. We have

$$\tilde{E}_r(\omega) = -\frac{ic^2}{\omega n^2}\frac{\partial^2\tilde{A}_z(\omega)}{\partial z\,\partial r}, \tag{7.60}$$

$$\tilde{E}_z(\omega) = -\frac{ic^2}{\omega n^2}\frac{\partial^2\tilde{A}_z(\omega)}{\partial z^2} - i\omega\tilde{A}_z(\omega), \tag{7.61}$$

and

$$\tilde{H}_\varphi(\omega) = -\frac{1}{\mu_0}\frac{\partial \tilde{A}_z(\omega)}{\partial r}. \tag{7.62}$$

The Poynting vector components are

$$\begin{aligned} S_r &= (E_\varphi H_z - E_z H_\varphi) = -E_z H_\varphi, \\ S_\varphi &= (E_z H_r - E_r H_z) = 0, \\ S_z &= (E_r H_\varphi - E_\varphi H_r) = E_r H_\varphi. \end{aligned} \tag{7.63}$$

The fact that there is no energy flux in the azimuthal direction is dictated by the propagation geometry: in vacuum, the Poynting vector is always parallel to the wavenumber, **k**. For Čerenkov radiation, this is in the direction of the cone, with a radial and an axial component for **k** but no azimuthal component.

The energy flux through a unit cylindrical surface is

$$\frac{dW}{d\ell} = 2\pi r\int_{-\infty}^{+\infty} S_r\, dt, \tag{7.64}$$

which is now recast as

$$\frac{dW}{d\ell} = -2\pi r\int_{-\infty}^{+\infty} E_z H_\varphi\, dt. \tag{7.65}$$

Equation 7.65 represents the Čerenkov radiation loss per unit distance; the time-dependent fields are given by the following asymptotic Fourier integrals:

$$E_z = \mu_0 e\int_{-\infty}^{+\infty}\left(1-\frac{1}{\beta^2 n^2}\right)\frac{\omega\, d\omega}{\sqrt{s(\omega) r}} \tag{7.66}$$

and

$$H_\varphi = -e\int_{-\infty}^{+\infty}\sqrt{\frac{s(\omega)}{r}}\cos\chi\, d\omega. \tag{7.67}$$

Using Equations 7.66 and 7.67 into Equation 7.65, we now have

$$\frac{dW}{d\ell} = \mu_0 e^2\int_{-\infty}^{+\infty}\int_{-\infty}^{+\infty}\left(1-\frac{1}{\beta^2 n^2}\right)\cos(\omega t+\alpha)\cos(\omega' t+\alpha')\sqrt{\frac{s(\omega)}{s(\omega')}}\,\omega' d\omega\, dt. \tag{7.68}$$

The delta-function representation,

$$\int_{-\infty}^{+\infty} \cos(\omega t + \alpha)\cos(\omega' t + \alpha')dt = \pi\delta(\omega' - \omega), \tag{7.69}$$

can then be used to obtain the important result, first derived by Tamm and Frank:

$$\frac{dW}{d\ell} = \mu_0 e^2 \int_{\beta n > 1} \left[1 - \frac{1}{\beta^2 n^2(\omega)}\right] \omega \, d\omega. \tag{7.70}$$

## 7.3 Fields and Inductions, Polarization and Nonlinear Susceptibilities

Having established the classical Tamm–Frank theory of Čerenkov radiation, we are now in a position to investigate the question of the relativistic transformation of the refractive index. In general, the interaction of electromagnetic waves with matter can be described according to two distinct theoretical formulations. On the one hand, the electromagnetic properties of the medium under consideration may be defined by introducing relations between the fields and the inductions; this approach is usually referred to as the Minkowski formulation. Generally, these relations, called constitutive relations, are complex, tensorial, and nonlinear. On the other hand, the other formulation describes the reaction of the medium to the electromagnetic waves in terms of an induced four-current density. As long as the theoretical analysis of the interaction of electromagnetic radiation with matter is performed in the rest frame of the medium, these two approaches are equivalent. However, whereas the four-vector current density formalism can lead to a covariant description of the electrodynamics of nonlinear media, the relations between fields and inductions become very complicated in any reference frame where the medium is moving relativistically. This is particularly true in the case of a nonlinear medium. Still, it should be noted that in the rest frame of the medium, the constitutive relations describing its electromagnetic properties, which are generally derived from quantum mechanics and group theory, directly reflect the underlying spatial symmetries of the medium and therefore are usually the preferred formulation in nonlinear optics. In the relativistic case, the difficulty arises from the fact that the Lorentz group conserves space–time symmetries rather than spatial symmetries. For example, it is possible to transform a tetragonal lattice into a cubic one through the Lorentz transform; a spin-polarized relativistic electron beam with the right energy will be sensitive to the magnetic phase transition corresponding to this relativistic symmetry effect, and spin-resonance phenomena could result from such experiments. Similarly, the Čerenkov radiation process in a linear, isotropic dielectric, which is studied in this chapter, can be viewed in the rest frame of the interacting electron as resulting from a singularity of the now anisotropic refractive index of the medium for electromagnetic waves propagating at the Čerenkov angle.

In this section, we review briefly the definition of the fields and inductions within the context of Maxwell's equations. The electromagnetic interaction is characterized, in the classical theory, by the electric field **E** and the magnetic field **H**. The corresponding electric and magnetic inductions are **D** and **B**, respectively. Maxwell's equations are conventionally separated in two groups. The first group, also called the source-free group, corresponds to

$$\begin{aligned} \nabla \times \mathbf{E} + \partial_t \mathbf{B} &= \mathbf{0}, \\ \nabla \cdot \mathbf{B} &= 0, \end{aligned} \tag{7.71}$$

and the second group is described by

$$\begin{aligned} \nabla \times \mathbf{H} - \partial_t \mathbf{D} &= \mathbf{j}, \\ \nabla \cdot \mathbf{D} &= \rho. \end{aligned} \tag{7.72}$$

Here, $j_\mu = (c\rho, \mathbf{j})$ is the four-vector current density.

Maxwell's equations combine the fields and inductions; the additional relations between the fields and inductions in vacuum are

$$\begin{aligned} \mathbf{D} &= \varepsilon_0 \mathbf{E}, \\ \mathbf{B} &= \mu_0 \mathbf{H}, \end{aligned} \tag{7.73}$$

where the permittivity, $\varepsilon_0$, and the permeability, $\mu_0$, of free space are related to the speed of light in vacuum through the well-known equation

$$\varepsilon_0 \mu_0 c^2 = 1. \tag{7.74}$$

It should be noted here that in classical electrodynamics, the vacuum is a linear, isotropic medium. In QED, vacuum nonlinearities appear near the Schwinger critical field for pair creation, introduced and briefly discussed in Chapter 1. In a medium, the most general relations are tensorial, nonlinear, and anisotropic and can be represented as

$$\begin{aligned} \mathbf{D} &= \mathbf{D}(\mathbf{E}, \mathbf{H}), \\ \mathbf{B} &= \mathbf{B}(\mathbf{E}, \mathbf{H}). \end{aligned} \tag{7.75}$$

Here, we allow the possibility of coupled nonlinear electric and magnetic effects, since relativity requires the treatment of electric and magnetic phenomena on an equal footing. In the low-field limit, one can expand the above expressions in a Taylor series and take into account lower-order nonlinearities only. The corresponding polynomial coefficients are the so-called nonlinear susceptibilities. In the rest frame of the medium, these constitutive relations are determined by the structure of the medium at atomic scale and by its spatial symmetries. The nonlinear susceptibilities can thus be derived from quantum mechanics and group theory. They are generally tensorial in character and describe the macroscopic electromagnetic properties of the nonlinear medium. The nonlinear susceptibilities are semiclassical in the sense that they are averaged over a large number of atomic systems, as optical wavelengths are generally long compared to typical lattice scales. Quantum effects do appear in certain optical nonlinear effects, as discussed at the end of Chapter 6; nevertheless, a large category of nonlinear media can be described adequately by the formalism outlined here.

A brief digression regarding the concept of nonlinear susceptibility will prove useful to better illustrate our discussion. In this brief introduction, we closely follow the presentation of Mandel and Wolf and use a similar notation, as it is widely adopted in nonlinear optics. We begin with the classical definition of the electromagnetic energy in a dielectric medium:

$$W = \iiint \frac{1}{2\mu_0}\mathbf{B}^2(\mathbf{x},t)d^3x + \iiint d^3x \int_0^{\mathbf{D}(\mathbf{x},t)} \mathbf{E}(\mathbf{x},t)\cdot d\mathbf{D}(\mathbf{x},t). \tag{7.76}$$

The first term takes a simple form because the medium has the permeability of vacuum: $\mathbf{B} = \mu_0\mathbf{H}$. The integral over the electric induction is generally difficult to perform in view of the complex relation between the electric field and induction in a nonlinear, anisotropic medium. On the other hand, in the case of vacuum, we have $\mathbf{D} = \varepsilon_0\mathbf{E}$, and Equation 7.76 reduces to the well-known equation for the electromagnetic energy density:

$$\frac{dW}{dv} = \frac{d^3W}{dx^3} = \frac{\varepsilon_0}{2}\mathbf{E}^2 + \frac{1}{2\mu_0}\mathbf{B}^2. \tag{7.77}$$

The relation between the electric field and induction can also be expressed in terms of the induced polarization, $\mathbf{P}(\mathbf{x}, t)$, by subtracting the vacuum induction or displacement:

$$\mathbf{D}(\mathbf{x},t) = \mathbf{D}[\mathbf{E}(\mathbf{x},t)] = \varepsilon_0\mathbf{E}(\mathbf{x},t) + \mathbf{P}[\mathbf{E}(\mathbf{x},t)]. \tag{7.78}$$

The polarization can then be Taylor-expanded as

$$P_i = \chi_{ij}^{(1)}E_j + \chi_{ijk}^{(2)}E_jE_k + \chi_{ijkl}^{(3)}E_jE_kE_l + \cdots, \tag{7.79}$$

which defines the linear susceptibility, $\chi_{ij}^{(1)}$, and its nonlinear counterparts, $\chi_{ij\ldots p}^{(n)}$. Note that the susceptibilities are tensors of rank $n + 1$. An interesting question, in view of the ideas expressed in Chapter 6 concerning the quantum nature of electromagnetic radiation, is the commutation of the various electric field components in Equation 7.79. Additionally, Equation 7.79 is valid only for a local and nondispersive medium. In frequency space, we can relate the polarization and the electric field as follows:

$$
\begin{aligned}
\boldsymbol{P}_i(\omega_1) = {}& \chi_{ij}^{(1)}(\omega_1;\omega_1)\boldsymbol{E}_j(\omega_1) + \chi_{ijk}^{(2)}(\omega_1;\omega_1-\omega_2,\omega_2)\boldsymbol{E}_j(\omega_1-\omega_2)\boldsymbol{E}_k(\omega_2) \\
& + \chi_{ijkl}^{(3)}(\omega_1;\omega_1-\omega_2-\omega_3,\omega_2,\omega_3)\boldsymbol{E}_j(\omega_1-\omega_2-\omega_3)\boldsymbol{E}_k(\omega_2)\boldsymbol{E}_l(\omega_3)+\cdots.
\end{aligned}
\tag{7.80}
$$

Equation 7.80 clearly indicates how various frequencies can be generated via nonlinear effects. The susceptibilities can be derived through various models of the medium under consideration, starting from the simple classical Thomson electron model, where electrons are subjected to a restoring force binding them to the atoms in a parabolic potential well, all the way to fully quantum mechanical descriptions of the radiation field-atom interactions involved. Deviations from the aforementioned parabolic potential well result in nonlinear interactions, as the restoring force depends on higher powers of the drive field. In closing, we note that with the definitions introduced above, the electromagnetic field energy in the nonlinear medium can now be expressed as

$$
W = \iiint d^3x\left[\frac{1}{2\mu_0}\mathbf{B}^2(\mathbf{x},t)+\frac{\varepsilon_0}{2}\mathbf{E}^2(\mathbf{x},t)+X_1(\mathbf{x})+X_2(\mathbf{x})+\cdots\right], \tag{7.81}
$$

where the linear component of the induced polarization contributes the term

$$
X_1(\mathbf{x}) = \frac{1}{2}\int d\omega\int d\omega'\chi_{ij}^{(1)}(\omega;\omega')E_i(\mathbf{x},\omega')E_j(\mathbf{x},\omega), \tag{7.82}
$$

while the nonlinear energy density associated with the $\chi^{(2)}$ susceptibility is of the form

$$
X_2(\mathbf{x}) = \frac{1}{3}\int d\omega\int d\omega'\int d\omega''\chi_{ijk}^{(2)}(\omega'';\omega-\omega',\omega')E_i(\mathbf{x},\omega'')E_j(\mathbf{x},\omega-\omega')E_k(\mathbf{x},\omega'). \tag{7.83}
$$

It is also important to emphasize that, in addition, the most general relations are nonlocal in character, as specified by the Kramer–Krönig dispersion theory, and can be described only through space–time integrals. In the following, we make the implicit assumption of steady-state, and we assume that the relations between the fields and the inductions can be satisfactorily described by quasilocal expressions.

We now consider the interaction of electromagnetic waves with a nonlinear medium, in the absence of external fields, except for the incident wave. Two equivalent descriptions are available. In the first approach, we consider

Maxwell's equations with no source term ($\rho = 0$, $\mathbf{j} = \mathbf{0}$), and describe the electromagnetic properties of the medium through its constitutive relations. We have the following set of equations:

$$\nabla \cdot \mathbf{D}(\mathbf{E}, \mathbf{H}) = 0, \tag{7.84}$$

$$\nabla \cdot \mathbf{B}(\mathbf{E}, \mathbf{H}) = 0, \tag{7.85}$$

$$\nabla \times \mathbf{E} + \partial_t \mathbf{B}(\mathbf{E}, \mathbf{H}) = \mathbf{0}, \tag{7.86}$$

and

$$\nabla \times \mathbf{H} - \partial_t \mathbf{D}(\mathbf{E}, \mathbf{H}) = \mathbf{0}. \tag{7.87}$$

Here, **E** and **H** represent the incoming electromagnetic wave, and **D** and **B** represent the reactions of the nonlinear medium; the sources are integrated into the inductions. Equations 7.84 to 7.87, together with the constitutive relations given in Equation 7.75, describe electromagnetic phenomena with the framework of the so-called Minkowski formulation.

In the second formulation, we consider Maxwell's equations in vacuum, and we describe the nonlinear reactions of the medium through source terms. The constitutive relations are those of vacuum, and we now have

$$\nabla \cdot \varepsilon_0 \mathbf{E} = \rho(\mathbf{E}, \mathbf{H}), \tag{7.88}$$

$$\nabla \cdot \mu_0 \mathbf{H} = 0, \tag{7.89}$$

$$\nabla \times \mathbf{E} + \mu_0 \partial_t \mathbf{H} = \mathbf{0}, \tag{7.90}$$

and

$$\nabla \times \mathbf{H} - \varepsilon_0 \partial_t \mathbf{E} = \mathbf{j}(\mathbf{E}, \mathbf{H}). \tag{7.91}$$

These two sets of equations constitute two alternative formulations of the electrodynamics of nonlinear media. However, their mathematical properties under transformations of the Lorentz group are quite different. As will be discussed in Section 7.7, the second formulation is covariant because the vacuum constitutive relations are invertible and because the induced source terms are described by a four-vector. In the next two sections, we will study the relativistic transform of the first set of equations in the case of a linear, isotropic, dielectric medium.

## 7.4 Transform of Linear Refractive Index: Minkowski Formulation

Here, we study the basic interaction of electromagnetic waves with a linear, isotropic, dielectric medium, within the Minkowski formulation. We thus make use of the relations between fields and inductions in the medium; in other words, we consider Maxwell's equations with no source terms and describe the electromagnetic properties of the scattering medium through its constitutive relations. In this section and in the remainder of the analysis, the primed variables refer to the rest frame of the medium. For the case of a linear, isotropic medium considered in this section, the constitutive relations are given, in the rest frame of the medium, by

$$\begin{aligned}\mathbf{D}' &= \varepsilon \mathbf{E}',\\ \mathbf{B}' &= \mu \mathbf{H}',\end{aligned} \tag{7.92}$$

and Maxwell's equations reduce to

$$\nabla' \cdot \varepsilon \mathbf{E}' = 0, \tag{7.93}$$

$$\nabla' \cdot \mu \mathbf{H}' = 0, \tag{7.94}$$

$$\nabla' \times \mathbf{E}' + \mu \partial_{t'} \mathbf{H}' = \mathbf{0}, \tag{7.95}$$

$$\nabla' \times \mathbf{H}' - \varepsilon \partial_{t'} \mathbf{E}' = \mathbf{0}. \tag{7.96}$$

At this point, we briefly review the dispersion of electromagnetic waves, as described in the rest frame of the medium. We can represent the electromagnetic wave by a space–time Fourier transform:

$$\mathbf{E}'(x'_\mu) = \frac{1}{(\sqrt{2\pi})^4} \int_{\mathbb{R}^4} \tilde{\mathbf{E}}'(k'_\mu) \exp(i k'_\mu x'^\mu) d^4k, \tag{7.97}$$

where $x'_\mu = (ct', \mathbf{x}')$ is the four-position in the rest frame of the medium, and $k'_\mu = (c^{-1}\omega', \mathbf{k}')$ is its conjugate, the four-wavenumber.

We then have the following operational equivalences in conjugate space:

$$\nabla' \equiv -i\mathbf{k}', \quad \partial_{t'} \equiv i\omega'. \tag{7.98}$$

Taking the curl of Equation 7.95, and making use of Equations 7.93 and 7.94, we obtain

$$(\varepsilon\mu \partial_{t'}^2 - \nabla'^2)\mathbf{E}' = \mathbf{0}; \tag{7.99}$$

combining Equation 7.99 with the operational equivalences defined above, we recover the usual dispersion relation, in the form

$$(\varepsilon\mu\omega'^2 - \mathbf{k}'^2)\tilde{\mathbf{E}}' = \mathbf{0}. \tag{7.100}$$

Finally, the refractive index $n'$ is defined as

$$n' = \left|\frac{c\mathbf{k}'}{\omega'}\right| = c\sqrt{\varepsilon\mu}. \tag{7.101}$$

We now study the same basic phenomenon, viewed from another reference frame. The invariance of Maxwell's equations under the Lorentz transform, which results directly from the principle of relativity, yields the transformation formulas for the fields and inductions. This is described in great detail in the monograph by Poincelot, which is listed in the references to this chapter. Here, we directly use the result of the derivation. We have

$$\mathbf{E}' = \gamma\left[\mathbf{E} - (1-\alpha)(\mathbf{E}\cdot\mathbf{v})\frac{\mathbf{v}}{v^2} + \mathbf{v}\times\mathbf{B}\right], \tag{7.102}$$

$$\mathbf{H}' = \gamma\left[\mathbf{H} - (1-\alpha)(\mathbf{H}\cdot\mathbf{v})\frac{\mathbf{v}}{v^2} - \mathbf{v}\times\mathbf{D}\right], \tag{7.103}$$

$$\mathbf{D}' = \gamma\left[\mathbf{D} - (1-\alpha)(\mathbf{D}\cdot\mathbf{v})\frac{\mathbf{v}}{v^2} + \frac{\mathbf{v}}{c^2}\times\mathbf{H}\right], \tag{7.104}$$

and

$$\mathbf{B}' = \gamma\left[\mathbf{B} - (1-\alpha)(\mathbf{B}\cdot\mathbf{v})\frac{\mathbf{v}}{v^2} - \frac{\mathbf{v}}{c^2}\times\mathbf{E}\right], \tag{7.105}$$

where $\mathbf{v}$ is the velocity of the medium relative to the reference frame under consideration, and $\gamma = 1/\sqrt{1-(\mathbf{v}/c)^2} = \alpha^{-1}$ is the relativistic factor.

Note that, as the transformation formulas result directly from the relativistic invariance of Maxwell's equations, they combine the fields and inductions. We can now rewrite the constitutive relations given in Equation 7.92 as follows:

$$\mathbf{D} - (1-\alpha)(\mathbf{D}\cdot\mathbf{v})\frac{\mathbf{v}}{v^2} + \frac{\mathbf{v}}{c^2}\times\mathbf{H} = \varepsilon\left[\mathbf{E} - (1-\alpha)(\mathbf{E}\cdot\mathbf{v})\frac{\mathbf{v}}{v^2} + \mathbf{v}\times\mathbf{B}\right], \tag{7.106}$$

$$\mathbf{B} - (1-\alpha)(\mathbf{B}\cdot\mathbf{v})\frac{\mathbf{v}}{v^2} - \frac{\mathbf{v}}{c^2}\times\mathbf{E} = \mu\left[\mathbf{H} - (1-\alpha)(\mathbf{H}\cdot\mathbf{v})\frac{\mathbf{v}}{v^2} - \mathbf{v}\times\mathbf{D}\right]. \tag{7.107}$$

Here, we have simply replaced the fields and inductions by the expressions obtained using Lorentz invariance. Taking the scalar product of Equations 7.106 and 7.107 with **v** yields the simple relations

$$\begin{aligned} \mathbf{D}\cdot\mathbf{v} &= \varepsilon\mathbf{E}\cdot\mathbf{v}, \\ \mathbf{B}\cdot\mathbf{v} &= \mu\mathbf{H}\cdot\mathbf{v}. \end{aligned} \tag{7.108}$$

Using these identities in Equations 7.106 and 7.107, we have

$$\mathbf{D}+\frac{\mathbf{v}}{c^2}\times\mathbf{H} = \varepsilon(\mathbf{E}+\mathbf{v}\times\mathbf{B}), \tag{7.109}$$

$$\mathbf{B}-\frac{\mathbf{v}}{c^2}\times\mathbf{E} = \mu(\mathbf{H}-\mathbf{v}\times\mathbf{D}). \tag{7.110}$$

Finally, after some straightforward calculations, we can eliminate **B** from Equation 7.109 and obtain the sought-after constitutive relations:

$$\mathbf{D}(\mathbf{E},\mathbf{H}) = \chi\frac{\varepsilon}{\gamma^2}\mathbf{E}+\chi\left(\varepsilon\mu-\frac{1}{c^2}\right)\mathbf{v}\times\mathbf{H}-\varepsilon\chi\left(\varepsilon\mu-\frac{1}{c^2}\right)\mathbf{v}(\mathbf{v}\cdot\mathbf{E}), \tag{7.111}$$

and

$$\mathbf{B}(\mathbf{E},\mathbf{H}) = \chi\frac{\mu}{\gamma^2}\mathbf{H}-\chi\left(\varepsilon\mu-\frac{1}{c^2}\right)\mathbf{v}\times\mathbf{E}-\mu\chi\left(\varepsilon\mu-\frac{1}{c^2}\right)\mathbf{v}(\mathbf{v}\cdot\mathbf{H}). \tag{7.112}$$

Here, we have defined the following dimensionless parameter:

$$\chi = \frac{1}{1-\varepsilon\mu\mathbf{v}^2}. \tag{7.113}$$

The constitutive relations can be recast in the following form:

$$\mathbf{D} = \xi\mathbf{E}+\eta\mathbf{v}\times\mathbf{H}-\varepsilon\eta\mathbf{v}(\mathbf{v}\cdot\mathbf{E}), \tag{7.114}$$

and

$$\mathbf{B} = \frac{\mu}{\varepsilon}\xi\mathbf{H}-\eta\mathbf{v}\times\mathbf{E}-\mu\eta\mathbf{v}(\mathbf{v}\cdot\mathbf{H}), \tag{7.115}$$

by introducing the coefficients

$$\xi = \varepsilon\frac{\chi}{\gamma^2},\quad \eta = \chi\left(\varepsilon\mu-\frac{1}{c^2}\right). \tag{7.116}$$

The new constitutive relations are still linear, but, as expected, they combine both electric and magnetic contributions, and they are obviously anisotropic, as the relative velocity **v** breaks the original spherical symmetry of the problem. In addition, the noncovariant character of the pseudoscalars $\varepsilon$ and $\mu$ appears very clearly.

## 7.5 Anomalous Refractive Index and Čerenkov Effect

At this point, it is possible to study the propagation of electromagnetic waves in a frame where the scattering medium is not at rest. We start from Maxwell's equations with no sources, and we make use of the constitutive relations derived above:

$$\nabla \times \mathbf{E} + \partial_t \left[ \frac{\mu}{\varepsilon} \xi \mathbf{H} - \eta \mathbf{v} \times \mathbf{E} - \mu\eta \mathbf{v}(\mathbf{v} \cdot \mathbf{H}) \right] = \mathbf{0}, \tag{7.117}$$

$$\nabla \times \mathbf{H} - \partial_t [\xi \mathbf{E} + \eta \mathbf{v} \times \mathbf{H} - \varepsilon\eta \mathbf{v}(\mathbf{v} \cdot \mathbf{E})] = \mathbf{0}. \tag{7.118}$$

Again, we represent the electromagnetic field by a four-dimensional Fourier transform,

$$\mathbf{E}(x_\mu) = \frac{1}{(\sqrt{2\pi})^4} \int_{\mathbb{R}^4} \tilde{\mathbf{E}}(k_\mu) \exp(ik_\mu x^\mu) d^4k, \tag{7.119}$$

which yields the usual operational equivalences,

$$\nabla \equiv -i\mathbf{k}, \quad \partial_t \equiv i\omega. \tag{7.120}$$

Combining Equations 7.117 and 7.118 to eliminate the magnetic field **H**, we obtain

$$(\mathbf{k} + \omega\eta\mathbf{v}) \times (\mathbf{k} + \omega\eta\mathbf{v}) \times \tilde{\mathbf{E}} + \omega^2 \frac{\mu}{\varepsilon} \xi^2 \tilde{\mathbf{E}} - \omega^2 \mu\xi\eta \mathbf{v}(\mathbf{v} \cdot \tilde{\mathbf{E}}) = \mathbf{0}. \tag{7.121}$$

We now define the transverse and parallel components of the following vectors:

$$\begin{aligned} \tilde{\mathbf{E}} &= \hat{z}\tilde{E}_\parallel + \hat{x}\tilde{E}_\perp, \\ \mathbf{k} &= \hat{z}k_\parallel + \hat{x}k_\perp, \end{aligned} \tag{7.122}$$

where we have defined the $z$-axis so that $\mathbf{v} = \hat{z}v$. Upon elimination of the amplitudes $\tilde{E}_{\parallel}$ and $\tilde{E}_{\perp}$ from Equation 7.121, which we write as

$$\begin{pmatrix} d_1 & d_2 \\ d_3 & d_4 \end{pmatrix}\begin{pmatrix} \tilde{E}_{\parallel} \\ \tilde{E}_{\perp} \end{pmatrix} = \begin{pmatrix} 0 \\ 0 \end{pmatrix}, \tag{7.123}$$

we end up with the following dispersion relation: $d_1 d_4 - d_2 d_3 = 0$, which yields

$$\omega^2 \chi^2 \frac{\varepsilon\mu}{\gamma^2} - (k_{\parallel} + \omega\eta v)^2 = k_{\perp}^2 \frac{\xi}{\xi - \varepsilon\eta v^2}. \tag{7.124}$$

Defining the propagation angle, $\theta$, the dispersion relation can be recast as

$$\omega^2 \chi^2 \frac{\varepsilon\mu}{\gamma^2} - \left[k\cos\theta + \omega v\chi\left(\varepsilon\mu - \frac{1}{c^2}\right)\right]^2 = k^2 \sin^2\theta \frac{\chi}{\gamma^2}, \tag{7.125}$$

which shows clearly the anisotropy of the medium acquired under the Lorentz transform.

We first check the relativistic invariance of the dispersion relation, which also defines the photon mass-shell. The Lorentz transformation of the four-wavenumber gives

$$\begin{aligned} \omega &= \gamma(\omega' + k'_{\parallel} v), \\ k_{\parallel} &= \gamma\left(k'_{\parallel} + \beta\frac{\omega'}{c}\right), \\ k_{\perp} &= k'_{\perp}. \end{aligned} \tag{7.126}$$

Here, $\beta = v/c$ is the normalized velocity.

We can use these expressions in the dispersion relation, Equation 7.124, to obtain, after some straightforward algebra,

$$(\omega'\sqrt{\varepsilon\mu} + k'_{\parallel})(-\omega'\sqrt{\varepsilon\mu} + k'_{\parallel}) = -k'^2_{\perp}, \tag{7.127}$$

which reduces to

$$k'^2_{\parallel} + k'^2_{\perp} = \omega'^2 \varepsilon\mu. \tag{7.128}$$

Equation 7.122 is identical to the dispersion relation of electromagnetic waves in a linear, isotropic medium, described by Equation 7.100.

We now derive the refractive index, $n = |c\mathbf{k}/\omega|$ from Equation 7.125 to obtain

$$n(\theta) = \frac{c\left(\frac{\mu}{\varepsilon}\xi^2 - \eta^2 v^2\right)}{\eta v \cos\theta + \sqrt{\cos^2\theta\,\xi^2\frac{\mu}{\varepsilon} + \frac{\sin^2\theta\,\xi\left(\frac{\mu}{\varepsilon}\xi^2 - \eta^2 v^2\right)}{\xi - \varepsilon\eta v^2}}}. \tag{7.129}$$

Despite its complexity, there are two simple limiting cases to the above equation. On the one hand, one may consider vacuum ($\varepsilon = \varepsilon_0$, $\mu = \mu_0$), in which case, one finds $n = 1$. The other limiting case is obtained by taking $v = 0$, $\gamma = 1$; we then verify that $n = n' = c\sqrt{\varepsilon\mu}$. The most interesting feature of Equation 7.129, however, is the fact that the index of refraction exhibits a singularity for

$$\tan^2\theta = -\frac{\xi - \varepsilon\eta v^2}{\xi} = -\frac{\gamma^2}{\chi}. \tag{7.130}$$

This means that we can expect a strong coupling of the radiation field to a static charge ($\omega = 0$) for this particular radiation angle, as a divergence of the refractive index corresponds to a finite wavenumber at zero frequency. We can translate this condition into the rest frame of the medium by noting that

$$\tan\theta' = \frac{k'_\perp}{k'_\parallel} = \frac{k_\perp}{\gamma\left(k_\parallel - \beta\frac{\omega}{c}\right)} = \frac{\tan\theta}{\gamma\left(1 - \frac{\beta}{n\cos\theta}\right)}. \tag{7.131}$$

At the singularity, where $n \to \infty$, we find

$$\tan^2\theta' = \frac{\tan^2\theta}{\gamma^2} = \varepsilon\mu v^2 - 1, \tag{7.132}$$

which finally yields

$$\cos\theta' = \frac{1}{n'\beta'}, \tag{7.133}$$

the well-known Čerenkov radiation threshold condition.

We have thus shown that the Čerenkov radiation condition to a singularity of the refractive index of the interacting medium, similar to that of an atomic transition, in the rest frame of the particle. It is particularly interesting to note that in the rest frame of the test particle, we only need to study the refractive index of the medium to infer the possibility of a radiation process, whereas in the rest frame of the medium, nothing in the dispersion relation indicates the possibility of Čerenkov radiation, and one has to solve entirely the field equations, following the Tamm–Frank method, to derive the Čerenkov threshold condition.

However, in the general case of a nonlinear medium, the constitutive relations now read

$$\begin{aligned} \mathbf{D}'(\mathbf{D},\mathbf{H}) &= \mathbf{D}'[\mathbf{E}'(\mathbf{E},\mathbf{B}),\ \mathbf{H}'(\mathbf{H},\mathbf{D})], \\ \mathbf{B}'(\mathbf{B},\mathbf{E}) &= \mathbf{B}'[\mathbf{E}'(\mathbf{E},\mathbf{B}),\ \mathbf{H}'(\mathbf{H},\mathbf{D})]. \end{aligned} \tag{7.134}$$

It is clear that for any complex nonlinear dependence of the inductions $\mathbf{D}'$ and $\mathbf{B}'$ on the fields $\mathbf{E}'$ and $\mathbf{H}'$, the inversion of the above equations will become analytically intractable. For the relativistic description of nonlinear media, the constitutive relation formalism proves to be inadequate, and there are no simple transformation formulae of the nonlinear susceptibilities. This is due to the incompatibility of the three-dimensional (spatial) tensorial character of the nonlinear susceptibilities with the fundamental four-dimensional character of the Lorentz transformation.

## 7.6 Linear Isotropic Medium: Induced-Source Formalism

In this section, we focus our attention on the induced-source formulation; in other words, we now consider Maxwell's equations in vacuum and describe the electromagnetic properties of the interacting medium through a source term, as prescribed in Equations 7.88 to 7.91. To illustrate this derivation, we first start from the basic example of a linear, isotropic medium. In this case, we have the following relation between the electromagnetic field and the induced current density, as expressed in the rest frame of the medium:

$$\mathbf{j}' = \varepsilon_0 \sigma \omega' \mathbf{E}' + \lambda \mathbf{k}' \times \mathbf{H}'. \tag{7.135}$$

The vectorial product for the magnetic field contribution results directly from the polar character of the vector **H**, as opposed to the axial character of **E**, and **j**; this, in turn, is related to the fact that the origin of magnetic properties in a material is determined by spin effects. We simplify matters further by considering a dielectric medium where, by definition, the charge density is null: $\rho' = 0$. Making use of the four-dimensional Fourier transform, we obtain the dispersion relation in the following form:

$$k'^2 = \frac{\omega'^2}{c^2}\left(\frac{1 - i\sigma}{1 - i\lambda}\right), \tag{7.136}$$

where the electric conductivity, $\sigma$, and its magnetic analog, $\lambda$, are defined as

$$\begin{aligned} 1 - i\sigma &= \frac{\varepsilon}{\varepsilon_0}, \\ 1 - i\lambda &= \frac{\mu_0}{\mu}. \end{aligned} \tag{7.137}$$

We can then identify Equation 7.136 to the usual dispersion relation for a linear, isotropic medium given by Equation 7.100.

We now consider the relativistic transformation of the source terms. Combining the dielectric condition and the Lorentz transform of the charge density, we have

$$\rho' = 0 = \gamma\left(\rho - \frac{\mathbf{v}}{c^2}\cdot\mathbf{j}\right). \tag{7.138}$$

Similarly, the relativistic transformation of the current density is given by

$$\mathbf{j}' = \mathbf{j} + \gamma\frac{\mathbf{v}}{v^2}[(1-\alpha)(\mathbf{j}\cdot\mathbf{v}) - \rho v^2]. \tag{7.139}$$

Using the dielectric condition, Equation 7.138 into Equation 7.139, we obtain a simplified expression of the current density:

$$\mathbf{j}' = \mathbf{j} + \frac{\mathbf{v}}{v^2}(1-\alpha)(\mathbf{j}\cdot\mathbf{v}). \tag{7.140}$$

Taking the scalar product of Equation 7.140 with **v** and making use of the relativistic transform of the four-wavenumber and electromagnetic fields, together with the expression of the induced linear current density, Equation 7.135 into Equation 7.138, we obtain the following expression for the charge density:

$$\rho = \frac{\gamma^2}{c^2}\{\varepsilon_0\sigma(\omega - \mathbf{v}\cdot\mathbf{k})(\mathbf{E}\cdot\mathbf{v}) + \lambda\mathbf{v}\cdot[\mathbf{k}\times(\mathbf{H} - \mathbf{v}\times\mathbf{D})]\}. \tag{7.141}$$

We now use the vacuum constitutive relations to rewrite the charge density transform as a function of the electromagnetic fields only:

$$\rho = \frac{\gamma^2}{c^2}\{\varepsilon_0\sigma(\omega - \mathbf{v}\cdot\mathbf{k})(\mathbf{E}\cdot\mathbf{V}) + \lambda\mathbf{v}\cdot[\mathbf{k}\times(\mathbf{H} - \varepsilon_0\mathbf{v}\times\mathbf{E})]\}. \tag{7.142}$$

Proceeding in the same way for the current density, we end up with

$$\begin{aligned}\mathbf{j} = {} & \gamma^2\varepsilon_0\sigma(\omega - \mathbf{v}\cdot\mathbf{k})(\mathbf{E} + \mu_0\mathbf{v}\times\mathbf{H}) + \gamma^2\lambda(1-\alpha)\frac{\mathbf{v}}{v^2}\{\mathbf{v}\cdot[\mathbf{k}\times(\mathbf{H} - \varepsilon_0\mathbf{v}\times\mathbf{E})]\} \\ & + \gamma\lambda\left[\!\left[\left\{\mathbf{k} + \gamma\frac{\mathbf{v}}{v^2}[(1-\alpha)(\mathbf{k}\cdot\mathbf{v}) - \beta^2\omega]\right\}\right.\right. \\ & \left.\left.\times(\mathbf{H} - \varepsilon_0\mathbf{v}\times\mathbf{E}) - (1-\alpha)(\mathbf{H}\cdot\mathbf{v})\frac{\mathbf{k}\times\mathbf{v}}{v^2}\right]\!\right].\end{aligned} \tag{7.143}$$

Here, again, $\mathbf{v}$ is the velocity of the medium relative to our frame of reference. We can now study the dispersion of electromagnetic waves in the linear, isotropic, dielectric medium considered here. We use the source term derived above to drive Maxwell's equations in vacuum:

$$\begin{aligned} \nabla \times \mathbf{E} + \mu_0 \partial_t \mathbf{H} &= \mathbf{0}, \\ \nabla \times \mathbf{H} - \varepsilon_0 \partial_t \mathbf{E} &= \mathbf{j}(\mathbf{E}, \mathbf{H}), \\ \nabla \cdot \varepsilon_0 \mathbf{E} &= \rho(\mathbf{E}, \mathbf{H}). \end{aligned} \tag{7.144}$$

Through four-dimensional Fourier analysis, we obtain, in the moving frame,

$$\tilde{\mathbf{E}}\left(\frac{\omega^2}{c^2} - \mathbf{k}^2\right) = i\left[\mu_0 \omega \mathbf{j}(\tilde{\mathbf{E}}, \tilde{\mathbf{H}}) - \frac{1}{\varepsilon_0}\rho(\tilde{\mathbf{E}}, \tilde{\mathbf{H}})\mathbf{k}\right], \tag{7.145}$$

and

$$\varepsilon_0 \mathbf{k} \cdot \tilde{\mathbf{E}} = \rho(\tilde{\mathbf{E}}, \tilde{\mathbf{H}}). \tag{7.146}$$

In the special case of a purely dielectric, nonmagnetic material ($\mu = \mu_0$, $\lambda = 0$), the expressions for the source term are greatly simplified:

$$\mathbf{j}(\tilde{\mathbf{E}}, \tilde{\mathbf{H}}) = \gamma^2 \varepsilon_0 \sigma(\omega - \mathbf{v} \cdot \mathbf{k})(\tilde{\mathbf{E}} + \mu_0 \mathbf{v} \times \tilde{\mathbf{H}}), \tag{7.147}$$

and

$$\rho(\tilde{\mathbf{E}}, \tilde{\mathbf{H}}) = \frac{\gamma^2}{c^2} \varepsilon_0 \sigma(\omega - \mathbf{v} \cdot \mathbf{k})(\tilde{\mathbf{E}} \cdot \mathbf{v}), \tag{7.148}$$

Making use of these expressions in Equation 7.145, we obtain the dispersion relation in the following form:

$$\frac{\omega^2}{c^2} - \mathbf{k}^2 = i\sigma \frac{\gamma^2}{c^2}(\omega - \mathbf{v} \cdot \mathbf{k})^2, \tag{7.149}$$

where we recognize the vacuum dispersion on the left-hand side and the usual Doppler-shifted beam mode on the right-hand side.

We note that the left-hand side of Equation 7.149 is a scalar representing the magnitude of the four-wavenumber and thus a relativistic invariant. We can then rewrite the dispersion relation as follows:

$$\left(\frac{\omega'^2}{c^2} - \mathbf{k}'^2\right) = i\sigma\omega'^2, \tag{7.150}$$

which is clearly identical to Equation 7.136 for $\lambda = 0$, thus demonstrating the relativistic invariance of the dispersive characteristics of the medium.

We now derive the expression of the refractive index, $n = |c\mathbf{k}/\omega|$, from Equation 7.149. We use the propagation angle $\theta$, previously defined, and the definition of $\sigma$, to recast Equation 7.149 as

$$\omega^2 - \mathbf{k}^2c^2 = \gamma^2\left(1 - \frac{\varepsilon}{\varepsilon_0}\right)(\omega - \beta ck\cos\theta)^2. \tag{7.151}$$

From this equation, we can easily solve for $k(\omega)$, and obtain the following result for the refractive index:

$$n(\theta) = \frac{\left(1 - \frac{\varepsilon}{\varepsilon_0}\right)\gamma^2\beta\cos\theta + \sqrt{\left(1 - \frac{\varepsilon}{\varepsilon_0}\right)\gamma^2(\beta^2\cos^2\theta - 1) + 1}}{\left(1 - \frac{\varepsilon}{\varepsilon_0}\right)\gamma^2\beta^2\cos^2\theta + 1}. \tag{7.152}$$

In the limiting case of vacuum ($\varepsilon = \varepsilon_0$), we easily find $n = 1$; in addition, for $\beta = 0$, $\gamma = 1$, we recover $n = n' = \sqrt{\varepsilon/\varepsilon_0}$. Again, the refractive index is clearly anisotropic, and it exhibits a singularity for waves propagating at an angle defined by the following equation:

$$\left(\frac{\varepsilon}{\varepsilon_0} - 1\right)\gamma^2\beta^2\cos^2\theta = 1. \tag{7.153}$$

To transform this condition on the propagation angle back to the rest frame of the medium, we use the relation between angles derived in Equation 7.131:

$$\tan^2\theta' = \frac{\tan^2\theta}{\gamma^2}, \tag{7.154}$$

which is valid at the singularity ($n \to \infty$). Using the trigonometric relation between $\tan^2\theta$ and $\cos^2\theta$, we then easily find

$$\frac{1}{\cos^2\theta'} = \beta^2 + \frac{1}{\gamma^2\cos^2\theta}. \tag{7.155}$$

Finally, the singularity in Equation 7.153 yields the following condition in the rest frame of the medium:

$$\frac{1}{\cos^2\theta'} = \beta^2\frac{\varepsilon}{\varepsilon_0}, \tag{7.156}$$

which is the Čerenkov radiation condition for a dielectric, nonmagnetic medium in the linear, isotropic case. We have thus shown the complete equivalence of the Minkowski formulation and the induced four-vector current density formalism in the linear regime.

## 7.7 Covariant Treatment of Nonlinear Effects

We now treat the full nonlinear problem. The general formalism is described in the following. In the rest frame of the nonlinear medium, the induced-source terms

$$\begin{aligned} \mathbf{j}' &= \mathbf{j}'(\mathbf{E}', \mathbf{H}'), \\ \rho' &= \rho'(\mathbf{E}', \mathbf{H}'), \end{aligned} \tag{7.157}$$

describe its nonlinear electromagnetic response. In addition, the relativistic transform of the electromagnetic field yields the relations

$$\begin{aligned} \mathbf{E}' &= \mathbf{E}'(\mathbf{E}, \mathbf{B}), \\ \mathbf{H}' &= \mathbf{H}'(\mathbf{H}, \mathbf{D}), \end{aligned} \tag{7.158}$$

as described explicitly in Equations 7.102 and 7.103.

The crucial point of this formulation is that the vacuum constitutive relations are relativistically invariant. Therefore, we have, within this formulation and in any Galilean frame,

$$\begin{aligned} \mathbf{D} &= \varepsilon_0 \mathbf{E}, \\ \mathbf{B} &= \mu_0 \mathbf{H}. \end{aligned} \tag{7.159}$$

We can thus transform the four-vector current density and the electromagnetic field, and make use of the vacuum constitutive relations to finally obtain the sought-after relativistic description of the nonlinear response of the medium:

$$\begin{aligned} \mathbf{j}(\mathbf{E}, \mathbf{H}) = {} & \mathbf{j}'[\mathbf{E}'(\mathbf{E}, \mu_0\mathbf{H}), \mathbf{H}'(\mathbf{H}, \varepsilon_0\mathbf{E})] \\ & - \gamma \frac{\mathbf{v}}{v^2} [\![ (1-\alpha)\{\mathbf{v} \cdot \mathbf{j}'[\mathbf{E}'(\mathbf{E}, \mu_0\mathbf{H}), \mathbf{H}'(\mathbf{H}, \varepsilon_0\mathbf{E})]\} \\ & - v^2 \rho'[\mathbf{E}'(\mathbf{E}, \mu_0\mathbf{H}), \mathbf{H}'(\mathbf{H}, \varepsilon_0\mathbf{E})] ]\!], \end{aligned} \tag{7.160}$$

$$\rho(\mathbf{E}, \mathbf{H}) = \gamma \left\{ \rho'[\mathbf{E}'(\mathbf{E}, \mu_0\mathbf{H}), \mathbf{H}'(\mathbf{H}, \varepsilon_0\mathbf{E})] + \frac{\mathbf{v}}{c^2} \cdot \mathbf{j}'[\mathbf{E}'(\mathbf{E}, \mu_0\mathbf{H}), \mathbf{H}'(\mathbf{H}, \varepsilon_0\mathbf{E})] \right\}. \tag{7.161}$$

We now address the same problem in a somewhat more detailed way. We consider a dielectric medium with electric nonlinearities, similar to those encountered in nonlinear optics and briefly discussed in Section 7.3. In its rest frame, we have the following expression of the induced nonlinear current density:

$$j'_i = \sum_{\ell=1}^{\infty} \sigma^{(\ell)}_{i;k,l,\ldots,p} E'_k E'_l \cdots E'_p. \tag{7.162}$$

Here, the italic indices refer to the three spatial coordinates, and repeated indices are summed over according to Einstein's convention. The integer $\ell$ refers to the order of the nonlinearity. Making use of Equation 7.160 together with the dielectric condition $(\rho' = 0)$, the relativistic transform of the current density yields

$$j_i = j'_i - \gamma \frac{v_i}{v^2}(1-\alpha) v_q j'_q. \tag{7.163}$$

Introducing the expression of the induced nonlinear current density in the rest frame of the medium, we obtain

$$j_i = \sum_{\ell=1}^{\infty} \left[ \sigma^{(\ell)}_{i;k,l,\ldots,p} E'_k E'_l \cdots E'_p + \gamma \frac{v_i}{v^2}(1-\alpha) v_q \sigma^{(\ell)}_{q;k,l,\ldots,p} E'_k E'_l \cdots E'_p \right]. \tag{7.164}$$

Finally, making use of the relativistic transform of the electric field and the vacuum constitutive relations, we find

$$j_i = \sum_{\ell=1}^{\infty} \left\{ \left[ \sigma^{(\ell)}_{i;k,l,\ldots,p} + \gamma \frac{v_i}{v^2}(1-\alpha) v_q \sigma^{(\ell)}_{q;k,l,\ldots,p} \right] \gamma^{\ell} \left[ E_k - (1-\alpha)\frac{v_k}{v^2} E_n v_n + (\mathbf{v} \times \mu_0 \mathbf{H})_k \right] \right.$$
$$\left. \times \left[ E_l - (1-\alpha)\frac{v_l}{v^2} E_n v_n + (\mathbf{v} \times \mu_0 \mathbf{H})_l \right] \cdots \left[ E_p - (1-\alpha)\frac{v_p}{v^2} E_n v_n + (\mathbf{v} \times \mu_0 \mathbf{H})_p \right] \right\}. \tag{7.165}$$

## 7.8 References for Chapter 7

Note: the numbers listed below refer to the main bibliography and reference sections at the end of this book.

1, 4, 8, 64, 71, 94, 116, 120, 121, 136, 153, 177, 214, 210, 220, 225, 248, 249, 506, 559, 565, 821, 908.

# 8

# *Three-Dimensional Waves in Vacuum, Ponderomotive Scattering, and Vacuum Laser Acceleration*

## 8.1 Introduction

In this chapter, we first review the theory of wave propagation in vacuum, including focusing and diffraction; the interaction of relativistic electrons with such waves is then studied in some detail, and the phenomenon of ponderomotive scattering is described. We also turn our attention to vacuum laser acceleration processes, including the inverse free-electron laser (IFEL) interaction and free-wave acceleration in the presence of a dephasing static magnetic field. Such novel acceleration techniques offer the potential to reach extremely high accelerating gradients, well in excess of 1 GeV/m, which represents an upper bound for conventional rf acceleration technology, as described in the proceedings of the 1998 High Energy Density Microwaves workshop, which is cited in the references to this chapter. We also note that there are concurrent plasma-based schemes pursued around the world, including plasma beatwave and wakefield acceleration that have demonstrated promising gradients; some of this work is also cited in the bibliography. The current state of the art in laser-plasma acceleration and other exotic schemes is ably summarized in the proceedings of the 1996, 1998, and 2000 Advanced Accelerator Concept (AAC) workshops and the 1997 and 1999 Particle Accelerator Conference (PAC), which are listed in the reference section. The proceedings of the workshop on quantum aspects of beam physics, will also prove very useful to the interested reader, and we have listed these as well.

The physics of laser–electron interactions changes dramatically at so-called relativistic intensities, where the transverse momentum of the charge, measured in electron units, exceeds one. Three fundamental processes are known to occur in this regime: nonlinear ponderomotive and Compton scattering, and high-intensity Kapitza–Dirac scattering. These multiphoton vacuum interactions correspond to the following geometries: collinear propagation, transverse or head-on collision, and electron diffraction in a laser standing wave, respectively.

An accurate description of the three-dimensional focus of a laser wave, in both the near-field and far-field regions, is required to describe properly the interaction of the electromagnetic field with charged leptons. In particular, the validity of the paraxial ray approximation, when used to model problems involving relativistic electrons copropagating with a laser wave over many Rayleigh ranges, must be firmly established. For applications involving ultrahigh-intensity and nonlinear Compton scattering, such as the proposed $\gamma$–$\gamma$ collider or focused x-ray sources, a detailed knowledge of the three-dimensional electromagnetic field distribution in the focal region is of paramount importance, since the axial component of the fields may play a major role in the electron dynamics. An accurate field distribution is also required to model properly experimental results. Two ultrahigh-intensity relativistic electron scattering experiments have been performed at the Stanford Linear Accelerator Center (SLAC) and at the Commissariat à l'Energie Atomique (CEA). In the first case, nonlinear (multiphoton) Compton backscattering was investigated using the SLAC 50 GeV beam and a tightly focused terawatt-class laser; at CEA, low-energy electrons were ponderomotively accelerated by a terawatt laser. In both instances, the three-dimensional nature of the focused laser pulse is an essential feature of the experiment and must be described accurately to interpret the resulting data correctly. In addition, considerable interest has been given recently to the detailed properties of laser focusing, partly because of potential applications such as the aforementioned plasmas and vacuum-based laser acceleration schemes. For example, super-Gaussian rings have been thoroughly studied. More in line with our motivation, the effect of the ponderomotive potential associated with an ultrahigh-intensity laser wave on the radial confinement of relativistic electrons copropagating with the pulse has been investigated by Moore. This analysis indicates that higher-order Gaussian modes can indeed confine the electrons through the focus because of the inward radiation pressure gradient. In this particular case, an accurate three-dimensional field distribution, satisfying both the vacuum wave equation and the gauge condition, is needed to demonstrate conclusively the validity of this approach.

An important goal of this chapter is to present a comprehensive theoretical and numerical description of the relativistic dynamics of a charged particle interacting with an external electromagnetic field propagating *in vacuo*. To accurately describe the focusing and diffraction of the drive laser wave in vacuum, the paraxial propagator approach is used, where the mass shell condition, or vacuum dispersion relation, is approximated by a quadratic Taylor expansion in the four-wavenumber. This approach proves extremely accurate for any realizable laser focus and yields analytical expressions for the fields. In addition, the gauge condition is satisfied exactly everywhere, thus yielding a proper treatment of the axial electromagnetic field components due to wavefront curvature. The electron phase is used as the independent variable, thus allowing for particle tracking over an arbitrarily large number of Rayleigh ranges, independent of the nonlinear slippage and relativistic Doppler shift due to radiation pressure. Ultrahigh-intensity ponderomotive scattering is studied as an example to demonstrate the relevance of this theoretical approach and the efficiency of the numerical algorithm developed in Section 8.6. We also note that the three-dimensional dynamics are different from earlier two-dimensional models cited in the references. In particular, the angular distribution of scattering energy no longer reflects canonical momentum invariance, as the light-cone variable is not invariant for focusing waves.

Before studying the full three-dimensional problem, we review briefly the interaction of a relativistic electron with an ultrahigh-intensity plane wave. We will return on this problem in considerable detail in Chapter 10.

The electron four-velocity and four-momentum are defined as

$$u_\mu = \frac{1}{c}\frac{dx_\mu}{d\tau} = \gamma(1, \beta), \quad p_\mu = m_0 c u_\mu, \quad u_\mu u^\mu = -1, \tag{8.1}$$

where $\tau$ is the proper time along the electron world line, $x_\mu(\tau)$.

In the absence of radiative corrections, which will be studied in Chapter 10, the energy–momentum transfer equations are governed by the Lorentz force:

$$\frac{du_\mu}{d\tau} = -\frac{e}{m_0 c}(\partial_\mu A_\nu - \partial_\nu A_\mu)u^\nu. \tag{8.2}$$

For plane waves, the four-vector potential of the laser pulse is simply given by

$$A_\mu(\phi) = [0, \mathbf{A}_\perp(\phi)], \quad \phi = -k_\mu x^\mu(\tau), \tag{8.3}$$

where $\phi$ is the relativistically invariant phase of the traveling wave along the electron trajectory.

Note that the temporal dependence of the wave is arbitrary. Choosing $k_\mu = \frac{\omega_0}{c}(1, 0, 0, 1)$, with the wave propagating in the $z$ direction, we have

$$\frac{d\phi}{d\tau} = \omega_0(\gamma - u_z) = \omega_0 \kappa, \tag{8.4}$$

which defines the light-cone variable, $\kappa$, and the four-momentum transfer equations read

$$\frac{d\mathbf{u}_\perp}{d\tau} = \omega_0 \kappa \frac{d}{d\phi}\left[\frac{e\mathbf{A}_\perp(\phi)}{m_0 c}\right], \tag{8.5}$$

and

$$\frac{du_z}{d\tau} = \frac{d\gamma}{d\tau} = \omega_0 \mathbf{u}_\perp \cdot \frac{d}{d\phi}\left[\frac{e\mathbf{A}_\perp(\phi)}{m_0 c}\right]. \tag{8.6}$$

In the above, $\omega_0$ is the characteristic laser frequency. Equation 8.6 shows that $\kappa$ is invariant:

$$\kappa = \kappa_0 = \gamma_0(1 - \beta_0). \tag{8.7}$$

Additionally, Equation 8.5 is readily integrated to yield the transverse momentum invariant,

$$\mathbf{u}_\perp(\phi) = \frac{e\mathbf{A}_\perp(\phi)}{m_0 c}, \tag{8.8}$$

and the energy and axial momentum are immediately obtained using the fact that the four-velocity is a unit four-vector; in other words, $\gamma^2 = 1 + \mathbf{u}_\perp^2 + u_z^2$,

$$u_z(\tau) = \gamma_0\left\{\beta_0 + \left[\frac{e\mathbf{A}_\perp(\phi)}{m_0 c}\right]^2\left(\frac{1+\beta_0}{2}\right)\right\}, \tag{8.9}$$

and

$$\gamma(\tau) = \gamma_0\left\{1 + \left[\frac{e\mathbf{A}_\perp(\phi)}{m_0 c}\right]^2\left(\frac{1+\beta_0}{2}\right)\right\}. \tag{8.10}$$

These results are quite general and hold as long as plane waves are considered. An important difference between polarization states immediately appears: the square of the vector potential varies adiabatically as the pulse envelope for circular polarization, while there is an extra modulation at the second harmonic, $2\omega_0$, for linear polarization. The transverse electron momentum depends linearly on the laser field, but the axial momentum is a quadratic function of that field, as it results from the nonlinear coupling of the transverse velocity to the laser magnetic field through the ponderomotive force, $\mathbf{v} \times \mathbf{B}$. This quadratic dependence of the energy and axial momentum on the four-vector potential, measured in electron units, distinguishes the relativistic scattering regime, where $|e\mathbf{A}_\perp/m_0 c| \geq 1$. In this regime, the ponderomotive force dominates the electron dynamics, yielding nonlinear slippage and Doppler shifts, as will be described in Chapter 10. Equation 8.10 also provides a scaling for the maximum energy in a plane wave, $\gamma^*/\gamma_0 \approx (e\mathbf{A}_\perp/m_0 c)^2$, for relativistic electrons. Finally, the electron position is given by

$$\mathbf{x}(\phi) = \mathbf{x}_0 + \frac{c}{\omega_0 \kappa_0}\int_0^\phi \mathbf{u}(\psi)\,d\psi. \tag{8.11}$$

Equation 8.10 also shows that there is no net energy gained by the electron after interacting with a plane: we have $\lim_{\phi \to \pm\infty}[\mathbf{A}_\perp(\phi)] = 0$, and therefore,

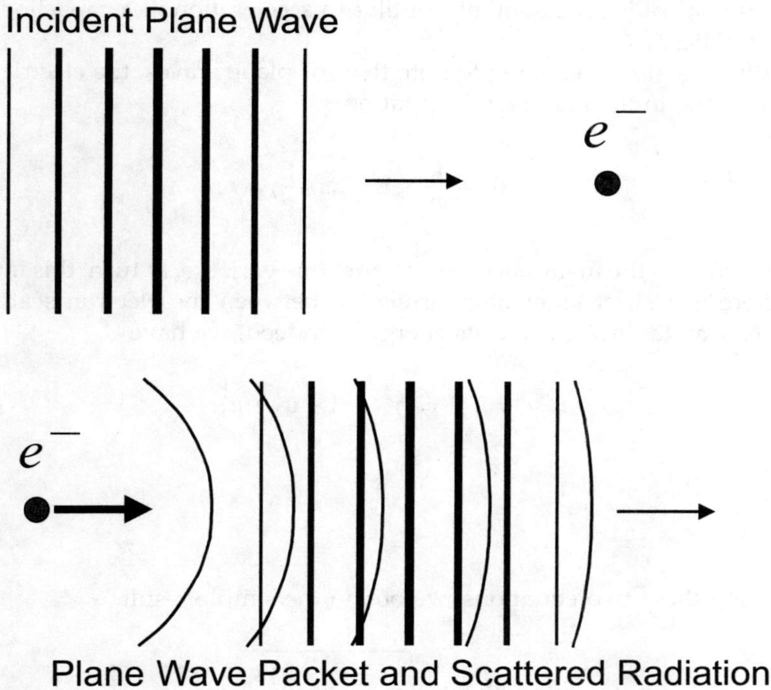


**FIGURE 8.1**
In the interaction of an electron with a plane wave and in the absence of radiative corrections the dipole radiation of the electron cannot interfere permanently with the incident plane wave; as a result, no net energy exchange is possible.

$\lim_{\phi \to \pm\infty} [\gamma(\phi)] = \gamma_0$. This is essentially the generalized version of the Lawson–Woodward theorem. The fact that a charged particle cannot exchange energy and momentum with an incident plane wave in vacuum can be understood easily: consider a frame where the electron is initially at rest, as illustrated in Figure 8.1. If the electron gains energy and momentum during the interaction, it is accelerated and therefore radiates. In the final state, the laser wave has been attenuated, which implies that there exists a permanent destructive interference between the laser wave and the wave radiated by the electron. This is the classical equivalent of photon annihilation in QED. However, the electron radiates waves whose energy decays like $r^{-2}$, and therefore no stable interference pattern can be obtained with a plane wave. In fact, in this case, any interference energy also decays like $r^{-2}$. This shows that, in the absence of radiative corrections (electron recoil), no net energy-momentum can be transferred from a plane wave to an electron in vacuum, in agreement with the generalized Lawson–Woodward theorem.

For certain pathological cases, such as so-called unipolar pulses, this rule can be violated; however, there is no practical merit to such approaches, as one cannot produce terawatt-class, unipolar optical pulses. In fact, acceleration by a unipolar pulse is essentially similar to acceleration in a capacitor with parallel plates.

Finally, it is also interesting to note that for plane waves, the electric field and magnetic induction obey the relation

$$\mathbf{B} = \frac{\mathbf{k}}{\omega_0} \times \mathbf{E} = \hat{z} \times \frac{\mathbf{E}}{c}, \tag{8.12}$$

which results in the invariance of the light-cone variable. In turn, this implies that there is a strict kinematic correlation between the electron scattering angle, $\theta = \arctan|\mathbf{u}_\perp / u_z|$, and its energy, $\gamma$. Indeed, we have

$$u_\mu u^\mu = -1 \Leftrightarrow \gamma^2 = 1 + \mathbf{u}_\perp^2 + u_z^2, \tag{8.13}$$

and

$$\gamma - u_z = \kappa; \tag{8.14}$$

combining these two equations, we obtain the simple result

$$\begin{aligned} \theta &= \arctan\sqrt{\frac{-1-\kappa_0^2+2\gamma\kappa_0}{\gamma^2+\kappa_0^2-2\gamma\kappa_0}} \\ &= \arctan\left[\frac{\sqrt{\left(\frac{2}{1+\beta_0}\right)\left(\frac{\gamma}{\gamma_0}-1\right)}}{\gamma-\gamma_0(1-\beta_0)}\right]. \end{aligned} \tag{8.15}$$

For focusing waves, however, we will see that the light-cone variable is no longer invariant, and the relation described in Equation 8.15 is not strictly valid anymore.

## 8.2 Exact Solutions to the Three-Dimensional Wave Equation in Vacuum

In vacuum, the wave equation takes the familiar form

$$\left[\nabla^2 - \frac{1}{c^2}\partial_t^2\right] A_\mu = [\partial_\nu \partial^\nu] A_\mu = 0, \tag{8.16}$$

where we recognize the four-gradient operator, $\partial_\mu = \partial/\partial x^\mu = (-c^{-1}\partial_t, \nabla)$, and the four-potential, $A_\mu = (c^{-1}\varphi, \mathbf{A})$, which is chosen to satisfy the Lorentz gauge condition,

$$\partial_\mu A^\mu = 0. \tag{8.17}$$

The electromagnetic field tensor is defined by

$$F_{\mu\nu} = \partial_\mu A_\nu - \partial_\nu A_\mu. \tag{8.18}$$

In vacuum, a general solution to the wave equation can be constructed as a Fourier superposition of wave packets of the form

$$A_\mu(x^\nu) = \frac{1}{\sqrt{2\pi}^4}\iiiint \tilde{A}_\mu(k_\nu)\exp(ik_\nu x^\nu)d^4k_\lambda, \tag{8.19}$$

where the notation $d^4k_\lambda = dk_0 dk_1 dk_2 dk_3 = c^{-1}d\omega d^3\mathbf{k}$, and where the four-wavenumber, $k_\mu = (c^{-1}\omega, \mathbf{k})$, satisfies the vacuum dispersion relation

$$k_\mu k^\mu = \mathbf{k}^2 - \frac{\omega^2}{c^2} = 0, \tag{8.20}$$

which is also the mass-shell condition for the photon field: $\hbar^2 k_\mu k^\mu = 0$. In Cartesian coordinates, this translates into

$$A_\mu(x, y, z, t) = \frac{1}{\sqrt{2\pi}^4}\iiiint \tilde{A}_\mu(\mathbf{k}, \omega)\exp[i(\omega t - \mathbf{k}\cdot\mathbf{x})]d^3\mathbf{k}\,d\omega. \tag{8.21}$$

In the case where the laser pulse characteristics are defined at focus ($z = 0$), or at any given plane, we can obtain the electromagnetic field distribution in any given $z$ plane by performing the following integral (i.e., by applying the propagation operator):

$$A_\mu(x, y, z, t) = \frac{1}{\sqrt{2\pi}^3}\iiint \tilde{A}_\mu(\mathbf{k}_\perp, \omega, z = 0) \times \exp\left[i\left(\omega t - k_x x + -k_y y - \sqrt{\frac{\omega^2}{c^2} - \mathbf{k}_\perp^2}\, z\right)\right]d^2\mathbf{k}_\perp d\omega. \tag{8.22}$$

This exact solution is easily interpreted: the temporal evolution of each wavepacket is described by the frequency spectrum, while the transverse profile of the laser wave is expressed as an integral over a continuous spectrum of transverse vacuum eigenmodes. The dispersion relation indicates how each transverse component of the wavepacket propagates, thus yielding wavefront curvature and transverse spreading, or diffraction, of the wavepacket. It should also be noted that the axial wavenumber can become purely imaginary, in which case the corresponding waves become evanescent modes. In Equation 8.22, we have introduced the frequency and transverse wavenumber spectral distributions at focus, which are determined by Fourier transforming the local field distribution according to

$$\tilde{A}_\mu(\mathbf{k}_\perp, \omega, z = 0) = \frac{1}{\sqrt{2\pi}^3}\iiint A_\mu(x, y, z = 0, t) \times \exp[-i(\omega t - k_x x - k_y y)]dx\,dy\,dt. \tag{8.23}$$

The next consideration is the gauge condition, which can be chosen to reduce to the Coulomb gauge, $\nabla \cdot \mathbf{A} = 0$, in a frame where the scalar potential is set to zero. Such a divergence-free vector potential can be generated by a vector field, **G**, defined such that $\mathbf{A} = \nabla \times \mathbf{G}$. As the curl and d'Alembertian operators commute, it is clear that if **G** satisfies the propagation equation, so will the vector potential. For an electromagnetic wave propagating along the $z$-axis, and linearly polarized in the $x$-direction, the generating vector field reduces to $\mathbf{G}(x_\mu) = \hat{y}G_y(x_\mu)$. For a Gaussian-elliptical focus, the generating field takes the form

$$G_y(x, y, z = 0, t) = \frac{A_\perp}{k_0}\exp\left[-\left(\frac{x}{w_{0x}}\right)^2 - \left(\frac{y}{w_{0y}}\right)^2\right]h(t), \tag{8.24}$$

where $w_{0x}$ refers to the beam waist along the $x$-axis, and $w_{0y}$ refers to the beam waist along the $y$-axis; $A_\perp$ is the amplitude of the vector potential at focus, $k_0 = \omega_0/c = 2\pi/\lambda_0$ corresponds to the central laser wavelength, and $h(t)$ is the temporal variation of the pulse, which can be arbitrary.

The corresponding focal spectral density is

$$\tilde{G}_y(\mathbf{k}_\perp, \omega, z = 0) = \frac{A_\perp}{2k_0}w_{0x}w_{0y}\tilde{h}(\omega)\exp\left[-\left(\frac{k_x x}{2}\right)^2 - \left(\frac{k_y y}{2}\right)^2\right], \tag{8.25}$$

as obtained by Fourier transforming Equation 8.24, as prescribed in Equation 8.23.

The propagation integral then takes the form

$$G_y(x, y, z, t) = \frac{1}{\sqrt{2\pi}^3}\iiint \tilde{G}_y(\mathbf{k}_\perp, z = 0, \omega) \times \exp\left[i\left(\omega t - k_x x - k_y y - \sqrt{\frac{\omega^2}{c^2} - \mathbf{k}_\perp^2}\,z\right)\right]d^2\mathbf{k}_\perp\,d\omega, \tag{8.26}$$

which is an exact solution to the three-dimensional wave equation in vacuum. The corresponding vector potential is obtained by taking the curl of the generating field **G**. In closing, we notice the Fourier-conjugate character of the transverse coordinates and wavenumber, $x$, $y$, and $\mathbf{k}_\perp$, and the time and frequency, $t$ and $\omega$.

## 8.3 The Paraxial Propagator

Although Equation 8.26 is an exact solution, it is somewhat impractical because we must perform a triple integral at any point in space–time to derive the corresponding electromagnetic field distribution. In this section, the paraxial propagator formalism is introduced and discussed. This approach has the virtue of yielding simple analytical solutions for the electromagnetic distribution. Although this formalism relies on an approximation, the field is reproduced with a very high degree of precision, sufficient in most practical situations. Moreover, the Coulomb gauge condition is still exactly satisfied, by construction.

The three-dimensional behavior of the laser electromagnetic field is now described within the context of the paraxial propagator formalism. Here, the photon mass-shell condition is approximated as a quadratic Taylor expansion; namely, Equation 8.26 is replaced by

$$G_y(x, y, z, t) \simeq \frac{1}{\sqrt{2\pi}^3}\iiint \tilde{G}_y(\mathbf{k}_\perp, z = 0, \omega) \times \exp\left\{i\left[\omega t - k_x x - k_y y - \left(\frac{\omega}{c} - \frac{\mathbf{k}_\perp^2}{2k_0}\right)z\right]\right\} d^2\mathbf{k}_\perp d\omega, \quad (8.27)$$

where the square root factor has been Taylor-expanded to second order around $\omega = \omega_0$, and $\mathbf{k}_\perp = \mathbf{0}$.

It is clear that the exact and Taylor-expanded axial phases differ only for large values of the transverse wavenumber, where the spectral density is vanishingly small.

The physical content of the paraxial approximation is illustrated in Figure 8.2. The Gaussian transverse wavenumber spectrum is shown for $w_0 k_0 = 20$, where $k_0 = 2\pi/\lambda_0$, and where we are considering a cylindrically symmetric focus, with $w_{0x} = w_{0y} = w_0$. Evanescent modes correspond to $|\mathbf{k}_\perp/k_0| > 1$. The axial wavenumber is also shown both for the exact dispersion relation and in the paraxial approximation. It is clear that for physically realizable foci, where the beam waist is significantly larger than the wavelength, the region of transverse wavenumber space where the paraxial phase differs significantly from the exact

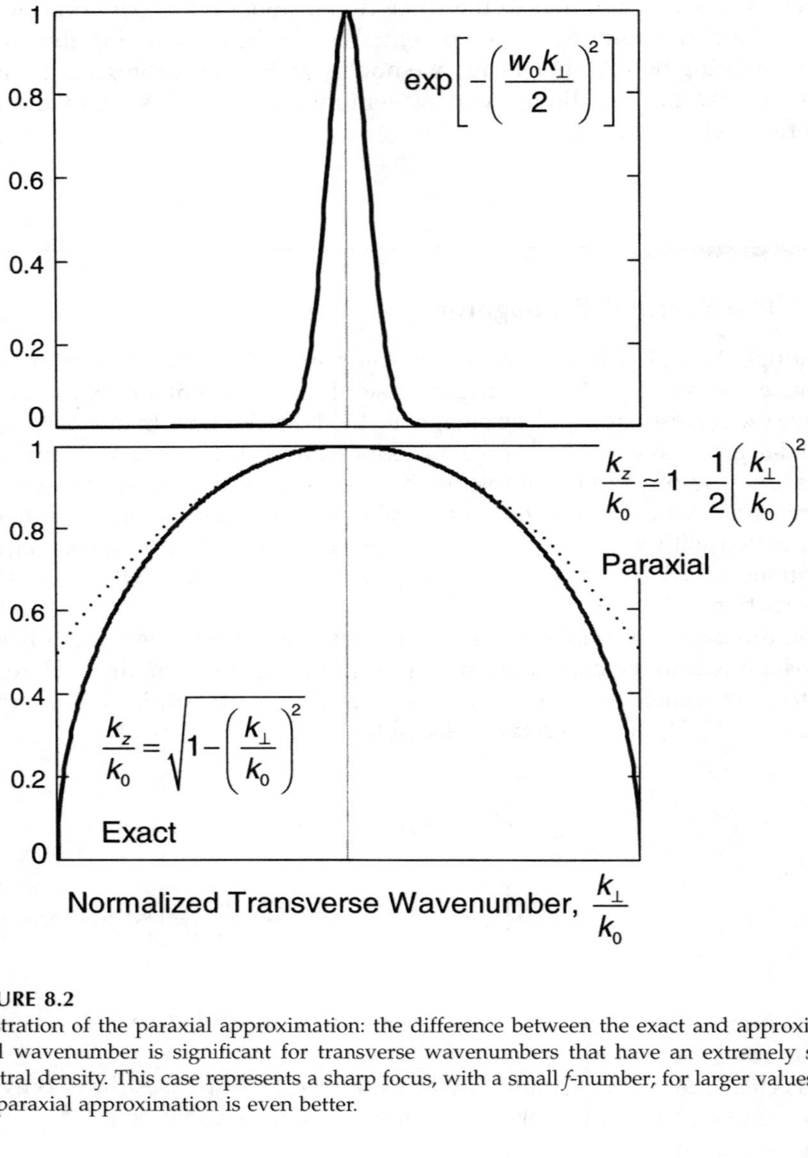


**FIGURE 8.2**
Illustration of the paraxial approximation: the difference between the exact and approximate axial wavenumber is significant for transverse wavenumbers that have an extremely small spectral density. This case represents a sharp focus, with a small *f*-number; for larger values of *f*, the paraxial approximation is even better.

value corresponds to very small spectral amplitudes. For larger values of the *f*-number, the transverse wavenumber spectrum is narrower, and the approximation is even better.

Inserting the Gaussian spectral density from Equation 8.25 into Equation 8.27, we see that the integral over the frequency $\omega$ is a simple Fourier transform of the frequency spectrum $\tilde{h}(\omega)$ and yields $h(t - c^{-1}z) = h(\phi/\omega_0)$. The integrals over the transverse momentum are also readily obtained, as they correspond to exponentials of quadratic complex polynomials:

$$G_y(x, y, z, t) = \frac{A_\perp}{k_0} h\left(\frac{\phi}{\omega_0}\right) \frac{1}{\sqrt[4]{1+\left(\frac{z}{z_{0x}}\right)^2}} \frac{1}{\sqrt[4]{1+\left(\frac{z}{z_{0y}}\right)^2}}$$

$$\times \exp\left\{-\left[\frac{x}{w_x(z)}\right]^2 - \left[\frac{y}{w_y(z)}\right]^2\right\}$$

$$\times \exp\left[\left[ i\left\{\frac{1}{2}\arctan\left(\frac{z}{z_{0x}}\right) + \frac{1}{2}\arctan\left(\frac{z}{z_{0y}}\right)\right.\right.\right.$$

$$\left.\left.\left. - \frac{z}{z_{0x}}\left[\frac{x}{w_x(z)}\right]^2 - \frac{z}{z_{0y}}\left[\frac{y}{w_y(z)}\right]^2\right\}\right]\right], \tag{8.28}$$

where

$$w_{x,y}(z) = w_{0x,y}\sqrt{1+\left(\frac{z}{z_{0x,y}}\right)^2}, \tag{8.29}$$

are the variable waist sizes of the Gaussian transverse distribution, and

$$z_{0x,y} = \frac{1}{2} k_0 w_{0x,y}^2, \tag{8.30}$$

represent the Rayleigh ranges for each *f*-number, which are defined by the relation

$$w_{0x,y} = \frac{\lambda_0}{\pi \arctan\left(\frac{1}{2f_{x,y}}\right)}. \tag{8.31}$$

Finally, taking the real part of the curl of the generating vector, we derive the vector potential:

$$\frac{\mathbf{A}(x_\mu)}{\mathcal{R}[G_y(x_\mu)]} = \hat{x}\left\{\frac{k_0}{g(\phi)}\frac{dg}{d\phi} + \frac{z}{2(z^2+z_{0x}^2)}\left[1-\left(\frac{2x}{w_x}\right)^2\right]\right.$$

$$\left.\times \frac{z}{2(z^2+z_{0y}^2)}\left[1-\left(\frac{2y}{w_x}\right)^2\right] + \tan(\Lambda)\frac{\partial\Lambda}{\partial z}\right\}$$

$$+\hat{z}\left[\frac{k_0 xz}{(z^2+z_{0x}^2)}\tan(\Lambda) - \frac{2x}{w_x^2}\right]. \tag{8.32}$$

Here, we have written the temporal laser pulse in terms of a slowly varying envelope, $h(\phi) = g(\phi)e^{i\phi}$.

In Equation 8.32, the total phase is given by

$$\Lambda = \phi + \frac{1}{2}\left[\arctan\left(\frac{z}{z_{0x}}\right) + \arctan\left(\frac{z}{z_{0y}}\right) - \frac{k_0 z x^2}{z^2 + z_{0x}^2} - \frac{k_0 x y^2}{z^2 + z_{0y}^2}\right], \tag{8.33}$$

which includes the Guoy phase, and the wavefront curvature terms in $x^2$ and $y^2$. Finally, the axial electromagnetic field component due to the three-dimensional effects appears clearly in Equation 8.32.

The vector potential of a focusing wave is shown in Figure 8.3. The fields are then derived from the vector potential in the usual fashion: $\mathbf{E} = -\partial_t \mathbf{A}$, $\mathbf{B} = \nabla \times \mathbf{A}$. We note that in the plane of polarization ($y = 0$), only $E_x$, $E_z$, and $B_y$ are nonzero; as a result, electrons seeded at $y = 0$, with no momentum in the $y$-direction ($p_y = 0$), will remain in this plane.

We have thus derived a general solution to the wave equation in vacuum, which reduces to the paraxial approximation in the limit of small transverse wavenumbers. In addition, the derivation of an analytical expression of the axial field component in the case of linear polarization, within this approximation, will prove quite useful to study the relativistic dynamics of electrons in ultrahigh-intensity laser fields. In particular, this derivation can be extended to higher-order Gaussian modes, which are thought to yield the particle confinement required for vacuum laser acceleration applications. We also note that circular polarization can be modeled in the same fashion, by adding a second component to the generating vector, $\mathbf{G}$:

$$\mathbf{G}(x, y, z = 0, t) = f(x, y)g(\phi)[\hat{x}\sin\phi \pm \hat{y}\cos\phi], \tag{8.34}$$

where $f(x, y)$ represents the transverse mode profile, while $g$ is the pulse envelope, and $\phi$ is the phase introduced earlier.

For a Gaussian-elliptical focus, we have

$$\mathbf{G}(x, y, z = 0, t) = \frac{A_\perp}{k_0} \exp\left[-\left(\frac{x}{w_{0x}}\right)^2 - \left(\frac{y}{w_{0y}}\right)^2\right] g(t)[\hat{x}\sin(\omega_0 t) \pm \hat{y}\cos(\omega_0 t)]. \tag{8.35}$$

Here again, $g(t)$ can be an arbitrary function of time.

Finally, Equation 8.28 takes a particularly simple form for a cylindrical focus:

$$G_y(x, y, z, t) = \frac{A_\perp}{k_0} h\left(\frac{\phi}{\omega_0}\right) \frac{\exp\left\{-\left[\frac{r}{w(z)}\right]^2\right\}}{\sqrt{1 + \left(\frac{z}{z_0}\right)^2}} \exp\left[\left[ i\left\{\arctan\left(\frac{z}{z_0}\right) - \left(\frac{z}{z_0}\right)\left[\frac{r}{w(z)}\right]^2\right\}\right]\right], \tag{8.36}$$

where we recognize the Rayleigh range, Guoy phase, and wavefront curvature.

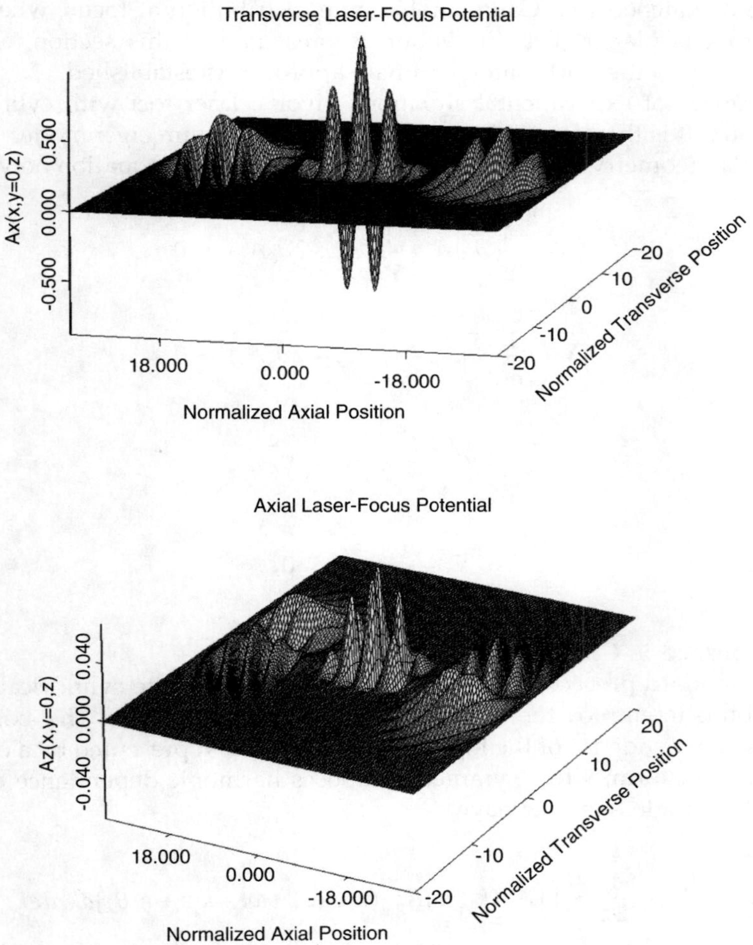


**FIGURE 8.3**
Top: snapshot of the transverse potential at three different times: before focus, at focus, and after focus. The wavefront curvature is clearly visible, as well as the increased amplitude and decreased transverse width of the laser pulse at focus. Bottom: axial potential component (note the vertical [amplitude] scale difference).

## 8.4 Bessel Functions and Hankel's Integral Theorem

This formalism can also be studied in cylindrical coordinates, where the radial dependence of the focusing wave is described as a continuous spectrum of Bessel functions and can be obtained by using Hankel's integral theorem. To define the boundary conditions for this problem, the beam profile is matched to a Gaussian–Hermite distribution at focus, where the wavefront is planar. This derivation is presented in this section, and its equivalence to the Cartesian coordinate approach is established.

A number of experimental situations involve laser foci with cylindrical symmetry. It is therefore of interest to study the vacuum eigenmodes of this particular geometry. For cylindrical symmetry, the wave equation now reads

$$\left[\nabla^2 - \frac{1}{c^2}\partial_t^2\right]A_r - \frac{1}{r^2}(A_r + 2\partial_\theta A_\theta) = 0, \tag{8.37}$$

$$\left[\nabla^2 - \frac{1}{c^2}\partial_t^2\right]A_\theta - \frac{1}{r^2}(A_\theta - 2\partial_\theta A_r) = 0, \tag{8.38}$$

$$\left[\nabla^2 - \frac{1}{c^2}\partial_t^2\right]A_z = 0, \tag{8.39}$$

$$\left[\nabla^2 - \frac{1}{c^2}\partial_t^2\right]\varphi = 0, \tag{8.40}$$

as established in Chapter 5.

The standard procedure to find a general solution to the cylindrical wave equation is to employ the method of separation of variables. The axial and temporal dependence of the four-vector potential is represented by a double Fourier transform, while symmetry imposes harmonic dependence on the azimuthal angle. We thus have

$$A_\mu(r, \theta, z, t) = \frac{1}{2\pi}\sum_m \iint \tilde{A}_m(k_\parallel, \omega) R_{\mu m}(r) \exp[i(\omega t - k_\parallel z + m\theta)] dk_\parallel d\omega. \tag{8.41}$$

Inserting Equation 8.41 into the cylindrical wave equation and using the orthogonality of complex exponentials, we obtain two sets of differential equations corresponding to two families of modes, TE and TM.

The TM modes are generated by the axial and temporal components of the four-potential. The corresponding wave equation admits solutions of the form

$$R_{zm}(r) = R_{\parallel m} \mathrm{J}_m(k_\perp r), \tag{8.42}$$

$$R_{tm}(r) = R_{\parallel m} \frac{kc^2}{\omega} \mathrm{J}_m(k_\perp r), \tag{8.43}$$

where the transverse eigenwavenumber is constrained by the vacuum dispersion relation, $\omega^2/c^2 = k_\perp^2 + k_\parallel^2$.

In Equations 8.42 and 8.43, the constants have been adjusted for the axial and temporal components in order to satisfy the Lorentz gauge condition.

On the other hand, the TE modes are generated by the radial and azimuthal components of the four-potential. The corresponding wave equation splits into two coupled differential equations, which admit solutions of the form

$$R_{rm}(r) = R_{\perp m}\frac{J_m(k_\perp r)}{k_\perp r}, \tag{8.44}$$

$$R_{\theta m}(r) = R_{\perp m}\frac{i}{m}J'_m(k_\perp r), \tag{8.45}$$

where the transverse eigenvalue is again constrained by the dispersion relation.

Note that here the gauge condition is satisfied automatically because the differential equations resulting from the wave equation are coupled.

In the case of a finite radial boundary at $r = a$ (cylindrical waveguide), we have seen in Chapter 5 that the eigenmode spectrum is discrete, and we have $k_\perp = \chi'_{mn}/a$ for TE modes and $k_\perp = \chi_{mn}/a$ for TM modes. Here, $\chi'_{mn}$ and $\chi_{mn}$ are the $n$-th zeros of $J'_m$ and $J_m$, respectively. However, in our case of interest, the radial boundary extends out to infinity, and the radial eigenmode spectrum is continuous. In addition, the distinction between TE and TM modes breaks down since focusing waves correspond to hybrid modes.

It is interesting to note that since Bessel functions are the eigenmodes of the cylindrical wave equation in vacuum, these modes can theoretically propagate as plane waves, without diffracting. However, the dispersion relation shows that these waves have a group velocity smaller than $c$, because of their nonzero cutoff frequency, thus indicating that such mode profiles cannot be maintained in vacuum without a waveguide boundary surface. In addition, in a waveguide, the energy flow is limited by the finite radial extent of the structure, whereas, in vacuum, the radial integral of a single unbounded Bessel node diverges.

We also note that, starting from these solutions, one can construct hybrid modes where the transverse components of the four-potential correspond to TE-like modes and the axial component is described by a TM-like profile. However, here the gauge condition is satisfied differently from guided waves. For propagation in vacuum, such hybrid modes are required: that is, in the case of a linearly polarized wave focusing in vacuum, a superposition of pure $TE_{11}$ modes will not yield an adequate description since they are not truly linearly polarized; therefore, an axial field component is required.

Now a general solution to the cylindrical wave equation can be expressed as a vacuum eigenmode expansion, but the continuous radial eigenvalue spectrum remains to be defined; namely, we have

$$\begin{aligned} A_\mu(r, \theta, z, t) = \frac{1}{2\pi}\sum_m \iiint A_m(k_\perp, k_\parallel, \omega) R_{\mu m}(k_\perp r) \\ \times \exp[i(\omega t - k_\parallel z + m\theta)]\, dk_\perp\, dk_\parallel\, d\omega, \end{aligned} \tag{8.46}$$

where

$$R_{\mu m}(k_\perp r) = R_{\perp m}\left[\hat{r}\frac{\mathrm{J}_m(k_\perp r)}{k_\perp r} + \hat{\theta}\frac{i}{m}\mathrm{J}'_m(k_\perp r)\right], \tag{8.47}$$

for the TE-like components, and

$$R_{\mu m}(r) = R_{\|m}\mathrm{J}_m(k_\perp r)\left[\hat{z} + \hat{t}\frac{k_\| c^2}{\omega}\right], \tag{8.48}$$

for the TM-like modes. The constraint between the radial eigenvalue and the frequency and wavenumber is given by the dispersion relation. This solution can easily be interpreted: the temporal evolution of the wavepacket is described by its frequency spectrum, while the radial profile of the laser wave is described by an integral over a continuous spectrum of transverse vacuum eigenmodes (Bessel functions). The dispersion relation indicates how each radial and temporal component of the wavepacket propagates, thus yielding wavefront curvature and diffraction of the wavepacket. The polarization state is described by $m$. If the temporal and radial spectral distribution are known at a given position along the propagation axis, as well as the polarization state, the pulse characteristics can be obtained at any other axial position by following the corresponding procedure outlined in the preceding section.

At this point, we need a mathematical procedure to determine the radial spectrum of the wavepacket. The most relevant case for practical applications corresponds to linearly and circularly polarized wavepackets, where the azimuthal number $|m| = 1$. For example, in the case of a circularly polarized hyperbolic secant laser pulse with a Gaussian radial profile at focus, we have

$$A_r(r, \theta, z = 0, t) = A\exp\left[-\left(\frac{r}{w_0}\right)^2\right]\frac{\cos(\omega_0 t - \theta)}{\cosh\left(\frac{t}{\Delta t}\right)}, \tag{8.49}$$

$$A_\theta(r, \theta, z = 0, t) = A\exp\left[-\left(\frac{r}{w_0}\right)^2\right]\frac{\sin(\omega_0 t - \theta)}{\cosh\left(\frac{t}{\Delta t}\right)}, \tag{8.50}$$

where $\Delta t$ is the pulse duration, $\omega_0$ is the laser frequency, and $w_0$ is the focal beam waist.

To express the Gaussian profile in terms of Bessel functions, one begins with Hankel's integral theorem, which reduces to Weber's integral in this case. For a Gaussian we have,

$$\int_0^\infty x e^{-a^2 x^2}\mathrm{J}_0(bx)\,dx = \frac{\exp\left(-\frac{b^2}{4a^2}\right)}{2a^2}. \tag{8.51}$$

Because of the polarization constraint, we now express $J_0$ in terms of $J_1$, using the recurrence relation

$$J_0(bx) = J_1'(bx) + \frac{J_1(bx)}{bx} \tag{8.52}$$

to obtain

$$\frac{\exp\left(-\frac{b^2}{4a^2}\right)}{2a^2} = \int_0^\infty x e^{-a^2x^2}\left[J_1'(bx) + \frac{J_1(bx)}{bx}\right]dx, \tag{8.53}$$

which is integrated by parts, yielding

$$\exp\left(-\frac{b^2}{4a^2}\right) = 2a^2\left\{\left[xe^{-a^2x^2}\frac{1}{b}J_1(bx)\right]_0^\infty + \int_0^\infty 2a^2x^2e^{-a^2x^2}\frac{J_1(bx)}{bx}dx\right\}. \tag{8.54}$$

The first term in the square brackets vanishes, and we are left with the sought-after Bessel transform of a Gaussian

$$\exp\left[-\left(\frac{r}{w_0}\right)^2\right] = \frac{w_0^4}{4}\int_0^\infty k_\perp^3 \exp\left[-\left(\frac{k_\perp w_0}{2}\right)^2\right]\frac{J_1(k_\perp r)}{k_\perp r}\,dk_\perp, \tag{8.55}$$

where we have expressed the various parameters in terms of physical quantities.

Performing the temporal Fourier transform of the circularly polarized hyperbolic secant pulse,

$$\frac{\cos(\omega_0 t - \theta)}{\cosh\left(\frac{t}{\Delta t}\right)} = \frac{1}{\sqrt{2\pi}}\int_{-\infty}^{+\infty}\sqrt{\frac{\pi}{8}}\,\Delta t\,\frac{\exp[i(\omega t \pm \theta)]}{\cosh\left[\frac{\pi}{2}(\omega \pm \omega_0)\Delta t\right]}\,d\omega, \tag{8.56}$$

the radial component of the four-potential can now be evaluated at any point along the propagation axis by performing the integral

$$A_r = \frac{\Delta t w_0^4}{16}\int_{-\infty}^{+\infty} d\omega \int_{-\infty}^{+\infty} dk_\perp \frac{k_\perp^3 \exp\left[-\left(\frac{k_\perp w_0}{2}\right)^2\right]}{\cosh\left[\frac{\pi}{2}(\omega \pm \omega_0)\Delta t\right]}\frac{J_1(k_\perp r)}{k_\perp r}$$

$$\times \exp\left[i\left(\omega t - \sqrt{\frac{\omega^2}{c^2} - k_\perp^2}\,z \pm \theta\right)\right], \tag{8.57}$$

where one must sum over the plus and minus signs.

The azimuthal component of the four-vector potential is obtained upon replacement of $\frac{J_1(k_\perp r)}{k_\perp r}$ by $iJ_1'(k_\perp r)$ in the integral. This procedure can be extended to Gaussian–Hermite profiles by noting that each term of the series has a Bessel transform given by Weber's integral. Specifically, we have

$$r^n \exp\left[-\left(\frac{r}{w_0}\right)^2\right] = \left(\frac{w_0^2}{2}\right)^{n+1} \int_0^\infty k_\perp^{n+1} \exp\left[-\left(\frac{w_0 k_\perp}{2}\right)^2\right] \mathrm{J}_n(k_\perp r) dk_\perp. \tag{8.58}$$

We then reduce the Bessel function order to 1, which is achieved by means of the recurrence relation

$$\mathrm{J}_{n-1}(k_\perp r) = n\left[\frac{\mathrm{J}_n(k_\perp r)}{k_\perp r}\right] + J_n'(k_\perp r), \tag{8.59}$$

and by integrating by parts. We have thus introduced a general mathematical procedure allowing for the exact description of the electromagnetic field distribution of a cylindrically symmetrical three-dimensional focus in vacuum, both in the near-field and in the far-field regions. The results derived here can also be obtained from the exact solution in rectangular coordinate by performing a coordinate transformation.

## 8.5 Plane Wave Dynamics: Lawson–Woodward Theorem

### 8.5.1 Canonical Invariants: Phase and Light-Cone Variable

We now return to the question of the relativistic dynamics of an electron in a plane wave and derive the full nonlinear theory in the case of a single plane wave electromagnetic eigenmode in vacuum, where the four-potential is an arbitrary function of the relativistically invariant phase:

$$A_\mu(x^\nu) = A_\mu(\phi), \quad \phi(x^\nu) = -k_\nu x^\nu. \tag{8.60}$$

In this case, the operation of the four-gradient upon the four-potential reduces to

$$\partial_\mu A_\nu = \frac{\partial \phi}{\partial x^\mu}\frac{dA_\nu}{d\phi} = -k_\mu \frac{dA_\nu}{d\phi}. \tag{8.61}$$

Our first task is to demonstrate the existence of two invariants: the canonical momentum, and the light-cone variable. In our analysis, charge is measured in units of $e$, mass in units of $m_0$, length is normalized to a reference wavelength, $k_0^{-1}$, while time is measured in units of the corresponding frequency,

$\omega_0^{-1} = (ck_0)^{-1}$. Neglecting radiative corrections, the electron motion is governed by the Lorentz force equation,

$$\frac{du_\mu}{d\tau} = -F_{\mu\nu}u^\nu = -(\partial_\mu A_\nu - \partial_\nu A_\mu)u^\nu. \tag{8.62}$$

Here, $u_\mu = \frac{dx_\mu}{d\tau}$ is the electron four-velocity along its world line, $x_\mu(\tau)$; $\tau$ is the electron proper time; $A_\mu$ is the four-potential from which the electromagnetic field derives; finally, $\partial_\mu = (-\partial_t, \nabla)$ is the four-gradient operator.

The electromagnetic field distribution considered here corresponds to a vacuum interaction; therefore, the four-potential satisfies the wave equation, $\Box A_\mu = [\partial_\nu \partial^\nu] A_\mu = 0$, and can be expressed as a superposition of plane waves, as described in Section 8.2. Furthermore, we choose to work in the Lorentz gauge, where $\partial_\mu A^\mu = 0$. Here, the symbol $\mathbb{0} = (0, \mathbf{0})$ represents the null four-vector.

Applying the result expressed in Equation 8.61 to the Lorentz force equation, we have

$$\frac{du_\mu}{d\tau} = k_\mu\left(u^\nu \frac{dA_\nu}{d\phi}\right) - (k_\nu u^\nu)\frac{dA_\mu}{d\phi} = k_\mu\left(u^\nu \frac{dA_\nu}{d\phi}\right) + \frac{d\phi}{d\tau}\frac{dA_\mu}{d\phi} = \frac{dA_\mu}{d\tau} + k_\mu\left(u^\nu \frac{dA_\nu}{d\phi}\right); \tag{8.63}$$

here, we recognize the canonical momentum, $\pi_\mu = u_\mu - A_\mu$.

We now consider the light-cone variable, $\kappa = d\phi/d\tau$, the evolution of this dynamical variable is described by

$$\frac{d\kappa}{d\tau} = -k_\mu \frac{du^\mu}{d\tau} = -k_\mu\left(k^\mu \frac{dA^\nu}{d\phi} - k^\nu \frac{dA^\mu}{d\phi}\right)u_\nu = -(k_\mu k^\mu)\left(\frac{dA^\nu}{d\phi}u_\nu\right) + (k^\nu u_\nu)\left(k_\mu \frac{dA^\mu}{d\phi}\right). \tag{8.64}$$

The first term in Equation 8.64 corresponds to the mass-shell condition, $k_\mu k^\mu = 0$, and can easily be derived by considering the propagation equation in vacuum,

$$\Box A_\mu = (\partial_\nu \partial^\nu) A_\mu = 0 = \frac{\partial\phi}{\partial x^\nu}\frac{\partial\phi}{\partial x_\nu}\frac{d^2 A_\mu}{d\phi^2} = (k_\nu k^\nu)\frac{d^2 A_\mu}{d\phi^2}; \tag{8.65}$$

while the second term in Equation 8.64 corresponds to the Lorentz gauge condition,

$$\partial_\mu A^\mu = 0 = \frac{\partial\phi}{\partial x^\mu}\frac{dA^\mu}{d\phi} = -k_\mu \frac{dA^\mu}{d\phi}. \tag{8.66}$$

With this, we see that the light-cone variable is a constant of the electron motion:

$$\frac{d\kappa}{d\tau} = 0. \tag{8.67}$$

We now return to Equation 8.63: from its structure, we can see that the solution must take the form $u_\mu = A_\mu + k_\mu g(\phi)$, where $g$ is a function of the electron phase to be determined. In addition, the nonlinear radiation pressure of the plane wave is proportional to $A_\mu A^\mu(\phi)$; finally, the solution must satisfy the condition $u^\mu u_\mu = -1$. Therefore, we consider

$$u_\mu = A_\mu + k_\mu \left[\frac{\varsigma + A_\nu A^\nu}{2\kappa}\right], \tag{8.68}$$

which has the appropriate structure, and where $\varsigma$ is a constant that will be determined from the normalization of the four-velocity. Deriving the trial solution with respect to proper time, we first have

$$\frac{du_\mu}{d\tau} = \frac{dA_\mu}{d\tau} + \frac{k_\mu}{2\kappa}\left(2A_\nu \frac{dA^\nu}{d\tau}\right) = \frac{dA_\mu}{d\tau} + \frac{k_\mu}{\kappa}\left(A_\nu \frac{dA^\nu}{d\phi}\frac{d\phi}{d\tau}\right) = \frac{dA_\mu}{d\tau} + k_\mu\left(A_\nu \frac{dA^\nu}{d\phi}\right). \tag{8.69}$$

We then use Equation 8.68 to replace the four-potential: $A_\nu = u_\nu - k_\nu \left[\frac{\varsigma + (A_\lambda A^\lambda)}{2\kappa}\right]$, and we find that

$$\frac{du_\mu}{d\tau} = \frac{dA_\mu}{d\tau} + k_\mu\left(u_\nu \frac{dA^\nu}{d\tau}\right) - k_\mu\left[\frac{\varsigma + (A_\lambda A^\lambda)}{2\kappa}\right]\left(k_\nu \frac{dA^\nu}{d\tau}\right), \tag{8.70}$$

which reduces exactly to the Lorentz force equation because the Lorentz gauge condition requires that $k_\nu \frac{dA^\nu}{d\tau} = 0$, as shown in Equation 8.66. In turn, the structure of the solution given in Equation 8.68 implies that the light-cone variable reduces to

$$\kappa = -k_\mu u^\mu = -k_\mu A^\mu - (k_\mu k^\mu)\left[\frac{\varsigma + A_\nu A^\nu}{2\kappa}\right] = -k_\mu A^\mu, \tag{8.71}$$

because of the mass-shell condition, $k_\mu k^\mu = 0$. We can also verify that the light-cone variable is, indeed, constant,

$$\frac{d\kappa}{d\tau} = -k_\mu \frac{dA^\mu}{d\tau} = -\left(k_\mu \frac{dA^\mu}{d\phi}\right)\frac{d\phi}{d\tau} = 0, \tag{8.72}$$

because of the Lorentz gauge condition. Using Equation 8.71, we can now rewrite the four-velocity as

$$u_\mu = A_\mu - k_\mu \left[ \frac{\varsigma + (A_\nu A^\nu)}{2k_\nu A^\nu} \right]; \tag{8.73}$$

finally, the constant $\varsigma$ is determined by taking the norm of the four-velocity,

$$u_\mu u^\mu = A_\mu A^\mu - 2k_\mu A^\mu \left[ \frac{\varsigma + (A_\nu A^\nu)}{2k_\nu A^\nu} \right] + (k_\mu k^\mu) \left[ \frac{\varsigma + (A_\nu A^\nu)}{2k_\nu A^\nu} \right]^2 = -\varsigma = -1, \tag{8.74}$$

where we have used the mass-shell condition again.

Thus, the fully covariant, nonlinear solution to the electron dynamics in a plane wave of arbitrary intensity and temporal dependence is found to be

$$u_\mu(x^\nu) = u_\mu(\phi) = A_\mu - k_\mu \left( \frac{1 + A_\nu A^\nu}{2k_\nu A^\nu} \right). \tag{8.75}$$

The question of the influence of the initial conditions on the electron trajectory can now be addressed: the four-potential can be regauged to incorporate the boundary conditions on the electron dynamics. Since we are shifting the four-potential by a constant four-vector, the Lorentz gauge condition is still satisfied, and the electromagnetic field tensor is unchanged. We have: $A^\mu \to A^\mu + u_0^\mu$, $\lim_{\phi \to \infty} A^\mu(\phi) = 0$, $\lim_{\phi \to \infty} u^\mu(\phi) = u_0^\mu$. With this, the invariant light-cone variable reads $\kappa = -k_\mu A^\mu = -k_\mu u_0^\mu$, and the four-velocity is given by

$$u^\mu = u_0^\mu + A^\mu - k^\mu \left( \frac{A_\nu A^\nu + 2A_\nu u_0^\nu}{2k_\nu u_0^\nu} \right), \tag{8.76}$$

where we have used the fact that $(A + u_0)_\mu (A + u_0)^\mu = A_\mu A^\mu + 2A_\mu u_0^\mu - 1$.

It should be emphasized that this nonlinear solution is fully covariant and makes explicit use of gauge invariance.

The nonlinear, covariant electron dynamics are thus fully determined; the Lawson–Woodward theorem is immediately recovered by considering the time-like component of Equation 8.76: $\lim_{\phi \to +\infty} \gamma(\phi) = \gamma_0$, as $\lim_{\phi \to +\infty} A_\mu(\phi) = 0$. In the linear regime, where the normalized four-potential satisfies the condition $|A_\mu A^\mu| \ll 1$, the electron motion is purely transverse, while at so-called relativistic intensities, ponderomotive effects dominate. In particular, the mass of the dressed electron can be derived immediately from the time-like component of Equation 8.76, in a frame where the electron is initially at rest:

$$\langle m \rangle = \langle \gamma \rangle = \left\langle 1 + \mathcal{V} + \frac{A_\mu A^\mu}{2} \right\rangle, \tag{8.77}$$

where $\mathcal{V}$ is the time-like component of the four-potential, or scalar potential.

It is also easy to recover more familiar expressions by rotating the coordinate system so that the four-wavenumber reduces to $k_\mu = (1, 0, 0, 1)$, and to introduce the four-potential as $A_\mu = (0, \mathbf{A}_\perp, 0)$; we then have $\kappa = \gamma - u_\parallel = \gamma_0 - u_{\parallel 0} = \kappa_0$, and

$$\begin{aligned}
\mathbf{u}_\perp(\phi) &= \mathbf{u}_{\perp 0} + \mathbf{A}_\perp(\phi), \\
u_\parallel(\phi) &= u_{\parallel 0} + \left[\frac{\mathbf{A}_\perp^2 2\mathbf{A}_\perp \cdot \mathbf{u}_{\perp 0}}{2(\gamma_0 - u_{\parallel 0})}\right], \\
\gamma(\phi) &= \gamma_0 + \left[\frac{\mathbf{A}_\perp^2 + 2\mathbf{A}_\perp \cdot \mathbf{u}_{\perp 0}}{2(\gamma_0 - u_{\parallel 0})}\right].
\end{aligned} \tag{8.78}$$

In particular, in the simple case where the reference frame is chosen such that $\mathbf{u}_{\perp 0} = \mathbf{0}$, we have $\gamma_0^2 = 1 + u_{\parallel 0}^2$, and Equation 8.78 reduces to the result previously derived,

$$\mathbf{u}_\perp = \mathbf{A}_\perp, \quad u_\parallel = \gamma_0\left[\beta_{\parallel 0} + \frac{\mathbf{A}_\perp^2}{2}(1 + \beta_{\parallel 0})\right], \quad \gamma = \gamma_0\left[1 + \frac{\mathbf{A}_\perp^2}{2}(1 + \beta_{\parallel 0})\right]. \tag{8.79}$$

### 8.5.2 Fluid Invariants

In this section, we explore the fluid equivalent of the nonlinear plane wave dynamics established in the previous section. We now consider the Lorentz force equation for a charged fluid,

$$\frac{du_\mu}{d\tau} = (u^\nu \partial_\nu) u_\mu = -(\partial_\mu A_\nu - \partial_\nu A_\mu) u^\nu, \tag{8.80}$$

and the charge conservation, or continuity, equation

$$\partial_\mu j^\mu = 0. \tag{8.81}$$

Here, $u_\mu(x_\nu)$ is the four-velocity field of the relativistic fluid, $A_\mu(x_\nu)$ is the four-potential of the laser pulse, and the total derivative with respect to proper time is to be considered as a convective operator, as indicated.

The four-current density of the relativistic fluid is given by

$$j_\mu(x_\nu) = -n(x_\nu)\frac{u_\mu(x_\nu)}{\gamma(x_\nu)}, \tag{8.82}$$

where $n(x_\nu)$ is the density. Space-charge and radiation reaction effects are neglected.

As the force equation is driven by the laser four-potential, which is a function of the fluid phase,

$$\phi(z, t) = t - z, \tag{8.83}$$

we seek a solution where the other fluid fields also depend on $\phi$; the convective derivative operator reduces to

$$(u^{\nu}\partial_{\nu})u_{\mu}(z,t) \equiv [\gamma\partial_t + u_z\partial_z]u_{\mu}(\phi) = \left(\gamma\frac{\partial\phi}{\partial t} + u_z\frac{\partial\phi}{\partial z}\right)\frac{du_{\mu}}{d\phi} = (\gamma - u_z)\frac{du_{\mu}}{d\phi}. \quad (8.84)$$

The Lorentz force equation now reads

$$(\gamma - u_z)\frac{d\mathbf{u}}{d\phi} = -(\gamma\mathbf{E} + \mathbf{u}\times\mathbf{B}); \quad (8.85)$$

in addition, energy conservation yields

$$(\gamma - u_z)\frac{d\gamma}{d\phi} = -\mathbf{u}\cdot\mathbf{E}. \quad (8.86)$$

Here, the electromagnetic field components are given by $F_{\mu\nu} = \partial_{\mu}A_{\nu} - \partial_{\nu}A_{\mu}$. The evolution of the momentum field can be separated into a transverse and an axial component:

$$(\gamma - u_z)\frac{d}{d\phi}(\mathbf{u}_{\perp} - \mathbf{A}_{\perp}) = \mathbf{0}, \quad \text{and} \quad (\gamma - u_z)\frac{du_z}{d\phi} = \mathbf{u}_{\perp}\cdot\frac{d\mathbf{A}_{\perp}}{d\phi} = (\gamma - u_z)\frac{d\gamma}{d\phi}. \quad (8.87)$$

We recover the transverse canonical momentum invariant, $\mathbf{u}_{\perp} - \mathbf{A}_{\perp}$; Equation 8.87 also shows that the light-cone variable is a fluid invariant. The sought-after fluid equilibrium is

$$u_z(z,t) = u_z(\phi) = u_{z0} + \kappa_0^{-1}\left[\mathbf{A}_{\perp}(\phi)\cdot\mathbf{u}_{\perp 0} + \frac{\mathbf{A}_{\perp}^2(\phi)}{2}\right], \quad (8.88)$$

and

$$\gamma(z,t) = \gamma(\phi) = \gamma_0 + \kappa_0^{-1}\left[\mathbf{A}_{\perp}(\phi)\cdot\mathbf{u}_{\perp 0} + \frac{\mathbf{A}_{\perp}^2(\phi)}{2}\right]. \quad (8.89)$$

To determine the density, we seek a solution to the charge conservation equation where the density field is a function of the phase. The continuity equation reads

$$\partial_t n(\phi) + \nabla\cdot[n(\phi)\beta(\phi)] = \frac{\partial\phi}{\partial t}\frac{dn}{d\phi} + \frac{\partial\phi}{\partial z}\frac{d}{d\phi}(n\beta_z) = \frac{d}{d\phi}[n(1-\beta_z)] = 0. \quad (8.90)$$

Multiplying and dividing the density by the energy, we can rewrite Equation 8.90 in terms of the light-cone variable,

$$\frac{d}{d\phi}\left[\frac{n}{\gamma}(\gamma-u_z)\right]=\frac{d}{d\phi}\left(\frac{n}{\gamma}\kappa\right)=\kappa_0\frac{d}{d\phi}\left(\frac{n}{\gamma}\right)=0: \tag{8.91}$$

$\frac{n(\phi)}{\gamma(\phi)}$ is a relativistic fluid invariant.

This is an important result, as it shows that the relativistic plasma frequency is invariant: the density modulation induced by the laser radiation pressure exactly compensates the variation of the fluid energy within the pulse. We finally find

$$n(z,t)=n(\phi)=n_0\left\{1+\kappa_0^{-1}\left[\mathbf{A}_\perp(\phi)\cdot\mathbf{u}_{\perp 0}+\frac{\mathbf{A}_\perp^2(\phi)}{2}\right]\right\}, \tag{8.92}$$

where $n_0$ is the initial electron beam density.

## 8.6 Ponderomotive Scattering

When the three-dimensional nature of the focused laser field is taken into account, it is found that electrons can be scattered by the laser and gain some energy in the process. Physically, this can be understood as follows: the relativistic electron copropagating with the laser is overtaken by the laser pulse close to the focus, and its energy is modulated, as it experiences the various phases of the electromagnetic field. In the focusing region, the laser field amplitude increases rapidly because of both spatial (focusing) and temporal effects (slippage towards the center of the pulse). As a result, the quiver amplitude increases nearly exponentially, but the particle is still confined around the propagation axis because the restoring force increases in step with the driving force. In close proximity to the focus, however, this situation becomes unstable because the quiver amplitude is of the order of or exceeds the focal spot radius. In this case, as the electron quivers through the spatial gradient of the Gaussian laser field, the restoring force decays exponentially, and the particle is scattered away from the focus with a high energy. The scattering effect is much more pronounced for electrons that are seeded into the laser wave prior to focus because these electrons are confined in the laser beam until they reach the focal region, where they experience the strongest electromagnetic fields. The maximum energy is obtained for those electrons that have the optimum phase when they are localized at the laser focus.

Using the analytic expression for the four-potential, the Lorentz force equation can be numerically integrated to yield the dynamics of a charge interacting with the three-dimensional laser fields. Typically, radiation reaction effects can be neglected in this situation, because the laser photons are Doppler downshifted in the electron frame, and thus behave as soft photons, which are well modeled by a classical continuous electromagnetic field. As the electrons are ponderomotively accelerated, this Doppler downshift increases further.

For the purpose of solving the Lorentz force equation for a three-dimensional laser field, one can employ the second-order Runge–Kutta method and utilize the axial electron phase as the variable of integration in order to handle both the nonlinear slippage and the large relativistic Doppler downshift. The normalized equations of motion are

$$\begin{aligned} \frac{d\mathbf{u}}{d\phi} &= -\frac{1}{\kappa}(\gamma\mathbf{e} + \mathbf{u}\times\mathbf{b}), \\ \frac{d\gamma}{d\phi} &= -\frac{1}{\kappa}\mathbf{u}\cdot\mathbf{e}, \end{aligned} \tag{8.93}$$

where we have introduced the normalized fields,

$$\mathbf{e} = \frac{e\mathbf{E}}{\omega_0 m_0 c}, \quad \text{and} \quad \mathbf{b} = \frac{e\mathbf{B}}{\omega_0 m_0}. \tag{8.94}$$

The inverse of the light-cone variable can be evaluated using the expression

$$\frac{1}{\kappa} = \frac{\gamma + u_z}{1 + \mathbf{u}_\perp^2}, \tag{8.95}$$

to avoid divergences. In addition, the evolution equation for the light-cone variable,

$$\frac{d\kappa}{d\phi} = -\frac{1}{\kappa}(\mathbf{u}\cdot\mathbf{e} - \gamma e_z - u_x b_y + u_y b_x), \tag{8.96}$$

may be used to randomize numerical noise and minimize the growth of numerical instabilities, by introducing the averaged quantities,

$$\begin{aligned} \langle\kappa\rangle &= \frac{1}{2}\left[\int_0^\phi \frac{d\kappa}{d\psi}d\psi + \frac{1 + \mathbf{u}_\perp^2}{\gamma + u_z}\right], \\ \langle\gamma\rangle &= \frac{1}{2}\left[\int_0^\phi \frac{d\gamma}{d\psi}d\psi + \sqrt{1 + \mathbf{u}^2}\right]. \end{aligned} \tag{8.97}$$

For more details, we refer the reader to the papers by Hartemann and co-authors, and by Quesnel and Mora, which are listed in the reference section for Chapter 8.

The typical results obtained for simulations of the dynamics of electrons in the field of a terawatt-class laser focused onto a micrometer-size spot, and reaching peak intensities well in excess of $10^{17}$ W/cm$^2$, are that the electrons are scattered with energies reaching a few MeVs, in a relatively narrow forward cone, in agreement with relativistic kinematics. One of the key difficulties with this process is that it proves very challenging to confine the electrons in the focusing laser beam long enough for them to gain the maximum ponderomotive energy available. Furthermore, the phase of the injected electrons with respect to the laser wave varies by large amounts, translating into a large phase space for the accelerated electrons. This last characteristic currently precludes the application of ponderomotive scattering to high-quality laser-driven electron acceleration.

Finally, we give a summarized version of the derivation of the relativistic expression of the ponderomotive force, as presented by Quesnel and Mora in the reference mentioned above. We start from the energy-momentum transfer equations:

$$\begin{aligned} d_t(\mathbf{p} - e\mathbf{A}) &= -(\nabla e\mathbf{A}) \cdot \mathbf{v}, \\ m_0 c^2 d_t \gamma &= e\mathbf{v} \cdot \partial_t \mathbf{A}. \end{aligned} \tag{8.98}$$

The derivation is performed in the Coulomb gauge, where $\nabla \cdot \mathbf{A} = 0$. We introduce the so-called "laser frame" coordinates, with the phase, $\phi = t - \frac{z}{c}$; Equation 8.98 becomes

$$\begin{aligned} [(1-\beta_z)\partial_\phi + \mathbf{v} \cdot \nabla](\mathbf{p} - e\mathbf{A}) &= -(\nabla e\mathbf{A}) \cdot \mathbf{v} + e(\beta \cdot \partial_\phi \mathbf{A})\hat{z}, \\ [(1-\beta_z)\partial_\phi + \mathbf{v} \cdot \nabla]\gamma m_0 c^2 &= e\mathbf{v} \cdot \partial_\phi \mathbf{A}. \end{aligned} \tag{8.99}$$

The momentum-transfer equation is now projected onto an axial and a transverse component, and we use the energy-transfer equation, to obtain

$$\begin{aligned} [(1-\beta_z)\partial_\phi + \mathbf{v} \cdot \nabla](\mathbf{p}_\perp - e\mathbf{A}_\perp) &= -(\nabla_\perp e\mathbf{A}) \cdot \mathbf{v}, \\ [(1-\beta_z)\partial_\phi + \mathbf{v} \cdot \nabla](p_z - eA_z - \gamma m_0 c) &= -e\mathbf{v} \cdot \partial_z \mathbf{A}. \end{aligned} \tag{8.100}$$

The next step consists in using the paraxial approximation and to separate all relevant quantities into fast and slow-varying components. Following the notation introduced by Quesnel and Mora, we write $f = \tilde{f} + \bar{f}$, where the first component varies on time-scales of order $\omega_0^{-1}$, while the time-scale for $\bar{f}$ is given by $G\Delta t = n^{-1}\omega_0^{-1}$, where $n$ represents the number of oscillation in the laser pulse of duration $\Delta t$: $n \approx \omega_0 \Delta t = 2\pi c \Delta t / \lambda_0$. Furthermore, a key assumption underlying this derivation is that the quantity $1 - \beta_z$ is of order zero; in other words, we require that

$$1-\beta_z \gg \frac{1}{k_0 w_0} \sim \frac{\lambda_0}{c\Delta t}. \tag{8.101}$$

The first parameter, $1/k_0 w_0$, corresponds to the paraxial approximation; the second parameter implies that the pulse has a slow-varying envelope.

Inspecting the second equality in Equation 8.100 at order zero, we see that

$$\partial_{\phi_0}(p_z - \gamma m_0 c) = 0, \tag{8.102}$$

which implies that this quantity varies slowly with respect to the phase of the laser pulse; in other words, it is an adiabatic invariant:

$$p_z - \gamma m_0 c \simeq \bar{p}_z - \bar{\gamma} m_0 c. \tag{8.103}$$

The same holds for the transverse canonical momentum:

$$\partial_{\phi_0}(\mathbf{p}_{\parallel} - e\mathbf{A}_{\perp}) = \mathbf{0}, \quad \mathbf{p}_{\perp} = e\tilde{\mathbf{A}}_{\perp} + \bar{\mathbf{p}}_{\perp}. \tag{8.104}$$

We now turn our attention to the first order component of the transverse momentum equation:

$$(\bar{\gamma} m_0 c - \bar{p}_z)\partial_{\phi_0}[\mathbf{p}_{\perp}^1 + \mathbf{p}_{\perp 1} + \mathbf{A}_{\perp 1}] + (\bar{\gamma} m_0 c - \bar{p}_z)\partial_{\phi_1}\bar{\mathbf{p}}_{\perp} + c(\mathbf{p}_{\perp} \cdot \nabla_{\perp})\bar{\mathbf{p}}_{\perp} = -c(\nabla_{\perp} e\mathbf{A}) \cdot \mathbf{p}. \tag{8.105}$$

In this equation, the superscript corresponds to the paraxial expansion order in $1/k_0 w_0$, while the subscript is related to the slow-varying envelope approximation, in $\lambda_0/c\Delta t$.

After averaging over the fast time-scale, Equation 8.105 reduces to

$$(\bar{\gamma} m_0 c - \bar{p}_z)\partial_{\phi_1}\bar{\mathbf{p}}_{\perp} + c(\bar{\mathbf{p}}_{\perp} \cdot \nabla_{\perp})\bar{\mathbf{p}}_{\perp} = -c\langle(\nabla_{\perp} e\mathbf{A}) \cdot \mathbf{p}\rangle. \tag{8.106}$$

Note that here, $\langle f \rangle = \bar{f}$.

The double expansion of the vector potential is now introduced, for a linearly polarized pulse:

$$\mathbf{A} \simeq (\tilde{A}_{\perp 0}^{0} + \tilde{A}_{\perp 1}^{0})\hat{x} + \tilde{A}_{z0}^{1}\hat{z}. \tag{8.107}$$

Here, we recognize the axial electromagnetic component introduced in our discussion of the paraxial approximation.

The average indicated in Equation 8.106 can now be performed, to yield the essential result:

$$\langle(\nabla_{\perp} e\mathbf{A})\cdot\mathbf{p}\rangle \simeq -\frac{1}{2}\nabla_{\perp}\langle|e\tilde{\mathbf{A}}_{\perp 0}^{0}|^{2}\rangle = -\frac{1}{2}\nabla_{\perp}\langle|e\tilde{\mathbf{A}}_{\perp}|^{2}\rangle. \tag{8.108}$$

Here, it is important to note that this quantity does not average to zero, since we are considering the square of the vector potential.

Proceeding similarly, we can obtain an expression for the relativistic factor. We start from the definition

$$\gamma^{2} = 1+\frac{1}{m_0^2c^2}\left[|\bar{\mathbf{p}}_{\perp}+e\tilde{\mathbf{A}}_{\perp}|^{2}+p_z^2\right]; \tag{8.109}$$

the axial momentum component is approximated using Equation 8.103,

$$p_z \simeq \bar{p}_z + m_0c(\gamma-\bar{\gamma}); \tag{8.110}$$

finally, we perform an average on the fast time-scale, which yields the sought-after result,

$$\bar{\gamma}^{2} \simeq 1+\frac{1}{m_0^2c^2}\left[|\bar{\mathbf{p}}_{\perp}|^{2}+\bar{p}_z^2+\langle|e\tilde{\mathbf{A}}_{\perp}|^{2}\rangle\right]. \tag{8.111}$$

The average velocity can now be defined as $\bar{\mathbf{v}} = \bar{\mathbf{p}}/\bar{\gamma}m_0 = \bar{\beta}c$, which allows us to recast Equation 8.106, with the help of the result given in Equation 8.108, as

$$[(1-\bar{\beta}_z)\partial_{\phi_1}+\bar{\mathbf{v}}_{\perp}\cdot\nabla_{\perp}]\bar{\mathbf{p}}_{\perp} = -\frac{1}{2m_0\bar{\gamma}}\nabla_{\perp}c\langle|e\tilde{\mathbf{A}}_{\perp}|^{2}\rangle. \tag{8.112}$$

For the axial motion, the averaging procedure outlined above first yields

$$[(\bar{\gamma}m_0c-\bar{p}_z)\partial_{\phi_1}+c(\bar{\mathbf{v}}_{\perp}\cdot\nabla_{\perp})]\bar{p}_z = [(\bar{\gamma}m_0c-\bar{p}_z)\partial_{\phi_1}+c(\bar{\mathbf{v}}_{\perp}\cdot\nabla_{\perp})]\bar{\gamma}m_0c. \tag{8.113}$$

On the other hand, deriving the average of the square of the relativistic factor, as expressed in Equation 8.111, with respect to the first-order phase results in the following identity:

$$\frac{1}{2}\partial_{\phi_1}\langle|e\tilde{\mathbf{A}}_{\perp}|^{2}\rangle = \bar{\gamma}m_0c\partial_{\phi_1}-\bar{p}_z\partial_{\phi_1}\bar{p}_z-\frac{1}{2}\partial_{\phi_1}|\bar{\mathbf{p}}_{\perp}|^{2}. \tag{8.114}$$

The last term in this equation can be obtained by using Equations 8.111 and 8.112: we multiply Equation 8.111 by the convective operator $\bar{\mathbf{p}}_{\perp}\cdot\nabla_{\perp}$, while

Equation 8.112 is multiplied by the average transverse momentum, $\bar{\mathbf{p}}_\perp$; with this we find that

$$-\frac{1}{2}\partial_{\phi_1}|\bar{\mathbf{p}}_\perp|^2 = c(\bar{\mathbf{p}}_\perp \cdot \nabla_\perp)\bar{\gamma}m_0c + \bar{p}_z\partial_{\phi_1}(\bar{p}_z - \bar{\gamma}m_0c). \qquad (8.115)$$

Finally, using the results expressed in Equations 8.114 and 8.115, the equation governing the averaged axial momentum component is derived:

$$[(\bar{\gamma}m_0c - \bar{p}_z)\partial_{\phi_1} + c(\bar{\mathbf{v}}_\perp \cdot \nabla_\perp)]\bar{p}_z = \frac{1}{2m_0c\bar{\gamma}}\partial_{\phi_1}\langle|e\tilde{\mathbf{A}}_\perp|^2\rangle. \qquad (8.116)$$

Equations 8.116 and 8.112 have the same structure: the motion is driven by the first-order phase derivative of $\langle|e\tilde{\mathbf{A}}_\perp|^2\rangle$. We can treat this quantity as an effective potential by adding the second-order terms $\bar{v}_z\partial_z$ and $-\frac{1}{2m_0\bar{\gamma}}\partial_z\langle|e\tilde{\mathbf{A}}_\perp|^2\rangle$ to both equations; we then find that

$$\frac{d\bar{\mathbf{p}}}{dt} = -\frac{1}{2m_0\bar{\gamma}}\nabla\langle|e\tilde{\mathbf{A}}_\perp|^2\rangle. \qquad (8.117)$$

The electrons behave as if subjected to an effective potential, the ponderomotive potential,

$$\mathcal{U} = \frac{1}{2m_0\bar{\gamma}}\langle|e\tilde{\mathbf{A}}_\perp|^2\rangle, \qquad (8.118)$$

and their averaged equation of motion is $\frac{d\bar{\mathbf{p}}}{dt} = -\nabla\mathcal{U}$.

## 8.7 Electron Dynamics in a Coherent Dipole Field

Continuing with our theoretical description of the interaction of relativistic electrons with high-intensity, coherent electromagnetic fields, we briefly outline the case of a dipole field. This field distribution is of particular interest because it satisfies Maxwell's equations exactly, as well as the gauge condition, and because it represents the lowest order in a multipole expansion, as discussed in Section 5.8. This is important, as some scientists have raised questions regarding the impact of the paraxial ray approximation upon such mechanisms as laser-driven particle acceleration. The coherent dipole field is an exact solution and can be used as a "toy model" to study these interactions. The fields also satisfy two important limits: in the vicinity of the dipole, the fields are similar to those of a diffracting laser pulse near focus;

in particular, wavefront curvature can be modeled exactly; in the far-field region, the fields tend to the plane-wave limit. Thus, one can study the transition from the regime where the Lawson–Woodward theorem applies, to the more realistic situation of diffracting dipole radiation and ponderomotive scattering.

In this section, radiation reaction effects are neglected; therefore, the natural units of length and time are given by the radiation wavepacket characteristic wavenumber, $k_0^{-1}$, and frequency, $\omega_0^{-1}$, respectively. In the case of an idealized oscillating electric dipole, the vector potential takes the form

$$\mathbf{A}(x_\mu) = A_0 \frac{f(\phi)}{r} \hat{x}, \tag{8.119}$$

where $r = \sqrt{x^2 + y^2 + z^2}$ is the distance from the dipole, $\hat{x}$ is the direction of polarization of the dipole (direction of the oscillating current), $\phi = t - r$ is the radial, expressed in normalized units, and $f(\phi)$ is an arbitrary function of the phase which corresponds to the temporal behavior of the dipole current.

Without loss of generality, we define $f(\phi) = g(\phi)\cos\phi$, where $g(\phi)$ is the temporal envelope of the dipole oscillatory motion. The scalar potential is obtained from the Lorentz gauge condition, $\partial_\mu A^\mu = 0$, with the result that

$$\varphi(x_\mu) = A_0 \frac{x}{r^2}\left[\frac{h(\phi)}{r} + f(\phi)\right], \tag{8.120}$$

where $h(\phi) = \int f(\phi)dt$.

At this point, it is important to verify that Maxwell's equations are completely satisfied by this form of the four-potential and the Lorentz gauge condition. The latter is satisfied by virtue of Equation 8.120; also, since we are using the four-potential, with $\mathbf{E} = -\nabla\varphi - \partial_t\mathbf{A}$ and $\mathbf{B} = \nabla \times \mathbf{A}$, Maxwell's source-free equations, $\nabla \cdot \mathbf{B} = 0$ and $\nabla \times \mathbf{E} + \partial_t\mathbf{B} = \mathbf{0}$, are automatically satisfied. Therefore, all we need to check is that $\nabla \cdot \mathbf{E} = 0$ and $\nabla \times \mathbf{B} - \partial_t\mathbf{E} = \mathbf{0}$, as we are considering the propagation of the dipole wave in vacuum. In terms of potential, these equations reduce to the wave equation,

$$\Box A_\mu = [\partial_\nu \partial^\nu] A_\mu = [\nabla^2 - \partial_t^2] A_\mu = 0, \tag{8.121}$$

which is indeed verified by the dipole four-potential, $A_\mu = (\varphi, A_x, 0, 0)$.

The Lorentz force components can easily be calculated from the four-potential. For the $y$- and $z$-components, this yields

$$\frac{du_y}{d\tau} = -u^\mu \partial_y A_\mu = -\frac{y}{r}(u^\mu \partial_r A_\mu), \tag{8.122}$$

and

$$\frac{du_z}{d\tau} = -u^{\mu}\partial_z A_{\mu} = -\frac{z}{r}(u^{\mu}\partial_r A_{\mu}), \tag{8.123}$$

where we recognize the fact that the partial derivatives operate only on the variable $r$ (when it occurs alone as a spatial derivative, and as part of the invariant phase).

Thus, we can rewrite the derivatives in terms of $r$, using the chain rule. For the $x$-component, the situation is similar, but we must proceed with care. There are two extra terms in the Minkowski force component along the polarization axis; the first is due to the multiplicative factor of the variable $x$ in the scalar potential, while the second comes from the fact that the vector potential has a component in that direction. Hence, the $x$-component of the Minkowski four-force can be expressed as

$$\frac{du_x}{d\tau} = -(u^{\mu}\partial_x A_{\mu} - u^{\mu}\partial_{\mu} A_x) = \gamma\frac{\varphi}{x} - \frac{x}{r}u^{\mu}\partial_r A_{\mu} + u^{\mu}\partial_{\mu} A_x. \tag{8.124}$$

Finally, the time-like component of the Lorentz force equation governs the evolution of the electron energy, and reads

$$\frac{d\gamma}{d\tau} = -(u^{\mu}\partial_t A_{\mu} - u^{\mu}\partial_{\mu}\varphi) = u_x\left(\frac{dA_x}{d\phi} + \frac{\varphi}{x}\right) + (\mathbf{u}\cdot\hat{r})\partial_r\varphi. \tag{8.125}$$

If one examines Equations 8.122 to 8.124, an interesting symmetry emerges: all of the spatial components of the Minkowski four-force depend upon a common term, namely $r^{-1}u^{\mu}\partial_r A_{\mu}$. Thus one can rearrange the terms in Equations 8.122 to 8.124 to obtain the following identities:

$$\frac{1}{y}\frac{du_y}{d\tau} = \frac{1}{z}\frac{du_z}{d\tau} = \frac{1}{x}\left(\frac{du_x}{d\tau} - u^{\mu}\partial_{\mu}A_x - \gamma\frac{\varphi}{x}\right) = -\frac{1}{r}u^{\mu}\partial_r A_{\mu}. \tag{8.126}$$

These identities are useful in checking the accuracy of the numerical code used to simulate the electron dynamics in the dipole field. The quantity to the right-hand side of Equation 8.126 can be expressed explicitly as

$$\begin{aligned} u^{\mu}\partial_r A_{\mu} &= u^{x}\partial_r A_x - \gamma\partial_r\varphi \\ &= \frac{A_0}{r}\left\{\gamma\frac{x}{r}\left[\frac{3}{r}\left(\frac{h}{r} + f\right) + \frac{df}{d\phi}\right] - u_x\left(\frac{f}{r} + \frac{df}{d\phi}\right)\right\}; \end{aligned} \tag{8.127}$$

additionally, we have

$$u^{\mu}\partial_{\mu}A_x = [(\mathbf{u}\cdot\nabla)+\gamma\partial_t]A_x = \frac{A_0}{r}\left[\gamma\frac{df}{d\phi}-u_r\left(\frac{f}{r}+\frac{df}{d\phi}\right)\right]. \quad (8.128)$$

The solution to the relativistic equations of motion is not analytically tractable, but in the limiting case where the distance from the dipole, as measured in units of $k_0^{-1}$, is a large number, we recover the plane wave dynamics discussed earlier in this chapter. For shorter distances, where the wavefront curvature and the axial electromagnetic components are significant, a numerical code can be used, and will be discussed shortly. We now derive the electromagnetic field corresponding to the dipole four-potential from the relation $F_{\mu\nu} = \partial_{\mu}A_{\nu} - \partial_{\nu}A_{\mu}$. The electric field of the ideal oscillating dipole is given by

$$E_x = \frac{A_0}{r}\left\{\left(\frac{x}{r}\right)^2\left[\frac{3}{r}\left(\frac{h}{r}+f\right)+\frac{df}{d\phi}\right]-\left[\frac{1}{r}\left(\frac{h}{r}+f\right)+\frac{df}{d\phi}\right]\right\}, \quad (8.129)$$

$$E_y = A_0\left(\frac{xy}{r^3}\right)\left[\frac{3}{r}\left(\frac{h}{r}+f\right)+\frac{df}{d\phi}\right], \quad (8.130)$$

and

$$E_z = A_0\left(\frac{xz}{r^3}\right)\left[\frac{3}{r}\left(\frac{h}{r}+f\right)+\frac{df}{d\phi}\right], \quad (8.131)$$

while the magnetic induction is given by

$$B_x = 0, \quad (8.132)$$

$$B_y = \frac{A_0}{r}\left(\frac{z}{r}\right)\left(\frac{f}{r}+\frac{df}{d\phi}\right), \quad (8.133)$$

and

$$B_z = \frac{A_0}{r}\left(\frac{y}{r}\right)\left(\frac{f}{r}+\frac{df}{d\phi}\right). \quad (8.134)$$

In the limit where the distance from the radiating dipole, as measured in units of the characteristic oscillation wavelength, $k_0^{-1} = \lambda_0/2\pi$, is large, it is easy to show that the electromagnetic field distribution reduces to that of a plane wave. We have

$$E_x \simeq -\frac{A_0}{r}\left\{\frac{df}{d\phi}\left[1-\left(\frac{x}{r}\right)^2\right]+O(r^{-1})\right\}+O(r^{-3})$$

$$\simeq -\frac{A_0}{r}\frac{df}{d\phi}\left[1-\left(\frac{x}{r}\right)^2\right], \tag{8.135}$$

$$E_y \simeq A_0\left(\frac{xy}{r^3}\right)\frac{df}{d\phi}+O(r^{-4}) \simeq 0, \tag{8.136}$$

and

$$E_z \simeq A_0\left(\frac{xz}{r^3}\right)\frac{df}{d\phi}+O(r^{-4}) \simeq 0, \tag{8.137}$$

for the electric field, and

$$B_x = 0, \tag{8.138}$$

$$B_y \simeq \frac{A_0}{r}\left(\frac{z}{r}\right)\left[\frac{df}{d\phi}+O(r^{-1})\right] \simeq \frac{A_0}{r}\left(\frac{z}{r}\right)\frac{df}{d\phi}, \tag{8.139}$$

and

$$B_z \simeq \frac{A_0}{r}\left(\frac{y}{r}\right)\left[\frac{df}{d\phi}+O(r^{-1})\right] \simeq \frac{A_0}{r}\left(\frac{y}{r}\right)\frac{df}{d\phi}, \tag{8.140}$$

for the magnetic induction. For $z = r$, we recover the plane wave relation $B_y = E_x$. It is also important to note that in order to compare the dipole field with a plane wave, diffraction must be taken into account by rescaling the amplitude of the four-potential as $A_0 \to A_0/r$. This is consistent as long as the relative displacement along the $z$-axis is small compared to the distance from the dipole, so that diffraction over the interaction length remains negligible.

Finally, we note that for a linearly polarized dipole wave, as described here, the electron trajectory remains in the plane of polarization; in other words, we can arbitrarily set $y = 0$ because of the azimuthal symmetry of the dipole radiation pattern (see Figure 8.4), and have $E_y = B_z = 0$. Because $B_x = 0$ as well, it is clear that there is no component of the Lorentz force in the $y$-direction; as a result, the electron trajectory remains two-dimensional and contained within the $x$-$z$ plane, or polarization plane.

According to classical electrodynamic theory, any electromagnetic wave can be described in terms of a multipole expansion, as studied in Chapter 5. The lowest-order moment is the dipole term, which we have employed here.

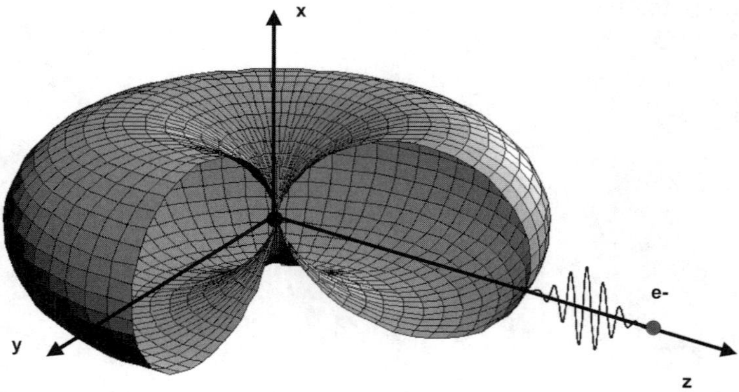


**FIGURE 8.4**
Schematic representation of the interaction of an electron with a dipole radiation wavepacket. The dipole is an exact solution of Maxwell's equations and satisfies the gauge condition.

Contributions from quadrupole, octupole, and other higher-order moments can be added to provide a full description of a laser focus in vacuum based on a multipole expansion.

To perform numerical simulations, the independent variable is chosen to be the radial phase, $\phi = t - r$. The evolution of the four-velocity is then described by

$$\frac{du_\mu}{d\phi} = \frac{du_\mu}{d\tau}\frac{d\tau}{d\phi} = a_\mu(\gamma - u_r)^{-1}, \tag{8.141}$$

where the four-acceleration, $a_\mu$, is described by Equations 8.122 to 8.125, and where the radial component of the four-velocity is given by

$$u_r = \frac{dr}{d\tau} = \mathbf{u}\cdot\hat{r} = \frac{\hat{x}u_x + \hat{y}u_y + \hat{z}u_z}{\sqrt{x^2+y^2+z^2}}. \tag{8.142}$$

The four-position of the electron is evaluated as

$$x_\mu(\phi) = x_\mu(\phi = 0) + \int_0^\phi \frac{dx_\mu}{d\psi}d\psi = x_{\mu 0} + \int_0^\phi \left[\frac{u_\mu}{\gamma - u_r}\right](\psi)\,d\psi. \tag{8.143}$$

Furthermore, in order to increase the numerical accuracy of the code, a second-order Runge–Kutta algorithm is used, where each dynamical variable, $w(\phi)$, is evaluated according to

$$w(\phi + \delta\phi) \simeq w(\phi) + \delta\phi\frac{dw}{d\phi}(\phi) + \frac{\delta\phi^2}{2}\frac{d^2w}{d\phi^2}(\phi). \tag{8.144}$$

The light-cone variable, $\kappa = \gamma - u_r$, is calculated using the identity $u_\mu u^\mu = -1$, with the result that

$$\kappa = \frac{1 + (\mathbf{u} \times \hat{r})^2}{\gamma + u_r}. \tag{8.145}$$

Also, the evolution equation for the light-cone variable,

$$\frac{d\kappa}{d\phi} = \frac{d}{d\phi}(\gamma - u_r) = \frac{1}{\gamma - u_r}\left[u_x\left(\frac{dA_x}{d\phi} + \frac{\varphi}{x}\right) + u_r\frac{\partial\varphi}{\partial r} - \left(\frac{\mathbf{u}^2 - u_r^2}{r}\right)\right] - \hat{r}\cdot\frac{d\mathbf{u}}{d\phi}, \tag{8.146}$$

is used to randomize the numerical noise and minimize the growth of numerical instabilities by introducing the averaged quantities

$$\langle\kappa\rangle = \frac{1}{2}\left[\kappa_0 + \int_0^\phi \frac{d\kappa}{d\psi}d\psi + \frac{1 + (\mathbf{u} \times \hat{r})^2}{\gamma + u_r}\right], \tag{8.147}$$

and

$$\langle\gamma\rangle = \frac{1}{2}\left[\gamma_0 + \int_0^\phi \frac{d\gamma}{d\psi}d\psi + \sqrt{1 + \mathbf{u}^2}\right]. \tag{8.148}$$

The convergence of the code is verified by comparing alternative calculated values of the energy, namely,

$$\gamma(\phi) = \gamma(\phi = 0) + \int_0^\phi \left[\frac{a_0}{\gamma - u_r}\right](\psi)\, d\psi, \tag{8.149}$$

and

$$\gamma(\phi) = \sqrt{1 + u_x^2(\phi) + u_y^2(\phi) + u_z^2(\phi)}. \tag{8.150}$$

The relative numerical error is obtained by dividing the difference between Equations 8.149 and 8.150 by the average value of the normalized energy, $\langle\gamma\rangle$.

The code can first be benchmarked against plane wave dynamics, as described earlier in this chapter. This is achieved by considering a region far from the dipole; in this case, the dipole wave is very close to a plane wave, and wavefront curvature is negligible. The electron dynamics agree precisely with the plane wave result, and there is no net energy exchange with the wave, as predicted by the Lawson–Woodward theorem. In terms of numerical precision, the maximum relative deviation observed between the dipole code and plane wave dynamics is $<10^{-10}$, for a normalized initial distance $z_0 = 10^6$.

Figure 8.5 shows a numerical simulation demonstrating ponderomotive scattering by a coherent dipole wave: the initial position of the electron is

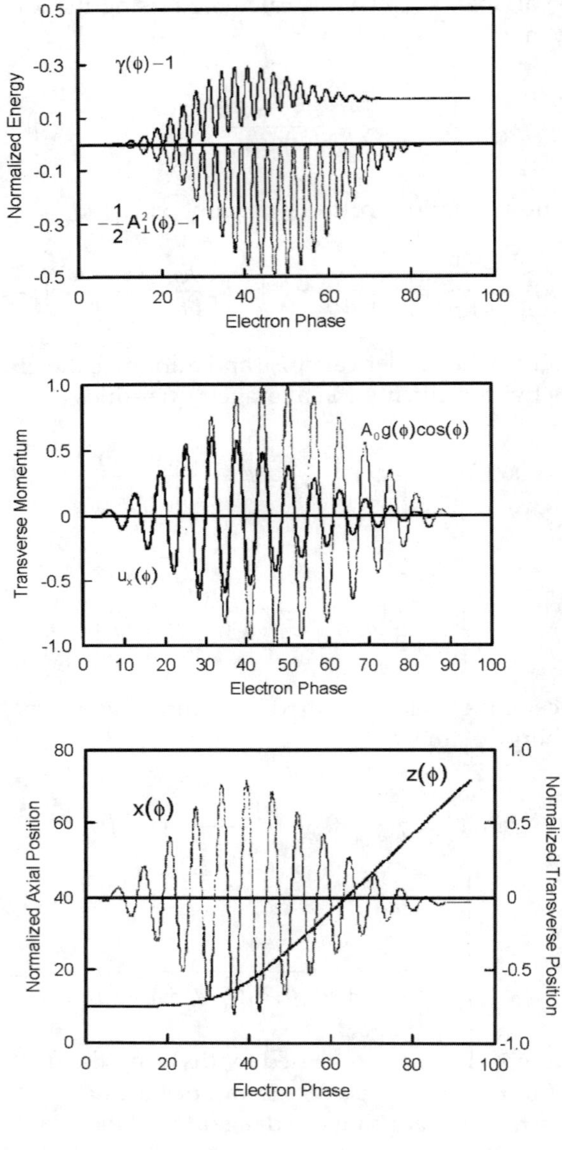


**FIGURE 8.5**
Top: electron energy as a function of phase and comparison with the plane wave case (no energy gain). Middle: transverse momentum and vector potential in the ponderomotive scattering regime. Bottom: typical electron trajectory for ponderomotive scattering.

$x_0 = y_0 = 0$, $z_0 = 10$; the amplitude of the four-potential is chosen so that $A_0/z_0 = 10$; the pulse envelope is modeled by the function

$$g(\phi) = \sin^2\left(\pi \frac{\phi}{\Delta\phi}\right), \tag{8.151}$$

where $\Delta\phi = \omega_0 \Delta t$ corresponds to a 10 fs pulse at a wavelength $\lambda_0 = 0.8$ $\mu$m; finally the electron is originally at rest, $\gamma_0 = 1$.

The net energy gain of the scattered electron is evident, and the trajectory is also shown, to better illustrate the physical mechanism underlying the scattering event. The conclusion of this study is that wavefront curvature and diffraction play a major role in ponderomotive scattering. It also helps emphasize the rather restrictive conditions under which the Lawson–Woodward is valid.

## 8.8 Chirped-Pulse Inverse FEL

The inverse free-electron laser (IFEL) interaction has been proposed as a viable vacuum laser acceleration process; the presence of a periodic magnetostatic field, or wiggler, clearly breaks the assumptions underlying the Lawson–Woodward theorem, and high-gradient acceleration is possible for such systems. Pioneering experimental work performed at Columbia University first demonstrated IFEL acceleration and was followed by experiments using a nanosecond-duration, GW $CO_2$ laser at Brookhaven National Laboratory (BNL). The IFEL is attractive for laser acceleration because the interaction is mediated in a vacuum, away from boundary conditions, thus allowing for a relatively large interaction region. In addition, the absence of a plasma as an accelerating medium resolves a number of potential problems, including plasma instabilities, nonlinear laser propagation, shot-to-shot reproducibility of the plasma, and the extremely small accelerating potential well, or "bucket," which characterizes laser-plasma acceleration schemes. Finally, the IFEL wiggler is suitable to provide good focusing and electron beam optics and transport, thus offering the potential for producing the high-quality electron beams required for advanced light sources, biomedical applications, and the next linear collider (NLC). However, one of the fundamental limitations of this acceleration scheme is the dephasing of the trapped electron with respect to the drive laser wave: as the electron energy increases, the free-electron laser (FEL) resonance condition can no longer be maintained, and the electron reaches a maximum energy given by the FEL interaction bandwidth. This problem can be addressed in two different ways: either the wiggler period and/or amplitude can be tapered, as successfully demonstrated in a high-efficiency FEL, or, equivalently, the drive laser pulse can be chirped. The latter approach will be discussed extensively in this section. Additionally, it should be noted that thus far, IFEL gradients were comparable to those possible with high-frequency rf systems (roughly up to 100 MeV/m).

By contrast, the main focus of this section is the theoretical and computational study of the IFEL interaction in a different regime, where we consider ultra-short, TW-class drive laser pulses which are now routinely generated by tabletop systems using chirped pulse amplification (CPA). For such femtosecond (fs) laser pulses, the IFEL interaction bandwidth is considerably wider than in the case of lower intensity drive pulses with durations in the picosecond to nanosecond range. Essentially, the well-known FEL resonance condition indicates that when the electron slips over one laser optical cycle, it also propagates over one wiggler period. Thus, for the fs pulses considered here, the wiggler interaction region is extremely short, and the IFEL resonance bandwidth is correspondingly wide. This translates directly into the fact that the electron energy can now vary significantly before the IFEL interaction detunes and saturates; in addition, the ultrahigh laser pulse intensity yields a high accelerating gradient. Therefore, the IFEL interaction physics is expected to change dramatically for broadband (fs) drive laser pulses near the so-called relativistic intensity regime ($>10^{17}$ W/cm$^2$ for optical wavelengths).

In this section, we also discuss how the use of a chirped laser pulse allows the FEL resonance condition to be maintained beyond the conventional dephasing limit, thus further improving the electron energy gain. Again, we note that such laser pulses are easily produced using the CPA technique which has led to the generation of femtosecond, multi-terawatt optical pulses, with tabletop laser systems operating at modest energies (in the joule range). We will show that the ultrashort, high-intensity laser pulses thus generated make it possible to design an IFEL with very high accelerating gradients (>1 GeV/m), in contrast with the longer pulse approaches previously considered. In fact, the need to alleviate electron dephasing for narrow-band laser pulses has motivated the development of period- and amplitude-tapered wigglers, which help maintain resonance throughout the IFEL interaction region, an approach which is similar and complementary to the concept presented here. Another practical limitation of IFEL accelerators is the diffraction of the drive laser pulse. In the conventional beam geometry, optical guiding cannot be used because the phase shift of the IFEL interaction has the opposite sign of the FEL phase shift, which results in the well-known guiding effect. We suggest how this can be alleviated by taking advantage of the ultra-wide optical bandwidth of the chirped laser pulse: negative dispersion focusing optics can be used to produce a chromatic line focus, where long wavelengths are focused first, while the shorter wavelengths required to maintain the FEL resonance condition at higher energies are focused further along the interaction region. Finally, it will be shown numerically that the accelerating IFEL bucket is very wide compared to plasma-based laser acceleration schemes, which are discussed briefly at the end of this chapter: We will show that for a 1-ps FWHM (full-width at half-maximum) Gaussian electron bunch, and a 1-cm period wiggler, the IFEL energy spread is <0.9%. This is extremely advantageous for a practical laser accelerator, as the device could be driven by a conventional rf photoinjector. This section is organized as follows: after a brief discussion of the motivation for the chirped-pulse IFEL concept, we provide a short presentation of the analytical theory of the chirped-pulse IFEL interaction. Next, a simple one-dimensional computer code is described, as well as the results of simulations demonstrating the relevance of the IFEL to high-gradient acceleration, and a short discussion of tapering and chirping as means to increase the IFEL interaction bandwidth. Finally, we focus on the study of the chirped-pulse IFEL output energy spectrum, where very low energy spreads are predicted for picosecond-duration electron bunches produced by a conventional rf system.

A brief discussion of this vacuum acceleration concept can be given by considering the well-known FEL resonance condition, $\lambda \approx \lambda_w(1 + A_w^2)/2\gamma^2$, where $\lambda$ is the drive laser wavelength, $\lambda_w$ is the wiggler period, $A_w = eB_w\lambda_w/2\pi m_0 c$ is the normalized vector potential of the wiggler, and $\gamma$ is the electron energy. It should be noted that in the above equation, the laser intensity is sufficiently small that its radiation pressure is negligible; in other words, the normalized vector potential of the laser satisfies the condition $A_0^2 = (eE_0/\omega_0 m_0 c)^2 \ll 1$. It is clear that, as the electron energy increases during the IFEL acceleration process, resonance cannot be maintained, thus saturating the interaction and limiting the energy gain. In particular, the higher the accelerating gradient, the faster dephasing occurs. However, for ultrashort laser pulses, this is balanced by the increased interaction bandwidth. To further alleviate this problem, it has been proposed to taper the wiggler amplitude, $B_w$, or the wiggler period, $\lambda_w$. This approach has proved extremely successful in the case of FELs, as exemplified by the results obtained by Orzechowski et al., at Lawrence Livermore National Laboratory (LLNL). However, one of the key requirements for the implementation of a tapered wiggler is the adiabaticity of the tapering, in order to preserve the electron beam quality. Clearly, for the very intense drive laser pulses necessary for high IFEL gradients, this requirement is difficult to meet, as the entire IFEL interaction now becomes strongly nonlinear and non-adiabatic. In contrast, with the technique discussed here, the drive laser wavelength, $\lambda$, decreases as the electron gains energy, thus maintaining resonance beyond the aforementioned saturation length. Such chirped, ultrahigh-intensity laser pulses are easily produced using CPA, and tabletop systems with high repetition rates (20–1000 Hz) are now available. We now present a short theoretical analysis of the chirped-pulse IFEL interaction. In this analysis, length is measured in units of the central laser wavelength, $1/k_0 = \lambda_0/2\pi$, time in units of the laser frequency, $1/\omega_0$, mass in units of $m_0$, and charge in units of $e$. For the sake of simplicity, plane waves are considered. Here, the combined four-potential of the drive laser pulse, $\mathbf{A}_\ell(\phi)$, and the wiggler magnetic field, $\mathbf{A}_w(z)$, is given by

$$A_\mu = (\varphi, \mathbf{A}_\perp, A_z), \quad \varphi = A_z = 0, \quad \mathbf{A}_\perp(x_\mu) = \mathbf{A}_\ell(\phi) + \mathbf{A}_w(z), \qquad (8.152)$$

where we have introduced the phase of the traveling laser wave, $\phi = k_\mu x^\mu = t - z$.

The corresponding electromagnetic field components are

$$\mathbf{E}_\perp = -\frac{\partial \mathbf{A}_\perp}{\partial t} = -\frac{d\mathbf{A}_\ell}{d\phi}, \quad \mathbf{B}_\perp = \nabla \times \mathbf{A}_\perp = \hat{z} \times \left( \mathbf{E}_\perp + \frac{d\mathbf{A}_w}{dz} \right). \tag{8.153}$$

Neglecting radiation reaction, the Lorentz force equation governs the evolution of the electron four velocity, $u_\mu = dx_\mu/d\tau$, where $\tau$ is the electron proper time, which, in our units, corresponds to the energy–momentum four vector. We thus have, for the transverse motion,

$$\frac{d\mathbf{u}_\perp}{d\tau} = -(\gamma \mathbf{E}_\perp + \hat{z} u_z \times \mathbf{B}_\perp) = (\gamma - u_z) \frac{d\mathbf{A}_\ell}{d\phi} + u_z \frac{d\mathbf{A}_w}{dz}, \tag{8.154}$$

where we recognize the light-cone variable, $\kappa = d\phi/d\tau = \gamma - u_z$. Equation 8.154 can be readily integrated to yield the transverse momentum invariant: $\mathbf{u}_\perp = \mathbf{A}_\ell(\phi) + \mathbf{A}_w(z)$. The evolution of the electron energy and axial momentum are given by

$$\frac{d\gamma}{d\tau} = -\mathbf{u}_\perp \cdot \mathbf{E}_\perp = \frac{1}{2} \frac{d\mathbf{A}_\ell^2}{d\phi} + \mathbf{A}_w \cdot \frac{d\mathbf{A}_\ell}{d\phi}, \tag{8.155}$$

and

$$\frac{du_z}{d\tau} = -\mathbf{u}_\perp \times \mathbf{B}_\perp = \frac{1}{2} \frac{d\mathbf{A}_\ell^2}{d\phi} + \frac{1}{2} \frac{d\mathbf{A}_w^2}{dz} - \mathbf{A}_\ell \cdot \frac{d\mathbf{A}_w}{dz} - \mathbf{A}_w \cdot \frac{d\mathbf{A}_\ell}{d\phi}, \tag{8.156}$$

where we recognize the laser radiation pressure (ponderomotive potential) $d\mathbf{A}_\ell^2/d\phi$, and the IFEL interaction terms, involving the nonlinear product of the laser and wiggler vector potentials.

In the case of a helically polarized wiggler with constant amplitude, $d\mathbf{A}_w^2/dz$ is identically zero. The IFEL acceleration gradient scaling is given by

$$\frac{d\gamma}{dz} = \frac{d\gamma}{d\tau}\frac{d\tau}{dz} = \frac{\mathbf{A}_w}{u_z} \cdot \frac{d\mathbf{A}_\ell}{d\phi}, \tag{8.157}$$

thus indicating that high acceleration gradients can be obtained by using ultrahigh-intensity drive laser pulses, where both the normalized laser and wiggler vector potentials approach unity. The equation governing the evolution of the light-cone variable is obtained by subtracting the axial force equation from the energy equation, to yield

$$\frac{d\kappa}{d\tau} = \frac{d}{d\tau}(\gamma - u_z) = -\mathbf{u}_\perp \cdot \mathbf{E}_\perp + \mathbf{u}_\perp \times \mathbf{B}_\perp = \mathbf{u}_\perp \cdot \frac{d\mathbf{A}_w}{dz}. \tag{8.158}$$

Using the result obtained for the transverse momentum, and for a helical wiggler, Equation 8.158 reduces to

$$\frac{d\kappa}{d\tau} = \mathbf{A}_\ell \cdot \frac{d\mathbf{A}_w}{dz}. \tag{8.159}$$

This equation is particularly simple, and it indicates that if either the wiggler or the laser field is zero, the light-cone variable is invariant. The axial momentum and energy can now be expressed in terms of the light-cone variable as

$$\gamma = \frac{1 + \mathbf{A}_\perp^2 + \kappa^2}{2\kappa}, \quad u_z = \frac{1 + \mathbf{A}_\perp^2 - \kappa^2}{2\kappa}. \tag{8.160}$$

Since $\lim_{\phi\to\pm\infty}|\mathbf{A}_\ell(\phi)| = \mathbf{0}$ and $\lim_{z\to\pm\infty}|\mathbf{A}_w(z)| = \mathbf{0}$, it is clear that the electron will have a net energy gain only if the light-cone variable is larger at the end of the interaction. We also note that, during the interaction, the energy can be increased locally by the laser radiation pressure or the IFEL interaction. This so-called "ponderomotive" acceleration process, which we have reviewed in Section 8.6, has been proposed as a laser acceleration scheme; however, a viable concept for the extraction of the accelerated electrons from the laser pulse proves extremely difficult to envision: the termination of the interaction necessarily involves a boundary condition which modifies the laser pulse characteristics, and it completely degrades the accelerated electron beam quality. It proves useful to consider the electron phase in the laser pulse as the independent variable; in this case, we have

$$\frac{d\kappa}{d\phi} = \frac{d\kappa}{d\tau}\frac{d\tau}{d\phi} = \frac{d}{d\tau}[\ln(\kappa)] = \frac{\mathbf{A}_\ell}{\kappa} \cdot \frac{d\mathbf{A}_w}{dz}, \tag{8.161}$$

and the electron position is given by

$$x_\mu(\phi) = \int_0^\phi \frac{u_\mu(\psi)}{\kappa(\psi)}\, d\psi. \tag{8.162}$$

Because we are considering ultrashort, chirped drive laser pulses, nonlinear slippage must be taken into account, as well as the diffraction (Rayleigh) length of the focused laser pulse. The nonlinear slippage length is given by the $z$ component of Equation 8.162. For such pulses, the slippage length can be much longer than the laser pulse itself, thus allowing long electron bunches to interact in the IFEL.

Based on the theoretical model described above, we have developed a simple, one-dimensional, nonlinear computer code, which follows the evolution of the light-cone variable as a function of the electron phase within the drive laser pulse. This one-dimensional analysis gives an adequate first-order description of the IFEL interaction physics provided that the drive laser pulse can be modeled by a plane wave, which implies a large *f* number and a long Rayleigh range, and as long as the IFEL electron trajectories are confined to a region where the transverse gradient of the wiggler and laser fields are small. Specifically, the vector potential of the laser is modeled by a circularly polarized plane wave with a linear chirp, $\alpha$,

$$\mathbf{A}_\ell(\phi) = A_0 g(\phi)[\hat{x}\sin(\phi+\alpha\phi^2)+\hat{y}\cos(\phi+\alpha\phi^2)]. \tag{8.163}$$

The pulse envelope is chosen as $g(\phi)=\sin^2[\phi/\Delta\phi]$, which closely approximates a Gaussian near the maximum and has a finite duration. Here, $\Delta\phi$ is the pulse FWHM, measured in units of $1/\omega_0$, and $A_0$ is the amplitude of the laser pulse measured in units of $m_0c/e$. The helical wiggler field is simply given by

$$\mathbf{A}_w(z) = A_w[\hat{x}\sin(k_w z)+\hat{y}\cos(k_w z)], \tag{8.164}$$

where the wiggler wave number is given in the units chosen here by $k_w = \lambda_0/\lambda_w$.

The position of the electron in the wiggler is given by Equation 8.162 as $z(\phi)=\int_0^\phi u_z(\psi)/\kappa(\psi)d\psi$. Finally, in order to increase the accuracy of the code, a second-order Runge–Kutta algorithm is used; for this purpose, we evaluate the second-order derivative of the light-cone variable with respect to phase,

$$\frac{d^2\kappa}{d\phi^2} = \frac{1}{\kappa}\left[\frac{d\mathbf{A}_\ell}{d\phi}\cdot\frac{d\mathbf{A}_w}{dz} - k_w^2\frac{dz}{d\phi}(\mathbf{A}_\ell\cdot\mathbf{A}_w) - \left(\frac{d\kappa}{d\phi}\right)^2\right]. \tag{8.165}$$

The accuracy of the code is verified by comparing two different values of the energy, namely,

$$\gamma(\phi) = \gamma(\phi=0) + \int_0^\phi \frac{1}{\kappa(\phi)}\left[\frac{1}{2}\frac{d\mathbf{A}_\ell^2(\psi)}{d\phi} + \mathbf{A}_w(\psi)\cdot\frac{d\mathbf{A}_\ell(\psi)}{d\phi}\right]d\psi, \tag{8.166}$$

and

$$\gamma(\phi) = \sqrt{1+\mathbf{u}_\perp^2(\phi)+u_z^2(\phi)}. \tag{8.167}$$

The relative numerical error is obtained by dividing the difference between Equations 8.166 and 8.167 by the average value of the energy. To demonstrate the relevance of the chirped-pulse IFEL concept, we run the code for the following parameters: the normalized vector potential of the drive laser pulse is $A_0 = 0.2$, the pulse duration is 10 fs, the central laser wavelength is $\lambda_0 = 0.8\ \mu$m, the wiggler field is $B_w = 2$ kG, the wiggler period is $\lambda_w = 1$ cm, and the injection energy is $\gamma_0 = 65$, while the resonant IFEL interaction energy

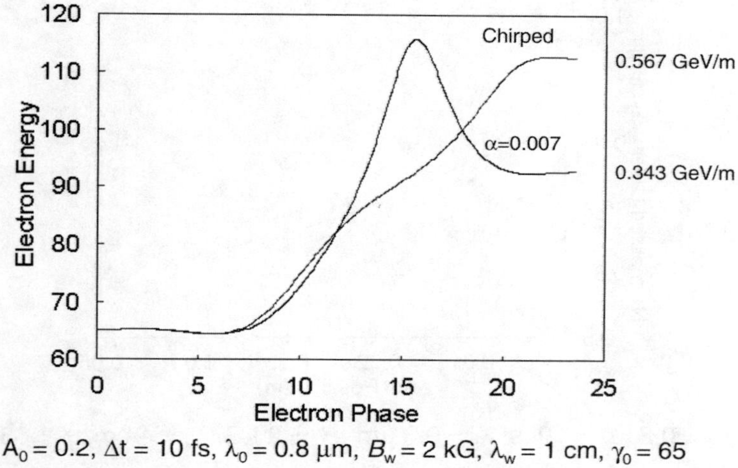


**FIGURE 8.6**
Electron energy as a function of its phase in the drive laser pulse, for an unchirped and a chirped pulse. The corresponding accelerating gradients are indicated, as well as the laser pulse and wiggler parameters. The injection energy corresponds to approximately 33 MeV, with an output energy of 57 MeV, over a 4-cm interaction region.

is $\gamma^* = 80$. The results are shown in Figure 8.6 for both an unchirped laser pulse, and an optimally chirped laser pulse, where $\alpha = 0.007$. In the first case, the electron clearly detunes away from the IFEL resonance and exits the interaction region with an energy gain corresponding to an average acceleration gradient of 0.343 GeV/m. This number is still quite high because we are using an ultrashort laser pulse with a high focused intensity. The intensity can be expressed as

$$I_0 = \left(\frac{A_0}{\lambda_0}\right)^2 \left(\frac{2\pi^2 \varepsilon_0 m_0^2 c^5}{e^2}\right) = \left(\frac{A_0}{\lambda_0}\right)^2 13.68 \text{ GW},$$

which corresponds to $I_0 = 8.55 \times 10^{16}$ W/cm$^2$ for the parameters given above. At this point, it is important to understand that attempts to decouple the electron at its peak energy will result in a severe loss of beam quality. In the second case, however, the IFEL resonance is maintained throughout the interaction region, and the electron energy now corresponds to an average gradient of 0.567 GeV/m, or a 65% increase over the unchirped case. More importantly, in this case, no extraction scheme is required, thus preserving the accelerated beam quality. The normalized wavelength of the chirped pulse can be given as a function of the electron position in the interaction region by plotting $1/(1 + \alpha\phi)$ versus $z(\phi)$, as shown in Figure 8.7. The relative optical bandwidth of the chirp is found to be 15%, which corresponds to 120 nm of

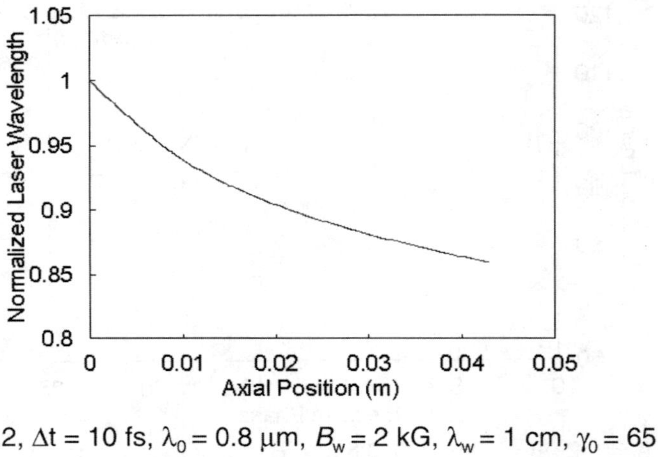


**FIGURE 8.7**
Resonant laser wavelength as a function of the distance in the wiggler; the bandwidth of the chirp is approximately 15%.

spectral width. The bandwidth of the 10 fs pulse, including the additional 15% chirp, is, therefore, approximately 200 nm, or 5 fs transform-limited duration. This value has been demonstrated experimentally for Ti:$Al_2O_3$ (titanium-doped sapphire) systems. The code has also been used to model a tapered wiggler IFEL; we find that using the same parameters, but without a chirped laser pulse, results in a similar gradient. This gradient is possible with an increase of the wiggler period from 1 to 1.3 cm over four periods because of the inherently large bandwidth of the 10 fs drive laser pulse; the amplitude of the wiggler remains constant over the interaction region. Therefore, tapering and chirping should be considered as complementary means to improve the IFEL gradient. The parameters discussed above can be translated into experimental parameters by assuming a long cylindrical focus for the drive laser pulse: using a Rayleigh range $z_0 = \pi(w_0^2/\lambda_0) = 2$ cm, the focal waist is $w_0 = 71.4$ $\mu$m. The corresponding peak laser power is $P_0 = I_0\pi w_0^2 = I_0\lambda_0 z_0 = 13.7$ TW, and the laser pulse energy takes the modest value of $W_0 = P_0\Delta t = 137$ mJ. This is considerably lower than the tens of joules commonly used in conventional rf systems to power accelerating sections of a few tens of MeVs. In fact, it is interesting to compare the energy in the drive laser pulse to the kinetic energy acquired by a 1 nC electron bunch in the IFEL: we find that the energy transferred to the beam is 22.7 mJ, or 16.6% of the laser energy. This shows that the IFEL can be significantly loaded, which is key to efficient acceleration. It is also interesting to evaluate the radius of the electron trajectory in the helical wiggler. We have $k_w r_\perp = u_\perp/u_\parallel \leq A_w/\gamma_0$, and the wiggler parameter $A_w = 0.18$; therefore, $r_\perp = 4.4$ $\mu$m $\ll w_0$, and the one-dimensional model should appropriately describe the IFEL interaction. Clearly, this represents a worst-case scenario, as a linearly polarized laser and wiggler and a slab focus could be used; in addition, it is important to note that the intensity of the laser along the focal region can be maximized by using an optical scheme that will provide a chromatic line focus. The long wavelength part of the pulse will be focused closer to the optics, with the shorter wavelengths focused further down the propagation axis. This can be accomplished in a number of ways, using either reflective, refractive, or diffractive optics. An interesting and simple approach is to use a Fresnel zone plate, which will produce the required $d\lambda/dz$ to match the computer calculation presented earlier. This approach could relax the drive laser energy requirement by one order of magnitude.

Detailed studies of the acceptance of the IFEL accelerating bucket can also be performed with the simple one-dimensional code introduced here. This is shown in Figure 8.8, where the IFEL energy gain is given as a function of the initial position in the wiggler, which is defined as the point where the laser pulse overtakes the electron [$z(\phi = 0)$]. The periodicity, which is clearly seen, results from the one-dimensional nature of our model. In the absence of diffraction, the IFEL is invariant by translation of $\lambda_w$. It is also important to note here that although both the drive laser pulse and the electron bunch are quite short, the nonlinear IFEL slippage length is a few centimeters, thus explaining why a picosecond electron bunch can be accelerated by a femtosecond laser pulse. In the present one-dimensional model, the phase of the ponderomotive force is determined by the initial electron position in the wiggler, as defined above; because the IFEL is a fast-wave device, the acceptance of the accelerating

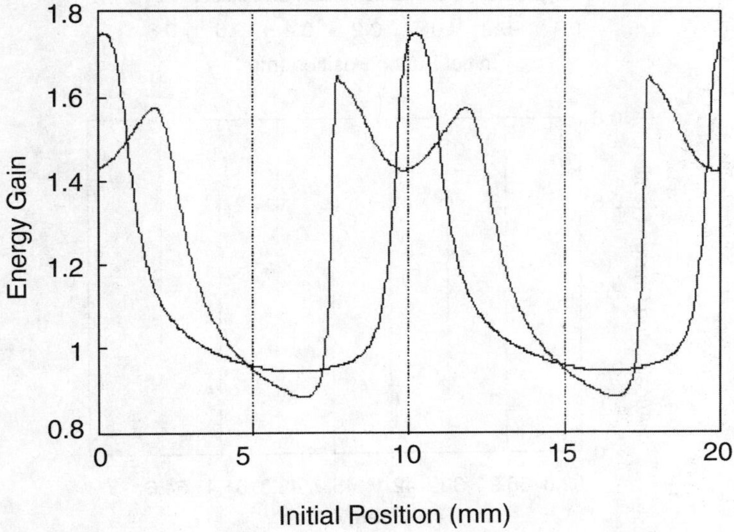


**FIGURE 8.8**
Relative energy gain for electron seeded at various positions (phases) in the laser and wiggler. The wiggler periodicity is evident, as well as the existence of a wide acceleration bucket, near 0 and 10 mm for the chirped pulse (smooth curve).

bucket scales with the wiggler period. This should not be confused with the bunching which occurs at the IFEL resonant wavelength. The width of the energy peak, in the chirped case, is seen to be approximately 2 mm FWHM. The fact that the IFEL bucket scale is given by the wiggler period directly results from the fast-wave nature of the IFEL interaction. This number is extremely large compared to the accelerating bucket in a laser-plasma system, which is roughly equal to a tenth of the plasma wavelength (typically, in the micron range). This number is also compatible with conventional rf injectors, as is demonstrated in Figure 8.9 (bottom). Here, we inject a 1-ps FWHM Gaussian electron bunch into the IFEL, at the optimum phase (0.2 mm) for maximum energy gain, as shown in Figure 8.9 (top); the output electron energy spectrum is then computed, and we find nearly 60% of the accelerated charge

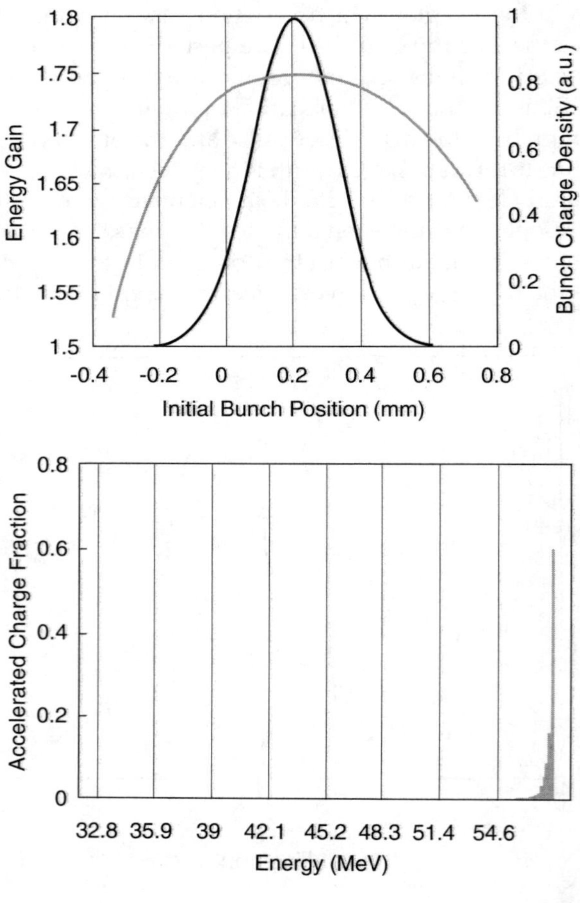


**FIGURE 8.9**
Top: 1 ps Gaussian electron bunch injected near the phase yielding the peak energy gain; resulting energy distribution after the IFEL, demonstrating the good acceptance of the accelerating bucket.

in a 0.125 MeV bin at 57.8 MeV. This represents an energy spread of < 0.2%. Furthermore, 90% of the accelerated charge is found within a 0.5 MeV energy interval, which corresponds to a spread smaller than 0.9%. The 32.8 MeV, 1-ps electron bunch injected in the IFEL could be produced by a modern rf photo-injector, as discussed in the introduction to Chapter 9. We also note that the IFEL resonance condition indicates that high-energy (TeV) scaling is possible, at least in principle: assuming that the wiggler magnetic field scales linearly with the wiggler wavelength, and using a value of $\chi = 2$ kG/cm for the scaling, which is easily achievable with current technology, the resonance condition can be recast as $\gamma \approx (\lambda_w/\lambda^*)^{5/2}$, where the length scale $\lambda^* = [2\pi(\lambda_0/r_0)m_0/\varepsilon_0\chi^2]^{1/5} = 3.4066$ mm. Here $r_0 = 2.8178 \times 10^{-15}$ m, is the classical electron radius, and we have used $\lambda_0 = 800$ nm. For $\lambda_w = 1$ m, the resonant IFEL energy is 0.754 TeV. Naturally, parasitic effects, such as synchrotron radiation losses, must be taken into account to further assess the high-energy scalability of the chirped-pulse IFEL. Finally, in the limit of high energies (GeV-TeV) and ultrahigh intensities, well into the relativistic regime ($A_0 > 1$), the IFEL concept has been studied theoretically by a few authors. Their results confirm that a static magnetic field can help dephase the accelerated electrons with respect to the drive laser pulse, thus yielding net energy gain even in the case of a plane wave interaction. Furthermore, these studies also show that the reversal of the static field allows one to maintain the accelerating gradient, as in an IFEL, and that for high gradients, the magnetic field should be tapered every half period. This approach is discussed in the next section.

## 8.9 Free-Wave Acceleration by Stimulated Absorption of Radiation

In this section, we briefly study a high-gradient laser acceleration process, whereby electrons are interacting in vacuum with a femtosecond laser operating at relativistic intensities, and a static magnetic field. The basic electrodynamical interaction underlying this concept can be viewed as free-wave acceleration by stimulated absorption of radiation (FASAR), where the magnetostatic field acts as a broadband wiggler, in a manner somewhat akin to the inverse free-electron laser (IFEL) mechanism. One of the main differences with conventional IFEL acceleration resides in the fact that we are considering a deeply nonlinear regime, where the drive laser pulse intensity is relativistic. The vector potential of the laser pulse, measured in units of $m_0c/e$, is much larger than one; for visible wavelengths, this corresponds to peak intensities exceeding $10^{18}$ W/cm$^2$. This interaction is best described as a three-stage process: first, the electron energy is boosted by the radiation pressure; at the peak energy, a transverse, static magnetic field is applied, which optimally dephases the electron and the wave; finally, the dephased electron is further accelerated in the second half of the pulse. The final energy, $\gamma_{+\infty}$, is shown to scale as $\frac{\gamma_{+\infty}}{\gamma_{-\infty}} \simeq (1 + A_0^2)^2$, where $\gamma_{-\infty}$ is the initial beam energy and $A_0$ is the normalized laser potential.

The first theoretical step towards the FASAR was independently taken by Appolonov and Kawata et al., who showed that a static electric field superimposed to a plane electromagnetic wave could dephase an electron and yield energy gain in a process closely paralleling the IFEL interaction. Further analysis of this process was performed in the case of a magnetic field, where net acceleration was also obtained; finally, the case of a high-intensity drive laser was first considered by Woodworth and Kerman. The present theory builds upon this pioneering work by considering the relativistic intensity regime, and offers a detailed analysis of the physics underlying the acceleration process; in particular, the energy gain scaling with laser intensity is derived for the first time.

The FASAR is attractive for laser acceleration because the interaction is mediated in a vacuum, away from boundary conditions, thus allowing for a relatively large interaction region. In addition, the absence of a plasma as an accelerating medium resolves a number of potential problems, including plasma instabilities, nonlinear laser propagation, shot-to-shot reproducibility, and the extremely small accelerating potential well, or "bucket," characterizing laser-plasma acceleration schemes. Finally, the FASAR interaction region is compatible with electron beam focusing optics, thus offering the potential for producing the high-quality electron beams required for advanced light sources, biomedical applications, and the next linear collider (NLC).

The size of the accelerating bucket is found to be considerably larger than that of proposed plasma-based laser accelerators, thus allowing the possibility of accelerating high-charge bunches and loading the laser pulse, which is key to efficient acceleration. It is also predicted that for a short Gaussian electron bunch, the FASAR energy spread will be quite small. This is extremely advantageous for a practical laser accelerator, as the device could be driven by a conventional rf linac, such as the NLCTA (NLC test accelerator).

In the present analysis, length is measured in units of the central laser wavelength, $k_0^{-1}$, time in units of the laser frequency, $\omega_0^{-1}$, mass in units of $m_0$, while charge is normalized to the electron charge, e. For the sake of simplicity, plane waves are considered. The combined four-potential of the drive laser pulse, $\mathbf{A}_\ell(\phi)$, and the static magnetic field, $\mathbf{A}_w(z)$, is given by

$$A_\mu = (\varphi, \mathbf{A}_\perp, A_z), \quad \varphi = A_z = 0, \quad \mathbf{A}_\perp(x_\mu) = \mathbf{A}_\ell(\phi) + \mathbf{A}_w(z), \quad (8.168)$$

where we have introduced the relativistically invariant phase of the drive laser pulse, $\phi = -k_\mu x^\mu = t - z$.

The four-potential satisfies the Lorentz gauge condition, $\partial_\mu A^\mu = 0$, where $\partial_\mu = (-\partial_t, \Delta)$ is the four-gradient. The corresponding electromagnetic field components are derived from the definition of the electromagnetic field tensor, $F_{\mu\nu} = \partial_\mu A_\nu - \partial_\nu A_\mu$:

$$\mathbf{E} = \mathbf{E}_\perp = -\partial_t \mathbf{A}_\perp = -\frac{d\mathbf{A}_\ell}{d\phi}, \tag{8.169}$$

and

$$\mathbf{B} = \mathbf{B}_\perp = \nabla \times \mathbf{A}_\perp = \hat{z} \times \left( \mathbf{E}_\perp + \frac{d\mathbf{A}_w}{dz} \right). \tag{8.170}$$

Radiation reaction effects can be neglected, as the drive laser photons are Doppler downshifted to very low energies in the electron frame; therefore the Lorentz force equation, $\frac{du_\mu}{d\tau} = -F_{\mu\nu}u^\nu$ governs the evolution of the electron four-velocity, $u_\mu = \frac{dx_\mu}{d\tau} = (\gamma, \mathbf{u}_\perp, u_z)$. Here, $\tau$ is the proper time along the electron world line, $x_\mu(\tau)$. Note that in our units, the four-velocity coincides with the four-momentum. The transverse component of the Lorentz force equation reads

$$\frac{d\mathbf{u}_\perp}{d\tau} = -(\gamma \mathbf{E}_\perp + \hat{z} u_z \times \mathbf{B}_\perp) = (\gamma - u_z)\frac{d\mathbf{A}_\ell}{d\phi} + u_z \frac{d\mathbf{A}_w}{dz}. \tag{8.171}$$

This equation can readily be integrated to yield the transverse momentum invariant: we introduce the light-cone variable,

$$\kappa = \gamma - u_z = \frac{d\phi}{d\tau}. \tag{8.172}$$

We can then recast Equation 8.171 as

$$\frac{d\mathbf{u}_\perp}{d\tau} = \frac{d\phi}{d\tau}\frac{d\mathbf{A}_\ell}{d\phi} + \frac{dz}{d\tau}\frac{d\mathbf{A}_w}{dz} = \frac{d}{d\tau}(\mathbf{A}_\ell + \mathbf{A}_w), \tag{8.173}$$

and obtain

$$\mathbf{u}_\perp - \mathbf{u}_{\perp -\infty} = \mathbf{A}_\ell + \mathbf{A}_w, \tag{8.174}$$

where $\mathbf{u}_{\perp -\infty}$ is the initial transverse momentum, which will be set equal to zero in the reminder of this derivation.

The dynamics of the axial momentum are then governed by

$$\frac{du_z}{d\tau} = -(\mathbf{u}_\perp \times \mathbf{B}_\perp)\cdot \hat{z} = \frac{1}{2}\frac{d\mathbf{A}_\ell^2}{d\phi} - \frac{1}{2}\frac{d\mathbf{A}_w^2}{dz} - \mathbf{A}_\ell \cdot \frac{d\mathbf{A}_w}{dz} + \mathbf{A}_w \cdot \frac{d\mathbf{A}_\ell}{d\phi}. \tag{8.175}$$

Finally, the energy evolves as

$$\frac{d\gamma}{d\tau} = -\mathbf{u}_\perp \cdot \mathbf{E}_\perp = \frac{1}{2}\frac{d\mathbf{A}_\ell^2}{d\phi} + \mathbf{A}_w \cdot \frac{d\mathbf{A}_\ell}{d\phi}. \tag{8.176}$$

In Equations 8.175 and 8.176, we recognize the laser radiation pressure (ponderomotive potential), $\frac{1}{2}\frac{d\mathbf{A}_\ell^2}{d\phi}$, and the FASAR/IFEL interaction terms, involving the nonlinear product of the laser and wiggler vector potentials. The energy and axial momentum can be expressed in terms of the light-cone variable and transverse momentum, using the norm of the four-velocity, $u_\mu u^\mu = -1$:

$$\gamma = \frac{1 + \mathbf{u}_\perp^2 + \kappa^2}{2\kappa}, \quad u_z = \frac{1 + \mathbf{u}_\perp^2 - \kappa^2}{2\kappa}. \tag{8.177}$$

In the regions where the magnetostatic field is zero, the motion of the electron can be described exactly in terms of plane wave dynamics; for $d\mathbf{A}_w/dz = \mathbf{0}$, Equations 8.172 and 8.173 can be combined to yield $d\kappa/d\tau = 0$, which shows that the light-cone variable is invariant, and Equation 8.177 yields the electron energy and axial momentum:

$$\gamma(\phi) = \gamma_{-\infty}\left[1 + \frac{\mathbf{A}_\perp^2(\phi)}{2}(1+\beta_{-\infty})\right], \quad u_z(\phi) = \gamma(\phi) = \gamma_{-\infty}\left[\beta_{-\infty} + \frac{\mathbf{A}_\perp^2(\phi)}{2}(1+\beta_{-\infty})\right], \tag{8.178}$$

where $\gamma_{-\infty}$ is the injection energy, and $\beta_{-\infty}$ is the initial velocity.

Equation 8.178 illustrates the Lawson–Woodward theorem: in the case of a laser pulse, $\lim_{\phi\to\pm\infty}[\mathbf{A}_\ell(\phi)] = \mathbf{0}$ and $\gamma_{+\infty} = \gamma_{-\infty}$; there is no net energy gain.

We now consider the three different stages of the FASAR interaction. The electron is first accelerated by the laser radiation pressure, in the absence of a magnetostatic field; the peak energy is reached when the potential of the drive pulse is maximized. We model $\mathbf{A}_\ell(\phi)$ as a linearly polarized plane wave:

$$\mathbf{A}_\ell(\phi) = A_0 g(\phi)\hat{x}\cos\phi, \tag{8.179}$$

where the normalized vector potential is given by

$$A_0 = \frac{e\lambda_0}{2\pi m_0 c^2}\sqrt{\frac{2I_0}{\varepsilon_0 c}}, \tag{8.180}$$

and where $\lambda_0$ is the laser wavelength, while $I_0$ is its focused intensity; $g(\phi)$ is the laser pulse envelope.

The phase is defined so that the envelope reaches its maximum at $\phi = 0$: $g(\phi = 0) = 1$; we then find that

$$\gamma(\phi = 0) = \gamma_0 = \gamma_{-\infty}\left[1 + \frac{A_0^2}{2}(1 + \beta_{-\infty})\right], \tag{8.181}$$

which can be approximated by $\gamma_0/\gamma_{-\infty} \simeq (1 + A_0^2)$, for highly relativistic electrons; furthermore, at $\phi = 0$, the electric field of the laser wave is null, since $\mathbf{E}_\perp = -\frac{d\mathbf{A}_\ell}{d\phi}$. This is an important condition, because in the next stage of the FASAR, a magnetostatic field is added to the interaction region. Within a sufficiently small phase interval around $\phi = 0$, the energy can be considered as a constant of the motion. With the addition of this magnetic field, the light-cone variable is no longer invariant, and starts evolving according to

$$\frac{d\kappa}{d\tau} = \frac{d}{d\tau}(\gamma - u_z) = -\mathbf{u}_\perp \cdot \mathbf{E}_\perp + (\mathbf{u}_\perp \times \mathbf{B}_\perp)\cdot\hat{z} = \mathbf{u}_\perp \cdot \frac{d\mathbf{A}_w}{dz} = \mathbf{A}_\ell \cdot \frac{d\mathbf{A}_w}{dz} + \frac{1}{2}\frac{d\mathbf{A}_w^2}{dz}. \tag{8.182}$$

Although Equation 8.182 is nonlinear and cannot be solved analytically, we can use the fact that the energy remains very nearly constant to evaluate the value of $\kappa$ over the magnetic interaction region:

$$\gamma_0 = \frac{1 + A_0^2 + \kappa_{-\infty}^2}{2\kappa_{-\infty}} \simeq \frac{1 + [A_0 + A_z(z)]^2 + \kappa^2(z)}{2\kappa(z)}, \tag{8.183}$$

where

$$\kappa_{-\infty} = \gamma_{-\infty} - u_{z-\infty} = \gamma_{-\infty}(1 - \beta_{-\infty}) = \frac{1}{\gamma_{-\infty}(1 + \beta_{-\infty})}, \tag{8.184}$$

is the initial value of the light-cone variable, and where the magnetostatic vector potential is given by

$$\mathbf{A}_w(z) = \hat{x}\int_0^z B_w(z')\,dz'. \tag{8.185}$$

Here, we have chosen the origin of the $z$-axis such that $z(\phi = 0) = 0$.

For highly relativistic electrons, we can use Equation 8.184 to find that $\kappa_{-\infty} \simeq 1/2\gamma_{-\infty} \ll 1$, and the applied magnetic field continuously decreases the value of $\kappa(z)$; therefore, we can further approximate Equation 8.183 to obtain

$$\kappa(z) \simeq \kappa_{-\infty}\frac{1 + [A_0 + \int_0^z B_w(z')\,dz']}{1 + A_0^2}. \tag{8.186}$$

The optimal electron dephasing condition is now obvious: the quantity in the brackets must be equal to zero at the end of the magnetostatic field. We then have

$$\int_0^{\Delta z} B_w(z)\,dz = -A_0, \quad \kappa(\Delta z) = \kappa_0 \simeq \frac{\kappa_{-\infty}}{1+A_0^2}. \tag{8.187}$$

We note that at the end of the magnetostatic field region, the transverse momentum of the electron is zero; in the third region, the plane wave dynamics apply again, and the light-cone variable is invariant again:

$$\mathbf{u}_\perp(\phi) = \mathbf{A}_\ell(\phi) + \hat{x}\int_0^{\Delta z} B_w(z)\,dz = \mathbf{A}_\ell(\phi) - \hat{x}A_0, \tag{8.188}$$

and

$$\kappa(\phi) = \kappa_0 \simeq \frac{\kappa_{-\infty}}{1+A_0^2} \simeq \frac{1}{2\gamma_{-\infty}(1+A_0^2)}. \tag{8.189}$$

Using Equation 8.177, we find that the energy evolves as

$$\gamma(\phi) = \frac{1+[\mathbf{A}_\ell(\phi) - \hat{x}A_0]^2 + \kappa_0^2}{2\kappa_0}. \tag{8.190}$$

At the end of the interaction, $\lim_{\phi\to+\infty}[\mathbf{A}_\ell(\phi)] = \mathbf{0}$, and we find the final energy:

$$\gamma_{+\infty} = \frac{1+A_0^2+\kappa_0^2}{2\kappa_0} \simeq \gamma_{-\infty}(1+A_0^2)^2. \tag{8.191}$$

This is a remarkable result, first derived by Kerman, as the energy gain is a quartic function of the vector potential; in the relativistic intensity regime, where $A_0 > 1$, considerable acceleration can be achieved.

Returning to the stationary phase condition, we must verify that

$$\Delta\phi = \int_0^{\Delta z} \frac{d\phi}{dz}\,dz \ll 2\pi; \tag{8.192}$$

changing variables, we have

$$\Delta\phi = \int_0^{\Delta z} \frac{d\phi}{d\tau}\frac{d\tau}{dz}\,dz = \int_0^{\Delta z} \frac{\kappa(z)}{u_z(z)}\,dz \simeq \int_0^{\Delta z} \frac{\kappa(z)}{\gamma_0 - \kappa(z)}\,dz \le \frac{\kappa_{-\infty}}{\gamma_0}\Delta z \simeq \frac{\Delta z}{2\gamma_{-\infty}^2(1+A_0^2)}. \tag{8.193}$$

For a useful comparison, we need to estimate the size of the interaction region required for optimal dephasing; we model the magnetic field profile as

$$B_w(z) = B_w \sin^2\left(\pi \frac{z}{\Delta z}\right). \tag{8.194}$$

With this, Equation 8.188 yields

$$\Delta z = -2\frac{A_0}{B_w} = -\frac{A_0}{\pi}\lambda_c, \tag{8.195}$$

where $\lambda_c$ is the cyclotron wavelength associated with the magnetic field, and is equal to 10.7 mm for 1 T. With this, Equation 8.194 can recast as

$$\Delta\phi \le \frac{A_0}{\gamma_{-\infty}^2(1+A_0^2)}\frac{\lambda_c}{\lambda_0}; \tag{8.196}$$

for $\gamma_{-\infty} = 10^3$, $A_0 = 3$, $B_w = 1$ T, and $\lambda_0 = 0.8\ \mu$m, we find that $\Delta\phi/2\pi \le 0.61 \times 10^{-3} \ll 1$, which satisfies the stationary phase assumption.

This brief presentation completes our introduction to vacuum laser acceleration. For the sake of completeness, plasma-based laser acceleration processes are briefly discussed in the next section, which is intended as a cursory overview.

## 8.10 Plasma-Based Laser Acceleration Processes

This section is intended as a brief introduction to the subject. This is, indeed, a rich field of research, which is growing in step with novel technologies for tabletop terawatt lasers, such as chirped-pulse amplification (CPA); compact electron injectors, including high-brightness photoinjectors; and sophisticated optical and x-ray diagnostics, including optical transition radiation (OTR), interferometry, and sub-picosecond streak cameras. Esarey and coauthors have well summarized the state of the field in their review articles, which are listed in the reference section, and the presentation outlined here follows the broad lines of these papers.

Although it is often difficult to pinpoint the exact origin of a new idea in applied physics, the seminal paper by Tajima and Dawson, listed in the bibliography for this chapter, can be regarded as an important steppingstone and introduction to the concept of plasma-based laser acceleration. The basic idea behind this concept is that plasmas can support extremely high fields, as the medium is already ionized. A cursory inspection of Poisson's equation also indicates that for sufficiently high densities, electrostatic fields comparable to the electric field of an intense laser wave, in the GV-TV/m ($10^9$–$10^{12}$ V $\times$ m$^{-1}$) range, can be achieved. In turn, such fields potentially translate into very high accelerating gradients for charged particles, particularly for leptons such as electrons and positrons: the 1–100 GeV/m range is the current goal for modern experiments, and some success has been achieved, as high gradients have been observed. However, in the current state of the art, the gradients are sustained over very small effective lengths, practically limiting the net energy gain to 1–100 MeV, and the phase space of the accelerated electron bunch typically resembles a two-temperature Maxwellian, rather than the well-defined, low axial and transverse emittance beams currently required for high-energy physics. Good transverse emittances have been measured, but this is mainly due to the fact that the transverse size of the accelerated plasma electron distribution is comparable to the focused laser spot size (a few $\mu$m). The angular divergence of the beam is still high, and its axial energy distribution contains a population of low-energy electrons (<1 MeV) that is many orders of magnitude higher than the desired, high-energy portion of the spectrum. Furthermore, plasma inhomogeneities and instabilities present a formidable challenge to researchers in the field, as well as the problem of diffraction due to the strong focusing of the drive laser pulse to reach the intensities required. Nonetheless, tremendous progress has been achieved, in terms of both experimental and computational physics, and the field is still growing rapidly, as can be assessed by consulting the list of participants to recent international conferences and workshops on advanced accelerator concepts. Perhaps the key factor here is the well-known gradient limitation of the conventional rf technology, lying somewhere between 0.1 and 1 GeV/m: new concepts are clearly required to reach the new high-energy frontier, beyond 10 TeV.

In laser-driven, plasma-based acceleration, one of the key parameters is the so-called wave-breaking field,

$$E_0 = \omega_p \frac{m_0 c}{e}, \tag{8.197}$$

where $\omega_p = \sqrt{n_0 e^2 / \varepsilon_0 m_0}$ is the nonrelativistic plasma frequency, and where we are considering electrons. Equation 8.197 can be recast as

$$\frac{E_0}{\sqrt{n_0}} \simeq 0.96 \ \text{V} \times \text{cm}^{-5/2}; \tag{8.198}$$

for example, $n_0 = 10^{18}$ cm$^{-3}$ yields $eE_0 \simeq 100$ GeV $\times$ m$^{-1}$. Relativistic effects affect this picture. In particular, the transverse quiver velocity of the electrons in the high-intensity drive laser field can approach the speed of light, thus changing the dynamics of the particles, and other important phenomena, such as relativistic self-focusing, laser filamentation, and nonlinear ponderomotive scattering are known to occur at these intensities. One of the key concepts for laser-driven, plasma-based acceleration is the plasma wakefield, which can be excited in two different ways: either the drive pulse is shorter than the plasma wavelength, and the ponderomotive pressure of the drive laser directly creates a significant charge imbalance by ejecting electrons from the plasma, or a longer laser pulse undergoes self-modulation by exciting a strong Raman instability in the plasma. In the later case, the growth or Raman sidebands, which can be observed by monitoring the spectrum of the laser light after the interaction, is a clear signature of the self-modulation process, which goes along with the aforementioned relativistic self-focusing. At this point, we also note that a strong plasma wakefield can be created by a short, relativistic charged particle bunch propagating through the plasma. In this case, the Coulomb field generates the charge imbalance giving rise to the a strong wake in the plasma. In either case, the plasma wakefield behaves like an accelerating structure, capable of trapping and accelerating electrons or positrons close to the speed of light; this is because the phase velocity of the wake is very close to $c$.

After initial, proof-of-principle experiments, which demonstrated the excitation of plasma wakefields and the presence of high-energy scattered electrons, more sophisticated laser-driven, plasma-based experiments are now being performed; for example, Milchberg and his group have been creating plasma channels capable of guiding the drive laser and the accelerated particles beyond the diffraction, or Rayleigh, length, as shown in Figure 8.10. Such new structures offer the potential of higher-quality, higher-energy accelerated electron distributions, with the ultimate goal of producing a real beam, occupying a small, well-defined fraction of phase space. With this in mind,

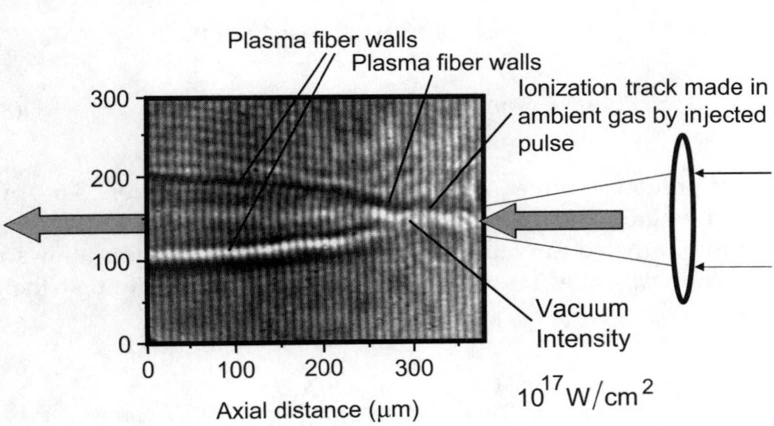


**FIGURE 8.10**
Production of plasma channel capable of guiding a high-intensity injected laser pulse over large distances. This structure is an excellent candidate for well-controlled laser-driven, plasma-based acceleration. (Courtesy of Howard Milchberg).

laser-injection schemes have also been proposed, where the goal is to selectively launch plasma electrons into the accelerating buckets of the large amplitude plasma wave excited by the drive laser. These are fairly challenging experiments and it will be interesting to see what results are produced in the near future.

Finally, we give a brief theoretical outline of the basic physics for the generation of a large amplitude plasma wave. In the following, we will use the normalized potential, $A_\mu = e(c^{-1}\phi, \mathbf{A})/m_0c$, which includes both the external laser field and the space-charge fields in the plasma. In the cold fluid limit, we have

$$(\partial_t + \mathbf{v}\cdot\nabla)\mathbf{p} = -e(\mathbf{E} + \mathbf{v}\times\mathbf{B}), \tag{8.199}$$

where the electric and magnetic fields of the drive laser pulse are given by

$$\begin{aligned} \mathbf{E} &= -\partial_t\mathbf{A}, \\ \mathbf{B} &= \nabla\times\mathbf{A}. \end{aligned} \tag{8.200}$$

In the one-dimensional limit, one can represent the drive laser pulse by a plane wave, with

$$\mathbf{A} = \mathbf{A}_\perp = A_0\cos(kz-\omega t)\hat{x}_\perp = A_0\cos\varphi\hat{x}_\perp, \tag{8.201}$$

where we recognize the invariant phase, $\varphi$. Furthermore, in the linear regime, where $A_0 \ll 1$, a simple linearization procedure yields the momentum perturbation equation

$$\begin{aligned} \frac{d}{dt}\delta\mathbf{p} &= -m_0[(\mathbf{v}_0\cdot\nabla)\mathbf{v}_0 + c\mathbf{v}_0\times(\nabla\times\mathbf{A}_\perp)] \\ &= -m_0c^2\nabla\left(\frac{\mathbf{A}_\perp^2}{2}\right), \end{aligned} \tag{8.202}$$

where the ponderomotive potential, introduced in Section 8.6 appears clearly, and where we have expanded the velocity as $\mathbf{v} = \mathbf{v}_0 + \delta\mathbf{v}$.

Furthermore, the conservation of the canonical momentum allows us to write $\mathbf{u}_\perp = \mathbf{p}_\perp/m_0c = \mathbf{A}_\perp$; in turn, this can be used to show that the axial force is given by

$$\frac{d}{dt}\delta p_z = -\frac{m_0c^2}{2\gamma}\frac{\partial\mathbf{A}_\perp^2}{\partial z}. \tag{8.203}$$

This is valid in the one-dimensional, linear regime; for a sufficiently large laser focal spot, such that $w_0 > \lambda_p \gg \lambda$, the transverse canonical momentum is still very nearly conserved, and we can use the laser phase, $\varphi$, as the independent variable, to write

$$\frac{\partial}{\partial \varphi}(\mathbf{u} - \mathbf{A}) \simeq \nabla(\gamma - \phi), \tag{8.204}$$

where $\phi$ is the scalar potential associated with the space-charge force.

The axial component of Equation 8.204 yields the light-cone variable,

$$\gamma - u_z = 1 + \phi - A_z, \tag{8.205}$$

and we can use relativistic kinematics to obtain

$$\gamma = \frac{1 + \langle \mathbf{u}_\perp \rangle^2 + \langle \mathbf{A}_\perp^2 \rangle + \langle 1 + \phi - A_z \rangle^2}{2\langle 1 + \phi - A_z \rangle}, \tag{8.206}$$

where the averaging produces the slow-varying component of the bracketed quantity, as discussed in Section 8.6.

In closing, for further details, we encourage the interested reader to consult the bibliography for this chapter, which contains a fairly extensive list of relevant publications in the field, as well as a series of review articles, including the works of Joshi, Leemans, Milchberg, Sprangle, and Umstadter, to name a few. Of particular interest is the concept of "braided light" introduced by Mori, who considered the mutual nonlinear interaction of two laser beams in a plasma. Also noteworthy is the proposed Linac Coherent Light Source (LCLS), an x-ray FEL operating on the self-amplified spontaneous radiation (SASE) mechanism pioneered by Pellegrini and Rosenzweig.

## 8.11 References for Chapter 8

Note: the numbers listed below refer to the main bibliography and reference sections at the end of this book.

1, 4, 17, 53, 56, 94, 157, 197, 214, 215, 250, 253, 259, 270, 276, 281, 290, 291, 292, 295, 296, 347, 350, 351, 352, 366, 367, 368, 369, 370, 381, 384, 397, 388, 409, 410, 411, 412, 413, 429, 431, 434, 439, 440, 441, 442, 443, 444, 445, 480, 512, 516, 525, 526, 544, 545, 546, 547, 548, 571, 576, 577, 578, 581, 602, 603, 604, 605, 609, 610, 619, 621, 622, 625, 626, 627, 628, 629, 632, 637, 644, 645, 647, 648, 650, 651, 652, 654, 662, 667, 670, 671, 672, 676, 679, 681, 685, 686, 711, 715, 716, 717, 718, 721, 722, 723, 724, 725, 727, 728, 729, 735, 736, 750, 754, 758, 762, 763, 768, 769, 770, 771, 772, 773, 774, 775, 776, 778, 784, 786, 788, 789, 791, 792, 810, 811, 813, 814, 815, 816, 817, 820, 821, 822, 823, 824, 832, 836, 837, 838, 839, 840, 842, 843, 845, 846, 847, 848, 849, 850, 883, 892, 893, 909, 912, 914, 915, 916, 922, 923, 925, 927.

# Part III

# Relativistic Electrons and Radiation

# 9

# *Coherent Synchrotron Radiation and Relativistic Fluid Theory*

## 9.1 Introduction

As discussed in depth in this book, one of the central problems of classical electrodynamics is the radiation of electromagnetic waves by an accelerated charge distribution. In this chapter, we first consider the theoretical description of the radiation for a high-charge electron bunch, such as those produced by a radio-frequency (rf) photoinjector, propagating on fixed trajectories through a helically polarized wiggler. The second topic described here is the self-consistent perturbation theory for a relativistic charged fluid, which is developed within the context of the so-called Eulerian formalism.

Photoinjectors are compact accelerators producing high-brightness, relativistic electron beams. The basic principle of operation of such a device is the following: the electrons are produced via the photoelectric effect by a short (typically, sub-picosecond) UV laser pulse striking a metal photocathode (4.66 eV work function for Cu); they are subsequently accelerated by a high-gradient (50 to 200 MeV/m) rf electric field in a high-Q cavity. Typically, S-band photoinjectors have been built, operating at 2.856 GHz and at 3.000 GHz; lower frequency devices, such as the Los Alamos National Laboratory (LANL) L-band photoinjector have also produced high-quality electron beams. Higher-frequency photoinjectors have also been studied experimentally, and Figure 9.1 shows an X-band version of such a device, which was operated at 8.547 GHz. The precise timing between the laser pulse and the rf fields in the photoinjector ensures the production of a short (few picosecond to sub-picosecond), relativistic (typically 2 to 5 MeV), high-quality electron bunch. The relative axial energy spread, $\Delta\gamma_\parallel/\gamma_\parallel$, is generally maintained near or below 1%, while the transverse normalized emittance, which is essentially a measure of the transverse phase space volume, lies in the range 1 to 10 $\pi$-mm $\cdot$ mrad for modern devices. These units are illustrated in Figure 9.2, where the transverse positions and momenta of the electrons are measured, and the phase space surface containing 90% of the electrons defines the 90% emittance. The factor of $\pi$ results from the fact that for an

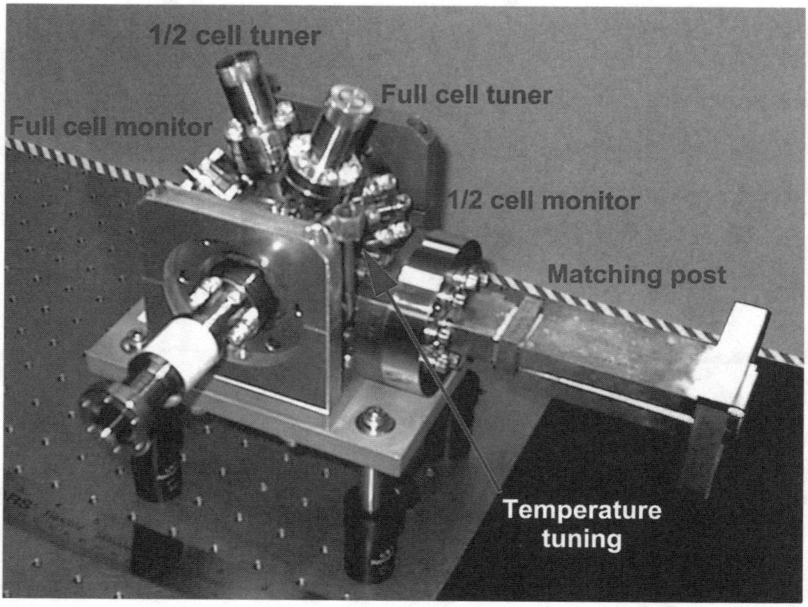


**FIGURE 9.1**
A high-frequency rf photoinjector; the entire structure is approximately 2 in. long and is capable of sustaining accelerating fields in excess of 100 MV/m.

ellipsoidal four-volume in phase space, defined by the equation

$$\left(\frac{x}{\Delta x}\right)^2 + \left(\frac{x'\Delta x}{\varepsilon_x}\right)^2 + \left(\frac{y}{\Delta y}\right)^2 + \left(\frac{y'\Delta y}{\varepsilon_y}\right)^2 = 1,$$

the four-volume is given by $\pi^2\varepsilon_x\varepsilon_y/2$. Moreover, the normalized emittance, defined as $\varepsilon_n = \varepsilon\beta\gamma$, is a relativistic invariant and compensates for the fact that, in an accelerator, as the axial momentum, $p_\parallel = m_0 c\gamma\beta_\parallel$ increases, the ratio $p_\perp/p_\parallel$ changes, as illustrated in Figure 9.2. Liouville's phase–space theorem governs the evolution of the normalized emittance, which is found to remain constant in a linear system. Finally, the brightness of the beam is defined as $B = dj/d\Omega \simeq 2I_b/\pi\varepsilon_x\varepsilon_y$, and is measured in $\text{A} \times \text{mm}^{-2} \times \text{mrad}^{-2}$; for the normalized brightness, one uses the normalized emittance. For more details on these important concepts in accelerator and beam physics, we refer the reader to the monographs by Reiser and Wiedemann listed in the reference section.

Photoinjectors have a number of important applications, as they produce high-quality relativistic electron beams. For example, they are used to drive free-electron lasers or as injectors for high-energy accelerators. In addition, the photoelectron bunches produced are very short in duration and can be focused tightly; thus, photoinjectors are also considered to produce prebunched beams, as discussed here, and to drive Compton scattering experiments

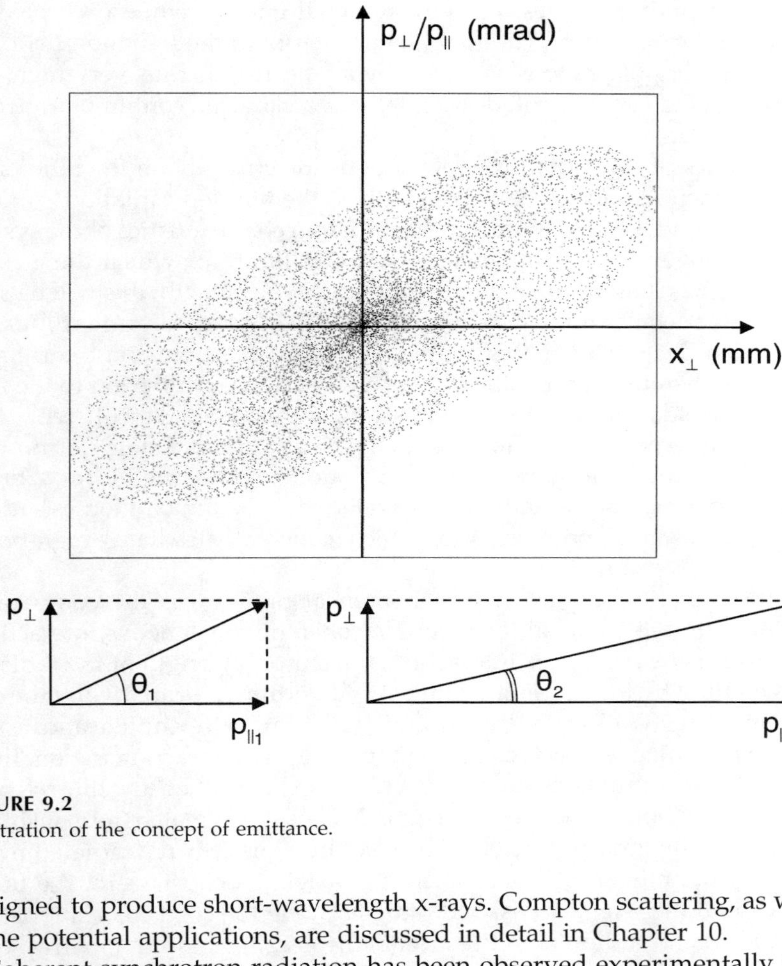


**FIGURE 9.2**
Illustration of the concept of emittance.

designed to produce short-wavelength x-rays. Compton scattering, as well as some potential applications, are discussed in detail in Chapter 10.

Coherent synchrotron radiation has been observed experimentally and is under investigation by several groups worldwide, as listed in the reference section. These preliminary experimental results are very encouraging for future applications, ranging from surface and solid-state physics and millimeter-wave and far infrared (FIR) photophysics and photochemistry to the next generation of ultrawideband radars. Other closely related experiments have been reported recently, where coherent radiation is emitted in a slow-wave structure by a train of photoelectron bunches. Such prebunched, relativistic electron beams offer the possibility of generating coherent pulses of electromagnetic radiation with very high peak powers, broad spectral bandwidth, and good conversion efficiency from the kinetic energy of the beam to the radiated energy.

The theoretical description of such electrodynamic systems can be constructed within the context of classical electrodynamics and special relativity. This framework is appropriate because we are considering millimeter waves, where each photon carries a relatively small energy, while each electron loses a significant part of its initial kinetic energy to the radiation field. The average number of photons scattered by an electron is thus very high, and the photon field can be treated essentially as a classical, continuous, electromagnetic field.

The physics of the radiation process depends critically on the ratio of the electron bunch length to the wavelength of the emitted radiation. We thus define two very different limits for the radiation characteristics of the system. First, the conventional spontaneous radiation limit, for which the electron bunch length is long compared to the radiation wavelength; this will be studied in greater details within the context of the stochastic electron gas formalism discussed in Chapter 10. In this regime, the electrons are not phase correlated, and the incoherent superposition of each electron's contribution to the radiation field yields very low power emission, with the usual incoherent phase noise statistics. In the other limit, which we call coherent synchrotron radiation, or the coherent super-radiant radiation regime, the electron bunch length is shorter that the radiation wavelength. The bunch then essentially behaves as a single point charge, which coherently radiates high-power millimeter-wave radiation.

Another essential aspect of our theoretical program is the comparative study of both free-space and waveguide coherent spontaneous interactions. In the case of the free-space interaction, the temporal length of the radiation pulses emitted by the accelerated bunch is determined primarily by the slippage between the electron bunch and the Doppler upshifted and downshifted electromagnetic waves through the wiggler interaction region. In the case of the waveguide interaction, which is exemplified by the so-called grazing interaction introduced in Section 5.6.7, one can match the axial bunch velocity and the group velocity of the electromagnetic wave radiated by the charge. In the reference frame where the axial velocity is zero, the bunch rotates about the wiggler axis, and radiates along the $z$-axis in a $\sin^2$ pattern similar to that of a dipole. In this frame, the grazing condition corresponds to cutoff. The electromagnetic power radiated by the accelerated charge is then confined around the electron bunch, and the resulting single compressed output pulse has unique spectral characteristics. Since, by definition, slippage is equal to zero at grazing, the question arises as to what physical mechanism determines the final output pulse duration in this regime. In addition, it will be shown that, at grazing, the radiation power level is considerably higher than that expected for the conventional coherent synchrotron radiation process, even without taking into account resonant radiation feedback.

It is also important to recognize that in the case of a single, point-like, transversally accelerated charge distribution, the physics of the radiation process can differ notably from the conventional free-electron laser (FEL) interaction. In particular, in a conventional FEL, the feedback between the electron beam dynamics in the wiggler and the radiation field (ponderomotive force) results in a density modulation (axial bunching), which in turn drives the radiation field and further bunches the beam, yielding the well-known exponential growth of the FEL instability. In the present case of an ultrashort electron pulse (or pre-bunched) FEL, it is not clear that the radiation can actually further bunch the beam; in fact, debunching space-charge forces generally dominate the bunch dynamics. Therefore, exponential gain is not expected for this interaction. Furthermore, away from grazing, the axial velocity of the photoelectron bunch and the group velocity of the electromagnetic waves are very different and can prevent the onset of the FEL super-radiant amplification mechanism. In the grazing limit only, if slippage is sufficiently small, the interaction can become similar to the super-radiant amplification of the coherent spontaneous radiation produced by the accelerated bunch in the first wiggler periods.

The next sections are organized as follows. In Section 9.2, we derive both the power and spectral radiation characteristics of a single point charge propagating on a fixed trajectory though a helically polarized wiggler magnetostatic field, with no boundary conditions (free-space interaction). We start from the Liénard–Wiechert potential for an accelerated point charge. We then consider a point charge propagating through a helical wiggler, and calculate its spatial and spectral power distribution. Finite-size effects are introduced in this derivation, and the transition between the coherent and incoherent radiation regimes can be approximately defined. We then derive the waveguide effects, including grazing, using a coupled-mode analysis presented in Section 9.3. This derivation yields a differential equation governing the evolution of the spectral density of each transverse electric (TE) cylindrical waveguide eigenmode as a function of position along the $z$-axis; we can thus calculate the output spectrum of this electrodynamic system, both in the free-space limit and at grazing and as a function of the bunch size.

## 9.2 Coherent Synchrotron Radiation in Free-Space

In this section, we drive both the power and spectral radiation characteristics of an axially extended charge distribution propagating on a fixed trajectory through a helically polarized wiggler magnetic field, with no boundary conditions (free-space interaction). This section is intended as a review of the basic features of coherent synchrotron radiation, and allows us to give a first definition of the coherence limit between conventional FEL spontaneous radiation and coherent synchrotron radiation. As mentioned earlier, the transition definition will be refined in Chapter 10.

For a single point charge, $e$, describing a prescribed trajectory $\mathbf{r}(t)$, the electric field at $x_\mu$ is obtained by deriving the Liénard–Wiechert four-vector potential. In MKSA units, we have

$$\mathbf{E}(\mathbf{x},t) = \frac{e}{4\pi\varepsilon_0}\left[\frac{\hat{n}-\beta}{\gamma^2(1-\beta\cdot\hat{n})R^2}\right]^- + \frac{e}{4\pi\varepsilon_0 c}\left[\frac{\hat{n}\times(\hat{n}-\beta)\times\dot{\beta}}{(1-\beta\cdot\hat{n})^3 R}\right]^-, \tag{9.1}$$

where the quantities in the brackets are evaluated at the retarded time $t^{-}$, such that

$$c(t-t^{-}) = R(t^{-}) = |\mathbf{x}-\mathbf{r}(t^{-})|. \tag{9.2}$$

Here, $\hat{n}$ is the unit vector in the direction of observation, and $\dot{\beta} = d\beta/dt$.

The first part of the electric field corresponds to the relativistic transform of the Coulomb electrostatic field. In the following, we will consider only the radiation field. The instantaneous power flux is given in terms of the Poynting vector, defined as

$$\mathbf{S} = \hat{n}\frac{dP}{R^2 d\Omega} = \mathbf{E}\times\mathbf{H}. \tag{9.3}$$

For the radiative field, we have $\mathbf{B} = \hat{n}\times\mathbf{E}/c = \mu_0\mathbf{H}$, and it is easily seen that we have

$$\mathbf{S} = \hat{n}\frac{\mathbf{E}^2}{\mu_0 c}. \tag{9.4}$$

In the instantaneous rest frame of the charge, where $\beta = \mathbf{0}$, but the acceleration can be nonzero, the radiative field reduces to

$$\mathbf{E} = \frac{e}{4\pi\varepsilon_0 c}\frac{\hat{n}\times(\hat{n}\times\dot{\beta})}{R}. \tag{9.5}$$

Let us call $\xi$ the angle between the acceleration and the direction of observation. We recover the familiar relation

$$\mathbf{S} = \frac{\hat{n}}{\mu_0 c}\left(\frac{e}{4\pi\varepsilon_0 c}\right)^2\frac{\dot{\beta}^2\sin^2\xi}{R^2}. \tag{9.6}$$

Here, we recognize the usual $\sin^2$ angular dependence of the radiation pattern in the rest frame of the charge, which is typical of a radiating dipole.

Integrating over all solid angles yields the total radiated power in the instantaneous rest frame of the charge. We have, for an appropriate choice of axis in spherical coordinates,

$$P = \frac{e^2}{16\pi^2\varepsilon_0 c}\dot{\beta}^2\int_0^{2\pi} d\theta\int_0^{\pi}\sin^3\xi\, d\xi, \tag{9.7}$$

which yields the following result for the total radiated power:

$$P = \frac{e^2}{6\pi\varepsilon_0 c}\dot{\beta}^2. \tag{9.8}$$

The radiated electromagnetic energy behaves as the time-like component of a four-vector under the Lorentz transformation. This fact, which will be established rigorously in Chapter 10, can be used to derive the total radiated power in the lab frame; we then obtain the Liénard formula

$$P = \frac{e^2}{6\pi\varepsilon_0 c}\gamma^6[\dot{\beta}^2 - (\beta \times \dot{\beta})^2]. \tag{9.9}$$

We now consider the motion of a charge propagating through a helically polarized wiggler magnetic field. In Cartesian coordinates, in the one-dimensional limit, the equilibrium helical trajectories are described by the well-known normalized velocity components

$$\beta = \beta_\perp[\hat{x}\cos(k_w z) + \hat{y}\sin(k_w z)] + \hat{z}\beta_\parallel, \tag{9.10}$$

where the perpendicular velocity is given in terms of the wiggler field amplitude, $B_w$, and wavenumber, $k_w = 2\pi/\ell_w$, by

$$\beta_\perp = \frac{eB_w}{\gamma_0 m_0 k_w c}, \tag{9.11}$$

and where energy conservation yields

$$\frac{1}{\gamma_0^2} = 1 - \beta_\perp^2 - \beta_\parallel^2 = 1 - \beta_0^2. \tag{9.12}$$

Here, $\gamma_0$ is the initial energy of the electron, and $\ell_w$ is the wiggler period.

Before proceeding further, we outline the method of derivation used to establish the results given in Equations 9.10, 9.11, and 9.12. We consider a slightly more general case, namely, that of an electron subjected to a helical wiggler field and a guide field. In this case, the external magnetic fields are described, in the one-dimensional limit, by

$$\mathbf{B} = \mathbf{B}_w + \mathbf{B}_\parallel = B_w[\hat{x}\cos(k_w z) + \hat{y}\sin(k_w z)] + \hat{z}B_\parallel. \tag{9.13}$$

The terminology "one-dimensional limit" corresponds to the fact the wiggler field depends only on the axial coordinate, $z$. It is clear, however, that although this field is divergence-free, it cannot be generated in vacuum because its curl is nonzero.

There are different approaches to solve for the electron trajectory in such a field configuration; for example, one could use the vector potential and the transverse canonical invariants derived in Chapter 8. Here, for the sake of illustration, we will directly solve the Lorentz force equation,

$$\frac{d\mathbf{p}}{dt} = m_0 c \frac{d\mathbf{u}}{dt} = -e\mathbf{v} \times \mathbf{B}, \tag{9.14}$$

where $\mathbf{u} = \gamma\beta = \gamma\mathbf{v}/c$ represents the spatial components of the normalized four-velocity.

In the configuration under consideration, Equation 9.14 explicitly reads

$$\frac{d\mathbf{u}}{dt} = -\frac{e}{m_0 c}\mathbf{v} \times \{B_w[\hat{x}\cos(k_w z) + \hat{y}\sin(k_w z)] + \hat{z}B_\parallel\}, \tag{9.15}$$

or, if we use the proper time, with $\gamma = dt/d\tau$,

$$\frac{d\mathbf{u}}{d\tau} = -\frac{e}{m_0}\mathbf{u} \times \{B_w[\hat{x}\cos(k_w z) + \hat{y}\sin(k_w z)] + \hat{z}B_\parallel\}. \tag{9.16}$$

Furthermore, the time-like component of the Lorentz force equation, which describes the evolution of the particle energy is identically zero:

$$\frac{d\gamma}{dt} = -\frac{e\mathbf{v} \cdot \mathbf{E}}{m_0 c^2} = 0, \tag{9.17}$$

or

$$\frac{d\gamma}{d\tau} = -\frac{e\mathbf{u} \cdot \mathbf{E}}{m_0 c} = 0. \tag{9.18}$$

In the absence of radiative corrections, the energy of the particle is conserved, as the magnetic forces do not produce work; therefore, we have

$$\gamma = \gamma_0. \tag{9.19}$$

Returning to Equation 9.16, we write each component by developing the cross-product:

$$\begin{aligned}
\frac{du_x}{d\tau} &= -\frac{e}{m_0}[u_y B_\parallel - u_z B_w \sin(k_w z)], \\
\frac{du_y}{d\tau} &= -\frac{e}{m_0}[u_z B_w \cos(k_w z) - u_x B_\parallel], \\
\frac{du_z}{d\tau} &= -\frac{e}{m_0}[u_x \sin(k_w z) - u_y \cos(k_w z)].
\end{aligned} \tag{9.20}$$

The nonrelativistic cyclotron frequencies, $\Omega_w = eB_w/m_0$, and $\Omega_\| = eB_\|/m_0$, can be introduced to simplify these equations; in fact, we can recast the general Lorentz force for a particle subjected to a magnetic field as

$$\frac{d\mathbf{u}}{dt} = -\frac{e}{m_0 c}\mathbf{v}\times\mathbf{B} = \Omega\times\frac{\mathbf{v}}{c} = \Omega\times\frac{\mathbf{u}}{\gamma}, \tag{9.21}$$

or

$$\frac{d\mathbf{u}}{d\tau} = -\frac{e}{m_0}\mathbf{u}\times\mathbf{B} = \Omega\times\mathbf{u}, \tag{9.22}$$

where we have introduced the cyclotron frequency vector, $\Omega = e\mathbf{B}/m_0$.

Energy conservation then takes the simple form

$$\mathbf{u}\cdot\frac{d\mathbf{u}}{d\tau} = \frac{1}{2}\frac{d\mathbf{u}^2}{d\tau} = \frac{1}{2}\frac{d}{d\tau}(\gamma^2 - 1) = \mathbf{u}\cdot(\Omega\times\mathbf{u}) = 0. \tag{9.23}$$

Returning to Equation 9.20, we now have:

$$\begin{aligned}
\frac{du_x}{d\tau} &= \Omega_w \sin(k_w z)u_z - \Omega_\| u_y,\\
\frac{du_y}{d\tau} &= \Omega_\| u_x - \Omega_w \cos(k_w z)u_z,\\
\frac{du_z}{d\tau} &= \Omega_w[u_y\cos(k_w z) - u_x\sin(k_w z)].
\end{aligned} \tag{9.24}$$

To solve the problem at hand, we recast Equation 9.24 by using the fact that $\mathbf{u} = d\mathbf{x}/cd\tau$: this first yields

$$\begin{aligned}
c\frac{du_x}{d\tau} &= \Omega_w \sin(k_w z)\frac{dz}{d\tau} - \Omega_\|\frac{dy}{d\tau},\\
c\frac{du_y}{d\tau} &= \Omega_\|\frac{dx}{d\tau} - \Omega_w\cos(k_w z)\frac{dz}{d\tau}.
\end{aligned} \tag{9.25}$$

The total derivatives with respect to the proper time can then be identified. We write

$$\begin{aligned}
c\frac{du_x}{d\tau} &= -\frac{d}{d\tau}\left[\frac{\Omega_w}{k_w}\cos(k_w z) + \Omega_\| y\right],\\
c\frac{du_y}{d\tau} &= \frac{d}{d\tau}\left[\Omega_\| x - \frac{\Omega_w}{k_w}\sin(k_w z)\right].
\end{aligned} \tag{9.26}$$

This yields two invariant quantities:

$$\begin{aligned} \frac{d}{d\tau}\left[u_x + \frac{\Omega_w}{k_w c}\cos(k_w z) + \frac{\Omega_\parallel}{c} y\right] &= 0, \\ \frac{d}{d\tau}\left[u_y + \frac{\Omega_w}{k_w c}\sin(k_w z) - \frac{\Omega_\parallel}{c} x\right] &= 0. \end{aligned} \tag{9.27}$$

These are related to the canonical momentum conservation, as we have

$$\begin{aligned} \mathbf{B}_w &= \nabla \times \mathbf{A}_w, \\ \mathbf{A}_w &= -\frac{B_w}{k_w}[\hat{x}\cos(k_w z) + \hat{y}\sin(k_w z)], \end{aligned} \tag{9.28}$$

for the wiggler field. In the absence of a guide field, we find that

$$\frac{d}{d\tau}\left[\mathbf{u}_\perp - \frac{e\mathbf{A}_w}{m_0 c}\right] = 0, \tag{9.29}$$

where we have introduced $\mathbf{u}_\perp = \gamma\beta_\perp = \gamma(\hat{x}\beta_x + \hat{y}\beta_y)$, and $u_\parallel = u_z$.

We then recover the result for equilibrium orbits in a helical wiggler:

$$\begin{aligned} \mathbf{u}_\perp &= \frac{e\mathbf{A}_w}{m_0 c}, \\ \mathbf{u}_\parallel &= \sqrt{\gamma_0^2 - 1 - \left(\frac{e\mathbf{A}_w}{m_0 c}\right)^2}. \end{aligned} \tag{9.30}$$

Using the fact that the energy is a constant, with $\gamma = \gamma_0$, we can translate this result in terms of the normalized velocity:

$$\begin{aligned} \mathbf{u}_\perp &= \frac{e\mathbf{A}_w}{m_0 c} = \gamma_0\beta_\perp = \gamma_0\beta_\perp[\hat{x}\cos(k_w z) + \hat{y}\sin(k_w z)], \\ \mathbf{u}_\parallel &= \sqrt{\gamma_0^2 - 1 - \beta_\perp^2}. \end{aligned} \tag{9.31}$$

On the other hand, when the wiggler field is equal to zero, we have

$$\frac{du_x}{d\tau} = -\Omega_\parallel u_y, \quad \frac{du_y}{d\tau} = \Omega_\parallel u_x, \tag{9.32}$$

and we can recover the well-known cyclotron motion. We first have

$$u_x\frac{du_x}{d\tau} + u_y\frac{du_y}{d\tau} = 2\frac{d\mathbf{u}_\perp^2}{d\tau} = 0; \tag{9.33}$$

furthermore,

$$\frac{d^2u_x}{d\tau^2} + \Omega_\parallel^2 u_x = 0, \quad \frac{d^2u_y}{d\tau^2} + \Omega_\parallel^2 u_y = 0, \tag{9.34}$$

yields

$$\frac{d^2\mathbf{u}_\perp}{d\tau^2} + \Omega_\parallel^2 \mathbf{u}_\perp = 0, \quad \frac{d^2\mathbf{u}_\perp}{dt^2} + \left(\frac{\Omega_\parallel}{\gamma_0}\right)^2 \mathbf{u}_\perp = 0, \tag{9.35}$$

and we find that

$$\mathbf{u}_\perp = u_{\perp 0}\left[\hat{x}\cos\left(\frac{\Omega_\parallel}{\gamma_0}t + \phi_0\right) + \hat{y}\sin\left(\frac{\Omega_\parallel}{\gamma_0}t + \phi_0\right)\right], \tag{9.36}$$

where the initial phase is obtained by solving the equation

$$\mathbf{u}_{\perp 0} = u_{\perp 0}[\hat{x}\cos\phi_0 + \hat{y}\sin\phi_0]. \tag{9.37}$$

When both the wiggler and guide fields are present, the equilibrium normalized four-velocity takes the form

$$\begin{aligned} u_\mu &= (\gamma_0, \mathbf{u}_\perp, \mathrm{u}_\parallel), \\ \mathbf{u}_\perp &= \gamma_0\beta_\perp\,[\hat{x}\cos(k_w z) + \hat{y}\sin(k_w z)], \\ u_\parallel &= \sqrt{\gamma_0^2 - 1 - \mathbf{u}_\perp^2} = \sqrt{\gamma_0^2(1-\beta_\perp^2) - 1} = \sqrt{\gamma_\parallel^2 - 1} = \gamma_0\beta_\parallel. \end{aligned} \tag{9.38}$$

To verify that this solution has the appropriate structure, we derive the four-velocity with respect to the proper time:

$$\begin{aligned} \frac{du_x}{d\tau} &= -\gamma_0\beta_\perp k_w \frac{dz}{d\tau}\sin(k_w z), \\ \frac{du_y}{d\tau} &= \gamma_0\beta_\perp k_w \frac{dz}{d\tau}\cos(k_w z), \\ \frac{du_z}{d\tau} &= 0. \end{aligned} \tag{9.39}$$

Since $cu_z = c\gamma_0\beta_\parallel = dz/d\tau$, we have

$$\begin{aligned} \frac{du_x}{d\tau} &= -\gamma_0^2\beta_\perp k_w \beta_\parallel c\sin(k_w z), \\ \frac{du_y}{d\tau} &= \gamma_0^2\beta_\perp k_w \beta_\parallel c\cos(k_w z). \end{aligned} \tag{9.40}$$

Returning to Equation 9.24, we now have

$$\begin{aligned}
-\gamma_0^2\beta_\perp\beta_\parallel k_w c\sin(k_w z) &= \Omega_w\sin(k_w z)\gamma_0\beta_\parallel - \Omega_\parallel\gamma_0\beta_\perp\sin(k_w z),\\
\gamma_0^2\beta_\perp\beta_\parallel k_w c\cos(k_w z) &= \Omega_\parallel\gamma_0\beta_\perp\cos(k_w z) - \Omega_w\cos(k_w z)\gamma_0\beta_\parallel,\\
0 &= \Omega_w[\gamma_0\beta_\perp\sin(k_w z)\cos(k_w z) - \gamma_0\beta_\perp\cos(k_w z)\sin(k_w z)].
\end{aligned} \tag{9.41}$$

The last equation is identically satisfied, and the transverse components yield the condition

$$\beta_\perp = \frac{\beta_\parallel\Omega_w}{\Omega_\parallel - \gamma_0 k_w\beta_\parallel c} \tag{9.42}$$

or

$$\beta_\perp = \frac{\beta_\parallel\dfrac{\Omega_w}{\gamma_0}}{\dfrac{\Omega_\parallel}{\gamma_0} - k_w\beta_\parallel c}. \tag{9.43}$$

The physics of this solution is easily understood. On the one hand, as shown in Equation 9.35, the transverse trajectory of the electron in the guide field corresponds to a harmonic motion at the eigenfrequency $\Omega_\parallel/\gamma_0 = eB_\parallel/\gamma_0 m_0$, which is the relativistic cyclotron frequency of the electron in the guide magnetic field. On the other hand, the wiggler drives the transverse motion of the electron at the frequency $k_w\beta_\parallel c$; when both frequencies are nearly equal, a resonant enhancement of the transverse motion occurs, as shown in Equation 9.43. Of course, the resonance is limited by energy conservation, as we must also satisfy the equation

$$\frac{1}{\gamma_0^2} = 1 - \beta_\perp^2 - \beta_\parallel^2. \tag{9.44}$$

This leads to the type I and type II trajectories in an FEL with a helical wiggler and a guide field, as discussed in great detail by Antonsen and Freund. The resonance condition can be written as

$$k_w\beta_\parallel c \approx \frac{\Omega_\parallel}{\gamma_0}, \quad \text{or} \quad \gamma_0 k_w\beta_\parallel c \approx \Omega_\parallel. \tag{9.45}$$

In the first case, we are considering the resonance in the lab frame, where the wiggler-induced motion has the characteristic frequency $k_w\beta_\parallel c = k_w v_\parallel$, and where the relativistic mass of the electron is $m = \gamma_0 m_0$, thus yielding the relativistic cyclotron frequency $eB_\parallel/m = eB_\parallel/\gamma_0 m_0 = \Omega_\parallel/\gamma_0$. The second equation corresponds to the electron frame, where the wiggler wavenumber is Lorentz transformed to yield the frequency $\gamma_0 k_w v_\parallel$, while there is no mass shift.

The helical motion of the electron yields an additional constraint on the radius of the trajectory: during the time taken by the electron to move by one wiggler period in the z-direction, it must also have described a full circle in the transverse plane. In other words, we have

$$c\beta_\perp\left(\frac{\ell_w}{c\beta_\parallel}\right) = 2\pi r_w, \tag{9.46}$$

where $r_w$ is the equilibrium radius of the helical trajectory.

This condition can be recast as

$$\beta_w = \frac{\beta_\perp}{\beta_\parallel} = \frac{2\pi r_w}{\ell_w} = k_w r_w, \tag{9.47}$$

which defines the pitch of the helix.

Finally, for one-dimensional systems such as the one considered here, it is often convenient to use the axial position, $z$, as the independent variable; this is easily achieved by noting that

$$\frac{d}{dz} = \frac{d\tau}{dz}\frac{d}{d\tau} = \frac{1}{cu_z}\frac{d}{d\tau}. \tag{9.48}$$

In the absence of a guide field, Equation 9.43 reduces to $\beta_\perp = \Omega_w/\gamma_0 k_w c = eB_w/\gamma_0 m_0 k_w c$, and we can calculate the total instantaneous power radiated by a single electron propagating through the wiggler by deriving Equation 9.10 with respect to time to obtain the acceleration

$$\begin{aligned}\dot{\beta} &= \frac{d}{dt}\{\beta_\perp[\hat{x}\cos(k_w z) + \hat{y}\sin(k_w z)] + \hat{z}\beta_\parallel\}\\ &= \beta_\perp k_w \frac{dz}{dt}[-\hat{x}\sin(k_w z) + \hat{y}\cos(k_w z)]\\ &= \beta_\perp k_w c\beta_\parallel[-\hat{x}\sin(k_w z) + \hat{y}\cos(k_w z)].\end{aligned} \tag{9.49}$$

We then find that

$$\dot{\beta}^2 = (\beta_\perp\beta_\parallel k_w c)^2, \tag{9.50}$$

and

$$\begin{aligned}\beta\times\dot{\beta} &= \{\beta_\perp[\hat{x}\cos(k_w z) + \hat{y}\sin(k_w z)] + \hat{z}\beta_\parallel\}\\ &\quad\times\{\beta_\perp k_w c\beta_\parallel[-\hat{x}\sin(k_w z) + \hat{y}\cos(k_w z)]\}\\ &= \hat{x}[-\beta_\parallel^2\beta_\perp k_w c\cos(k_w z)] + \hat{y}[-\beta_\parallel^2\beta_\perp k_w c\sin(k_w z)]\\ &\quad + \hat{z}\{\beta_\parallel\beta_\perp^2 k_w c[\cos^2(k_w z) + \sin^2(k_w z)]\}\\ &= \hat{x}[-\beta_\parallel^2\beta_\perp k_w c\cos(k_w z)] + \hat{y}[-\beta_\parallel^2\beta_\perp k_w c\sin(k_w z)] + \hat{z}(\beta_\parallel\beta_\perp^2 k_w c).\end{aligned} \tag{9.51}$$

This result can be used to calculate that

$$\begin{aligned}(\beta\times\dot{\beta})^2 &= (\beta_\parallel^2\beta_\perp k_w c)^2[\cos^2(k_w z)+\sin^2(k_w z)]+(\beta_\parallel\beta_\perp^2 k_w c)^2\\ &= (\beta_\parallel\beta_\perp k_w c)^2(\beta_\parallel^2+\beta_\perp^2),\end{aligned} \tag{9.52}$$

and we can employ Equation 9.50 to obtain

$$\begin{aligned}\gamma^6[\dot{\beta}^2-(\beta\times\dot{\beta})^2] &= \gamma_0^6[(\beta_\perp\beta_\parallel k_w c)^2-(\beta_\parallel\beta_\perp k_w c)^2(\beta_\parallel^2+\beta_\perp^2)]\\ &= \gamma_0^6(\beta_\perp\beta_\parallel k_w c)^2(1-\beta_\parallel^2+\beta_\perp^2)\\ &= \gamma_0^4(\beta_\perp\beta_\parallel k_w c)^2.\end{aligned} \tag{9.53}$$

The power is then given by applying Equation 9.9:

$$P=\frac{e^2}{6\pi\varepsilon_0 c}\gamma^6[\dot{\beta}^2-(\beta\times\dot{\beta})^2]=\frac{e^2}{6\pi\varepsilon_0 c}\gamma_0^4(\beta_\perp\beta_\parallel k_w c)^2. \tag{9.54}$$

We now derive in a more general way the distribution in frequency and solid angle of the energy radiated by an axially extended charge distribution propagating through a helical wiggler. First, we briefly review the basic steps of the derivation given by Jackson in the case of a point charge.

The distribution of energy radiated per unit solid angle, per unit frequency is defined as

$$\frac{d^2 I(\omega,\hat{n})}{d\omega\, d\Omega}=\frac{e^2}{8\pi^2\varepsilon_0 c}|\tilde{\Lambda}(\omega,\hat{n})|^2, \tag{9.55}$$

where the vector $\tilde{\Lambda}(\omega,\hat{n})$ is the Fourier transform of the vector

$$\Lambda(t,\hat{n})=\left[\frac{\hat{n}\times(\hat{n}-\beta)\times\dot{\beta}}{(1-\hat{n}\cdot\beta)^3}\right], \tag{9.56}$$

which, in turn, is defined in terms of the instantaneous solid angle power distribution by

$$\frac{e^2}{16\pi^2\varepsilon_0 c}|\Lambda(t,\hat{n})|^2=\frac{dP(t,\hat{n})}{d\Omega}. \tag{9.57}$$

Using Parseval's theorem and closely following the steps of the theoretical derivation given by Jackson, we find

$$\frac{d^2 I(\omega,\hat{n})}{d\omega\, d\Omega}=\frac{e^2\omega^2}{16\pi^3\varepsilon_0 c}\left|\int_{-\infty}^{+\infty}\hat{n}\times(\hat{n}\times\beta)\exp\left\{i\omega\left[t-\frac{\hat{n}\cdot\mathbf{r}(t)}{c}\right]\right\}dt\right|^2. \tag{9.58}$$

This result does not involve the acceleration because we have integrated by parts; it essentially states that the spectral intensity is the Fourier transform of the trajectory followed by the accelerated charge, provided that the relativistic Doppler effect is properly taken into account, as expressed by the phase term, $\omega[t-(\hat{\boldsymbol{n}}\cdot\mathbf{r}(t)/c)]$, in the complex exponential.

In the particular example of interest here, of a point charge propagating on stable orbits through a helically polarized wiggler field, we have

$$\mathbf{r}(t) = c\int_0^t \beta(t')dt' = r_w[\hat{x}\sin\phi(t) + y\cos\phi(t)] + \hat{z}\frac{\phi(t)}{k_w}, \tag{9.59}$$

and

$$\beta(\tau) = \beta_\perp[\hat{x}\cos(k_w z) + \hat{y}\sin(k_w z)] + \hat{z}\beta_\parallel. \tag{9.60}$$

Here, we have introduced the phase factor

$$\phi(t) = k_w c\beta_\parallel t = k_w z(t), \tag{9.61}$$

and the quiver radius of the helical trajectory, as defined in Equation 9.47.

For $N_w$ wiggler periods, we have to perform the following integral:

$$\frac{d^2 l(\omega,\hat{\boldsymbol{n}})}{d\omega\, d\Omega} = \frac{e^2}{16\pi^3\varepsilon_0 c}\left(\frac{\omega}{k_w c\beta_\parallel}\right)^2\left|\int_{-\infty}^{+\infty}\hat{\boldsymbol{n}}\times[\hat{\boldsymbol{n}}\times\beta(\phi)]\exp\left\{i\omega\left[\frac{\phi}{k_w c\beta_\parallel} - \frac{\hat{\boldsymbol{n}}\cdot\mathbf{r}(\phi)}{c}\right]\right\}d\phi\right|^2. \tag{9.62}$$

In Equation 9.61, we have changed variables, and used the phase as the independent variable.

The radiation emitted by the charge in the wiggler is circularly polarized, and the radiation pattern is azimuthally symmetric around the $z$-axis. The most interesting case is the radiation spectrum emitted on-axis, where most of the power is radiated and where we obtain the maximum relativistic Doppler upshift. Setting $\hat{\boldsymbol{n}} = \hat{z}$, the integral can easily be performed and the complex amplitude calculated, to yield

$$\frac{d^2 l(\omega,\hat{\boldsymbol{n}})}{d\omega\, d\Omega} = \frac{e^2}{16\pi^3\varepsilon_0 c}\left(\frac{\omega}{k_w c\beta_\parallel}\right)^2\beta_\perp^2\left[\frac{1-\cos u(\omega)\chi}{u^2(\omega)} + \frac{1-\cos v(\omega)\chi}{v^2(\omega)}\right], \tag{9.63}$$

where we have defined the following parameters:

$$\begin{aligned} u(\omega) &= \frac{\omega(1-\beta_{\|})}{k_w c\beta_{\|}} + 1, \\ v(\omega) &= \frac{\omega(1-\beta_{\|})}{k_w c\beta_{\|}} - 1, \end{aligned} \tag{9.64}$$

and $\chi = 2\pi N_w$.

This spectrum has a maximum at the usual FEL Doppler upshifted frequency,

$$\omega^{+} = \gamma_{\|}^{2}\beta_{\|}k_w c(1+\beta_{\|}), \tag{9.65}$$

such that $v(\omega^{+}) = 0$.

The spectral width varies like $N_w^{-1}$, as the Fourier transform becomes better defined as the electron undergoes more wiggler oscillations. Satellite lines are also produced because the power spectrum is a $\text{sinc}^2$ structure.

At $\omega = \omega^{+}$, the amplitude of the main spectral line is

$$\frac{d^2 I(\omega^{+},\hat{z})}{d\omega\, d\Omega} = \frac{e^2}{8\pi\varepsilon_0 c}\left(\frac{1}{1-\beta_{\|}}\right)^2 \beta_{\perp}^2 N_w^2, \tag{9.66}$$

which exhibits the quadratic scaling in terms of charge, acceleration, and interaction length characterizing coherent electromagnetic radiation processes.

To investigate the case of an axially extended or finite-sized charge distribution propagating through the wiggler, we can use Equation 9.58, provided that we make the following substitution:

$$\beta \exp\left[i\omega\frac{\hat{\boldsymbol{n}}\cdot\mathbf{r}(t)}{c}\right] \to \frac{1}{ec}\int d^3\mathbf{x}\,\mathbf{j}(\mathbf{x},t)\exp\left(i\omega\frac{\hat{\boldsymbol{n}}\cdot\mathbf{x}}{c}\right). \tag{9.67}$$

Here, the fluid helical current density, $\mathbf{j}(\mathbf{x}, t)$, is defined in terms of bunch charge density, $\rho$, as

$$\mathbf{j}(z,t) = \rho(z,t)c\beta(z), \tag{9.68}$$

where $\beta(z)$ is the one-dimensional helical velocity field given in Equation 9.60, and where we model the bunch axial charge density with a Gaussian linear density distribution of width $\Delta z$, and total charge $q$:

$$\rho(z,t) = \frac{q}{\sqrt{\pi}\Delta z}\exp\left[-\left(\frac{z-\beta_{\|}ct}{\Delta z}\right)^2\right]. \tag{9.69}$$

We note in passing that Equations 9.68 and 9.69 satisfy the continuity equation. The explicit expression for the one-dimensional helical current density is then

$$\mathbf{j}(z,t) = \frac{qc}{\sqrt{\pi}\Delta z}\{\beta_\perp[\hat{x}\cos(k_w z) + \hat{y}\sin(k_w z)] + \hat{z}\beta_\parallel\}\exp\left[-\left(\frac{z-\beta_\parallel ct}{\Delta z}\right)^2\right]. \quad (9.70)$$

Making the substitution described in Equation 9.67, and using the current density given in Equation 9.70, we find the on-axis energy spectral distribution after integrating over space and time with the appropriate boundaries. The result is

$$\begin{aligned}\frac{d^2 I(\omega,\hat{z})}{d\omega\, d\Omega} &= \frac{q^2\beta_\perp^2}{16\pi^3\varepsilon_0 c}\left(\frac{\omega}{k_w\beta_\parallel c}\right)\\ &\times\left|\left(\frac{\hat{x}}{2} - i\frac{\hat{y}}{2}\right)\exp\left[-\frac{\Delta z^2}{4}\left(\frac{\omega}{c} - k_w\right)^2\right]\frac{\cos u(\omega)\chi + i\sin u(\omega)\chi - 1}{u(\omega)}\right|^2\\ &\times\left|\left(\frac{\hat{x}}{2} + i\frac{\hat{y}}{2}\right)\exp\left[-\frac{\Delta z^2}{4}\left(\frac{\omega}{c} + k_w\right)^2\right]\frac{\cos v(\omega)\chi + i\sin v(\omega)\chi - 1}{v(\omega)}\right|^2.\end{aligned} \quad (9.71)$$

At maximum Doppler upshift, this expression assumes the value

$$\frac{d^2 I(\omega^+,\hat{z})}{d\omega\, d\Omega} = \frac{q^2\beta_\perp^2 N_w^2}{8\pi\varepsilon_0 c}\left(\frac{1}{1-\beta_\parallel}\right)^2\exp\left[-\frac{2\pi^2}{\beta_\parallel^2}\left(\frac{\Delta z}{\lambda}\right)^2\right]. \quad (9.72)$$

Here, $\lambda$ is the wavelength associated to the upshifted FEL spectral line.

This expression is identical to that derived in the case of a point charge and presented in Equation 9.66, except for the exponential coherence function, which describes how the energy density in the main spectral line decreases as the bunch length to wavelength ratio increases. This exponential suppression factor, while describing the transition from coherence to incoherence in an approximately correct fashion, does not adequately model the scaling of the radiated power in the incoherent regime. This important question will be fully resolved in Chapter 10, where a stochastic electron gas model will allow us to take into account the discrete, point-like nature of electrons, and its implications in terms of coherence. For our current purpose, however, this model is quite adequate.

## 9.3 Coherent Synchrotron Radiation in a Waveguide

In this section, we study the coupling between an axially extended charge distribution describing fixed trajectories in a helically polarized wiggler, and TE and TM cylindrical waveguide eigenmodes, in the absence of radiation feedback. The TM modes couple to the space-charge density distribution and to the axial component of the current density, while the TE modes are excited by the wiggler-induced radial and azimuthal current density components. The mathematical analysis presented in the following is based on a coupled-mode theory similar to that used by Haus and Islam to derive the characteristics of synchrotron radiation in a rectangular waveguide; it is also in line with the Green function analysis in a cylindrical waveguide discussed in Chapter 5.

The major differences between the present analysis and the aforementioned theoretical approaches are the following. First, we are treating the problem in cylindrical geometry, with a helical wiggler field. Next, we are specifically interested in the coherent synchrotron radiation limit; it is therefore essential to model an axially extended charge distribution to determine the transition between this limit and the usual incoherent spontaneous radiation studied in the above-mentioned reference. Finally, we use a temporal Fourier analysis only; the spatial evolution of the respective eigenmode amplitudes is governed by a differential equation in $z$.

### 9.3.1 Four-Current in a Helical Wiggler

For the sake of simplicity, we consider a one-dimensional wiggler field approximation, given by

$$\mathbf{B} = B_w[\hat{r}\cos(k_w z - \theta) + \hat{\theta}\sin(k_w z - \theta)], \tag{9.73}$$

in cylindrical coordinates. The connection with the Cartesian coordinates is described in detail in Section 5.6.7. In Equation 9.73, $B_w$ is the wiggler amplitude, and $k_w = 2\pi/\ell_w$ is the wiggler wavenumber, with $\ell_w$ its period. The equilibrium trajectories in the wiggler are expressed, in cylindrical coordinates, as

$$\beta = \beta_\perp[\hat{r}\cos(k_w z - \theta) + \hat{\theta}\sin(k_w z - \theta)] + z\beta_\|, \tag{9.74}$$

where the perpendicular and axial velocities are given by Equations 9.11 and 9.12.

A simple model of the axially extended bunch charge density can be given in terms of a Gaussian distribution moving along the $z$-axis with the axial velocity $v_\| = \beta_\| c$:

$$\rho(r,\theta,z,t) = \frac{q}{\sqrt{\pi}\Delta z \pi r_\perp^2}\exp\left[-\left(\frac{z-\beta_\| ct}{\Delta z}\right)^2\right], \quad (r \le r_\perp), \tag{9.75}$$

where $q$ is the total bunch charge, and $\Delta z$ is its characteristic axial length scale.

Note that, as the bunch is azimuthally symmetric, the charge density does not depend on $\theta$. The effective radius, $r_\perp$, in Equation 9.75 is defined in terms of the input beam radius, $r_b$, as

$$r_\perp \simeq \sqrt{r_b^2 + r_w^2} = \sqrt{r_b^2 + \left(\frac{\beta_w}{k_w}\right)^2}. \tag{9.76}$$

This model is valid when $r_b > r_w = \beta_w / k_w$, $\beta_w$ being the wiggler pump parameter. The corresponding current density in the helical wiggler is then given by $\mathbf{j} = \rho \beta c$, which yields the explicit expression

$$\mathbf{j}(r, \theta, z, t) = \frac{qc}{\sqrt{\pi}\Delta z \pi r_\perp^2}\{\beta_\perp [\hat{r}\cos(k_w z - \theta) + \hat{\theta}\sin(k_w z - \theta)] + z\beta_\parallel\} \times \exp\left[-\left(\frac{z - \beta_\parallel ct}{\Delta z}\right)^2\right]. \tag{9.77}$$

One can easily verify that the charge density distribution described in Equation 9.75, together with the current density given by Equation 9.77 satisfies the continuity equation

$$\partial_t \rho + \nabla \cdot \mathbf{j} = 0, \tag{9.78}$$

everywhere, except at the beam edge.

We now proceed to derive the temporal Fourier transform of the radial component of the current density, which will be needed in the next section to drive the coupled-mode wave equation for TE modes. We have

$$j_r(r, \theta, z, t) = \frac{1}{\sqrt{2\pi}} \int_{-\infty}^{+\infty} \tilde{j}_r(r, \theta, z, \omega) e^{-i\omega t} d\omega, \tag{9.79}$$

where, by definition, the Fourier transform is given by

$$\tilde{j}_r(r, \theta, z, \omega) = \frac{1}{\sqrt{2\pi}} \int_{-\infty}^{+\infty} \frac{qc\beta_\perp}{\pi^{3/2}\Delta z r_\perp^2} \cos(k_w z - \theta) \exp\left[-\left(\frac{z - \beta_\parallel ct}{\Delta z}\right)^2\right] e^{i\omega t} dt. \tag{9.80}$$

This integral can be performed easily, to yield

$$\tilde{j}_r(r, \theta, z, \omega) = \frac{q\beta_w}{\sqrt{2}\pi^{3/2} r_\perp^2} \cos(k_w z - \theta) \exp\left(i\frac{\omega z}{\beta_\parallel c}\right) \exp\left[-\left(\frac{\omega \Delta z}{2\beta_\parallel c}\right)^2\right]. \tag{9.81}$$

The term $\omega z / \beta_\parallel c$ will be shown to give rise to beam modes, while the second exponential corresponds to the coherence factor introduced earlier.

### 9.3.2 Coupling to Cylindrical Waveguide Modes

In cylindrical coordinates, the wave equation for the four-vector potential is

$$\begin{aligned}
&\Box A_r - \frac{1}{r^2}(A_r + 2\partial_\theta A_\theta) + \mu_0 j_r = 0,\\
&\Box A_\theta - \frac{1}{r^2}(A_\theta - 2\partial_\theta A_r) + \mu_0 j_\theta = 0,\\
&\Box A_r + \mu_0 j_z = 0,\\
&\Box \phi + \frac{1}{\varepsilon_0}\rho = 0,
\end{aligned} \tag{9.82}$$

where

$$\Box \equiv \left[\Delta - \frac{1}{c^2}\partial_t^2\right] \equiv \left[\frac{1}{r}\partial_r(r\partial_r) + \frac{1}{r^2}\partial_\theta^2 + \partial_z^2 - \frac{1}{c^2}\partial_t^2\right]. \tag{9.83}$$

It is easily seen that in the case of a spatially extended charge propagating in a helical wiggler, the TE modes, represented by $A_r$ and $A_\theta$, couple to the wiggler-induced motion, as the current components $j_r$ and $j_\theta$ contain the term $\beta_\perp$, which is proportional to the wiggler field amplitude, $B_w$. By contrast, the TM modes, represented by $A_z$ and $\phi$ are driven by the uniform motion of the space-charge distribution in the cylindrical waveguide. Therefore, in the remainder of this derivation, we will focus on the TE modes. We now expand the radial and azimuthal components of the four-vector potential in terms of TE cylindrical eigenmodes, which yield the following expressions:

$$A_r(r,\theta,z,t) = \frac{1}{\sqrt{2\pi}}\sum_m\sum_n\int_{-\infty}^{+\infty}\tilde{A}_{mn}(z,\omega)\frac{\mathrm{J}_m\left(\chi'_{mn}\frac{r}{a}\right)}{\chi'_{mn}\frac{r}{a}}e^{i(m\theta-\omega t)}d\omega, \tag{9.84}$$

and

$$A_\theta(r,\theta,z,t) = \frac{1}{\sqrt{2\pi}}\sum_m\sum_n\int_{-\infty}^{+\infty}\tilde{A}_{mn}(z,\omega)\frac{i}{m}\mathrm{J}'_m\left(\chi'_{mn}\frac{r}{a}\right)e^{i(m\theta-\omega t)}d\omega, \tag{9.85}$$

where $\chi'_{mn}$ is the $n$-th zero of $\mathrm{J}'_m$.

The main difference between this mode expansion and the one used in Chapter 5 is the fact that we have not Fourier-transformed the axial dependence; instead, we shall seek a differential equation in $z$, describing the evolution of the spectral density of each TE vacuum eigenmode, as driven by the pre-bunched FEL current. Within this context, the term $\tilde{A}_{mn}(z,\omega)$

represents the spectral density of each $TE_{mn}$ mode as a function of position along the waveguide axis.

Inserting Equations 9.84 and 9.85 into the radial component of the wave equation, together with the temporal Fourier transform of the radial component of the current density, we obtain, after some straightforward algebra,

$$\sum_m \sum_n \left[\frac{\omega^2}{c^2} - \frac{\chi_{mn}'^2}{a^2} + \partial_z^2\right] \tilde{A}_{mn}(\omega, z) \frac{J_m\left(\chi_{mn}' \frac{r}{a}\right)}{\chi_{mn}' \frac{r}{a}} e^{im\theta} + \mu_0 \tilde{j}_r(\theta, z, \omega) = 0. \quad (9.86)$$

Here, we have used the fact that the complex exponential functions are orthogonal, which equivalent to implicitly diagonalizing the time-dependence. The azimuthal wave equation is essentially identical to the radial equation, which is sufficient to uniquely determine the spectral amplitude $\tilde{A}_{mn}(z,\omega)$. This derives from the fact that the modal expansions given in Equations 9.84 and 9.85 satisfy the Lorentz gauge condition term by term.

Using Equation 9.81, Equation 9.86 can be written explicitly as

$$\sum_m \sum_n \left[\frac{\omega^2}{c^2} - \frac{\chi_{mn}'^2}{a^2} + \partial_z^2\right] \tilde{A}_{mn}(\omega, z) \frac{J_m\left(\chi_{mn}' \frac{r}{a}\right)}{\chi_{mn}' \frac{r}{a}} e^{im\theta}$$

$$+ \frac{\mu_0 q \beta_w}{\sqrt{2}\pi^{3/2} r_\perp^2} \cos(k_w z - \theta) \exp\left(i\frac{\omega z}{\beta_\| c}\right) \exp\left[-\left(\frac{\omega \Delta z}{2\beta_\| c}\right)^2\right] = 0. \quad (9.87)$$

We now diagonalize this set of coupled differential equations using the orthogonality of the TE cylindrical vacuum eigenmodes. To this purpose, we apply the following operator,

$$\int_0^{2\pi} d\theta\, e^{-ip\theta} \int_0^a r^2 dr J_p\left(\chi_{pq}' \frac{r}{a}\right), \quad (9.88)$$

to Equation 9.87. Here, $a$ is the waveguide radius. The operator described in Equation 9.88 diagonalized the left-hand side of Equation 9.87, while it projects the source term on a particular $TE_{pq}$ cylindrical waveguide mode. We then have

$$\sum_m \sum_n \left[\frac{\omega^2}{c^2} - \frac{\chi_{mn}'^2}{a^2} + \partial_z^2\right] \tilde{A}_{mn}(\omega, z) \int_0^{2\pi} e^{i\theta(m-p)} d\theta \int_0^a \frac{a}{\chi_{mn}'} J_p\left(\chi_{pq}' \frac{r}{a}\right) J_m\left(\chi_{mn}' \frac{r}{a}\right) r\, dr$$

$$+ \frac{\mu_0 q \beta_w}{\sqrt{2}\pi^{3/2} r_\perp^2} \exp\left(i\frac{\omega z}{\beta_\| c}\right) \exp\left[-\left(\frac{\omega \Delta z}{2\beta_\| c}\right)^2\right]$$

$$\times \int_0^{2\pi} \cos(k_w z - \theta) e^{-ip\theta} d\theta \int_0^{r_\perp} J_p\left(\chi_{pq}' \frac{r}{a}\right) r^2 dr = 0. \quad (9.89)$$

We now use the fact that

$$\int_0^{2\pi} e^{i\theta(m-p)} d\theta = 2\pi\delta_{mp}, \tag{9.90}$$

and we introduce the normalized radius, $x = r/a$. With this, Equation 9.89 reduces to

$$2\pi\sum_n \left[\frac{\omega^2}{c^2} - \frac{\chi_{pn}'^2}{a^2} + \partial_z^2\right]\tilde{A}_{pn}(\omega, z)\frac{1}{\chi_{pn}'}\int_0^1 J_p(\chi_{pq}'x)J_p(\chi_{pn}'x)x\,dx$$
$$+\frac{\mu_0 q\beta_w}{\sqrt{2}\pi^{3/2}r_\perp^2}\exp\left(i\frac{\omega z}{\beta_\| c}\right)\exp\left[-\left(\frac{\omega\Delta z}{2\beta_\| c}\right)^2\right]\int_0^{2\pi}\cos(k_w z-\theta)e^{-ip\theta}d\theta$$
$$\times\int_0^{r_\perp/a} J_p(\chi_{pq}'x)x^2 dx = 0,$$

since all the terms in the summation over the azimuthal number $m$ are zero except for $m = p$. Next, we use the following orthogonality relation for the radial mode profile:

$$\int_0^1 J_p(\chi_{pq}'x)J_p(\chi_{pn}'x)x\,dx = \frac{1}{2}\left[1-\left(\frac{p}{\chi_{pq}'}\right)^2\right]J_p^2(\chi_{pq}')\delta_{nq}, \tag{9.91}$$

which results in a single differential equation for a given $TE_{pq}$ mode,

$$2\pi\left[\frac{\omega^2}{c^2} - \frac{\chi_{pq}'^2}{a^2} + \partial_z^2\right]\tilde{A}_{pq}(\omega, z)\frac{1}{\chi_{pn}'}\frac{1}{2}\left[1-\left(\frac{p}{\chi_{pn}'}\right)^2\right]J_p^2(\chi_{pn}')$$
$$+\frac{\mu_0 q\beta_w}{\sqrt{2}\pi^{3/2}r_\perp^2}\exp\left(i\frac{\omega z}{\beta_\| c}\right)\exp\left[-\left(\frac{\omega\Delta z}{2\beta_\| c}\right)^2\right]$$
$$\times\int_0^{2\pi}\cos(k_w z-\theta)e^{-ip\theta}d\theta\int_0^{r_\perp/a} J_p(\chi_{pn}'x)x^2 dx = 0. \tag{9.92}$$

We now consider the wiggler helicity integral,

$$\int_0^{2\pi}\cos(k_w z-\theta)e^{-ip\theta}d\theta = 2\pi\left\{\frac{\exp[\pm(-ik_w z)]}{2}\right\}, \tag{9.93}$$

for $p = \pm 1$. This integral is equal to zero for all other values of $p$. Because of the wiggler helicity, the wiggler-induced current can only couple to $TE_{\pm 1q}$ modes, a fact which is mathematically represented by the selection rule derived above. These modes correspond to right-hand-circularly polarized (RHCP) and left-hand-circularly polarized (LHCP) waves propagating through the cylindrical waveguide. From here on, we consider the $p = 1$ modes only. We can then calculate the radial integral in the source term, with the result that

$$\int_0^{r_\perp/a} J_1(\chi'_{1q}x)x^2 dx = \frac{1}{\chi'_{1q}}\left(\frac{r_\perp}{a}\right)^2 J_2\left(\chi'_{1q}\frac{r_\perp}{a}\right). \tag{9.94}$$

Using Equations 9.93 and 9.94, we obtain the sought-after differential equations governing the evolution of the $TE_{1q}$ cylindrical waveguide mode spectral density along the $z$-axis, driven by an axially extended charge distribution propagating through a helical wiggler:

$$\left[\frac{\omega^2}{c^2} - \frac{\chi'^2_{1q}}{a^2} + \partial_z^2\right]\tilde{A}_{1q}(\omega, z) + \frac{\mu_0 q \beta_w J_2\left(\chi'_{1q}\frac{r_\perp}{a}\right)}{\sqrt{2}\pi^{3/2} a^2 J_1^2(\chi'^2_{1q})\left[1 - \left(\frac{1}{\chi'_{1q}}\right)^2\right]}$$
$$\times \exp\left[i\left(\frac{\omega}{\beta_\parallel c} - k_w\right)z\right]\exp\left[-\left(\frac{\omega \Delta z}{2\beta_\parallel c}\right)^2\right] = 0. \tag{9.95}$$

To obtain a clear picture of the physics involved, we can introduce two different wavenumbers. First,

$$k_1(\omega) = \sqrt{\frac{\omega^2}{c^2} - \frac{\chi'^2_{1q}}{a^2}}, \tag{9.96}$$

which corresponds to the propagation of the $TE_{1q}$ mode in the cylindrical waveguide, and

$$k_2(\omega) = \frac{\omega}{\beta_\parallel c} - k_w, \tag{9.97}$$

which corresponds to the usual FEL wiggler-shifted beam mode. Within this context, the differential equation 9.95 can be recast in the simple form

$$f''(z) + k_1^2 f(z) = Ce^{ik_2 z}, \tag{9.98}$$

where we have defined $f(z) = \tilde{A}_{1q}(\omega, z)$, and where the coupling coefficient is

$$C = -\frac{\mu_0 q \beta_w}{\sqrt{2}\pi^{3/2} a^2} \frac{J_2\left(\chi'_{1q}\frac{r_\perp}{a}\right)}{J_1^2(\chi'_{1q})\left[1-\left(\frac{1}{\chi'_{1q}}\right)^2\right]} \exp\left[-\left(\frac{\omega \Delta z}{2\beta_\| c}\right)^2\right]. \tag{9.99}$$

Equation 9.98 describes a harmonic oscillator, with eigenfrequency, $k_1$, corresponding to the electromagnetic mode, driven harmonically at $k_2$, by the beam mode. This system is driven resonantly when

$$k_1(\omega) = k_2(\omega), \tag{9.100}$$

which corresponds to the two roots $\omega = \omega^{\pm}$, where

$$\omega^{\pm} = \gamma_\|^2 \beta_\| k_w c\left[1 \pm \beta_\| \sqrt{1-\left(\frac{\omega_c}{\gamma_\| \beta_\| k_w c}\right)^2}\right], \tag{9.101}$$

are the well-known waveguide FEL Doppler upshifted and downshifted interaction frequencies. Here, $\omega_c = \chi'_{1q} c/a$ is the cutoff frequency of the $TE_{1q}$ mode under consideration.

The solution to Equation 9.98 corresponding to forward propagating modes is given by

$$f(z) = A\exp(ik_1 z) + \frac{C}{k_1^2 - k_2^2}\exp(ik_2 z). \tag{9.102}$$

The boundary condition is

$$f(z=0) = 0: \tag{9.103}$$

no power is radiated into the TE modes before the wiggler entrance. With this, we have

$$\tilde{A}_{1q}(\omega, z) = -\frac{\mu_0 q \beta_w}{\sqrt{2}\pi^{3/2} a^2} \frac{J_2\left(\chi'_{1q}\frac{r_\perp}{a}\right)}{J_1^2(\chi'_{1q})\left[1-\left(\frac{1}{\chi'_{1q}}\right)^2\right]} \exp\left[-\left(\frac{\omega \Delta z}{2\beta_\| c}\right)^2\right] \times \frac{\exp[ik_2(\omega)z] - \exp[ik_1(\omega)z]}{k_1^2(\omega) - k_2^2(\omega)}. \tag{9.104}$$

One can easily show that

$$k_1^2(\omega) - k_2^2(\omega) = -\frac{1}{(\beta_\| \gamma_\| c)^2}(\omega - \omega^-)(\omega - \omega^+); \tag{9.105}$$

we can then insert this result into Equation 9.104, and find, after some simple trigonometry,

$$\tilde{A}_{1q}(\omega, z) = \frac{\sqrt{2} i q \beta_w \beta_\parallel^2 \gamma_\parallel^2}{\sqrt{\pi^3} \varepsilon_0 a^2} \frac{J_2\left(\chi'_{1q} \frac{r_\perp}{a}\right)}{J_1^2(\chi'_{1q})\left[1 - \left(\frac{1}{\chi'_{1q}}\right)^2\right]} \exp\left[-\left(\frac{\omega \Delta z}{2\beta_\parallel c}\right)^2\right]$$
$$\times \frac{\sin\left[\left(\sqrt{\omega^2 - \omega_c^2} - \frac{\omega}{\beta_\parallel} + k_w c\right)\frac{z}{2c}\right]}{(\omega - \omega^-)(\omega - \omega^+)} \exp\left[i\left(\sqrt{\omega^2 - \omega_c^2} + \frac{\omega}{\beta_\parallel} - k_w c\right)\frac{z}{2c}\right]. \tag{9.106}$$

Using the relation $\mathbf{E} = -\nabla \phi - \partial_t \mathbf{A}$, yields the Fourier transform of the electric field in the FEL interaction region:

$$\tilde{E}_{1q}(\omega, z) = \sqrt{\frac{2}{\pi^3}} \frac{q \beta_w}{\varepsilon_0 a^2} g_{1q} \omega \exp\left[-\left(\frac{\omega \Delta z}{2\beta_\parallel c}\right)^2\right] \frac{\sin\left[\Delta k(\omega)\frac{z}{2}\right]}{\Delta k(\omega)} \frac{\exp\left\{i\left[k_1(\omega) + k_2(\omega)\frac{z}{2}\right]\right\}}{k_1(\omega) + k_2(\omega)}, \tag{9.107}$$

where we have defined the radial overlap integral of the bunch over the eigenmode,

$$g_{1q} = \frac{J_2\left(\chi'_{1q} \frac{r_\perp}{a}\right)}{J_1^2(\chi'_{1q})\left[1 - \left(\frac{1}{\chi'_{1q}}\right)^2\right]}. \tag{9.108}$$

The main features of this solution are the following. First, the spectral amplitude of the four-vector potential and electric field are proportional to the bunch charge, $q$, and acceleration $\beta_w = \beta_\perp / \beta_\parallel$; this yields the usual quadratic scalings for the power spectrum. The next factor is the overlap integral, $g_{1q}$, of the bunch transverse distribution with the $TE_{1q}$ mode. The exponential factor describes the degree of coherence of the radiation; its argument is a quadratic function of the bunch length to wavelength ratio. The next factor, which appears in the form of a modified sinc function, is the envelope of the frequency spectrum. Here, the difference between the waveguide and free-space interactions appear clearly, as the waveguide structure allows the possibility of operating at grazing, where

$$\omega^+ = \omega^- = \omega^* = \gamma_\parallel \omega_c. \tag{9.109}$$

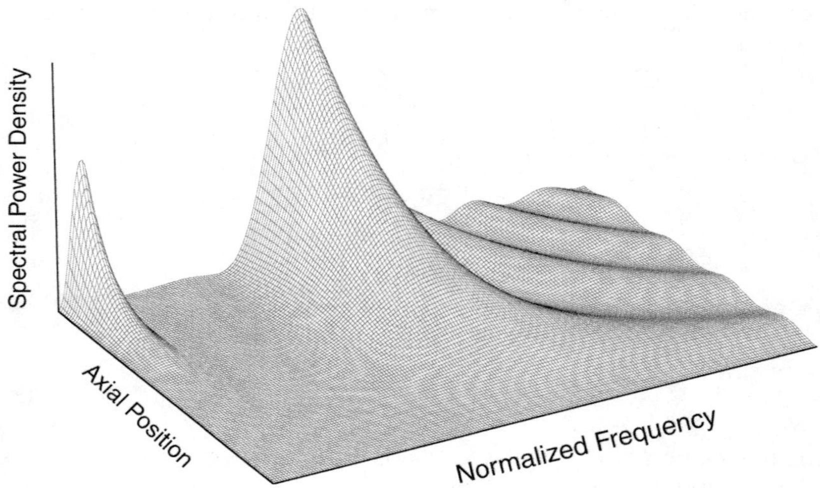


**FIGURE 9.3**
Spectral power density for coherent synchrotron radiation as a function of distance in a prebunched FEL. The quadratic increase of power with distance is evident, as well as the two Doppler upshifted and downshifted frequencies. The narrowing of the spectral bandwidth as the number of wiggler periods increases is also clear.

In the case of well-separated Doppler upshifted and downshifted interaction frequencies, the denominator of the modified sinc term tends to zero linearly. By contrast, in the case of grazing, the denominator has a double pole, as $\omega^+ = \omega^-$, yielding a quadratic behavior and a maximum interaction bandwidth. Finally, the spectral phase information, including frequency-chirping, is the described by the argument of the complex exponential. Figures 9.3 and 9.4 show the normalized power spectral density, which is proportional to $|\tilde{E}_{1q}(\omega, z)|^2$, as a function of both position in the wiggler and frequency, for two different cases, corresponding to well-separated Doppler upshifted and downshifted frequencies, and near grazing, where the interaction bandwidth is seen to be maximal. In both cases, the power increases quadratically in the wiggler, a result which is typical of coherent radiation processes; at the same time, the Fourier bandwidth narrows with increasing $z$, as it scales like $N_w^{-1}$.

In the reference frame where the axial velocity is zero, the bunch rotates about the wiggler axis and radiates along the $z$-axis in a $\sin^2$ pattern similar to that of a dipole. In this frame, the grazing condition corresponds to cutoff, as can be seen by Lorentz-transforming the dispersion relation for the beam and electromagnetic modes. The electromagnetic power radiated by the accelerated charge is then confined around the electron bunch, and the resulting single compressed output pulse is chirped over the full interaction bandwidth. This is schematically illustrated in Figure 9.5, where the radiation process is seen from the lab frame and from the bunch frame.

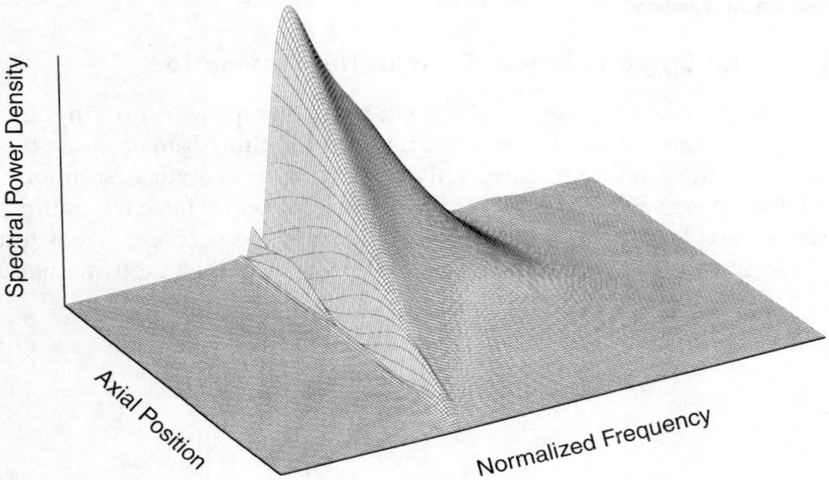


**FIGURE 9.4**
Identical to Figure 9.3, but the parameters are adjusted to yield a grazing interaction: only one spectral peak is visible; the bandwidth is extremely wide.

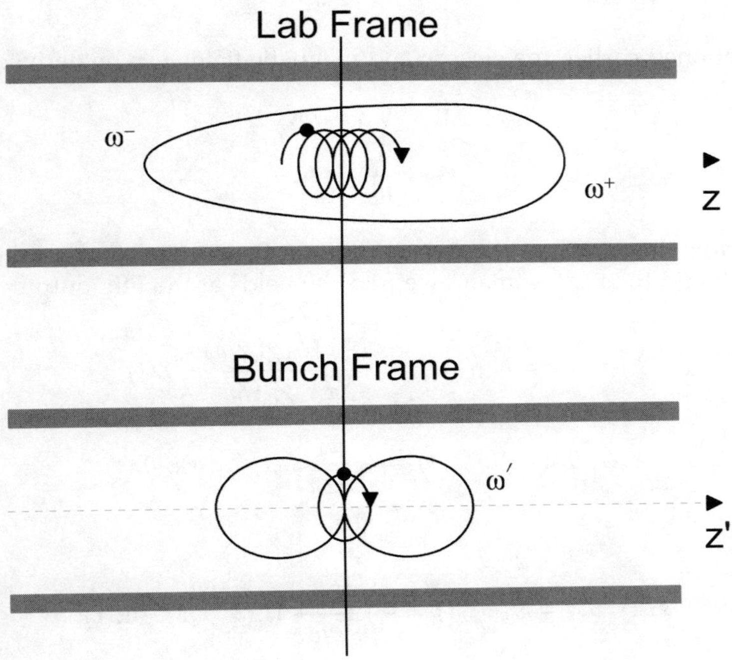


**FIGURE 9.5**
The coherent synchrotron radiation process, as seen in the bunch and lab frames.

## 9.4 Instantaneous Power Flow in the Waveguide

We now proceed to derive the instantaneous power flowing in the waveguide. The Fourier transform back into the time domain must be calculated on the computer, since only approximate analytical solutions are available; these will be discussed in the next section. Here, we outline the various steps leading to a numerical solution of the problem. We begin by normalizing the relevant parameters with respect to the cutoff frequency:

$$w = \frac{\omega}{\omega_c}, \quad x = \frac{ck_w}{\omega_c}, \quad \xi = \frac{N_w \ell_w \omega_c}{2c}, \quad \delta = \left(\frac{\omega_c \Delta z}{2c}\right)^2, \tag{9.110}$$

and

$$w^{\pm} = \frac{\omega^{\pm}}{\omega_c} = \gamma_{\|}^2 \beta_{\|} x\left[1 \pm \sqrt{1 - \left(\frac{1}{\gamma_{\|}\beta_{\|}x}\right)^2}\right]. \tag{9.111}$$

In addition, we introduce the electric field

$$E_0 = i\sqrt{\frac{2}{\pi^3}} \frac{q\beta_w \beta_{\|}^2 \gamma_{\|}^2}{\varepsilon_0 a^2} g_{1q}. \tag{9.112}$$

As mentioned earlier, the electric and magnetic fields can be derived from

$$\begin{aligned} \mathbf{E} &= \nabla\phi - \partial_t \mathbf{A}, \\ \mathbf{H} &= \frac{1}{\mu_0}\nabla \times \mathbf{A}. \end{aligned} \tag{9.113}$$

Performing the space and time derivatives in Fourier space, and transforming back into the time domain, we find the fields at the FEL output:

$$\begin{aligned} E_r(r,\theta,z = N_w\ell_w,t) &= \frac{iE_0}{\sqrt{2\pi}} \frac{J_{1q}(\chi'_{1q}r/a)}{\chi'_{1q}r/a} e^{i\theta}\Sigma(t), \\ E_\theta(r,\theta,z = N_w\ell_w,t) &= -\frac{E_0}{\sqrt{2\pi}} J'_{1q}(\chi'_{1q}r/a)e^{i\theta}\Sigma(t), \end{aligned} \tag{9.114}$$

and

$$\begin{aligned} H_r(r,\theta,z = N_w\ell_w,t) &= -\frac{iE_0}{\sqrt{8\pi}\mu_0 c} J'_{1q}(\chi'_{1q}r/a)e^{i\theta}\,\Xi(t), \\ H_\theta(r,\theta,z = N_w\ell_w,t) &= \frac{E_0}{\sqrt{8\pi}\mu_0 c} \frac{J_{1q}(\chi'_{1q}r/a)}{\chi'_{1q}r/a} e^{i\theta}\,\Xi(t). \end{aligned} \tag{9.115}$$

Here, we have introduced the Fourier integral

$$\Sigma(t) = \int_{-\infty}^{+\infty} \exp\left[-\delta\left(\frac{w}{\beta_\parallel}\right)^2\right] \frac{w \sin[a(w)\xi]}{(w - w^+)(w - w^-)} \exp\{ib(w)\xi - w\tau\} dw, \quad (9.116)$$

and

$$\Xi(t) = \int_{-\infty}^{+\infty} \exp\left[-\delta\left(\frac{w}{\beta_\parallel}\right)^2\right] \frac{\{a(w)\cos[a(w)\xi] + ib(w)\sin[a(w)\xi]\}}{(w - w^+)(w - w^-)} \times \exp\{ib(w)\xi - w\tau\} dw; \quad (9.117)$$

in addition, the following parameters are used in Equations 9.116 and 9.117:

$$a(w) = \sqrt{w^2 - 1} - \frac{w}{\beta_\parallel} + x,$$
$$b(w) = \sqrt{w^2 - 1} + \frac{w}{\beta_\parallel} - x, \quad (9.118)$$

and $\tau = \omega_c t$ is the normalized time.

We now use Poynting's theorem to calculate the instantaneous power flowing through the waveguide:

$$P(t) = \int_0^{2\pi} d\theta \int_0^a \{[\mathbf{E}(t) \times \mathbf{H}^*(t)] \cdot \hat{z}\} r\, dr. \quad (9.119)$$

Performing the integral over the waveguide cross-section, we finally find

$$P(t) = \frac{q^2 \beta_\perp^2 \beta_\parallel^2 \gamma_\parallel^4}{a^2} \frac{c}{\varepsilon_0} \left\{ \frac{J_2\left(\chi'_{1q} \frac{r_\perp}{a}\right)}{J_1^2 \chi'_{1q}\left[1 - \left(\frac{1}{\chi'_{1q}}\right)^2\right]} \right\}^2 |\Sigma(t)||\Xi(t)|. \quad (9.120)$$

In the simple case where the beam fills the waveguide, when $r_\perp = a$, this reduces to

$$P(t) = \frac{q^2 \beta_\perp^2 \beta_\parallel^2 \gamma_\parallel^4}{a^2} \frac{c}{\varepsilon_0} \frac{1}{\chi'^2_{1q}} |\Sigma(t)||\Xi(t)|. \quad (9.121)$$

## 9.5 Time-Dependent Chirped Wavepacket

In this section, we consider the temporal characteristics of the wavepacket generated by the coherent synchrotron radiation process in a waveguide FEL. While approximate analytical expressions for the temporal behavior of the pulse can be obtained in a straightforward manner in the case where

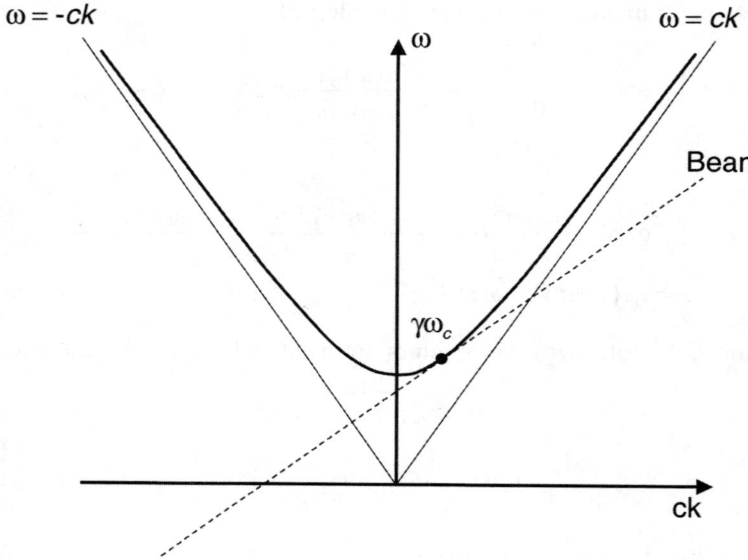


**FIGURE 9.6**
FEL grazing interaction.

slippage is the dominant broadening process, as shown by Jerby and Gover, a derivation in the case of grazing was first derived by Hartemann and coauthors. The focus of this section is a detailed derivation of such an analytical expression for the time-dependent chirped wavepacket generated at grazing. In addition, it will be shown that the duration of the pulses generated in the FEL structure at grazing can be extremely short, and it will be compared to the theoretical Fourier transform limit corresponding to the ultra-wideband interaction bandwidth.

The FEL dispersion relation at grazing is shown in Figure 9.6, for normalized units, defined as

$$x = w - 1 = \frac{\omega}{\omega^*} - 1, \quad w_c = \frac{\omega_c}{\omega^*} = \frac{1}{\gamma_\parallel}, \quad q_w = \frac{ck_w}{\omega^*} = \frac{1}{\gamma_\parallel^2 \beta_\parallel}, \tag{9.122}$$

in which case the dispersion of the waveguide mode is given by

$$q_1(w) = \sqrt{w^2 - \frac{1}{\gamma_\parallel^2}}, \tag{9.123}$$

while the beam mode disperses according to

$$q_2(w) = \frac{w}{\beta_\parallel} - q_w. \tag{9.124}$$

In these units, the following relations hold at grazing:

$$w^* = 1, \quad x^* = 0, \quad q_1(w^*) = q_2(w^*) = \beta_\parallel. \tag{9.125}$$

In Figure 9.4, the axial beam velocity $\beta_\parallel = 0.85$, and the other parameters are adjusted to satisfy the grazing condition.

The time-dependent electric field in the FEL interaction region has the general form

$$E_{1q}(z,t) = \frac{\sqrt{2}q\beta_w}{\pi^2 \varepsilon_0 a^2} g_{1q} \int_{-\infty}^{+\infty} \omega \exp\left[-\left(\frac{\omega \Delta z}{2\beta_\parallel c}\right)\right] \times \frac{\sin\left[(\Delta)k\frac{z}{2}\right]}{\Delta k(k_1+k_2)} \exp\left\{ i\left[(k_1+k_2)\frac{z}{2} - \omega t\right]\right\} d\omega, \tag{9.126}$$

where the poles of Equation 9.126 correspond to the resonant interaction frequencies.

To obtain the temporal characteristics of the radiation pulse(s), one must perform the Fourier transform in Equation 9.126. In general, it is not possible to derive an analytical expression for the corresponding time-dependent wavepacket. However, the Fourier integral can be simplified considerably by using a Taylor expansion of the dispersion relation around a given resonant frequency $\omega^*$:

$$\Delta k(\omega - \omega^*) \simeq (\omega - \omega^*)\left(\frac{1}{v_g} - \frac{1}{v_\parallel}\right) - \frac{1}{2}\left(\frac{\omega - \omega^*}{v_g}\right)^2 \frac{dv_g}{d\omega} + \frac{1}{6}\left(\frac{\omega - \omega^*}{v_g}\right)^3 \left[2\left(\frac{dv_g}{d\omega}\right)^2 - v_g \frac{d^2 v_g}{d\omega^2}\right] + \cdots, \tag{9.127}$$

where we have introduced the group velocity, $v_g(\omega^*)$, the group velocity dispersion, and higher-order dispersive effects.

Note that the linear term corresponds to slippage. For a given type of interaction, e.g., slippage dominated, grazing, zero-dispersive grazing, etc., corresponding to a given minimal order of the Taylor expansion, the wavenumber mismatch takes the general form

$$\begin{aligned} \Delta k(\omega - \omega^*) &\simeq a_n(\omega - \omega^*)^n, \\ a_1 &= \frac{1}{v_g} - \frac{1}{v_\parallel}, \\ a_2 &= -\frac{1}{2}\frac{v_g'}{v_g^2}, \\ a_3 &= -\frac{1}{6}\frac{v_g''}{v_g^2}, \ldots . \end{aligned} \tag{9.128}$$

In the case of interest for this section, the linear term is zero (no slippage), and the time-dependent electric field takes the form

$$E_{1q}(z,t) \simeq \frac{\sqrt{2}q\beta_w}{\pi^2\varepsilon_0 a^2} g_{1q} \frac{\omega^* \exp\left[-\left(\frac{\omega^*\Delta z}{2\beta_\| c}\right)^2\right]}{\left(\frac{v_g'}{v_g^2}\right)\left(\frac{\omega^*}{\beta_\| c} - k_w\right)} \exp\left\{i\left[\omega^*\left(\frac{z}{\beta_\| c} - t\right) - k_w z\right]\right\} F, \quad (9.129)$$

where $F$ is the Fourier integral

$$F = \int_{-\infty}^{+\infty} \frac{\sin \alpha x^2}{x^2} \exp[i(\delta x - \alpha x^2)]dx, \quad (9.130)$$

and where we have approximated the nonresonant terms by their value at the double pole $\omega^*$, and introduced

$$x = \omega - \omega^*, \quad \alpha(z) = \frac{v_g' z}{v_g^2 4}, \quad \text{and} \quad \delta(z,t) = \frac{z}{\beta_\| c} - t. \quad (9.131)$$

The parameter $\alpha$ governs the pulse duration and is directly relates to group velocity dispersion, while $\delta$ corresponds to the group velocity delay.

In the case of a waveguide FEL operating at grazing, the following relations hold:

$$\begin{aligned} v_g &= \frac{d\omega}{dk}(\omega = \omega^*) = c\beta_\|, \\ v_g' &= \frac{dv_g}{d\omega}(\omega = \omega^*) = \frac{c}{\omega^* \gamma_\|^2 \beta_\|}, \end{aligned} \quad (9.132)$$

and the pulse duration parameter is given by

$$\alpha(z) = \frac{z}{4\omega^* c \gamma_\|^2 \beta_\|^3}. \quad (9.133)$$

We now focus on the evaluation of the Fourier integral given in Equation 9.130. We first eliminate the antisymmetrical terms in the integrand, to obtain

$$F = \int_{-\infty}^{+\infty} \frac{\sin(\alpha x^2)}{x^2} e^{-i\alpha x^2} \cos(\delta x) dx. \quad (9.134)$$

Using simple trigonometry, the real part of $F$ is shown to take the form

$$\frac{1}{2}\int_{-\infty}^{+\infty} \sin(2\alpha x^2)\cos(\delta x)\frac{dx}{x^2}, \quad (9.135)$$

and can be directly evaluated in terms of the Fresnel sine and cosine integrals, $S$ and $C$. We use the following formula:

$$\int_0^{\infty} \sin(ax^2)\cos(bx)\frac{dx}{x^2} = \frac{b\pi}{2}\left[S\left(\frac{b}{2\sqrt{a}}\right) - C\left(\frac{b}{2\sqrt{a}}\right)\right] + \sqrt{a\pi}\sin\left(\frac{b^2}{4a} + \frac{\pi}{4}\right), \quad (9.136)$$

to obtain

$$\Re(F) = \frac{\pi}{2}\left(\frac{z}{v_g} - t\right)\left[S\left(\frac{z - v_g t}{\sqrt{2v'_g z}}\right) - C\left(\frac{z - v_g t}{\sqrt{2v'_g z}}\right)\right] + \frac{1}{\sqrt{2}}\frac{\sqrt{\pi v'_g z}}{v_g}\sin\left[\frac{(z - v_g t)^2}{2v'_g z} + \frac{\pi}{4}\right]. \quad (9.137)$$

The argument of the sine clearly shows the chirped nature of the pulse through the quadratic dependence on the group delay parameter. The pulse duration is governed by group velocity dispersion through $2v'_g z$, and is seen to increase with the square root of this quantity as the pulse propagates down the interaction region.

We now consider the imaginary part of $F$. The first step consists in integrating by parts and using simple trigonometry to obtain

$$\Im(F) = \left[\frac{1}{x}\sin^2(\alpha x^2)\cos(\delta x)\right]_{-\infty}^{+\infty} - 2\alpha\int_{-\infty}^{+\infty}\sin(2\alpha x^2)\cos(\delta x)dx$$
$$+ \delta\int_{-\infty}^{+\infty}\sin^2(\alpha x^2)\sin(\delta x)\frac{dx}{x}. \quad (9.138)$$

The first term vanishes, and the first remaining integral can readily be evaluated by using the following formula:

$$\int_0^{\infty}\sin(ax^2)\cos(bx)dx = \frac{1}{2}\sqrt{\frac{\pi}{a}}\cos\left(\frac{b^2}{4a} + \frac{\pi}{4}\right). \quad (9.139)$$

Using simple trigonometry, the second integral can be separated into two terms to yield

$$\frac{1}{2}\int_{-\infty}^{+\infty}[1 - \cos(2\alpha x^2)]\frac{\sin(\delta x)}{x}dx = \frac{\pi}{2}\text{sgn}(d) - \frac{1}{2}\int_{-\infty}^{+\infty}\cos(2\alpha x^2)\sin(\delta x)\frac{dx}{x}. \quad (9.140)$$

To evaluate the last remaining integral, we shall employ Cauchy's residue theorem. After a change of variable, and using symmetry considerations, we have

$$\frac{1}{2}\int_{-\infty}^{+\infty}\cos(2\alpha x^2)\sin(\delta x)\frac{dx}{x} = \text{sgn}(\delta)\Re\left[\frac{1}{i}\int_0^{\infty}\exp\left(i\frac{2\alpha}{\delta^2}x^2\right)\frac{e^{ix}}{x}dx\right]. \quad (9.141)$$

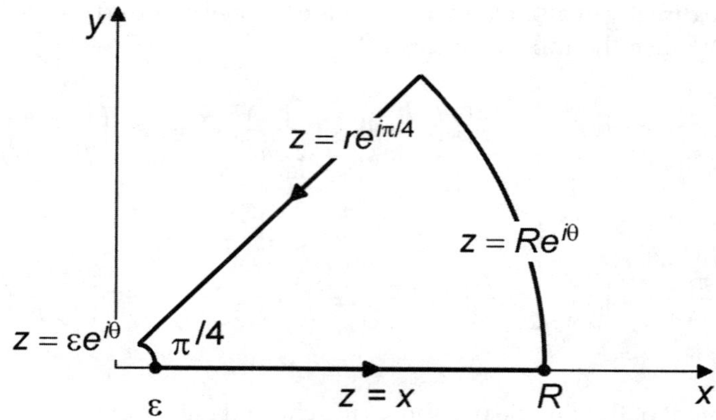


**FIGURE 9.7**
Contour used to apply Cauchy's residue theorem.

The contour $K$ used to apply Cauchy's theorem is shown in Figure 9.7. We have

$$\oint_K \exp\left(i\frac{2\alpha}{\delta^2}z^2\right)\frac{e^{iz}}{z}dz = 0, \tag{9.142}$$

since this complex function has no poles inside $K$. This results in the following identity:

$$\begin{aligned} 0 = &\int_\varepsilon^R \exp\left(i\frac{2\alpha}{\delta^2}x^2\right)\frac{e^{ix}}{x}dx + \int_0^{\pi/4}\exp\left(i\frac{2\alpha}{\delta^2}R^2e^{2i\theta}\right)\frac{e^{i\mathrm{Re}^{i\theta}}}{\mathrm{Re}^{i\theta}}i\mathrm{Re}^{i\theta}d\theta \\ &+ \int_R^\varepsilon \exp\left(i\frac{\alpha}{\delta^2}r^2e^{i\pi/2}\right)\frac{e^{ire^{i\pi/4}}}{re^{i\pi/4}}e^{i\pi/4}dr + \int_{\pi/4}^0 \exp\left(i\frac{2\alpha}{\delta^2}\varepsilon^2e^{2i\theta}\right)\frac{e^{i\varepsilon e^{i\theta}}}{\varepsilon e^{i\theta}}i\varepsilon e^{i\theta}d\theta. \end{aligned} \tag{9.143}$$

We now take the limit where $\varepsilon \to 0$ and $R \to \infty$. One has to pay special attention to the two circular segments on the contour $K$. Taking the limit inside these integrals, we first have

$$\lim_{\varepsilon\to 0}\int_{\pi/4}^0 \exp\left(i\frac{2\alpha}{\delta^2}\varepsilon^2e^{2i\theta}\right)\frac{e^{i\varepsilon e^{i\theta}}}{\varepsilon e^{i\theta}}i\varepsilon e^{i\theta}d\theta = -i\frac{\pi}{4}, \tag{9.144}$$

and we can show that

$$\left|\int_0^{\pi/4}\exp\left(i\frac{2\alpha}{\delta^2}R^2e^{2i\theta}\right)\frac{e^{i\mathrm{Re}^{i\theta}}}{\mathrm{Re}^{i\theta}}i\mathrm{Re}^{i\theta}d\theta\right| \le \int_0^{\pi/4}\exp\left\{-\theta\left[\frac{8(\alpha/\delta^2)R^2+2R}{\pi}\right]\right\}d\theta, \tag{9.145}$$

by considering the inequality $\sin x \geq 2x/\pi$, which holds on the interval $[0,\pi]$. The integral on the right-hand side of Equation 9.145 is easily seen to vanish as $R \to \infty$. We are then left with

$$\int_0^{\infty} \exp\left(i\frac{2\alpha}{\delta^2}x^2\right)\frac{e^{ix}}{x}dx = \int_0^{\infty} \exp\left(-\frac{2\alpha}{\delta^2}r^2 - \frac{\sqrt{2}}{2}r\right)\left[\cos\left(\frac{\sqrt{2}}{2}r\right) + i\sin\left(\frac{\sqrt{2}}{2}r\right)\right]\frac{dr}{r} + i\frac{\pi}{4}. \tag{9.146}$$

Using Equation 9.141, we now have

$$\frac{1}{2}\int_{-\infty}^{+\infty} \cos(2\alpha x^2)\sin(\delta x)\frac{dx}{x} = \operatorname{sgn}(\delta)\left[\int_0^{\infty} \exp\left(-\frac{2\alpha}{\delta^2}r^2 - \frac{\sqrt{2}}{2}r\right)\sin\left(\frac{\sqrt{2}}{2}r\right)\frac{dr}{r} + \frac{\pi}{4}\right], \tag{9.147}$$

where the integral on the right-hand side can now be evaluated in terms of parabolic cylinder functions, $D_\mu$, by using the formula

$$\int_0^{\infty} x^{-1}e^{-ax-bx^2}\sin(ax)dx = -\frac{i}{2}\left\{\exp\left(-i\frac{a^2}{4b}\right)D_0\left[\frac{a(1-i)}{\sqrt{2b}}\right] - \exp\left(i\frac{a^2}{4b}\right)D_0\left[\frac{a(1+i)}{\sqrt{2b}}\right]\right\}. \tag{9.148}$$

Combining these results together yields the sought-after expression for the imaginary part of $F$:

$$\begin{aligned}\Im(F) = -\sqrt{2\alpha\pi}\cos\left(\frac{\delta^2}{8\alpha}+\frac{\pi}{4}\right) - |\delta|\left[\left[\frac{i}{2}\left\{\exp\left(i\frac{\delta^2}{16\alpha}\right)D_0\left[\frac{\delta}{\sqrt{8\alpha}}(1+i)\right]\right.\right.\right.\\ \left.\left.\left. - \exp\left(-i\frac{\delta^2}{16\alpha}\right)D_0\left[\frac{\delta}{\sqrt{8\alpha}}(1-i)\right]\right\} - \frac{\pi}{4}\right]\right].\end{aligned} \tag{9.149}$$

We have thus derived a complete analytical expression for the Fourier integral $F$ that represents the evolution of the time-dependent chirped wave-packet propagating along the FEL interaction region.

Considerably simpler expression can be obtained by approximating the generalized sinc function by the superposition of two Gaussians of appropriate width and spacing. We then have

$$\sin\left(\frac{\alpha x^2}{x^2}\right) \simeq f(x) = \frac{\alpha}{2e^{-\chi^2}}\exp\left[-\alpha\left(x \pm \frac{\chi}{\sqrt{\alpha}}\right)^2\right], \tag{9.150}$$

where we have adjusted the parameters $\chi$ to 0.73, so that $f(0) = f(\pm\chi/\sqrt{\alpha})$.

The corresponding Fourier transform can then easily be performed to obtain

$$F \simeq \frac{1}{2e^{-\chi^2}}\sqrt{\frac{\alpha\pi}{\sqrt{2}}}\exp\left[-\frac{\left(\frac{\delta}{2}\pm\chi\sqrt{\alpha}\right)^2}{2\alpha}\right]\exp\left[-i\left(\frac{-\frac{\delta^2}{4}\pm\delta\chi\sqrt{\alpha}+\chi^2\alpha}{2\alpha}+\frac{\pi}{8}\right)\right]. \quad (9.151)$$

Taking the sum of the squares of the real and imaginary parts, the pulse envelope can be expressed as

$$|F|^2 \simeq \frac{\alpha\pi e^{\chi^2}}{2\sqrt{2}}\left[\cosh\left(\frac{\delta\chi}{\sqrt{\alpha}}\right)+\cos\left(\frac{\delta\chi}{\sqrt{\alpha}}\right)\right]\exp\left(-\frac{\delta^2}{4\alpha}\right). \quad (9.152)$$

The corresponding full width at half-maximum (FWHM) of the pulse intensity is then proportional to

$$\Delta t \propto \sqrt{\alpha(z)} = \frac{1}{v_g}\sqrt{v_g' z} = \frac{1}{\gamma_\parallel}\sqrt{\frac{z}{\omega^* c\beta_\parallel^3}} = \sqrt{\frac{z}{\omega^* c\gamma_\parallel^2\beta_\parallel^3}}. \quad (9.153)$$

The square root of the group velocity dispersion parameter, $\alpha(z)$, clearly determines the pulse duration in the FEL, and increases as the pulse propagates down the interaction region, while the quadratic dependence of the imaginary argument of the exponential on the group delay parameter, $\delta(z, t) = zv_g^{-1} - t$, in Equation 9.151 corresponds to the chirp. This is illustrated in Figure 9.8, where the temporal structure of a wavepacket generated at grazing in a short (10-period) wiggler is shown. The chirped nature of the pulse appears clearly, and the comparison with a direct numerical simulation shows good agreement. In particular, the group delay is well modeled by the Taylor expansion, as well as the rate of chirping.

At this point, it is important to note that the pulses generated by the FEL are actually very close to their Fourier transform limit, which is obtained for monochromatic pulses with no higher-order chirp. This is due to the intimate connection between the interaction bandwidth and the rate of chirping in the FEL structure, as evidenced by Equation 9.130, for example. Here, the spectral envelope is governed by the modified sinc function, $\sin(\alpha x^2)/x^2$, while the chirp is described by the quadratic term in the argument of the complex exponential, $\exp[i(\delta x - \alpha x^2)]$. As the spectral width decreases when the wavepacket radiated by the electrons propagates through the interaction region because GVD disperses the pulse, the rate of chirping increases correspondingly, yielding a strong correlation between the spectrum envelope and its phase. The resulting time-dependent wavepackets are extremely short, close to the Fourier transform limit. In addition, we note that the spectral width of the wavepacket also decreases as the pulse propagates down the FEL interaction region because the wiggler wavenumber becomes better defined as the electrons perform an increasing number of oscillations in the wiggler.

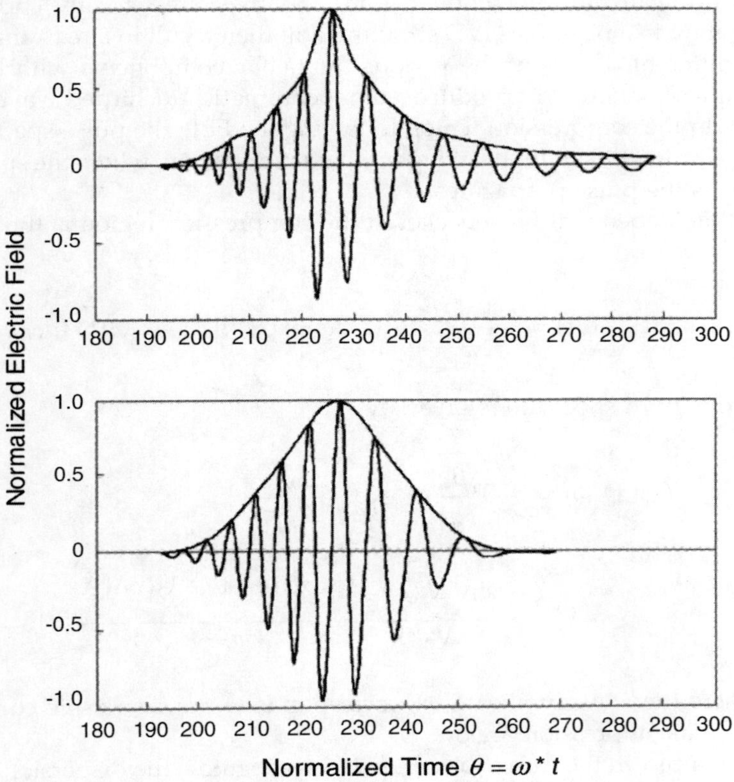


**FIGURE 9.8**
Chirped coherent synchrotron radiation wavepackets: computer calculation (top), and analytical model (bottom).

## 9.6 Propagation in a Negative GVD Structure

We now briefly study the propagation of chirped pulses in negative GVD guiding structure, such as a corrugated waveguide, as studied in Section 5.6.8. Starting from the chirped pulse input spectrum, the pulse shape at the output of the negative GVD section is obtained by Fourier transforming back after taking into account the different phase shifts accumulated by the various frequency components of the pulse. In the case of wavepackets that have no specific correlation between the rate of chirping and the spectral bandwidth, the pulse duration reaches its Fourier transform limit when the negative chirp accumulated through the compression region exactly compensates the initial chirp. In the case of the chirped pulses generated by the FEL, both the spectral width and chirp at $z = z_0 = N_w \ell_w$ (wiggler exit) are due to GVD in the interaction region, and they are strongly correlated. Propagation in a negative GVD structure will then result in a rearrangement of the different phases of the various frequency components with little or no compression, but will produce a monochromatic, unchirped wavepacket instead. In the compression region following the FEL, the pulse spectrum is "frozen," in contrast with the wiggler interaction region, where the spectrum evolves as the pulse propagates.

The time-dependent wavepacket in the compression region is thus given by

$$E_{1q}(z = z_0 + \ell, t) = \frac{1}{\sqrt{2\pi}} \int_{-\infty}^{+\infty} \tilde{E}_{1q}(z_0, \omega) \exp\{-i[\omega t - k(\omega)\ell]\} d\omega, \quad (9.154)$$

where the input spectrum is given by

$$\tilde{E}_{1q}(z_0, \omega) = \frac{\sqrt{2} q \beta_w}{\pi^{3/2} \varepsilon_0 a^2} g_{1q} \exp\left[-\left(\frac{\omega \Delta z}{2\beta_{\|} c}\right)^2\right] \times \frac{\sin\left[\Delta k(\omega)\frac{z_0}{2}\right]}{\Delta k(\omega)} \frac{\exp\left\{i[k_1(\omega) + k_2(\omega)]\frac{z_0}{2}\right\}}{k_1(\omega) + k_2(\omega)}, \quad (9.155)$$

and where the wavenumber, $k(\omega)$, corresponds to the dispersive characteristics of the compression region.

For example, in the case of a corrugated waveguide, the dispersion can be approximated by

$$\frac{\omega}{\omega^*} = w_0 + \Delta w\left[1 - \cos\left(\pi \frac{q}{q_0}\right)\right], \quad q(w) = \frac{ck(\omega)}{\omega^*}, \quad (9.156)$$

where we have used normalized units, and where the parameters $\Delta w$ and $q_0$ correspond to the bandpass (corrugation depth) and period of the waveguide, respectively.

This type of structure can be analyzed using Floquet's theorem, as discussed in Section 5.6.8. The Fourier integral in Equation 9.154 can be either calculated numerically or evaluated by means of a quadratic Taylor expansion. We then have

$$q(w) \simeq q(w^*) + (w - w^*)\frac{1}{\beta_g} - \frac{1}{2}(w - w^*)^2 \frac{\beta_g'}{\beta_g^2}, \quad (9.157)$$

where the group velocity and GVD are given by

$$\beta_g = \frac{\pi}{q_0 \Delta w} \sin\left(\pi \frac{q^*}{q_0}\right), \quad (9.158)$$

and

$$\beta'_g = \frac{d\beta_g}{dw} = \frac{1}{\beta_g}\frac{d\beta_g}{dq} = \frac{\pi}{q_0}\Delta w \cot\left(\pi\frac{q^*}{q_0}\right), \tag{9.159}$$

and where $w^* = 1$ is the FEL resonant frequency at grazing; in this case, $q^* = q(w^*)$.

Since we are using a quadratic Taylor expansion for the wavenumber, the mathematical formalism described in the previous section can be directly applied here. In particular, one can use the expressions given in Equations 9.137 and 9.149, and essentially replace the group delay and GVD parameters by

$$\delta(z,t) = \frac{z_0}{\beta_\| c} + \frac{\ell}{\beta_g c} - t = \delta_0 + \frac{\ell}{\beta_g c}, \tag{9.160}$$

and

$$\alpha(\ell) = \frac{z_0}{4\omega^* c \gamma_\|^2 \beta_\|^3} + \frac{1}{2}\frac{v'_g}{v_g^2}\ell = \alpha_0\left(1 - \frac{\ell}{\ell^*}\right), \tag{9.161}$$

respectively. Equation 9.161 shows that the chirp vanishes after a compression distance

$$\ell^* = -2\alpha_0 \frac{v_g^2}{v'_g} = -z_0 \frac{\beta_g^2}{2\gamma_\|^2 \beta_\|^3 \beta'_g}, \tag{9.162}$$

such that $\alpha(\ell^*) = 0$. At this point, the full spectrum of the pulse has the proper phase coherence and the pulse duration reaches its Fourier transform limit. In addition, Equation 9.161 shows that if the GVD in the FEL interaction region is comparable to the negative GVD in the compression region, the propagation length necessary to obtain the shortest pulse duration will be of the same order as the wiggler length. Finally, using Equation 9.159, we can recast the expression for the compression distance in terms of the corrugated waveguide characteristics:

$$\frac{\ell^*}{z_0} = -\frac{1}{2\gamma_\|^2 \beta_\|^3}\frac{\pi}{q_0}\Delta w \frac{\sin^3[\pi(q^*/q_0)]}{\cos[\pi(q^*/q_0)]}. \tag{9.163}$$

## 9.7 Relativistic Eulerian Fluid Perturbation Theory

A large class of beam–wave interaction problems involve electromagnetic energies that are small compared to the particles' kinetic energy, and perturbation theory is appropriate to describe such linear beam–wave interactions. This category of problem will be the focus of our attention in the present section. Different formal mathematical descriptions of this type of interaction are possible, such as the Maxwell-Vlasov kinetic theory, or the Maxwell-Euler fluid model. In this section, we consider the latter theoretical framework, which involves the manipulation of fields for both the electromagnetic and dynamical quantities, and of operators, such as the electromagnetic wave propagator (d'Alembertian operator), or the fluid convective derivation, providing a compact and elegant mathematical framework to study these interactions.

One of the main objects of our discussion is to show that, starting from the standard set of eight equations in the four-vector potential, $A_\mu$, and the four-vector current density, $j_\mu$, we can obtain a canonical system of four coupled partial differential equations (PDEs) describing the evolution of the electromagnetic perturbation, $\delta A_\mu$, by linearizing the interaction equations. The compact set of PDEs derived in this manner involves the perturbed electromagnetic four-potential, and the equilibrium, or unperturbed, fluid field components. Within this framework, different specific problems are characterized by different fluid equilibria and boundary conditions for $\delta A_\mu$ and its derivatives. The initial set of eight equations consist of the four Maxwell equations with source describing the evolution of the four-potential, the Lorentz gauge condition, which is equivalent here to the conservation of charge, or to the continuity equation, and the three components of the fluid equation of motion.

At this level, two main formal approaches can be used to solve this linear system of PDEs. On the one hand, one can expand $\delta A_\mu$ into known eigenmodes satisfying the appropriate boundary conditions, and study the excitation of these modes through the coupled PDEs. The other approach consists in solving directly these equations, then using the boundary conditions to determine the actual eigenvalues and eigenfunctions of the problem.

This section is organized as follows: in subsection 9.7.1, we linearize the standard system of eight interaction equations in $A_\mu$ and $j_\mu$, and reduce it to a canonical system in the four-potential perturbation, $\delta A_\mu$. Subsection 9.7.2 is focused on the study of space–charge waves supported by an electron beam confined by an axial guide magnetic field and is intended as an example of the use of the general formalism and equations derived in subsection 9.7.1. It will also expand our understanding of plasma waves, which were introduced in Chapters 2 and 5.

### 9.7.1 Covariant Linearized Fluid Theory

The purpose of this subsection is to give a detailed account of the formal mathematical description of the interaction between a relativistic electron beam and electromagnetic fields, within the frame of classical electrodynamics. A very large number of methods have been described in the literature, and there is, sometimes, some confusion about which equations and which variables should be used. For example, it is well known that the gauge condition, the conservation of charge, and the continuity equation are equivalent. Here, our objective is to reduce the linearized interaction equations to the canonical system of four equations in the four-potential perturbation, $\delta A_\mu(x_\nu)$.

We first briefly review the equations relevant to the problem. The interaction of charged particles with electromagnetic fields can be described, in the classical limit, by two sets of equations. On the one hand, there are Maxwell's two group of equations, governing the fields,

$$\begin{aligned} \nabla \times \mathbf{E} + \partial_t \mathbf{B} &= \mathbf{0}, \\ \nabla \cdot \mathbf{B} &= 0, \end{aligned} \tag{9.164}$$

and the group with sources,

$$\begin{aligned} \nabla \cdot \mathbf{E} &= \frac{1}{\varepsilon_0}\rho, \\ \nabla \times \mathbf{B} - \frac{1}{c^2}\partial_t \mathbf{E} &= \mu_0 \mathbf{j}. \end{aligned} \tag{9.165}$$

On the other hand, there are the equations governing the particle dynamics, which are given by the expression of the Lorentz force,

$$d_t \mathbf{p} = -e(\mathbf{E} + \mathbf{v} \times \mathbf{B}), \tag{9.166}$$

and the continuity, or charge conservation, equation:

$$\partial_t \rho + \nabla \cdot \mathbf{j} = 0. \tag{9.167}$$

Here, $j_\mu = (c\rho, \mathbf{j}) = -en(c, \mathbf{v})$ is the four-vector current density, with $n$ the particle density, and $\mathbf{v} = c\beta$ their velocities.

The particles' momentum is given by $\mathbf{p} = \gamma m_0 \mathbf{v}$, and their normalized energy by $\gamma = 1/\sqrt{1-\beta^2}$.

At this point, it is important to remember that Maxwell's first group of equations suggests the introduction of the four-vector potential, $A_\mu = (c^{-1}\phi, \mathbf{A})$, defined such that

$$\begin{aligned} \mathbf{E} &= -\nabla\phi - \partial_t \mathbf{A}, \\ \mathbf{B} &= \nabla \times \mathbf{A}. \end{aligned} \tag{9.168}$$

As a result, Equation 9.164 is automatically satisfied. If, in addition, we impose that the four-vector potential satisfies the Lorentz gauge condition,

$$\frac{1}{c^2}\partial_t \phi + \nabla \cdot \mathbf{j} = 0, \tag{9.169}$$

we see that the second group of equations is equivalent to

$$\begin{aligned}\left[\nabla^2 - \frac{1}{c^2}\partial_t^2\right]\phi + \frac{1}{\varepsilon_0}\rho &= 0,\\ \left[\nabla^2 - \frac{1}{c^2}\partial_t^2\right]\mathbf{A} + \mu_0\mathbf{j} &= \mathbf{0}.\end{aligned} \tag{9.170}$$

It should also be noted that the gauge condition is equivalent to the continuity equation, within the formalism developed here.

The equation of momentum transfer 9.166 implicitly satisfies energy conservation, as can be seen by taking the dot product of Equation 9.166 by $\mathbf{p}$, to obtain

$$d_t\gamma = -\frac{e\mathbf{E}\cdot\mathbf{v}}{m_0c^2}. \tag{9.171}$$

Finally, using the definitions, Equation 9.166 can be transformed to read, within the framework of a relativistic fluid model,

$$(\partial_t + \mathbf{v}\cdot\nabla)\mathbf{v} = -\frac{e}{m_0}\sqrt{1-\frac{v^2}{c^2}}\left[-\nabla\phi - \partial_t\mathbf{A} + \mathbf{v}\times(\nabla\times\mathbf{A}) + \frac{\mathbf{v}}{c^2}(\nabla\phi + \partial_t\mathbf{A})\cdot\mathbf{v}\right], \tag{9.172}$$

where we have used Equation 9.171, and the fact that

$$d_t\mathbf{p} = m_0d_t(\gamma\mathbf{v}) = \gamma m_0d_t\mathbf{v} + m_0\mathbf{v}d_t\gamma = \gamma m_0d_t\mathbf{v} - \mathbf{v}\left(\frac{e\mathbf{E}\cdot\mathbf{v}}{c^2}\right). \tag{9.173}$$

We thus obtain a closed system of eight equations, with eight unknowns, $A_\mu$, $n$, and $\mathbf{v}$:

$$\begin{aligned}\left[\nabla^2 - \frac{1}{c^2}\partial_t^2\right]\phi &= \frac{1}{\varepsilon_0}en,\\ \left[\nabla^2 - \frac{1}{c^2}\partial_t^2\right]\mathbf{A} &= \mu_0en\mathbf{v},\\ \frac{1}{c^2}\partial_t\phi + \nabla\cdot\mathbf{A} &= 0,\end{aligned} \tag{9.174}$$

together with Equation 9.172.

We now focus on the linear analysis of the beam–field interaction. Any fluid field component, $f(x_\nu)$, is written as $f = f_0 + \delta f$. The quantity $f_0$ refers to the beam self-consistent equilibrium in the external fields, while $\delta f$ corresponds to the electrodynamic perturbation. We assume that for all fluid field components, we have $|\delta f| \ll |f_0|$. We can then linearize the equations presented above, with the result that

$$\left[\nabla^2 - \frac{1}{c^2}\partial_t^2\right]\delta\phi = \frac{1}{\varepsilon_0}e\,\delta n,$$

$$\left[\nabla^2 - \frac{1}{c^2}\partial_t^2\right]\delta\mathbf{A} = \mu_0 e(n_0\delta\mathbf{v} + \mathbf{v}_0\delta_n), \tag{9.175}$$

$$\frac{1}{c^2}\partial_t\delta\phi + \nabla\cdot\delta\mathbf{A} = 0,$$

and

$$\begin{aligned}(\partial_t + \mathbf{v}_0\cdot\nabla)\delta\mathbf{v} + (\delta\mathbf{v}\cdot\nabla)\mathbf{v}_0 \\ = -\frac{e}{\gamma_0 m_0}\Big\{-\nabla\delta\phi - \partial_t\delta\mathbf{A} + \delta\mathbf{v}\times\mathbf{B}_0 + \mathbf{v}_0\times(\nabla\times\delta\mathbf{A}) \\ -\frac{\mathbf{v}_0}{c^2}[\mathbf{E}_0\cdot\delta\mathbf{v} - (\nabla\delta\phi + \partial_t\delta\mathbf{A})\cdot\mathbf{v}_0] - \frac{\delta\mathbf{v}}{c^2}(\mathbf{E}_0\cdot\mathbf{v}_0) \\ -\frac{\gamma_0^2}{c^2}(\mathbf{v}_0\cdot\delta\mathbf{v})\left[\mathbf{E}_0 + \mathbf{v}_0\times\mathbf{B}_0 - \frac{\mathbf{v}_0}{c^2}(\mathbf{E}_0\cdot\mathbf{v}_0)\right]\Big\}.\end{aligned} \tag{9.176}$$

Here, the equilibrium electric field and magnetic induction are $\mathbf{E}_0(x_v)$ and $\mathbf{B}_0(x_v)$, respectively.

We shall now reduce this system by considering

$$n_0\delta\mathbf{v} = \frac{1}{\mu_0 e}\left(\Box\,\delta\mathbf{A} - \beta_0\Box\frac{\delta\phi}{c}\right), \tag{9.177}$$

where

$$\Box \equiv \partial_\mu\partial^\mu \equiv \Delta - \frac{1}{c^2}\partial_t^2, \tag{9.178}$$

is the d'Alembertian operator, or electromagnetic wave propagator. Equation 9.177 is obtained as follows. We have

$$\frac{1}{\mu_0 e}\Box\,\delta\mathbf{A} = n_0\delta\mathbf{v} + \mathbf{v}_0\delta n, \tag{9.179}$$

and

$$\mathbf{v}_0\frac{e}{\varepsilon_0}\Box\,\delta\phi = \mathbf{v}_0\delta n; \tag{9.180}$$

subtracting Equation 9.180 from Equation 9.179 and using the fact that $\varepsilon_0\mu_0 c^2 = 1$, yields the sought-after result. On the other hand, we have

$$(\partial_t + \mathbf{v}_0 \cdot \nabla)(n_0 \delta\mathbf{v}) = n_0(\partial_t + \mathbf{v}_0 \cdot \nabla)\delta\mathbf{v} + \delta\mathbf{v}(\partial_t + \mathbf{v}_0 \cdot \nabla)n_0, \quad (9.181)$$

and, after Equation 9.177,

$$(\partial_t + \mathbf{v}_0 \cdot \nabla)(n_0 \delta\mathbf{v}) = \frac{1}{\mu_0 e}(\partial_t + \mathbf{v}_0 \cdot \nabla)\left(\Box\, \delta\mathbf{A} - \beta_0 \Box \frac{\delta\phi}{c}\right). \quad (9.182)$$

The first term on the right-hand side of Equation 9.181 is given by Equation 9.176:

$$\begin{aligned} & n_0(\partial_t + \mathbf{v}_0 \cdot \nabla)\delta\mathbf{v} + n_0(\delta\mathbf{v} \cdot \nabla)\mathbf{v}_0 \\ & = -n_0 \frac{e}{\gamma_0 m_0}\left\{ -\nabla\delta\phi - \partial_t \delta\mathbf{A} + \delta\mathbf{v} \times \mathbf{B}_0 + \mathbf{v}_0 \times (\Delta \times \delta\mathbf{A}) \right. \\ & \quad - \frac{\mathbf{v}_0}{c^2}[\mathbf{E}_0 \cdot \delta\mathbf{v} - (\nabla\delta\phi + \partial_t \delta\mathbf{A}) \cdot \mathbf{v}_0] - \frac{\delta\mathbf{v}}{c^2}(\mathbf{E}_0 \cdot \mathbf{v}_0) \\ & \quad \left. - \frac{\gamma_0^2}{c^2}(\mathbf{v}_0 \cdot \delta\mathbf{v})\left[\mathbf{E}_0 + \mathbf{v}_0 \times \mathbf{B}_0 - \frac{\mathbf{v}_0}{c^2}(\mathbf{E}_0 \cdot \mathbf{v}_0)\right] \right\}, \end{aligned} \quad (9.183)$$

while the second term can be derived from the equilibrium continuity equation,

$$\begin{aligned} & \partial_t n_0 + \nabla \cdot (n_0 \mathbf{v}_0) = 0 = \partial_t n_0 + n_0(\nabla \cdot \mathbf{v}_0) + \mathbf{v}_0 \cdot \nabla n_0 \\ & \Rightarrow (\partial_t + \mathbf{v}_0 \cdot \nabla)n_0 = -n_0(\nabla \cdot \mathbf{v}_0). \end{aligned} \quad (9.184)$$

We thus have

$$(\partial_t + \mathbf{v}_0 \cdot \nabla + \nabla \cdot \mathbf{v}_0)\left(\Box\, \delta\mathbf{A} - \beta_0 \Box \frac{\delta\phi}{c}\right) = \mu_0 e n_0(\partial_t + \mathbf{v}_0 \cdot \nabla)\delta\mathbf{v}. \quad (9.185)$$

We now use Equation 9.183 to obtain

$$\begin{aligned} & (\partial_t + \mathbf{v}_0 \cdot \nabla + \nabla \cdot \mathbf{v}_0)\left(\Box\, \delta\mathbf{A} - \beta_0 \Box \frac{\delta\phi}{c}\right) + \left[\left(\Box\, \delta\mathbf{A} - \beta_0 \Box \frac{\delta\phi}{c}\right) \cdot \nabla\right]\mathbf{v}_0 \\ & = -\mu_0 e n_0 \left\{ -\nabla\delta\phi - \partial_t \delta\mathbf{A} + \delta\mathbf{v} \times \mathbf{B}_0 + \mathbf{v}_0 \times (\nabla \times \delta\mathbf{A}) \right. \\ & \quad - \frac{\mathbf{v}_0}{c^2}[\mathbf{E}_0 \cdot \delta\mathbf{v} - (\nabla\delta\phi - \partial_t \delta\mathbf{A}) \cdot \mathbf{v}_0] - \frac{\delta\mathbf{v}}{c^2}(\mathbf{E}_0 \cdot \mathbf{v}_0) \\ & \quad \left. - \frac{\gamma_0^2}{c^2}(\mathbf{v}_0 \cdot \delta\mathbf{v})\left[\mathbf{E}_0 + \mathbf{v}_0 \times \mathbf{B}_0 - \frac{\mathbf{v}_0}{c^2}(\mathbf{E}_0 \cdot \mathbf{v}_0)\right] \right\}. \end{aligned} \quad (9.186)$$

At this point, we define the following parameters:

$$\Omega_0(x_\nu) = \frac{e\mathbf{B}_0(x_\nu)}{\gamma_0(x_\nu)m_0}, \quad \omega_p^2(x_\nu) = \frac{n_0(x_\nu)e^2}{\gamma_0(x_\nu)m_0\varepsilon_0},$$
$$\beta_0(x_\nu) = \frac{\mathbf{v}_0(x_\nu)}{c}, \quad \Lambda(x_\nu) = \frac{e\mathbf{E}_0(x_\nu)}{\gamma_0(x_\nu)m_0c}, \tag{9.187}$$

which are, respectively, the vectorial relativistic cyclotron frequency in the equilibrium magnetic induction, the relativistic beam plasma frequency, the normalized fluid equilibrium velocity field, and the normalized equilibrium electric field, governing the energy time scale. The formalism described here includes the most general case, where the dynamical quantities describing the fluid equilibrium state are functions of both space and time.

Upon replacement of every quantity $n_0\delta\mathbf{v}$ appearing on the right-hand side of Equation 9.186 by the value defined in Equation 9.177, we end up with the sought-after canonical system of four equations in the four-potential vector perturbation, $\delta A_\mu = (c^{-1}\delta\phi, \delta\mathbf{A})$:

$$\{\partial_t + \mathbf{v}_0 \cdot \nabla + \nabla \cdot \mathbf{v}_0 - \Lambda_0 \cdot \beta_0 - \beta_0\Lambda_0 \cdot -\Omega_0 \times -\gamma_0^2[\Lambda_0 - \beta_0(\Lambda_0 \cdot \beta_0)$$
$$+ \beta_0 \times \Omega_0]\,\beta_0 \cdot\}\left(\square\,\delta\mathbf{A} - \beta_0\square\frac{\delta\phi}{c}\right) + \left[\left(\square\,\delta\mathbf{A} - \beta_0\square\frac{\delta\phi}{c}\right)\cdot\nabla\right]\mathbf{v}_0$$
$$+ \frac{\omega_p^2}{c^2}[-\nabla\delta\phi - \partial_t\delta\mathbf{A} + \mathbf{v}_0 \times (\nabla \times \delta\mathbf{A}) + \beta_0(\nabla\delta\phi + \partial_t\delta\mathbf{A})\cdot\beta_0] = 0, \tag{9.188}$$

$$\frac{1}{c^2}\partial_t\delta\phi + \nabla\cdot\delta\mathbf{A} = 0. \tag{9.189}$$

Note that we can easily identify the different terms in Equation 9.188 as a beam-mode type operator, coupled to an electromagnetic wave propagator, and a beam coupling term proportional to the beam density, $\omega_p^2(x_\nu)$, and containing the ponderomotive force. Equation 9.188 is written in terms of operators acting on the electromagnetic four-potential; the dot and vectorial product symbols appearing there apply either explicitly to a vector (e.g., $\Lambda_0 \cdot \beta_0$), or implicitly to the four-potential perturbation (e.g., $\beta_0\Lambda_0\cdot$), as do the space–time derivative.

At this point, different beam–wave interactions are characterized by different fluid equilibria and different boundary conditions for $\delta A_\mu$. On the one hand, one can expand $\delta A_\mu$ into known eigenmodes satisfying the appropriate boundary conditions, and study the excitation of these modes through the coupled PDEs describing the evolution of the four-vector potential perturbation. The other approach consists in solving directly these equations, then using the boundary conditions to determine the actual eigenvalues and eigenfunctions of the problem.

The reduction in the number of unknown field components to a minimum of four, corresponding to the four degrees of freedom of space–time, clearly warrants the terminology "canonical system." However, it should be noted that the degree of the system derived above, namely third-order PDEs, is higher than that of the initial system of second-order PDEs described by Equations 9.172 and 9.174. In addition, we wish to remark that the mathematical formalism exposed here generalizes a procedure which is virtually of universal use for the study of any particular beam–wave interaction within the framework of a linear relativistic fluid model. The equations presented above form a compact system of PDEs describing the evolution of the four-vector potential perturbation. Starting the electromagnetic stability analysis of a given beam–wave interaction from this system allows one to avoid the lengthy algebraic manipulations generally involved in the conventional mathematical procedure, therefore reducing the risk of errors to a minimum; the mathematical consistency of our perturbation formalism is also guaranteed by these equations.

Equation 9.188 is not written in a manifestly covariant manner, but some of its terms can easily be recast within the tensorial notation of relativity. For example, the fluid convective derivative, $\partial_t + \mathbf{v} \cdot \nabla$, takes the form

$$\frac{d}{d\tau} = \frac{dt}{d\tau}\frac{d}{dt} = \gamma\frac{d}{dt} = \gamma(\partial_t + \mathbf{v} \cdot \nabla) = u^\mu \partial_\mu, \tag{9.190}$$

which gives rise to two different terms:

$$u_0^\mu \partial_\mu, \quad \text{and} \quad \delta u^\mu \partial_\mu, \tag{9.191}$$

where $u_0^\mu(x_\nu) = \gamma_0(c, \mathbf{v}_0)$ is the equilibrium fluid four-velocity, and $\delta u_\mu = (\delta\gamma, \gamma_0\delta\mathbf{v} + \mathbf{v}_0\delta\gamma)$ is the fluid four-velocity perturbation. Similarly, the terms $\Omega_0$ and $\Lambda_0$ are related to the equilibrium electromagnetic tensor,

$$F^0_{\mu\nu} = \partial_\mu A^0_\nu - \partial_\nu A^0_\mu, \tag{9.192}$$

where $A^0_\mu = (c^{-1}\phi_0, \mathbf{A}_0)$ is the equilibrium four-potential.

The factor $1/\gamma_0$ simply corresponds to a change of variable between the laboratory time and the proper time.

In the next subsection, which is intended as an example of the use of these equations, we study the propagation of space–charge waves supported by an electron beam confined by an axial guide magnetic field in a cylindrical waveguide.

### 9.7.2 Cylindrical Waveguide Electrostatic Modes

Here, we study waves supported by an electron beam, which is confined by an axial guide magnetic field, $\hat{z}B_\parallel$. We start from the master system of equations governing the evolution of the four-potential perturbation $\delta A_\mu$,

Equations 9.188 and 9.189. Since this subsection is intended as an example of the use of the equations derived above, the fluid equilibrium studied here is the simplest possible in this case; in particular, we neglect the equilibrium self-fields produced by the beam space–charge. The fluid field equilibrium assumptions are then

$$\mathbf{B}_0 = \hat{z}B_{\parallel}, \tag{9.193}$$

which corresponds to the externally applied guide magnetic field, and

$$\beta_0 = \hat{z}\beta_{\parallel} = \hat{z}\frac{v_{\parallel}}{c}, \tag{9.194}$$

which describes the axial drift velocity of the electron beam. In addition, there is no externally applied electric field, therefore $\mathbf{E}_0 = \mathbf{0}$. The radial beam density profile is

$$n_0(r) = \begin{cases} n_0, & \text{for } r < r_b, \\ 0, & \text{for } a < r < r_b, \end{cases} \tag{9.195}$$

where $r_b$ is the beam radius and $a$ is the waveguide radius.

Note that because the equilibrium fluid velocity field is purely axial, the continuity equation is satisfied everywhere, except at the beam edge. The equations of evolution for the four-vector potential perturbation are, in cylindrical coordinates,

$$(\partial_t + v_{\parallel}\partial_z)\left[\Box\,\delta A_r - \frac{1}{r^2}(\delta A_r + 2\partial_\theta \delta A_\theta)\right] + \Omega_{\parallel}\left[\Box\,\delta A_\theta - \frac{1}{r^2}(\delta A_\theta - 2\partial_\theta\,\delta A_r)\right]$$
$$+ \frac{\omega_{p0}^2}{c^2}[-(\partial_t + v_{\parallel}\partial_z)\delta A_r + \partial_r(v_{\parallel}\delta A_z - \delta\phi)] = 0; \tag{9.196}$$

for the radial component,

$$(\partial_t + v_{\parallel}\partial_z)\left[\Box\,\delta A_\theta - \frac{1}{r^2}(\delta A_\theta - 2\partial_\theta\,\delta A_r)\right] - \Omega_{\parallel}\left[\Box\,\delta A_r - \frac{1}{r^2}(\delta A_r + 2\partial_\theta \delta A_\theta)\right]$$
$$+ \frac{\omega_{p0}^2}{c^2}\left[-(\partial_t + v_{\parallel}\partial_z)\delta A_\theta + \frac{1}{r}\partial_\theta(v_{\parallel}\delta A_z - \delta\phi)\right] = 0; \tag{9.197}$$

for the azimuthal component,

$$(\partial_t + v_{\parallel}\partial_z)\Box\left(\delta A_z - \beta_{\parallel}\frac{\delta\phi}{c}\right) - \frac{\omega_{p0}^2}{\gamma_{\parallel}^2 c^2}(\partial_z\,\delta\phi + \partial_t\,\delta A_z) = 0; \tag{9.198}$$

for the axial component, and finally, the Lorentz gauge condition,

$$\frac{1}{c^2}\partial_t \delta\phi + \frac{1}{r}\partial_r(r\delta A_r) + \frac{1}{r}\partial_\theta \delta A_\theta + \partial_z \delta A_z = 0. \tag{9.199}$$

Here, $\omega_{p0}^2 = n_0 e^2/\gamma_0 \varepsilon_0 m_0$ is the equilibrium relativistic beam plasma frequency, and $\gamma_0 = 1/\sqrt{1-\beta_0^2} = \gamma_\parallel$.

We shall demonstrate that this system of four coupled differential equations in the four-potential vector $\delta A_\mu$ admits solutions of the following form:

$$\delta A_r(r,\theta,z,t) = \left[\mathcal{A}\mathrm{J}_{m-1}(\chi r) + \mathcal{B}\frac{\mathrm{J}_m(\chi r)}{\chi r}\right]\exp[i(\omega t - kz + m\theta)], \tag{9.200}$$

$$\delta A_\theta(r,\theta,z,t) = i\left[\mathcal{A}\mathrm{J}_{m-1}(\chi r) + \mathcal{B}\frac{\mathrm{J}_m'(\chi r)}{m}\right]\exp[i(\omega t - kz + m\theta)], \tag{9.201}$$

$$\delta A_z(r,\theta,z,t) = \mathcal{C}\mathrm{J}_m(\chi r)\exp[i(\omega t - kz + m\theta)], \tag{9.202}$$

$$\delta\phi(r,\theta,z,t) = c\mathcal{D}\mathrm{J}_m(\chi r)\exp[i(\omega t - kz + m\theta)]. \tag{9.203}$$

Note that the terms in $\mathcal{B}$ are similar to the usual vacuum transverse electric (TE) modes, while $\delta A_z$ and $\delta\phi$ are similar to the usual vacuum transverse magnetic (TM) modes. In the first case, the boundary condition at the waveguide wall ($r = a$) would yield $\chi = \chi_{mn}'/a$, and in the second case, $\chi = \chi_{mn}/a$, where $\chi_{mn}'$ and $\chi_{mn}$ are the $n$-th zeros of $\mathrm{J}_m'$ and $\mathrm{J}_m$, respectively. The axial magnetic creates hybrid modes resulting from the mixing of vacuum TE and TM eigenmodes.

In order to demonstrate that Equations 9.200 to 9.203 describe the new eigenmodes of the system, we start by inspecting the gauge equation, which now reads

$$\frac{i\omega}{c}\mathcal{D}\mathrm{J}_m(\chi r) + \frac{1}{r}d_r\left\{r\left[\mathcal{A}\mathrm{J}_{m-1}(\chi r) + \mathcal{B}\frac{\mathrm{J}_m(\chi r)}{\chi r}\right]\right\}$$
$$-\frac{m}{r}\left[\mathcal{A}\mathrm{J}_{m-1}(\chi r) + \mathcal{B}\frac{\mathrm{J}_m'(\chi r)}{\chi r}\right] - ik\mathcal{C}\mathrm{J}_m(\chi r) = 0. \tag{9.204}$$

In Equation 9.204, we have factored out the complex exponential, to keep only the radial dependence. The terms in $\mathcal{B}$ cancel out, and Equation 9.204 reduces to

$$i\left[\frac{\omega}{c}\mathcal{D} - k\mathcal{C}\right]\mathrm{J}_m(\chi r) + \frac{\mathcal{A}}{r}[d_r(r) - m]\mathrm{J}_{m-1}(\chi r) = 0. \tag{9.205}$$

At this point, we use the following formula:

$$\mathrm{J}_{m-1}(\chi r) = \mathrm{J}'_m(\chi r) - m\frac{\mathrm{J}_m(\chi r)}{\chi r}, \tag{9.206}$$

which yields

$$i\left[\frac{\omega}{c}\mathcal{D} - kC\right]\mathrm{J}_m(\chi r) + \mathcal{A}\left\{\frac{1}{r}d_r[r\mathrm{J}'_m(\chi r)] - \frac{1}{\chi}\frac{m^2}{r^2}\mathrm{J}_m(\chi r)\right\} = 0. \tag{9.207}$$

The terms in $\mathcal{A}$ now correspond to the usual Bessel differential equation, and we end up with

$$\left\{i\left[\frac{\omega}{c}\mathcal{D} - kC\right] - \chi\mathcal{A}\right\}\mathrm{J}_m(\chi r) = 0. \tag{9.208}$$

Since Equation 9.208 must hold for any value of the radius, we obtain the condition

$$i\left[\frac{\omega}{c}\mathcal{D} - kC\right] - \chi\mathcal{A} = 0. \tag{9.209}$$

We now examine the $z$-component of the coupled wave equation, Equation 9.202. Remembering the definition of the d'Alembertian operator in cylindrical coordinates,

$$\square \equiv \left[\frac{1}{r}\partial_r(r\partial_r) + \frac{1}{r^2}\partial_\theta^2 + \partial_z^2 - \frac{1}{c^2}\partial_t^2\right], \tag{9.210}$$

and using the conventional Bessel differential equation, we easily obtain

$$\left(\frac{\omega}{c} - \beta_\parallel k\right)\left(\frac{\omega^2}{c^2} - k^2 - \chi^2\right)(C - \beta_\parallel \mathcal{D}) - \frac{\omega_{p0}^2}{\gamma_\parallel^2 c^2}\left(\frac{\omega}{c}C - k\mathcal{D}\right) = 0. \tag{9.211}$$

At this point, we need to evaluate the quantities

$$\square_r = \square\,\delta A_r - \frac{1}{r^2}(\delta A_r + 2\partial_\theta \delta A_\theta), \tag{9.212}$$

and

$$\square_\theta = \square\,\delta A_\theta - \frac{1}{r^2}(\delta A_\theta - 2\partial_\theta \delta A_r). \tag{9.213}$$

We have, by definition,

$$\square_r = \left\{\left[\frac{1}{r}d_r(rd_r) + \frac{\omega^2}{c^2} - k^2 - \frac{m^2+1}{r^2}\right]\left[\mathcal{A}\mathrm{J}_{m-1}(\chi r) + \mathcal{B}\frac{\mathrm{J}_m(\chi r)}{\chi r}\right]\right.$$

$$\left. + \frac{2m}{r^2}\left[\mathcal{A}\mathrm{J}_{m-1}(\chi r) + \mathcal{B}\frac{\mathrm{J}'_m(\chi r)}{\chi r}\right]\right\}\exp[i(\omega t - kz + m\theta)] = 0. \quad (9.214)$$

We group the terms in $\mathcal{A}$ and $\mathcal{B}$ together, to obtain

$$\square_r = \exp[i(\omega t - kz + m\theta)]\left[\!\!\left[\mathcal{A}\left[\frac{1}{r}d_r(rd_r) + \frac{\omega^2}{c^2} - k^2 - \frac{m^2-1}{r^2}\right]\mathrm{J}_{m-1}(\chi r)\right.\right.$$

$$\left.\left. + \mathcal{B}\left\{\frac{1}{r}d_r\left[rd_r\frac{\mathrm{J}_m(\chi r)}{\chi r}\right] + \left(\frac{\omega^2}{c^2} - k^2 - \frac{m^2+1}{r^2}\right)\frac{\mathrm{J}_m(\chi r)}{\chi r} + \frac{2}{r^2}\mathrm{J}'_m(\chi r)\right\}\right]\!\!\right]. \quad (9.215)$$

We now use the identity

$$\frac{1}{r^2}\mathrm{J}'_m(\chi r) = \frac{1}{r}d_r\left[\frac{\mathrm{J}_m(\chi r)}{\chi r}\right] + \frac{1}{r^2}\left[\frac{\mathrm{J}_m(\chi r)}{\chi r}\right], \quad (9.216)$$

to find

$$\square_r = \exp[i(\omega t - kz + m\theta)]\left[\!\!\left[\left(\frac{\omega^2}{c^2} - k^2 - \chi^2\right)\mathcal{A}\mathrm{J}_{m-1}(\chi r)\right.\right.$$

$$\left.\left. + \mathcal{B}\left\{d_r^2\left[\frac{\mathrm{J}_m(\chi r)}{\chi r}\right] + \frac{3}{r}\left[\frac{\mathrm{J}_m(\chi r)}{\chi r}\right] + \left(\frac{\omega^2}{c^2} - k^2 + \frac{1-m^2}{r^2}\right)\left[\frac{\mathrm{J}_m(\chi r)}{\chi r}\right]\right\}\right]\!\!\right]. \quad (9.217)$$

Finally, we make use of the following Bessel differential equation:

$$y'' + \frac{1-2\alpha}{x}y' + \left(\beta^2 + \frac{\alpha^2 - p^2}{x^2}\right)y = 0, \quad (9.218)$$

which is satisfied for $y(x) = x^{\alpha}Z_p(\beta x)$, where $Z_p$ represents an arbitrary Bessel function. In our case, we make the following identifications: $p \equiv m$, $\alpha \equiv -1$, $Z_p \equiv \mathrm{J}_m$. We then end up with the eigenvalue equation:

$$\square_r = \left(\frac{\omega^2}{c^2} - k^2 - \chi^2\right)\left[\mathcal{A}\mathrm{J}_{m-1}(\chi r) + \mathcal{B}\frac{\mathrm{J}_m(\chi r)}{\chi r}\right]\exp[i(\omega t - kz + m\theta)]. \quad (9.219)$$

After similar algebraic manipulations, we also obtain

$$\Box_\theta = i\left(\frac{\omega^2}{c^2} - k^2 - \chi^2\right)\left[\mathcal{A}J_{m-1}(\chi r) + \mathcal{B}\frac{J'_m(\chi r)}{m}\right]\exp[i(\omega t - kz + m\theta)]. \quad (9.220)$$

We can now proceed with the $r$- and $\theta$-components of the wave equation. Making use of Equations 9.219 and 9.220, the $r$-component of the wave equation now reads

$$i(\omega - v_\| k)\left(\frac{\omega^2}{c^2} - k^2 - \chi^2 - \frac{\omega_{p0}^2}{c^2}\right)\left[\mathcal{A}J_{m-1}(\chi r) + \mathcal{B}\frac{J_m(\chi r)}{\chi r}\right] + i\Omega_\|\left(\frac{\omega^2}{c^2} - k^2 - \chi^2\right)$$

$$\times\left[\mathcal{A}J_{m-1}(\chi r) + \mathcal{B}\frac{J'_m(\chi r)}{m}\right] + \frac{\omega_{p0}^2}{c^2}(v_\| C - c\mathcal{D})\chi J'_m(\chi r) = 0, \quad (9.221)$$

while the $\theta$-component is given by

$$(\omega - v_\| k)\left(\frac{\omega^2}{c^2} - k^2 - \chi^2 - \frac{\omega_{p0}^2}{c^2}\right)\left[\mathcal{A}J_{m-1}(\chi r) + \mathcal{B}\frac{J'_m(\chi r)}{m}\right] + \Omega_\|\left(\frac{\omega^2}{c^2} - k^2 - \chi^2\right)$$

$$\times\left[\mathcal{A}J_{m-1}(\chi r) + \mathcal{B}\frac{J_m(\chi r)}{\chi r}\right] - \frac{\omega_{p0}^2}{c^2}(v_\| C - c\mathcal{D})\frac{im}{r}J_m(\chi r) = 0. \quad (9.222)$$

In the $r$-component of the wave equation, we use the identity

$$\frac{J_m(\chi r)}{\chi r} = \frac{1}{m}[J_{m-1}(\chi r) - J'_m(\chi r)], \quad (9.223)$$

to obtain two equations,

$$(\omega - v_\| k)\left(\frac{\omega^2}{c^2} - k^2 - \chi^2 - \frac{\omega_{p0}^2}{c^2}\right)\left(\mathcal{A} + \frac{\mathcal{B}}{m}\right) + \Omega_\|\left(\frac{\omega^2}{c^2} - k^2 - \chi^2\right)\mathcal{A} = 0, \quad (9.224)$$

and

$$\left[(\omega - v_\| k)\left(\frac{\omega^2}{c^2} - k^2 - \chi^2 - \frac{\omega_{p0}^2}{c^2}\right) - \Omega_\|\left(\frac{\omega^2}{c^2} - k^2 - \chi^2\right)\right]\mathcal{B} + i\frac{\omega_{p0}^2}{c^2}(v_\| C - c\mathcal{D})m\chi = 0, \quad (9.225)$$

corresponding to terms in $J_{m-1}(\chi r)$ and $J'_m(\chi r)$, respectively. The second equation, Equation 9.225, is of particular interest since it couples the quasi-TE modes ($\mathcal{A}$, $\mathcal{B}$) to the quasi-TM modes ($\mathcal{C}$, $\mathcal{D}$) through the beam plasma frequency, $\omega_{p0}$. Proceeding in the same manner with the $\theta$-component of the wave equation 9.222 and using the identity Equation 9.223 to express $J'_m(\chi r)$ as a function of $J_{m-1}(\chi r)$ and $J_m(\chi r)/\chi r$, we obtain two equations that are identical to Equations 9.224 and 9.225.

The dispersion relation, $D(\omega, k, \chi) = 0$, for waves propagating inside the electron beam is obtained by taking the determinant of the system of four equations, 9.209, 9.211, 9.224, and 9.225, in the four unknown amplitudes, $\mathcal{A}$, $\mathcal{B}$, $\mathcal{C}$, and $\mathcal{D}$. We obtain

$$\left(\omega - kv_\parallel + \frac{\omega_{p0}}{\gamma_\parallel}\sqrt{\frac{\frac{\omega^2}{c^2} - k^2}{\frac{\omega^2}{c^2} - k^2 - \chi^2}}\right)\left(\omega - kv_\parallel - \frac{\omega_{p0}}{\gamma_\parallel}\sqrt{\frac{\frac{\omega^2}{c^2} - k^2}{\frac{\omega^2}{c^2} - k^2 - \chi^2}}\right)$$
$$\times\left[\omega - kv_\parallel + \Omega_\parallel\left(\frac{\frac{\omega^2}{c^2} - k^2 - \chi^2}{\frac{\omega^2}{c^2} - k^2 - \chi^2 - \frac{\omega_{p0}^2}{c^2}}\right)\right]\left[\omega - kv_\parallel - \Omega_\parallel\left(\frac{\frac{\omega^2}{c^2} - k^2 - \chi^2}{\frac{\omega^2}{c^2} - k^2 - \chi^2 - \frac{\omega_{p0}^2}{c^2}}\right)\right]$$
$$+\chi^2\frac{\omega_{p0}^2}{\gamma_\parallel^2}\left[\frac{(\omega - kv_\parallel)^2}{\frac{\omega^2}{c^2} - k^2 - \chi^2}\right] = 0, \tag{9.226}$$

where the four different types of space–charge waves appear clearly: the fast and slow ordinary modes couple to the fast and slow extraordinary or cyclotron modes through the term proportional to $\chi^2(\omega_{p0}^2/\gamma_\parallel^2)$.

In the limiting case where $B_\parallel \to 0$, this result is similar to that derived by Trotman. We now outline the mathematical procedure followed to obtain the complete solution to this problem. At this point, it is important to note that the transverse wavenumber, $\chi$, remains unknown. To determine $\chi$, we must also solve the vacuum wave equation and match the solution at the beam edge, where $r = r_b$. Outside the beam ($a > r > r_b$), the plasma frequency is zero, and we have

$$\begin{aligned}
&\square\,\delta A_r - \frac{1}{r^2}(\delta A_r + 2\partial_\theta \delta A_\theta) = 0,\\
&\square\,\delta A_\theta - \frac{1}{r^2}(\delta A_\theta - 2\partial_\theta \delta A_r) = 0,\\
&\square\,\delta A_z = 0,\\
&\square\,\delta\phi = 0,
\end{aligned} \tag{9.227}$$

together with the Lorentz gauge condition,

$$\frac{1}{c^2}\partial_t\delta\phi + \frac{1}{r}\partial_r(r\delta A_r) + \frac{1}{r}\partial_\theta\delta A_\theta + \partial_z\delta A_z = 0. \tag{9.228}$$

The general solution is:

$$\delta A_\mu(r, \theta, z, t) = \delta A_\mu(r)\exp[i(\omega t - kz + m\theta)], \tag{9.229}$$

with

$$\begin{aligned} \delta A_r(r) &= \frac{1}{\chi r}[\mathcal{W}\mathrm{J}_m(\chi r) + \mathcal{X}\mathrm{Y}_m(\chi r)], \\ \delta A_\theta(r) &= \frac{1}{m}[\mathcal{W}\mathrm{J}'_m(\chi r) + \mathcal{X}\mathrm{Y}'_m(\chi r)], \end{aligned} \tag{9.230}$$

which are similar to vacuum TE modes, and

$$\begin{aligned} \delta A_z(r) &= \frac{\omega}{ck}[\mathcal{U}\mathrm{J}_m(\chi r) + \mathcal{V}\mathrm{Y}_m(\chi r)], \\ \delta\phi(r) &= c[\mathcal{U}\mathrm{J}_m(\chi r) + \mathcal{V}\mathrm{Y}_m(\chi r)], \end{aligned} \tag{9.231}$$

which correspond to vacuum TM modes. Note that in this region of space, the modified Bessel functions of the first kind, $\mathrm{Y}_m(\chi r)$, must be included in the general solution because the symmetry axis is excluded: $r > r_b \neq 0$. In addition, $\omega$, $k$, and $\chi$ are constrained by the vacuum dispersion relation,

$$\frac{\omega^2}{c^2} - k^2 - \chi^2 = 0. \tag{9.232}$$

At this point, we have found a general solution to the four-dimensional wave equation in two distinct regions of space: region 1, outside the electron beam ($a > r > r_b$), and region 2, inside the electron beam ($r_b > r$). The corresponding solutions are

$$\delta A_{\mu 1}(r, \theta, z, t) = \delta A_{\mu 1}(r)\exp[i(\omega_1 t - k_1 z + m_1\theta)], \tag{9.233}$$

where the radial dependence of $\delta A_{\mu 1}$ is described by Equations 9.230 and 9.231, for $\chi \equiv \chi_1$, and $\omega_1$, $k_1$, and $\chi_1$ must satisfy the vacuum dispersion relation,

$$\frac{\omega_1^2}{c^2} - k_1^2 - \chi_2^2 = 0, \tag{9.234}$$

and, inside the beam,

$$\delta A_{\mu 2}(r, \theta, z, t) = \delta A_{\mu 2}(r) \exp[i(\omega_2 t - k_2 z + m_2 \theta)], \tag{9.235}$$

where the $\delta A_{\mu 2}(r)$ are given by Equations 9.200 to 9.203, for $\chi \equiv \chi_2$, together with the beam dispersion relation, Equation 9.226, which we summarize by writing

$$D(\omega_2, k_2, \chi_2) = 0. \tag{9.236}$$

The boundary conditions are the following: at the beam edge, where $r = r_b$, all the components of the four-vector potential must be continuous, except the radial component,

$$\begin{aligned} \Delta \delta A_\theta(r = r_b) &= 0, \\ \Delta \delta A_z(r = r_b) &= 0, \\ \Delta \delta \phi(r = r_b) &= 0, \end{aligned} \tag{9.237}$$

to avoid infinite field components. In addition, because the cylindrical distribution of surface charges and currents cannot contribute to the discontinuity of the following field components, we have

$$\begin{aligned} \Delta \delta E_\theta(r = r_b) &= 0, \\ \Delta \delta E_z(r = r_b) &= 0, \\ \Delta \delta B_z(r = r_b) &= 0. \end{aligned} \tag{9.238}$$

We note, however, that the first two conditions in Equation 9.238 are automatically satisfied if Equation 9.237 is satisfied. Finally, at the waveguide wall, where $r = a$, the tangential electric field and the perpendicular magnetic induction must be zero:

$$\begin{aligned} \delta E_\theta(r = a) &= 0, \\ \delta E_z(r = a) &= 0, \\ \delta B_r(r = a) &= 0. \end{aligned} \tag{9.239}$$

Again, we note that the condition on the magnetic induction at the waveguide wall is redundant, as it will be automatically satisfied if the first two conditions in Equation 9.239 are met. At this level, we have two series of independent boundary conditions, four at the beam edge and two at the waveguide wall, while eight amplitudes characterize the system ($\mathcal{A}$, $\mathcal{B}$, $\mathcal{C}$, and $\mathcal{D}$ inside the beam, and $\mathcal{U}$, $\mathcal{V}$, $\mathcal{W}$, and $\mathcal{X}$ outside the beam). Therefore, we require two extra boundary conditions to solve the problem completely. The remaining two boundary conditions are obtained by considering the surface charge and current densities at the beam edge that generate discontinuities in $\delta E_r$ and $\delta B_\theta$. We have

$$\Delta\delta E_r(r = r_b) = -\frac{1}{\varepsilon_0} e n_0 \delta r,$$
$$\Delta\delta B_\theta(r = r_b) = -\mu_0 e n_0 v_\parallel \delta r, \tag{9.240}$$

where $\delta r(\theta, z, t)$ is the beam edge perturbation induced by the electromagnetic waves.

This quantity can be evaluated by considering

$$n_0 \delta v_r = \frac{1}{\mu_0 e}\left[\Box\,\delta A_r - \frac{1}{r^2}(\delta A_r + 2\partial_\theta \delta A_\theta)\right]. \tag{9.241}$$

We have, by definition,

$$\delta v_r(r = r_b) = (\partial_t + \mathbf{v}_0 \cdot \nabla)\delta r, \tag{9.242}$$

which yields

$$\delta r(\theta, z, t) = -\frac{i\left(\frac{\omega^2}{c^2} - k^2 - \chi_2^2\right)}{\mu_0 e n_0(\omega - k v_\parallel)}\left[\mathcal{A} J_{m-1}(\chi_2 r_b) + \mathcal{B}\frac{J_m(\chi_2 r_b)}{\chi_2 r_b}\right]\exp[i(\omega t - kz + m\theta)]. \tag{9.243}$$

Here, we have used the fact that the continuity of $\delta A_\theta$, $\delta A_z$, and $\delta\phi$ at the beam edge, as required in Equation 9.237, for any value of $z$, $\theta$, and $t$, immediately yields the following relations:

$$k_1 = k_2 = k,$$
$$m_1 = m_2 = m, \tag{9.244}$$
$$\omega_1 = \omega_2 = \omega.$$

We now have eight equations, and we can eliminate the amplitudes to obtain a relation between $\omega$, $k$, $\chi_1$, and $\chi_2$, of the form

$$B(\omega, k, \chi_1, \chi_2) = 0, \tag{9.245}$$

which includes the geometrical factors of the problem, such as the beam equilibrium radius, $r_b$, and the waveguide radius, $a$. Equation 9.245 and the two dispersion relations in vacuum, Equation 9.234, and inside the beam,

Equation 9.236, form a system of three nonlinear equations in $k$, $\chi_1$, and $\chi_2$. For a given value of the frequency, $\omega$, we can determine the wavenumber, $k$, and the radial profile of the electromagnetic waves propagating along the electron beam in the cylindrical waveguide. In general, because of the waveguide wall boundary conditions, a discrete spectrum of eigenmode will emerge. Finally, we note that from the form of the beam dispersion relation, there are generally two distinct values of $\chi_2$ allowed, reflecting the birefringence of the magnetized electron beam.

In closing, we wish to emphasize that the systematic method discussed in this section can be generalized to treat a wide variety of linear beam–wave interactions, including complex boundary conditions and beam geometries. In particular, the boundary condition matching technique presented here is used widely in the design of electron tubes for high-power, coherent microwave generation, including traveling-wave tube amplifiers (TWTAs), free-electron masers (FEMs), and gyro-devices. Knowledge of the dispersion relation allows for the detailed study of the stability of the electromagnetic modes supported by the system, including unstable modes that can lead to amplification. Moreover, the dispersion relation also indicates the effects of propagation on the phase of the signal and can be used as a starting point to analyze the phase noise characteristics of the device, an important characteristic for modern radar and communication systems.

## 9.8 References for Chapter 9

Note: the numbers listed below refer to the main bibliography and reference sections at the end of this book.

1, 4, 53, 86, 94, 192, 197, 244, 245, 248, 262, 268, 270, 295, 296, 308, 331, 332, 333, 334, 335, 336, 337, 338, 339, 340, 346, 411, 413, 501, 543, 559, 566, 569, 570, 574, 577, 597, 598, 621, 622, 655, 684, 727, 728, 729, 736, 750, 757, 777, 784, 789, 808, 848, 850, 912, 917, 918, 919, 920, 921.

# 10

# *Compton Scattering, Coherence, and Radiation Reaction*

## 10.1 Introduction

In this final chapter, a number of important concepts are introduced and discussed in depth. We begin with the classical theory of Compton scattering, where the Lorentz force equation is used to describe the covariant dynamics of an electron in a plane wave and where radiation reaction effects are neglected. The classical differential Compton scattering cross-section is obtained, and it closely matches its QED equivalent in the limit where the incident photon energy, as observed in the rest frame of the electron, is much smaller than the rest energy of the electron. In this regime, recoil is negligible, and the classical description proves useful, as the coherent field of a laser, for example, can be treated as a classical, continuous electromagnetic field. Using the differential Compton scattering cross-section, the influence of the phase space of an ensemble of electrons upon the scattered radiation can then be studied in detail. In the case of a relativistic electron beam, the transverse emittance and the axial energy spread can be modeled analytically using Gaussian distributions, and incoherent summations can be performed, leading to analytical expressions for the scattered radiation.

In the case of a three-dimensional electromagnetic wave, an important radiation theorem is derived using a representation of the field in terms of a superposition of plane waves. This theorem is quite general, as plane waves are the eigenmodes for the d'Alembertian operator, or photon propagator, in vacuum. The electron trajectory is linearized, and the scattered radiation field is derived, leading to the central result that each vacuum eigenmode gives rise to a single Doppler-shifted classical dipole excitation. For illustration, the three-dimensional analytical theory is supplemented by computer simulations, for a compact x-ray source based on Compton scattering.

In the nonlinear regime, where the radiation pressure of the drive laser field is high, on-axis harmonic can be generated, and the cases of both linear and circular polarization are discussed. The mechanism for off-axis harmonic generation is also briefly outlined, while for more detail, the reader is referred to the series of comprehensive analyses performed by Esarey and co-authors, listed in the reference section. The question of coherence is then addressed in some detail, within the context of a stochastic electron gas model. In particular, the transition from coherent radiation to spontaneous emission is described, as well as the influence of the point-like nature of electrons upon incoherent radiation. Comparisons with a fluid model are also briefly discussed.

Finally, two fundamental questions are addressed within the context of classical electrodynamics, radiation reaction and Dirac monopoles. In the first case, the derivation of the Dirac–Lorentz equation is presented in detail, including a number of useful mathematical digressions, yielding a thorough analysis of the problem. Furthermore, a number of important ancillary questions are addressed, including the problem of electromagnetic mass renormalization in classical electrodynamics, acausal effects, runaway solutions, and the relation between the Schott term of the Dirac–Lorentz equation, the radiation damping force, and the Maxwell stress tensor. Dirac monopoles and dyons are also discussed within the context of symmetrized electrodynamics, where the electromagnetic field tensor and its dual are used to construct a higher-level, symmetrized object. Next, radiation reaction is treated within the context of electric–magnetic charge-symmetrized electrodynamics. We note that, because the derivation of the symmetrized Dirac–Lorentz equation includes the special case of point electric charges, we have chosen to present the general case, instead of repeating the detailed calculation. However, some care is taken to indicate which steps are specific to the symmetrized problem and which apply to the case of a point electron. Furthermore, a brief overview of the derivation and properties of the Dirac–Lorentz equation is given as an introduction in Section 10.7. Both problems are intended to stimulate further inquiries, rather than giving the reader the impression that the field of electrodynamics is fully understood. For example, despite periodic claims to the contrary, the question of whether the Dirac–Lorentz equation represents the classical limit to quantum electrodynamics has not been fully elucidated at the time of this writing; indeed, even the question of what constitutes the correct definition for such a limit is unclear. Nonrelativistic treatments of the problem do yield the Schott term, within a factor of 4/3, but one of the key problems remains: the quantum and the classical scales differ by a factor of $\alpha = e^2/2\varepsilon_0 hc$, the fine structure constant, and this term diverges when $h \rightarrow 0$. In other words, classical electrodynamics has a nonzero scale and its action unit is $\hbar/\alpha = m_0 c r_0 \neq 0$.

## 10.2 Classical Theory of Compton Scattering

Remarkable advances in ultrashort-pulse laser technology based on chirped-pulse amplification and the recent development of high-brightness, relativistic electron sources allow the design of novel, compact, monochromatic, tunable, femtosecond x-ray sources using Compton scattering. Such new light sources are expected to have a major impact in a number of important fields of research, including the study of fast structural dynamics, advanced biomedical imaging, and x-ray protein crystallography. However, the quality of both the electron and laser beams is of paramount importance in achieving the peak and average x-ray spectral brightness required for such applications. For a fairly comprehensive bibliography of the field, we refer the reader to the reference section for this chapter.

One of the primary purposes of this section is to establish a theoretical formalism capable of describing fully the three-dimensional nature of the laser-electron beam interaction, as well as the influence of the electron and laser beam phase space topologies upon the x-ray spectral brightness. In addition, the radiation theorem that is demonstrated and used in this section is of a general nature and represents a useful tool for the study of classical electrodynamics. Finally, analytical expressions of the x-ray spectral brightness including the effects of emittance and energy spread are obtained in the one-dimensional limit.

### 10.2.1 The HLF Radiation Theorem

Our first task is to demonstrate the following theorem, due to Hartemann and Le Foll: in the linear regime, where the four-potential amplitude satisfies the condition $eA/m_0c \ll 1$, and in the absence of radiative corrections, where the frequency cutoff is $\omega \ll m_0c^2/\hbar$, as measured in the electron frame, the spectral photon number density scattered by an electron interacting with an *arbitrary* electromagnetic field distribution *in vacuum* is given by the momentum space distribution of the incident vector potential at the Doppler-shifted frequency:

$$\frac{d^2N_x(k_\mu^s)}{d\omega_s d\Omega} = \frac{\alpha}{(2\pi)^4}\frac{1}{\gamma_0^2\omega_s}\left|\mathbf{k}_s \times \int_{\mathbb{R}^3}\left[1+\left(\frac{\mathbf{k}}{\kappa_s}\right)\mathbf{u}_0\cdot\right]\right.$$

$$\left.\times\tilde{\mathbf{A}}\left[\omega_s - \frac{\mathbf{u}_0}{\gamma_0}\cdot(\mathbf{k}_s-\mathbf{k}), \mathbf{k}\right]\exp(i\mathbf{k}\cdot\mathbf{x}_0)d^3\mathbf{k}\right|^2. \tag{10.1}$$

Here, $k_\mu^s = (\omega_s, \mathbf{k}_s) = \omega_s(1, \hat{\mathbf{n}})$ is the four-wavenumber of the wave scattered in the observation direction $\hat{\mathbf{n}}$, at the frequency $\omega_s$; $\alpha = e^2/2\varepsilon_0hc \simeq 1/137.036$ is the fine structure constant; $u_\mu^0 = (\gamma_0, \mathbf{u}_0)$ is the electron initial four-velocity; $x_\mu^0 = (0, \mathbf{x}_0)$ is its initial four-position, and we have introduced the scattered light-cone variable, $\kappa_s = -u_\mu^0 k_\mu^s = \gamma_0\omega_s - \mathbf{u}_0 \cdot \mathbf{k}_s$.

The term $[1 + (\mathbf{k}/\kappa_s)\mathbf{u}_0\cdot]$ is to be considered as an operator acting on the Fourier transform of the spatial components of the four-potential, $A_\mu = (\mathcal{V}, \mathbf{A})$,

$$\tilde{A}_\mu(k_\nu) = \frac{1}{\sqrt{2\pi}^4}\int_{\mathbb{R}^4} A_\mu(x^\nu)\exp(ik_\nu x^\nu)d^4k_\nu, \tag{10.2}$$

while the term $\exp(i\mathbf{k}\cdot\mathbf{x}_0)$ will be shown to give rise to the coherence factor.

This theorem, which we refer to as the HLF theorem, is then applied to the specific case of Compton scattering in a three-dimensional Gaussian-elliptical focus. The effects of the electron beam phase space topology are also included, in the form of energy spread and emittance.

In our analysis, charge is measured in units of $e$, mass in units of $m_0$, length is normalized to a reference wavelength, $k_0^{-1}$, while time is measured in units of the corresponding frequency, $\omega_0^{-1} = (ck_0)^{-1}$. Neglecting radiative corrections, the electron motion is governed by the Lorentz force equation $du_\mu/d\tau = -(\partial_\mu A_\nu - \partial_\nu A_\mu)u^\nu$. Here, $u_\mu = dx_\mu/d\tau$ is the electron four-velocity along its world line, $x_\mu(\tau)$; $\tau$ is the electron proper time; $A_\mu$ is the four-potential from which the electromagnetic field derives; finally, $\partial_\mu = (-\partial_t, \nabla)$ is the four-gradient operator. The electromagnetic field distribution considered here corresponds to a vacuum interaction; therefore, the four-potential satisfies the wave equation, $\square A_\mu = [\partial_\nu A^\nu]A_\mu = \mathbb{0}$. and can be expressed as a superposition of plane waves, as described in Equation 10.2. Furthermore, we choose to work in the Lorentz gauge, where $\partial_\mu A^\mu = 0$. Here, the symbol $\mathbb{0} = (0, \mathbf{0})$ represents the null four-vector.

### 10.2.2 Covariant Linearization

Introducing the maximum amplitude of the four-potential, $A$, we can linearize the Lorentz force equation, provided that $A \ll 1$, a condition which is typically satisfied in most experimental situations. For example, in the case of an ultrahigh intensity laser focus, this condition translates into a maximum intensity below $10^{17}$ W/cm$^2$ for visible wavelengths. We then write $u_\mu = u_\mu^0 + u_\mu^1 + u_\mu^2 \ldots$, where $u_\mu^n \propto A^n$, and the Lorentz force equation yields, to first order,

$$\frac{du_\mu^1}{d\tau} \simeq -(\partial_\mu A_\nu - \partial_\nu A_\mu)u_0^\nu. \tag{10.3}$$

To solve Equation 10.3, we Fourier transform the first-order four-velocity perturbation into momentum space:

$$u_\mu^1(x^\nu) = \frac{1}{\sqrt{2\pi}^4}\int_{\mathbb{R}^4} d^4k_\nu \tilde{u}_\mu^1(k_\nu)\exp(ik_\nu x^\nu); \tag{10.4}$$

because of the orthogonality of complex exponentials, Equation 10.3 takes the form

$$\tilde{u}_\mu^1 ik_\nu \frac{dx^\nu}{d\tau} \simeq -u_0^\nu(ik_\mu \tilde{A}_\nu - ik_\nu \tilde{A}_\mu), \tag{10.5}$$

and the term $dx^\nu/d\tau = u^\nu$ must be approximated by $u_0^\nu$ within the context of the linear theory presented here; we then find that $\tilde{u}_\mu^1 \simeq \tilde{A}_\mu - k_\mu(\tilde{A}_\nu u_0^\nu / k_\nu u_0^\nu)$.

Fourier transforming this result back into space–time, we finally obtain

$$u_\mu(x^\nu) \simeq u_\mu^0 + \frac{1}{\sqrt{2\pi}^4}\int_{\mathbb{R}^4} d^4k_\nu\left[\tilde{A}_\mu(k_\nu) - k_\mu\frac{\tilde{A}_\nu(k_\nu)u_0^\nu}{k_\nu u_0^\nu}\right]\exp(ik_\nu x^\nu). \quad (10.6)$$

The linearization procedure used here is manifestly covariant.

### 10.2.3 Nonlinear Plane Wave Dynamics

It is interesting to compare the result given in Equation 10.6 with the full nonlinear theory in the case of a single plane wave electromagnetic eigenmode in vacuum, where the four-potential is an arbitrary function of the relativistically invariant phase: $A_\mu(x^\nu) = A_\mu(\phi)$, $\phi(x^\nu) = -k_\nu x^\nu$. This derivation was first presented in Chapter 8, and the reader can skip to Section 10.2.4, if already familiar with these results.

In the case of a plane wave, the operation of the four-gradient upon the four-potential reduces to:

$$\partial_\mu A_\nu = \frac{\partial\phi}{\partial x^\mu}\frac{dA_\nu}{d\phi} = -k_\mu\frac{dA_\nu}{d\phi}. \quad (10.7)$$

Applying this result to the Lorentz force equation, we have

$$\frac{du_\mu}{d\tau} = k_\mu\left(u^\nu\frac{dA_\nu}{d\phi}\right) - (k_\nu u^\nu)\frac{dA_\mu}{d\phi} = k_\mu\left(u^\nu\frac{dA_\nu}{d\phi}\right) + \frac{d\phi}{d\tau}\frac{dA_\mu}{d\phi} = \frac{dA_\mu}{d\tau} + k_\mu\left(u^\nu\frac{dA_\nu}{d\phi}\right). \quad (10.8)$$

Here, we recognize the canonical momentum, $\pi_\mu = u_\mu - A_\mu$. We now consider the light-cone variable, $\kappa = d\phi/d\tau$; the evolution of this dynamical variable is described by

$$\frac{d\kappa}{d\tau} = -k_\mu\frac{du^\mu}{d\tau} = -k_\mu\left(k^\mu\frac{dA^\nu}{d\phi} - k^\nu\frac{dA^\mu}{d\phi}\right)u_\nu = -(k_\mu k^\mu)\left(\frac{dA^\nu}{d\phi}u_\nu\right) + (k^\nu u_\nu)\left(k_\mu\frac{dA^\mu}{d\phi}\right). \quad (10.9)$$

The first term in Equation 10.9 corresponds to the mass-shell condition, $k_\mu k^\mu = 0$, and can be derived easily by considering the propagation equation in vacuum,

$$\Box A_\mu = (\partial_\nu\partial^\nu)A_\mu = 0 = \frac{\partial\phi}{\partial x^\nu}\frac{\partial\phi}{\partial x_\nu}\frac{d^2A_\mu}{d\phi^2} = (k_\nu k^\nu)\frac{d^2A_\mu}{d\phi^2}; \quad (10.10)$$

while the second term in Equation 10.9 corresponds to the Lorentz gauge condition,

$$\partial_\mu A^\mu = 0 = \frac{\partial\phi}{\partial x^\mu}\frac{dA^\mu}{d\phi} = -k_\mu\frac{dA^\mu}{d\phi}. \quad (10.11)$$

With this, we see that the light-cone variable is a constant of the electron motion, $d\kappa/d\tau = 0$.

We now return to Equation 10.8: from its structure, we can see that the solution must take the form $u_\mu = A_\mu + k_\mu g(\phi)$, where $g$ is a function of the electron phase to be determined; in addition, the nonlinear radiation pressure of the plane wave is proportional to $A_\mu A^\mu(\phi)$. Finally, the solution must satisfy the condition $u^\mu u_\mu = -1$. Therefore, we consider

$$u_\mu = A_\mu + k_\mu \left[ \frac{\varsigma + A_\nu A^\nu}{2\kappa} \right], \tag{10.12}$$

which has the appropriate structure, and where $\varsigma$ is a constant, which will be determined from the normalization of the four-velocity. Deriving the trial solution with respect to proper time, we first have

$$\frac{du_\mu}{d\tau} = \frac{dA_\mu}{d\tau} + \frac{k_\mu}{2\kappa}\left(2A_\nu \frac{dA^\nu}{d\tau}\right) = \frac{dA_\mu}{d\tau} + \frac{k_\mu}{\kappa}\left(A_\nu \frac{dA^\nu}{d\phi}\frac{d\phi}{d\tau}\right) = \frac{dA_\mu}{d\tau} + k_\mu\left(A_\nu \frac{dA^\nu}{d\phi}\right). \tag{10.13}$$

We then use Equation 10.12 to replace the four-potential: $A_\nu = u_\nu - k_\nu[\varsigma + (A_\lambda A^\lambda)/2\kappa]$, and we find that

$$\frac{du_\mu}{d\tau} = \frac{dA_\mu}{d\tau} + k_\mu\left(u_\nu \frac{dA^\nu}{d\tau}\right) - k_\mu\left[\frac{\varsigma + (A_\lambda A^\lambda)}{2\kappa}\right]\left(k_\nu \frac{dA^\nu}{d\tau}\right), \tag{10.14}$$

which reduces exactly to the Lorentz force equation because the Lorentz gauge condition requires that $k_\nu (dA^\nu/d\tau) = 0$, as shown in Equation 10.11. In turn, the structure of the solution given in Equation 10.12 implies that the light-cone variable reduces to

$$\kappa = -k_\mu u^\mu = -k_\mu A^\mu - (k_\mu k^\mu)\left[\frac{\varsigma + A_\nu A^\nu}{2\kappa}\right] = -k_\mu A^\mu, \tag{10.15}$$

because of the mass-shell condition, $k_\mu k^\mu = 0$. We can also verify that the light-cone variable is, indeed, constant: $d\kappa/d\tau = -k_\mu(dA^\mu/d\tau) = -[k_\mu(dA_\mu/d\phi)]d\phi/d\tau = 0$, because of the Lorentz gauge condition. Using Equation 10.15, we can now rewrite the four-velocity as $u_\mu = A_\mu - k_\mu[\varsigma + (A_\nu A^\nu)/2k_\nu A^\nu]$; finally, the constant $\varsigma$ is determined by taking the norm of the four-velocity:

$$u_\mu u^\mu = A_\mu A^\mu - 2k_\mu A^\mu \left[\frac{\varsigma + (A_\nu A^\nu)}{2k_\nu A^\nu}\right] + (k_\mu k^\mu)\left[\frac{\varsigma + (A_\nu A^\nu)}{2k_\nu A^\nu}\right]^2 = -\varsigma = -1, \tag{10.16}$$

where we have used the mass-shell condition again.

Thus, the fully covariant, nonlinear solution to the electron dynamics in a plane wave of arbitrary intensity and temporal dependence is found to be

$$u_\mu(x^\nu) = u_\mu(\phi) = A_\mu - k_\mu\left(\frac{1 + A_\nu A^\nu}{2k_\nu A^\nu}\right). \tag{10.17}$$

The question of the influence of the initial conditions on the electron trajectory can now be addressed: the four-potential can be regauged to incorporate the boundary conditions on the electron dynamics. Since we are shifting the four-potential by a constant four-vector, the Lorentz gauge condition is still satisfied, and the electromagnetic field tensor is unchanged. We have $A^\mu \to A^\mu + u_0^\mu$, $\lim_{\phi\to\infty} A^\mu(\phi) = \mathbf{0}$, $\lim_{\phi\to\infty} u^\mu(\phi) = u_0^\mu$. With this, the invariant light-cone variable reads $\kappa = -k_\mu A^\mu = k_\mu u_0^\mu$, and the four-velocity is given by

$$u^\mu = u_0^\mu + A^\mu - k^\mu\left(\frac{A_\nu A^\nu + 2A_\nu u_0^\nu}{2k_\nu u_0^\nu}\right), \tag{10.18}$$

where we have used the fact that $(A + u_0)_\mu (A + u_0)^\mu = A_\mu A^\mu + 2A^\mu u_0^\mu - 1$.

It should be emphasized that this nonlinear solution is fully covariant and makes explicit use of gauge invariance.

The nonlinear, covariant electron dynamics are thus fully determined; in particular, the mass of the dressed electron can be derived immediately from the time-like component of Equation 10.18, in a frame where the electron is initially at rest:

$$\langle m \rangle = \langle \gamma \rangle = \left\langle 1 + \mathcal{V} + \frac{A_\mu A^\mu}{2} \right\rangle, \tag{10.19}$$

where $\mathcal{V}$ is the time-like component of the four-potential, or scalar potential.

It is also easy to recover more familiar expressions by rotating the coordinate system so that the four-wavenumber reduces to $k_\mu = (1, 0, 0, 1)$ and to introduce the four-potential as $A_\mu = (0, \mathbf{A}_\perp, 0)$; we then have $\kappa = \gamma - u_\parallel = \gamma_0 - u_{\parallel 0} = \kappa_0$, and

$$\begin{aligned} \mathbf{u}_\perp(\phi) &= \mathbf{u}_{\perp 0} + \mathbf{A}_\perp(\phi), \\ u_\parallel(\phi) &= u_{\parallel 0} + \left[\frac{\mathbf{A}_\perp^2 + 2\mathbf{A}_\perp \cdot \mathbf{u}_{\perp 0}}{2(\gamma_0 - u_{\parallel 0})}\right], \\ \gamma(\phi) &= \gamma_0 + \left[\frac{\mathbf{A}_\perp^2 + 2\mathbf{A}_\perp \cdot \mathbf{u}_{\perp 0}}{2(\gamma_0 - u_{\parallel 0})}\right]. \end{aligned} \tag{10.20}$$

In particular, in the simple case where the reference frame is chosen such that $\mathbf{u}_{\perp 0} = \mathbf{0}$, we have $\gamma_0^2 = 1 + u_{\parallel 0}^2$, and Equation 10.20 reduces to the well-known result,

$$\mathbf{u}_\perp = \mathbf{A}_\perp, \quad u_\parallel = \gamma_0\left[\beta_{\parallel 0} + \frac{\mathbf{A}_\perp^2}{2}(1 + \beta_{\parallel 0})\right], \quad \gamma = \gamma_0\left[1 + \frac{\mathbf{A}_\perp^2}{2}(1 + \beta_{\parallel 0})\right]. \tag{10.21}$$

Furthermore, if we linearize Equation 10.18 and consider the expression of the four-potential in momentum space, the result given in Equation 10.6 becomes intuitively clear.

### 10.2.4 Radiation

We now consider the second half of the demonstration of the HLF theorem. The electromagnetic radiation scattered by the accelerated charge is described by the number of photons radiated per unit frequency, per unit solid angle, which is determined by Fourier transforming the electron trajectory into momentum space:

$$\frac{d^2N_x(k_\mu^s)}{d\omega_s d\Omega} = \frac{\alpha\omega_s}{4\pi^2}\left|\hat{\mathbf{n}}\times\int_{-\infty}^{+\infty}\mathbf{u}(\tau)\exp[-ik_\mu^s x^\mu(\tau)]d\tau\right|^2. \tag{10.22}$$

Here, we have used the fact that $\hat{\mathbf{n}}$ is a unit vector to simplify the double cross product.

The spatial component of the electron four-velocity is replaced by the linearized solution given in Equation 10.6; with this, and now using the Coulomb gauge, we have

$$\frac{d^2N_x(k_\mu^s)}{d\omega_s d\Omega} = \frac{\alpha}{(4\pi^2)^3}\frac{1}{\omega_s} \times\left|\mathbf{k}_s\times\int_{-\infty}^{+\infty}d\tau\int_{\mathbb{R}^4}d^4k_\mu\left[\tilde{\mathbf{A}}(k_\mu)+\frac{\mathbf{k}}{\kappa}\mathbf{u}_0\cdot\tilde{\mathbf{A}}(k_\mu)\right]\exp[i(k_\mu-k_\mu^s)x^\mu(\tau)]\right|^2. \tag{10.23}$$

The electron four-position is now approximated by $x_\mu(\tau) \simeq x_\mu^0 + u_\mu^0\tau$; this corresponds to the lowest-order convective term due to the ballistic component of the electron motion and excludes harmonic production mechanisms. This approximation is valid for high Doppler-shift scattering, where the transverse oscillation scale is given by $A\lambda_0/\gamma_0 \ll \lambda_0$. Using the light-cone variables, $\kappa$ and $\kappa_s$, we find that

$$\frac{d^2N_x(k_\mu^s)}{d\omega_s d\Omega} = \frac{\alpha}{(4\pi^2)^3}\frac{1}{\omega_s}\left|\mathbf{k}_s\times\int_{\mathbb{R}^4}d^4k_\mu\left(\tilde{\mathbf{A}}+\frac{\mathbf{k}}{\kappa}\mathbf{u}_0\cdot\tilde{\mathbf{A}}\right)\right. \left.\times\exp[i(k_\mu-k_\mu^s)x_0^\mu]\int_{-\infty}^{+\infty}d\tau\exp[i(\kappa-\kappa_s)\tau]\right|^2. \tag{10.24}$$

The integral over proper time yields a $\delta$-function: $\int_{-\infty}^{+\infty}d\tau\exp[i(\kappa-\kappa_s)\tau] = 2\pi\delta(\kappa-\kappa_s)$. We now perform the change of variable $d^4k_\mu = |\partial q_\nu/\partial k_\mu|^{-1}d^4q_\nu$, where we have introduced the four-vector $q_\nu = [\kappa(k_\mu), \mathbf{k}] = (u_0^\mu k_\mu, \mathbf{k}) = (\gamma_0\omega - \mathbf{u}_0\cdot\mathbf{k}, \mathbf{k})$, which allows us to perform the integral over the $\delta$-function. We then find that the Doppler condition derives from the equality of $\kappa_s = -u_0^\mu k_\mu^s = \gamma_0\omega_s - \mathbf{u}_0 \cdot \mathbf{k}_s$, and $\kappa = -u_0^\mu k_\mu = \gamma_0\omega - \mathbf{u}_0 \cdot \mathbf{k}$: this yields the well-known Compton scattering relation $\omega_s = (\gamma_0\omega - \mathbf{u}_0 \cdot \mathbf{k})/(\gamma_0 - \mathbf{u}_0 \cdot \hat{\mathbf{n}})$, in the limit where recoil is negligible. Finally, using the Jacobian of the transform, $|\partial q_\nu/\partial k_\mu|^{-1} = \gamma_0^{-1}$, we obtain the sought-after result, as presented in the Hartemann–Le Foll theorem.

## 10.3 Electron Beam Phase Space

### 10.3.1 Classical Compton Scattering Differential Cross-Section

To demonstrate the usefulness of the HLF theorem, we first consider the case of a linearly polarized plane wave with an arbitrary temporal profile. The four-potential is $A_\mu(\phi) = \hat{\mathbf{x}} A_0 g(\phi) e^{-i\phi}$, where $\phi = -k_\mu^0 x^\mu$, and $k_\mu^0 = (1, 0, 0, 1)$, for a wave propagating along the $z$-axis. Introducing the temporal Fourier transform of the pulse envelope, $\tilde{g}(\omega) = \int_{-\infty}^{+\infty} g(t) e^{-i\omega t} dt/\sqrt{2\pi}$, we have

$$\tilde{A}_\mu(k_\nu) = \hat{\mathbf{x}}\sqrt{2\pi}^3 A_0 \delta(k_x)\delta(k_y)\delta(\omega - k_z)\tilde{g}(1-\omega), \tag{10.25}$$

where $\delta(\omega - k_z)$ corresponds to the pulse propagation, and $\tilde{g}(1-\omega)$ is the spectrum of the pulse, centered around the normalized frequency $\omega_0 = 1$, in our units.

Applying the HLF theorem, we immediately find

$$\frac{d^2N_x}{d\omega_s d\Omega} = \frac{\alpha}{2\pi}\omega_s \frac{A_0^2}{\kappa_0^2}\left|\hat{\mathbf{n}} \times \left(\hat{\mathbf{x}} + \frac{u_{0x}}{\kappa_0}\hat{\mathbf{z}}\right)\right|^2 \tilde{g}^2\left(1 - \frac{\kappa_0^s}{\kappa_0}\right), \tag{10.26}$$

where $\kappa_0 = \gamma_0 - \mathbf{u}_0 \cdot \hat{\mathbf{z}} = \gamma_0 - u_{0z}$, and $\kappa_0^s = \omega_s(\gamma_0 - \mathbf{u}_0 \cdot \hat{\mathbf{n}})$. Introducing the normalized Doppler-shifted frequency $\chi = \kappa_0^s/\kappa_0 = \omega_s(\gamma_0 - \mathbf{u}_0 \cdot \hat{\mathbf{n}})/(\gamma_0 - u_{0z})$, and the differential scattering cross-section, or radiation pattern, $f = |\hat{\mathbf{n}} \times (\kappa_0\hat{\mathbf{x}}_0 + u_{0x}\hat{\mathbf{z}}_0)/\kappa_0^2|^2$, this result can be recast as

$$\frac{d^2N_x}{d\omega_s d\Omega} = \frac{\alpha}{2\pi}A_0^2\omega_s f\tilde{g}^2(1-\chi). \tag{10.27}$$

In the case of a Gaussian pulse envelope, where $g(t) = e^{-t^2/\Delta t^2}$, and for the interaction geometry shown in Figure 10.1, Equation 10.27 takes the familiar form

$$\frac{d^2N_x}{d\omega_s d\Omega} = \frac{\alpha}{4\pi}A_0^2\Delta\phi^2\omega_s \exp\left\{-\frac{\Delta\phi^2}{2}[\chi(\omega_s, \gamma_0, \theta, \varphi) - 1]^2\right\}\frac{[\gamma_0\cos(\theta+\varphi) - u_0\cos\theta]^2}{[\gamma_0 - u_0\cos\varphi]^4}. \tag{10.28}$$

Here, $\varphi$ is the incidence angle between the initial electron velocity and the

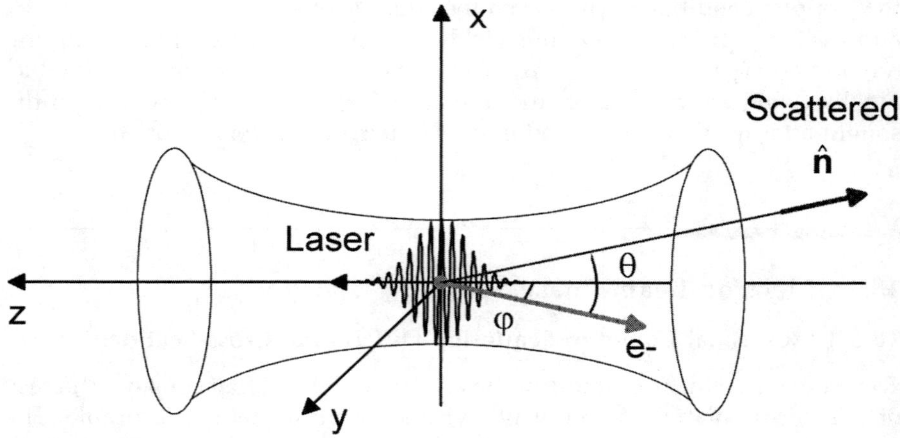


**FIGURE 10.1**
Three-dimensional Compton scattering geometry.

direction of propagation of the plane wave, and is the scattering angle, measured with respect to the electron initial velocity. Equation 10.28 clearly shows that the scattering spectral density is proportional to the incident photon number density, as represented by the laser intensity $A_0^2\Delta\phi$, and that the cold spectral bandwidth of the x-rays is given by that of the incident laser pulse, $\Delta\phi^{-1} = 1/\omega_0\Delta t$. Equation 10.28 also indicates that the peak intensity is radiated near the Doppler-shifted frequency, where $\chi(\omega_x, \gamma_0, \theta, \varphi) \simeq 1$; this yields $\hbar\omega_x(\gamma_0, \theta, \varphi) \simeq \hbar\omega_0[(\gamma_0 - u_0\cos\varphi)/(\gamma_0 - u_0\cos\theta)]$. For a head-on collision, where $\varphi = \pi$, the frequency radiated on-axis, for $\theta = 0$, is the same as the well-known free-electron laser (FEL) frequency for an electromagnetic wiggler. For ultra-relativistic (UR) electrons, we recover the well-known relation, $\omega_x = \gamma^2(1+\beta)^2 \simeq 4\gamma^2$.

For a number of radiating electrons, $N_e$, the various initial positions must be taken into account, as the coherence factor now appears as a sum of phasors, $\left|\sum_{n=1}^{N_e}\exp(i\mathbf{k}\cdot\mathbf{x}_{0n})\right|^2$. This forms the basis of the stochastic electron gas (SEG) theory of coherence in Compton scattering, which is presented in Section 10.5. Here, incoherent summations over the electron beam phase space distribution will be used to study the influence of the phase space topology on the scattered radiation.

The angular x-ray energy distribution can be mapped by considering the position of the spectral peak, where $\omega_s = \omega_x$, and $\chi = 1$. We then find that

$$\frac{d^2N_x}{d\omega d\Omega}(\omega_x, \hat{n}) = \frac{\alpha}{4\pi}A_0^2\Delta\phi^2\frac{[\gamma\cos(\varphi+\theta) - u(\gamma)\cos\theta]^2}{[\gamma - u(\gamma)\cos\theta][\gamma - u(\gamma)\cos\varphi]^3}; \quad (10.29)$$

in the particular case of a head-on collision ($\varphi = \pi$), the angular behavior reduces to

$$\frac{\cos^2\theta}{[\gamma - u(\gamma)\cos\theta][\gamma + u(\gamma)]} \simeq \frac{\gamma - u(\gamma)}{\gamma - u(\gamma)\cos\theta}, \tag{10.30}$$

where the approximation holds for small angles.

The full width at half maximum (FWHM) of the x-ray cone can be derived by further simplifying Equation 10.30 for UR electrons and small angles, where we can use the following approximations: $u(\gamma) \simeq \gamma - (1/2\gamma)$, and $\cos\theta \simeq 1 - (\theta^2/2)$, respectively. With this, the angular energy distribution is described by a Lorentzian, $1/[1 + (\gamma\theta)^2]$, which has an angular FWHM equal to $2/\gamma$. This well-known behavior of the x-ray frequency-integrated cone is illustrated in Figure 10.2.

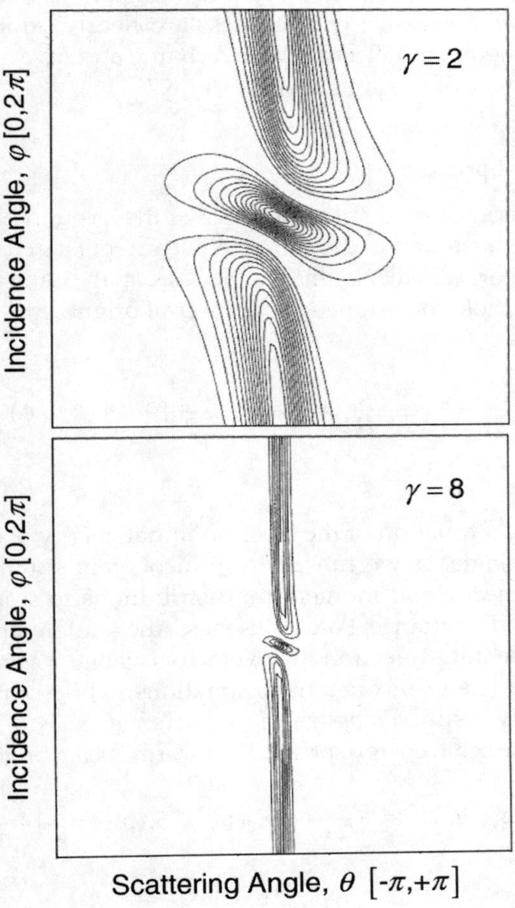


**FIGURE 10.2**
Angular x-ray distribution, showing the $1/\gamma$ cone for the frequency-integrated radiated energy.

Before studying the effect of energy spread and emittance, we also note that the cold, average on-axis brightness of the x-ray source can be estimated by multiplying the spectral brightness by the normalized average electron bunch current $\langle I_b \rangle = q\rho$, where $\rho$ is the repetition rate of the system. By considering a 1 mrad$^2$ solid angle, $\Delta\Omega = 10^{-6}$, and a 0.1% fractional bandwidth, $\Delta\omega = \omega_x \times 10^{-3}$, and by normalizing the source size to 1 mm$^2$ we obtain

$$\langle B_x \rangle = \frac{\alpha}{4\pi} \frac{\langle I_b \rangle}{\pi r_b^2} A_0^2 \Delta\phi^2 \omega_x \times 10^{-15}, \tag{10.31}$$

where $\langle B_x \rangle$ is expressed in units of photons/0.1% bandwidth/mrad$^2$/mm$^2$/s, or synchrotron units, and $r_b$ is the electron beam spot size, which we assume to be equal to the laser spot size. The normalized vector potential is given by $A_0 = (e/\omega_0 m_0 c)\sqrt{(2/\varepsilon_0 c)(\mathcal{W}_0/\pi w_0^2 \Delta t)}$, as expressed in terms of the laser pulse energy $\mathcal{W}_0$, duration $\Delta t$, frequency $\omega_0$, and focal spot size $w_0$. With this, the main scaling laws for the x-ray brightness are clearly exhibited, bilinear in the laser pulse energy and electron bunch charge and inversely proportional to the fourth power of the source size, $1/w_0^2 r_b^2$.

### 10.3.2 Energy Spread

The formalism used to model the influence of the electron beam phase space topology is now illustrated in the case of a linearly polarized plane wave with an arbitrary temporal profile; in this simple case, analytical results are derived. We introduce the cold, one-dimensional spectral brightness,

$$S_0(\omega, \gamma, \theta, \varphi) = \frac{4\pi}{\alpha A_0^2 \Delta\phi^2} \frac{d^2 N_x}{d\omega d\Omega} = \omega \exp\left\{ -\frac{\Delta\phi^2}{2} [\chi(\omega, \gamma, \theta, \varphi) - 1]^2 \right\} f(\gamma, \theta, \varphi). \tag{10.32}$$

Note that as $S_0$ is a function of the electron initial energy, $\gamma$, scattering angle, $\theta$, and incident angle, $\varphi$, we can perform incoherent summations over the electron initial energy and momentum distributions to study the effects of energy spread and emittance. For conciseness, the scattered frequency is now labeled $\omega$, and the initial electron four-velocity is labeled as $u_\mu^0 = (\gamma, u)$, where $u = \sqrt{\gamma^2 - 1}$. The use of incoherent summations, while intuitively obvious, will be rigorously justified *a posteriori* in Section 10.5.

We start with the beam energy spread. The "warm" beam brightness is given by

$$\begin{aligned} S_\gamma(\omega, \gamma_0, \Delta\gamma, \theta, \varphi) &\simeq \frac{1}{\sqrt{\pi}\Delta\gamma} \int_1^\infty S_0(\omega, \gamma, \theta, \varphi) \exp\left[ -\left( \frac{\gamma - \gamma_0}{\Delta\gamma} \right)^2 \right] d\gamma, \\ &\simeq \frac{\omega f(\gamma_0, \theta, \varphi) \exp\left( \frac{u^2}{v} - w \right)}{\sqrt{1 + \frac{1}{2}\left( \Delta\phi \frac{\Delta\gamma}{\gamma_0} \right)^2 \left[ \frac{\omega}{\gamma_0^2} \frac{\cos\varphi - \cos\theta}{(1 - \cos\varphi)^2} \right]^2}}, \end{aligned} \tag{10.33}$$

where we have used a Gaussian distribution to model the beam longitudinal phase space.

Note that as $\lim_{\Delta\gamma\to 0}(1/\sqrt{\pi}\Delta\gamma)\exp[-(\gamma-\gamma_0/\Delta\gamma)^2] = \delta(\gamma-\gamma_0)$, the cold brightness is automatically recovered for a monoenergetic electron beam.

The analytical result in Equation 10.33 is obtained by Taylor-expanding to second order around the central electron energy, $\gamma_0$. The normalization constant is given by

$$\int_1^\infty \exp\left[-\left(\frac{\gamma-\gamma_0}{\Delta\gamma}\right)^2\right]d\gamma \simeq \sqrt{\pi}\,\Delta\gamma, \tag{10.34}$$

an excellent approximation for $\gamma_0 \gg 1$ and $\Delta\gamma/\gamma_0 \ll 1$. Here, $\Delta\gamma$ refers to the energy spread; in addition, $a = (\omega/\gamma_0^3)(\cos\varphi - \cos\theta)/(1-\cos\varphi)^2$, $b = \chi(\omega, \gamma_0, \theta, \varphi) - 1$, $u = (1/\Delta\gamma^2)[1 + a/2(\Delta\phi\Delta\gamma)^2]$, $v = (\Delta\phi^2/2)ab$, and $w = (\Delta\phi^2/2)b^2$. Since $v$ and $w$ are both linear functions of $b$, which is equal to zero at the peak of the x-ray spectrum, the exponential is equal to one for $\omega = \omega_x$. In addition, the factor $[\Delta\phi(\Delta\gamma/\gamma_0)]^2$ in the square root shows that the relative energy spread must be compared to the normalized laser pulse duration, which is equivalent to the number of electromagnetic wiggler periods. This indicates that to increase the x-ray spectral brightness by lengthening the drive laser pulse, the requirement on the electron beam energy spread becomes increasingly stringent. Figure 10.3 illustrates the effects of energy spread, which are seen to broaden symmetrically the scattered x-ray spectrum and lower the peak

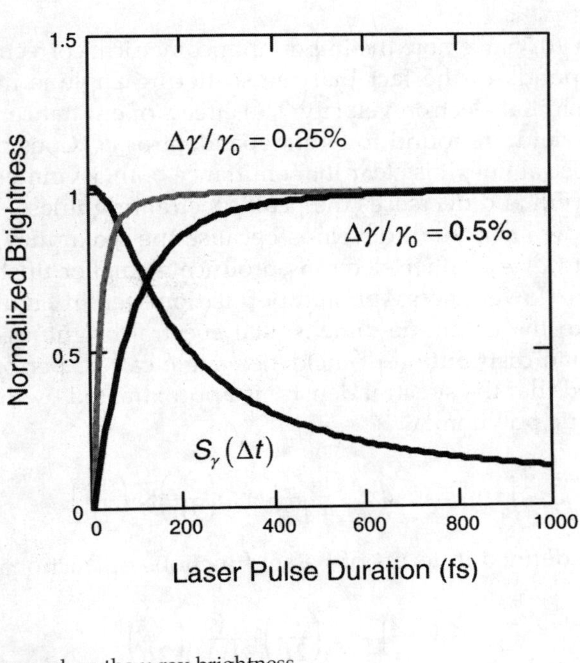


**FIGURE 10.3**
Effect of energy spread on the x-ray brightness.

intensity. Figure 10.3 shows the aforementioned saturation of the spectral brightness, as the laser pulse duration is increased, for a given value of the energy spread, and a fixed value of the normalized potential. We see that for 0.5% relative energy spread and $\gamma_0 = 50$, a 200 fs laser pulse will yield the optimum spectral brightness with a minimum energy, longer pulses requiring more energy without any beneficial effects on the x-ray spectral width.

### 10.3.3 Emittance

We now turn our attention to the influence of the electron beam emittance:

$$S_\varepsilon(\omega, \gamma_0, \Delta\gamma, \theta, \varphi_0, \Delta\varphi) \simeq \frac{1}{\sqrt{\pi}\Delta\varphi}\int_0^{2\pi} S_\gamma(\omega, \gamma_0, \Delta\gamma, \theta - \delta, \varphi_0 + \delta) \times \exp\left[-\left(\frac{\delta}{\Delta\varphi}\right)^2\right] d\delta, \tag{10.35}$$

where the spread of incidence angle is given in terms of the beam emittance $\varepsilon$, and radius $r_b$, by $\Delta\varphi = \varepsilon/\gamma_0 r_b$, and where $\varphi_0$ is the mean incidence angle, defined by the laser and electron beams. Again, the normalization constant is given by

$$\int_0^{2\pi} \exp\left[-\left(\frac{\delta}{\Delta\varphi}\right)^2\right] d\delta \simeq \sqrt{\pi}\Delta\varphi, \tag{10.36}$$

provided that $\varphi_0 \ll 1$.

In Equation 10.35, we note the important geometrical correction term, $\theta - \delta$ which corresponds to the fact that the scattering angle is measured with respect to the initial electron velocity. The effects of emittance are illustrated in Figure 10.4 and are found to be independent of $\varphi_0$. Considering the on-axis x-ray spectral line, it is clear that emittance both asymmetrically broadens the spectrum and decreases the peak spectral brightness. Near head-on collisions, a low energy tail develops because the maximum Doppler-shift corresponds to $\delta \simeq 0$: other electrons produce a smaller upshift, thus contributing to the lower energy photon population seen in Figure 10.4.

Returning to the cold, one-dimensional spectral brightness, the integral over a Gaussian distribution of incidence angle can be performed analytically, provided that the spectral density is approximated by the exponential of a biquadratic polynomial:

$$\int_0^{\infty} e^{-\mu x^4 - 2\nu x^2} dx = \frac{1}{4}\sqrt{\frac{2\nu}{\mu}} \exp\left(\frac{\nu^2}{2\mu}\right) \bar{K}_{\frac{1}{4}}\left(\frac{\nu^2}{2\mu}\right), \tag{10.37}$$

where $\bar{K}_{1/4}$ is defined in terms of Bessel functions of fractional order,

$$\bar{K}_{\frac{1}{4}}\left(\frac{\nu^2}{2\mu}\right) = I_{\frac{1}{4}}\left(\frac{\nu^2}{2\mu}\right) - \frac{\nu}{|\nu|} I_{\frac{1}{4}}\left(\frac{\nu^2}{2\mu}\right). \tag{10.38}$$

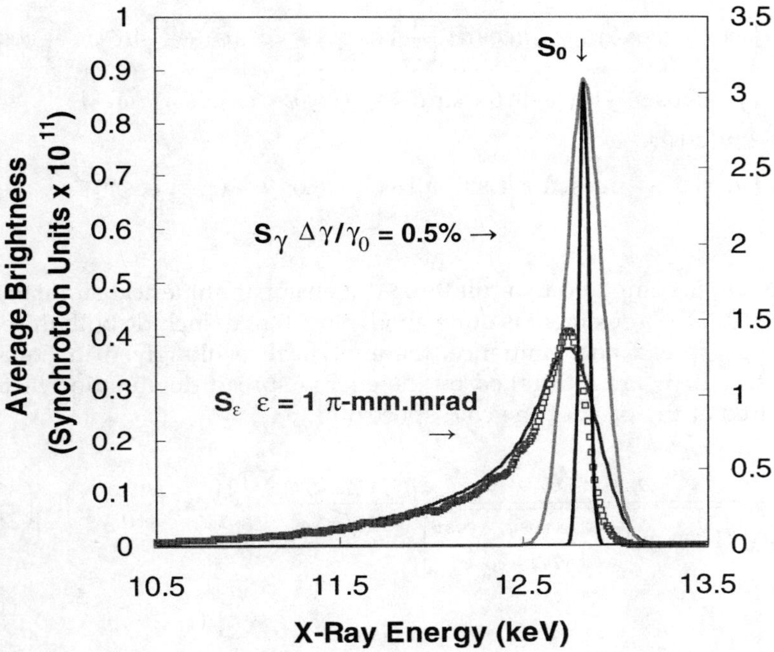


**FIGURE 10.4**
Combined effects of energy spread and emittance on the x-ray brightness. $S_0$ corresponds to the right-hand scale, and a cold beam, while $S_\gamma$ includes energy spread, and $S_\varepsilon$ combines energy spread and emittance. The asymmetric spectrum is typical of emittance. The squares are from the three-dimensional code and confirm the validity of the analytical theory.

Since $\omega f(\gamma, \theta, \varphi)$ is a slow-varying function of the incidence angle, we can seek an approximate expression for the cold spectral density of the form:

$$S_0(\omega, \gamma, \theta-\delta, \varphi+\delta) \simeq \omega f(\gamma, \theta, \varphi)\exp[-\mu(\omega, \gamma, \theta, \varphi)\delta^4 - 2\nu(\omega, \gamma, \theta, \varphi)\delta^2 + \lambda(\omega, \gamma, \theta, \varphi)]. \quad (10.39)$$

The constant term is obtained by taking $\delta = 0$: $\lambda(\omega, \gamma, \theta, \varphi) = -(\Delta\phi^2/2)[\chi(\omega, \gamma, \theta, \varphi) - 1]^2$; the other coefficients are derived using $\delta \simeq 1 - (\delta^2/2!) + (\delta^4/4!)$ and $\sin\delta \simeq \delta - (\delta^3/3!)$. We then find that $\mu = (\Delta\phi^2/2)[(\mu_1 + \mu_2)/(\gamma - u\cos\varphi)^4]$, and $2\nu = (\Delta\phi^2/2)[(\nu_1 - \nu_2)/(\gamma - u\cos\varphi)^4]$, with

$$\mu_1 = (\gamma - u\cos\varphi)^2\left\{\frac{u}{12}(\cos\varphi - \cos\theta)[\gamma(\omega-1) + u(\cos\varphi - \omega\cos\theta)] - \frac{u^2}{3}(\omega\sin\theta + \sin\varphi)^2 + \frac{u^2}{4}(\cos\varphi - \omega\cos\theta)^2\right\},$$

$$\mu_2 = [\omega(\gamma - u\cos\theta) - \gamma + u\cos\varphi]^2\left[\frac{u}{12}\cos\varphi(\gamma - u\cos\varphi) + \frac{u^2}{3}\sin^2\varphi - \frac{u^2}{4}\cos^2\varphi\right],$$
$$\nu_1 = (\gamma - u\cos\varphi)^2\{u^2(\omega\sin\theta + \sin\theta)^2 - u(\cos\varphi - \omega\cos\theta)[\gamma(\omega - 1) + u(\cos\varphi - \omega\cos\theta)]\}$$
$$\nu_2 = [u\cos\varphi(\gamma - u\cos\varphi) + u^2\sin^2\varphi][\omega(\gamma - u\cos\theta) - (\gamma - u\cos\varphi)]^2. \quad (10.40)$$

This result is compared to a full three-dimensional numerical simulation on Figure 10.4; the agreement is quite good. Note that to include both the effects of energy spread and emittance, the analytical results given in Equations 10.37 and 10.39 are multiplied by the energy spread degradation factor, as measured at the peak of the cold spectrum:

$$S_\varepsilon \simeq \frac{\omega_x f(\gamma_0, \theta, \varphi_0)}{\sqrt{1 + \frac{1}{2}\left(\Delta\phi\frac{\Delta\gamma}{\gamma_0}\right)^2\left[\frac{\omega_x\cos\varphi_0 - \cos\theta}{\gamma_0^2(1 - \cos\varphi_0)^2}\right]^2}} \frac{\omega f(\gamma_0, \theta, \varphi_0)}{2\sqrt{\pi}\Delta\varphi}\sqrt{\frac{2\nu}{\mu}}\exp\left(\frac{\nu^2}{2\mu} + \lambda\right)\bar{K}_{\frac{1}{4}}\left(\frac{\nu^2}{2\mu}\right). \quad (10.41)$$

At this point, the combined effects of energy spread and emittance can be further studied by varying the bunch charge and modeling the behavior of the electron beam phase space as follows:

$$\frac{\Delta\gamma}{\gamma_0}(q) \simeq \sqrt{\left[\frac{\gamma_0}{2}(\omega_{rf}\Delta\tau)^2\right]^2 + \left(\frac{e}{m_0c^2}\frac{q}{2\pi\varepsilon_0 c\Delta\tau}\right)^2}, \quad (10.42)$$

where the first term is the spread due to the finite duration of the bunch in the rf accelerating bucket of frequency $\omega_{rf}/2\pi$, while the second term corresponds to space–charge; for the emittance, an empirical linear scaling with charge is chosen, with $\varepsilon(q) \simeq \sigma q$, and $\sigma = 1\pi$-mm · mrad/nC. This results in the brightness curve shown in Figure 10.5, where the brightness first scales linearly with the charge, reaches a maximum near 0.5 nC, and starts degrading thereafter under the combined influences of energy spread and emittance. This optimum value of the charge is quite interesting as it very nearly corresponds to the state-of-the-art for high-brightness photoinjectors.

## 10.4 Three-Dimensional Theory of Compton Scattering

The Hartemann–Le Foll theorem is now applied to the case of a three-dimensional laser focus. The transverse laser profile is specified at the focal plane and propagated using the method discussed in Chapter 8, where the vector potential derives from a generating function: $\mathbf{A} = \nabla \times \mathbf{G}$. In this

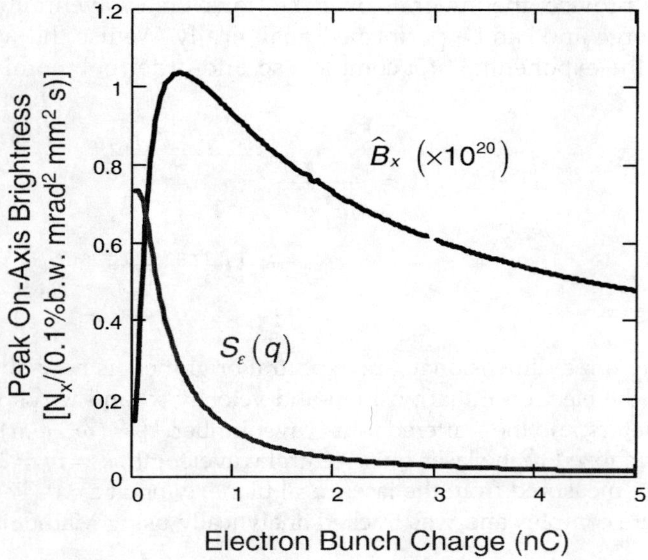


**FIGURE 10.5**
X-ray brightness degradation under the influence of space–charge, which increases both the electron beam energy spread and emittance.

manner, the Coulomb gauge condition, $\nabla \cdot \mathbf{A} = 0$, is automatically satisfied. For a linearly polarized Gaussian-elliptical focus, with focal waists $w_{0x}$ and $w_{0y}$ and a monochromatic wave at the central frequency $\omega_0 = 1$, with a Gaussian envelope of duration $\omega_0 \Delta t = \Delta\phi$, the four-potential is represented in momentum-space by

$$\tilde{A}_\mu(k_\nu) = \frac{\sqrt{\pi}}{2} A_0 w_{0x} w_{0y} \Delta t \exp\left\{-\left[\frac{w_{0x}k_x}{2}\right]^2 - \left[\frac{w_{0y}k_y}{2}\right]^2 - \left[\frac{\Delta t(\omega-1)}{2}\right]^2\right\}$$
$$\times\, \delta\left(k_z - \sqrt{\omega^2 - k_x^2 - k_y^2}\right)[(i - \hat{\mathbf{x}}k_z + \hat{\mathbf{z}}k_x)]. \tag{10.43}$$

Here, we recognize the $\mathbf{k}_\perp$-spectrum, the frequency spectrum, the propagator, $\delta(k_\mu k^\mu)$, and the curl operator, as expressed in momentum space.

The scattered radiation can now be determined by using the HLF theorem. To obtain an analytical result than can be further exploited to include the phase space topology of the electron beam interacting with the laser pulse using the method outlined above, the paraxial propagator formalism is used. The phase function $\delta(k_z - \sqrt{\omega^2 - k_x^2 - k_y^2})$ is replaced by $\delta[k_z - \omega + (k_x^2/2k_0) + (k_y^2/2k_0)]$. The accuracy of the paraxial approximation has been studied in detail and found to be extremely good over a wide range of parameters; however, in the case of Compton scattering, the following conditions must also be satisfied: $\kappa_0^2 w_{0x,y}^2 > \Delta\phi^2[u_z(\kappa_s - \kappa_0) - u_{x,y}^2]$.

With this proviso, the integrals over the transverse wavenumber components converge and can be performed analytically. We use the well-known integral of the exponential of a complex, second-order polynomial,

$$\int_{-\infty}^{+\infty} e^{-(ax^2+2bx+c)} e^{i(px^2+2qx+r)} dx = \frac{\sqrt{\pi}}{\sqrt[4]{a^2+p^2}} e^{\frac{a(b^2-ac)-(aq^2-2bq+cp^2)}{a^2+p^2}} \times e^{i\left[\frac{1}{2}\arctan\left(\frac{p}{a}\right)-\frac{p(q^2-pr)-(b^2p-2abq+a^2r)}{a^2+p^2}\right]}, \tag{10.44}$$

and the fully three-dimensional x-ray spectral brightness is now obtained as a function of the electron initial position and velocity, $\mathbf{x}_0$ and $\mathbf{u}_0$, as well as the laser parameters, and the scattered four-wavenumber, $k_\mu^s = (\omega_s, \omega_s\hat{\mathbf{n}})$. Frequencies are normalized to the laser pulse central wavelength, $k_0 = \omega_0 = 1$, and axial positions are measured from the laser focal plane, lying at $z = 0$. The complete result is quite complex and was tracked analytically using Mathematica.

### 10.4.1 The Cold Three-Dimensional Spectral Density

Writing the cold (single electron) three-dimensional brightness as

$$\frac{\alpha A_0^2 \Delta\phi^2 w_{0x}^2 w_{0y}^2}{(4\pi)^3} S_0(k_\mu^s, x_\mu^0, u_\mu^0) = \frac{d^2 N_x}{d\omega d\Omega} = \frac{\alpha}{(4\pi)^3} \frac{\omega}{\kappa^2} A_0^2 w_{0x}^2 w_{0y}^2 \Delta t^2 \times \left| \hat{\mathbf{n}} \times \iint_{\mathbb{R}^2} (\alpha k_x^2 + \beta k_y^2 + \gamma k_x k_y + \delta k_x + \varepsilon k_y + \zeta) e^{-(ak_x^2+bk_y^2+ck_xk_y+dk_x+ek_y+f)} \times e^{i(pk_x^2+qk_y^2+rk_x+sk_y+t)} dk_x dk_y \right|^2, \tag{10.45}$$

we can introduce the following coefficients for the exponentials:

$$a = \frac{1}{4}\left(\frac{\Delta t^2(u_x^2 - u_z(\kappa_s - \kappa))}{\kappa^2} + w_{0x}^2\right), \quad b = \frac{1}{4}\left(\frac{\Delta t^2(u_y^2 - u_z(\kappa_s - \kappa))}{\kappa^2} + w_{0y}^2\right),$$

$$c = \frac{\Delta t^2 u_x u_y}{2\kappa^2}, \quad d = \frac{\Delta t^2 u_x(\kappa_s - \kappa)}{2\kappa^2}, \quad e = \frac{\Delta t^2 u_y(\kappa_s - \kappa)}{2\kappa^2}, \quad f = \frac{\Delta t^2(\kappa_s - \kappa)^2}{4\kappa^2},$$

$$p = -\frac{\gamma z_0}{2\kappa}, \quad q = -\frac{\gamma z_0}{2\kappa}, \quad r = x_0 + \frac{u_x z_0}{\kappa}, \quad s = y_0 + \frac{u_y z_0}{\kappa}, \quad t = \frac{\kappa_s z_0}{\kappa}. \tag{10.46}$$

Furthermore, the curl operator, expressed in transverse momentum space, yields the following vector components:

$$\begin{aligned}
&\alpha_x = \frac{\gamma\kappa_s - 2u_x^2 + 2\kappa u_z}{2\kappa\kappa_s}, \quad \alpha_y = 0, \quad \alpha_z = \frac{\kappa u_x u_z - u_x^3 + u_x\gamma\kappa_s}{\kappa^2\kappa_s},\\
&\beta_x = \frac{\gamma}{2\kappa}, \quad \beta_y = -\frac{u_x u_y}{\kappa\kappa_s}, \quad \beta_z = \frac{-u_x u_y^2 + u_x\gamma\kappa_s}{\kappa^2\kappa_s},\\
&\gamma_x = -\frac{u_x u_y}{\kappa\kappa_s}, \quad \gamma_y = \frac{\kappa u_z - u_x^2}{\kappa\kappa_s}, \quad \gamma_z = \frac{\kappa u_y u_z - 2u_x^2 u_y}{\kappa^2\kappa_s},\\
&\delta_x = -2\frac{u_x}{\kappa}, \quad \delta_y = 0, \quad \delta_z = 1 + \frac{u_z}{\kappa} - 2\frac{u_x^2}{\kappa^2},\\
&\varepsilon_x = -\frac{u_y}{\kappa}, \quad \varepsilon_y = -\frac{u_x}{\kappa}, \quad \varepsilon_z = -2\frac{u_x u_y}{\kappa\,\kappa},\\
&\zeta_x = -\frac{\kappa_s}{\kappa}, \quad \zeta_y = 0, \quad \zeta_z = -\frac{\kappa_s u_x}{\kappa^2}.
\end{aligned} \tag{10.47}$$

These terms are functions of the laser parameters, as well as the scattering four-wavenumber, $k_\mu^s$, and the electron initial position, $x_\mu^0$, and velocity, $u_\mu^0$; therefore the six-dimensional electron phase space now appears explicitly in the three-dimensional spectral density of the scattered x-rays. Finally, the incident and scattered light-cone variables are

$$\kappa = \gamma - u_z, \quad \kappa_s = \omega(\gamma - \hat{\mathbf{n}} \times \mathbf{u}). \tag{10.48}$$

The advantage of a fully analytical treatment of the problem, afforded by the use of the paraxial propagator formalism, resides in the much shorter computing time required to map the radiation produced by each electron in the beam.

A much simplified form of the general result is obtained by considering a centered electron, where $\mathbf{x}_0 = \mathbf{0}$. In this case,

$$\begin{aligned}
\frac{d^2N(k_\mu^s)}{d\omega_s d\Omega} &= \frac{\alpha\omega_s A_0^2}{4\pi\kappa_0^2}\eta_x\eta_y\eta_t\left|\hat{\mathbf{n}} \times v(u_\nu^0, x_\nu^0, k_\mu^s)\right|^2\\
&\times \exp\left\{-\frac{\Delta\phi^2(\chi-1)^2 + \mathcal{F}(u_\nu^0, x_\nu^0, k_\mu^s)}{2\left[1 + \eta_x\Delta\phi^2\left(\frac{u_x}{\bar{w}_x}\right)^2 + \eta_y\Delta\phi^2\left(\frac{u_y}{\bar{w}_y}\right)^2\right]}\right\}.
\end{aligned} \tag{10.49}$$

Here, we have defined $\bar{w}_{x,y} = \kappa_0 w_{0x,y}$, and the normalization constants

$$\eta_{x,y}(k_\mu^s) = \frac{\bar{w}_{x,y}^2}{\bar{w}_{x,y}^2 + \Delta\phi^2 u_z(\kappa_0 - \kappa_s)},$$

and

$$\eta_t(k_\mu^s) = \left[\Delta\phi^{-2} + \eta_x\left(\frac{u_x}{\varpi_x}\right)^2 + \eta_y\left(\frac{u_y}{\varpi_y}\right)^2\right]^{-1}; \tag{10.50}$$

we have also introduced the normalized Doppler-shifted frequency $\chi(u_\nu^0, k_\mu^s) = \kappa_s/\kappa_0$. Equation 10.50 indicates that the minimum x-ray spectral width is given by the laser pulse duration; it also shows that three-dimensional effects change the x-ray spectrum because of the convective terms due to the electron crossing the laser focus. The vector $\mathbf{v}(u_\nu^0, x_\nu^0, k_\mu^s)$, and the function $\mathcal{F}(u_\nu^0, x_\nu^0, k_\mu^s)$ both depend on the initial position of the electron; when $\mathbf{x}_\nu^0 = \mathbb{0}$, $\mathcal{F} = 0$. In addition, in the specific cases of transverse or axial collisions, $|\hat{\mathbf{n}} \times \mathbf{v}|^2$ takes a relatively simpler form:

$$\left|\hat{\mathbf{n}} \times \mathbf{v}(\hat{\mathbf{z}}u_z, \mathbb{0}, k_\mu^s)\right|^2 = |\hat{\mathbf{n}} \times \hat{\mathbf{x}}|^2\left[\frac{\eta_x\kappa_0(2\kappa_0 u_z + \gamma\kappa_s)}{\varpi_x^2\kappa_s} + \frac{\eta_y\kappa_0\gamma}{\varpi_y^2} - \frac{\kappa_s}{\kappa_0}\right]^2, \tag{10.51}$$

$$\frac{\left|\hat{\mathbf{n}} \times \mathbf{v}(\hat{\mathbf{y}}u_z, \mathbb{0}, k_\mu^s)\right|^2}{|\hat{\mathbf{n}} \times \hat{\mathbf{x}}|^2} = \left\{\frac{\kappa_0\gamma}{\varpi_x^2} - \frac{\kappa_s}{\kappa_0} - \frac{\Delta\phi^2 u_y^2(\kappa_0 - \kappa_s) - \dfrac{\gamma\kappa_0^2 u_y^2}{\Delta\phi^2 u_y^2 + \varpi_y^2}\left[\Delta\phi^2 + \dfrac{\Delta\phi^4}{2}\left(1 - \dfrac{\kappa_s}{\kappa_0}\right)^2 + \dfrac{\varpi_y^2}{u_y^2}\right]}{\kappa_0(\Delta\phi^2 u_y^2 + \varpi_y^2)}\right\}^2, \tag{10.52}$$

$$\frac{\left|\hat{\mathbf{x}} \times \mathbf{v}[\hat{\mathbf{x}}u_x, \mathbb{0}, \omega_s(1, \hat{\mathbf{x}})]\right|^2}{|\hat{\mathbf{x}} \times \hat{\mathbf{z}}|^2} = \left\{\frac{\kappa_0\gamma}{\varpi_y^2} - \frac{\kappa_s}{\kappa_0} - \frac{2\Delta\phi^2 u_x^2(\kappa_0 - \kappa_s) + \dfrac{\gamma\kappa_s - 2u_x^2}{2\kappa_s}[\Delta\phi^4 u_x^2(\kappa_0 - \kappa_s)^2 + 2\kappa_0^2(\Delta\phi^2 u_x^2 + \varpi_x^2)]}{\kappa_0(\Delta\phi^2 u_x^2 + \varpi_x^2)}\right\}^2. \tag{10.53}$$

### 10.4.2 Three-Dimensional Effects

The physics underlying these results can be summarized as follows. First, the one-dimensional kernel $(\alpha\omega_s A_0^2\Delta\phi^2/4\pi\kappa_0^2)\exp\{-\Delta(\phi^2/2)[\chi(\omega, \gamma, \theta, \varphi) - 1]^2\}$ always appears, indicating that the scattered radiation spectrum peaks near the Doppler-shifted frequency $\chi = 1$; furthermore, the scattering is proportional to the laser pulse energy $A_0^2\Delta\phi$, and the minimum spectral width is given by that of the laser, $\Delta\phi^{-1}$. Second, the interaction geometry modifies the spectral width, as illustrated in Figure 10.6. In the case of a transverse interaction, wider bandwidths can be obtained for narrow focal spots, corresponding to the aforementioned shorter effective interaction time due to

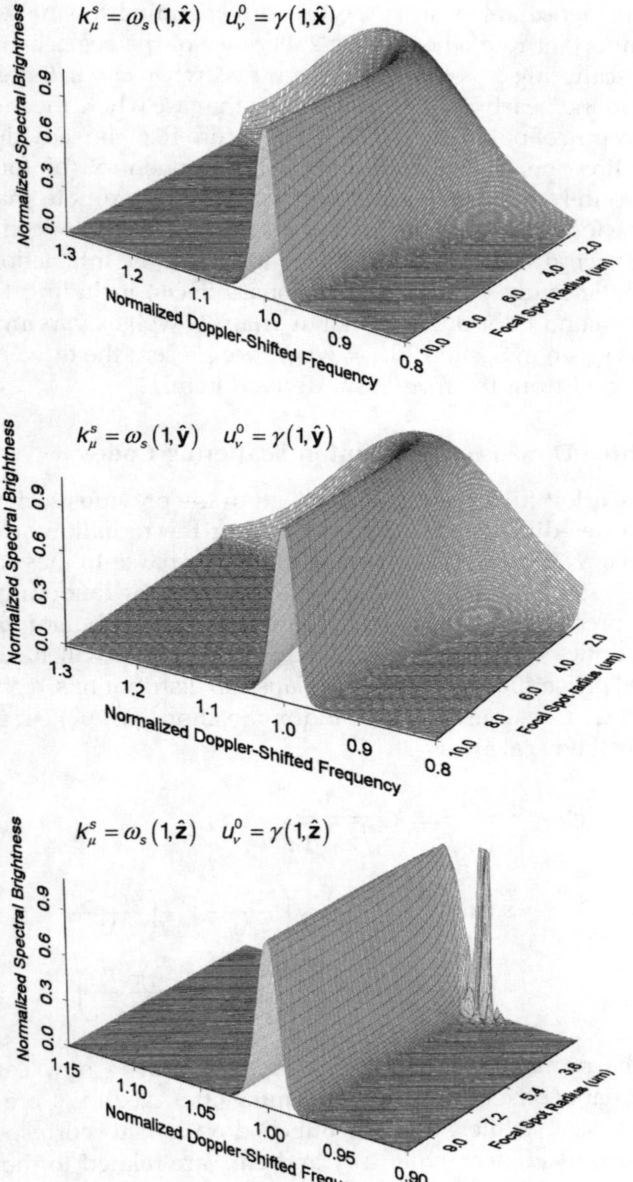


**FIGURE 10.6**
On-axis spectral brightness as a function of the Doppler-shifted frequency and focal radius for transverse interactions; each spectrum is normalized to clearly exhibit the broadening and frequency shift. Here, $\gamma = 50$, $\lambda_0 = 800$ nm, and $\Delta t = 50$ fs. The spurious peak on the bottom graph corresponds to a regime where the integral convergence condition is violated (see text).

the convective motion of the electron through the focus. In addition, the spread of incidence angle $\Delta\varphi \simeq \pi\lambda_0/w_0$ corresponding to smaller spot sizes plays an important role when $\varphi_0 = \pi/2$, because of the $\cos\varphi_0$ behavior of the differential scattering cross-section. By contrast, when $\varphi_0 = \pi$, these corrections are quadratic and nearly negligible. Third, in the case where the initial velocity is in the direction of polarization, shown in Figure 10.6, the radiation observed along that direction results from the axial component of the four-potential, which is a purely three-dimensional effect. Finally, we note that for larger values of the focal spot size, around $w_0 = 100\ \mu$m, the spectra converge to the minimum spectral width of the laser, regardless of the interaction geometry.

A systematic study of three-dimensional effects, including timing jitter, electron beam and laser pulse spatial overlap, as well as varying interaction geometry, is given in Section 10.4.3, where we present the three-dimensional code developed from the formalism derived here.

### 10.4.3 Three-Dimensional Compton Scattering Code

In order to exploit fully the results derived in the previous sections, one can develop a three-dimensional code describing the radiation scattered by a distribution of $N_e$ point charges having the same charge-to-mass ratio as electrons. The $6N_e$-dimensional phase space is generated by randomly loading the particles in prescribed statistical distribution, or can be the output of an electron beam optics code, such as PARMELA. Here, we will focus on a $6N_e$-dimensional phase space model using Gaussian distributions: a given particle, numbered $1 \leq i \leq N_e$, is assigned a random position, $(\mathbf{x}_i^0, \mathbf{u}_i^0)$, in phase space. Its charge is then scaled as

$$q_i = \frac{1}{\sqrt{1+\beta_x^2 z^2}} \times \frac{1}{\sqrt{1+\beta_y^2 z^2}} \times \exp\left[-\left(\frac{x_i^0 - x_0}{\Delta x}\right)^2 - \left(\frac{y_i^0 - y_0}{\Delta y}\right)^2 - \left(\frac{z_i^0 - z_0}{\Delta z}\right)^2 - \left(\frac{u_{xi}^0 - u_{x0}}{\Delta u_x}\right)^2 - \left(\frac{u_{yi}^0 - u_{y0}}{\Delta u_y}\right)^2 - \left(\frac{u_{zi}^0 - u_{z0}}{\Delta u_z}\right)^2\right], \tag{10.54}$$

to reflect the probability distribution. Here, $x_0$ and $y_0$ now correspond to transverse spatial offsets, $z_0$ represents timing jitter, $\Delta x$ and $\Delta y$ are the electron bunch waist size, while $\Delta z$ is the bunch duration; $\mathbf{u}_0$ corresponds to the electron beam incidence, while $\Delta u_x$ and $\Delta u_y$ are related to the beam horizontal and vertical emittance. Finally, $\Delta u_z$ represents energy spread, while the normalization factors, $(1+\beta_{x,y}^2 z^2)^{-1/2}$ are related to the fact that, in close analogy with the laser beam focusing and diffraction, the electron beam inverse $\beta$-functions, which are defined as $\beta_{x,y} = (1/\Delta x, y)\tan(\varepsilon_{x,y}/\gamma_0 \Delta x, y)$, indicate how the beam envelope varies with axial position, while the current through a given z-plane is conserved. Here, $\varepsilon_{x,y}$ represent the horizontal and vertical emittance, and $\gamma_0$ is the average beam energy. Note that far from head-on collisions, one needs to perform the appropriate projections to relate correctly the quantities discussed above to the conventional parameters describing the electron beam phase space. We also note that the random Gaussian loading technique presented here does not include correlations found in typical beams modeled by PARMELA. For our current purpose, however, this model proves entirely sufficient as it allows us to verify the accuracy of the one-dimensional theory of emittance and energy spread developed in Section 10.3 and to benchmark the code in the plane wave limit. It also allows for the systematic study of the x-ray spectral brightness degradation under a variety of conditions closely approaching experimental constraints. The code keeps track of the total charge, $\sum_{i=1}^{N_e} q_i$, and after performing the incoherent summation over all particles, the charge is rescaled to the desired value.

The terms in Equations 10.46 and 10.47 are easily tracked by the code, and Formula 10.44 is used twice to perform the integrals over the $k_x$ and $k_y$ components of the transverse wavenumber, as prescribed in Equation 10.45. In running the code initially, it can be determined that good statistical convergence is obtained for $N_e \geq 3 \times 10^4$; generally, we have used 50,000 particles in the results presented here. Two different types of output data files are created by the code: spectral brightness measured at a prescribed scattering angle, or angular maps at a specified x-ray frequency; for angular maps, the code also integrates the flux over the map, in a small (typically 1 eV) x-ray photon energy interval. This last result is important for x-ray protein crystallography and other applications.

We first systematically study the effects of energy spread, emittance, electron beam focal spot size, bunch duration, and timing jitter. For the sake of clarity, we choose cylindrical foci: $\Delta x = \Delta y$, and $w_{0x} = w_{0y}$. To distinguish between the various effects, we can vary each time one of the parameters listed above, while maintaining all other parameters equal to zero, and the radiation frequency $\chi = 1$; furthermore, to help compare the various effects, we use the normalized brightness. The results are given in Figure 10.7 and very clearly show that the various degradation mechanisms studied yield similar curves, roughly scaling as $1/\sqrt{1+\xi^2}$ where $\xi$ represents the degradation parameter, properly scaled. For example, in the case of energy spread, $\xi \simeq (\Delta\phi/\sqrt{2})(\Delta\gamma/\gamma_0)(\omega/\gamma_0^2)[(\cos\varphi - \cos\theta)/(1-\cos\varphi)^2]$. From an experimental point of view, the most stringent requirements are clearly on emittance and energy spread. Finally, the optimum electron beam size does not correspond to a match with the laser mode. Rather, the beam should be focused as tightly as possible within the laser focal spot. This can easily be understood if we consider the laser pulse as an electromagnetic wiggler, which has maximal field strength, or photon density, on-axis. For high-energy beams, where emittance and space–charge effects are very small and can therefore be focused over a few tens of nm, this is a potentially important result. We also note that this conclusion does not hold if the laser beam depletion becomes important.

The case of a realistic beam is presented in Figures 10.8 and 10.9. The laser parameters are as follows: $\lambda_0 = 800$ nm, $w_0 = 10$ $\mu$m, $\Delta t = 200$ fs, and a pulse

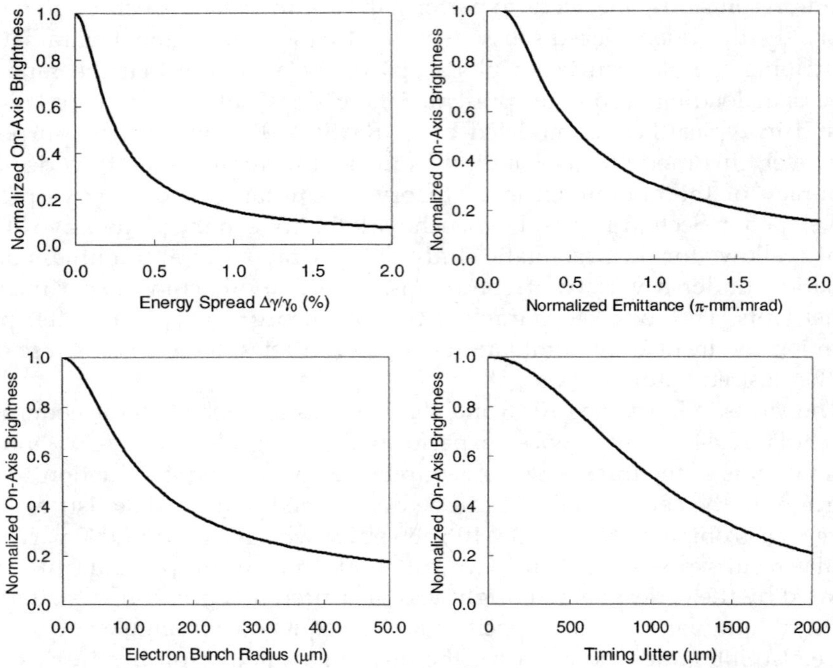


**FIGURE 10.7**
Degradation of the on-axis x-ray spectral brightness as a function of energy spread (top left), emittance (top right), electron bunch radial overlap (bottom left), and timing jitter (bottom right). The other parameters are maintained equal to zero, and the beam energy is 22.75 MeV, the bunch charge is 0.5 nC, its duration is 1 ps, the laser wavelength is 800 nm, the focal spot size is 10 $\mu$m, and the laser pulse energy is 50 mJ.

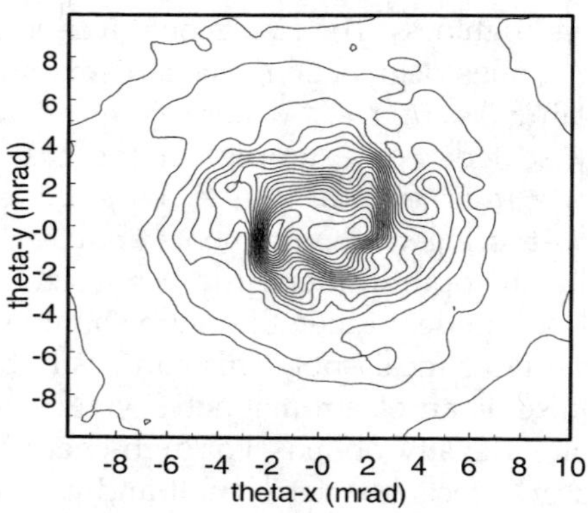


**FIGURE 10.8**
Angular map of the x-ray spectral brightness; the parameters are as in Figure 10.7, and the x-ray energy is 12.66 eV.

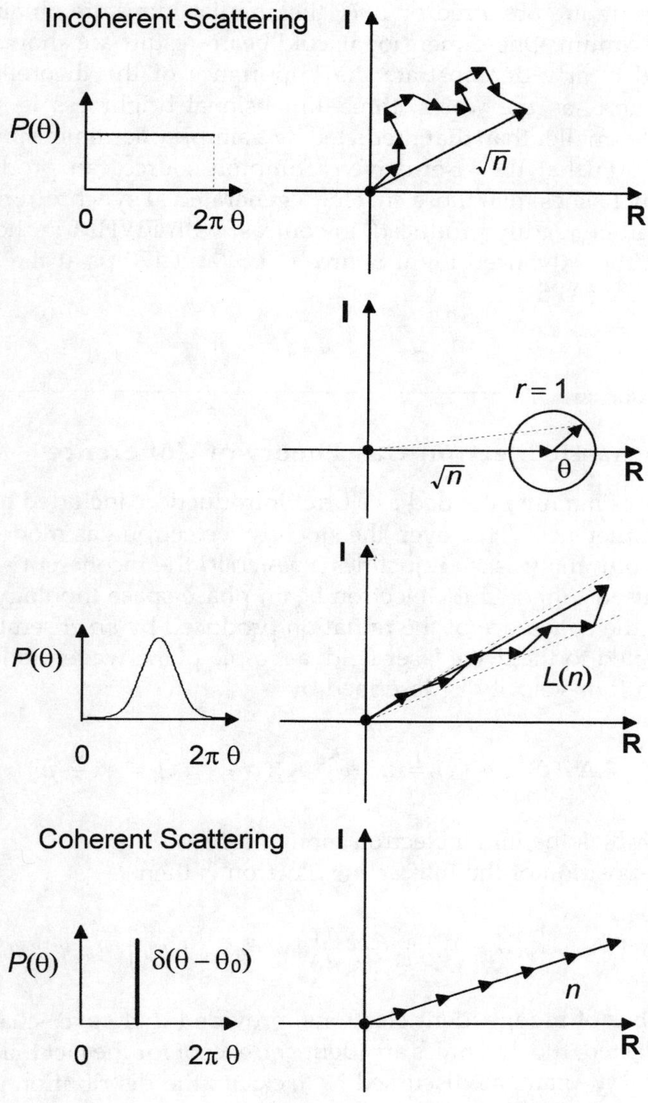


**FIGURE 10.9**
Phasor summation and phase angle probability density for (from top to bottom) incoherent, partially coherent, and fully coherent scattering.

energy of 50 mJ. The electron bunch energy is 22.75 MeV, its charge is 0.5 nC, the relative energy spread is $\Delta\gamma/\gamma_0 = 0.5\%$, the beam normalized emittance is 1 $\pi$-mm · mrad, and the focal spot size matches the laser focal distribution. A repetition rate of 1 kHz is used to scale the average spectral brightness of the source; the maximum brightness compares well with that produced by bend magnets on synchrotron beamlines. The angle of incidence is 180°, and the spectra are observed on-axis; the angular maps are obtained at the spectral maximum. One-dimensional, cold beam results are shown for comparison and clearly demonstrate the importance of the theoretical model developed here, as the warm, three-dimensional brightness is seen to be considerably smaller than that predicted by a simple one-dimensional theory.

We also note that the laser-driven Compton source can produce much shorter x-ray flashes than those currently generated at synchrotrons: sub-100 fs pulses will be readily produced, in contrast with FWHM in the 35 to 100 ps range at the Advanced Light Source (ALS) and 170 ps at the Advanced Photon Source (APS).

## 10.5 Stochastic Electron Gas Theory of Coherence

This section is mainly intended as a brief introduction, included here for the sake of completeness. However, the stochastic electron gas model is highly relevant to our analysis, as it justifies *a posteriori* the incoherent summation technique used to model the electron beam phase space topology.

To study the coherence of the radiation produced by an ensemble of electrons subjected to the drive laser field, a simple plane-wave model suffices: the electron four-velocity is described by

$$\mathbf{u}_\perp(\tau) = \mathbf{A}_\perp(\phi), \quad u_z(\tau) = u_0 + \frac{\kappa_0^{-1}}{2}\mathbf{A}_\perp^2(\phi), \quad \gamma(\tau) = \gamma_0\frac{\kappa_0^{-1}}{2}\mathbf{A}_\perp^2(\phi). \tag{10.55}$$

Here, $u_0 = \gamma_0\beta_0$ is the initial electron momentum.

The four-position of the interacting electron is then

$$x_\mu(\phi) = x_\mu(\phi = 0) + \int_0^\phi \frac{dx_\mu}{d\tau}\frac{d\tau}{d\phi}d\psi = x_{\mu 0} + \kappa_0^{-1}\int_0^\phi u_\mu(\psi)d\psi. \tag{10.56}$$

For a number of independent electrons, provided that space-charge effects can be neglected, the dynamics are identical, except for the fact that the initial positions vary. Again, as discussed by Jackson, the distribution of photons radiated per unit solid angle, per unit frequency is

$$\frac{d^2N_x(\omega, \hat{\mathbf{n}})}{d\omega d\Omega} = \frac{\alpha}{4\pi^2}\frac{\omega}{\kappa_0^2}\left|\sum_{n=1}^{N_e}\int_{-\infty}^{+\infty}\hat{\mathbf{n}}\times[\hat{\mathbf{n}}\times\mathbf{u}_n(\phi)]\exp\{i\omega[\phi + z_n(\phi) - \hat{\mathbf{n}}\cdot\mathbf{x}_n(\phi)]\}d\phi\right|^2, \tag{10.57}$$

where $N_x$ is the photon number, $\hat{\mathbf{n}}$ is the unit vector in the direction of observation, $\alpha$ is the fine-structure constant, $N_e$ is the electron number, and where we have used the phase as the independent variable, as well as the invariance of $\kappa = \kappa_0$.

As discussed above, $\mathbf{u}_n(\phi)$ can be replaced by the single-electron result obtained in Equation 10.55.

In the case of Compton backscattered radiation ($\hat{\mathbf{n}} = -\hat{z}$), Equation 10.57 reduces to

$$\frac{d^2N_x(\omega,-\hat{z})}{d\omega\, d\Omega} = \frac{\alpha}{4\pi^2}\frac{\omega}{\kappa_0^2}\left|\sum_{n=1}^{N_e}\int_{-\infty}^{+\infty}\mathbf{A}_\perp(\phi)\exp\{i\omega[\phi+2z_n(\phi)]\}d\phi\right|^2. \quad (10.58)$$

The axial position of each radiating electron must now be specified:

$$z_n(\phi) = z_n + \frac{u_0}{\kappa_0}\phi + \frac{1}{2\kappa_0^2}\int_0^\phi \mathbf{A}_\perp^2(\psi)d\psi, \quad (10.59)$$

where $z_n$ is the initial position of the $n$-th electron. $\int_0^\phi \mathbf{A}_\perp^2(\psi)d\psi$ represents the relativistic mass correction of the "dressed" electron within the high intensity laser pulse. The integral over phase and the sum over electrons are now separated:

$$\frac{d^2N_x(\omega,-\hat{z})}{d\omega\, d\Omega} = \frac{\alpha}{4\pi^2}\chi\left|\int_{-\infty}^{+\infty}\mathbf{A}_\perp(\phi)\exp\left\{i\chi\left[\phi+\int_0^\phi \mathbf{A}_\perp^2(\psi)d\psi\right]\right\}\right|^2\left|\sum_{n=1}^{N_e}\exp(i2\omega z_n)\right|^2. \quad (10.60)$$

Here, $\chi = \omega[(1+\beta)/(1-\beta_0)]$ is the normalized Doppler-shifted frequency and $|\sum_{n=1}^{N_e}\exp(i2\omega z_n)|^2$ is the coherence factor.

In the case of a uniform initial electron distribution with random phase, illustrated in Figure 10.9 (top), the coherence factor is simply the amplitude of a sum of phasors, each with unit length and a random angle. To show that the average length of the sum is given by $\sqrt{N_e}$, one can simply use a proof by recurrence. Let us first assume that the average length of the first $n$ phasors is $\sqrt{n}$; we now add a vector of unit length with random orientation, as shown in Figure 10.9 (middle), and the new vector has a length given by $\sqrt{(\sqrt{n}+\cos\theta)^2+\sin^2\theta} = \sqrt{n+1+2\sqrt{n}\cos\theta}$. To obtain the average length of the new vector, we perform the integral over the random angle $\theta$:

$$\left\langle\sqrt{|(\sqrt{n}+\cos\theta)\hat{x}+\hat{y}\sin\theta|^2}\right\rangle = \left[n+1+\frac{\sqrt{n}}{\pi}\int_0^{2\pi}\cos\theta\, d\theta\right]^{1/2} = \sqrt{n+1}, \quad (10.61)$$

which proves the recurrence.

We have thus shown that $\langle|\sum_{n=1}^{N_e}\exp(i\theta_n)|^2\rangle = N_e$; this result is independent of the radiation frequency, because there are no boundary conditions to define a length scale. In a realistic situation, the initial electron distribution naturally defines a transition from coherent to incoherent radiation for a given wavelength. For electron distributions much shorter than that wavelength, the radiation process is expected to be coherent, while for longer bunches one should obtain incoherent radiation.

To model this situation properly, the derivation presented above must be modified. The key point in the derivation is to replace the average over the random phase angle performed in Equation 10.61 by a weighted average including the probability density of the initial phase. As derived previously, the length $\mathcal{L}(n+1)$ of $n+1$ phasors is given by

$$\mathcal{L}^2(n+1) = [\mathcal{L}(n) + \cos\theta]^2 + \sin^2\theta = \mathcal{L}^2(n) + 1 + 2\mathcal{L}(n)\cos\theta. \quad (10.62)$$

Averaging Equation 10.62 over the random phase angle and taking the limit where $n \gg 1$, we find that

$$\mathcal{L}(n) \simeq \sqrt{n(1-\langle\cos\theta\rangle^2)} + n\langle\cos\theta\rangle, \quad (10.63)$$

with a relative error equal to $1/\sqrt{n}$. The accuracy of this solution is illustrated in Figure 10.10 (top), where the behavior of $\mathcal{L}(n)$, calculated exactly with a computer, is shown as a function of $\langle\cos\theta\rangle$, and compared to Equation 10.63 for $n = 10^6$; the precision is excellent.

For incoherent radiation, $\langle\cos\theta\rangle = 0$, and we recover the linear scaling of the radiated power with $n$; the case of coherent radiation corresponds to $\langle\cos\theta\rangle = 1$, for which the power scales as $n^2$. At this point, the averaging over the random phase angle must be specified. We have

$$\langle\cos\theta\rangle = \int_{-\infty}^{+\infty} P(\theta)\cos[\theta(z)]d\theta, \quad (10.64)$$

where $P(\theta)$ is the probability density for the initial phase of the electron.

$P(\theta)$ is directly related to the initial electron distribution by the relation $\theta = 2\omega z$ and is normalized: $\int_{-\infty}^{+\infty} P(\theta)d\theta = 1$. Here, a Gaussian bunch of width $\Delta z$ is considered. The probability density takes the form $P[\theta(z)] = (1/\sqrt{\pi}2\omega\Delta z)\exp[-(z/\Delta z)^2]$, and the average over the phase angle becomes

$$\langle\cos(\theta)\rangle = \eta(a) = \sqrt{\frac{a}{\pi}}\int_{-\infty}^{+\infty} e^{-a\theta^2}\cos\theta\, d\theta, \quad (10.65)$$

where $a = 1/(2\omega z)^2$. The integral in Equation 10.65 can be performed analytically to obtain

$$\langle\cos\theta\rangle = \exp(-\omega^2\Delta z^2) = e^{-1/2a}. \quad (10.66)$$

In the case where the electron distribution is much longer than the radiation wavelength, $a \ll 1$, and $\eta(a) \to 0$; the coherence exponent approaches one half, and the radiation is incoherent. When the electron distribution is much shorter than the radiation wavelength, the probability density distribution approaches a Dirac delta-function, with $\lim_{a\to\infty}(\sqrt{a/\pi}e^{-a\theta^2}) = \delta(\theta)$, and the radiated power scales as $N_e^2$. This is illustrated in Figure 10.9. For incoherent scattering (top), each phasor angle has equal probability; $\langle\cos\theta\rangle = 0$, and the

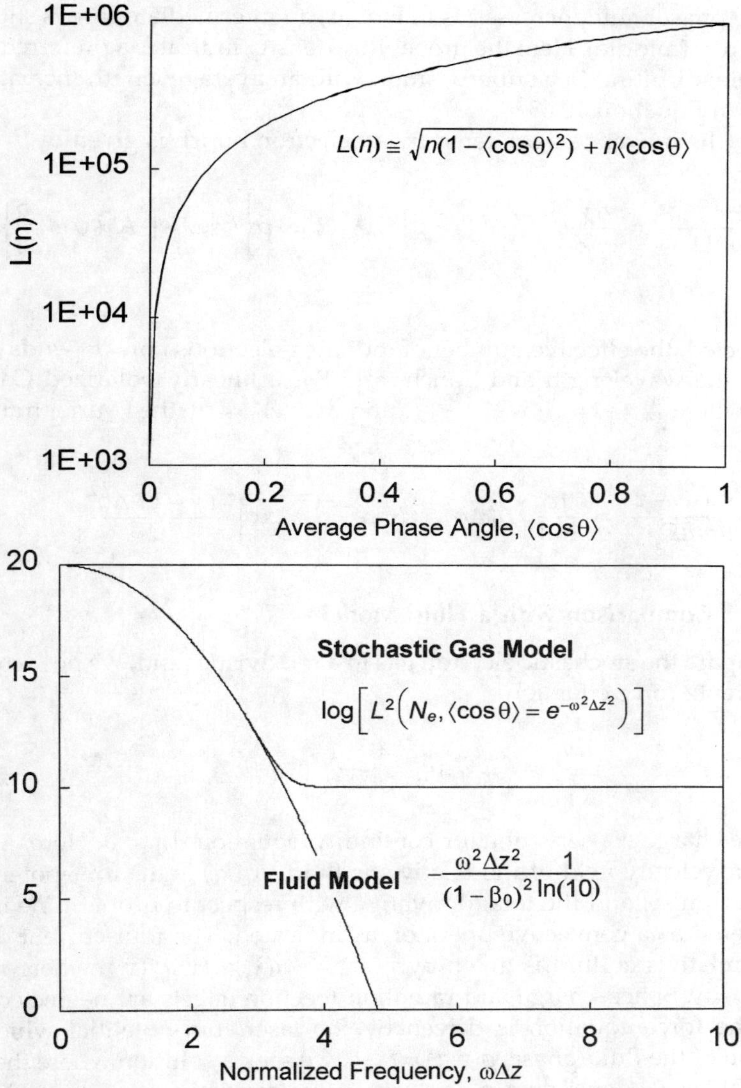


**FIGURE 10.10**
Top: comparison between the average length of $n = 10^6$ phasors, as calculated exactly with a computer and as derived analytically. Bottom: logarithm of the effective number of radiating electrons as a function of the normalized bunch length, for both the stochastic electron gas and fluid models.

resulting superposition increases as $\sqrt{n}$. The case of fully coherent radiation is shown in Figure 10.9 (bottom); here the angular probability density is a Dirac delta-function, where $P(\theta) = \delta(\theta - \theta_0)$, the interference term, $\langle\cos\theta\rangle = 1$, and the phasor sum increases as $n$. Finally, an intermediate case is shown in Figure 10.9 (middle). Here the probability density indicates a preferred angular range, resulting in a superposition with an average length increasing as shown in Equation 10.63.

Thus, the power backscattered by an electron bunch is given by

$$\frac{d^2N_x(\omega,-\hat{z})}{d\omega d\Omega} = \frac{\alpha\chi}{4\pi^2}\mathcal{L}^2(N_e, e^{-\omega^2\Delta z^2})\left|\int_{-\infty}^{+\infty}\mathbf{A}_\perp(\phi)\exp\left\{i\chi\left[\phi+\int_0^{\phi}\mathbf{A}_\perp^2(\psi)d\psi\right]\right\}d\phi\right|^2; \tag{10.67}$$

as expected, the effective number of radiating electrons now depends explicitly on the wavelength and bunch size. For a linearly polarized Gaussian pulse, where $\mathbf{A}_\perp(\phi) = A_0\hat{x}e^{-i\phi-(\phi/\Delta\phi)^2}$, and with $A_0^2 \ll 1$, the Fourier transform yields

$$\frac{d^2N_x(\omega,-\hat{z})}{d\omega d\Omega} = \frac{\alpha}{16\pi}\chi A_0^2\Delta\phi^2\mathcal{L}^2(N_e, e^{-\omega^2\Delta z^2})\exp\left[-\frac{(\chi-1)^2\Delta\phi^2}{2}\right]. \tag{10.68}$$

### 10.5.1 Comparison with a Fluid Model

To compare the stochastic electron gas to a relativistic fluid, we now consider the Lorentz force equation

$$\frac{du_\mu}{d\tau} = (u^\nu\partial_\nu)u_\mu = -(\partial_\mu A_\nu - \partial_\nu A_\mu)u^\nu, \tag{10.69}$$

and the charge conservation, or continuity, equation $\partial_\mu j^\mu = 0$. Here, $u_\mu(x_\nu)$ is the four-velocity field of the relativistic fluid, $A_\mu(x_\nu)$ is the four-potential of the laser pulse, and the total derivative with respect to proper time is to be considered as a convective operator, as indicated. The four-current density of the relativistic fluid is given by $j_\mu(x_\nu) = -n(x_\nu)u_\mu(x_\nu)/\gamma(x_\nu)$, where $n(x_\nu)$ is the density. Space–charge and radiation reaction effects are neglected.

As the force equation is driven by the laser four-potential which is a function of the fluid phase, $\phi(z, t) = t - z$, we seek a solution where the other fluid fields also depend on $\phi$. The convective derivative operator reduces to

$$(u^\nu\partial_\nu)u_\mu(z,t) \equiv [\gamma\partial_t + u_z\partial_z]u_\mu(\phi) = \left(\gamma\frac{\partial\phi}{\partial t} + u_z\frac{\partial\phi}{\partial z}\right)\frac{du_\mu}{d\phi} = (\gamma - u_z)\frac{du_\mu}{d\phi}. \tag{10.70}$$

The Lorentz force equation now reads $(\gamma - u_z)(d\mathbf{u}/d\phi) = -(\gamma\mathbf{E} + \mathbf{u}\times\mathbf{B})$; in addition, energy conservation yields $(\gamma - u_z)(d\gamma/d\phi) = -\mathbf{u}\cdot\mathbf{E}$. The electromagnetic field components are given by $F_{\mu\nu} = \partial_\mu A_\nu - \partial_\nu A_\mu$. The evolution of the momentum field can be separated into a transverse and an axial component:

$$(\gamma - u_z)\frac{d}{d\phi}(\mathbf{u}_\perp - \mathbf{A}_\perp) = \mathbf{0}, \quad \text{and} \quad (\gamma - u_z)\frac{du_z}{d\phi} = \mathbf{u}_\perp \cdot \frac{d\mathbf{A}_\perp}{d\phi} = (\gamma - u_z)\frac{d\gamma}{d\phi}. \tag{10.71}$$

We recover the transverse canonical momentum invariant, $\mathbf{u}_\perp - \mathbf{A}_\perp$; Equation 10.71 also shows that the light-cone variable is a fluid invariant. The sought-after fluid equilibrium is

$$u_z(z,t) = u_z(\phi) = u_{z0} + \kappa_0^{-1}\left[\mathbf{A}_\perp(\phi)\cdot\mathbf{u}_{\perp 0} + \frac{\mathbf{A}_\perp^2(\phi)}{2}\right], \tag{10.72}$$

and

$$\gamma(z,t) = \gamma(\phi) = \gamma_0 + \kappa_0^{-1}\left[\mathbf{A}_\perp(\phi)\cdot\mathbf{u}_{\perp 0} + \frac{A_\perp^2(\phi)}{2}\right]. \tag{10.73}$$

To determine the density, we seek a solution to the charge conservation equation where the density field is a function of the phase. The continuity equation reads

$$\partial_t n(\phi) + \nabla\cdot[n(\phi)\beta(\phi)] = \frac{\partial\phi}{\partial t}\frac{dn}{d\phi} + \frac{\partial\phi}{\partial z}\frac{d}{d\phi}(n\beta_z) = \frac{d}{d\phi}[n(1-\beta_z)] = 0. \tag{10.74}$$

Multiplying and dividing the density by the energy, we can rewrite Equation 10.74 in terms of the light-cone variable,

$$\frac{d}{d\phi}\left[\frac{n}{\gamma}(\gamma - u_z)\right] = \frac{d}{d\phi}\left(\frac{n}{\gamma}\kappa\right) = \kappa_0\frac{d}{d\phi}\left(\frac{n}{\gamma}\right) = 0: \tag{10.75}$$

$n(\phi)/\gamma(\phi)$ is a relativistic fluid invariant. This is an important result, as it shows that the relativistic plasma frequency is invariant. The density modulation induced by the laser radiation pressure exactly compensates the variation of the fluid energy within the pulse. We finally find

$$n(z,t) = n(\phi) = n_0\left\{1 + \kappa_0^{-1}\left[\mathbf{A}_\perp(\phi)\cdot\mathbf{u}_{\perp 0} + \frac{\mathbf{A}_\perp^2(\phi)}{2}\right]\right\}, \tag{10.76}$$

where $n_0$ is the initial electron beam density.

Having defined the fluid dynamical variables, we can derive the distribution of energy radiated by the fluid per unit solid angle, per unit frequency, by Fourier transforming the four-current into momentum space:

$$\frac{d^2N_x(\omega,\hat{\mathbf{n}})}{d\omega d\Omega} = \frac{\alpha}{4\pi^2}\omega\left|\int_{\mathbb{R}^4} d^4x_\mu \hat{\mathbf{n}}\times[\hat{\mathbf{n}}\times\mathbf{j}(x_\mu)]\exp[i\omega(t - \hat{\mathbf{n}}\cdot\mathbf{x})]\right|^2, \tag{10.77}$$

where $\mathbf{j}(x_\mu)$ is given by $\mathbf{j} = -n\beta = -n_0(\gamma/\gamma_0)\beta = -(n_0/\gamma_0)\mathbf{u}$.

The initial density distribution of the unmodulated beam has the form $n_0(x_\mu) = \rho f(\zeta)$, where we have defined the electron beam phase $\zeta(z, t) = z - \beta_0 t$, and where $f(\zeta)$ is the normalized axial envelope of the bunch, which propagates with the axial velocity $\beta_0$. The normalization constant $\rho$ is defined by the total charge in the bunch: $\rho \int_{-\infty}^{+\infty} dz\, f(\zeta) = N_e$. Note that with the initial density field used here, the charge conservation equation is automatically satisfied. This model for the background fluid density is valid as long as its spatial and temporal gradients are small compared to $k_0$ and $\omega_0$. The radiated photon number density now takes the form

$$\frac{d^2N_x(\omega, \hat{\mathbf{n}})}{d\omega d\Omega} = \frac{\alpha}{4\pi^2}\frac{\omega}{\gamma_0^2}\rho^2\left|\int_{\mathbb{R}^2} dz\, dt\, f(\zeta)\,\hat{\mathbf{n}}\times[\hat{\mathbf{n}}\times\mathbf{u}(x_\mu)]\exp[i\omega(t-\hat{\mathbf{n}}\cdot\mathbf{x})]\right|^2. \tag{10.78}$$

We use a Gaussian profile to obtain an analytical result for the radiated spectra. The calculations presented here can easily be generalized to other distributions. We have $f(\zeta) = \exp[-(\zeta/\Delta z)^2]$, and $\rho = N_e/\sqrt{\pi}\Delta z$, where $\Delta z$ is the axial scale of the electron bunch;

$$\frac{d^2N_x(\omega, -\hat{z})}{d\omega d\Omega} = \frac{\alpha}{4\pi^2}\frac{\omega}{\gamma_0^2}\rho^2 A_0^2\left|\int_{\mathbb{R}^2} dz\, dt\, f(\zeta)\, g(\phi)\exp\{i[\omega(t+z)-\phi]\}\right|^2. \tag{10.79}$$

Here, a linearly polarized laser pulse, with temporal envelope $g(\phi)$, is considered.

The integrals over axial position and time can be separated by using $\zeta(z,t) = z - \beta_0 t$, and $\phi(z, t) = t - z$, as independent variables. The product of the differential elements is obtained using the Jacobian of the transform:

$$dz\, dt = \begin{vmatrix} \dfrac{\partial z}{\partial \phi} & \dfrac{\partial z}{\partial \zeta} \\ \dfrac{\partial t}{\partial \phi} & \dfrac{\partial t}{\partial \zeta} \end{vmatrix} d\phi\, d\zeta = -\frac{d\phi\, d\zeta}{1-\beta_0}. \tag{10.80}$$

Equation 10.79 now takes the simple form

$$\frac{d^2N_x(\omega, -\hat{z})}{d\omega d\Omega} = \alpha\chi\rho^2 A_0^2\left|\int_{-\infty}^{+\infty} d\zeta\, \exp\left[i\frac{2\chi\zeta}{1+\beta_0} - \frac{\zeta^2}{\Delta z^2}\right]\int_{-\infty}^{+\infty} d\phi \exp\left[i(\chi-1)\phi - \frac{\phi^2}{\Delta\phi^2}\right]\right|^2 \tag{10.81}$$

The first integral can be identified as the fluid coherence factor, while the second integral corresponds to the spectral density of the Doppler-shifted Compton backscattered line. Performing the integrals over $\phi$ and $\zeta$, we finally obtain

$$\frac{d^2N_x(\omega,-\hat{z})}{d\omega d\Omega} = \frac{\alpha}{16\pi}\chi A_0^2\Delta\phi^2 N_e^2 \exp\left[-2\left(\frac{\chi\Delta z}{1+\beta_0}\right)^2\right]\exp\left[-\frac{(\chi-1)^2\Delta\phi^2}{2}\right]. \quad (10.82)$$

The physics of this solution can be understood as follows: the photon number density radiated on-axis scales linearly with the laser pulse energy $A_0^2\Delta\phi$, and quadratically with the number of electrons in the bunch, while the coherence factor scales exponentially with the inverse electron bunch duration squared, measured in units of the Doppler upshifted wavelength. This indicates that long electron bunches radiate incoherently at short wavelengths; also note that the fluid incoherent radiation scaling is quite different from the simple $N_e$ scaling. It is a function of frequency and shows very strong (i.e., exponential) suppression of short wavelengths. Finally, the Compton backscattered spectral linewidth is determined by the laser pulse linewidth $\sqrt{2}/\Delta\phi$, and its frequency is obtained for $\chi = 1$. This corresponds to the frequency $\omega = [\gamma_0(1-\beta_0)]^2$; for $\beta_0 \to -1$, we recover the result $\omega \simeq 4\gamma_0^2$.

The stochastic electron gas and relativistic fluid models are compared by inspecting Equations 10.68 and 10.82, in a frame where the initial electron distribution is at rest ($\beta_0 = 0$). The difference between these theoretical models is shown in Figure 10.10 (bottom), where the number of electrons has been chosen as $N_e = 10^{10}$. The logarithm of both coherence factors is calculated as a function of the normalized electron distribution axial scale length, $\omega\Delta z$. In the case of a perfectly coherent radiation process, where $\omega\Delta z \to 0$, both models yield the well-known $N_e^2$ scaling. When the electron distribution becomes long compared to the radiation wavelength, the stochastic electron gas model correctly predicts the linear scaling with $N_e$. The fluid model yields a very different result: the coherence factor continues to decrease exponentially, as shown by the parabolic curve in Figure 10.10 (bottom). This is due to the fact that the Fourier transform of the Gaussian fluid distribution is a Gaussian with an argument proportional to the product $\omega\Delta z$. For arbitrarily short wavelengths, the fluid four-current yields a vanishingly small Fourier component. The fluid model introduces an unphysical cutoff scale given by the length of the electron distribution. Thus, the fundamental difference between the stochastic electron gas approach and the relativistic fluid model resides in the fact that, for any number of incoherently phased point electrons, the four-current contains Fourier components at arbitrarily short wavelengths, whereas the fluid model introduces an unphysical cutoff scale. Therefore, the discrete and point-like nature of electric charge is shown to play a fundamental role in the physics of incoherent radiation processes.

## 10.6 Harmonics and Nonlinear Radiation Pressure

In Compton scattering, harmonic production is directly related to a modulation of the Doppler-shift term in the radiation equation. As discussed earlier, the spectral brightness for a single electron is given by

$$\frac{d^2N_x(\omega,\hat{\mathbf{n}})}{d\omega d\Omega} = \frac{\alpha}{4\pi^2}\frac{\omega}{\kappa_0^2}\left|\int_{-\infty}^{+\infty}\hat{\mathbf{n}}\times[\hat{\mathbf{n}}\times\mathbf{u}(\phi)]\exp\{i\omega[\phi+z(\phi)-\hat{\mathbf{n}}\cdot\mathbf{x}(\phi)]\}d\phi\right|^2. \tag{10.83}$$

Therefore, the position of the radiating electron is required to determine the spectral brightness of the scattered light, as shown in the Doppler shift term, corresponding to the argument of the exponential in Equation 10.83; this is obtained by performing the following integral: $x_\mu(\phi) = x_{\mu 0} + \kappa_0^{-1}\int_0^\phi u_\mu(\vartheta)d\vartheta$. At this point, a number of important remarks are in order. First, for a number of radiating electrons, the various initial positions must be taken into account. This forms the basis of the stochastic electron gas (SEG) theory of coherence in Compton scattering presented in Section 10.5. Second, the linear motion of the electron in the laser field yields a modulation of the Doppler shift, which is at the origin of harmonics radiated off-axis, as discussed in detail by Esarey and co-authors. The origin of this effect is the projected term in the Doppler shift: for $\hat{\mathbf{n}}\neq\pm\hat{z}$, the argument of the complex exponential, $\hat{\mathbf{n}}\cdot\mathbf{x}(\phi) = z(\phi)\cos\theta + x_\perp(\phi)\sin\theta$, now contains a term which is linearly proportional to the laser four-potential, as $x_\perp(\phi) = x_{\perp 0} + \kappa_0^{-1}\int_0^\phi u_\perp(\psi)d\psi = x_{\perp 0} + \kappa_0^{-1}\int_0^\phi A_\perp(\psi)d\psi$. This term modulates the off-axis Doppler shift, yielding off-axis harmonic generation. This mechanism is quite similar to the Bessel function modulation of synchrotron radiation due to the finite displacement of the radiating electron bunch in a bend magnet. Finally, at higher intensities, in the so-called relativistic intensity regime, the ponderomotive force modulates the axial motion of the electron. This result in the radiation of harmonics on-axis, as extensively discussed by Hartemann and Kerman. This nontrivial effect, which reflects the modulation of the electron proper time by the laser pressure, has recently been measured experimentally by Umstadter and his group. For details, we refer the reader to the references at the end of this chapter.

In the case of circular polarization, the dimensionless vector potential takes the form $\mathbf{A}_\perp(\phi) = A_0 g(\phi)[\hat{x}\sin\phi + \hat{y}\cos\phi]$, which implies that the magnitude of the four-vector potential varies adiabatically as the pulse intensity envelope: $A_\mu A^\mu = \mathbf{A}_\perp^2(\phi) = A_0^2 g^2(\phi)$. Furthermore, for a hyperbolic secant pulse, the full nonlinear spectrum can also be determined analytically. With $g(\phi) = \cosh^{-1}(\phi/\Delta\phi)$, the electron's axial position is

$$\int_{-\infty}^\phi \mathbf{A}_\perp^2(\psi)d\psi = \int_{-\infty}^\phi \frac{A_0^2}{\cosh^2\left(\frac{\psi}{\Delta\phi}\right)}d\psi = A_0^2\Delta\phi\left[1+\tanh\left(\frac{\phi}{\Delta\phi}\right)\right], \tag{10.84}$$

and the nonlinear backscattered spectrum is now proportional to

$$\chi\left|A_0 e^{i\chi A_0^2\Delta\phi}\int_{-\infty}^{+\infty}\frac{\hat{x}\cos\phi+\hat{y}\sin\phi}{\cosh\left(\frac{\phi}{\Delta\phi}\right)}\exp\left\{i\chi\Delta\phi\left[\frac{\phi}{\Delta\phi}+A_0^2\tanh\left(\frac{\phi}{\Delta\phi}\right)\right]\right\}d\phi\right|^2. \quad (10.85)$$

The Fourier transform can be evaluated analytically by performing two changes of variables: namely, we first set $y = e^{\phi/\Delta\phi}$, then $x = (y^2 - 1)/(y^2 + 1)$, with the result that

$$\frac{d^2N_x(\omega,-\hat{z})}{d\omega d\Omega} = \frac{\alpha}{8}\chi A_0^2\Delta\phi^2\left\{\sum_{\pm}\left|\frac{\Phi(\mu_\pm,1,2iA_0^2\chi\Delta\phi)}{\cosh\left[\frac{\pi}{2}\Delta\phi(\chi\pm 1)\right]}\right|^2\right\}. \quad (10.86)$$

Here $\Phi$ is the degenerate (confluent) hypergeometric function, and the argument $\mu_\pm = (1/2)\left[1 + i\Delta\phi(\chi \pm 1)\right]$.

The downshift of the Compton backscattered line due to radiation pressure is determined by considering the nonlinear phase in the Fourier integral. The argument of the complex exponential takes the form $\Lambda(\phi) = (\chi - 1)\phi + \chi A_0^2\Delta\phi\tanh(\phi/\Delta\phi)$. We Taylor expand this nonlinear phase around $\phi = 0$ by using the fact that $\tanh x \simeq x$, to obtain $\Lambda(\phi) \simeq [\chi(1 + A_0^2) - 1]\phi + O(\phi^3)$. The frequency of the main Compton backscattered line is obtained by canceling the linear coefficient of the expansion, which yields $\chi = (1 + A_0^2)^{-1}$. This can be recast in the more familiar form $\omega = 1/(1 + A_0^2)[(1 - \beta_0)/(1 + \beta_0)] \simeq 4\gamma_0^2/(1 + A_0^2)$, where the last equality holds for $\beta_0 \to -1$. This is then well-known radiation frequency for an FEL with an electromagnetic wiggler.

Finally, we note that temporal shaping of the drive laser pulse can increase the contrast between the main radiated line and the satellites produced by the transient during the rise and fall of the pulse. We have demonstrated theoretically that square optical pulses, such as those studied experimentally by Weiner et al., could be used to filter efficiently the spectral content radiated on-axis, in the case of a circularly polarized laser pulse.

## 10.7 Radiative Corrections: Overview

This section is intended as a brief introduction to the Dirac–Lorentz equation and its main properties; for a considerably more detailed discussion, we refer the reader to Sections 10.8 through 10.14.

Radiative correction corresponds to the electron interaction with the electromagnetic fields it scatters and can be treated using quantum electrodynamics (QED), or classically, using the Dirac–Lorentz equation. The latter equation describes the covariant dynamics of a classical point electron, including the radiation reaction effects due to the electron self-interaction. The main steps of Dirac's derivation are briefly outlined here; for conciseness, we now use the classical electron radius, $r_0$, to measure length, and the corresponding time unit, $r_0/c$. In these units, the vacuum permittivity is $\varepsilon_0 = 1/4\pi$, and its permeability is $\mu_0 = 4\pi$. The electron four-current density is

$$j^s_\mu(x_\lambda) = -\int_{-\infty}^{+\infty} u_\mu(x'_\lambda)\delta_4(x_\lambda - x'_\lambda)d\tau', \tag{10.87}$$

and the corresponding self-electromagnetic field $F^s_{\mu\nu} = \partial_\mu A^s_\nu - \partial_\nu A^s_\mu$ satisfies the driven wave equation, $\square A^s_\mu(x_\lambda) = -4\pi j^s_\mu(x_\lambda)$, which can be solved in terms of Green functions as $A^s_\mu(x_\lambda) = 4\pi\int_{-\infty}^{+\infty} u_\mu(x'_\lambda)G(x_\lambda - x'_\lambda)d\tau'$.

The self-force on the electron can now be evaluated as

$$F^s_\mu = -(\partial_\mu A^s_\nu - \partial_\nu A^s_\mu)u^\nu = -\int_{-\infty}^{+\infty} u^\nu(x_\lambda)[u_\nu(x'_\lambda)\partial_\mu - u_\mu(x'_\lambda)\partial_\nu]G(x_\nu - x'_\nu)d\tau'. \tag{10.88}$$

The advanced and retarded Green functions both depend on the space–time interval $s^2 = (x - x')_\mu(x - x')^\mu$: $G^\pm = -\delta(s^2)[1 \mp (x_0 - x'_0)/(|x_0 - x'_0|)]$. As a result, the partial derivatives can now be replaced by the operator $\partial_\mu \equiv 2(x_\mu - x'_\mu)(\partial/\partial s^2)$, and Equation 10.88 reads

$$F^s_\mu = -2\int_{-\infty}^{+\infty} u^\nu(x_\lambda)[u_\nu(x'_\lambda)(x_\mu - x'_\mu) - u_\mu(x'_\lambda)(x_\nu - x'_\nu)]\frac{\partial G}{\partial s^2}d\tau'. \tag{10.89}$$

At this point, the new variable $\tau'' = \tau - \tau'$ is introduced, so that the range of integration explicitly includes the electron (singular point $\tau'' = 0$). To evaluate the integral in Equation 10.89, one can now use Taylor–McLaurin expansions in powers of $\tau''$:

$$\begin{aligned} x_\mu - x'_\mu &= \tau'' u_\mu - \frac{1}{2}\tau''^2 a_\mu + \frac{1}{6}\tau''^3 d_\tau a_\mu + \cdots \\ u_\mu(x'_\lambda) &= u_\mu(\tau - \tau'') = u_\mu - \tau'' a_\mu + \frac{1}{2}\tau''^2 d_\tau a_\mu + \cdots, \end{aligned} \tag{10.90}$$

where the four-acceleration $a_\mu = d_\tau u_\mu$. Using the above expansions, one first finds that $s^2 \simeq \tau''^2$, which yields $\partial G/\partial s^2 \simeq -(1/2\tau'')(\partial G/\partial \tau'')$. With this, the expression for the self-electromagnetic force now reads

$$F^s_\mu \simeq \int_{-\infty}^{+\infty}\left\{-\frac{\tau''}{2}a_\mu + \frac{\tau''^2}{3}\left[\frac{da_\mu}{d\tau} - u_\mu(a_\nu a^\nu)\right]\right\}\frac{\partial G}{\partial \tau''}d\tau''. \tag{10.91}$$

This equation can be integrated by parts; using the retarded (causal) Green function, one finds

$$F^{s}_{\mu} \simeq -\frac{1}{2}a_{\mu}\int_{-\infty}^{+\infty}\frac{\delta(\tau'')}{|\tau''|}d\tau'' + \frac{2}{3}\left[\frac{da_{\mu}}{d\tau} - u_{\mu}(a_{\nu}a^{\nu})\right]. \tag{10.92}$$

The corresponding four-momentum transfer equation now reads

$$\left[1+\frac{1}{2}\int_{-\infty}^{+\infty}\frac{\delta(\tau'')}{|\tau''|}d\tau''\right]a_{\mu} = -F_{\mu\nu}u^{\nu} + \tau_0\left[\frac{da_{\mu}}{d\tau} - u_{\mu}(a_{\nu}a^{\nu})\right], \tag{10.93}$$

where $\tau_0 = 2/3$ is the Compton time scale, in the units of $r_0/c$ used here.

The divergent integral on the left-hand side of the equation is the infinite electromagnetic mass of the point electron, which multiplies the four-acceleration. Dirac first proposed to renormalize this term away by using the time symmetrical Green function $G = 1/2(G^{+} - G^{-})$; with this one obtains the Dirac–Lorentz equation,

$$a_{\mu} = -F_{\mu\nu}u^{\nu} + \tau_0\left[\frac{da_{\mu}}{d\tau} - u_{\mu}(a_{\nu}a^{\nu})\right]. \tag{10.94}$$

One of the conceptual difficulties associated with this equation is the existence of unphysical, runaway solutions. Contracting Equation 10.94 with $a^{\mu}$, it is easily seen that, in the absence of an external field, it reduces to $a_{\mu}a^{\mu} = (\tau_0/2)(d/d\tau)(a_{\mu}a^{\mu})$, which admits the runaway solution $[a_{\mu}a^{\mu}](\tau) = [a_{\mu}a^{\mu}]_0 \exp[2(\tau/\tau_0)]$. To avoid these unphysical solutions, one must require that the Dirac–Rohrlich (DR) asymptotic condition be satisfied: $\lim_{\tau\to\pm\infty} a_{\mu}(\tau) = 0$.

Also note that the second radiative correction term corresponds to the radiated four-momentum, $H_{\mu}$. Thus we can rewrite Equation 10.94 as

$$a_{\mu} = -F_{\mu\nu}u^{\nu} + \tau_0\frac{da_{\mu}}{d\tau} - \frac{dH_{\mu}}{d\tau}. \tag{10.95}$$

In the case of an external electric field deriving from a static potential, the time-like component of the Dirac–Lorentz equation, which describes energy conservation, takes the simple form

$$\frac{d\gamma}{d\tau} = \mathbf{u}\cdot\nabla\varphi + \tau_0\frac{d^2\gamma}{d\tau^2} - \frac{dH_0}{d\tau} = \frac{d}{d\tau}\left[\varphi + \tau_0\frac{d\gamma}{d\tau} - H_0\right], \tag{10.96}$$

and can be formally integrated to yield the conservation law

$$\Delta(\gamma - \varphi + H_0) = \tau_0\left[\frac{d\gamma}{d\tau}\right]_{-\infty}^{+\infty}, \tag{10.97}$$

which indicates that, provided the DR asymptotic condition $\lim_{\tau\to\pm\infty}[d\gamma/d\tau] = 0$ is satisfied, the electron potential energy is converted into kinetic energy and radiation.

In the case of nonlinear Compton scattering, the Dirac–Lorentz equation can be given as

$$\begin{aligned} a_\mu &= L_\mu + \tau_0\left[\frac{da_\mu}{d\tau} - u_\mu(a_\nu a^\nu)\right], \\ L_\perp &= \kappa \mathbf{E}_\perp, \quad L_z = L_0 = \mathbf{u}_\perp \cdot \mathbf{E}_\perp, \end{aligned} \tag{10.98}$$

where we recognize the light-cone variable $\kappa = \gamma - u_z$, and the laser transverse electric field. $\tau_0 = (2/3)(r_0/c) = 0.626 \times 10^{-23}$ s is the Compton time-scale. Subtracting the axial component of Equation 10.98 from the temporal component, we obtain an equation governing the evolution of $\kappa$,

$$\frac{d\kappa}{d\tau} = \tau_0\left[\frac{d^2\kappa}{d\tau^2} - \kappa(a_\nu a^\nu)\right]. \tag{10.99}$$

Introducing the small parameter $\varepsilon = \omega_0\tau_0$, which measures the Doppler-shifted laser wavelength in units of $r_0$, and noting that $\mathbf{E}_\perp = \varepsilon(d\mathbf{A}_\perp/d\phi)$, we also obtain an equation governing the evolution of the canonical momentum:

$$\frac{d}{d\tau}(\mathbf{u}_\perp - \mathbf{A}_\perp) = \tau_0\left[\frac{d^2\mathbf{u}_\perp}{d\tau^2} - \mathbf{u}_\perp(a_\nu a^\nu)\right]. \tag{10.100}$$

Now using the laser phase, $\phi$, as the independent variable, Equation 10.99 reads

$$\frac{d\kappa}{d\phi} = \varepsilon\left[\frac{d^2}{d\phi^2}\left(\frac{\kappa^2}{2}\right) - \kappa^2\left(\frac{du_\mu}{d\phi}\frac{du^\mu}{d\phi}\right)\right]. \tag{10.101}$$

Since the right-hand side of Equation 10.101 is at least of order $\varepsilon$, we can replace the terms in the brackets with their zeroth-order (Lorentz dynamics) approximation. In this case, we obtain a simple differential equation for the light-cone variable perturbation

$$\frac{d}{d\phi}\left[\frac{1}{\kappa(\phi)}\right] \simeq \varepsilon A_0^2 g^2(\phi), \tag{10.102}$$

where we recognize the envelope of the circularly polarized laser pulse. Equation 10.102 can easily be integrated to yield

$$\frac{1}{\kappa(\phi)} = \frac{1}{\kappa_0} + \varepsilon A_0^2 \int_{-\infty}^{\phi} g^2(\psi)\,d\psi. \tag{10.103}$$

This last equation describes the electron recoil from the coherent laser field; it is clear that at sufficient intensities and short wavelengths, the relative radiative energy loss becomes significant.

Finally, the Dirac–Lorentz equation can be integrated backward in time to avoid runaways due to the electromagnetic mass renormalization, and the Compton backscattered spectrum is obtained by evaluating

$$\frac{\partial^2 N_x(\varpi, -\hat{z})}{d\varpi d\Omega} = \frac{\alpha}{4\pi^2}\varpi \left| \int_{-\infty}^{+\infty} \frac{u_\perp(\phi)}{\kappa(\phi)} \exp\left\{ i\varpi o \left[ \phi + 2\int_{-\infty}^{\phi} \frac{u_z(\psi)}{\kappa(\psi)} d\psi \right] \right\} \right|^2. \quad (10.104)$$

## 10.8 Symmetrized Electrodynamics: Introduction

Recently, close attention has been paid to the concept of duality in quantum field theories, as summarized by Witten. In particular, recent work in superstring theory has resulted in the convergence of four main themes: electromagnetic duality in four dimensions, the symmetries of supergravity, dualities in superstring theory, and gauge theory dynamics in four dimensions.

The concept of duality in electrodynamics results from the symmetry between the electric and magnetic components of the field tensor. The source-free equations of the Maxwell set are symmetrical in vacuum under the transformation $\mathbf{E} \rightarrow \mathbf{B}$ and $\mathbf{B} \rightarrow -\mathbf{E}$ in addition, the symmetry can be maintained in the presence of four-currents, provided that both electric and magnetic monopoles are introduced, thus suggesting a deeper hidden symmetry. Since Dirac's brilliant insight on charge quantization, the role and importance of magnetic monopoles and duality in electrodynamics have taken on a much more profound significance. Feynman and Wheeler first demonstrated the deep connection between time reversal and charge conjugation, while Schwinger proposed to associate the electric and magnetic charge in a single electrically charged monopole, referred to as a dyon. Such a particle should exhibit the full symmetries of the electromagnetic interaction. A comprehensive list of references on these subjects is given in the bibliography for this chapter.

Although magnetic monopoles have never been observed, it can be argued that the apparent quantization of electric charge might represent indirect evidence for the existence of magnetic charge. The argument, originally put forth by Dirac and later simplified by Saha, can be summarized as follows. If a magnetic field with nonzero divergence is added to an electric field with the same property, the total field has nonzero angular momentum, even in the static case. The field angular momentum turns out to be proportional to the product of the charges of the electric and magnetic sources involved and independent of distance. Since angular momentum is quantized and assuming that the amount of magnetic charge in the universe is finite, it follows that electric charge must be quantized. Although other explanations have been proposed for the observed quantization of charge, Dirac's argument remains the most elegant. In addition, in 1977, Montonen and Olive showed that in a limiting case of electroweak interaction theory, a particle of electric charge $q$ and magnetic charge $g$ acquires a mass $m = \langle\psi\rangle\sqrt{q^2+g^2}$ under spontaneous symmetry breaking, where $\langle\psi\rangle$ is a constant measuring the gauge symmetry breaking.

Within this theoretical context, there exists a beautiful and compelling case for studying fully symmetrized versions of classical and quantum electrodynamics (CED and QED). In addition, these theories might also provide the correct approach to demonstrating that CED indeed represents the classical limit of QED, a problem that is still unresolved. This is because the duality of fully symmetrized QED implies that if the electric fine structure constant $\alpha = e^2/2\varepsilon_0 hc \simeq 1/137.036$ and its magnetic counterpart, $1/\alpha$, are exchanged, electric and magnetic phenomena will appear to be switched for a classical observer. This, in turn, can be related to the notion of running coupling constants, used in gauge theory dynamics, where two important limiting cases might shed some light on the exact relation between QED and CED, the case where $\alpha \to \infty$, and quantum effects disappear, and the case where the full symmetry between electricity and magnetism is restored, with $\alpha = 1$.

Thus, the main thrust of this section is to present a classical derivation of radiation reaction for electric and magnetic monopoles, as well as dyons. The approach used here includes a generalization of Dirac's derivation of classical radiation reaction for a point charge from general principles, including gauge invariance and Lorentz covariance, where the double four-potential introduced by Cabbibo and Ferrari is further extended by introducing a complex electromagnetic tensor unifying the conventional electromagnetic tensor and its dual. In addition, the electric and magnetic four-currents are unified into a single complex four-vector, and the connection with electromagnetic duality is now completely explicit. Electric and magnetic charges can be rotated into one another, while preserving the global invariance of the symmetrized form of Maxwell's equations. One advantage of the complex four-potential formalism over Dirac's well-known model of magnetic monopoles is the absence of string-like singularities; in addition, within this framework, one can readily derive a Hamiltonian for dyon–dyon interactions. Finally, our complex notation proves extremely compact, allowing for an elegant derivation of the symmetrized Dirac–Lorentz equation.

## 10.9 Symmetrized Electrodynamics: Complex Notation

Here and throughout the remainder of this chapter, we use electron units, where length is measured in units of the classical electron radius, $r_0 = e^2/4\pi\varepsilon_0 m_0 c^2$, while time is measured in units of $r_0/c$, mass is measured in units of $m_0$, electric charge is measured in units of $e$, and magnetic charge in units of $\hbar/e$. In these units, $\varepsilon_0 = 1/4\pi$, $\mu_0 = 4\pi$, and for a particle of mass $m_0$, the four-momentum is equal to the four-velocity, $p_\mu = u_\mu = dx_\mu/d\tau$, where $x_\mu(\tau)$ is the world line of the particle, and $\tau$ is its proper time.

We now focus on the problem of a dyon, having both electric and magnetic charges $q$ and $g$, respectively. If magnetic sources are allowed, Maxwell's equations become symmetrized as follows:

$$\partial_\nu F^{\mu\nu} = 4\pi j^\mu, \quad \partial_\nu \tilde{F}^{\mu\nu} = 4\pi g^\mu, \tag{10.105}$$

where $j^\mu$ and $g^\mu$ correspond to the electric and magnetic four-current densities, and where $\tilde{F}^{\mu\nu}$ is the dual electromagnetic tensor. Here, the four-gradient operator is defined as usual: $\partial^\mu = (-\partial_t, \nabla)$.

Now introducing the electric four-potential $A^\mu = (\phi, \mathbf{A})$, and its magnetic counterpart, $V^\mu = (\varphi, \mathbf{V})$, the field tensor and its dual may be written as follows:

$$F^{\mu\nu} = \partial^\mu A^\nu - \partial^\nu A^\mu - \varepsilon^{\mu\nu\alpha\beta}\partial_\alpha V_\beta, \tag{10.106}$$

and

$$\tilde{F}^{\mu\nu} = \varepsilon^{\mu\nu\alpha\beta}\partial_\alpha A_\beta + \partial^\mu V^\nu - \partial^\nu V^\mu, \tag{10.107}$$

where $\varepsilon^{\mu\nu\alpha\beta}$ is the completely antisymmetrical Levi–Civita tensor. Applying the Lorentz gauge condition to the four-potentials, we have $\partial_\mu A^\mu = 0$, and $\partial_\mu V^\mu = 0$. The symmetrized version of Maxwell's equations become

$$\partial_\nu \partial^\nu A_\mu = -4\pi j_\mu, \tag{10.108}$$

and

$$\partial_\nu \partial^\nu V_\mu = -4\pi g_\mu. \tag{10.109}$$

At this point, we note that Equations 10.108 and 10.109 are invariant under the dual transformation

$$A'_\mu = A_\mu \cos\theta + V_\mu \sin\theta, \quad V'_\mu = V_\mu \cos\theta - A_\mu \sin\theta, \tag{10.110}$$

provided that the charges and four-currents are also transformed:

$$j'_\mu = j_\mu \cos\theta + g_\mu \sin\theta, \quad g'_\mu = g_\mu \cos\theta - j_\mu \sin\theta. \tag{10.111}$$

By analogy with the idea that Lorentz invariance becomes manifest in covariant notation, dual-invariance can be expressed clearly in complex notation. Since the dual transformation is essentially a rotation in the complex charge plane, this suggests the notation $\bar{q} = q + ig$ which yields the complex four-current density $\bar{j}_\mu = j_\mu + ig_\mu$; similarly, the complex four-potential is defined as $\bar{A}_\mu = A_\mu + iV_\mu$, from which the complex electromagnetic field tensor is derived as

$$\bar{F}_{\mu\nu} = F_{\mu\nu} + i\tilde{F}_{\mu\nu} = \partial_\mu \bar{A}_\nu - \partial_\nu \bar{A}_\mu + i\varepsilon_{\mu\nu\alpha\beta}\partial^\alpha \bar{A}^\beta. \tag{10.112}$$

Within this context, the dual transform reduces to

$$\bar{F}'_{\mu\nu} = \bar{F}_{\mu\nu} e^{-i\theta}, \quad \bar{j}'_\mu = \bar{j}_\mu e^{-i\theta}. \tag{10.113}$$

This notation also proves extremely compact: the symmetrized form of Maxwell's equations takes the form

$$\partial_\mu \bar{F}^{\mu\nu} = -\partial_\nu \partial^\nu \bar{A}^\mu = 4\pi \bar{j}^\mu, \tag{10.114}$$

and it is now obvious that Equations 10.112 and 10.113 are dual-invariant.

Within this context, the dual-invariant Lorentz force equation takes the form

$$F_\mu = (qF_{\mu\nu} + g\tilde{F}_{\mu\nu})u^\nu = \Re(q\bar{F}^*_{\mu\nu} u^\nu). \tag{10.115}$$

Note that whenever a product between any two complex electromagnetic quantities defined previously is taken, one of the quantities must be complex-conjugated so that its magnetic component changes sign. This operation is analogous to raising and lowering an index in covariant notation, when contracting tensors, so that the sign of the time-like component is reversed. In both cases, the sign reversal ensures the invariance of the product under the respective transform. The duality and Lorentz transforms are both rotations, in two- and four-dimensional spaces, respectively.

## 10.10 Symmetrized Dirac–Lorentz Equation

To derive the radiation force, two different approaches can be used. First, one can follow Dirac's treatment and derive the radiation reaction from general principles including gauge invariance and Lorentz covariance; this is the focus of the present section. The second type of derivation relies on a careful study of the conservation of the four-momentum of the electromagnetic field during the interaction.

In the case of a classical point dyon, which possesses both an electric charge, $q$, and a magnetic charge, $g$, the complex four-current is given by

$$\bar{j}_\mu(x_\nu) = \bar{q}\int_{-\infty}^{+\infty} u_\mu(x'_\nu)\delta_4(x_\nu - x'_\nu)d\tau', \tag{10.116}$$

where the integral over proper time allows the use of the invariant four-dimensional Dirac delta-function, as discussed in Section 10.16.

Now assuming, as Dirac did, that a particle acts on itself through the Lorentz force but using the symmetrized expression thereof, the self-force is

$$F_\mu^s = \Re[\bar{q}(\partial^\mu \bar{A}_s^\nu - \partial^\nu \bar{A}_s^\mu + i\varepsilon^{\mu\nu\alpha\beta}\partial_\alpha \bar{A}_\beta^s)^* u_\nu], \tag{10.117}$$

where the complex self-potential satisfies the driven wave equation

$$\partial_\nu \partial^\nu \bar{A}_\mu^s = \Box \bar{A}_\mu^s = -4\pi\bar{q}\int_{-\infty}^{+\infty} u_\mu(x'_\nu)\delta_4(x_\nu - x'_\nu)d\tau'. \tag{10.118}$$

Equation 10.117 can also be written as

$$F_\mu^s = \Re[\bar{q}(\partial^\mu \bar{A}_s^{\nu^*} - \partial^\nu \bar{A}_s^{\nu^*})u_\nu - i\bar{q}\varepsilon^{\mu\nu\alpha\beta}\partial_\alpha \bar{A}_\beta^{s^*} u_\nu], \tag{10.119}$$

while the driven wave equation 10.118 can be solved in terms of Green functions:

$$\bar{A}_\mu^s(x_\lambda) = -4\pi\bar{q}\int_{-\infty}^{+\infty} u_\mu(x'_\lambda)G(x_\lambda - x'_\lambda)d\tau', \tag{10.120}$$

where $G$ is the Green function formally defined as $G(x_\lambda - x'_\lambda) \equiv -\delta_4(x_\lambda - x'_\lambda)/\Box$. Using this Green function solution in Equation 10.119, it becomes clear that the last term in the square brackets is purely imaginary: $-i\bar{q}\,\varepsilon^{\mu\nu\alpha\beta}\partial_\alpha \bar{A}_\beta^{s*}\,u_\nu \propto i\bar{q}\bar{q}^* = i|\bar{q}|^2$; thus, the self-force reduces to

$$F_\mu^s = \Re[\bar{q}(\partial^\mu \bar{A}_s^{\nu^*} - \partial^\nu \bar{A}_s^{\mu^*})u_\nu]. \tag{10.121}$$

Lest this manipulation appears as sleight of hand, we depart momentarily from our elegant shorthand to elucidate the reason for the vanishing of the last term in Equation 10.119. In terms of real quantities, the self-force reads

$$F_s^\mu = q(\partial^\mu A_s^\nu - \partial^\mu A_s^\nu - \varepsilon^{\mu\nu\alpha\beta}\partial_\alpha V_\beta^s)u_\nu + g(\varepsilon^{\mu\nu\alpha\beta}\partial_\alpha A_\beta^s + \partial^\mu V_s^\nu - \partial^\nu V_s^\mu)u_\nu, \tag{10.122}$$

where the electric and magnetic self-potentials are driven by the dyon electric and magnetic four-currents, with

$$\begin{bmatrix} A \\ V \end{bmatrix}_\mu^s (x_\lambda) = -4\pi\begin{bmatrix} q \\ g \end{bmatrix}\int_{-\infty}^{+\infty} u_\mu(x'_\lambda)G(x_\lambda - x'_\lambda)d\tau', \tag{10.123}$$

which implies that $V_\mu^s = (g/q)A_\mu^s$. This last relation allows some cancellation in Equation 10.122, yielding the simpler expression

$$F_s^\mu = q(\partial^\mu A_s^\nu - \partial^\mu A_s^\nu)u_\nu + g(\partial^\mu V_s^\nu - \partial^\nu V_s^\mu)u_\nu, \tag{10.124}$$

in agreement with Equation 10.121.

Physically, the disappearance of the cross-terms involving the action of the magnetic self-potential on the electric charge and that of the electric self-potential on the magnetic charge are due to the fact that the corresponding ponderomotive self-forces exactly cancel out. This decoupling of the radiation reaction forces is to be expected because the polarization of the radiation generated by the dyon's electric charge is always orthogonal to that radiated by the magnetic charge; thus, there is no interference between the electric and magnetic components of the dyon self-electromagnetic field.

Using the explicit form of the Green function in the force equation, we have

$$F_\mu^s(x_\lambda) = -4\pi|\bar{q}|^2\int_{-\infty}^{+\infty} u^\nu[u'_\nu\partial_\mu - u'_\mu\partial_\nu]G(x_\lambda - x'_\lambda)d\tau', \qquad (10.125)$$

where we have used the notation $u'_\mu = u_\mu(x'_\lambda)$.

We now apply Dirac's procedure for finding the self-force in the point limit. The Green function in Equation 10.125 depends on the space–time interval $s^2 = (x - x')_\mu(x - x')^\mu$; using $s^2$ as the independent variable, the four-gradient operator reads $\partial_\mu \equiv 2\,(x_\mu - x'_\mu)\partial/\partial s^2$, and the self-force is

$$F_\mu^s = -8\pi|\bar{q}|^2\int_{-\infty}^{+\infty} d\tau' u^\nu[u'_\nu(x_\mu - x'_\mu) - u'_\mu(x_\nu - x'_\nu)]\frac{\partial G}{\partial s^2}. \qquad (10.126)$$

At this point, we introduce the new variable $\tau'' = \tau - \tau'$ so that the range of integration explicitly includes the electron (singular point at $\tau'' = 0$). To evaluate the integral in Equation 10.126, we can now use Taylor–McLaurin expansions in powers of $\tau''$. We first have

$$x_\mu - x'_\mu = x_\mu(\tau) - x_\mu(\tau - \tau'') = \tau'' u_\mu - \frac{1}{2}\tau''^2 a_\mu + \frac{1}{6}\tau''^3\frac{da_\mu}{d\tau} + \cdots, \qquad (10.127)$$

where we have used the four-velocity and four-acceleration. For the four-velocity, we have

$$u'_\mu = \mu_\mu(\tau - \tau'') = u_\mu - \tau'' a_\mu + \frac{1}{2}\tau''^2\frac{da_\mu}{d\tau} + \cdots. \qquad (10.128)$$

Using expansions 10.127 and 10.128, and factoring, we have

$$\begin{aligned}
&\frac{u^\nu}{\tau''}[u'_\nu(x_\mu - u'_\mu) - u'_\mu(x_\nu - u'_\nu)] \\
&= (u^\nu u_\nu)\left(u_\mu - \frac{\tau''}{2}a_\mu + \frac{\tau''^2}{6}\frac{da_\mu}{d\tau}\right) + \frac{\tau''^2}{2}\left(u^\nu\frac{da_\nu}{d\tau}\right)u_\mu \\
&\quad - \left(u_\mu - \tau'' a_\mu + \frac{\tau''^2}{2}\frac{da_\mu}{d\tau}\right)(u^\nu u_\nu) - \frac{\tau''^2}{6}\left(u^\nu\frac{da_\nu}{d\tau}\right)u_\mu + O(\tau''^3). \qquad (10.129)
\end{aligned}$$

We now use the Lorentz invariant $u^\mu u_\mu = -1$. Differentiating this equation with respect to the proper time $\tau$, we first find that $u_\mu(du^\mu/d\tau) = 0 = u_\mu a^\mu$; this result corresponds to the fact that the derivative of a vector with fixed length is orthogonal to the original vector. The four-acceleration is always perpendicular to the four-velocity. Differentiating a second time with respect to $\tau$, we also have $a_\mu a^\mu = -u_\mu(da^\mu/d\tau)$.

Grouping terms, we finally obtain the important result

$$u^\nu[u'_\nu(x_\mu - x'_\mu) - u'_\mu(x_\nu - x'_\nu)] = \tau''^2\left\{-\frac{1}{2}a_\mu + \frac{\tau''}{3}\left[\frac{da_\mu}{d\tau} - u_\mu(a_\nu a^\nu)\right]\right\} + O(\tau''^4). \tag{10.130}$$

The relation between the space–time interval, $s^2$, and the proper time difference, $\tau''$, can be expanded as well: $s^2 = \tau''^2(u_\mu u^\mu) - (\tau''^3/2)(u_\mu a^\mu + a_\mu u^\mu) + O(\tau''^4)$. Using the orthogonality of the four-velocity and four-acceleration, this reduces to $s^2 = -\tau''^2 + O(\tau''^4)$, and we have $\partial G/\partial s^2 = [-1/2\tau'' + O(\tau'')]\partial G/\partial\tau''$.

With this, the expression for the self-force reads

$$F_\mu^s = 4\pi|\bar{q}|^2\int_{-\infty}^{+\infty} d\tau''\left\{-\frac{\tau''}{2}a_\mu + \frac{\tau''^2}{3}\left[\frac{da_\mu}{d\tau} - u_\mu(a_\nu a^\nu)\right] + O(\tau''^3)\right\}\frac{\partial G}{\partial\tau''}. \tag{10.131}$$

We can integrate Equation 10.131 by parts, according to $\int d\tau'' f(\tau'')\partial G/\partial\tau'' = -\int d\tau''(\partial f/\partial\tau'')G(\tau'')$, and obtain

$$F_\mu^s = 4\pi|\bar{q}|^2\int_{-\infty}^{+\infty} d\tau''\left\{-\frac{1}{2}a_\mu + \frac{2}{3}\left[\frac{da_\mu}{d\tau} - u_\mu(a_\nu a^\nu)\right] + O(\tau''^2)\right\}G(\tau''). \tag{10.132}$$

We now use the retarded (causal) Green function; Equation 10.132 reads

$$F_\mu^s = 4\pi|\bar{q}|^2\int_{-\infty}^{+\infty} d\tau''\left\{-\frac{1}{2}a_\mu + \frac{2}{3}\left[\frac{da_\mu}{d\tau} - u_\mu(a_\nu a^\nu)\right] + O(\tau''^2)\right\}\frac{\delta(\tau'')}{|\tau''|}\left(1 + \frac{\tau''}{|\tau''|}\right) \tag{10.133}$$

where we have used and $x_0 - x'_0/|x_0 - x'_0| = \tau''/|\tau''|$ and $\delta(s^2) = \delta(-\tau''^2) = \delta(\tau'')/|\tau''|$. This last identity has to be defined mathematically with care, as is discussed in Section 5.2.

We now proceed with the integration of Equation 10.133 to obtain

$$F_\mu^s = -\frac{1}{2}|\bar{q}|^2\left[\int_{-\infty}^{+\infty}\frac{\delta(\tau'')}{|\tau''|}d\tau''\right]a_\mu + \frac{2}{3}|\bar{q}|^2\left[\frac{da_\mu}{d\tau} - u_\mu(a_\nu a^\nu)\right], \tag{10.134}$$

which is the sought-after expression for the self-force. Note that we have dropped the antisymmetrical terms in $\tau''\delta(\tau'')/|\tau''|$ and $\delta(\tau'')/\tau''$, and that this expression is exact because all the higher-order terms in the expansion integrate out. This fact is rarely properly appreciated in the literature. The momentum transfer equation, including the radiation reaction, now reads

$$\left[m+\frac{1}{2}|\bar{q}|^2\int_{-\infty}^{+\infty}\frac{\delta(\tau'')}{|\tau''|}d\tau''\right]a_\mu = \Re(\bar{q}\bar{F}^*_{\mu\nu}u^\nu)+\frac{2}{3}|\bar{q}|^2\left[\frac{da_\mu}{d\tau}-u_\mu(a_\nu a^\nu)\right]. \quad (10.135)$$

Equation 10.135 clearly exhibits the infinite electromagnetic mass, in the form of the divergent integral multiplying the four-acceleration.

## 10.11 Conceptual Difficulties: Electromagnetic Mass Renormalization, Runaways, Acausal Effects

In this section, the main conceptual problems associated with the classical Dirac–Lorentz electron model are reviewed and discussed. The Dirac–Rohrlich asymptotic condition is then introduced to determine the physical solutions of the Dirac–Lorentz equation.

As shown in Equation 10.135, the mass term contains an infinite contribution from the self-electromagnetic fields of the point dyon. There are two different ways to circumvent this difficulty. First, we can consider that the infinite potential energy associated with a point charge model must be balanced by an infinite binding energy $-W$, such as that produced by the Poincaré stress tensor, so that the finite observed rest mass of the dyon is given in units of $m_0$ by $m = (^1\!/_2)|\bar{q}|^2 \int d\tau''\delta(\tau'')/|\tau''| - W$. This procedure is essentially equivalent to mass renormalization in QED. The divergent electromagnetic mass, which is produced by the singular part of the Green function can also be removed by considering the time-symmetrical Green function $G = (^1\!/_2)(G^- - G^+)$, as first proposed by Dirac [9]; here $G^\pm$ represents the retarded and advanced Green functions, as studied in detail in Section 5.4.2. There is little doubt that the removal of the infinite self-energy of the (nonradiative) Coulomb field is connected deeply to the charge conjugation and time reversal properties of electrodynamics, as exemplified by the Wheeler–Feynman electrodynamics; however, the connection is not entirely clear.

Using either approach to renormalize the electromagnetic mass, we finally obtain the complete equation of motion for a particle with arbitrary electric and magnetic charge:

$$ma_\mu = \Re(\bar{q}\bar{F}^*_{\mu\nu}u^\nu)+\frac{2}{3}|\bar{q}|^2\left[\frac{da_\mu}{d\tau}-u_\mu(a_\nu a^\nu)\right], \quad (10.136)$$

where $m$ is the renormalized dyon mass.

It is manifest that Equation 10.136, like the generalized form of Maxwell's equations, is invariant under a duality transform. In the case of an electron,

$m/m_0 = 1$ and $\bar{q} = -1$, which yields the well-known Dirac–Lorentz equation,

$$a_\mu = -F_{\mu\nu}u^\nu + \tau_0\left[\frac{da_\mu}{d\tau} - u_\mu(a_\nu a^\nu)\right], \tag{10.137}$$

where $\tau_0 = 2/3$ is the Compton time-scale, expressed in the units of $r_0/c$ used here.

In mksa units, $\tau_0 = \mu_0 e^2/6\pi m_0 c = 0.626 \times 10^{-23}$ s. The first term on the right-hand side is the Lorentz force, while the radiation reaction contain the Schott term and the radiation damping force.

A very important property of the Dirac–Lorentz equation is the fact that it satisfies energy–momentum conservation, as is easily seen by contracting Equation 10.137 with the four-velocity. We then have

$$u^\mu a_\mu = 0 = u^\mu F_{\mu\nu}u^\nu + \frac{2}{3}\left[u^\mu \frac{da_\mu}{d\tau} - (u^\mu u_\mu)(a_\nu a^\nu)\right], \tag{10.138}$$

which is satisfied by virtue of the antisymmetry of the electromagnetic field tensor $F_{\mu\nu}$ and the orthogonality $u_\mu$ and $a_\mu$.

We now briefly review some of the conceptual difficulties associated with the Dirac–Lorentz equation itself. First, it is easily seen that, in the absence of an external field, Equation 10.137 can be contracted with $a^\mu$ to obtain $a^\mu a_\mu = (\tau_0/2)d_\tau(a^\mu a_\mu)$, which admits the so-called "runaway" solution $[a_\mu a^\mu](\tau) = [a_\mu a^\mu]_{\tau=0} \exp(2\tau/\tau_0)$.

Note that this self-excited motion implies that $[a_\mu a^\mu]_{\tau=0} \neq 0$, and can be eliminated through the use of the appropriate asymptotic conditions, $\lim_{\tau\to\pm\infty} a_\mu(\tau) = 0$, as first suggested by Dirac and then by Rohrlich. This type of boundary condition on the electron motion also satisfies the law of inertia: the electron velocity remains constant when no external force is applied. A detailed analysis of Equation 10.137 also reveals the existence of acausal or "preacceleration" solutions. This is directly connected to the implicit electromagnetic mass renormalization underlying the Dirac–Lorentz equation. The self-force can be explicitly derived by using the time-symmetrical Green function $G = (1/2)(G^- - G^+)$, as first noted by Dirac. As a result, although the electron is modeled as a point charge, it can interact electromagnetically with external fields localized within its classical radius. To show the implicit acausality of the Dirac–Rohrlich solution, we recast the Dirac–Lorentz equation in the form

$$a_\mu - \tau_0\frac{da_\mu}{d\tau} = K_\mu, \quad K_\mu = -F_{\mu\nu}u^\nu - \tau_0 u_\mu(a_\nu a^\nu). \tag{10.139}$$

Multiplication by the integrating factor $e^{-\tau/\tau_0}$ yields

$$\frac{d}{d\tau}\left[a_\mu(\tau)e^{-\tau/\tau_0}\right] = -\frac{1}{\tau_0}e^{-\tau/\tau_0}K_\mu(\tau); \tag{10.140}$$

Equation 10.140 can now be formally integrated to obtain

$$a_\mu(\tau) = e^{\tau/\tau_0}\int_{-\infty}^{\tau} e^{-\tau'/\tau_0}\left[\frac{1}{\tau_0}u^\nu F_{\mu\nu} + u_\mu(a_\nu a^\nu)\right](\tau')d\tau'. \qquad (10.141)$$

The structure of this formal solution, which implicitly satisfies the Dirac–Rohrlich asymptotic condition, clearly exhibits the acausal convolution integral operator, $\int_{-\infty}^{\tau} e^{-\tau'/\tau_0}d\tau'$, which "weighs" the externally applied electromagnetic field exponentially within a characteristic space–time interval equal to the classical electron radius. This type of solution does not run away because the preacceleration of the electron over the Compton time-scale "launches" it on a stable trajectory. In other words, the preacceleration exactly compensates the runaway instability, and when the external field is applied, the electron executes a motion which conserves the total four-momentum, including the pump and scattered fields, and asymptotically satisfies the law of inertia.

## 10.12 Schott Term

Here, we consider the exchange of four-momentum between the electron, the external field, and the scattered field. An elementary treatment of this problem can be given in the instantaneous rest frame of the particle, as discussed by Jackson, where one can balance to zero the time-averaged work produced by the radiation force on the particle with the time-averaged radiated electromagnetic energy, to obtain the Schott term of the Abraham–Lorentz force. The Schott term depends on the second time derivative of the particle velocity. However, it should be noted here that, strictly speaking, in the instantaneous rest frame ($\beta = \mathbf{0}$) where, by definition, both the particle velocity and kinetic energy are equal to zero, the infinitesimal variation of the work of the damping force, $dW = \mathbf{F}\cdot\beta dt$, must also be zero. In fact, it will be shown that in that frame, the dipole radiation pattern of the scattered field is symmetrical and that there is no momentum exchanged between the charge and the radiated wave. The method of derivation used here consists of evaluating the instantaneous variation of the energy–momentum of the radiated field first. This can be done either by integrating the Poynting vector flux and the radiation pressure of the scattered field on a sphere of finite radius, then taking the limit where the radius tends to zero, assuming no internal particle structure or by generalizing results obtained in the instantaneous rest frame in a covariant way.

For a point charge moving along a world line, $x_\mu(\tau)$, with three-velocity, $\beta = d\mathbf{x}/dt$, and three-acceleration, $\dot{\beta} = d\beta/dt$, the radiative electric field at $r_\mu$ is obtained by deriving the Liénard–Wiechert four-potential. In electron units, we have for the radiative field

$$\mathbf{E}(r_\mu) = -\left[\frac{\hat{\mathbf{n}}\times(\hat{\mathbf{n}}-\beta)\times\dot{\beta}}{(1-\beta\cdot\hat{\mathbf{n}})^3 R}\right], \tag{10.142}$$

where the quantities in the bracket are evaluated at the retarded time $t^-$ such that the retardation condition, $t - t^- = R(t^-) = |\mathbf{r} - \mathbf{x}(t^-)|$, is satisfied, and where $\hat{\mathbf{n}}$ is the unit vector in the direction of observation.

The instantaneous electromagnetic momentum flux is given in terms of the Maxwell stress tensor, defined as

$$T_{ij} = \frac{1}{4\pi}\left[E_iE_j + B_iB_j - \frac{1}{2}(\mathbf{E}^2+\mathbf{B}^2)\delta_{ij}\right]. \tag{10.143}$$

The total radiation pressure force applied to a sphere of radius $R$, corresponding to the momentum recoil of the photons emitted by the particle at $t^-$ is given by $\iint T_{ij}n_jR^2d\Omega$, where $n_j$ is the $j$-th component of $\hat{\mathbf{n}}$. The instantaneous variation of the momentum of the scattered field can be expressed as

$$\frac{d\mathbf{G}}{dt} = -\lim_{R\to 0}\left[\iint(\hat{\mathbf{n}}\circ\mathbf{T})R^2d\Omega\right], \tag{10.144}$$

where $\circ$ denotes tensorial contraction.

The details of the derivation are given in Section 10.13; the covariant form of the instantaneous variation of the scattered wave four-momentum is found to be

$$\frac{dG_\mu}{d\tau} = \frac{2}{3}(a_\nu a^\nu)u_\mu. \tag{10.145}$$

The corresponding radiation damping force acting on the charge is essentially a relativistic effect. Indeed, if we first consider the instantaneous rest frame of the particle, we see that this force vanishes, as indicated by Equation 10.145. This is due to the symmetry of the dipole radiation pattern in this particular frame, as shown in Figure 10.11 (top). Although electromagnetic energy is radiated by the particle, there is no net recoil force because for each photon radiated in a given direction of space there is a photon with the same momentum radiated in the opposite direction. In any other frame, as shown in Figure 10.11 (bottom), the relativistic Doppler effect breaks this symmetry: the photons radiated in the forward direction are blue-shifted and carry more momentum than their backscattered counterparts, resulting in a net radiation force opposite to the direction of motion. In the instantaneous rest frame, the electron merely mediates the transfer of energy from the external field to the radiated wave by scattering the incident photons. This physical picture is in agreement with the fact that in that frame the electron has no free energy to yield and that the work of any force acting on the electron must be zero. It also clearly indicates that in that frame, energy is directly exchanged

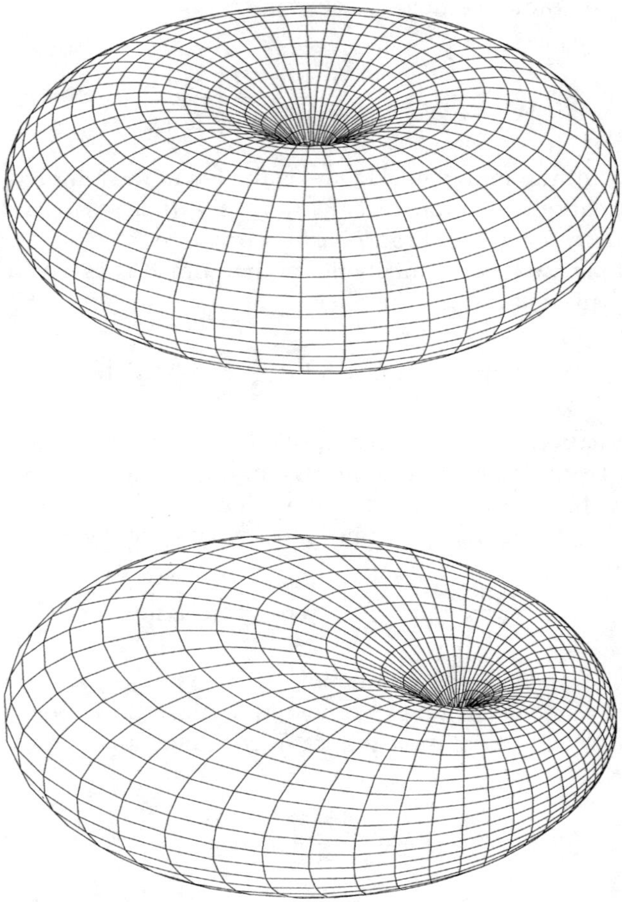

**FIGURE 10.11**
Top: dipole radiation pattern, as observed in the electron instantaneous rest frame, where $\gamma = 1$. Bottom: relativistic Doppler effect, for $\gamma = 1.01$, showing the asymmetry introduced by relativistic effects; this asymmetry is at the origin of the radiation damping force.

between the external field and the scattered wave. With this in mind, we now need to carefully investigate the conservation of the energy–momentum of the three interacting bodies.

The covariant energy–momentum transfer equation between the charge and the electromagnetic field now takes the form $a_\mu = d_\tau p_\mu = -F_{\mu\nu}u^\nu - d_\tau G_\mu - d_\tau H_\mu$, where the first term is the usual Lorentz force expressed in terms of the electromagnetic tensor, while the second term corresponds to the four-momentum radiated away by the scattered wave as derived above and where we have introduced a third term corresponding to the instantaneous variation of the energy–momentum of the external field resulting from the interaction.

Within this context, the radiation force is defined as $F^s_\mu = -d_\tau(G_\mu + H_\mu)$; here, we have also used the principle of action and reaction which holds as long as we consider the instantaneous interaction of a point particle. In that case, both the space-like and time-like intervals are zero and there is no propagation delay to consider.

We now use the relations between the four-velocity and its successive derivatives; using Equation 10.145, and contracting the four-momentum transfer equation with the four-velocity, we first have

$$u^\mu a_\mu = 0 = -u^\mu F_{\mu\nu} u^\nu + \frac{2}{3}\left(u_\nu \frac{da^\nu}{d\tau}\right)(u^\mu u_\mu) - u^\mu \frac{dH_\mu}{d\tau}. \tag{10.146}$$

The first term on the right-hand side is equal to zero, since the electromagnetic tensor is antisymmetrical; in the second term, we use $u^\mu u_\mu = -1$ to obtain

$$\frac{2}{3} u_\nu \frac{da^\nu}{d\tau} = -u^\mu \frac{dH_\mu}{d\tau}. \tag{10.147}$$

As noted by Pauli, the general solution is

$$\frac{dH_\mu}{d\tau} = -\frac{2}{3}\frac{da_\mu}{d\tau} + \kappa u^\nu K_{\mu\nu}, \tag{10.148}$$

where we have introduced the antisymmetrical tensor,

$$K_{\mu\nu} = \frac{2}{3}\left[u_\mu \frac{da_\nu}{d\tau} - u_\nu \frac{da_\mu}{d\tau}\right], \tag{10.149}$$

and where $\kappa$ is an arbitrary constant.

It is clear that $\kappa = 0$ yields the Dirac–Lorentz equation; in that case, we can identify the variation of the four-momentum in the external field with the Schott term:

$$\frac{dH_\mu}{d\tau} = -\frac{2}{3}\frac{da_\mu}{d\tau}. \tag{10.150}$$

With this, the manifestly covariant expression for the radiation reaction becomes

$$F_\mu = \frac{2}{3}\left[\frac{da_\mu}{d\tau} - u_\mu (a_\nu a^\nu)\right], \tag{10.151}$$

and it is easily seen that $F_\mu = u^\nu K_{\mu\nu}$. In addition, the antisymmetrical character of the tensor $K_{\mu\nu}$ guarantees that $u^\mu F_\mu = 0$. For completeness, we give the corresponding expression of the radiation reaction force in vector form, as expressed in electron units where the force is normalized to $m_0c^2/r_0$:

$$\mathbf{F} = \tau_0\gamma^2\{\ddot{\beta}+3\gamma^2\dot{\beta}(\beta\cdot\dot{\beta})+\gamma^2\beta[\beta\cdot\ddot{\beta}+3\gamma^2(\beta\cdot\dot{\beta})^2]\}. \qquad (10.152)$$

It is easily verified that the variation of the electron energy due to the radiative effects (time-like component of the radiation force) satisfies the equation

$$\frac{d\gamma}{dt} = \tau_0\gamma^4[\beta\cdot\ddot{\beta}+3\gamma^2(\beta\cdot\dot{\beta})^2] = \beta\cdot\mathbf{F}. \qquad (10.153)$$

Equation 10.151 corresponds exactly to the covariant expression of the Abraham–Lorentz force. The self-interaction nature of the radiation force is evident, as the expression derived scales with the square of the particle charge: $\tau_0 = \mu_0 e^2/6\pi m_0 c$. In the first term of Equation 10.151, we recover the Schott term which depends on the second time derivative of the particle velocity and which is identified here with the depletion of energy–momentum from the pump (accelerating) field, while we recover the quadratic scaling with acceleration for the second term corresponding to the radiation damping force. As indicated by Equation 10.151, the total radiation force can be attributed to two distinct effects. On the one hand, energy–momentum is radiated away by the scattered wave, as described by Equation 10.145. The asymmetry of the Doppler-shifted dipole radiation pattern in any frame where the particle is not instantaneously at rest gives rise to this force, which dominates in the ultrarelativistic limit. It also has a nonzero value for a particle submitted to a constant acceleration, as opposed to the Schott term. On the other hand, the second term in Equation 10.151 is attributed to the energy–momentum exchanged between the scattered wave and the external field. This term allows for the local simultaneous conservation of energy and momentum during the radiation process. The physics of the interaction can be illustrated by considering the process shown in Figure 10.12. Here, we consider the total energy and momentum of the electrodynamical system initially comprising a high intensity, short wavelength incoming laser pulse (pump) and an electron at rest. In general, after the interaction, the electron has gained some energy and momentum (in the minimal case, the electron would be left precisely at rest after the scattering) and is now moving at relativistic velocity, while the scattered wave carries energy and momentum in all spatial directions. In this case, it is clear that all the energy and momentum gained by both the electron and the scattered wave come at the expense of the external field. It is equally clear that, in such a process, the radiated electromagnetic power and the variation of the electron energy cannot be equal, therefore invalidating any theoretical model based on the local conservation of four-momentum between the electron and the radiated field only. We also note that while the backscattered radiation does not interfere with the laser pulse, the forward scattered radiation, which has the same spectral characteristics as the pump and copropagates in the positive $z$-direction, does interfere destructively

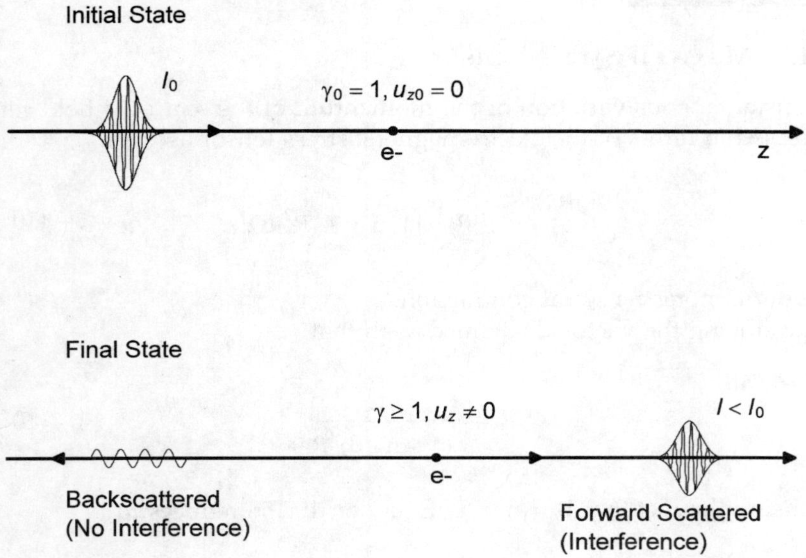


**FIGURE 10.12**
Scattering of a laser pulse by an electron initially at rest.

with the laser pulse and lowers its energy and momentum, yielding pump-field depletion.

Finally, in the case of an external electric field deriving from a static potential, $\varphi(r)$, the time-like component of the Dirac–Lorentz equation, which describes energy conservation, takes the simple form

$$\frac{d\gamma}{d\tau} = \mathbf{u} \cdot \nabla\varphi + \frac{2}{3}\frac{d^2\gamma}{d\tau^2} - \frac{dG_0}{d\tau} = \frac{d}{d\tau}\left[\varphi + \frac{2}{3}\frac{d\gamma}{d\tau} - G_0\right], \tag{10.154}$$

and can formally be integrated to yield the conservation law

$$\nabla(\gamma - \varphi + G_0) = \frac{2}{3}\left[\frac{d\gamma}{d\tau}\right]_{-\infty}^{+\infty}, \tag{10.155}$$

which indicates that, provided the Dirac–Rohrlich asymptotic condition, $\lim_{\tau\to\pm\infty}[d\gamma/d\tau] = 0$, is satisfied, the electron potential energy is converted to kinetic energy and radiation.

Within this context, the small value of the fine structure constant, which corresponds to the ratio of the classical to quantum electron scale (classical electron radius divided by the electron Compton wavelength), guarantees that the acausal effects related to the electromagnetic mass renormalization will be smeared by quantum fluctuations before the strong classical radiative correction regime is reached, thus preventing "naked acausalities." If magnetic charges are considered, however, the radiation reaction dominate over the quantum effects because the effective coupling constant is now $1/\alpha$, which is a large number.

## 10.13 Maxwell Stress Tensor

The instantaneous variation of the momentum of the scattered field can be expressed in terms of the electromagnetic stress tensor as

$$\frac{d\mathbf{G}}{dt} = -\lim_{R\to 0}\left[\iint(\hat{\mathbf{n}} \circ \mathbf{T})R^2 d\Omega\right], \tag{10.156}$$

where $\circ$ denotes tensorial contraction.

Introducing the vector $\xi$, defined such that

$$\mathbf{E} = \frac{\dot{\beta}\xi}{(1-\hat{\mathbf{n}}\cdot\beta)^3 R}, \tag{10.157}$$

and using the fact that $\mathbf{B} = \hat{\mathbf{n}} \times \mathbf{E}$, Equation 10.156 reduces to

$$\frac{dG_i}{dt} = \frac{1}{4\pi}\iint n_j \left\{\frac{\dot{\beta}^2[\delta_{ij}\xi^2 - \xi_i\xi_j - (n_j\xi_k - n_k\xi_j)(n_k\xi_i - n_i\xi_k)]}{(1-\hat{\mathbf{n}}\cdot\beta)^5}\right\}_{t=t^-} d\Omega. \tag{10.158}$$

Following Sommerfeld, we change variables and express the variation of momentum as a function of the retarded time. After some straightforward vector calculations, we obtain

$$\frac{d\mathbf{G}}{dt^-} = -\frac{1}{4\pi}\dot{\beta}^2\iint\frac{[\xi\times(\xi\times\hat{\mathbf{n}})]^2}{(1-\hat{\mathbf{n}}\cdot\beta)^5}d\Omega \tag{10.159}$$

which can be further reduced to

$$\frac{d\mathbf{G}}{dt^-} = -\frac{1}{4\pi}\iint\hat{\mathbf{n}}\frac{\{\hat{\mathbf{n}}\times[(\hat{\mathbf{n}}-\beta)\times\dot{\beta}]\}^2}{(1-\hat{\mathbf{n}}\cdot\beta)^5}d\Omega, \tag{10.160}$$

by noting that $\xi\times(\xi\times\hat{\mathbf{n}}) = (\hat{\mathbf{n}}\cdot\xi)\xi - \xi^2\hat{\mathbf{n}}$, and $\hat{\mathbf{n}}\cdot\xi = 0$. It is interesting to notice that Equation 10.160 can also be derived directly by using the Poynting vector $\mathbf{S} = \mathbf{E}\times\mathbf{H}$ in the simpler equation, $\lim_{R\to 0}(\iint \mathbf{S}R^2 d\Omega)$. To evaluate the integral in Equation 10.160, we expand the numerator using spherical coordinates:

$$\begin{aligned}\{\hat{\mathbf{n}}\times[(\hat{\mathbf{n}}-\beta)\times\dot{\beta}]\}^2 = \dot{\beta}^2[&(\sin\alpha\sin\theta\cos\phi + \cos\alpha\cos\theta)^2(\beta^2-1)\\ &+(1-\beta\cos\theta)^2 + 2\beta\cos\alpha(1-\beta\cos\theta)\\ &\times(\sin\alpha\sin\theta\cos\phi+\cos\alpha\cos\theta)].\end{aligned} \tag{10.161}$$

Here, we have chosen the axis of the Galilean frame $L$ such that we have

$$\begin{aligned}\beta &= \hat{z}\beta,\\ \dot{\beta} &= \dot{\beta}(\hat{z}\cos\alpha + \hat{x}\sin\alpha),\\ \hat{\mathbf{n}} &= \hat{x}(\sin\theta\cos\phi) + \hat{y}(\sin\theta\sin\phi) + \hat{z}\cos\theta.\end{aligned}$$

The integral over all solid angles is

$$\frac{d\mathbf{G}}{dt^-} = \frac{1}{4\pi}\int_0^{2\pi} d\phi \int_0^{\pi} \hat{\mathbf{n}} \frac{\{\hat{\mathbf{n}} \times [(\hat{\mathbf{n}} - \beta) \times \dot{\beta}]\}^2}{(1 - \hat{\mathbf{n}} \cdot \beta)^5} \sin\theta\, d\theta, \tag{10.162}$$

where the explicit dependence of the numerator on $\theta$ and $\phi$ is given by Equation 10.161.

The integral corresponding to the $y$-component averages to zero over $\phi$, and the integral corresponding to the $x$-component averages to zero over $\theta$. We are left with

$$\frac{d\mathbf{G}}{dt^-} = \hat{z}\frac{1}{4\pi}\dot{\beta}^2\left[\frac{8}{3}\pi\beta\gamma^6(1 - \beta^2 + \beta^2\cos^2\alpha)\right]. \tag{10.163}$$

At this point it is important to note that, as the sphere radius tends to zero, the retarded time tends to the instantaneous interaction time; Equation 10.163 is easily shown to reduce to

$$\frac{d\mathbf{G}}{dt} = \frac{2}{3}\beta\gamma^4[\dot{\beta}^2 + \gamma^2(\beta \cdot \dot{\beta})^2]. \tag{10.164}$$

The instantaneous variation of the energy of the scattered wave can be derived in the same way by integrating the Poynting vector flux over all solid angles, and taking the limit where $R$ tends to zero, to recover the Liénard formula

$$\frac{dW}{dt} = \frac{2}{3}\gamma^4[\dot{\beta}^2 + \gamma^2(\beta \cdot \dot{\beta})^2]. \tag{10.165}$$

Moreover, the velocity-dependent term in Equations 10.164 and 10.165 can be expressed in terms of the four-acceleration as

$$\gamma^4[\dot{\beta}^2 + \gamma^2(\beta \cdot \dot{\beta})^2] = a_\mu a^\mu. \tag{10.166}$$

The covariant generalization of Equations 10.164 and 10.165 then becomes quite straightforward. Following Becker, we combine Equations 10.164 and 10.165 to obtain the sought-after covariant form of the instantaneous variation of the energy–momentum of the scattered wave:

$$\frac{dG_\mu}{d\tau} = \frac{dG_\mu}{dt}\frac{dt}{d\tau} = \gamma\frac{dG_\mu}{dt} = \frac{2}{3}(a_\nu a^\nu)u_\mu. \tag{10.167}$$

Here, we have used the definition of the four-velocity, $u_\mu = dx_\mu/d\tau = \gamma dx_\mu/dt = \gamma(1, \beta)$.

## 10.14 Hamiltonian Formalism

It is also quite instructive to consider the dynamics of a point electron within the context of Hamiltonian formalism. It can be shown that, in the temporal gauge, the Hamiltonian

$$\mathcal{H} = -\sqrt{(\pi - \mathrm{A})^2 + \mu^2} + \frac{1}{16\pi}\iiint(\mathbf{E}^2 + \mathbf{B}^2)d^3\mathbf{x}, \tag{10.168}$$

yields the covariant Dirac–Lorentz equation in an external field, $F_{\mu\nu}$, provided that the mass term in Equation 10.168 satisfies the condition

$$\mu = \left[\frac{1}{2}\int\frac{\delta(\tau)}{|\tau|}d\tau\right] - 1, \tag{10.169}$$

which corresponds to the mass renormalization previously introduced. In Equation 10.168, $\pi$ is the particle canonical momentum, and $\mathbf{A}(q)$ is the vector-potential at the position of the particle. Note that in Equation 10.168, there is a negative sign in front of the term usually associated with the kinetic energy. This readily explains the existence of runaway solutions. It is equally important to notice that the normalized electron mass in the Dirac–Lorentz equation has the usual value of one, and not the value of $\mu$ given in the Hamiltonian (Equation 10.168). In this sense, it is clear that the (dissipative) Dirac–Lorentz equation cannot be derived from a conventional Hamiltonian (as expected).

One should also recognize that, by neglecting the radiation reaction terms in the Dirac–Lorentz equation, which then reduces to the usual Lorentz force equation, one can recover the standard Hamiltonian in an external potential,

$$\mathcal{H} = \sqrt{(\pi - \mathbf{A})^2 + 1} + \phi, \tag{10.170}$$

which defines a positive particle kinetic energy, as exemplified by the plus sign in front of the square root.

Furthermore, in the case where the external force is derived from a potential, it is possible to integrate the time-like component of the Dirac–Lorentz equation to obtain

$$\Delta\gamma = \Delta\varphi - W + \frac{2}{3}\left[\frac{d\gamma}{d\tau}\right]_{-\infty}^{+\infty}, \tag{10.171}$$

where the usual balance between the kinetic, potential, and radiated energy is realized, as long as $\lim_{\tau\to+\infty}[d\gamma/d\tau] = \lim_{\tau\to-\infty}[d\gamma/d\tau]$, which the Dirac–Rohrlich asymptotic condition obviously satisfies. Because of the implied nonlocality of the radiation, this balance is generally not realized differentially.

Finally, it is worth noting that the runaway solutions of the Dirac–Lorentz equation correspond to (unphysical) trajectories that minimize the particle energy associated with the Hamiltonian given in Equation 10.168 by making it tend to negative infinity. In this sense, there is an interesting analogy between the Dirac–Rohrlich asymptotic conditions (law of inertia), which are assumed by Dirac to yield the only physical solution to the Dirac–Lorentz equation, and the assumption made in QED that the negative energy states are entirely occupied by electrons in order to prevent transitions from positive to negative energies. The general problem of the classical limit of QED remains an outstanding difficulty in classical electrodynamics at high field strengths. For example, using the path-integral formulation of QED, the photon coordinates can be functionally integrated out, but the radiation reaction terms yield divergences on the electron world line that are exactly analogous to those introduced in Section 10.8. However, the Dirac–Lorentz equation, coupled to the prescription that all runaway solutions must be excluded, offers a simple and economical classical electron model, which yields a consistent electrodynamics that includes the usual Maxwell–Lorentz theory and gives a reasonable description of such phenomena as nonlinear Compton scattering, which can now be studied experimentally at energies in the 50 GeV range and laser intensities exceeding $10^{18}$ W/cm$^2$. Finally, we note that the Hamiltonian formalism can also be generalized within the framework of symmetrized electrodynamics. The symmetrized Hamiltonian including radiation reaction is simply given by

$$\begin{aligned}\mathcal{H} = &-\left(\sqrt{(\pi - e\mathbf{A} - g\mathbf{V})^2 + \mu^2} + e\phi + g\varphi\right)\\ &+ \frac{1}{8\pi}\iiint\left[\frac{\mathbf{E}^2+\mathbf{B}^2}{2} - \mathbf{V}\cdot(\nabla\times\mathbf{E}) + \mathbf{A}\cdot(\nabla\times\mathbf{B}) - \varphi(\nabla\cdot\mathbf{E}) - \phi(\nabla\cdot\mathbf{B})\right]d^3\mathbf{x}.\end{aligned} \tag{10.172}$$

while the Lorentz force Hamiltonian is

$$\mathcal{H} = \sqrt{(\pi - e\mathbf{A} - g\mathbf{V})^2 + m^2} + e\phi + g\varphi \tag{10.173}$$

Of course, there is no Hamiltonian for the Dirac–Lorentz equations following Equation 10.172.

## 10.15 Symmetrized Electrodynamics in the Complex Charge Plane and the Running Fine Structure Constant

At this point, the connection between duality and the fully symmetrized version of electrodynamics can be discussed within the context of a dynamical gauge theory, where the fine structure constant is now a running coupling constant. We start from the Dirac–Schwinger charge quantization condition for electric and magnetic monopoles:

$$\mathbf{q}_1 \times \mathbf{q}_2 = \Im(\bar{q}_1 \bar{q}_2^*)\hat{z} = n\alpha^{-1}\hat{z}, \quad n \in \mathbb{N}. \tag{10.174}$$

In Equation 10.174 the $z$-axis corresponds to angular momentum; this is schematically illustrated on Figure 10.13 (top), where two different charge state vectors are shown in the complex charge plane for a positron, with $\mathbf{q}_1 = [\hat{x}\Re(\bar{q}_1) + \hat{y}\Im(\bar{q}_1)] = \hat{x}$, along the electric axis, and a magnetic monopole, $\mathbf{q}_2 = [\hat{x}\Re(\bar{q}_2) + \hat{y}\Im(\bar{q}_2)] = \alpha^{-1}\hat{y}$, along the magnetic axis. The $\pi/2$ angle between both charge states corresponds to the orthogonality of the electric and magnetic axes. The total angular momentum of the system is now represented by the cross product of $\mathbf{q}_1$ and $\mathbf{q}_2$, and is quantized according to Equation 10.174. It is clear that a duality transform simply rotates the electric and magnetic axes, as shown in Figure 10.13 (top); however, the cross product remains unchanged, as the relative angle between the monopole charge states and their length are preserved by this rotation. Therefore, to symmetrize electrodynamics fully, one needs to take $\alpha = 1$, as first observed by Dirac, in which case the distinction between electric and magnetic charges disappears. In this case, radiation reaction are equal for an electric or a magnetic point charge interacting with external fields, and the full symmetry of electrodynamics is realized, as illustrated in Figure 10.13 (bottom). One of the deepest questions associated with this theory is the exact connection with spin and the Dirac equation of QED.

In conclusion, the basic electrodynamic equations for a dyon have been presented within the context of a covariant formalism in the complex charge plane. A double-potential formalism has been introduced which facilitates symmetrization of the calculations. An expression for the general self-force of a dyon has been derived, and it has been found that this expression is proportional to Dirac's expression for the self-force on an electron, differing only by a factor involving the electric and magnetic charge. Dirac's procedure for taking the point limit of the self-force has been applied, and the complete electrodynamic equation of motion for a dyon has been obtained. Finally, the connection with electromagnetic duality has been outlined.

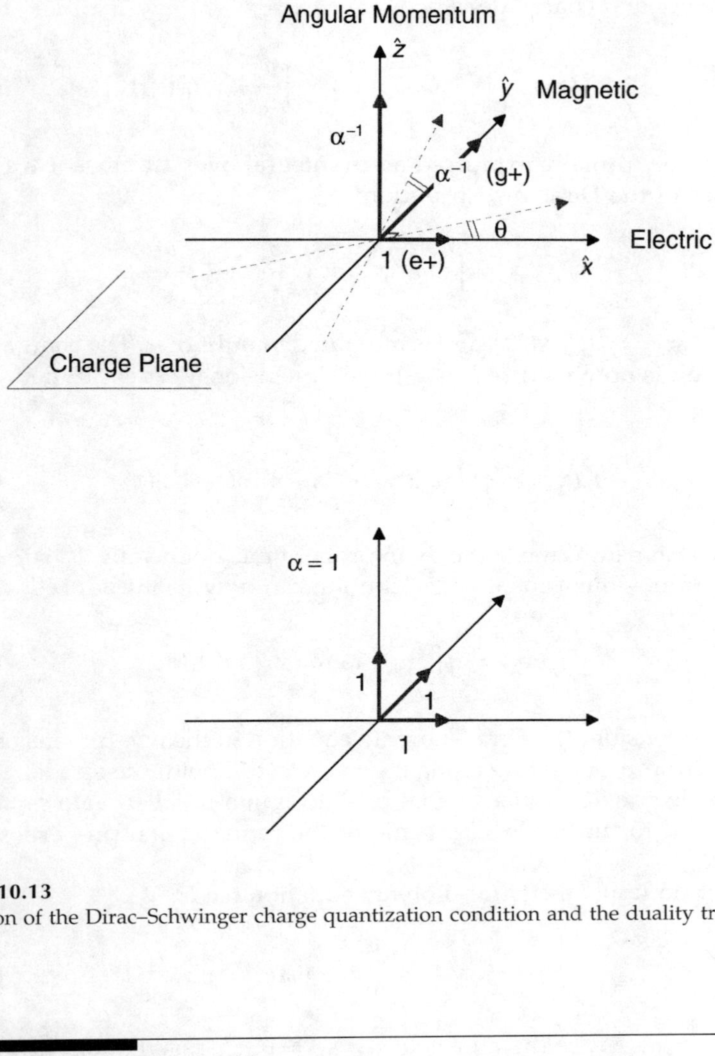


**FIGURE 10.13**
Illustration of the Dirac–Schwinger charge quantization condition and the duality transform.

## 10.16 Notes to Chapter 10

In these notes, we first derive the dyon four-current required to establish the expression of radiative corrections in symmetrized electrodynamics; we also consider the one-dimensional Dirac–Lorentz equation in the presence of an electric field with constant direction, which helps exemplify some of its basic properties.

In Section 10.10, the dyon four-current is modeled by the integral over the dyon proper time of a four-dimensional delta function; here, we show how to go from a three-dimensional point charge model to an invariant delta function. In general, the four-current density can be expressed in terms of four-velocity and charge density as

$$j_\mu(x_\lambda) = \left[\frac{u_\mu}{\gamma}\right](x_\lambda)\rho(x_\lambda) = \left[\frac{u_\mu}{u_0}\right](x_\lambda)\rho(x_\lambda), \tag{10.175}$$

which can be formally expressed as an integral over all times if we use the properties of the Dirac $\delta$-distribution:

$$j_\mu(x_\lambda) = \int_{-\infty}^{+\infty} u_\mu(x'_\lambda)\rho(x'_\lambda)\delta(t-t')\frac{dt'}{\gamma'}. \tag{10.176}$$

Here, $x'_\lambda = x_\lambda(t') = [t', \mathbf{x}(t')]$, and is measured in units of $r_0$. The charge density of the dyon is now modeled by a three-dimensional $\delta$-distribution, and we have

$$j^s_\mu(x_\lambda) = \bar{q}\int_{-\infty}^{+\infty} u_\mu(x'_\lambda)\delta_3(\mathbf{x}-\mathbf{x}')\delta(t-t')d\tau', \tag{10.177}$$

where we have introduced the dyon proper time, defined by $dt' = \gamma' d\tau'$. The invariant four-dimensional $\delta$-distribution can now be introduced, to yield

$$j^s_\mu(x_\lambda) = \bar{q}\int_{-\infty}^{+\infty} u_\mu(x'_\lambda)\delta_4(x_\lambda - x'_\lambda)d\tau'. \tag{10.178}$$

We now consider the Dirac–Lorentz equation in the one-dimensional case of an electron propagating along a given electric field; the special case of a constant electric field, such as that produced in a parallel-plate capacitor, is then treated to further exhibit some of the fundamental properties of this equation.

In electron units, the Dirac–Lorentz equation reads:

$$\frac{du_\mu}{d\tau} = F_\mu + \frac{2}{3}[a_\mu - u_\mu(a_\nu a^\nu)], \tag{10.179}$$

where $u_\mu = dx_\mu/d\tau$ is the four-velocity, $a_\mu = du_\mu/d\tau$ is the four-acceleration of the charge along its world line, $x_\mu(\tau)$, and $F_\mu$ is the force applied to the particle. In the one-dimensional case, Eq. (10.179) reduces to

$$\dot{u} = E(\tau)\sqrt{1+u^2} + \frac{2}{3}\left(\ddot{u} - u\frac{\dot{u}^2}{1+u^2}\right), \tag{10.180}$$

where we have introduced the notation $\dot{u} = du/d\tau$, $\ddot{u} = d^2u/d\tau^2$. To solve Eq. (10.180), let us introduce the temporal component of the four-velocity,

$$\gamma = u_0 = \sqrt{1+u^2} \tag{10.181}$$

and show that the square of the four-acceleration take the simple form used in Eq. (10.180): we first have

$$a_\mu a^\mu = \left(\frac{d\mathbf{u}}{d\tau}\right)^2 - \left(\frac{d\gamma}{d\tau}\right)^2 = \left(\frac{du}{d\tau}\right)^2 - \left(\frac{d\gamma}{d\tau}\right)^2; \tag{10.182}$$

the length of the normalized four-velocity is $u_\mu u^\mu = -1$, which is equivalent to $\gamma^2 = 1 + u^2$ in the one-dimensional case; deriving this last equality with respect to the proper time, we now have

$$2\gamma\frac{d\gamma}{d\tau} = 2u\frac{du}{d\tau}, \qquad \frac{d\gamma}{d\tau} = \frac{u}{\gamma}\frac{du}{d\tau}, \tag{10.183}$$

and we can express Eq. (10.182) as

$$a_\mu a^\mu = \left(1 - \frac{u^2}{\gamma^2}\right)\left(\frac{du}{d\tau}\right)^2 = \frac{1}{\gamma^2}\left(\frac{du}{d\tau}\right)^2 = \frac{\dot{u}^2}{1+u^2}. \tag{10.184}$$

Although Eq. (10.180) is highly nonlinear, an inspired change of variable allows us to resolve it: let

$$u(\tau) = \sinh[w(\tau)]; \tag{10.185}$$

we then have

$$\dot{u} = \dot{w}\cosh(w), \qquad \ddot{u} = \ddot{w}\cosh(w) + \dot{w}^2\sinh(w); \tag{10.186}$$

now remembering that $\cosh^2(w) - \sinh^2(w) = 1$, we find that

$$\sqrt{1+u^2} = \sqrt{1+\sinh^2(w)} = \cosh(w), \tag{10.187}$$

and we can recast Eq. (10.180) as

$$\dot{w}\sinh(w) = E\cosh(w) + \frac{2}{3}\left[\ddot{w}\cosh(w) + \dot{w}^2\sinh(w) - \sinh(w)\frac{\dot{w}^2\cosh^2(w)}{\cosh(w)}\right], \tag{10.188}$$

which yields, remarkably,

$$\dot{w} = E(\tau) + \frac{2}{3}\ddot{w}. \tag{10.189}$$

In the general case, this equation can be formally integrated to yield

$$\begin{aligned} \dot{w}(\tau) &= \mathcal{A}e^{\tau/\tau_0} + \int_\tau^{+\infty} e^{-(\tau'-\tau)/\tau_0} E(\tau') \frac{d\tau'}{\tau_0} \\ w(\tau) &= \mathcal{B} + \mathcal{A}\tau_0 e^{\tau/\tau_0} + \int_\tau^{+\infty} \left[e^{-(\tau'-\tau)/\tau_0} - 1\right] E(\tau')\, d\tau'. \end{aligned} \tag{10.190}$$

To verify that Eq. (10.190) constitutes the sought-after solution, we recall the definition of the integral of an arbitrary function, $f(x)$:

$$\int_a^b f(x)dx = F(b) - F(a), \qquad \frac{dF}{dx} = f(x). \tag{10.191}$$

With this, we can derive the acceleration with respect to proper time:

$$\begin{aligned} \ddot{w} = \frac{d\dot{w}}{d\tau} &= \frac{\mathcal{A}}{\tau_0} e^{\tau/\tau_0} + \frac{d}{d\tau}\left[e^{\tau/\tau_0} \int_\tau^{+\infty} e^{-\tau'/\tau_0} E(\tau') \frac{d\tau'}{\tau_0}\right] \\ &= \frac{\mathcal{A}}{\tau_0} e^{\tau/\tau_0} + \frac{e^{\tau/\tau_0}}{\tau_0} \int_\tau^{+\infty} e^{-\tau'/\tau_0} E(\tau') \frac{d\tau'}{\tau_0} - e^{\tau/\tau_0}\left[e^{-\tau/\tau_0} E(\tau) \frac{1}{\tau_0}\right] \\ &= \frac{\mathcal{A}}{\tau_0} e^{\tau/\tau_0} + \frac{1}{\tau_0} \int_\tau^{+\infty} e^{-(\tau'-\tau)/\tau_0} E(\tau') \frac{d\tau'}{\tau_0} - \frac{E(\tau)}{\tau_0} \\ &= \frac{1}{\tau_0}[\dot{w}(\tau) - E(\tau)]; \end{aligned} \tag{10.192}$$

multiplying this result by $\tau_0$, we find that Eq. (10.190) is, indeed, identically satisfied.

In Eq. (10.190), we have used the time-scale $\tau_0 = 2/3$, in electron units, and the runaway solution discussed in Section 10.11 appears clearly as the term $\mathcal{A}e^{\tau/\tau_0}$; setting the integration constant $\mathcal{A} = 0$ alleviates this problem. The integral term in Eq. (10.190) corresponds to the so-called pre-acceleration of the charge, which occurs over a time-scale of order $\tau_0$. Since $\tau_0$ is a very small quantity, the corresponding photon energies are quite high, as the equivalent wavelength is of order $c\tau_0 = 2r_0/3 = 2\alpha\lambda\!\!\!^{-}_c/3$, which yields $hc/e\lambda = 3m_0c^2/4\pi\alpha e =$ 16.72 MeV. Therefore, as strong QED effects will appear at lower energies, the reality of the acausal nature of the Dirac–Lorentz equation cannot be probed directly. Also note that a specific choice of the integration constants $\mathcal{A}$ and $\mathcal{B}$ allows us to eliminate pre-acceleration: if we set

$$\begin{aligned} \mathcal{A} &= \int_{-\infty}^{+\infty} e^{-\tau'/\tau_0} E(\tau')\, d\tau'/\tau_0, \\ \mathcal{B} &= \lim_{\tau\to-\infty}[w(\tau)] + \int_{-\infty}^{+\infty} E(\tau')\, d\tau', \end{aligned} \tag{10.193}$$

we see that the solution for the four-velocity takes the form,

$$\dot{w}(\tau) = -\int_{-\infty}^{\tau} e^{(\tau-\tau')/\tau_0} E(\tau') \frac{d\tau'}{\tau_0}$$

$$w(\tau) = \lim_{\tau\to-\infty} [w(\tau)] - \int_{-\infty}^{\tau} \left[e^{(\tau-\tau')/\tau_0} - 1\right] E(\tau')\, d\tau'. \tag{10.194}$$

We have eliminated both the pre-acceleration and the runaway motion, but the acceleration takes a non-zero value after the force has been applied; in other words, if $E(\tau) = 0$ for $\tau > \tau^*$, runaway motion will occur after $\tau^*$.

Returning to the solution obtained in Eq. (10.190), we can recast the acceleration as

$$\dot{w}(\tau) = e^{\tau/\tau_0}\left[\mathcal{A} + \int_{\tau}^{+\infty} e^{-\tau'/\tau_0} E(\tau') \frac{d\tau'}{\tau_0}\right]$$

$$= e^{\tau/\tau_0}\left[\mathcal{A} + \int_{-\infty}^{+\infty} e^{-\tau'/\tau_0} E(\tau') \frac{d\tau'}{\tau_0} - \int_{-\infty}^{\tau} e^{-\tau'/\tau_0} E(\tau') \frac{d\tau'}{\tau_0}\right]. \tag{10.195}$$

Let us now assume that the domain where $E(\tau) \neq 0$ is bounded; in other words, there exist two values of the proper time, $\tau_2 > \tau_1$, such that $E(\tau) = 0$ for $\tau < \tau_1$ or $\tau > \tau_2$; the Dirac–Rohrlich asymptotic conditions are then satisfied: we first have

$$\dot{w}(\tau) = e^{\tau/\tau_0}\left[\mathcal{A} + \int_{\tau_1}^{\tau_2} e^{-\tau'/\tau_0} E(\tau') \frac{d\tau'}{\tau_0} - \int_{\tau_1}^{\tau} e^{-\tau'/\tau_0} E(\tau') \frac{d\tau'}{\tau_0}\right], \tag{10.196}$$

and

$$\lim_{\tau\to-\infty} [\dot{w}(\tau)] = \lim_{\tau\to-\infty} \left\{ e^{\tau/\tau_0}\left[\mathcal{A} + \int_{\tau_1}^{\tau_2} e^{-\tau'/\tau_0} E(\tau') \frac{d\tau'}{\tau_0}\right]\right\} = 0 \tag{10.197}$$

moreover, if we choose $\mathcal{A} = 0$, we have

$$\lim_{\tau\to+\infty} [\dot{w}(\tau)] = \lim_{\tau\to+\infty} \left\{ e^{\tau/\tau_0}\left[\int_{\tau_1}^{\tau_2} e^{-\tau'/\tau_0} E(\tau') \frac{d\tau'}{\tau_0} - \int_{\tau_1}^{\tau_2} e^{-\tau'/\tau_0} E(\tau') \frac{d\tau'}{\tau_0}\right]\right\} = 0. \tag{10.198}$$

Therefore, the solution with pre-acceleration is physically more appealing, as it asymptotically satisfies the law of inertia, and deviates significantly from the classical behavior in a small region of space–time, where pre-acceleration occurs; furthermore, the solution correctly accounts for the energy radiated by the charge while it is accelerated, in contradistinction with the Lorentz force. One can think of the pre-acceleration phase as launching the electron on a trajectory exactly compensating runaways.

We now turn to the special case of a constant electric field in a parallel-plate capacitor, where $E(\tau) = E_0$ for $0 < \tau < \tau^*$, and $E(\tau) = 0$ otherwise; applying the derivation performed above, we find

$$\dot{w}(\tau) = e^{\tau/\tau_0}\left[\frac{E_0}{\tau_0}\int_0^{\tau^*} e^{-\tau'/\tau_0} d\tau' - \int_0^{\tau} e^{-\tau'/\tau_0} E(\tau') \frac{d\tau'}{\tau_0}\right]. \tag{10.199}$$

Here, $\tau = 0$ corresponds to the time when the electron enters the space between the plates, while $\tau^*$ marks the proper time when the charge leaves the constant field region.

In order to obtain a more pedagogical approach to this specific situation, we return to Eq. (10.189): we now have

$$\begin{aligned} \dot{w} &= E_0 + \tau_0 \ddot{w}, \quad 0 < \tau < \tau^* \\ \dot{w} &= \tau_0 \ddot{w}, \qquad \text{otherwise.} \end{aligned} \tag{10.200}$$

Outside the capacitor, one simply obtains the exponential solution,

$$\begin{aligned} \dot{w}(\tau) &= \mathcal{A}^{\pm} e^{\tau/\tau_0}, \\ w(\tau) &= \tau_0 \mathcal{A}^{\pm} e^{\tau/\tau_0} + \mathcal{B}^{\pm}, \end{aligned} \tag{10.201}$$

where the superscript refers to $\tau < 0$ for the minus sign, and $\tau > \tau^*$ for the plus sign. Inside the capacitor, we have

$$\begin{aligned} \dot{w}(\tau) &= E_0 + \mathcal{C} e^{\tau/\tau_0}, \\ w(\tau) &= E_0 \tau + \tau_0 \mathcal{C} e^{\tau/\tau_0} + \mathcal{D}. \end{aligned} \tag{10.202}$$

To determine the integration constants, we require that the four-velocity and four-acceleration be continuous at the capacitor boundaries, and that the four-acceleration satisfies the Dirac–Rohrlich asymptotic condition. The latter implies that $\lim_{\tau \to \pm\infty}[\dot{w}(\tau)] = 0$, which is satisfied if $\mathcal{A}^+ = 0$. Examining the boundary at $\tau = 0$, we now have

$$\begin{aligned} \mathcal{A}^- &= E_0 + \mathcal{C}, \\ \tau_0 \mathcal{A}^- + \mathcal{B}^- &= \tau_0 \mathcal{C} + \mathcal{D}; \end{aligned} \tag{10.203}$$

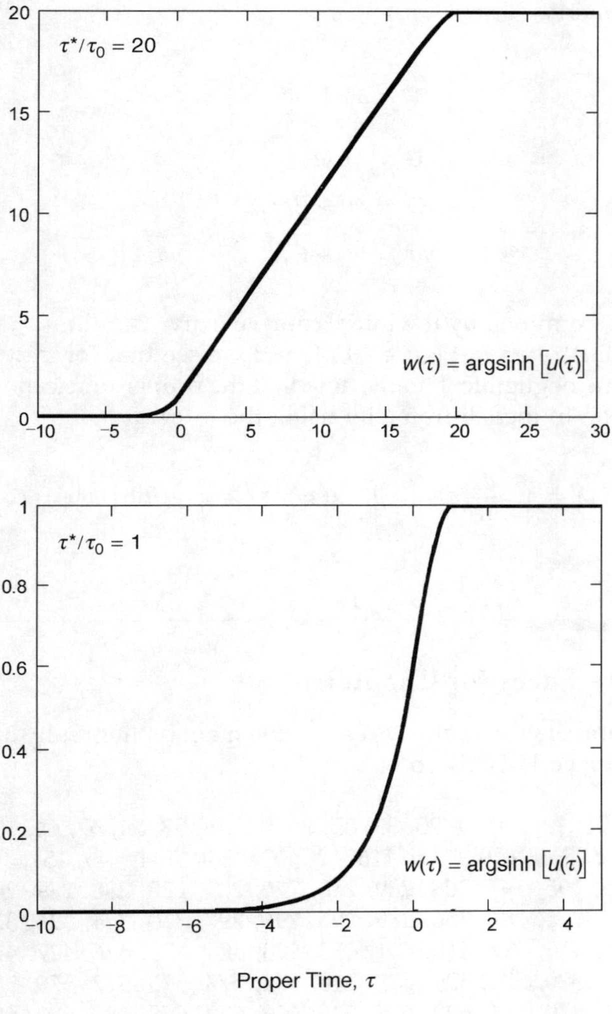


**FIGURE 10.14**
Top: solution to the one-dimensional Dirac–Lorentz equation for a constant unit electric field, $E_0 = 1$, for $0 < \tau < \tau^*$, with $\tau^* = 20\tau_0$; this is very close to the solution of the Lorentz force equation. Bottom: here, $\tau^* = \tau_0$, and the acausal pre-acceleration appears clearly. In both cases, the radiation emitted by the charge during its acceleration is properly modeled by the Dirac–Lorentz solution: the work of the electric field on the charge is equal to the sum of the kinetic energy gained by the particle and the electromagnetic energy radiated during the process.

while at $\tau = \tau^*$, we find

$$\begin{aligned} &E_0 + Ce^{\tau^*/\tau_0} = 0, \\ &E_0\tau^* + \tau_0 Ce^{\tau^*/\tau_0} + \mathcal{D} = \mathcal{B}^+. \end{aligned} \tag{10.204}$$

We easily obtain

$$\begin{aligned} \mathcal{A}^- &= E_0(1 - e^{-\tau^*/\tau_0}), \\ \mathcal{C} &= -E_0 e^{\tau^*/\tau_0}, \\ \mathcal{D} &= \mathcal{B}^- + E_0\tau_0, \\ \mathcal{B}^+ &= \mathcal{B}^- + E_0\tau^*, \end{aligned} \tag{10.205}$$

while $\mathcal{B}^-$ is determined by the initial four-velocity: $\mathcal{B}^- = \lim_{\tau\to-\infty}[w(\tau)]$. This solution is illustrated in Figure 10.14, and we see that for $\tau^* \gg \tau_0$, acausal effects become negligible. Finally, note that the proper time can be connected to the position in the lab frame by using the integral

$$x(\tau) = \int_{-\infty}^{\tau} \frac{dx}{d\tau'} d\tau' = \int_{-\infty}^{\tau} u(\tau')d\tau' = \int_{-\infty}^{\tau} \sinh[w(\tau')]d\tau'. \tag{10.206}$$

## 10.17 References for Chapter 10

Note: the numbers listed below refer to the main bibliography and reference sections at the end of this book.

1, 4, 6, 8, 9, 11, 12, 13, 19, 23, 33, 35, 36, 44, 46, 53, 54, 57, 64, 69, 71, 73, 74, 87, 92, 94, 96, 97, 99, 102, 105, 116, 120, 121, 140, 144, 149, 157, 159, 165, 170, 172, 182, 185, 195, 197, 208, 209, 210, 220, 221, 225, 238, 239, 240, 246, 250, 262, 268, 270, 275, 285, 286, 289, 295, 296, 297, 310, 324, 326, 327, 348, 353, 363, 411, 413, 416, 417, 418, 421, 422, 423, 424, 425, 426, 427, 428, 435, 436, 437, 438, 479, 520, 521, 529, 539, 540, 572, 573, 575, 577, 579, 580, 588, 597, 598, 600, 601, 620, 621, 622, 635, 636, 669, 677, 678, 680, 682, 683, 688, 689, 714, 720, 727, 728, 729, 736, 745, 746, 747, 748, 749, 750, 765, 779, 780, 781, 783, 784, 789, 790, 798, 807, 812, 825, 835, 848, 850, 882, 893, 899, 900, 901, 912, 913, 914, 915, 916, 923, 924, 925, 926, 927.

# Bibliography


1. Abramowitz and Stegun, *Handbook of Mathematical Functions,* Dover, New York, 1965.
2. Alibek, *Biohazard,* Random House, New York, 1999.
3. Andreadis, *To Seek Out New Life: The Biology of Star Trek,* Three Rivers Press, New York, 1998.
4. Arfken and Weber, *Mathematical Methods for Physicists,* 4th ed., Academic Press, San Diego, 1995.
5. Asimov, *The Collapsing Universe: The Story of Black Holes,* Walker Publishing, New York, 1977.
6. Atwood, *Soft X-Rays and Extreme Ultraviolet Radiation,* Cambridge University Press, Cambridge, 1999.
7. Barrow, *The Origin of the Universe,* Basic Books, Harper Collins, New York, 1994.
8. Barut, *Electrodynamics and Classical Theory of Fields and Particles,* Dover, New York, 1980.
9. Becker, *Electromagnetic Fields and Interactions,* Dover, New York, 1964.
10. Beckmann, *A History of Pi,* St. Martin's Press, New York, 1971.
11. Berestetskii, Lifshitz, and Pitaevskii, *Quantum Electrodynamics,* 2nd ed., Pergamon Press, Oxford, 1980.
12. Blackburn and Gait, *Nucleic Acids in Chemistry and Biology,* 2nd ed., Oxford University Press, Oxford, 1996.
13. Bleistein and Handelsman, *Asymptotic Expansions of Integrals,* Dover, New York, 1986.
14. Bloom, *The Lucifer Principle: A Scientific Expedition into the Forces of History,* Atlantic Monthly Press, New York, 1995.
15. Bohm, *Quantum Theory,* Dover, New York, 1979.
16. Bohr, *Physique Atomique et Connaissance Humaine,* Gauthier-Villars, Paris, 1972.
17. Born and Wolf, *Principles of Optics,* 7th ed., Cambridge University Press, Cambridge, 1999.
18. Boslough, *Stephen Hawking's Universe,* Avon Books, New York, 1985.
19. Bowman, *Introduction to Bessel Functions,* Dover, New York, 1958.
20. Brennan, *Heisenberg Probably Slept Here,* John Wiley & Sons, New York, 1997.
21. Cantor, *Contributions to the Founding of the Theory of Transfinite Numbers,* Dover, New York, 1955.
22. Capra, *The Taos of Physics,* 2nd ed., Bantam Books, New York, 1984.
23. Carroll, *Perspectives on Structure and Mechanism in Organic Chemistry,* Brooks/Cole Publishing, Pacific Grove, CA, 1998.
24. Casti, *The Cambridge Quintet,* Helix Books, Perseus Books, Reading, MA, 1998.
25. Cavelos, *The Science of the X-Files,* Berkley Boulevard Books, New York, 1998.
26. Cavelos, *The Science of Star Wars,* St. Martin's Press, New York, 1999.

27. Chandrasekhar, *The Mathematical Theory of Black Holes,* Oxford Science Publications, Oxford, 1992.
28. Charon, *J'ai Vécu Quinze Milliards d'Années,* Albin Michel, Paris, 1982.
29. Charon, *L'Esprit et la Relativité Complexe,* Albin Michel, Paris, 1983.
30. Cline, *Men Who Made a New Physics,* University of Chicago Press, Chicago, 1987.
31. Cohen-Tannoudji, Diu, and Laloe, *Quantum Mechanics,* Vol. 1, John Wiley & Sons, New York, 1978.
32. Cohen-Tannoudji, Diu, and Laloe, *Quantum Mechanics,* Vol. 2, John Wiley & Sons, New York, 1978.
33. Collins, *Mathematical Methods for Physicists and Engineers,* 2nd ed., Dover, Mineola, NY, 1999.
34. Costa de Beauregard, *Le Second Principe de la Science du Temps,* Seuil, Paris, 1963.
35. Courant and Hilbert, *Methods of Mathematical Physics,* Vol. 1, John Wiley & Sons, New York, 1953.
36. Courant and Hilbert, *Methods of Mathematical Physics,* Vol. 2, John Wiley & Sons, New York, 1962.
37. Cronin, *Microwave and Optical Waveguides,* Institute of Physics Publishing, Bristol, England, 1995.
38. Davidson, *Theory of Nonneutral Plasma,* W. A. Benjamin, Reading, MA, 1974.
39. Davies, *The Accidental Universe,* Cambridge University Press, Cambridge, 1982.
40. Davies, *Superforce,* Simon and Schuster, New York, 1984.
41. Davies, *The Last Three Minutes,* Basic Books, Harper Collins, New York, 1994.
42. Davies, *Are We Alone?,* Basic Books, Harper Collins, New York, 1995.
43. Davies and Brown, *Superstrings: A Theory of Everything?,* Cambridge University Press, Cambridge, 1988.
44. Davis, *Fourier Series and Orthogonal Functions,* Dover, New York, 1963.
45. Davis and Park, *No Way: The Nature of the Impossible,* W. H. Freeman, New York, 1987.
46. Dawkins, *The Blind Watchmaker,* W. W. Norton, New York, 1996.
47. De Broglie, *La Physique Nouvelle et les Quanta,* Champs, Flammarion, Paris, 1937.
48. Delachet, *Le Calcul Tensoriel,* Presses Universitaires de France, Paris, 1974.
49. Delcroix, *Physique des Plasmas,* Vols. 1 and 2, Dunod, Paris, 1966.
50. Dennett, *Consciousness Explained,* Back Bay Books, Boston, MA, 1991.
51. D'Espagnat, *A la Recherche du Réel,* 2nd ed., Gauthier-Villars, Paris, 1981.
52. D'Espagnat, *Un Atome de Sagesse,* Seuil, Paris, 1982.
53. DeVries, *A First Course in Computational Physics,* John Wiley & Sons, New York, 1994.
54. DiMauro, Murmane, and L'Huillier, *Applications of High-Field and Short Wavelength Sources,* Plenum Press, New York, 1998.
55. Dirac, *The Principles of Quantum Mechanics,* 4th ed., Oxford University Press, New York, 1958.
56. Drake, *Atomic, Molecular, and Optical Physics Handbook,* AIP, New York, 1996.
57. Drenth, *Principles of Protein X-Ray Crystallography,* 2nd ed., Springer-Verlag, New York, 1999.
58. Dyson, *Disturbing the Universe,* Basic Books, Harper Collins, New York, 1979.
59. Eddington, *Space, Time and Gravitation,* Cambridge University Press, Cambridge, 1920.
60. Einstein, *Quatre Conférences sur la Théorie de la Relativité,* Gauthier-Villars, Paris, 1971.
61. Einstein, *Ideas and Opinions,* Crown Publishers, New York, 1982.

62. Einstein and Born, *Correspondence 1916–1955,* Seuil, Paris, 1972.
63. Einstein, Lorentz, Weyl, and Minkowski, *The Principle of Relativity,* Dover, New York, 1952.
64. Eyges, *The Classical Electromagnetic Field,* Dover, New York, 1972.
65. Ferguson, *Stephen Hawking: Quest for a Theory of Everything,* Bantam Books, New York, 1992.
66. Fermi, *Thermodynamics,* Dover, New York, 1936.
67. Ferris, *The Whole Shebang,* Simon and Schuster, New York, 1997.
68. Ferris, *The World Treasury of Physics, Astronomy, and Mathematics,* Back Bay Books, Little, Brown, and Co., Boston, 1991.
69. Feynman, *Quantum Electrodynamics,* Addison-Wesley, Reading, MA, 1961.
70. Feynman, Leighton, and Sands, *The Feynman Lectures on Physics,* Vol. 1, Addison-Wesley, Reading, MA, 1963.
71. Feynman, Leighton, and Sands, *The Feynman Lectures on Physics,* Vol. 2, Addison-Wesley, Reading, MA, 1964.
72. Feynman, Leighton, and Sands, *The Feynman Lectures on Physics,* Vol. 3, Addison-Wesley, Reading, MA, 1965.
73. Feynman, *QED: The Strange Theory of Light and Matter,* Princeton University Press, Princeton, 1985.
74. Feynman and Weinberg, *Elementary Particles and the Laws of Physics,* Cambridge University Press, New York, 1987.
75. Feynman, Morinigo, and Wagner, *Feynman Lectures on Gravitation,* Addison-Wesley, Reading, MA, 1995.
76. Feynman, *The Character of Physical Law,* Modern Library, New York, 1994.
77. Feynman, *Six Easy Pieces,* Helix Books, Addison-Wesley, Reading, MA, 1995.
78. Feynman, *Six Not-So-Easy Pieces,* Helix Books, Addison-Wesley, Reading, MA, 1997.
79. Feynman, *The Pleasure of Finding Things Out,* Helix Books, Perseus Books, Cambridge, 1999.
80. Feynman, *Surely You're Joking, Mr. Feynman!,* W. W. Norton, New York, 1985.
81. Feynman, *What Do You Care What Other People Think?,* W. W. Norton, New York, 1988.
82. Feynman, *The Meaning of It All,* Helix Books, Addison-Wesley, Reading, MA, 1998.
83. Fowles, *Introduction to Modern Optics,* 2nd ed., Dover, New York, 1975.
84. Frege, *The Foundations of Arithmetic,* Northwestern University Press, Evanston, IL, 1980.
85. French and Kennedy, *Niels Bohr: A Centenary Volume,* Harvard University Press, Cambridge, 1985.
86. Freund and Antonsen, *Principles of Free-Electron Lasers,* 2nd ed., Chapman and Hall, New York, 1996.
87. Gallo, *Virus Hunting, AIDS, Cancer, and the Human Retrovirus: A Story of Scientific Discovery,* Basic Books, Perseus Books, Reading, MA, 1991.
88. Gamow, *Mr. Tompkins in Paperback,* Cambridge University Press, Cambridge 1990.
89. Gamow, *Thirty Years that Shook Physics,* Dover, New York, 1966.
90. Gell-Mann, *The Quark and the Jaguar,* W. H. Freeman, New York, 1994.
91. Gleick, *Genius: The Life and Science of Richard Feynman,* Vintage Books, Random House, New York, 1992.
92. Goldstein, *Classical Mechanics,* 2nd ed., Addison-Wesley, Reading, MA, 1980.

93. Goodstein and Goodstein, *Feynman's Lost Lecture,* W. W. Norton, New York, 1996.
94. Gradshteyn and Ryzhik, *Table of Integrals, Series, and Products,* 4th ed., Academic Press, Orlando, FL, 1980.
95. Green, Schwarz, and Witten, *Superstring Theory,* Vols. 1 and 2, Cambridge University Press, Cambridge, 1987.
96. Greiner, Muller, and Rafelski, *Quantum Electrodynamics of Strong Fields,* Springer-Verlag, New York, 1985.
97. Greiner, *Relativistic Quantum Mechanics: Wave Equations,* Springer-Verlag, New York, 1990.
98. Greiner and Schafer, *Quantum Chromodynamics,* Springer-Verlag, New York, 1994.
99. Greiner and Reinhardt, *Quantum Electrodynamics,* 2nd ed., Springer-Verlag, New York, 1994.
100. Greiner, *Quantum Mechanics: An Introduction,* 3rd ed., Springer-Verlag, New York, 1994.
101. Greiner, *Quantum Mechanics: Symmetries,* 2nd ed., Springer-Verlag, New York, 1994.
102. Greiner, *Classical Electrodynamics, Springer-Verlag,* New York, 1998.
103. Gribbin, *Schrödinger's Kittens and the Search for Reality,* Back Bay Books, Little, Brown, and Co., Boston, MA, 1995.
104. Gribbin and Gribbin, *Richard Feynman: A Life in Science,* Dutton, New York, 1997.
105. Griffiths, *Introduction to Electrodynamics,* 2nd ed., Prentice Hall, Englewood Cliffs, NJ, 1989.
106. Guillen, *Bridges to Infinity,* Jeremy P. Tarcher, Houghton Mifflin, Boston, MA, 1983.
107. Guitton, *Une Recherche de Dieu,* Editions du Félin, Paris, 1996.
108. Gullberg, *Mathematics from the Birth of Numbers,* W. W. Norton, New York, 1997.
109. Guth, *The Inflationary Universe,* Addison-Wesley, Reading, MA, 1997.
110. Hanley, *The Physics of Star Trek,* Basic Books, Harper Collins, New York, 1997.
111. Hawking, *A Brief History of Time,* Bantam Books, Toronto, Canada, 1988.
112. Hawking and Penrose, *The Nature of Space and Time,* Princeton University Press, Princeton, 1996.
113. Heisenberg, *The Physical Principles of Quantum Theory,* Dover, New York, 1949.
114. Heisenberg, *Tradition in Science,* Seabury Press, New York, 1983.
115. Heisenberg, *Encounters with Einstein,* Princeton University Press, Princeton, 1983.
116. Heitler, *The Quantum Theory of Radiation,* 3rd ed., Dover, New York, 1984.
117. Herbert, *Faster than Light: Superluminal Loopholes in Physics,* Plume Books, Penguin Group, New York, 1988.
118. Hofstadter, *Metamagical Themas: Questing for the Essence of Mind and Pattern,* Basic Books, New York, 1985.
119. Horgan, *The End of Science,* Helix Books, Addison-Wesley, Reading, MA, 1995.
120. Jackson, *Classical Electrodynamics,* 3rd ed., John Wiley, New York, 1999.
121. Jahnke and Emde, *Tables of Functions,* Dover, New York, 1945.
122. Jenkins and Jenkins, *The Biology of Star Trek,* Boxtree, Harper Collins, New York, 1998.
123. Johnson, *Strange Beauty: Murray Gell-Mann and the Revolution in Twentieth-Century Physics,* Vintage Books, Random House, New York, 1999.
124. Johnson, Jr. and Gabriel, *Laser Light Scattering,* Dover, New York, 1981.

125. Kaku, *Quantum Field Theory,* Oxford University Press, New York, 1993.
126. Kaku, *Hyperspace,* Anchor Books, Doubleday, New York, 1994.
127. Kaku, *Visions,* Anchor Books, Doubleday, New York, 1997.
128. Kaku and Thompson, *Beyond Einstein,* Anchor Books, Doubleday, New York, 1995.
129. Krauss, *The Physics of Star Trek,* Basic Books, Harper Collins, New York, 1995.
130. Krauss, *Beyond Star Trek,* Basic Books, Harper Collins, New York, 1997.
131. Kuhn, *The Structure of Scientific Revolutions,* University of Chicago Press, Chicago, IL, 1970.
132. Kuhn, *Black-Body Theory and the Quantum Discontinuity,* 1894–1912, University of Chicago Press, Chicago, IL, 1978.
133. Kane, *The Particle Garden,* Helix Books, Addison-Wesley, Reading, MA, 1995.
134. Kurzweil, *The Age of Spiritual Machines,* Viking, Penguin Group, New York, 1999.
135. Landau and Lifshitz, *The Classical Theory of Fields,* 4th ed., Pergamon Press, Oxford, 1975.
136. Landau, Lifshitz, and Pitaevskii, *Electrodynamics of Continuous Media,* 2nd ed., Pergamon Press, Oxford, 1984.
137. Landau and Lifshitz, *Quantum Mechanics (Non-Relativistic Theory),* 3rd ed., Pergamon Press, Oxford, 1977.
138. Landau and Lifshitz, *Fluid Mechanics,* 2nd ed., Pergamon Press, Oxford, 1987.
139. Landsberg, *The Enigma of Time,* Adam Hilger, Bristol, England, 1982.
140. Lebedev, *Special Functions and Their Applications,* Dover, New York, 1972.
141. Lederman with Teresi, *The God Particle,* Delta Books, Dell Publishing, New York, 1993.
142. Leighton, *Tuva or Bust!,* W. W. Norton, New York, 1991.
143. Levi-Civita, *The Absolute Differential Calculus,* Dover, New York, 1977.
144. Lide, *CRC Handbook of Chemistry and Physics,* 75th ed., CRC Press, Boca Raton, FL, 1994.
145. Lightman, *Einstein's Dreams,* Warner Books, New York, 1993.
146. Lightman, *Dance for Two,* Pantheon Books, New York, 1996.
147. Lindley, *The End of Physics,* Basic Books, Harper Collins, New York, 1993.
148. Loudon, *The Quantum Theory of Light,* 2nd ed., Oxford University Press, New York, 1983.
149. Low, *Classical Field Theory. Electromagnetism and Gravitation,* John Wiley & Sons, New York, 1997.
150. Mach, *Space and Geometry,* Open Court Publishing, La Salle, IL, 1906.
151. Mandel and Wolf, *Optical Coherence and Quantum Optics,* Cambridge University Press, New York, 1995.
152. Maxwell, *A Treatise on Electricity and Magnetism,* Vols. 1 and 2, Dover, New York, 1954.
153. Marshall, *Free-Electron Lasers,* Macmillan, New York, 1985.
154. Menzel, *Fundamental Formulas of Physics,* Vol. 1, Dover, New York, 1960.
155. Menzel, *Fundamental Formulas of Physics,* Vol. 2, Dover, New York, 1960.
156. Mehra, *The Beat of a Different Drum,* Oxford University Press, New York, 1994.
157. Milonni and Eberly, *Lasers,* John Wiley & Sons, New York, 1988.
158. Misner, Thorne, and Wheeler, *Gravitation,* W. H. Freeman, San Francisco, 1973.
159. Montagnier, *Virus,* W. W. Norton, New York, 2000.
160. Morris, *Time's Arrow,* Simon and Schuster, New York, 1985.
161. Nagel and Newman, *Gödel's Proof,* New York University Press, New York, 1958.

162. Newton, *Principia, Vol. 1: The Motion of Bodies,* and *Vol. 2: The System of the World,* University of California Press, Berkeley, CA, 1962.
163. Omnès, *L'Univers et ses Métamorphoses,* Collection Savoir, Hermann, Paris, 1973.
164. Papas, *Theory of Electromagnetic Wave Propagation,* Dover, New York, 1988.
165. Pauli, *Lectures on Physics, Vol. 1: Electrodynamics,* MIT Press, Cambridge, 1973.
166. Pauli, *Lectures on Physics, Vol. 2: Optics and the Theory of Electrons,* MIT Press, Cambridge, 1973.
167. Pauli, *Lectures on Physics, Vol. 3: Thermodynamics and the Kinetic Theory of Gases,* MIT Press, Cambridge, 1973.
168. Pauli, *Lectures on Physics, Vol. 4: Statistical Mechanics,* MIT Press, Cambridge, 1973.
169. Pauli, *Lectures on Physics, Vol. 5: Wave Mechanics,* MIT Press, Cambridge, 1973.
170. Pauli, *Lectures on Physics, Vol. 6: Selected Topics in Field Quantization,* MIT Press, Cambridge, 1973.
171. Pauli, *General Principles of Quantum Mechanics,* Springer-Verlag, New York, 1980.
172. Pauli, *Theory of Relativity,* Dover, New York, 1981.
173. Pauli, *Writings on Physics and Philosophy,* Springer-Verlag, Berlin, 1994.
174. Pauling and Wilson, *Introduction to Quantum Mechanics with Applications to Chemistry,* Dover, New York, 1963.
175. Peebles, *Principles of Physical Cosmology,* Princeton University Press, Princeton, 1993.
176. Peebles, *The Large-Scale Structure of the Universe,* Princeton University Press, Princeton, 1980.
177. Penfield and Haus, *Electrodynamics of Moving Media,* MIT Press, Cambridge, 1967.
178. Penrose, *The Emperor's New Mind,* Penguin Books, London, 1989.
179. Penrose, *Shadows of the Mind,* Oxford University Press, Oxford, 1994.
180. Penrose, with Shimony, Cartwright, and Hawking, *The Large, the Small, and the Human Mind,* Cambridge University Press, Cambridge, 1997.
181. Peskin and Schroeder, *An Introduction to Quantum Field Theory,* Addison-Wesley, Reading, MA, 1995.
182. Peters and Olshaker, *Virus Hunter,* Anchor Books, Doubleday, New York, 1997.
183. Petit Bois, *Tables of Indefinite Integrals,* Dover, New York, 1961.
184. Poincaré, *La Science et l'Hypothèse,* Champs, Flammarion, Paris, 1968.
185. Poincelot, *Principes et Applications Usuelles de la Relativité,* Editions de la Revue d'Optique Theorique et Instrumentale, Paris, 1968.
186. Polkinghorne, *Rochester Roundabout. The Story of High Energy Physics,* Longman, Burnt Mill, Harlow, England, 1989.
187. Polkinghorne, *Belief in God in an Age of Science,* Yale University Press, New Haven, CT, 1998.
188. Prigogine, *Les Lois du Chaos,* Flammarion, Paris, 1994.
189. Prigogine and Stengers, *La Nouvelle Alliance,* Folio Essais, Gallimard, Paris, 1979, 1986.
190. Regis, *Who Got Einstein's Office?,* Addison-Wesley, Reading, MA, 1987.
191. Reichenbach, *Philosophic Foundations of Quantum Mechanics,* Dover, New York, 1944.
192. Reiser, *Theory and Design of Charged Particle Beams,* John Wiley & Sons, New York, 1994.
193. Reitz, Milford and Christy, *Foundations of Electromagnetic Theory,* 3rd ed., Addison-Wesley, Reading, MA, 1979.

194. Rojansky, *Electromagnetic Fields and Waves*, Dover, New York, 1979.
195. Rohrlich, *Classical Charged Particles*, Addison-Wesley, Reading, MA, 1965.
196. Rosebury, *Handbook of Electron Tube and Vacuum Techniques*, Addison-Wesley, Reading, MA, 1965.
197. Rulliere, *Femtosecond Laser Pulses*, Springer-Verlag, New York, 1998.
198. Russel, *The Will to Doubt*, Philosophical Library, New York, 1958.
199. Russel, *The ABC of Relativity*, 4th ed., Mentor Books, New American Library, 1985.
200. Russel, *Principles of Mathematics*, W. W. Norton, New York, 1903.
201. Sagan, *The Cosmic Connection, an Extraterrestrial Perspective*, Anchor Press, Doubleday, New York, 1973.
202. Sagan, *The Dragons of Eden*, Ballantine Books, New York, 1977.
203. Sagan, *The Demon-Haunted World*, Random House, New York, 1996.
204. Sagan, *Billions and Billions*, Ballantine Books, New York, 1997.
205. Schrödinger, *What is Life? And Mind and Matter*, Cambridge University Press, Cambridge, 1967.
206. Schrödinger, *My View of the World*, Ox Bow Press, Woodbridge, CT, 1983.
207. Schwartz, *Principles of Electrodynamics*, Dover, New York, 1972.
208. Schweber, *QED and the Men Who Made It*, Princeton University Press, Princeton, 1994.
209. Schwinger, *Quantum Electrodynamics*, Dover, New York, 1958.
210. Schwinger, DeRaad, Jr., Milton, and Tsai, *Classical Electrodynamics*, Perseus Books, Reading, MA, 1998.
211. Segrè, *From Falling Bodies to Radio Waves*, W. H. Freeman, New York, 1984.
212. Segrè, *From X-Rays to Quarks*, W. H. Freeman, New York, 1980.
213. Shadowitz, *The Electromagnetic Field*, Dover, New York, 1975.
214. Shen, *The Principles of Nonlinear Optics*, John Wiley & Sons, New York, 1984.
215. Siegman, *Lasers*, University Science Books, Sausalito, CA, 1986.
216. Simon, *The Real Science Behind the X-Files*, Simon and Schuster, New York, 1999.
217. Smoot and Davidson, *Wrinkles in Time*, Avon Books, New York, 1993.
218. Smythe, *Static and Dynamic Electricity*, 3rd ed., Hemisphere Publishing, New York, 1989.
219. Sneddon, *Fourier Transforms*, Dover, New York, 1995.
220. Sokolov and Ternov, *Radiation from Relativistic Electrons*, AIP Translation Series, New York, 1986.
221. Sommerfeld, *Electrodynamics*, Academic Press, New York, 1964.
222. Spiegel, *Complex Variables*, McGraw-Hill, New York, 1993.
223. Sugimoto, *Albert Einstein: A Photographic Biography*, Schocken Books, New York, 1989.
224. Sykes, *No Ordinary Genius: The Illustrated Richard Feynman*, W. W. Norton, New York, 1994.
225. Teplitz, *Electromagnetism, Paths to Research*, Plenum Press, New York, 1982.
226. Thaller, *The Dirac Equation*, Springer-Verlag, New York, 1992.
227. Thorne, *Black Holes and Time Warps: Einstein's Outrageous Legacy*, W. W. Norton, New York, 1994.
228. Tipler, *The Physics of Immortality*, Anchor Books, Doubleday, New York, 1994.
229. Tomonaga, *The Story of Spin*, University of Chicago Press, Chicago, 1997.
230. Trefil, *101 Things You Don't Know About Science and No One Else Does Either*, Mariner Books, Houghton Mifflin, Boston, 1996.
231. Trigg, *Landmark Experiments in Twentieth Century Physics*, Dover, New York, 1975.

232. Valentin, *Subatomic Physics: Nuclei and Particles,* Elsevier Science, Amsterdam, 1982.
233. van Holde, Johnson, and Ho, *Principles of Physical Biochemistry,* Prentice Hall, Upper Saddle River, NJ, 1998.
234. Wald, *General Relativity,* University of Chicago Press, Chicago, 1984.
235. Wald, *Quantum Field Theory in Curved Spacetime and Black Hole Thermodynamics,* University of Chicago Press, Chicago, 1994.
236. Watson, *Dark Nature: A Natural History of Evil,* Harper Perennial, Harper Collins, New York, 1995.
237. Weinberg, *The First Three Minutes: A Modern View of the Origin of the Universe,* Basic Books, New York, 1977.
238. Weinberg, *The Quantum Theory of Fields, Vol. 1: Foundations,* Cambridge University Press, New York, 1995.
239. Weinberg, *The Quantum Theory of Fields, Vol. 2: Modern Applications,* Cambridge University Press, New York, 1996.
240. Weinberg, *The Quantum Theory of Fields, Vol. 3: Supersymmetry,* Cambridge University Press, New York, 2000.
241. Weinberg, *Dreams of a Final Theory,* Vintage Books, Random House, New York, 1992, 1993.
242. Wheeler and Ford, Geons, *Black Holes, and Quantum Foam,* W. W. Norton, New York, 1998.
243. Wheeler and Zurek, *Quantum Theory and Measurement,* Princeton University Press, Princeton, 1983.
244. Wiedemann, *Particle Accelerator Physics 1: Basic Principles and Linear Beam Dynamics,* 2nd ed., Springer-Verlag, Berlin, 1999.
245. Wiedemann, *Particle Accelerator Physics 2: Nonlinear and Higher-Order Beam Dynamics,* 2nd ed., Springer-Verlag, Berlin, 1999.
246. Yaghjian, *Relativistic Dynamics of a Charged Sphere,* Springer-Verlag, Berlin, 1992.
247. Yariv, *Quantum Electronics,* 2nd ed., John Wiley & Sons, New York, 1975.
248. Zotter and Kheifets, *Impedance and Wakes in High-Energy Particle Accelerators,* World Scientific, Singapore, 1998.
249. Zrelov, *Cherenkov Radiation in High-Energy Physics,* Israel Program for Scientific Translations, Jerusalem, 1970.
250. Zwillinger, *Standard Mathematical Tables and Formulae,* 30th ed., CRC Press, Boca Raton, FL, 1996.

## References

251. Al-Abawi et al., Electron dynamics in a free-electron laser, *Phys. Rev.,* A16, 666, 1977.
252. Abramovich et al., High spectral coherence in long-pulse and continuous free-electron laser: measurements and theoretical limitations, *Phys. Rev. Lett.,* 82, 5257, 1999.
253. Adolphsen et al., Measurement of wake-field suppression in a detuned X-band accelerator structure, *Phys. Rev. Lett.,* 74, 2475, 1995.
254. Alberti et al., Experimental study of a 28 GHz high-power long-pulse cyclotron autoresonance maser oscillator, *Phys. Rev. Lett.,* 71, 2018, 1993.
255. Antonsen, Jr., Stability of bound eigenmode solutions for the collisionless universal instability, *Phys. Rev. Lett.,* 41, 33, 1978.

256. Antonsen, Jr., Density and deceleration limits in tapered free electron lasers, *Phys. Rev. Lett.*, 58, 211, 1987.
257. Antonsen, Jr. and Levush, Mode competition and control in free-electron laser oscillators, *Phys. Rev. Lett.*, 62, 1488, 1989.
258. Antonsen, Jr. and Ott, Multifractal power spectra of passive scalars convected by chaotic fluid flows, *Phys. Rev.*, A44, 851, 1991.
259. Antonsen, Jr. and Mora, Self-focusing and Raman scattering of laser pulses in tenuous plasmas, *Phys. Rev. Lett.*, 69, 2204, 1992.
260. Antonsen, Jr. and Bondeson, Effect of trapped thermal particles on the $n = 1$ internal kink mode in tokamaks, *Phys. Rev. Lett.*, 71, 2046, 1993.
261. Antonsen, Jr. et al., Statistics of wave-function scars, *Phys. Rev.*, E51, 111, 1995.
262. Arthur et al., The LCLS: a fourth generation light source using the SLAC linac, *Rev. Sci. Instrum.*, 66, 1967, 1995.
263. Aspect, Proposed experiment to test the nonseparability of quantum mechanics, *Phys. Rev.*, D14, 1944, 1976.
264. Aspect et al., Time correlations between the two sidebands of the resonance fluorescence triplet, *Phys. Rev. Lett.*, 45, 617, 1980.
265. Aspect, Grangier, and Roger, Experimental test of realistic local theories via Bell's theorem, *Phys. Rev. Lett.*, 47, 460, 1981.
266. Aspect, Grangier, and Roger, Experimental realization of Einstein–Podolsky–Rosen *Gedankenexperiment*: a new violation of Bell's inequalities, *Phys. Rev. Lett.*, 49, 91, 1982.
267. Aspect, Dalibard and Roger, Experimental test of Bell's inequalities using time-varying analyzers, *Phys. Rev. Lett.*, 49, 1804, 1982.
268. Atwood et al., Tunable coherent radiation in the soft x-ray and extreme ultraviolet spectral regions, *IEEE J. Quantum Electron.*, QE35, 709, 1999.
269. Audretsch and Müller, Radiation from a uniformly accelerated particle detector: energy, particles, and the quantum measurement process, *Phys. Rev.*, D49, 6566, 1994.
270. Backus et al., High power ultrafast lasers, *Rev. Sci. Instrum.*, 69, 1207, 1998.
271. Bagnato and Kleppner, Bose-Einstein condensation in low-dimensional traps, *Phys. Rev.*, A44, 7439, 1991.
272. Bagrov et al., Stimulated emission of radiation by electrons moving in the field of a plane electromagnetic wave, Soviet Physics, *J. Exp. Th. Phys. (JETP)*, 30, 503, 1970.
273. Baier et al., Contribution to the theory of quantum processes in the field of an intense electromagnetic wave, Soviet Physics, *J. Exp. Th. Phys. (JETP)*, 42, 400, 1975.
274. Baier and Milstein, To the theory of a free-electron laser, *Phys. Lett.*, A65, 319, 1978.
275. Baier and Milstein, Radiative effects in a plane wave moving along a magnetic field, *J. Phys.*, A11, 297, 1978.
276. Baker et al., Thomson scattering measurements of the Langmuir wave spectra resulting from stimulated Raman scattering, *Phys. Rev. Lett.*, 77, 67, 1996.
277. Baldis et al., Second-harmonic light from the interaction of a nanosecond $CO_2$ laser pulse with the plasma produced from polyethylene sheet, *Phys. Rev. Lett.*, 35, 37, 1975.
278. Baldis et al., Two-plasmon decay and profile modification produced by 10.6 $\mu$m radiation at quarter-critical density, *Phys. Rev. Lett.*, 41, 1719, 1978.

279. Baldis and Corkum, Self-focusing of 10.6-$\mu$m radiation in an underdense plasma, *Phys. Rev. Lett.*, 45, 1260, 1980.
280. Baldis and Walsh, Experimental observations of nonlinear saturation of the two-plasmon decay instability, *Phys. Rev. Lett.*, 47, 1658, 1981.
281. Baldis et al., Competition between the stimulated Raman and Brillouin scattering instabilities in 0.35-$\mu$m irradiated CH foil targets, *Phys. Rev. Lett.*, 62, 2829, 1989.
282. Baldis et al., Resonant seeding of stimulated Brillouin scattering by crossing laser beams, *Phys. Rev. Lett.*, 77, 2957, 1996.
283. Baldwin et al., Pulse-stimulated emission from plasma columns, *Phys. Rev. Lett.*, 20, 314, 1968.
284. Balkcum et al., Theory and design of a high-harmonic gyrofrequency multiplier, *IEEE Trans. Plasma Sci.*, 22, 913, 1994.
285. Balter, "New hope in HIV disease," *Science*, 274, 1988, 1996.
286. Bamber et al., Studies of nonlinear QED in collisions of 46.6 GeV electrons with intense laser pulses, *Phys. Rev.*, D60, 092004, 1999.
287. Bambini et al., Classical theory of the free-electron laser in a moving frame, *Phys. Rev.*, A19, 2013, 1979.
288. Bambini and Stenholm, Quantum description of free electrons in the laser, *Opt. Comm.*, 30, 391, 1979.
289. Baranger, Bethe, and Feynman, Relativistic correction to the Lamb shift, *Phys. Rev.*, 92, 482, 1953.
290. Baranova et al., Acceleration of charged particles by laser beams, *J. Exp. Th. Phys. (JETP)*, 78, 249, 1994.
291. Barov and Rosenzweig, Propagation of short electron pulses in underdense plasmas, *Phys. Rev.*, E49, 4407, 1994.
292. Barrow et al., Strongly coupled stimulated Raman backscatter from subpicosecond laser-plasma interactions, *Phys. Rev. Lett.*, 69, 442, 1992.
293. Barty et al., 12.8-eV laser in neutral cesium, *Phys. Rev. Lett.*, 61, 2201, 1988.
294. Barty et al., Studies of a 96.9-nm laser in neutral cesium, *Phys. Rev.*, A46, 4286, 1992.
295. Barty et al., Generation of 18-fs, multiterawatt pulses by regenerative shaping and chirped-pulse amplification, *Opt. Lett.*, 21, 668, 1996.
296. Barty et al., Multiterawatt 30-fs Ti:sapphire laser system, *Opt. Lett.*, 19, 1442, 1994.
297. Barut and Bornzin, $SO(4,2)$-formulation of the symmetry breaking in relativistic Kepler problems with or without magnetic charges, *J. Math. Phys.*, 12, 841, 1971.
298. Baylis and Yao, Relativistic dynamics of charges in electromagnetic fields: an eigenspinor approach, *Phys. Rev.*, A60, 785, 1999.
299. Beck et al., Measurement of collisional anisotropic temperature relaxation in a strongly magnetized pure electron plasma, *Phys. Rev. Lett.*, 68, 317, 1992.
300. Becker and Mitter, Quantum theory of a free electron laser, *Z. Phys.*, B35, 399, 1979.
301. Bekefi and Brown, Collision cross section and energy loss of slow electrons in hydrogen, *Phys. Rev.*, 112, 159, 1958.
302. Bekefi et al., Incoherent microwave radiation from plasmas, *Phys. Rev.*, 116, 1051, 1959.
303. Bekefi et al., Cyclotron emission from plasmas with non-Maxwellian distributions, *Phys. Rev.*, 122, 1037, 1961.
304. Bekefi et al., Microwave emission and absorption at cyclotron harmonics of a warm plasma, *Phys. Rev. Lett.*, 9, 6, 1962.
305. Bekefi et al., Microwave emission form $n$-type indium antimonide at 4.2 and 77°K, *Phys. Rev. Lett.*, 19, 24, 1967.

306. Bekefi and Orzechowski, Giant microwave bursts emitted from a field-emission, relativistic-electron-beam magnetron, *Phys. Rev. Lett.*, 37, 379, 1976.
307. Bekefi et al., Free-electron laser radiation induced by a periodic dielectric medium, *Phys. Rev.*, A34, 1228, 1986.
308. Bekefi et al., Temporal evolution of beam emittance from a field-emission electron gun, *J. Appl. Phys.*, 62, 1564, 1987.
309. Bekenstein, Black holes and entropy, *Phys. Rev.*, D7, 2333, 1973.
310. Belinfante, On the part played by scalar and longitudinal photons in ordinary electromagnetic fields, *Phys. Rev.*, 76, 226, 1949.
311. Benerofe et al., 116-nm $H_2$ laser pumped by a traveling-wave photoionization electron source, *Phys. Rev. Lett.*, 66, 3136, 1991.
312. Bernstein and Hirshfield, Theory of a free-electron laser, *Phys. Rev. Lett.*, 40, 761, 1978.
313. Bhattacharjee et al., Observation of optical guiding in a Raman free-electron laser, *Phys. Rev. Lett.*, 60, 1254, 1988.
314. Bluem et al., Demonstration of a new free-electron-laser harmonic interaction, *Phys. Rev. Lett.*, 67, 824, 1991.
315. Boivin et al., Analytical solution to the quantum field theory of self-phase modulation with a finite response time, *Phys. Rev. Lett.*, 73, 240, 1994.
316. Bondeson et al., Quasiperiodically forced damped pendula and Schrödinger equations with quasiperiodic potentials: implications of their equivalence, *Phys. Rev. Lett.*, 55, 2103, 1985.
317. Bonifacio et al., Spectrum, temporal structure, and fluctuations in a high-gain free-electron laser starting from noise, *Phys. Rev. Lett.*, 73, 70, 1994.
318. Boyer, Derivation of the blackbody radiation spectrum without quantum assumptions, *Phys. Rev.*, 185, 1374, 1969.
319. Boyer, Asymptotic retarded van der Waals forces derived from classical electrodynamics with classical electromagnetic zero-point radiation, *Phys. Rev.*, A5, 1799, 1972.
320. Boyer, Van der Waals forces and zero-point energy for dielectric and permeable materials, *Phys. Rev.*, A9, 2078, 1974.
321. Boyer, Random electrodynamics: the theory of classical electrodynamics with classical electromagnetic zero-point radiation, *Phys. Rev.*, D11, 790, 1975.
322. Boyer, Thermal effects of acceleration through random classical radiation, *Phys. Rev.*, D21, 2137, 1980.
323. Briggs and Lau, Localization of trapped-particle instabilities, *Phys. Rev. Lett.*, 28, 1248, 1972.
324. Brown and Feynman, Radiative corrections to Compton scattering, *Phys. Rev.*, 85, 231, 1952.
325. Bucksbaum et al., Nonresonant above-threshold ionization by circularly polarized subpicosecond pulses, *Phys. Rev.*, A41, 4119, 1990.
326. Bula et al., Observation of nonlinear effects in Compton scattering, *Phys. Rev. Lett.*, 76, 3116, 1996.
327. Burke et al., Positron production in multiphoton light-by-light scattering, *Phys. Rev. Lett.*, 79, 1626, 1997.
328. Burnett et al., The theory of Bose–Einstein condensation of dilute gases, *Physics Today*, 52 (12), 37, 1999.
329. Caponi and Davidson, Influence of ion-resonance broadening on the anomalous heating and momentum transfer in a current-carrying plasma, *Phys. Rev. Lett.*, 31, 86, 1973.

330. Carmel et al., Demonstration of a two-stage backward-wave-oscillator free-electron laser, *Phys. Rev. Lett.*, 51, 566, 1983.
331. Carlsten, New photoelectric injector design for the Los Alamos National Laboratory XUV FEL accelerator, *Nucl. Instrum. Meth. Phys. Res.*, A285, 313, 1989.
332. Carlsten and Raubenheimer, Emittance growth of bunched beams in bends, *Phys. Rev.*, E51, 1453, 1995.
333. Carlsten and Russel, Subpicosecond compression of 0.1-1 nC electron bunches with a magnetic chicane at 8 MeV, *Phys. Rev.*, E53, R2072, 1996.
334. Carlsten, Calculation of the noninertial space-charge force and the coherent synchrotron radiation force for short electron bunches in circular motion using the retarded Green's function technique, *Phys. Rev.*, E54, 838, 1996.
335. Carlsten, Centrifugal space-charge force of an electron beam in a focusing element, *Phys. Rev.*, E55, R4893, 1997.
336. Carlsten, Growth rate of nonthermodynamic emittance of intense electron beams, *Phys. Rev.*, E58, 2489, 1998.
337. Carlsten, Emittance growth due to radial density variations of an emittance-dominated electron beam in a channel with continuous acceleration and focusing, *Phys. Plasmas*, 5, 1148, 1998.
338. Carlsten, Thermalization of an intense, space-charge-dominated electron beam in a long focusing channel, *Phys. Rev.*, E60, 2280, 1999.
339. Carlsten, Long-term, correlated emittance decrease in intense, high-brightness induction linacs, *Phys. Plasmas*, 6, 3615, 1999.
340. Carlsten and Palmer, Enhanced emittance compensation in a high-frequency rf photoinjector using rf radial focusing, *Nucl. Instrum. Meth. Phys. Res.*, A425, 37, 1999.
341. Carlsten et al., Collisional redistribution and saturation of near-resonance scattered light, *Phys. Rev.*, A15, 1029, 1977.
342. Carmel et al., From linearity towards chaos: basic studies of relativistic backward-wave oscillators, *Phys. Rev. Lett.*, 69, 1652, 1992.
343. Casimir, On Onsager's principle of microscopic reversibility, *Rev. Mod. Phys.*, 17, 343, 1945.
344. Casimir and Polder, The influence of retardation on the London–van der Waals forces, *Phys. Rev.*, 73, 360, 1948.
345. Caves et al., Comment on the quantitative limit on the ability of a Maxwell demon to extract work form heat, *Phys. Rev. Lett.*, 65, 1387, 1990.
346. Chen and Marshall, Thomson backscattering from a relativistic electron beam as a diagnostic for parallel velocity spread, *Phys. Rev. Lett.*, 52, 425, 1984.
347. Chen et al., Evolution of a plasma waveguide created during relativistic–ponderomotive self-channeling of an intense laser pulse, *Phys. Rev. Lett.*, 80, 2610, 1998.
348. Chen et al., Experimental observation of relativistic nonlinear Thomson scattering, *Nature*, 396, 653, 1998.
349. Chen and Renard, Evolution condition for electroweak interactions in composite models, *Phys. Rev.*, D28, 1758, 1983.
350. Chen et al., Acceleration of electrons by the interaction of a bunched electron beam with a plasma, *Phys. Rev. Lett.*, 54, 693, 1985.
351. Chen and Yokoya, Field-gradient effect in quantum beamstrahlung, *Phys. Rev. Lett.*, 61, 1101, 1988.
352. Chen et al., Final focusing and enhanced disruption from an underdense plasma lens in a linear collider, *Phys. Rev.*, D40, 923, 1989.

353. Chen and Telnov, Coherent pair creation in linear colliders, *Phys. Rev. Lett.*, 63, 1796, 1989.
354. Chen, Differential luminosity under multiphoton beamstrahlung, *Phys. Rev.*, D46, 1186, 1992.
355. Chen et al., Hadron production in gamma gamma collisions as a background for e+-e- linear colliders, *Phys. Rev.*, E50, 3209, 1994.
356. Chen, Resonant photon–graviton conversion and cosmic microwave background fluctuations, *Phys. Rev. Lett.*, 74, 634, 1995.
357. Chen and Tajima, Testing Unruh radiation with ultraintense lasers, *Phys. Rev. Lett.*, 83, 256, 1999.
358. Chen and Davidson, Chaotic electron dynamics for relativistic-electron-beam propagation through a planar wiggler magnetic field, *Phys. Rev.*, A42, 5041, 1990.
359. Chen and Davidson, Chaotic particle dynamics in free-electron lasers, *Phys. Rev.*, A43, 5541, 1991.
360. Chen and Davidson, Nonlinear resonances and chaotic behavior in a periodically focused intense charged-particle beam, *Phys. Rev. Lett.*, 72, 2195, 1994.
361. Chen and Davidson, Nonlinear properties of the Kapchinskij–Vladimirskij equilibrium and envelope equation for an intense charged-particle beam in a periodic focusing field, *Phys. Rev.*, E49, 5679, 1994.
362. Chen et al., Rigid-rotor Vlasov equilibrium for an intense charged-particle beam propagating through a periodic solenoidal magnetic field, *Phys. Rev. Lett.*, 79, 1997.
363. Chin et al., Ultrafast structural dynamics in InSb probed by time-resolved x-ray diffraction, *Phys. Rev. Lett.*, 83, 336, 1999.
364. Christenson and Lau, One-dimensional modulational instability in a crossed-field gap, *Phys. Rev. Lett.*, 76, 3324, 1996.
365. Chu et al., Single-mode operation of a Bragg free-electron maser oscillator, *Phys. Rev. Lett.*, 72, 2391, 1994.
366. Cicchitelli et al., Longitudinal field components for laser beams in vacuum, *Phys. Rev.*, A41, 3727, 1990.
367. Clark and Milchberg, Time- and space-resolved density evolution of the plasma waveguide, *Phys. Rev. Lett.*, 78, 2373, 1997.
368. Clark and Milchberg, Frequency selective tunnel coupling to the plasma fiber, *Phys. Rev. Lett.*, 81, 357, 1998.
369. Clayton et al., Relativistic plasma-wave excitation by collinear optical mixing, *Phys. Rev. Lett.*, 54, 2343, 1985.
370. Clayton et al., Ultrahigh-gradient acceleration of injected electrons by laser-excited relativistic electron plasma waves, *Phys. Rev. Lett.*, 70, 37, 1993.
371. Cohen and Feynman, Theory of inelastic scattering of cold neutrons from liquid helium, *Phys. Rev.*, 107, 13, 1957.
372. Cohen-Tannoudji et al., Detection of the static magnetic field produced by the oriented nuclei of optically pumped $^3$He gas, *Phys. Rev. Lett.*, 22, 758, 1969.
373. Cohen–Tannoudji and Dupont-Roc, Experimental study of Zeeman light shifts in weak magnetic fields, *Phys. Rev.*, A5, 968, 1972.
374. Cohen-Tannoudji et al., Comment on the momentum-translation approximation, *Phys. Rev.*, A8, 2747, 1973.
375. Cole, Classical electrodynamic system interacting with classical electromagnetic random radiation, *Found. Phys.*, 20, 225, 1990.
376. Coleman and Witten, Chiral-symmetry breakdown in large-*N* chromodynamics, *Phys. Rev. Lett.*, 45, 100, 1980.

377. Colombant and Lau, Nonlinear beam loading and dynamical limiting currents in a high-power microwave gap, *Phys. Rev. Lett.*, 64, 2320, 1990.
378. Colombant and Lau, Maximum microwave conversion efficiency from a modulated intense relativistic electron beam, *Phys. Rev.*, A45, R2179, 1992.
379. Conde and Bekefi, Experimental study of a 33.3-GHz free-electron-laser amplifier with a reversed axial guide magnetic field, *Phys. Rev. Lett.*, 67, 3082, 1991.
380. Couchman et al., Quantum chaos in systems with ray splitting, *Phys. Rev.*, A46, 6193, 1992.
381. Courant et al., High-energy inverse free-electron-laser accelerator, *Phys. Rev.*, A32, 2813, 1985.
382. Courtney et al., Classical, semiclassical, and quantum dynamics in the Lithium Stark system, *Phys. Rev.*, A51, 3604, 1995.
383. Courtney and Kleppner, Core-induced chaos in diamagnetic lithium, *Phys. Rev.*, A53, 178, 1996.
384. Coverdale et al., Propagation of intense subpicosecond laser pulses through underdense plasmas, *Phys. Rev. Lett.*, 74, 4659, 1995.
385. Danly et al., Direct spectral measurements of a quasi-cw free-electron laser oscillator, *Phys. Rev. Lett.*, 65, 2251, 1990.
386. Danly et al., Long-pulse millimeter-wave free-electron laser and cyclotron autoresonance maser experiments, *Phys. Fluids*, B4, 2307, 1992.
387. Darrow et al., Saturation of beat-excited plasma waves by electrostatic mode coupling, *Phys. Rev. Lett.*, 56, 2629, 1986.
388. Darrow et al., Strongly coupled stimulated Raman backscatter from subpicosecond laser-plasma interactions, *Phys. Rev. Lett.*, 69, 442, 1992.
389. Davidovich et al., Quantum switches and nonlocal microwave fields, *Phys. Rev. Lett.*, 71, 2360, 1993.
390. Davidovich et al., Teleportation of an atomic state between two cavities using nonlocal microwave fields, *Phys. Rev.*, A50, R895, 1994.
391. Davidson and Goldman, Diffusion of electron plasma oscillations in a large-amplitude ion-wave background, *Phys. Rev. Lett.*, 21, 1671, 1968.
392. Davidson and Krall, Vlasov description of an electron gas in a magnetic field, *Phys. Rev. Lett.*, 22, 833, 1969.
393. Davidson and Goldman, Diffusion of electron plasma oscillations in a large-amplitude ion-wave background, *Phys. Rev. Lett.*, 22, 159, 1968.
394. Davidson et al., Electron heating by electron-ion beam instabilities, *Phys. Rev. Lett.*, 24, 579, 1970.
395. Davidson et al., Influence of finite-beta effects on the lower-hybrid-drift instability in post-implosion theta pinches, *Phys. Rev. Lett.*, 37, 750, 1976.
396. Davidson and McMullin, Stochastic particle instability for electron motion in combined helical wiggler, radiation, and longitudinal wave fields, *Phys. Rev.*, A26, 410, 1982.
397. Davidson and McMullin, Higher harmonic emission by a relativistic electron beam in a longitudinal magnetic field, *Phys. Rev.*, A26, 1997, 1982.
398. Davidson and McMullin, Detrapping stochastic particle instability for electron motion in combined longitudinal wiggler and radiation wave fields, *Phys. Rev.*, A29, 791, 1984.
399. Davidson and Yin, Influence of finite radial geometry on the generation of coherent radiation by a relativistic electron beam in a longitudinal magnetic wiggler, *Phys. Rev.*, A30, 3078, 1984.

400. Davidson et al., Radially confined relativistic-electron-beam equilibria for longitudinal wiggler free-electron-laser applications, *Phys. Rev.*, A30, 1984.
401. Davidson et al., Collective instabilities driven by anode plasma ions and electrons in a nonrelativistic cylindrical diode with applied magnetic field, *Phys. Rev.*, A32, 1044, 1985.
402. Davidson and Uhm, Kinetic stability properties of nonrelativistic non-neutral electron flow in a planar diode with applied magnetic field, *Phys. Rev.*, A32, 3554, 1985.
403. Davidson et al., Kinetic analysis of the sideband instability in a helical wiggler free-electron laser for electrons trapped near the bottom of the ponderomotive potential, *Phys. Rev.*, A34, 3063, 1986.
404. Davidson et al., Nonlinear traveling waves in a helical free-electron laser, *Phys. Rev.*, A34, 392, 1986.
405. Davidson and Yoon, Stabilization of the cyclotron autoresonance maser instability by axial momentum spread, *Phys. Rev.*, A39, 2534, 1989.
406. Davies, Scalar particle production in Schwarzschild and Rindler metric, *J. Phys. Math. Gen.*, A8, 609, 1975.
407. Davies et al., Energy-momentum tensor near an evaporating black hole, *Phys. Rev.*, D13, 2720, 1976.
408. Davies and Unruh, Perturbation techniques for quantum fields in curved space, *Phys. Rev.*, D20, 388, 1979.
409. Dawson et al., Damping of large-amplitude plasma waves propagating perpendicular to the magnetic field, *Phys. Rev. Lett.*, 50, 1455, 1983.
410. Decker and Mori, Group velocity of large amplitude electromagnetic waves in a plasma, *Phys. Rev. Lett.*, 72, 490, 1994.
411. Delfyett, High-power ultrafast laser diodes, *IEEE J. Quantum Electron.*, QE28, 2203, 1992.
412. Derbenev et al., Proposal for a novel two-beam accelerator, *Phys. Rev. Lett.*, 72, 3025, 1994.
413. Derickson et al., Comparison of timing jitter in external and monolithic cavity mode-locked semiconductor lasers, *Appl. Phys. Lett.*, 59, 3372, 1991.
414. Deutsch and Bekefi, Stark-broadening calculations of the Lyman-alpha line in a turbulent plasma, *Phys. Rev.*, A14, 854, 1976.
415. Dicke, Coherence in spontaneous radiation processes, *Phys. Rev.*, 93, 99, 1954.
416. Dirac, Classical theory of radiating electrons, *Proc. R. Soc. London*, Ser. A167, 148, 1938.
417. Dirac, Quantum theory of localizable dynamical systems, *Phys. Rev.*, 73, 1092, 1948.
418. Dirac, The theory of magnetic poles, *Phys. Rev.*, 74, 817, 1948.
419. Dirac, Energy of the gravitational field, *Phys. Rev. Lett.*, 2, 368, 1959.
420. Dirac, Fixation of coordinates in the Hamiltonian theory of gravitation, *Phys. Rev.*, 114, 924, 1959.
421. Dirac, Quantum electrodynamics without dead wood, *Phys. Rev.*, 139, B684, 1965.
422. Ditmire et al., Energy-yield and conversion-efficiency measurements of high-order harmonic radiation, *Phys. Rev.*, A51, R902, 1995.
423. Ditmire et al., Amplification of xuv harmonic radiation in a Gallium amplifier, *Phys. Rev.*, A51, R4337, 1995.
424. Ditmire et al., Strong x-ray emission from high-temperature plasmas produced by intense irradiation of clusters, *Phys. Rev. Lett.*, 75, 3122, 1995.

425. Ditmire et al., Interaction of intense laser pulses with atomic clusters, *Phys. Rev.*, A53, 3379, 1996.
426. Donnelly et al., Dynamics of optical-field-ionized plasmas for x-ray lasers, *Phys. Rev.*, A51, R2691, 1995.
427. Donnelly et al., High-order harmonic generation in atom clusters, *Phys. Rev. Lett.*, 76, 2472, 1996.
428. Drake et al., Observation of stimulated Compton scattering from resonant electrons in a laser-produced plasma, *Phys. Rev. Lett.*, 64, 423, 1990.
429. Durfee III and Milchberg, Light pipe for high intensity laser pulses, *Phys. Rev. Lett.*, 71, 2409, 1993.
430. Ebrahim et al., Bernstein-mode quasioptical maser experiment, *Phys. Rev. Lett.*, 49, 1556, 1982.
431. Ehrlich et al., Guiding of high intensity laser pulses in straight and curved plasma channel experiments, *Phys. Rev. Lett.*, 77, 4186, 1996.
432. Einstein et al., The gravitational equations and the problem of motion, *Ann. Math.*, 20, 65, 1937.
433. Esarey et al., Stabilization of the tearing mode in low-density tokamak plasmas by turbulent electron diffusion, *Phys. Rev. Lett.*, 50, 583, 1983.
434. Esarey and Ting, Comment on cascade focusing in the beat-wave accelerator, *Phys. Rev. Lett.*, 65, 1961, 1990.
435. Esarey et al., Frequency shifts induced in laser pulses by plasma waves, *Phys. Rev.*, A42, 3526, 1990.
436. Esarey et al., Frequency up-shifting of laser pulses by copropagating ionization fronts, *Phys. Rev.*, A44, 3908, 1991.
437. Esarey and Sprangle, Generation of stimulated backscattered harmonic radiation from intense-laser interactions with beams and plasmas, *Phys. Rev.*, A45, 5872, 1992.
438. Esarey et al., Nonlinear Thomson scattering of intense laser pulses from beams and plasmas, *Phys. Rev.*, E48, 3003, 1993.
439. Esarey et al., Envelope analysis of intense laser pulse self-modulation in plasmas, *Phys. Rev. Lett.*, 72, 2887, 1994.
440. Esarey et al., Laser acceleration of electrons in vacuum, *Phys. Rev.*, E52, 5443, 1995.
441. Esarey et al., Overview of plasma-based accelerator concepts, *IEEE Trans. Plasma Sci.*, 24, 252, 1996.
442. Esarey et al., Electron injection into plasma wakefields by colliding laser pulses, *Phys. Rev. Lett.*, 79, 2682, 1997.
443. Esarey et al., Nonlinear theory of nonparaxial laser pulse propagation in plasma channels, *Phys. Rev. Lett.*, 84, 3081, 2000.
444. Everett et al., Evolution of stimulated Raman into stimulated Compton scattering of laser light via wave breaking of plasma waves, *Phys. Rev. Lett.*, 74, 1355, 1995.
445. Everett et al., Coupling between high-frequency plasma waves in laser-plasma interactions, *Phys. Rev. Lett.*, 74, 2236, 1995.
446. Fajans and Rosen, Specular reflection of phonons in superfluid helium at 0.33°K, *Phys. Rev. Lett.*, 13, 328, 1964.
447. Fajans et al., Spectral measurements from a tunable, Raman, free-electron laser, *Phys. Rev. Lett.*, 53, 246, 1984.
448. Fajans et al., Off-axis electron orbits in realistic helical wigglers for free-electron laser applications, *Phys. Rev.*, A32, 3448, 1985.

449. Fajans et al., Nonlinear power saturation and phase (wave refractive index) in a collective free-electron laser amplifier, *Phys. Rev. Lett.*, 57, 579, 1986.
450. Fakir and Unruh, Improvement on cosmological chaotic inflation through non-minimal coupling, *Phys. Rev.*, D41, 1783, 1990.
451. Fakir and Unruh, Induced-gravity inflation, *Phys. Rev.*, D41, 1792, 1990.
452. Falcone et al., Observation of laser-induced inelastic collisions, *Phys. Rev.*, A15, 1333, 1977.
453. Fedorov and McIver, Multiphoton stimulated Compton scattering, *Opt. Comm.*, 32, 179, 1980.
454. Feynman, Forces in molecules, *Phys. Rev.*, 56, 340, 1939.
455. Feynman, Relativistic cut-off for quantum electrodynamics, *Phys. Rev.*, 74, 1430, 1948.
456. Feynman, A relativistic cut-off for classical electrodynamics, *Phys. Rev.*, 74, 939, 1948.
457. Feynman et al., Equations of state of elements based on the generalized Fermi–Thomas theory, *Phys. Rev.*, 75, 1561, 1949.
458. Feynman, Space-time approach to quantum electrodynamics, *Phys. Rev.*, 76, 769, 1949.
459. Feynman, The theory of positrons, *Phys. Rev.*, 76, 749, 1949.
460. Feynman, Mathematical formulation of the quantum theory of electromagnetic interaction, *Phys. Rev.*, 80, 440, 1950.
461. Feynman, An operator calculus having applications in quantum electrodynamics, *Phys. Rev.*, 84, 108, 1951.
462. Feynman, The lambda-transition in liquid helium, *Phys. Rev.*, 90, 1116, 1953.
463. Feynman, Atomic theory of liquid helium near absolute zero, *Phys. Rev.*, 91, 1301, 1953.
464. Feynman, Atomic theory of the lambda transition in helium, *Phys. Rev.*, 91, 1291, 1953.
465. Feynman, Atomic theory of the two-fluid model of liquid helium, *Phys. Rev.*, 94, 262, 1954.
466. Feynman and Speisman, Proton–neutron mass difference, *Phys. Rev.*, 94, 500, 1954.
467. Feynman, Slow electrons in a polar crystal, *Phys. Rev.*, 97, 660, 1955.
468. Feynman and Cohen, Energy spectrum of the excitations in liquid helium, *Phys. Rev.*, 102, 1189, 1956.
469. Feynman and Gell-Mann, Theory of the Fermi interaction, *Phys. Rev.*, 109, 193, 1958.
470. Feynman et al., Mobility of slow electrons in a polar crystal, *Phys. Rev.*, 127, 1004, 1962.
471. Feynman, Gell-Mann, and Zweig, Group $U(6) \times U(6)$ generated by current components, *Phys. Rev. Lett.*, 13, 678, 1964.
472. Feynman, Very high-energy collisions of hadrons, *Phys. Rev. Lett.*, 23, 1415, 1969.
473. Feynman et al., Some comments on baryonic states, *Phys. Rev.*, D2, 1267, 1970.
474. Feynman et al., Current matrix elements from a relativistic quark model, *Phys. Rev.*, D3, 2706, 1971.
475. Feynman et al., Quantum-chromodynamic approach for the large-transverse-momentum production of particle jets, *Phys. Rev.*, D18, 3320, 1978.
476. Feynman and Kleinert, Effective classical partition functions, *Phys. Rev.*, A34, 5080, 1986.

477. Field and Feynman, Quark elastic scattering as a source of high-transverse-momentum mesons, *Phys. Rev.*, D15, 2590, 1977.
478. Fischer et al., Phase locking, amplification, and mode selection in an 85 GHz quasioptical gyroklystron, *Phys. Rev. Lett.*, 72, 2395, 1994.
479. Fitzgerald, Phase-sensitive x-ray imaging, *Phys. Today*, 53 (7), 23, 2000.
480. Forslund et al., Two-dimensional simulations of single-frequency and beat-wave laser-plasma heating, *Phys. Rev. Lett.*, 54, 558, 1985.
481. Freeman et al., Argon-induced hyperfine frequency shift in potassium, *Phys. Rev.*, A13, 907, 1976.
482. Freeman and Cooke, Argon-induced *g*-factor shift in potassium, *Phys. Rev.*, A13, 1692, 1976.
483. Freeman and Kleppner, Core polarization and quantum defects in high-angular-momentum states of alkali atoms, *Phys. Rev.*, A14, 1614, 1976.
484. Freeman and Bjorklund, Effects of electric fields upon autoionizing states of Sr, *Phys. Rev. Lett.*, 40, 118, 1978.
485. Freeman and Haas, Periodic structure in the heavy-ion reactions $^{18}O + ^{12}C$ and $^{18}O + ^{16}O$, *Phys. Rev. Lett.*, 40, 927, 1978.
486. Freeman et al., Observation of electric-field-induced resonances above the ionization limit in a one-electron atom, *Phys. Rev. Lett.*, 41, 1463, 1978.
487. Freeman and Economou, Electric field dependence of the photoionization cross-section of Rb, *Phys. Rev.*, A20, 2356, 1979.
488. Freeman et al., Isotope shift in the $2^3P$-$3^3D$ transition in helium, *Phys. Rev.*, A22, 1510, 1980.
489. Freeman et al., Four-wave parametric mixing in optically inverted barium ions, *Phys. Rev.*, A26, 3029, 1982.
490. Freeman et al., Ponderomotive effects on angular distributions of photoelectrons, *Phys. Rev. Lett.*, 57, 3156, 1986.
491. Freeman et al., Above-threshold ionization with subpicosecond laser pulses, *Phys. Rev. Lett.*, 59, 1092, 1987.
492. Freund et al., Spontaneous emission near electron plasma frequency in a plasma with a runaway electron tail, *Phys. Rev. Lett.*, 40, 1563, 1978.
493. Freund et al., Coherent and incoherent radiation from free-electron lasers with an axial guide field, *Phys. Rev.*, A24, 1965, 1981.
494. Freund et al., Effect of fluctuating space-charge fields on sideband instabilities in free-electron lasers, *Phys. Rev.*, A25, 3121, 1982.
495. Freund et al., Collective effects on the operation of free-electron lasers with an axial guide field, *Phys. Rev.*, A26, 2004, 1982.
496. Freund, Nonlinear analysis of free-electron laser amplifiers with axial guide fields, *Phys. Rev.*, A27, 1977, 1983.
497. Freund and Ganguly, Three-dimensional theory of the free-electron laser in the collective regime, *Phys. Rev.*, A28, 3438, 1983.
498. Freund and Sprangle, Unstable electrostatic beam modes in free-electron laser systems, *Phys. Rev.*, A28, 1835, 1983.
499. Freund and Gold, Efficiency enhancement in free-electron lasers using a tapered axial guide field, *Phys. Rev. Lett.*, 52, 926, 1984.
500. Freund and Ganguly, Nonlinear analysis of efficiency enhancement in free-electron laser amplifiers, *Phys. Rev.*, A33, 1060, 1986.
501. Freund and Ganguly, Effect of beam quality on the free-electron maser, *Phys. Rev.*, A34, 1242, 1986.

502. Freund et al., Linear gain of a free-electron laser with an electromagnetic wiggler and an axial-guide magnetic field, *Phys. Rev.*, A34, 2007, 1986.
503. Freund et al., Three-dimensional nonlinear analysis of free-electron-laser amplifiers with planar wigglers, *Phys. Rev.*, A36, 2182, 1987.
504. Freund et al., Harmonic generation in free-electron lasers, *Phys. Rev.*, A36, 3218, 1987.
505. Freund, Multimode nonlinear analysis of free-electron laser amplifiers in three dimensions, *Phys. Rev.*, A37, 3371, 1988.
506. Freund, Nonlinear analysis of high-power Cherenkov masers, *Phys. Rev. Lett.*, 65, 2993, 1990.
507. Freund and Chang, Effect of lower beat wave on optical guiding in planar wiggler free-electron lasers, *Phys. Rev.*, A42, 6737, 1990.
508. Freund and Jackson, Self-consistent analysis of wiggler-field errors in free-electron lasers, *Phys. Rev.*, A45, 7488, 1992.
509. Freund, Nonlinear theory of short-wavelength free-electron lasers, *Phys. Rev.*, E52, 5401, 1995.
510. Freund and Neil, Free-electron lasers: vacuum electronic generators of coherent radiation, *Proc. IEEE*, 87, 782, 1999.
511. Friedland and Hirshfield, Free-electron laser with a strong axial magnetic field, *Phys. Rev. Lett.*, 44, 1456, 1980.
512. Friedman et al., Electron accelerators driven by modulated intense relativistic electron beams, *Phys. Rev. Lett.*, 63, 2468, 1989.
513. Frolov et al., Gravitational fields of straight and circular cosmic strings: relation between gravitational mass, angular deficit, and internal structure, *Phys. Rev.*, D39, 1084, 1989.
514. Fulling, Nonuniqueness of canonical field quantization in Riemannian space–time, *Phys. Rev.*, D7, 2850, 1973.
515. Fulling and Davies, Radiation from a moving mirror in two dimensional space–time. Conformal anomaly, *Proc. R. Soc. London*, A348, 393, 1976.
516. Gai et al., Experimental demonstration of wake-field effects in dielectric structures, *Phys. Rev. Lett.*, 61, 2756, 1988.
517. Ganguly and Freund, Nonlinear analysis of free-electron-laser amplifiers in three dimensions, *Phys. Rev.*, A32, 2275, 1985.
518. Ganguly and Freund, Nonlinear theory of the orbitron maser in three dimensions, *Phys. Rev.*, A36, 2199, 1987.
519. Ganguly and Hirshfield, Nonlinear theory of gyroharmonic radiation from spatiotemporally modulated electron beams, *Phys. Rev. Lett.*, 70, 291, 1993.
520. Garber, The next wave of the genomics business, *Tech. Rev.*, 103 (4), 46, 2000.
521. Gea-Banacloche et al., Soft x-ray free-electron laser with a laser undulators, *IEEE J. Quantum Electron.*, QE23, 1558, 1987.
522. Gibson and Freeman, Verification of the dominant role of resonant enhancement in short-pulse multiphoton ionization, *Phys. Rev. Lett.*, 69, 1904, 1992.
523. Gibson et al., Excitation and ionization dynamics in short-pulse multiphoton ionization, *Phys. Rev.*, A49, 3870, 1994.
524. Glauber, The quantum theory of optical coherence, *Phys. Rev.*, 130, 2529, 1963.
525. Glover et al., Subpicosecond Thomson scattering measurements of optically ionized helium plasmas, *Phys. Rev. Lett.*, 73, 78, 1994.
526. Glover et al., Measurement of velocity distributions and recombination kinetics in tunnel-ionized helium plasmas, *Phys. Rev. Lett.*, 75, 445, 1995.

527. Gold and Nusinovich, Review of high-power microwave source research, *Rev. Sci. Instrum.*, 68, 3945, 1997.
528. Gold et al., X-band magnicon amplifier for the Next Linear Collider, *Phys. Plasmas*, 4, 1900, 1997.
529. Golden and Lemonick, The race is over, *Time*, 156 (1), 19, 2000.
530. Goldstein et al., Focused-flow model of relativistic diodes, *Phys. Rev. Lett.*, 33, 1471, 1974.
531. Gover et al., Time and frequency domain analysis of superradiant coherent synchrotron radiation in a waveguide free-electron laser, *Phys. Rev. Lett.*, 72, 1192, 1994.
532. Goy et al., Observation of cavity-enhanced single-atom spontaneous emission, *Phys. Rev. Lett.*, 50, 1903, 1983.
533. Grangier, Aspect, and Vigue, Quantum interference effect for two atoms radiating a single photon, *Phys. Rev. Lett.*, 54, 418, 1985.
534. Grishchuk et al., Generation of squeezed radiation from vacuum in the cosmos and the laboratory, *Phys. Rev.*, D46, 1440, 1992.
535. Grove, On an inertial observer's interpretation of the detection of radiation by linearly accelerated particle detectors, *Class. Quantum Grav.*, 3, 801, 1986.
536. Guo and Drake, Multiphoton ionization in circularly polarized standing waves, *Phys. Rev.*, A45, 6622, 1992.
537. Guo et al., Schrödinger-like equation for a nonrelativistic electron in a photon field of arbitrary intensity, *Phys. Rev.*, A58, 521, 1998.
538. Guo et al., Relativistic electron moving in a multimode quantized radiation field, *Phys. Rev.*, A54, 1087, 1996.
539. Guo, Theory of the Kapitza–Dirac effect in strong radiation fields, *Phys. Rev.*, A53, 4311, 1996.
540. Gupta, Theory of longitudinal photons in quantum electrodynamics, *Proc. Phys. Soc.*, 63, 46, 1950.
541. Guss et al., Sideband mode competition in a gyrotron oscillator, *Phys. Rev. Lett.*, 69, 3727, 1992.
542. Guth, Inflationary universe: a possible solution to the horizon and flatness problems, *Phys. Rev.*, D23, 347, 1981.
543. Hafizi and Roberson, Role of beam quality in free-electron lasers, *Phys. Plasma*, 3, 2156, 1996.
544. Hafizi et al., Vacuum beat wave acceleration, *Phys. Rev.*, E55, 5924, 1997.
545. Hafizi et al., Laser-driven acceleration with Bessel beams, *Phys. Rev.*, E55, 3539, 1997.
546. Hairapetian et al., Experimental demonstration of dynamic focusing of a relativistic electron bunch by an overdense plasma lens, *Phys. Rev. Lett.*, 72, 2403, 1994.
547. Hamster et al., Subpicosecond, electromagnetic pulses from intense laser-plasma interaction, *Phys. Rev. Lett.*, 71, 2725, 1993.
548. Hamster et al., Short-pulse teraHertz radiation from high-intensity-laser-plasma interaction, *Phys. Rev.*, E49, 671, 1994.
549. Hansen and Fajans, Dynamic and Debye shielding and antishielding in magnetized, collisionless plasmas, *Phys. Rev. Lett.*, 74, 4209, 1995.
550. Hanson et al., Influence of finite wavelength on the quantum kicked rotator in the semiclassical regime, *Phys. Rev.*, A29, 819, 1984.
551. Haroche et al., Modified Zeeman hyperfine spectra observed in H1 and Rb87 ground states interacting with a nonresonant rf field, *Phys. Rev. Lett.*, 24, 861, 1970.

552. Haroche and Cohen-Tannoudji, Resonant transfer of coherence in nonzero magnetic field between atomic levels of different *g* factors, *Phys. Rev. Lett.*, 24, 974, 1970.
553. Haroche and Hartmann, Theory of saturated-absorption line shapes, *Phys. Rev.*, A6, 1280, 1972.
554. Haroche, Paisner, and Schawlow, Hyperfine quantum beats observed in Cs vapor under pulsed dye laser excitation, *Phys. Rev. Lett.*, 30, 948, 1973.
555. Haroche et al., Observation of fine-structure quantum beats following stepwise excitation in sodium *D* states, *Phys. Rev. Lett.*, 33, 1063, 1974.
556. Haroche and Kleppner, Cavity quantum electrodynamics, *Physics Today*, 42 (1), 24, 1989.
557. Haroche and Raimond, Cavity quantum electrodynamics, *Sci. Am.*, 268, 26, 1993.
558. Hartemann et al., The rippled-field magnetron (cross-field free-electron laser), *IEEE Trans. Plasma Sci.*, PS13, 484, 1985.
559. Hartemann and Bekefi, Time-resolved studies of intense, relativistic electron beams with a subnanosecond Čerenkov electro-optic shutter, *Appl. Phys. Lett.*, 49, 1680, 1986.
560. Hartemann et al., Wave-profile modification (optical guiding) induced by the free-electron laser interaction, *Phys. Rev. Lett.*, 59, 1177, 1987.
561. Hartemann and Bekefi, Microwave radiation from a tunable circular free-electron laser, *Phys. Fluids*, 30, 3283, 1987.
562. Hartemann et al., Generation of short pulses of coherent electromagnetic radiation in a free-electron laser amplifier, *IEEE J. Quantum Electron.*, QE24, 105, 1988.
563. Hartemann et al., Novel adiabatic bifilar helical wiggler entrance for free-electron laser applications, *Appl. Phys. Lett.*, 53, 631, 1988.
564. Hartemann et al., Pulse compression in a free electron laser amplifier, *Nucl. Instrum. Meth. Phys. Res.*, A272, 125, 1988.
565. Hartemann and Toffano, Relativistic electrodynamics of continuous media, *Phys. Rev.*, A41, 5066, 1990.
566. Hartemann, Eulerian formalism of linear beam-wave interactions, *Phys. Rev.*, A42, 2906, 1990.
567. Hartemann et al., Anomalous (stimulated) refraction induced by the free-electron laser interaction, *Phys. Lett.*, A147, 495, 1990.
568. Hartemann, Fonctionnement en régime Raman et optimisation du rendement d'un laser à electrons libres opérant dans la gamme des ondes millimétriques, *Rev. Tech.* Thomson-CSF, Edited by Gauthier-Villars, 22, 270, 1990.
569. Hartemann et al., Coherent synchrotron radiation in a cylindrical waveguide with a helical wiggler, *Phys. Plasmas*, 1, 1306, 1994.
570. Hartemann et al., Coherent photoelectron bunch generation and quantum efficiency enhancement in a photocathode optical resonator, *Appl. Phys. Lett.*, 65, 2404, 1994.
571. Hartemann et al., Nonlinear ponderomotive scattering of relativistic electrons by an intense laser field at focus, *Phys. Rev.*, E51, 4833, 1995.
572. Hartemann and Luhmann, Classical electrodynamical derivation of the radiation damping force, *Phys. Rev. Lett.*, 74, 1107, 1995.
573. Hartemann and Kerman, Classical theory of nonlinear Compton scattering, *Phys. Rev. Lett.*, 76, 624, 1996.
574. Hartemann et al., Transform-limited coherent synchrotron radiation wavepackets in a chirped pulse free-electron laser, *Phys. Plasmas*, 3, 2446, 1996.

575. Hartemann et al., Spectral analysis of the nonlinear relativistic Doppler shift in ultrahigh intensity Compton scattering, *Phys. Rev.*, E54, 2956, 1996.
576. Hartemann et al., Three-dimensional relativistic electron scattering in an ultrahigh-intensity laser focus, *Phys. Rev.*, E58, 5001, 1998.
577. Hartemann, High-intensity scattering processes of relativistic electrons in vacuum, *Phys. Plasmas*, 5, 2037, 1998.
578. Hartemann et al., The chirped-pulse inverse free-electron laser: a high-gradient vacuum laser accelerator, *Phys. Plasmas*, 6, 4104, 1999.
579. Hartemann, Stochastic electron gas theory of coherence in laser-driven synchrotron radiation, *Phys. Rev.*, E61, 972, 2000.
580. Hartemann et al., Three-dimensional theory of emittance in Compton scattering and x-ray protein crystallography, *Phys. Rev.*, E63, 016501, 2001.
581. Hartman and Rosenzweig, Ponderomotive focusing in axisymmetric rf linacs, *Phys. Rev.*, E47, 2031, 1993.
582. Hassam et al., Spontaneous poloidal spin-up of tokamaks and the transition to the *H* mode, *Phys. Rev. Lett.*, 66, 309, 1991.
583. Haus and Mullen, Quantum noise in linear amplifiers, *Phys. Rev.*, 128, 2407, 1962.
584. Haus, Noise in free-electron laser amplifier, *IEEE J. Quantum Electron.*, QE17, 1427, 1981.
585. Haus and Islam, Synchrotron radiation of wiggled electron beam in rectangular waveguide, *J. Appl. Phys.*, 54, 4784, 1983.
586. Haus and Yamamoto, Quantum noise of an injection-locked laser oscillator, *Phys. Rev.*, A29, 1261, 1984.
587. Haus and Islam, Theory of the soliton laser, *IEEE J. Quantum Electron.*, QE21, 1172, 1985.
588. Haus, On the radiation from point charges, *Am. J. Phys.*, 54, 1126, 1986.
589. Haus and Yamamoto, Theory of feedback-generated squeezed states, *Phys. Rev.*, A34, 270, 1986.
590. Haus and Kärtner, Quantization of the nonlinear Schrödinger equation, *Phys. Rev.*, A46, R1175, 1992.
591. Haus and Kärtner, Optical quantum nondemolition measurements and the Copenhagen interpretation, *Phys. Rev.*, A53, 3785, 1996.
592. Heiderich and Unruh, Spin-two fields, general covariance, and conformal invariance, *Phys. Rev.*, D38, 490, 1988.
593. Heiderich and Unruh, Nonlinear, non-covariant spin-two theories, *Phys. Rev.*, D42, 2057, 1990.
594. Hinterleitner, Inertial and accelerated particle detectors with back-reaction in flat space–time, *Ann. Phys.* 226, 165, 1993.
595. Hirshfield and Watchel, Electron cyclotron maser, *Phys. Rev. Lett.*, 12, 533, 1964.
596. Hirshfield and Park, Electron-beam cooling by stimulated synchrotron emission and absorption, *Phys. Rev. Lett.*, 66, 2312, 1991.
597. Hirshfield, Synchrotron-radiation laser, *Phys. Rev. Lett.*, 68, 792, 1992.
598. Hogan et al., Measurements of gain larger than $10^5$ at 12 $\mu$m in a self-amplified spontaneous-emission free-electron laser, *Phys. Rev. Lett.*, 81, 4867, 1998.
599. Horne and Witten, Conformal gravity in three dimensions as a gauge theory, *Phys. Rev. Lett.*, 62, 501, 1988.
600. Huang et al., Radiation reaction in a continuous focusing channel, *Phys. Rev. Lett.*, 74, 1759, 1995.
601. Huang and Ruth, Laser-electron storage ring, *Phys. Rev. Lett.*, 80, 976, 1998.

602. Hui and Lau, Injection and extraction of a relativistic electron beam in a modified betatron, *Phys. Rev. Lett.*, 53, 2024, 1984.
603. Hui et al., Hose-induced current enhancement associated with beam propagation, *Phys. Rev. Lett.*, 55, 87, 1985.
604. Hussein et al., Theory of free-wave acceleration, *Phys. Rev.*, A46, 3562, 1992.
605. Hussein and Pato, Nonlinear amplification of inverse-Bremsstrahlung electron acceleration, *Phys. Rev. Lett.*, 68, 1136, 1992.
606. Jerby et al., Observations of periodic intensity bursts during the start-up phase of a free-electron-laser oscillator, *Phys. Rev. Lett.*, 66, 2068, 1991.
607. Jerby and Bekefi, Cyclotron-maser experiments in a periodic waveguide, *Phys. Rev.*, E48, 4637, 1993.
608. Johnston et al., Resonant operation of the cross-field free-electron laser: kinetic and fluid equilibria, *Phys. Rev.*, A38, 1309, 1988.
609. Joshi et al., Forward Raman instability and electron acceleration, *Phys. Rev. Lett.*, 47, 1285, 1981.
610. Joshi et al., Resonant self-focusing of laser light in a plasma, *Phys. Rev. Lett.*, 48, 874, 1982.
611. Kaku, Soliton 'dictionary' for massive quantum electrodynamics, *Phys. Rev.*, D12, 2330, 1975.
612. Kaku and Senjanovic, Functional measure for quantum gravity, *Phys. Rev.*, D15, 1019, 1977.
613. Kaku, Gauge theory on a random supersymmetric lattice, *Phys. Rev. Lett.*, 50, 1893, 1983.
614. Kaku, Strong-coupling approach to the quantization of conformal gravity, *Phys. Rev.*, D27, 2819, 1983.
615. Kaku, Super lattices and gauge theory, *Phys. Rev.*, D31, 1992, 1985.
616. Kaku, Geometric derivation of string field theory from first principles: closed strings and modular invariance, *Phys. Rev.*, D38, 3052, 1988.
617. Kaku, Nonpolynomial closed-string field theory, *Phys. Rev.*, D41, 3734, 1990.
618. Kaku, Subcritical closed string theory in less than 26 dimensions, *Phys. Rev.*, D49, 5364, 1994.
619. Kapetanos et al., Destructive instabilities in hollow intense relativistic electron beams, *Phys. Rev. Lett.*, 30, 1303, 1973.
620. Kapitza and Dirac, The reflection of electrons from standing light waves, *Proc. Cambridge Phil. Soc.*, 29, 297, 1933.
621. Kapteyn et al., Observation of a short-wavelength laser pumped by Auger decay, *Phys. Rev. Lett.*, 57, 2939, 1986.
622. Kapteyn and Falcone, Auger-pumped short-wavelength lasers in xenon and krypton, *Phys. Rev.*, A37, 2033, 1988.
623. Kärtner and Haus, Quantum-nondemolition measurements and the 'collapse of the wave function', *Phys. Rev.*, A47, 4585, 1993.
624. Kärtner and Haus, Quantum-mechanical stability of solitons and the correspondence principle, *Phys. Rev.*, A48, 2361, 1993.
625. Katsouleas and Dawson, Unlimited electron acceleration in laser-driven plasma waves, *Phys. Rev. Lett.*, 51, 392, 1983.
626. Katsouleas and Dawson, Unlimited electron acceleration in laser-driven plasma waves, *Phys. Rev. Lett.*, 51, 846, 1983.
627. Katsouleas and Dawson, Katsouleas and Dawson respond, *Phys. Rev. Lett.*, 53, 1027, 1984.

628. Katsouleas et al., Comment on free-electron laser and laser electron acceleration based on the megagauss magnetic fields in laser-produced plasmas, *Phys. Rev. Lett.*, 57, 1960, 1986.
629. Katsouleas and Mori, Wave-breaking amplitude of relativistic oscillations in a thermal plasma, *Phys. Rev. Lett.*, 61, 90, 1988.
630. Katsouleas and Mori, Comment on packet spreading, stabilization, and localization in superstrong fields, *Phys. Rev. Lett.*, 70, 1561, 1993.
631. Kaw et al., Nonlinear 1D laser pulse solitons in a plasma *Phys. Rev., Lett.*, 68, 3172, 1992.
632. Kawata et al., Inverse-Bremsstrahlung electron acceleration, *Phys. Rev. Lett.*, 66, 2072, 1991.
633. Ketterle, Experimental studies of Bose–Einstein condensation, *Physics Today*, 52 (12), 30, 1999.
634. Kim, Quantum field and uniformly accelerated oscillator, *Phys. Rev.*, D59, 064024, 1999.
635. Kim, The equation of motion of an electron: a debate in classical and quantum physics, *Nucl. Instrum. Meth. Phys. Res.*, A429, 1, 1999.
636. Kimel and Elias, The role of radiation reaction in Lienard–Wiechert description of FEL interaction, *Nucl. Instrum. Meth. Phys. Res.*, A375, 565, 1996.
637. Kimura et al., First staging of two laser accelerators, *Phys. Rev. Lett.*, 86, 4041, 2001.
638. Kishek and Lau, Interaction of multipactor discharge and rf circuit, *Phys. Rev. Lett.*, 75, 1218, 1995.
639. Kitagawa et al., Demonstration of collisionally enhanced degenerate four-wave mixing in a plasma, *Phys. Rev. Lett.*, 62, 151, 1989.
640. Kleppner, Goldenberg, and Ramsey, Theory of the hydrogen maser, *Phys. Rev.*, 126, 603, 1962.
641. Kleppner et al., Hydrogen-maser principles and techniques, *Phys. Rev.*, 138, A972, 1965.
642. Kleppner, Inhibited spontaneous emission, *Phys. Rev. Lett.*, 47, 233, 1981.
643. Kleppner, Quantum chaos and laser spectroscopy, *At. Phys.*, 13, 1993.
644. Krall et al., Vlasov simulations of very-large-amplitude-wave generation in the plasma wake-field accelerator, *Phys. Rev.*, A44, 6854, 1991.
645. Krall et al., Enhanced acceleration in a self-modulated-laser wake-field accelerator, *Phys. Rev.*, E48, 2157, 1993.
646. Kreischer and Temkin, Single-mode operation of high-power, step-tunable gyrotrons, *Phys. Rev. Lett.*, 59, 547, 1987.
647. Krushelnick et al., Second harmonic generation of stimulated Raman scattered light in underdense plasmas, *Phys. Rev. Lett.*, 75, 3681, 1995.
648. Krushelnick et al., Plasma channel formation and guiding during high intensity short pulse laser plasma experiments, *Phys. Rev. Lett.*, 78, 4047, 1997.
649. Kulin et al., Coherent manipulation of atomic wave packets by adiabatic transfer, *Phys. Rev. Lett.*, 78, 4185, 1997.
650. Labaune et al., Upshifted Raman light produced by coupling between stimulated Raman and Brillouin scattering, *Phys. Rev. Lett.*, 69, 285, 1992.
651. Labaune et al., Large-amplitude ion acoustic waves in laser-produced plasma, *Phys. Rev. Lett.*, 75, 248, 1995.
652. Labaune et al., Location of ion-acoustic waves from back and side stimulated Brillouin scattering, *Phys. Rev. Lett.*, 76, 3727, 1996.

653. Lai et al., Demonstration of microwave generation from a static field by a relativistic ionization front in a capacitor array, *Phys. Rev. Lett.*, 77, 4764, 1996.
654. Lai et al., Transient filamentation of a laser beam in a thermal force dominated plasma, *Phys. Rev. Lett.*, 78, 670, 1997.
655. Landahl et al., Phase noise reduction and photoelectron acceleration in a high-$Q$ rf gun, *IEEE Trans. Plasma Sci.*, 26, 814, 1998.
656. Landragin et al., Measurement of the van der Waals force in an atomic mirror, *Phys. Rev. Lett.*, 77, 1727, 1996.
657. Lane and Davidson, Nonlinear traveling-wave equilibria for free-electron laser applications, *Phys. Rev.*, A27, 2008, 1983.
658. Lau, Transient growth in a current-carrying plasma, *Phys. Rev. Lett.*, 42, 779, 1979.
659. Lau and Chu, Electron-cyclotron maser instability driven by a loss-cone distribution, *Phys. Rev. Lett.*, 50, 243, 1983.
660. Lau and Chernin, Stabilization of the negative mass instability in a rotating electron beam, *Phys. Rev. Lett.*, 52, 1425, 1984.
661. Lau, Collective interaction klystron, *Phys. Rev. Lett.*, 53, 1984.
662. Lau, Classification of beam breakup instabilities in linear accelerators, *Phys. Rev. Lett.*, 63, 1141, 1989.
663. Lau et al., Quantum extension of the Child-Langmuir law, *Phys. Rev. Lett.*, 66, 1446, 1991.
664. Lawson, Some attributes of real and virtual photons, *Contemp. Phys.*, 11, 575, 1970.
665. Leahy and Unruh, Angular dependence of neutrino emission from rotating black holes, *Phys. Rev.*, D19, 3509, 1979.
666. Leahy and Unruh, Effects of a lambda Phi$^4$ interaction on black-hole evaporation in two dimensions, *Phys. Rev.*, D28, 694, 1983.
667. Le Blanc et al., Temporal characterization of a self-modulated laser wakefield, *Phys. Rev. Lett.*, 77, 5381, 1996.
668. Leemans et al., Detection of trapped magnetic fields in a theta-pinch plasma using a relativistic electron beam, *Rev. Sci. Instrum.*, 59, 1641, 1988.
669. Leemans et al., Stimulated Compton scattering from preformed underdense plasmas, *Phys. Rev. Lett.*, 67, 1434, 1991.
670. Leemans et al., Experiments and simulations of tunnel-ionized plasmas, *Phys. Rev.*, A46, 1091, 1992.
671. Leemans et al., Plasma physics aspects of tunnel-ionized gases, *Phys. Rev. Lett.*, 68, 321, 1992.
672. Leemans et al., Non-linear dynamics of driven relativistic electron plasma waves, *Phys. Rev.*, A46, 5112, 1992.
673. Leemans et al., A technique for single-pulse spectrum and pulse width measurements for an IR-FEL, *Nucl. Instrum. Meth. Phys. Res.*, 331, 615, 1993.
674. Leemans et al., Measurement of single pulse spectra of an infrared FEL, *Nucl. Instrum. Meth. Phys. Res.*, A341, 473, 1994.
675. Leemans et al., Time-resolved study of sideband generation and transition to chaos on an infrared FEL, *Nucl. Instrum. Meth. Phys. Res.*, A358, 208, 1995.
676. Leemans et al., Plasma guiding and wakefield generation for second generation experiments, *IEEE Trans. Plasma Sci.*, PS24, 331, 1996.
677. Leemans et al., X-ray based subpicosecond electron bunch characterization using 90° Thomson scattering, *Phys. Rev. Lett.*, 77, 4182, 1996.

678. Leemans et al., Interaction of relativistic electrons with ultrashort laser pulses: generation of femtosecond x-rays and microprobing of electron beams, *IEEE J. Quantum Electron.*, QE11, 1925, 1997.
679. Leemans et al., Laser driven plasma based accelerators: wakefield excitation, channel guiding, and laser triggered particle injection, *Phys. Plasmas*, 5, 1615, 1998.
680. Leemans et al., Femtosecond x-ray generation through relativistic electron beam-laser interaction, *Les Comptes Rendus de l'Académie des Sciences*, Série IV, 1, 279, 2000.
681. Leemans et al., Gamma-neutron activation experiments using laser wakefield accelerators, *Phys. Plasmas*, 8, 2510, 2001.
682. Lemoff et al., Demonstration of a 10-Hz femtosecond-pulse-driven XUV laser at 41.8 nm in Xe IX, *Phys. Rev. Lett.*, 74, 1574, 1995.
683. Lemonick, The genome is mapped. Now what?, *Time* 156 (1), 24, 2000.
684. Le Sage et al., Theory and design of a photoinjector-driven chirped pulse free-electron maser, *IEEE Trans. Plasma Sci.*, 24, 781, 1996.
685. Lidow et al., Inelastic collision induced by intense optical radiation, *Phys. Rev. Lett.*, 36, 462, 1976.
686. Lidow et al., Inelastic collision induced by intense optical radiation, *Phys. Rev. Lett.*, 37, 1590, 1976.
687. Lin and Marshall, Solitary wave spikes emitted from a microwave free-electron laser, *Phys. Rev. Lett.*, 70, 2403, 1993.
688. Lindenberg et al., Time-resolved x-ray diffraction from coherent phonons during laser-induced phase transition, *Phys. Rev. Lett.*, 84, 111, 2000.
689. Litvinenko et al., Gamma-ray production in a storage ring free-electron laser, *Phys. Rev. Lett.*, 78, 4569, 1997.
690. Lou et al., New modes in a plasma with periodic boundaries: the origin of the dense spectrum, *Phys. Rev. Lett.*, 67, 2481, 1991.
691. Luginsand et al., Two-dimensional Child–Langmuir law, *Phys. Rev. Lett.*, 77, 4668, 1996.
692. Lutwak et al., Circular states of atomic hydrogen, *Phys. Rev.*, A56, 1443, 1997.
693. McDermott et al., High-power free-electron laser based on stimulated Raman backscattering, *Phys. Rev. Lett.*, 41, 1368, 1978.
694. McDermott et al., Prebunched high-harmonic FEL for short-pulse millimeter wave emission, *IEEE Trans. Plasma Sci.*, 24, 808, 1996.
695. McDonald and Suen, Membrane viewpoint on black holes: dynamical electromagnetic fields near the horizon, *Phys. Rev.*, D32, 848, 1985.
696. McMullin and Bekefi, Stimulated emission from relativistic electrons passing through a spatially periodic longitudinal magnetic field, *Phys. Rev.*, A25, 1826, 1982.
697. McMullin and Davidson, Low-gain free-electron laser near cyclotron resonance, *Phys. Rev.*, A25, 3130, 1982.
698. McMullin et al., Stimulated emission from a relativistic electron beam in a variable-parameter longitudinal magnetic wiggler, *Phys. Rev.*, A28, 517, 1983.
699. McNally et al., High performance, 70 kV third-harmonic smooth-bore gyroklystron amplifier, *IEEE Trans. Plasma Sci.*, 22, 932, 1994.
700. Mahaffey et al., Rigid rotation and surface envelopes of nonneutral-plasma columns, *Phys. Rev. Lett.*, 35, 1439, 1975.
701. Marable et al., Vlasov theory of wiggler field errors and radiation growth in a free-electron laser, *Phys. Rev.*, A42, 3006, 1990.

702. Markovic and Unruh, Vacuum for a massless scalar field outside a collapsing body in de Sitter spacetime, *Phys. Rev.*, D43, 332, 1991.
703. Massar et al., On the problem of the uniformly accelerated oscillator, *Class. Quantum Grav.*, 10, 385, 1993.
704. Masud et al., Gain measurements from start-up and spectrum of a Raman free-electron laser oscillator, *Phys. Rev. Lett.*, 56, 1567, 1986.
705. Masud et al., Sideband control in a millimeter-wave free-electron laser, *Phys. Rev. Lett.*, 58, 763, 1987.
706. Mathur et al., Hyperfine separation of tritium, *Phys. Rev.*, 158, 14, 1967.
707. Matsas, Rindler and Minkowski particles relationships revisited, *Phys. Lett.*, B380, 24, 1996.
708. Mazenko et al., Does a phase transition in the early universe produce the conditions needed for inflation?, *Phys. Rev.*, D31, 273, 1985.
709. Mendel, Jr. et al., Stability of magnetically insulated electron flow, *Phys. Rev.*, A32, 1091, 1985.
710. Mendelsohn et al., Emission spectra of quasimetastable levels of alkali-metal atoms, *Phys. Rev.*, A35, 2095, 1987.
711. Menge et al., Experimental reduction of beam-breakup instability growth by external cavity coupling in long-pulse electron-beam transport, *Phys. Rev. Lett.*, 69, 2372, 1992.
712. Merecerau and Feynman, Physical conditions for ferromagnetic resonance, *Phys. Rev.*, 104, 63, 1956.
713. Meyer, Note on accelerated detectors, *Phys. Rev.*, D18, 609, 1976.
714. Meyerhofer, High-intensity-laser-electron scattering, *IEEE J. Quantum Electron.*, QE33, 1935, 1997.
715. Milchberg et al., Resistivity of a simple metal from room temperature to $10^6$ °K, *Phys. Rev. Lett.*, 61, 2364, 1988.
716. Milchberg and Freeman, Expansion-induced Doppler shifts from ultrashort-pulse laser-produced plasmas, *Phys. Rev.*, A41, 2211, 1990.
717. Mima et al., Magnetic field generation by the Rayleigh-Taylor instability, *Phys. Rev. Lett.*, 41, 1715, 1978.
718. Mohideen et al., Interaction of free electrons with an intense focused laser pulse in Gaussian and conical axicon geometries, *J. Opt. Soc. Am.*, B9, 2190, 1992.
719. Mohideen et al., High intensity above-threshold ionization of He, *Phys. Rev. Lett.*, 71, 509, 1993.
720. Moore et al., Observation of the transition from Thomson to Compton scattering in multiphoton interactions with low-energy electrons, *Phys. Rev. Lett.*, 74, 2439, 1995.
721. Moore et al., Electron trapping in self-modulated laser wakefields by Raman backscatter, *Phys. Rev. Lett.*, 79, 3909, 1997.
722. Mora and Antonsen, Jr., Electron cavitation and acceleration in the wake of an ultraintense, self-focused laser pulse, *Phys. Rev.*, E53, R2068, 1996.
723. Mori et al., Evolution of self-focusing of intense electromagnetic waves in plasma, *Phys. Rev. Lett.*, 60, 1298, 1988.
724. Mori and Katsouleas, Ponderomotive force of a uniform electromagnetic wave in a time varying dielectric medium, *Phys. Rev. Lett.*, 69, 3495, 1992.
725. Mori et al., Raman forward scattering of short-pulse high-intensity lasers, *Phys. Rev. Lett.*, 72, 1482, 1994.
726. Mori et al., Conversion of dc fields in a capacitor array to radiation by a relativistic ionization front, *Phys. Rev. Lett.*, 74, 542, 1995.

727. Morgner et al., Sub-two-cycle pulses from a Kerr-lens mode-locked Ti:sapphire laser, *Opt. Lett.*, 24, 411, 1999.
728. Mourou et al., Ultrahigh-intensity lasers: physics of the extreme on a tabletop, *Phys. Today,* 51 (1), 22, 1998.
729. Murmane et al., High density plasmas produced by ultrafast laser pulses, *Phys. Rev. Lett.*, 62, 155, 1989.
730. Mussetto et al., Ion confinement in pulsed laser-produced plasmas, *Phys. Rev. Lett.*, 40, 321, 1978.
731. Myint et al., Absolute value of the proton *g* factor, *Phys. Rev. Lett.*, 17, 405, 1966.
732. Namenson et al., Fractal dimension fluctuations for snapshot attractors of random maps, *Phys. Rev.*, E53, 2287, 1996.
733. Nantel et al., Pressure ionization and line merging in strongly coupled plasmas produced by 100-fs laser pulses, *Phys. Rev. Lett.*, 80, 4442, 1998.
734. Nappi and Witten, Wess–Zumino–Witten model based on a nonsemisimple group, *Phys. Rev. Lett.*, 71, 3751, 1993.
735. Nikitin et al., High efficiency coupling and guiding of intense femtosecond laser pulses in preformed plasma channels in an elongated gas jet, *Phys. Rev.*, E59, R3839, 1999.
736. Nisoli et al., Compression of high-energy laser pulses below five fs, *Opt. Lett.*, 22, 522, 1997.
737. Olson and Unruh, Conversion of electromagnetic to gravitational radiation by scattering from a charged black hole, *Phys. Rev. Lett.*, 33, 1116, 1974.
738. Opat and Unruh, Theory of an earth-bound clock comparison experiment as test of the principle of equivalence, *Phys. Rev.*, D44, 3342, 1991.
739. Orzechowski et al., High-efficiency extraction of microwave radiation from a tapered-wiggler free-electron laser, *Phys. Rev. Lett.*, 57, 2172, 1986.
740. O'Shea, Free-electron lasers: bright light sources for the future, *Opt. Photon. News,* 47, May 1998.
741. Ott et al., Effect of noise on time-dependent quantum chaos, *Phys. Rev. Lett.*, 53, 2187, 1984.
742. Ott and Antonsen, Jr., Chaotic fluid convection and the fractal nature of passive scalar gradients, *Phys. Rev. Lett.*, 61, 2839, 1988.
743. Ott and Antonsen, Jr., Fractal measures of passively convected vector fields and scalar gradients in chaotic fluid flows, *Phys. Rev.*, A39, 3660, 1989.
744. Parker et al., Axial magnetic-field effects in a collective-interaction free-electron laser at millimeter wavelengths, *Phys. Rev. Lett.*, 48, 238, 1982.
745. Perry et al., Nonresonant multiphoton ionization of noble gases: theory and experiment, *Phys. Rev. Lett.*, 60, 1270, 1988.
746. Perry et al., Multiphoton ionization of the noble gases by an intense $10^{14}$-W/$cm^2$ dye laser, *Phys. Rev.*, A37, 747, 1988.
747. Perry and Landen, Resonantly enhanced multiphoton ionization of krypton and xenon with intense ultraviolet laser radiation, *Phys. Rev.*, A38, 2815, 1988.
748. Perry et al., Resonantly enhanced above-threshold ionization of helium, *Phys. Rev. Lett.*, 63, 1058, 1989.
749. Perry and Crane, High-order harmonic emission from mixed fields, *Phys. Rev.*, A48, R4051, 1993.
750. Perry and Mourou, Terawatt to petawatt subpicosecond lasers, *Science,* 264, 917, 1994.
751. Piran et al., Filtering of spacecraft Doppler tracking data and detection of gravitational radiation, *Phys. Rev.*, D34, 984, 1986.

752. Prange et al., Smoothed density of states for problems with ray splitting, *Phys. Rev.*, E53, 207, 1996.
753. Pritchard et al., Atomic differential spin-exchange scattering, *Phys. Rev. Lett.*, 19, 1363, 1967.
754. Pukhov et al., Particle acceleration in relativistic laser channels, *Phys. Plasmas*, 6, 2847, 1999.
755. Qian et al., Halo formation induced by density nonuniformities in intense ion beams, *Phys. Rev.*, E51, R5216, 1995.
756. Qian and Davidson, Nonlinear dynamics of intense ion beam envelopes, *Phys. Rev.*, E53, 5349, 1996.
757. Qiu et al., Demonstration of emittance compensation through the measurement of the slice emittance of a 10-ps electron bunch, *Phys. Rev. Lett.*, 76, 3723, 1996.
758. Quesnel and Mora, Theory and simulation of the interaction of ultraintense laser pulses with electrons in vacuum, *Phys. Rev.*, E58, 3719, 1998.
759. Quigg, Top-ology, *Phys. Today*, 50 (5), 20, 1997.
760. Raimond, Brune, and Haroche, Reversible decoherence of a mesoscopic superposition of field states, *Phys. Rev. Lett.*, 79, 1964, 1997.
761. Raine et al., Does a uniformly accelerated quantum oscillator radiate?, *Proc. R. Soc. London*, A435, 205, 1991.
762. Reiche et al., Experimental confirmation of transverse focusing and adiabatic damping in a standing wave linear accelerator, *Phys. Rev.*, E56, 3572, 1997.
763. Renard et al., Detailed characterization of electron plasma waves produced by stimulated Raman scattering, *Phys. Rev. Lett.*, 77, 3807, 1996.
764. Reynaud et al., Experimental evidence for compensation of Doppler broadening by light shifts, *Phys. Rev. Lett.*, 42, 756, 1979.
765. Ride et al., Thomson scattering of intense lasers from electron beams at arbitrary interaction angles, *Phys. Rev.*, E52, 5425, 1995.
766. Roberson et al., Electrostatic wave reflection from a plasma density gradient, *Phys. Rev. Lett.*, 31, 1041, 1973.
767. Roberson and Sprangle, A review of free-electron lasers, *Phys. Fluids*, B1, 3, 1989.
768. Robin et al., Quasi-isochronous storage rings, *Phys. Rev.*, E48, 2149, 1993.
769. Rosenzweig, Nonlinear plasma dynamics in the plasma wake-field accelerator, *Phys. Rev. Lett.*, 58, 555, 1987.
770. Rosenzweig et al., Experimental observation of plasma wake-field acceleration, *Phys. Rev. Lett.*, 61, 98, 1988.
771. Rosenzweig, Trapping, thermal effects, and wave breaking in the nonlinear plasma wake-field accelerator, *Phys. Rev.*, A38, 3634, 1988.
772. Rosenzweig, Multiple-fluid models for plasma wake-field phenomena, *Phys. Rev.*, A40, 5249, 1989.
773. Rosenzweig and Chen, Beam optics of a self-focusing plasma lens, *Phys. Rev.*, D39, 2039, 1989.
774. Rosenzweig et al., Experimental measurements of nonlinear plasma wake fields, *Phys. Rev.*, A39, 1586, 1989.
775. Rosenzweig et al., Acceleration and focusing of electrons in two-dimensional nonlinear plasma wake fields, *Phys. Rev.*, A44, R6189, 1991.
776. Rosenzweig and Chen, Transverse equilibria and luminosity enhancement in linear collider beam–beam collisions, *Phys. Rev.*, E50, 526, 1994.
777. Rosenzweig and Serafini, Transverse particle motion in radio-frequency linear accelerators, *Phys. Rev.*, E49, 1599, 1994.

778. Rosenzweig et al., A proposed dielectric-loaded resonant laser accelerator, *Phys. Rev. Lett.*, 74, 2467, 1995.
779. Rose-Petruck et al., Ultrafast electron dynamics and inner-shell ionization in laser driven clusters, *Phys. Rev.*, A55, 1182, 1997.
780. Rose-Petruck et al., Picosecond-milliAngstrom lattice dynamics measured by ultrafast x-ray diffraction, *Nature*, 398, 310, 1999.
781. Saeta et al., Ultrafast electronic disordering during femtosecond laser melting of GaAs, *Phys. Rev. Lett.*, 67, 1023, 1991.
782. Sakawa et al., Nonlinear evolution of the modified Simon–Hoh instability via a cascade of sideband instabilities in a weak beam plasma system, *Phys. Rev. Lett.*, 69, 85, 1992.
783. Salamin, Forward and backscattered ultrahigh-intensity Compton spectra in a uniform magnetic field, *Phys. Rev.*, A60, 3276, 1999.
784. Salin et al., Thermal eigenmode amplifiers for diffraction-limited amplification of ultrashort pulses, *Opt. Lett.*, 23, 718, 1998.
785. Salour and Cohen-Tannoudji, Observation of Ramsey's interference fringes in the profile of Doppler-free two-photon resonances, *Phys. Rev. Lett.*, 38, 757, 1977.
786. Sarkisov et al., Self-focusing, channel formation, and high-energy ion generation in interaction of an intense short laser pulse with a He jet, *Phys. Rev.*, E59, 7042, 1999.
787. Saubaméa et al., Direct measurement of the spatial correlation function of ultracold atoms, *Phys. Rev. Lett.*, 79, 3146, 1997.
788. Savage et al., Frequency upconversion of electromagnetic radiation upon transition into an ionization front, *Phys. Rev. Lett.*, 68, 946, 1992.
789. Schnürer et al., Coherent 0.5-keV x-ray emission from helium driven by a sub-10-fs laser, *Phys. Rev. Lett.*, 80, 3236, 1998.
790. Schoenlein et al., Femtosecond x-ray pulses at 0.4 Å generated by 90° Thomson scattering: a tool for probing the structural dynamics of materials, *Science*, 274, 236, 1996.
791. Schroeder et al., Generation of ultrashort electron bunches by colliding laser pulses, *Phys. Rev.*, E59, 6037, 1999.
792. Schroeder et al., Multimode analysis of the hollow plasma channel wakefield accelerator, *Phys. Rev. Lett.*, 82, 1177, 1999.
793. Schwinger, On quantum-electrodynamics and the magnetic moment of the electron, *Phys. Rev.*, 73, 416, 1947.
794. Schwinger, Quantum electrodynamics. I. A covariant formulation, *Phys. Rev.*, 74, 1439, 1948.
795. Schwinger, Quantum electrodynamics. II. Vacuum polarization and self-energy, *Phys. Rev.*, 75, 651, 1948.
796. Schwinger, Quantum electrodynamics. III. The electromagnetic properties of the electron-radiative correction to scattering, *Phys. Rev.*, 76, 790, 1949.
797. Schwinger, On radiative corrections to electron scattering, *Phys. Rev.*, 75, 898, 1949.
798. Schwinger, On the classical radiation of accelerated electrons, *Phys. Rev.*, 75, 1912, 1949.
799. Schwinger, On gauge invariance and vacuum polarization, *Phys. Rev.*, 82, 664, 1951.
800. Schwinger, The theory of quantized fields. I, *Phys. Rev.*, 82, 914, 1951.
801. Schwinger, The theory of quantized fields. II, *Phys. Rev.*, 91, 713, 1953.

802. Schwinger, The theory of quantized fields. III, *Phys. Rev.*, 91, 728, 1953.
803. Schwinger, The theory of quantized fields. IV, *Phys. Rev.*, 92, 1283, 1953.
804. Schwinger, The theory of quantized fields. V, *Phys. Rev.*, 93, 615, 1954.
805. Schwinger, The theory of quantized fields. VI, *Phys. Rev.*, 94, 1362, 1954.
806. Schwinger, Gauge invariance and mass, *Phys. Rev.*, 125, 397, 1962.
807. Schwinger, Classical radiation of accelerated electrons. II. A quantum viewpoint, *Phys. Rev.*, D7, 1696, 1973.
808. Serafini and Rosenzweig, Envelope analysis of intense relativistic quasilaminar beams in rf photoinjectors: a theory of emittance compensation, *Phys. Rev.*, E55, 7565, 1997.
809. Shkvarunets et al., Plasma influence on the dispersion properties of finite-length, corrugated waveguides, *Phys. Rev.*, E53, R2045, 1996.
810. Shvets et al., Superradiant amplification of an ultrashort laser pulse in a plasma by a counterpropagating pump, *Phys. Rev. Lett.*, 81, 4879, 1998.
811. Shvets et al., Electromagnetically induced guiding of counterpropagating lasers in plasmas, *Phys. Rev.*, E59, 1033, 1999.
812. Siders et al., Detection of nonthermal melting by ultrafast x-ray diffraction, *Science*, 286 (5443), 1340, 1999.
813. Sprangle et al., Nonlinear theory of intense laser-plasma interactions, *Phys. Rev. Lett.*, 64, 2011, 1990.
814. Sprangle et al., Nonlinear interaction of intense laser pulses in plasmas, *Phys. Rev.*, A41, 4463, 1990.
815. Sprangle and Hafizi, Comments on nondiffracting beams, *Phys. Rev. Lett.*, 66, 837, 1991.
816. Sprangle and Esarey, Stimulated backscattered harmonic generation from intense laser interactions with beams and plasmas, *Phys. Rev. Lett.*, 67, 2021, 1991.
817. Sprangle et al., Propagation and guiding of intense laser pulses in plasmas, *Phys. Rev. Lett.*, 69, 2200, 1992.
818. Sprangle et al., Methods for conditioning electron beams in free-electron lasers, *Phys. Rev. Lett.*, 70, 2896, 1993.
819. Sprangle et al., Hose-modulation instability of laser pulses in plasmas, *Phys. Rev. Lett.*, 73, 3544, 1994.
820. Sprangle et al., Laser driven electron acceleration in vacuum, gases, and plasmas, *Phys. Plasma*, 3, 2183, 1996.
821. Sprangle et al., Ionization and pulse lethargy effects in inverse Cherenkov accelerators, *Appl. Phys. Lett.*, 70, 2198, 1997.
822. Sprangle et al., Self-guiding and stability of intense optical beams in gases undergoing ionization, *Phys. Rev.*, E54, 4211, 1996.
823. Sprangle et al., Propagation and stability of intense laser pulses in partially stripped plasmas, *Phys. Rev.*, E56, 5894, 1997.
824. Sprangle et al., Intense laser pulse propagation and stability in partially stripped plasmas, *Phys. Rev. Lett.*, 79, 1046, 1997.
825. Springer et al., Enhanced interaction energy shifts in the x-ray spectrum of $^{163}$Ho, *Phys. Rev.*, A31, 1965, 1985.
826. Springer et al., Measurement of the neutrino mass using the inner bremsstrahlung emitted in the electron-capture decay of $^{163}$Ho, *Phys. Rev.*, A35, 679, 1987.
827. Springer et al., Spectroscopic absorption measurements of an iron plasma, *Phys. Rev. Lett.*, 69, 3735, 1992.
828. Swegle and Ott, Instability of the Brillouin-flow equilibrium in magnetically insulated structures, *Phys. Rev. Lett.*, 46, 929, 1981.

829. Swegle et al., Backward wave oscillator with rippled wall resonators: analytic theory and numerical simulation, *Phys. Fluids*, 28, 2882, 1985.
830. Tajima et al., Alfvén ion-cyclotron instability: its physical mechanism and observation in computer simulation, *Phys. Rev. Lett.*, 39, 201, 1977.
831. Tajima et al., Double-layer forward shocks in a magnetohydrodynamic fluid, *Phys. Rev. Lett.*, 40, 652, 1978.
832. Tajima and Dawson, Laser electron accelerator, *Phys. Rev. Lett.*, 43, 267, 1979.
833. Tajima and Cavenago, Crystal x-ray accelerator, *Phys. Rev. Lett.*, 59, 1440, 1987.
834. Taylor et al., Timing jitter in mode-locked and gain-switched InGaAsP injection lasers, *Appl. Phys. Lett.*, 49, 681, 1986.
835. Ting et al., Demonstration experiments of a laser synchrotron source for tunable, monochromatic x-rays at 500 eV, *Nucl. Instrum. Meth. Phys. Res.*, A375, ABS68, 1996.
836. Ting et al., Temporal evolution of self-modulated laser wakefields measured by coherent Thomson scattering, *Phys. Rev. Lett.*, 77, 5377, 1996.
837. Ting et al., Backscattered supercontinuum emission from high-intensity laser–plasma interactions, *Opt. Lett.*, 21, 1096, 1996.
838. Ting et al., Plasma wakefield generation and electron acceleration in a self-modulated laser wakefield accelerator experiment, *Phys. Plasma*, 4, 1889, 1997.
839. Tremaine et al., Electromagnetic wake fields and beam stability in slab-symmetric dielectric structures, *Phys. Rev.*, E56, 7204, 1997.
840. Troha et al., Vacuum electron acceleration by coherent dipole radiation, *Phys. Rev.*, E60, 926, 1999.
841. Tsang and Davidson, Macroscopic cold-fluid equilibrium properties of relativistic non-neutral electron flow in a cylindrical diode, *Phys. Rev.*, A33, 4284, 1986.
842. Tzeng et al., Anomalous absorption and scattering of short-pulse high-intensity lasers in underdense plasma, *Phys. Rev. Lett.*, 76, 3332, 1996.
843. Tzeng et al., Electron beam characteristics from laser-driven wave breaking, *Phys. Rev. Lett.*, 79, 5258, 1997.
844. Uhm and Davidson, Kinetic equilibrium properties of relativistic non-neutral electron flow in a cylindrical diode with applied magnetic field, *Phys. Rev.*, A31, 2556, 1985.
845. Umstadter et al., Observation of steepening in electron plasma waves driven by stimulated Raman backscattering, *Phys. Rev. Lett.*, 59, 292, 1987.
846. Umstadter et al., Nonlinear plasma waves resonantly driven by optimized laser pulse trains, *Phys. Rev. Lett.*, 72, 1224, 1994.
847. Umstadter et al., Resonantly laser-driven plasma waves for electron acceleration, *Phys. Rev.*, E51, 3484, 1995.
848. Umstadter, Terawatt lasers produce faster electron acceleration, *Laser Focus World*, 1996.
849. Umstadter et al., Laser injection of ultrashort electron pulses into wakefield plasma waves, *Phys. Rev. Lett.*, 76, 2073, 1996.
850. Umstadter et al., Tabletop, ultrahigh-intensity lasers: dawn of nonlinear relativistic optics, *Opti. Photon. News*, 9 (7), 41, 1998.
851. Unruh, Separability of the neutrino equations in a Kerr background, *Phys. Rev. Lett.*, 31, 1265, 1973.
852. Unruh, Second quantization in the Kerr metric, *Phys. Rev.*, D10, 3194, 1974.
853. Unruh, Notes on black-hole evaporation, *Phys. Rev.*, D14, 870, 1976.
854. Unruh, Absorption cross section of small black holes, *Phys. Rev.*, D14, 3251, 1976.

855. Unruh, Origin of the particles in black-hole evaporation, *Phys. Rev.*, D15, 365, 1977.
856. Unruh, Quantum nondemolition measurements and coherent states, *Phys. Rev.*, D17, 1180, 1978.
857. Unruh, Analysis of quantum-nondemolition measurements, *Phys. Rev.*, D18, 1764, 1978.
858. Unruh, Quantum nondemolition and gravity-wave detection, *Phys. Rev.*, D19, 2888, 1979.
859. Unruh, Experimental black-hole evaporation?, *Phys. Rev. Lett.*, 46, 1351, 1981.
860. Unruh and Wald, Acceleration radiation and the generalized second law of thermodynamics, *Phys. Rev.*, D25, 942, 1982.
861. Unruh and Wald, Entropy bounds, acceleration radiation, and the generalized second law, *Phys. Rev.*, D27, 2271, 1983.
862. Unruh and Wald, What happens when an accelerating observer detects a Rindler particle, *Phys. Rev.*, D29, 1047, 1984.
863. Unruh and Weiss, Acceleration radiation in interacting field theories, *Phys. Rev.*, D29, 1656, 1984.
864. Unruh, Collapse of radiating fluid spheres and cosmic censorship, *Phys. Rev.*, D31, 2693, 1985.
865. Unruh and Wald, Damping mechanisms for coherent oscillations of axions, *Phys. Rev.*, D32, 831, 1985.
866. Unruh, Accelerated monopole detector in odd spacetime dimensions, *Phys. Rev.*, D34, 1222, 1986.
867. Unruh et al., Cosmic-string loops are straight, *Phys. Rev. Lett.*, 62, 2897, 1989.
868. Unruh, Unimodular theory of canonical quantum gravity, *Phys. Rev.*, D40, 1048, 1989.
869. Unruh, Quantum coherence, wormholes, and the cosmological constant, *Phys. Rev.*, D40, 1053, 1989.
870. Unruh and Weiss, Relationship between the usual formulation of a massless scalar field theory and its formulation in terms of a two-form potential, *Phys. Rev.*, D40, 1064, 1989.
871. Unruh and Zurek, Reduction of a wave packet in quantum Brownian motion, *Phys. Rev.*, D40, 1071, 1989.
872. Unruh and Wald, Time and the interpretation of canonical quantum gravity, *Phys. Rev.*, D40, 2598, 1989.
873. Unruh, Thermal bath and decoherence of Rindler spacetimes, *Phys. Rev.*, D46, 3271, 1992.
874. Unruh, Straight strings and Friedmann–Robertson–Walker spacetimes, *Phys. Rev.*, D46, 3265, 1992.
875. Unruh and Newbury, Solution to 2+1 gravity in the dreibein formalism, *Phys. Rev.*, D48, 2686, 1993.
876. Unruh, Reality and the measurement of the wave function, *Phys. Rev.*, A50, 882, 1994.
877. Unruh, Maintaining coherence in quantum computers, *Phys. Rev.*, A51, 992, 1995.
878. Unruh, Sonic analog of black holes and the effects of high frequencies on black hole evaporation, *Phys. Rev.*, D51, 2827, 1995.
879. Unruh and Wald, Evolution laws taking pure states to mixed states in quantum field theory, *Phys. Rev.*, D52, 2176, 1995.

880. Vafa and Witten, Parity conservation in quantum chromodynamics, *Phys. Rev. Lett.*, 53, 535, 1984.
881. Vallarta and Feynman, The scattering of cosmic rays by the stars of a galaxy, *Phys. Rev.*, 55, 506, 1939.
882. Van Meter et al., Radiative corrections in symmetrized classical electrodynamics, *Phys. Rev.*, E62, 8640, 2000.
883. Van Steenbergen et al., Observation of energy gain at the BNL inverse free-electron laser accelerator, *Phys. Rev. Lett.*, 77, 2690, 1996.
884. Vollick and Unruh, Light deflection caused by a cosmic string carrying a wave pulse, *Phys. Rev.*, D42, 2621, 1990.
885. Vollick and Unruh, Gravitational lensing properties of curved cosmic strings, *Phys. Rev.*, D44, 2388, 1991.
886. Wagner et al., Electron acceleration by a laser wakefield in a relativistically self-guided channel, *Phys. Rev. Lett.*, 78, 3125, 1997.
887. Wang et al., Efficient co-generation of seventh-harmonic radiation in cyclotron autoresonance acceleration, *Phys. Rev. Lett.*, 77, 3819, 1996.
888. Watchel and Hirshfield, Interference beats in pulse-stimulated cyclotron radiation, *Phys. Rev. Lett.*, 17, 348, 1966.
889. Watchel and Hirshfield, Negative electron cyclotron resonance absorption due to collisions, *Phys. Rev. Lett.*, 19, 293, 1967.
890. Weiner et al., High-resolution femtosecond pulse shaping, *J. Opt. Soc. Am.*, B5, 1563, 1988.
891. Welch et al., Positive energy structure of the Rydberg diamagnetic spectrum, *Phys. Rev. Lett.*, 62, 1975, 1989.
892. Wilks et al., Frequency up-conversion of electromagnetic radiation with use of an overdense plasma, *Phys. Rev. Lett.*, 61, 337, 1988.
893. Wilks et al., Photon accelerator, *Phys. Rev. Lett.*, 62, 2600, 1989.
894. Willison et al., Emission spectra of core excited even-parity $^2P$ states of neutral lithium, *Phys. Rev. Lett.*, 44, 1125, 1980.
895. Willison et al., Laser spectroscopy of metastable extreme-ultraviolet levels in lithium atoms and ions, *Phys. Rev. Lett.*, 47, 1827, 1981.
896. Witten, Some exact multipseudoparticle solutions of classical Yang–Mills theory, *Phys. Rev. Lett.*, 38, 121, 1977.
897. Witten, Some inequalities among hadron masses, *Phys. Rev. Lett.*, 51, 2351, 1983.
898. Witten, Space–time and topological orbifolds, *Phys. Rev. Lett.*, 61, 670, 1988.
899. Witten, Reflections on the fate of spacetime, *Physics Today*, 49 (4), 24, 1996.
900. Witten, Duality, spacetime and quantum mechanics, *Physics Today*, 50 (5), 28, 1997.
901. Workman et al., Control of bright picosecond x-ray emission from intense subpicosecond laser-plasma interactions, *Phys. Rev. Lett.*, 75, 2324, 1995.
902. Yoder et al., Energy-gain measurements from a microwave inverse free-electron laser accelerator, *Phys. Rev. Lett.*, 86, 1765, 2001.
903. Yoon and Davidson, Closed-form analytical model of the electron whistler and cyclotron maser instabilities in relativistic plasma with arbitrary energy anisotropy, *Phys. Rev.*, A35, 2619, 1987.
904. Yoon and Davidson, Exact analytical model of the classical Weibel instability in a relativistic anisotropic plasma, *Phys. Rev.*, A35, 2718, 1987.
905. Yablonovitch, Accelerating reference frame for electromagnetic waves in a rapidly growing plasma: Unruh–Davies–Fulling–De Witt radiation and the nonadiabatic Casimir effect, *Phys. Rev. Lett.*, 62, 1742, 1989.

906. Yamamoto and Haus, Commutation relations and laser linewidth, *Phys. Rev.*, A41, 5164, 1990.
907. Yamamoto and Haus, Effect of electrical partition noise on squeezing in semiconductor lasers, *Phys. Rev.*, A45, 6596, 1992.
908. Yoshii et al., Radiation from Čerenkov waves in a magnetized plasma, *Phys. Rev. Lett.*, 79, 4194, 1997.
909. Yu et al., Ponderomotive acceleration of electrons at the focus of high intensity lasers, *Phys. Rev.*, E61, R2220, 2000.
910. Zhang et al., A millimeter wave FEL driven by a photocathode rf linac, *Nucl. Instrum. Meth. Phys. Res.*, A341, 67, 1994.
911. Zhang and Marshall, Possibility of generating a high-power self-similar radiation pulse from a free-electron laser, *Phys. Rev. Lett.*, 74, 916, 1995.
912. Zhou et al., Amplification of 26-fs, 2-TW pulses near the gain-narrowing limit in Ti:sapphire, *Opt. Lett.*, 20, 64, 1995.
913. Zholents and Zolotorev, Femtosecond x-ray pulses of synchrotron radiation, *Phys. Rev. Lett.*, 76, 912, 1996.
914. *Advanced Accelerator Concepts*, 7th Workshop, Edited by Chattopadhyay, McCullough and Dahl, AIP Conferences Proceedings 398, American Institute of Physics, Woodbury, NY, 1997.
915. *Advanced Accelerator Concepts*, 8th Workshop, Edited by Lawson, Bellamy and Brosius, AIP Conferences Proceedings 472, American Institute of Physics, Woodbury, NY, 1999.
916. *Advanced Accelerator Concepts*, 9th Workshop, Edited by Colestock and Kelley, AIP Conferences Proceedings 569, American Institute of Physics, Woodbury, NY, 2001.
917. *Free-Electron Lasers '87*, Edited by Sprangle, Tang and Walsh, North-Holland, Amsterdam, 1988.
918. *Free-Electron Lasers '89*, Edited by Elias and Kimel, North-Holland, Amsterdam, 1990.
919. *Free-Electron Lasers '90*, Edited by Buzzi and Ortega, North-Holland, Amsterdam, 1991.
920. *Free-Electron Lasers '91*, Edited by Goldstein and Newnam, North-Holland, Amsterdam, 1992.
921. *Free-Electron Lasers '93*, Edited by Van Amersfoort, Van der Slot and Witteman, North-Holland, Amsterdam, 1994.
922. *High Energy Density Microwaves*, Edited by Phillips, AIP Conference Proceedings 474, American Institute of Physics, Woodbury, NY, 1999.
923. *High-Field Science*, Edited by Tajima, Mima and Baldis, Kluwer Academic/Plenum Publishers, New York, 2000.
924. *Physics Potential and Development of Muon Colliders and Neutrino Factories*, Edited by Cline, AIP Conference Proceedings 542, American Institute of Physics, Woodbury, NY, 2000.
925. *Proceedings of the 1997 Particle Accelerator Conference*, Edited by Comyn, Craddock, Reiser, and Thomson, IEEE, Piscataway, NJ, 1998.
926. *Proceedings of the 1999 Particle Accelerator Conference*, Edited by Luccio and MacKay, IEEE, Piscataway, NJ, 1999.
927. *Quantum Aspects of Beam Physics*, Edited by Chen, World Scientific, Singapore, 1999.

# Index

## A



## B



## C

## D



## E



## F

## K



## L



## M



## N

## O



## P



## Q



## R

## U



## V



## W



## X



## Y



## Z